CHEMISTRY

IN THE COMMUNITY

Sixth Edition

AMERICAN CHEMICAL SOCIETY
Chief Editor: Angela Powers
Revision Team: Laurie Langdon, Thomas Pentecost, Cece Schwennsen
ACS: Mary Kirchhoff (Director, ACS Education Division), Marta Gmurczyk,
Karen Kaleuati, Michael Mury, Terri Taylor
Chemistry at Work: Christen Brownlee
ACS Committee on Chemical Safety: Harry J. Elston
Ancillary Materials: Susan Cooper, Michael Dianovsky, Sara Marchlewicz,
Stephanie Ryan
Editorial Advisory Board: Steven Long (Chair), Henry Heikkinen, Cathy Middlecamp,
Barbara Sitzman, Michael Tinnesand, Don Wink
Teacher Reviewers and Pilot Testers: Bonnie Bloom, Patricia Deibert, Pamela Diaz,
Regis Goode, Jennifer Kieffer-Gerckens, John Novak, Deborah Pusateri,
Barbara Sitzman, Linda Tilton

W. H. FREEMAN/BFW
Executive Editor: Ann Heath
Assistant Editor: Dora Figueiredo
Development Editor: Don Gecewicz
Media Editor: Dave Quinn
Executive Marketing Manager: Cindi Weiss
Photo Editor: Bianca Moscatelli
Photo Researcher: Jacqui Wong
Project Editor: Vivien Weiss
Design Manager: Blake Logan
Text Designer: Rae Grant Design
Cover Image: Per Eriksson/The Image Bank/Getty Images, illustration by ACS
Illustrations: Network Graphics
Illustrations Coordinator: Janice Donnola
Production Manager: Susan Wein
Composition: MPS Content Services, A Macmillan Company; Rae Grant Design
Printing and Binding: Quad Graphics

This material is based upon work supported by the National Science Foundation
under Grant No. SED-88115424 and Grant No. MDR-8470104. Any opinions,
findings, and conclusions or recommendations expressed in this publication are
those of the authors and do not necessarily reflect the views of the National Science
Foundation. Any mention of trade names does not imply endorsement by the
National Science Foundation.

Library of Congress Control Number: 2011928734

ISBN-13: 978-1-4292-1952-5
ISBN-10: 1-4292-1952-1

Copyright © 2012, 2006, 2002, 1998, 1993, 1988 by American Chemical Society

Second printing

Printed in the United States of America

CHEMISTRY
IN THE COMMUNITY
Sixth Edition

CHEMCOM
A PROJECT OF THE
AMERICAN CHEMICAL SOCIETY

bfw
Freeman

W. H. FREEMAN AND COMPANY/ BFW

NEW YORK

Important Notice

Chemistry in the Community (ChemCom) is intended for use by high school students in the classroom laboratory under the direct supervision of a qualified chemistry teacher. The experiments described in this book involve substances that may be harmful if they are misused or if the procedures described are not followed. Read cautions carefully and follow all directions. Do not use or combine any substances or materials not specifically called for in carrying out investigations. Other substances are mentioned for educational purposes only and should not be used by students unless the instructions specifically so indicate.

The materials, safety information, and procedures contained in this book are believed to be reliable. This information and these procedures should serve only as a starting point for good laboratory practices, and they do not purport to specify minimal legal standards or to represent the policy of the American Chemical Society. No warranty, guarantee, or representation is made by the American Chemical Society as to the accuracy or specificity of the information contained herein, and the American Chemical Society assumes no responsibility in connection therewith. The added safety information is intended to provide basic guidelines for safe practices. It cannot be assumed that all necessary warnings and precautionary measures are contained in the document or that other additional information and measures may not be required.

Safety and Laboratory Activity

In *ChemCom*, you will frequently complete laboratory investigations. While no human activity is completely risk free, if you use common sense, as well as chemical sense, and follow the rules of laboratory safety, you should encounter no problems. Chemical sense is just an extension of common sense. Sensible laboratory conduct won't happen by memorizing a list of rules, any more than a perfect score on a written driver's test ensures an excellent driving record. The true "driver's test" of chemical sense is your actual conduct in the laboratory.

You will find Rules of Laboratory Conduct on pages 8–10 in Unit 0.

BRIEF CONTENTS

CONTENTS

UNIT 4

WATER: EXPLORING SOLUTIONS   388

UNIT 5

**INDUSTRY: APPLYING CHEMICAL
REACTIONS**   502

UNIT **6**

ATOMS: NUCLEAR INTERACTIONS   584

PREFACE

It is appropriate that the publication of the sixth edition of *Chemistry in the Community (ChemCom)* coincides with the International Year of Chemistry (IYC) 2011, which is a global celebration of chemistry and its contributions to the world around us. The theme chosen for IYC 2011 has been central to *ChemCom* for more than 25 years: Chemistry—our life, our future.

ChemCom is written for high school students taking their first chemistry course. *ChemCom* aims to develop chemistry-literate and science-literate citizens by focusing on chemistry for life and citizenship. Each unit is centered on a chemistry-related societal issue or challenge, which provides a "need-to-know" for learning chemical principles. Laboratory, skill-building, modeling, and decision-making activities are integrated into the text as students progress toward a culminating project that addresses the unit's societal issue or challenge.

ABOUT CHEMCOM

Developed by the American Chemical Society (ACS) with funding from the National Science Foundation and input from hundreds of teachers, university educators, scientists, and social-science consultants, *ChemCom* has been used successfully by more than 2.2 million students and teachers throughout the United States and the world.

The world's largest scientific society, ACS is a congressionally chartered independent membership organization that represents professionals at all degree levels and in all fields of chemistry and related sciences. Since its founding in 1876, ACS has promoted excellence in science education and community outreach.

The original eight *ChemCom* units were developed by teams of professors and high school teachers. The topics included water, mineral resources, petroleum, food, nuclear chemistry, air and climate, health, and the chemical industry. Over the past thirty years, *ChemCom* has continued to develop to meet the needs of students and teachers. One constant has been its organization around real-world issues, particularly issues of sustainability.

Development of the sixth edition of *ChemCom* was influenced by feedback from teachers and students, as well as a comprehensive review by BSCS using the Analyzing Instructional Materials (AIM) process. Starting from this information, the revision team—comprised of high school and university educators—drew upon their own experiences in classrooms and chemical education research, as well as those of the sixth edition *ChemCom* Editorial Advisory Board, to create a text designed for twenty-first-century learners. Beyond the BSCS review, significant influences on this edition include Wiggins and McTighe's *Understanding by Design*, The National Research Council's *How People Learn*, and Hand and Greenbowe's Science Writing Heuristic.

NEW TO THE SIXTH EDITION

The most apparent changes in this edition are the reordering of units and the addition of Unit 0. Unit 0 offers a brief introduction to the study of chemistry, the idea of community, safety in the laboratory, and *ChemCom* itself.

Units 1 through 4, designed to be studied in order and thus sometimes referred to as the sequential units, have been reorganized and reordered. *ChemCom* now begins with the study of metals, which allows for early introduction of key chemistry topics including atomic structure, the periodic table and periodicity, the mole concept, and chemical equations. Unit 2 addresses gases and the atmosphere; explicitly introduces ideas about scientific inquiry; and challenges students to design their own investigations, building skills and knowledge that will be beneficial throughout the rest of the course. Unit 3 on petroleum maintains its position, but the water unit is now the final unit in this sequence, allowing students to use all the concepts and skills they have developed to understand ideas about solution chemistry and solve the fish-kill mystery.

A new feature in this edition is the Concept Check. Concept Checks serve as a formative assessment tool by eliciting student ideas to make them more apparent both to students and to teachers. In each set of questions, the initial queries probe concepts or skills that students have just learned to help both students and teachers monitor progress. At least one question in each Concept Check asks students about their ideas on a topic they are about to encounter. This allows students to articulate their existing ideas from everyday experiences and provides important information about students' prior knowledge.

The final key change in this edition involves the Investigating Matter activities. In each unit, and in several sections, an introductory investigation was added to provide students with hands-on experience before they tackle a new topic. Even more critically, each Investigating Matter activity was revised to more clearly reflect the components of scientific inquiry. Students begin by asking questions and preparing to investigate, then gather evidence, analyze and interpret that evidence to make scientific claims, and finally reflect upon what they did and what they learned.

The following table shows how each key revision goal was accomplished.

Key Goals	How *CHEMCOM* 6e Accomplishes These Goals
Make interaction with phenomena central • Provide a common set of experiences as a basis for student construction of understanding • Develop habits of mind consistent with the nature of science • Model scientific ways of knowing	• Begins each unit with structured opportunities to explore matter or data, such as *Investigating Matter* or *Developing Skills* activities. • Reorganizes the *sequential units* of the text to emphasize characteristics of matter, scientific inquiry, and building problem-solving skills. • Introduces *Unit 0* to help students develop their ideas about chemistry and community and become familiar with the format and features of the text.
Emphasize scientific inquiry • Identify and practice skills necessary for scientific investigation • Explicitly address understandings about inquiry	• Restructures *Investigating Matter* activities to more explicitly reflect an inquiry focus. • Uses new subheadings within investigations that articulate expectations for students, including: *Asking Questions, Preparing to Investigate, Making Predictions, Gathering Evidence, Analyzing Evidence, Interpreting Evidence, Making Claims*, and *Reflecting on the Investigation.* • Supports students' learning of inquiry by creating a progression of activities from proscriptive to open-ended throughout the text. • Provides *culminating projects* that include opportunities to share and refine claims and construct scientific explanations.
Enhance student construction of understanding • Create an essential flow of concepts and skills within each unit and throughout the sequential units • Address advances in learning and cognitive science, including the importance of prior conceptions and metacognition	• Adds *Unit 0* and reorders the *sequential units* so that they build on each other and provide the foundation necessary for understanding and extending challenging concepts in the Water unit, now Unit 4. • Poses a *section question* to focus student attention on the essential concept or "big idea" to be learned in that section. • States *learning goals* at the beginning of each section. • Organizes the *Section Summary* to correspond with the stated learning goals and prompts students to summarize their understanding of the section's essential ideas in a written response to the section question. • Introduces *Concept Checks* so students can make connections to prior learning as well as express initial ideas about concepts they are about to learn. They are designed to be formative assessment tools for students and teachers. • Reframes *Making Decisions* activities as regular checkpoints for applying chemistry principles to the unit's central issue or challenge as presented in the unit opener. This allows students to make consistent progress toward the unit's culminating activity, *Putting It All Together.* • Provides at least one *sample problem* in every *Developing Skills* activity. • Enhances *Modeling Matter* activities to build more connections between particulate models and macroscopic phenomena, especially those phenomena encountered in previous *Investigating Matter* activities.

HOW TO GET THE MOST FROM *CHEMCOM* 6e

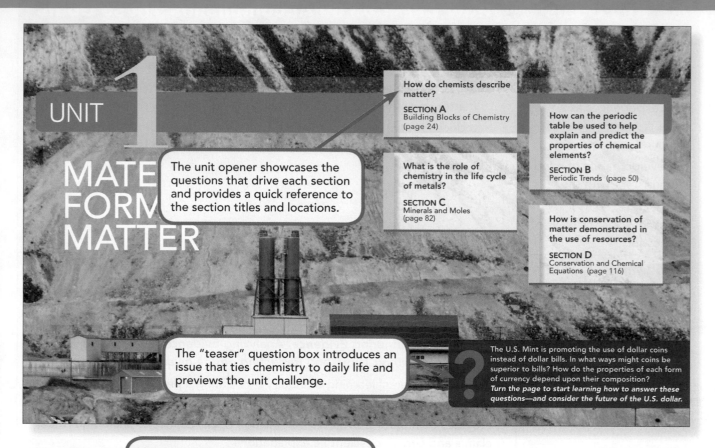

UNIT 1

MATE FORM MATTER

The unit opener showcases the questions that drive each section and provides a quick reference to the section titles and locations.

How do chemists describe matter?

SECTION A
Building Blocks of Chemistry (page 24)

What is the role of chemistry in the life cycle of metals?

SECTION C
Minerals and Moles (page 82)

How can the periodic table be used to help explain and predict the properties of chemical elements?

SECTION B
Periodic Trends (page 50)

How is conservation of matter demonstrated in the use of resources?

SECTION D
Conservation and Chemical Equations (page 116)

The "teaser" question box introduces an issue that ties chemistry to daily life and previews the unit challenge.

?

The U.S. Mint is promoting the use of dollar coins instead of dollar bills. In what ways might coins be superior to bills? How do the properties of each form of currency depend upon their composition? *Turn the page to start learning how to answer these questions—and consider the future of the U.S. dollar.*

The unit challenge invites students to address an interesting real-world problem. Making an informed decision about the problem will require chemistry concepts and scientific inquiry, providing a "need-to-know" that propels the unit.

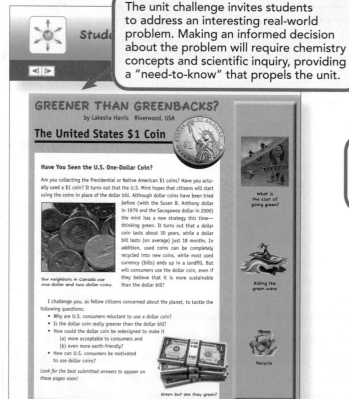

Stud

GREENER THAN GREENBACKS?
by Lakesha Harris   Riverwood, USA

The United States $1 Coin

Have You Seen the U.S. One-Dollar Coin?

Are you collecting the Presidential or Native American $1 coins? Have you actually used a $1 coin? It turns out that the U.S. Mint hopes that citizens will start using the coins in place of the dollar bill. Although dollar coins have been tried before (with the Susan B. Anthony dollar in 1979 and the Sacagawea dollar in 2000) the mint has a new strategy this time— thinking green. It turns out that a dollar coin lasts about 30 years, while a dollar bill lasts (on average) just 18 months. In addition, used coins can be completely recycled into new coins, while most used currency (bills) ends up in a landfill. But will consumers use the dollar coin, even if they believe that it is more sustainable than the dollar bill?

Our neighbors in Canada use one-dollar and two-dollar coins.

I challenge you, as fellow citizens concerned about the planet, to tackle the following questions:

- Why are U.S. consumers reluctant to use a dollar coin?
- Is the dollar coin really greener than the dollar bill?
- How could the dollar coin be redesigned to make it
  (a) more acceptable to consumers and
  (b) even more earth-friendly?
- How can U.S. consumers be motivated to use dollar coins?

Look for the best submitted answers to appear on these pages soon!

What is the cost of going green?

Riding the green wave

Recycle

Green but are they green?

143

PUTTING IT ALL TOGETHER

MAKING THE CASE FOR CURRENCY

You probably realize that new products are not easily created and that even the design process requires multiple stages. One common step in designing a new product is soliciting proposals from several individuals or teams. Once the proposals are submitted, a panel of experts or

The **Putting It All Together** feature appears at the end of each unit and provides a framework for students' completion of the unit challenge.

clearly confirm your recommendations for printing and minting dollars in the United States.

- Should both the dollar bill and dollar coin continue to be produced?
- Should one or the other be discontinued?
- Should the bill or the coin (or both) be made from different materials or otherwise redesigned?

Rationale
Support your recommendations with evidence, including:

- The factors that most influenced your recommendations and why they are important.
- Descriptions of the materials that make up the forms of currency that you recommend.
- Details about the raw materials needed to produce the currency, including
  a. for metals:
    ◦ Where major deposits are located.
    ◦ Important ores of the chosen raw materials.
  b. for all materials:
    ◦ How the materials are mined, collected, or harvested.
    ◦ How the materials are processed for use or production.
    ◦ Your rationale for choosing the selected materials.
    ◦ An analysis of both necessary and desirable properties of the selected materials.

Goals are provided at the start of each section to highlight key skills and ideas that students should know, understand, or be able to do by the end of the section.

Concept Checks elicit student ideas to make them more apparent to students and teachers. Some questions relate to concepts or skills that students have previously encountered and help students to monitor their own progress. At least one question addresses a topic students are about to encounter.

Each section opener highlights a question designed to stimulate and guide students' learning within that section.

The section question is revisited in the **Section Summary** to provide students an opportunity to recap the section, pull together the important ideas, and demonstrate their understanding of the chemistry they have learned.

Connecting the Concepts encourages students to synthesize ideas from one or more sections or units.

The boxes within **Reviewing the Concepts** highlight key ideas—linked to the learning goals at the beginning of the section—and organize the end-of-section questions.

Extending the Concepts provides opportunities for more in-depth exploration of related chemistry and science concepts.

---

50   Unit 1 Materials: Formulating Matter

SECTION **B**

PERIODIC TRENDS

How can the periodic table be used to help explain and predict the properties of chemical elements?

More than 100 chemical elements have been discovered and explored. Of those, a small number are considered highly prized and used for coins, jewelry, and art. Why are some more valuable than others? How do scientists classify and organize the characteristics of the elements? In this section, you will examine properties of some elements and construct statements about groups of elements and their reactivity. You will also be introduced to the periodic table, one of a chemist's most important tools. As you study the ideas in this section, think about how this tool could help you to choose or explain the use of certain elements in forms of currency.

GOALS

- Use the periodic table to
  a. predict physical and chemical properties of an element,
  b. identify elements by their atomic numbers, and
  c. locate periods and groups (families) of elements.
- Recognize and distinguish characteristics of the basic subatomic particles: protons, neutrons, and electrons.
- Describe what constitutes an ion. Indicate the electrical charge of an ion containing a specified number of protons and electrons.
- Use the basic structure of the atom to explain the organization of the periodic table.
- Collect, organize, and represent data.
- Explore periodic trends of groups of elements.
- Write the formula and name of an ionic compound, given the compound's anion and cation names and electrical charges.

concept check 3

1. How can you distinguish between chemical and physical properties?
2. What information does a chemical formula contain about a compound?
3. Draw and label a diagram that illustrates the structure of an atom.

---

78   Unit 1 Materials: Formulating Matter

Elements are arranged in the periodic table based on their properties. Elements with similar chemical properties are placed in the same columns. Physical properties vary in predictable patterns across rows and down columns.

9. Give another term for these features of the periodic table:
   a. Row   b. Column
10. Give the names and symbols of two elements other than lithium in the alkali metal family.
11. Consider the noble gas family.

13. The melting points of sulfur (S) and tellurium (Te) are 115 °C and 450 °C, respectively. Estimate the melting point of selenium (Se).
14. Would you expect the boiling point of chlorine to be higher or lower than that of iodine? Explain.

The properties of an element are determined largely by the number and arrangement of electrons in its atoms.

15. Are atoms of metallic or non-metallic elements more likely to lose one or more electrons?
16. Predict whether each element would be more likely to form an anion or a cation. (Note that anions are *negatively* charged; cations are *positively* charged.)
   a. sodium   d. copper   g. tin
   b. calcium   e. oxygen   h. iodine
   c. fluorine   f. lithium

17. Noble gas elements rarely lose or gain electrons. What does this indicate about their chemical reactivity?

Ionic compounds are composed of positively and negatively charged ions (atoms that have lost or gained electrons), combined so that the compound has no net electrical charge.

. Classify each of these as an electrically neutral atom, an anion, or a cation.
   a. O^{2-} b. Li c. Cl d. Ag^+ e. Hg^{2+}
19. For each particle in Question 18, indicate whether the electrical charge or lack of charge resulted from an atom ns, losing electrons, or

ol and show the electrical n the following atoms or ions:
th 1 proton and 1 electron
h 11 protons and 10 electrons
h 17 protons and 18 electrons
with 13 protons and
s

21. Write the name and formula for the ionic compound that can be formed from these cations and anions:
   a. K^+ and I^- d. Ba^{2+} and OH^-
   b. Ca^{2+} and S^{2-} e. NH_4^+ and PO_4^{3-}
   c. Fe^{3+} and Br^- f. Al^{3+} and O^{2-}

Tables, graphs, and models are all used to represent scientific data and illustrate scientific ideas so that they are easier to analyze, interpret, and understand

22. Why was the periodic table crea
23. What type of graph should be u to represent discontinuous data Why?
24. How is a data table helpful
   a. before an investigation?
   b. during an investigation?
   c. after an investigation?

25. Think about the ion cards yo used in Modeling Matter B.11.
   a. How are they helpful in arning about ions and ionic compounds?
   b. How are they imperfect models of ions?

Metals react with one an her in predictable patterns acc rding to their reactivities.

26. Which of these reactio s is more likely to occur? Why? (Refer to able 1.3, page 75.)
   a. Calcium metal wi h chromium(III) chloride solution.
   b. Chromium metal with calcium chloride solution.
27. Consider these two equations. Which represents a reaction that is more likely to occur? Why?
   a. $Zn^{2+}(aq) + 2\ Ag(s) \longrightarrow Zn(s) + 2\ Ag^+(aq)$
   b. $2\ Ag^+(aq) + Zn(s) \longrightarrow 2\ Ag(s) + Zn^{2+}(aq)$
28. a. Why would t be a poor idea to stir a solution of ead(II) nitrate with an iron spoon? (Se Table 1.3, page 75.)
   b. Write a ch mical equation to support your answer.

How an the periodic table be used to help explain and predict the properties of chemical elements?

Throughout Section B, you have observed and studied elements
k about
answe

our
as in

ving
of
elements, subatomic particles, periodicity and trends, and reactivity.

SECTION B Section Summary   **79**

Connecting the Concepts

29. Which pair is more similar chemically? Defend your choice.
   a. copper metal and copper(II) ions
   or
   b. oxygen with mass number 16 and oxygen with mass number 18
30. The diameter of a magnesium ion (Mg^{2+}) is 156 pm (picometers, where 1 pm = 10^{-12} m); the diameter of a strontium ion (Sr^{2+}) is 254 pm. Estimate the diameter of a calcium ion (Ca^{2+}).
31. Identify each element in the periodic table described by the statements below:
   a. This element is a nonmetal. It forms anions with a 1– charge. It is in the same period as the metals used in a penny.
   b. This element is a metalloid. It is in the same period as the elements found in table salt.
32. Mendeleev arranged elements in his periodic table in order of their atomic masses. In the modern periodic table, however, elements are arranged in order of their atomic numbers. Cite two examples from the periodic table for which these two schemes would produce a different ordering of adjacent elements.
33. In building ships, common practice is to attach a piece of magnesium to the hull to act as a "sacrificial anode." In terms of metal activity, explain how this helps to prevent the corrosion of other metal parts of the ship.

Extending the Concepts

34. Construct a graph of the price per gram of an element versus its atomic number for each of the first 20 elements. Can the current cost of those elements be regarded as a periodic property? Explain. (*Hint:* Use a chemical supply catalog or the Web to locate the current price of each element.)
35. Although aluminum is a more reactive metal than iron, it is often used for outdoor products. Investigate why this makes sense.

■ INVESTIGATING MATTER

B.12 RELATIVE REACTIVITIES OF METALS

Each **Investigating Matter** activity has been reframed to provide explicit instruction in the concepts and skills of scientific inquiry. The new subheadings reflect the process of scientific investigation as it is actually conducted by scientists.

Preparing to Investigate

In this investigation, you will observe the reactions of the metals copper, magnesium, and zinc with four different solutions. Each solution contains a particular cation. The solutions you will use are copper(II) nitrate, $Cu(NO_3)_2$ (containing Cu^{2+}); magnesium nitrate, $Mg(NO_3)_2$ (containing Mg^{2+}); zinc nitrate, $Zn(NO_3)_2$ (containing Zn^{2+}); and silver nitrate, $AgNO_3$ (containing Ag^+).

Before you begin, read *Gathering Evidence* to learn what you will need to do and note safety precautions. Devise a systematic procedure that allows you to observe the reaction (if any) between each metal and each of the four ionic solutions. You will conduct each reaction in a separate well of your well plate using 10 drops of the specified solution and a small strip of

added to indicate copper's ionic charge. Copper(I) oxide is Cu_2O because it contains Cu^+ ions. Copper(II) oxide is CuO; it contains Cu^{2+} ions.

Scientific American Illustrations are information-rich visuals intended to help students deepen their conceptual understanding of an idea or develop a working knowledge of a device related to the unit concept.

A **ChemQuandary** is a puzzling chemistry-related question or situation that is designed to stimulate thinking and decision making. It can be used as a think-pair-share, to begin a discussion, or as a prompt for a journal entry.

Scientific American Conceptual Illustration

Figure 1.20 *Model of a lithium (Li) atom. Each electrically neutral lithium atom contains three protons, three electrons, and either three or four neutrons. In this model, the protons and three neutrons reside in the Li nucleus (center). Electrons are not depicted as individual particles; they occupy "electron clouds" around the nucleus. Two electrons (yellow cloud) remain close to the nucleus; the third electron (fainter cloud) is relatively farther away. Electron clouds occupy most of an atom's volume; the nucleus accounts for nearly all the atom's mass. Note that an atom does not have a distinctly defined "outer edge," but, instead, has a rather fuzzy outer region.*

CHEM**QUANDARY**

FIVE CENTS' WORTH

A U.S. nickel is composed of an alloy of nickel and copper. Based on your familiarity with the appearance of that common five-cent coin, you might be surprised to learn it is composed of more copper than nickel! Specifically, each U.S. five-cent coin contains only 25% nickel and 75% copper by mass. What does this suggest about the difference between an alloy and a simple **mixture** of powdered copper and powdered nickel?

Developing Skills activities focus on problem-solving skills, including real-world situations and practice problems. One or more sample problems are provided at the beginning of each Developing Skills section to model how to reason through, set up, and solve similar problems.

Making Decisions activities guide students in applying their understanding of chemical principles to real-life situations. Students collect and/or analyze data for underlying patterns, identify and assess benefits and risks of particular decisions, and evaluate claims. Most activities relate to the unit's organizing theme and help students progress toward the **Putting It All Together** activity.

■ DEVELOPING SKILLS

B.14 TRENDS IN METAL REACTIVITY

Sample Problem: Will Pb metal react with Ag⁺ ions?

Yes; according to Table 1.3, lead is a more reactive metal than silver, so lead metal will cause silver ions to change to silver metal.

Use Table 1.3 and the periodic table (page 60) to answer these questions

1. a. What trend in metallic reactivity is found as you move from le to right across a horizontal row (period) of the periodic table?

■ MAKING DECISIONS

C.15 LIFE CYCLE OF A COIN

So far in this unit you have considered properties that ar desirable for coins and bank notes. You have also learned ab uses, and properties of metals, and have begun to explore products and materials. Now you will use this knowledge life cycle of a particular metal product, the current U.S. doll ure 1.35 (page 88) and Figure 1.49 to answer the followin

In **Modeling Matter** activities, students create and critique visual representations of matter, often making connections between what they sense at the macroscopic level and what happens at the molecular level. Students also engage in formulating and revising scientific explanations, proposing and evaluating analogies, and using physical molecular models to better understand connections between molecular structure and physical and chemical properties of substances.

Thousands of people use chemistry every day. **Chemistry at Work** interviews introduce students to some of these individuals and show how they apply chemical processes and principles in their careers.

■ MODELING MATTER

D.2 REPRESENTING REACTIONS

In Modeling Matter A.8 (page 38), you saw how chemical can be represented with formulas and pictures. As you m Section D, you will again be asked to model elements, compound tions, and translate among symbolic, particulate, and macroscop

CHEMISTRY *AT WORK*
Q&A

John Conkling, Pyrotechnic Chemist at Washington College in Chestertown, Maryland

Fireworks are one of the most striking parts of every Independence Day. But have you ever wondered how fireworks, well, work? Pyrotechnic chemists are experts on how these and other types of explosives function. They study how to make fireworks and other pyrotechnic devices safe to use and better for the environment. Have a blast reading about one pyrotechnic chemist and his work!

Q. What is pyrotechnic chemistry?

A. Pyrotechnic chemistry is the chemistry of producing heat from chemical reactions and using that heat to produce color and light and audible effects. We use pyrotechnic chemistry not only for entertainment, but also for practical purposes like emergency signaling and military applications.

Q. How did you get into this field?

A. While I was teaching undergraduate chemistry, I was approached by a fireworks company that wanted to hire me for a side project developing chemical compositions for fireworks that made them safe to transport and store. I got really interested in the chemistry of fireworks, so I wrote some articles about it. One ended up being the cover story for a chemistry magazine! Once that published, the army called and wanted to work with me on some military pyrotechnic applications, and my pyrotechnic chemistry career really shot off. Nowadays, I do training seminars for people interested in anything that explodes: from people who design and manufacture fireworks to people who dispose of bombs.

Q. How do fireworks work?

A. Every fireworks mixture needs at least one chemical that's oxygen rich and one chemical that acts as a fuel. By choosing these compounds carefully, we can determine how much heat will be produced once a firework's fuse is lit, how fast a reaction will take place, and what solid and gas products will be produced. To make fireworks more attractive, we include compounds that color the flames produced when they explode. Different elements produce different colors when they burn. For example, strontium compounds burn bright red. Barium compounds produce a green light. Sodium compounds have a yellow-orange flame.

ACS
Chemistry for Life®

Dear Learners and Educators,

It is my pleasure to welcome you to the *Chemistry in the Community* (*ChemCom*) family. First published in 1988, this innovative text introduces high school students to chemistry on a "need-to-know" basis in the context of real-world issues. *ChemCom* was developed with initial funding from the National Science Foundation and has received ongoing support from the American Chemical Society (ACS) and its publisher, W. H. Freeman.

The goals of *Chemistry in the Community* have remained constant throughout all six editions. *ChemCom* is designed to help students:

- Develop an understanding of chemistry
- Cultivate problem-solving and critical-thinking skills related to chemistry
- Apply chemistry knowledge to decision making about scientific and technological issues
- Recognize the importance of chemistry in daily life
- Understand benefits, as well as limitations, of science and technology

The writing team, led by Editor-in-Chief Angela Powers, has drawn upon the expertise of *ChemCom* teachers, insights gained through education research, and advances in new areas of chemistry in revising the text. The sixth edition of *ChemCom* features a stronger emphasis on sustainability and green chemistry, topics of increasing importance in applying scientific solutions to global challenges. This focus aligns with the mission of the American Chemical Society, which is "To advance the broader chemistry enterprise and its practitioners for the benefit of Earth and its people."

I am confident that *Chemistry in the Community* will provide you with an exceptional foundation in chemistry, one that will increase your appreciation for and understanding of the role of chemistry in everyday life, while preparing you for further study in the chemical sciences. *ChemCom* truly embodies the vision of the ACS, "Improving people's lives through the transforming power of chemistry," and I am pleased that your introduction to high school chemistry begins with *Chemistry in the Community*.

Sincerely,

Mary Kirchhoff
Education Division
American Chemical Society

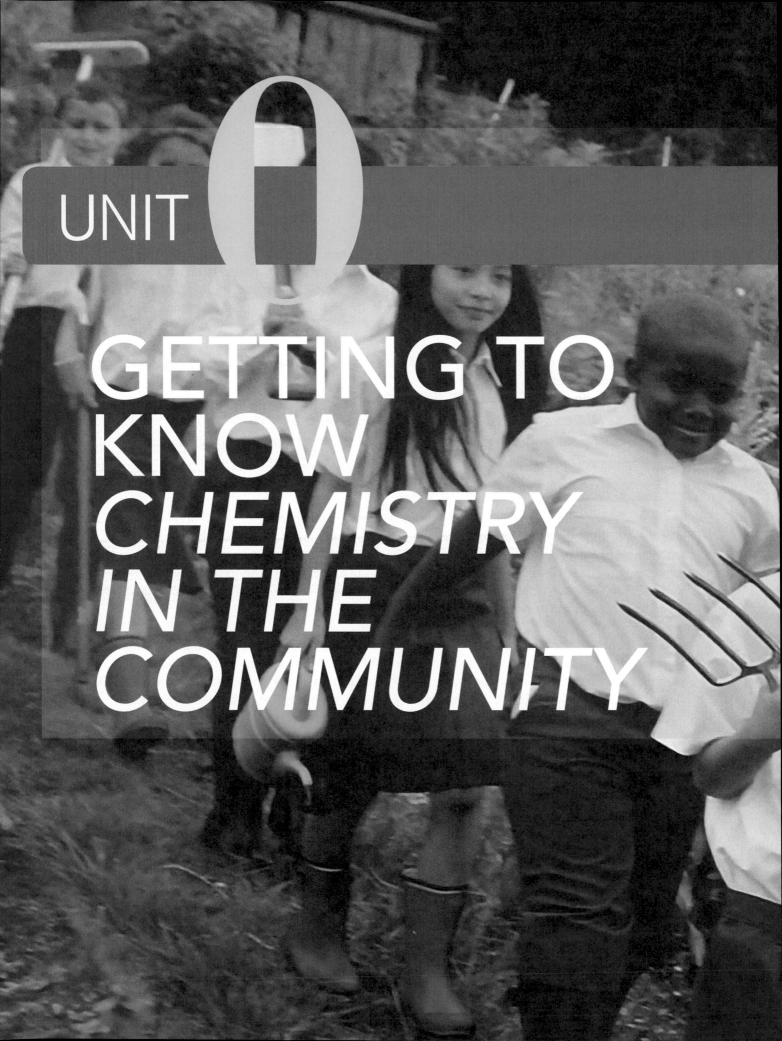

UNIT 0

GETTING TO KNOW CHEMISTRY IN THE COMMUNITY

What is chemistry?

SECTION A
The Central Science (page 4)

How do chemists investigate?

SECTION C
Inquiry and Investigation (page 8)

What is community?

SECTION B
Living Within Communities (page 7)

How is *ChemCom* designed to help you learn and apply chemistry?

SECTION D
Learning and Applying Chemistry (page 15)

?

You are about to begin the study of chemistry, which may change the way you look at the world around you, including the products and resources you use, the food you eat, and the air you breathe. What is chemistry? How is it important in your life and in your community?

Turn the page to begin your exploration of Chemistry in the Community.

SECTION A

THE CENTRAL SCIENCE
What is chemistry?

GOALS

Goals highlight important skills and concepts that you should master while studying each section of the text. (In other units, you will find goals for each section. The goals here are for all of Unit 0.) Content and activities within each section support and lead to the section goals. Goals can also help you to organize and monitor your learning as you progress through the course.

- Define chemistry and recognize its presence all around you.
- Develop a concept of community.
- Understand the roles and responsibilities required when working in a group.
- Know and apply safety guidelines in the laboratory and be able to recognize safety concerns in an investigation.
- Begin to become familiar with the structure of investigations within *ChemCom*.
- Identify characteristics and expectations of features within *ChemCom*.
- Begin to use tools and strategies to assess your learning in chemistry.

✓ concept check 1

Concept Checks will help you draw out knowledge you already have about a topic. You may be asked to refer to topics that you studied in previous courses or in previous units. At least one question in each Concept Check will ask about something you have not yet studied in this course, but about which you may have some initial ideas.

1. What motivated you to study chemistry?
2. What topics did you expect to study when you enrolled in this course?
3. How do you use chemistry in your daily life?
4. How would you define chemistry?

A.1 WHAT IN THE WORLD IS CHEMISTRY?

You are likely taking a chemistry course for the first time. Even if you have not realized it, you have been immersed in chemistry all of your life. Chemistry is the study of matter and its changes. Matter is the "stuff" all around you that makes up your home, your vehicle, what you wear—and, in fact, even you. Common examples of changes in matter include digesting food, burning fuel, making synthetic fabrics, and producing medicines. As someone wryly observed, you simply cannot ignore chemistry, because chemistry will not ignore you.

Each unit in this textbook introduces a chemistry-related concern that affects your life or your community. You will complete laboratory investigations and other activities that encourage you to apply your chemistry knowledge and skills to a particular issue or problem. You will seek solutions and weigh consequences of decisions that you and your classmates propose.

To get ready for these challenges, look at the images in Figure 0.1. Which of the images evoke the concept of chemistry as you currently understand it? Why? Discuss your thoughts with a classmate or group of classmates. Now, as a group, look at the images that you did not choose. What aspects of the object or situation actually do involve chemistry? Keep these ideas in mind as you preview the issues and related chemistry you will address in this course.

Figure 0.1 *How do these seemingly unrelated images—canyons, fabric, bacteria, people—relate to the study of chemistry?*

MAKING DECISIONS

A.2 WHY STUDY CHEMISTRY?

Making Decisions activities give you experience with real-life decision-making strategies—many related to the unit challenge. Each Making Decisions activity asks you to gather and analyze data, then propose a solution, ask further questions, or make a supported claim.

Part I:

Work with a partner or group to identify benefits of knowing chemistry in a variety of settings. Discuss each situation with your partner or group and make a list of your answers.

1. You might have already begun thinking about your life after high school. List some career options that you have considered. For each option that your group lists, explain how an understanding of chemistry would be necessary or useful.

2. Consider a situation in which you are living in a town that must decide whether a garbage incinerator should be built nearby. How would an understanding of chemistry help you make an informed decision about how to vote?

3. How is knowledge of chemistry useful to a consumer? List and explain at least 10 specific examples.

Part II:

In this part of the activity, you will talk to family members or friends who are not enrolled in chemistry. Discuss each of the questions in Part I with two or three people (at least one should not be a student) and compile their responses to share with the class.

Garbage incinerators reduce the volume of solid waste (trash) by burning it at high temperature, resulting in the formation of ash, gases, and heat.

SECTION B LIVING WITHIN COMMUNITIES

What is community?

CHEM**QUANDARY**

VISIONS OF COMMUNITY

The term *community* probably brings some images to mind. Some of those pictures may be the same for you and your classmates or similar to those on this page. Others may be uniquely your own. An in-class activity will help you and your classmates explore your ideas about community.

*A **ChemQuandary** is a puzzling, chemistry-related question or situation designed to stimulate your thinking. A ChemQuandary often results in more questions than answers and rarely has a single "correct" answer.*

SECTION C INQUIRY AND INVESTIGATION

How do chemists investigate?

One of the most important tools available to chemists is the process of inquiry. You may already be familiar with inquiry, which can be generally defined as a seeking of information through questioning. Chemists and other scientists use scientific inquiry to guide their investigations of natural phenomena.

As a student of chemistry, you will practice the skills and abilities necessary to do scientific inquiry. You will also develop your own understanding of scientific inquiry, so that you can explain how scientific knowledge develops and changes; that is, how we know what we know.

Investigations are an integral part of *Chemistry in the Community* and provide opportunities to learn and practice scientific inquiry. This initial investigation will help you become familiar with the structure of scientific inquiry in this text, as well as some of the equipment in your school's chemistry laboratory.

C.1 INVESTIGATING SAFELY

Although no human activity is completely risk-free, if you use common sense, as well as chemical sense, and follow the rules of laboratory safety, you should encounter no safety problems in the laboratory. Chemical sense is just an extension of common sense. Sensible laboratory conduct will not happen by memorizing a list of rules any more than a perfect score on a written driver's test ensures an excellent driving record. The true "driver's test" of chemical sense is your actual conduct in the laboratory.

The following safety pointers apply to all laboratory activity. For your personal safety and that of your classmates, make adherence to these guidelines second nature in the laboratory. Your teacher will point out any special safety guidelines that apply to each investigation. Two safety icons appear in your textbook. They appear at the beginning of the laboratory procedure, but apply to the entire investigation.

When you see the goggle icon, you should put on your protective goggles and continue to wear them until you are completely finished in the laboratory.

The caution icon means there are substances or procedures requiring special care. See your teacher for specific information on these cautions.

Rules of Laboratory Conduct

1. Do laboratory work only when your teacher is present. Unauthorized or unsupervised laboratory experimentation is not allowed.

2. Your concern for safety should begin even before the first laboratory investigation. Before starting any laboratory work, always read and think about the details of your investigation.

3. Know the location and procedures for use of all safety equipment in your laboratory. These should include the safety shower, eye wash, first-aid kit, fire extinguisher, fire blanket, exits (and evacuation routes), and emergency warning system.

4. Wear a laboratory coat or apron and impact/splash-proof goggles for all laboratory work. Wear closed shoes (rather than sandals or open-toed shoes), preferably constructed of leather or similar water-impervious materials, and tie back loose hair. Shorts or short skirts must not be worn. See Figure 0.2.

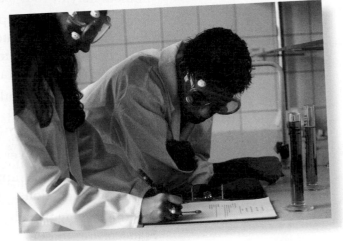

Figure 0.2 *These students are wearing appropriate clothing for laboratory work. Note that the lab bench is clear of miscellaneous clutter.*

5. Clear your bench top of all unnecessary material, such as books and clothing, before starting your work.

6. Check chemistry labels twice to ensure that you have the correct substance and the correct solution concentration. Some chemical formulas and names differ by only a letter or a number.

7. You may be asked to transfer some chemical substances from a supply bottle or jar to your own container. Do not return any excess material to its original container unless authorized by your teacher, as you may contaminate the supply bottle.

8. Avoid unnecessary movement and talk in the laboratory.

9. Never taste any laboratory materials. Do not bring gum, food, or drinks into the laboratory. Do not put fingers, pens, or pencils in your mouth while in the laboratory.

10. If you are instructed to smell something, do so by fanning some of the vapor toward your nose. Do not place your nose near the opening of the container. Your teacher will show you the correct technique.

11. Never look directly down into a test tube; view the contents from the side. Never point the open end of a test tube toward yourself or your neighbor. Never directly heat a test tube in a Bunsen burner flame.

12. Any laboratory accident, however small, should be reported immediately to your teacher.

13. In case of a chemical spill on your skin or clothing, rinse the affected area with plenty of water. If your eyes are affected, rinse with water immediately and continue for at least 10 to 15 minutes. Professional assistance must be obtained.

14. Minor skin burns should be placed under cold, running water.

15. When discarding or disposing of used materials, carefully follow all provided instructions. Waste chemical substances usually are not permitted in the sewer system.

16. Return equipment, supplies, aprons, and protective goggles to their designated locations.

17. Before leaving the laboratory, make sure that gas lines and water faucets are shut off.

18. Wash your hands before leaving the laboratory.

19. If you are uncertain or confused about proper safety procedures, ask your teacher for clarification. If in doubt, ask!

DEVELOPING SKILLS

C.2 SAFETY IN THE LABORATORY AND EVERYDAY LIFE

Developing Skills activities reinforce the skills, concepts, and processes discussed and demonstrated in the preceding section. Each question has a specific answer or set of answers. Developing Skills activities usually begin with a sample question to model expected responses.

If you understand the reasons behind them, the safety rules listed in Section C.1 will be easy to remember and to follow. To become more familiar with the safety rules, complete the following activities.

Sample Problem: Identify a rule similar to Safety Rule #1 that applies in everyday life.

In most U.S. states, inexperienced drivers must have a licensed, adult driver in the vehicle with them while they are learning to drive. This is similar to requiring an experienced, adult scientist in the laboratory while learning to investigate.

1. For Safety Rules 2–10, identify a similar rule or precaution that applies in everyday life—for example, in cooking, repairing or driving a car, or playing a sport.
2. For Safety Rules 11–19, briefly describe possible harmful consequences if the rule is not followed.
3. Look again at Figure 0.2. Which safety rules are illustrated in the image?

INVESTIGATING MATTER
C.3 DENSITY OF SOLIDS AND LIQUIDS

Asking Questions

Scientific investigations usually begin with a question to be answered through data gathering and experimentation. Sometimes this question will be provided, while other times you will be asked to develop the question with your laboratory partners or classmates.

Your teacher will demonstrate the behavior of several solids and liquids with water. Your goal for this initial investigation is to be able to determine in advance—to predict—whether solids and liquids will float or sink when placed atop water. The question for this investigation could be phrased as: How can I predict whether a solid or liquid will sink or float when I add it to water?

Preparing to Investigate

Before you begin experiments, it is important to clearly outline a procedure for gathering evidence that includes identifying the data to be collected and the steps to be followed. In some cases, a complete or partial procedure will be included in the investigation, but many times you will devise all or part of the procedure with your laboratory partners or classmates

Whether the procedure is provided or devised, you will need to study it completely before beginning. You will also need to create a system—usually a data table—for recording the observations and measurements you will make during the investigation.

Before you begin, read *Gathering Evidence* to learn what you will need to do and note safety precautions. *Gathering Evidence* also provides guidance about when you should collect and record data. Construct a data table appropriate for recording the data you will collect. In your data table, create six columns: one column will be used to list the solids and liquids tested, and the other columns will be used to record each result for dimensions, volume, mass, density, and sinking/floating behavior.

Making Predictions

In some investigations, you will predict what you think will happen as you gather evidence. These predictions should be based on your prior experience and will not be evaluated for correctness, but you may be asked to reflect upon them after the investigation.

Since you have some experience with the behavior of solids and liquids in water from everyday observations and prior science courses, make a prediction about the differences in properties that you expect to find between materials that float in water and those that sink. Write this prediction on the page containing your data table.

Gathering Evidence

Gathering Evidence is the core of the investigation. It contains directions, steps, or guidance for collecting data and observations.

Part I: Investigating Liquids

1. Before you begin, put on your goggles, and wear them properly throughout the investigation.

2. Collect individual samples of solids and liquids as instructed by your teacher.

3. Record the name of each sample in your data table. Label another row of your data table "water."

4. Find the mass of a clean, dry, 10-mL graduated cylinder to the nearest 0.1 g. Record the mass in your data table.

5. Dispense 8 to 9 mL of your first liquid sample into the graduated cylinder whose mass you just measured.

6. Find the mass of the graduated cylinder containing the liquid sample.

7. Measure and record the volume of the liquid sample to the nearest 0.1 mL. See Figure 0.3.

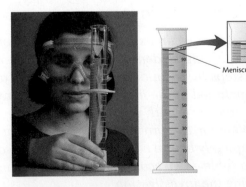

Figure 0.3 *To find the volume of liquid in a graduated cylinder, read the scale at the bottom of the curved part of the liquid (meniscus).*

8. Place ~100 mL water into a 250-mL beaker.

9. Carefully pour the contents of the graduated cylinder atop the water in the beaker. See Figure 0.4.

> The symbol "~" means "approximately" or "about."

10. Note whether the liquid floats or sinks, then dispose of the contents of the beaker as directed by your teacher.

11. Repeat Steps 4–10 for each remaining liquid sample, then repeat Steps 4–7 for water.

Part II: Investigating Solids

12. Find the mass of the first solid sample to the nearest 0.1 g. Record the mass in your data table.

13. Measure the dimensions—height, length, and width—of the solid sample in centimeters.

14. Place ~100 mL water into a 250-mL beaker.

15. Gently place the solid sample atop the water in the beaker.

16. Note whether the solid floats or sinks, then remove the solid from the beaker and dry off any excess water.

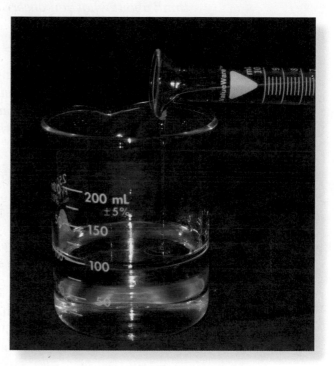

Figure 0.4 *Pouring a liquid onto water.*

17. Repeat Steps 12–16 for each remaining solid sample.

18. Return the solid samples and clean and replace all equipment as directed by your teacher.

19. Wash your hands thoroughly before leaving the laboratory.

Analyzing Evidence

The evidence gathered in some investigations requires further processing before it is useful in answering questions. Guidance is often provided to facilitate calculations and other analysis.

Recall that the formula for density is mass divided by volume.

1. Calculate the mass of each liquid sample, including water. Record the answer in your data table.

2. Calculate the volume of each solid sample by multiplying width by length by height for each sample. Record the answer in your data table. Be sure to include units.

3. Calculate the density of each sample. Record the answer in your data table. Be sure to include units.

(Note: The Interpreting Evidence, Making Claims, and Reflecting on the Investigation sections contain the results of your investigation and the conclusions you draw about the results. The questions in these sections are numbered sequentially to indicate that answers build on each other as you think about what you observed in the investigation.)

Interpreting Evidence

The next step after analyzing evidence is to ask, "What does the evidence mean?" Answering this question allows you to propose explanations for scientific phenomena. Questions within this section are designed to help you think about implications of the evidence and connect it to the purpose of the investigation.

1. What patterns do you notice in the data?

Making Claims

Once data have been analyzed and interpreted, an answer to the initial question can be proposed. This answer often comes in the form of a scientific claim. Such claims must be supported by evidence from the investigation.

2. What claim can you make about whether a solid or liquid will sink or float when you add it to water? (*Note:* Make sure that your claim can predict sinking or floating in water.)
3. What evidence from the investigation supports your answer to Question 2?
4. If a sample of olive oil floats atop a sample of vinegar, what can you conclude about their relative densities?

Reflecting on the Investigation

The final task in most investigations is to reflect on what was done, think about how your understanding has developed, and apply what was determined to other situations.

5. Consider the prediction that you made before the investigation.
   a. How does your answer to Question 2 compare to your prediction?
   b. If your prediction was accurate, explain why you were able to make an accurate prediction. If your prediction was not accurate, describe what you were thinking when you made the prediction.
6. Consider a solid object that sinks in water. If you cut the object in half, will it now sink or float? How do you know?
7. How could you predict the sinking/floating behavior of a solid sample in isopropyl (rubbing) alcohol?
8. How could you predict the sinking/floating behavior of a spherical object?

SECTION D LEARNING AND APPLYING CHEMISTRY

How is *ChemCom* designed to help you learn and apply chemistry?

Your entire life has been a journey in learning and applying new knowledge and skills. If you reflect on how you have learned, you will likely identify several people, tools, resources, and strategies that helped you, including skilled teachers, parents, books, Internet articles, hands-on activities, debates, group discussions, writing, and exploring ideas and objects. Through it all, though, *you* are the most important part of the learning process. When you are motivated to learn something, you feel pride in your efforts because you know that you own that knowledge—it is yours to use forever!

You already know some chemistry, whether it is from taking a previous science course or from your life experiences to this point. The *ChemCom* textbook is an important tool (but not your only tool) for learning more chemistry. It has been designed with you in mind—your motivations, interests, possible ideas you already have about chemistry, and activities that will engage you and help you begin thinking more like a chemist. In this section, you will learn more about how specific features within *ChemCom* are structured to help your learning.

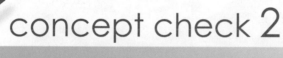

concept check 2

1. Why is it important to study chemistry?
2. What is the purpose of laboratory investigations?
3. Reflect on your past learning experiences. Describe three activities or strategies that seem to really help you learn.
4. Describe how you use textbooks to help you learn.

D.1 LEARNING ACTIVITIES IN A *CHEMCOM* UNIT

You have already encountered several of the main titled features in *ChemCom*. In Section A, you completed a Making Decisions activity. In Section C, you completed a Developing Skills activity, which required you to apply your knowledge of laboratory safety. In Section C, you also performed an Investigating Matter laboratory activity. Along the way, you completed two Concept Checks and a ChemQuandary.

Figure 0.5 uses a "club sandwich" analogy to illustrate the overall structure of each *ChemCom* unit. Examine the figure and then refer back to it as you examine a *ChemCom* unit in Developing Skills D.2.

Figure 0.5 *A club sandwich analogy for the structure of a* ChemCom *unit.*

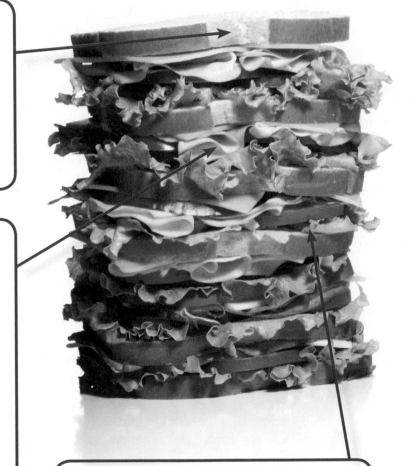

The bread: Each *ChemCom* unit is framed and driven by **chemistry-related issues or problems** embedded within community, regional, national, or global settings. The unit begins by framing the issue; the issue is revisited throughout the unit; and the issue is the basis for "putting it all together" at the end.

The meat, cheese, and veggies: Regular features that appear throughout the unit. ***Developing Skills*** activities give you practice applying chemistry ideas and problem-solving skills. ***Investigating Matter*** activities provide opportunities to interact with matter and to develop your inquiry skills. ***Modeling Matter*** activities make abstract chemical ideas easier to grasp and require you to interpret and draw visual representations of matter. ***Making Decisions*** activities give you experience with real-life decision-making strategies—many related to the unit's organizing theme. (This may make them more like the "bread" in the middle of the unit.) ***Section Summary*** questions appear at the end of each section and help you to review, connect, and extend what you have learned.

The condiments: *Goals, Concept Check* questions, *Chemistry at Work* features, and *ChemQuandary* puzzles appear less often (though very regularly) throughout the text and help hold the activities and concepts together—and make the sandwich tastier.

DEVELOPING SKILLS

D.2 EXPLORING THE STRUCTURE OF A *CHEMCOM* UNIT

The following questions will help you become more familiar with the *ChemCom* text. The first question addresses the chemistry-related challenges that drive each unit, whereas the remaining questions require you to find, describe, and use various textbook features within *ChemCom*.

Sample Problem: What is the challenge that you will address in Unit 1? What key chemistry ideas will you need to learn more about to complete this challenge? What is the final product that you will be expected to create to demonstrate your knowledge of chemistry and your solution to the problem?

To address these questions, look at the opening page of Unit 1, the Making Decisions activities at the end of each section, and the Putting It All Together at the end of the unit.

The opening Web page identifies the Unit 1 challenge—to decide whether a dollar coin is "greener" than a dollar bill and examine the many issues affecting use of the dollar coin. With your class, identify and discuss some of the key chemistry ideas that you will encounter while addressing this challenge and the final product that you will create.

Your teacher will divide your class into six groups. Each group will be assigned a unit.

1. Look at the opening page of your unit, the Making Decisions activities at the end of each section, and the Putting It All Together at the end of your unit, then answer the following questions.

    a. What is the challenge that you will address or problem you will solve in the unit?

    b. What are the key chemistry ideas that you will need to learn more about to complete the challenge?

    c. How might your solution to the challenge or problem impact your life or the lives of people around you?

    d. What is the final product that you will be expected to create to demonstrate your knowledge of chemistry and your solution to the problem?

    e. Refer back to Figure 0.5. Describe how the challenge and its components are represented by the bread in the club sandwich image.

2. Some major features within *ChemCom* are Developing Skills, Investigating Matter, Modeling Matter, and Making Decisions. Look at these activities within your assigned unit and answer the following questions:

   a. How does each of the following activities help you complete the unit challenge?

      i. Developing Skills

      ii. Investigating Matter

      iii. Modeling Matter

      iv. Making Decisions

   b. What do you think is the primary purpose of the Investigating Matter feature?

   c. How could Modeling Matter help you understand new chemistry ideas?

   d. What role do these activities play in the club sandwich analogy? Explain this in your own words.

3. The beginning of each section lists goals for concepts and skills you should develop within that section. Examine at least one of these lists in your unit. (*Note:* You are not expected to know and be able to do everything in this list before beginning the section.)

   a. How could you use the goals to help monitor your learning while you are working through a section?

   b. How could you use the goals to check your understanding once you've completed a section?

   c. How are Concept Checks and Section Summary questions related to section goals?

   d. How are Goals, Concept Checks, ChemQuandaries, Chemistry at Work features, and Section Summary questions

      i. important to the structure of a *ChemCom* unit?

      ii. important to your learning?

4. Figure 0.5 uses an analogy to illustrate the structure of a *ChemCom* unit. You will encounter several analogies in other figures and in Modeling Matter sections.

   a. What is your definition of an analogy?

   b. How are analogies useful when learning a new idea?

   c. How can analogies be confusing or incomplete?

   d. Suggest a different analogy for the structure of a *ChemCom* unit. Describe your analogy.

PUTTING IT ALL TOGETHER

WELCOME TO CHEMISTRY

A **Putting It All Together** activity concludes each of the seven units in ChemCom. In these culminating activities, you will sum up, review, and apply knowledge gained through your study of the unit. In each case, you will produce a performance and a product to communicate and defend a position on a science-related community issue.

A LETTER TO YOURSELF

Through this introductory unit, you have had the opportunity to consider what it means to be enrolled in a chemistry course using the *ChemCom* textbook. Reflect upon the activities in this unit, then compose a letter to yourself, welcoming you to this course. Be sure that your letter addresses the following questions:

- What does "success" mean for you in this course?
- What will you do in order to be successful in chemistry?
- How will you know whether you are successful in chemistry?
- How will you contribute to the overall success of your classroom community?
- How is your success important for the well-being of the larger community as you have defined it in this unit?

Your letter is a confidential communication between you and your teacher. It will be evaluated for thoughtfulness and the degree to which you address each question with the knowledge you have gained in this unit. Your answers may differ from those of your classmates.

Good luck, and welcome to chemistry!

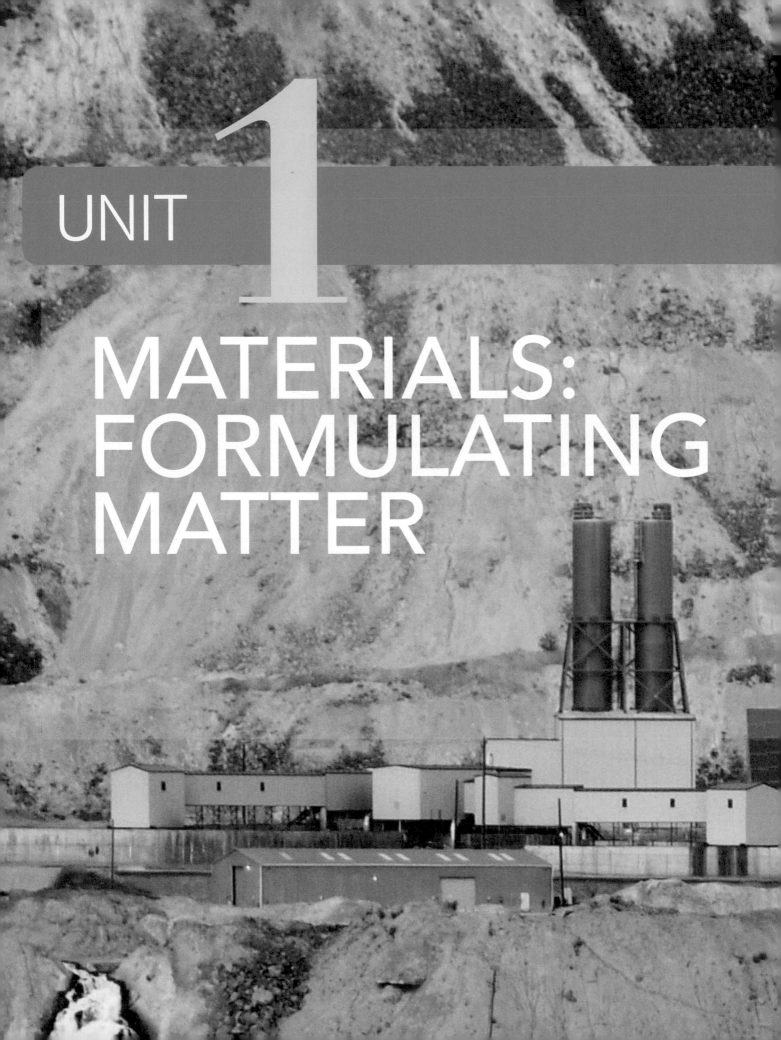

UNIT 1

MATERIALS: FORMULATING MATTER

How do chemists describe matter?

SECTION A
Building Blocks of Chemistry
(page 24)

How can the periodic table be used to help explain and predict the properties of chemical elements?

SECTION B
Periodic Trends (page 50)

What is the role of chemistry in the life cycle of metals?

SECTION C
Minerals and Moles
(page 82)

How is conservation of matter demonstrated in the use of resources?

SECTION D
Conservation and Chemical Equations (page 116)

The U.S. Mint is promoting the use of dollar coins instead of dollar bills. In what ways might coins be superior to bills? How do the properties of each form of currency depend upon their composition?
Turn the page to start learning how to answer these questions—and consider the future of the U.S. dollar.

◀ ▶
HOME   PROJECTS   LINKS

GREENER THAN GREENBACKS?

by Lakesha Harris   Riverwood, USA

The United States $1 Coin

Have You Seen the U.S. One-Dollar Coin?

Are you collecting the Presidential or Native American $1 coins? Have you actually used a $1 coin? It turns out that the U.S. Mint hopes that citizens will start using the coins in place of the dollar bill. Although dollar coins have been tried

Our neighbors in Canada use one-dollar and two-dollar coins.

before (with the Susan B. Anthony dollar in 1979 and the Sacagawea dollar in 2000) the mint has a new strategy this time— thinking green. It turns out that a dollar coin lasts about 30 years, while a dollar bill lasts (on average) just 18 months. In addition, used coins can be completely recycled into new coins, while most used currency (bills) ends up in a landfill. But will consumers use the dollar coin, even if they believe that it is more sustainable than the dollar bill?

I challenge you, as fellow citizens concerned about the planet, to tackle the following questions:

- Why are U.S. consumers reluctant to use a dollar coin?
- Is the dollar coin really greener than the dollar bill?
- How could the dollar coin be redesigned to make it
    - (a) more acceptable to consumers and
    - (b) even more earth-friendly?
- How can U.S. consumers be motivated to use dollar coins?

Look for the best submitted answers to appear on these pages soon!

Green but are they green?

What is the cost of going green?

Riding the green wave

Recycle

Nearly every choice that you make affects other people, and many choices impact Earth and our environment—even selection of the form of money that you use. How can you become well informed when making these choices? What role does chemistry play in making everyday decisions?

In this unit, you will consider the benefits and drawbacks of using a dollar coin. You will gather information on the opinions of others and evaluate claims made in the opening Web page. Your final products will include recommendations for dollar production and use, as well as strategies to convince U.S. consumers to adopt your choice of **currency**.

As you consider the pros and cons of coins and bills, you will learn about Earth's mineral resources and how nations use those resources. You will learn why certain materials are used for particular new products, including currency, and how those materials are developed from available resources. Your decisions and recommendations will be guided by chemistry concepts and knowledge you will learn through this unit, including chemical and physical properties of substances, relationships among structure and properties of materials, and how to account for atoms as they are transferred from one substance to another. In order to successfully communicate your conclusions, you will learn how to use representations of atoms and molecules to describe the chemical composition of your chosen materials. Throughout this unit, keep in mind how such chemical knowledge can help guide your decision making.

Currency refers to circulating money. In this unit we will use it to include both coins and bills (banknotes).

SECTION A BUILDING BLOCKS OF CHEMISTRY

How do chemists describe matter?

Every human-produced object, old or new, is made of materials selected for their specific properties. What makes a particular material best for a particular use? You can begin to answer this question by exploring some properties of materials.

In this unit, you will be considering the design of something that you use every day—money. Throughout history, people have used many different items as money: beads, stones, printed paper, and precious metals, to name a few (see Figure 1.1). What characteristics make a material suitable (useful) as money? How important is appearance? Cost? Sustainability? What other characteristics or properties can you suggest?

As you examine designs and properties of currency, you will gather information as scientists do in order to consider properties that make materials appropriate for specific applications. You will also begin to describe matter using a language shared by chemists and other scientists around the world. Think about the ways in which chemists describe matter as you investigate the money that you use and make recommendations for the currency of the future.

Figure 1.1 *What properties should be considered when designing currency?*

GOALS

- Make predictions and observations of chemical and physical changes. Record observations in organized data tables.

- Distinguish between chemical and physical properties and between chemical and physical changes.

- Classify specific examples as either chemical or physical properties. Classify specific examples as either chemical or physical changes.

- Recognize chemical symbols and formulas that represent elements and compounds. Use chemical symbols and formulas to describe the composition of materials.

- Interpret and create models that represent elements and compounds at the particulate level.

- Classify selected elements as metals, nonmetals, or metalloids based on observations of chemical and physical properties.

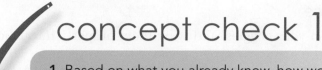

concept check 1

1. Based on what you already know, how would you describe *matter*?
2. Read the Section A Goals. You may already be familiar with the concepts and terminology in some of the goals; others might be less familiar to you. (Don't worry—you're not expected to be able to achieve the goals at this point.)
   a. For those concepts or terms that seem familiar to you, write down what you think they mean at this point. Give a specific example to illustrate your understanding.
   b. Identify concepts and terms that you do not know. You will learn more about these ideas in Section A.

■ INVESTIGATING MATTER

A.1 EXPLORING PROPERTIES OF MATTER

Preparing to Investigate

One of the most important decisions that designers of coins and bills must make is the choice of material (or materials) that will make up the currency. In order to select or evaluate currency materials, they must investigate and predict properties of many different materials. In this investigation, you will predict and observe properties and changes of several materials (most of which will already be familiar to you). Focus on making careful observations; even if you have not practiced this skill in the laboratory before, you have already developed a sense for properties of some materials and have encountered hundreds of chemical reactions in your daily life. You will refer back to the observations you make within this investigation as you develop your understanding of chemistry ideas throughout this unit.

Before you begin, read *Gathering Evidence* to learn what you will need to do and note safety precautions. *Gathering Evidence* also provides guidance about when you should collect and record data.

Making Predictions

Predict what you think will happen in each of the activities, and write down your predictions. A sample data table is provided here.

DATA TABLE		
Investigation number	Predictions	Observations
1		
2		

Gathering Evidence

Before you begin, put on your goggles, and wear them properly throughout the investigation.

Six stations, A through F, have been set up around the laboratory. At each station, you will complete the investigations indicated for that station. The stations can be completed in any order; that is, work at Station D can be completed before Station B activities, and so on. When working at a particular station, you must complete the investigations at that station in order. Follow these general instructions:

- Take note of what the station looks like and how it is set up. You will be expected to reset it after you are finished.

- Reread the procedure and safety reminders.

- Review your predictions.

- Complete the investigation.

  ○ Record your complete data and observations. Quantitative **data** may include masses or volumes measured using tools such as a balance or graduated cylinder. **Observations** refer to data you can collect using your senses. Thus record what you see, hear, feel, or smell. (**Caution:** *Never taste anything in the laboratory.*)

  ○ At this point, do not try to go beyond observations to infer what is happening. For instance, suppose you add two liquids together and a solid forms. Your observation would be, "When I poured liquid A into liquid B, it became cloudy. After a few minutes, I saw some white powder on the bottom of the test tube, with clear, colorless liquid on top." Later you may decide that the observation is evidence that a chemical reaction occurred; however, your conclusion about the reaction is an **inference** about what happened, not an observation. While you are engaged in the investigations, simply focus on making accurate, detailed observations.

- Restore the station to its original condition.

> An *inference* is a conclusion based on analysis of data and observations.

When you have completed your work at all six stations, answer the questions that follow Investigation 13 to analyze and interpret evidence, and reflect upon the investigation.

Station A: Paper

Investigation 1

1. Tear a small piece of paper into smaller pieces and place the pieces on a watch glass.
2. Record your observations.

Investigation 2

1. Place the watch glass and pieces of paper on a heat-resistant ceramic pad.

2. Light the pieces of paper with a match and allow them to burn completely.

3. Record your observations.

4. Discard the burned paper as directed by your teacher, clean the watch glass, and reset the station.

Station B: Solutions

Investigation 3

1. Place two clean, dry test tubes into a test tube rack.

2. Dispense one drop of Universal Indicator into one of the clean, dry test tubes.

3. Add 10 drops of ammonia solution to the test tube containing the Universal Indicator.

4. Record your observations. Keep the test tube and its contents for Investigation 4.

5. Dispense one drop of Universal Indicator into a second clean, dry test tube.

6. Add 10 drops of vinegar to the second test tube.

7. Record your observations. Keep the test tube and its contents for Investigation 4.

Investigation 4

1. Carefully pour the contents of the test tube containing ammonia solution from Investigation 3 into the test tube containing vinegar from Investigation 3.

2. Touch the outside of the test tube at the level of the combined solutions.

3. Record your complete observations.

4. Discard the test tube contents as directed by your teacher, clean both test tubes, and reset the station.

Station C: Blue Crystals

Investigation 5

1. Use a spatula to place a blue crystal into the mortar. See Figure 1.2.

2. Use the pestle to grind the blue crystal.

3. Record your observations.

4. Put the powder into a crucible to use for Investigation 6.

5. Clean and reset the station as instructed by your teacher.

Figure 1.2 *A blue crystal is transferred into a mortar. Note the pestle, which will be used to grind the crystal in Step 2.*

Investigation 6

1. Turn on the hot plate to a setting of "high." (**Caution:** *Do not touch the hot plate surface. It may already be hot.*)

2. Measure and record the mass of the crucible and blue powder from Investigation 5.

3. Place the crucible on the hot plate and heat for 3 minutes, stirring gently with the glass stirring rod. Use tongs to hold the crucible as you stir. See Figure 1.3.

The appearance of a hot plate surface does not change when it is hot. Always exercise caution when using hot plates.

Figure 1.3 *Heating and stirring blue powder in Investigation 6.*

4. Use tongs to remove the crucible from the hot plate and allow it to cool for several minutes.

5. Measure and record the mass of the crucible and its contents.

6. Record your data and observations.

7. Put the powder into a waste container as directed by your teacher.

8. Turn off the hot plate and reset the station.

Station D: Tea lights

Investigation 7

1. Turn on the hot plate to a setting of "high." (**Caution:** *Do not touch the hot plate surface. It may already be hot.*)

2. Use tongs to place the tea light (in the metal holder, without a wick) on the hot plate.

3. Observe for three to five minutes.

4. Use tongs to carefully remove the tea light and place it on a heat-resistant ceramic pad.

5. Turn off the hot plate.

6. Record your observations.

Investigation 8

1. Light the wick on the second tea light with a match.

2. Observe for three to five minutes.

3. Carefully extinguish the flame.

4. Record your observations.

5. Reset the station.

Station E: Baking soda

Investigation 9

1. Place a small amount of solid baking soda into a clean well in a 24-well plate (see Figure 1.4).
2. Add 5 drops of vinegar to the baking soda in the well.
3. Make careful observations and record your results.

Investigation 10

1. Dispense 10 drops of baking soda solution from a dropper bottle or pipet into a clean well.
2. Add 5 drops of vinegar to the baking soda solution in the well.
3. Make careful observations and record your results.

Investigation 11

1. Dispense 10 drops of baking soda solution from a dropper bottle or pipet into a clean well.
2. Add 1 drop of silver nitrate solution to the baking soda solution in the well. **(Caution:** *Silver nitrate solution can stain your skin or clothing. Handle with care.)*
3. Make careful observations and record your results.
4. Discard the well plate contents as directed by your teacher, clean the well plate, and reset the station.

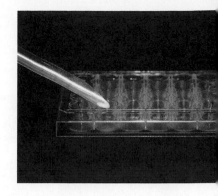

Figure 1.4 *A spatula is used to transfer baking soda into a 24-well plate in Investigation 9.*

Station F: Metals

Investigation 12

1. Count and use 10 post-1982 pennies to make the measurements that follow.
2. Make sure the 10 pennies are completely dry. Then measure and record the mass of the 10-penny sample.
3. Pour ~50 mL of water into a 100-mL graduated cylinder. Accurately measure and record the volume of water in the cylinder. See Figure 1.5, which demonstrates how to use a graduated cylinder to measure volume.
4. Carefully place your 10 pennies into the graduated cylinder. Accurately measure and record the volume of water (plus objects).
5. Remove the pennies from the cylinder and dry them.
6. Repeat steps 1–5 using a different set of metal objects (10 nails, 10 pre-1982 pennies, or other available metal samples).
7. When you have completed this procedure for at least two metal samples, reset the station.

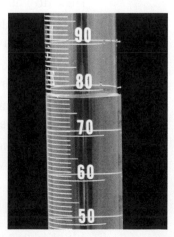

Figure 1.5 *To find the volume of a liquid in a graduated cylinder, read the scale at the bottom of the curved part of the liquid (meniscus).*

Investigation 13

1. Fill a clean test tube to a height of 2 to 3 cm with blue solution.

2. Use forceps to carefully place an iron nail into the test tube so that it is partially, but not totally, immersed in the blue solution.

3. Record your observations over several minutes.

4. Remove the nail from the solution and record any additional observations you have.

5. Dispose of the solution and the nail as directed by your teacher, clean the test tube, and reset the station.

Analyzing Evidence

1. In which investigations did you collect quantitative data?

> Recall that *density* refers to the mass of a material within a given volume.

2. The **density** of solid objects is often reported using units of g/cm³. A cubic centimeter (cm³) is equal to 1 milliliter (mL). See Figure 1.6. How can you use your data from Investigation 12 to determine the density of post-1982 pennies?

   a. Calculate the density of post-1982 pennies.

   b. Calculate the density of the other metal object(s) you tested in Investigation 12.

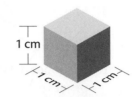

Figure 1.6 *One cubic centimeter (shown actual size). 1 cm³ = 1 mL.*

Interpreting Evidence and Making Claims

1. Did either treatment of the paper at Station A create a new material? How do your observations support your thinking?

2. When you added the ammonia and vinegar solutions together in Investigation 4, did you form a new substance? How do you know?

3. Did either treatment of the blue crystals at Station C create a new material? Why do you think that?

4. Was the burning tea light different from the melting tea light at Station D? Describe the observations that led to your conclusion.

5. Did any of the investigations of baking soda at Station E create a new material? How do your observations support your thinking?

Reflecting on the Investigation

6. Identify one investigation in which your observations closely matched your predictions. Why do you think your prediction was so accurate?

7. Identify one investigation in which your observations were different from your predictions.

   a. Describe what you were thinking when you made the prediction.

   b. Write one question you have about the investigation now that your observations did not match your prediction.

8. In your own words,
    a. describe what it means to make observations during investigations in the laboratory.
    b. identify types of evidence you will look for as you make observations in future investigations.

A.2 PROPERTIES MAKE THE DIFFERENCE

Every substance has characteristic properties that distinguish it from other substances, thus allowing it to be identified. These characteristic properties include **physical properties** such as color, density, and odor—properties that can be determined without altering the chemical makeup of the material. Physical properties and the ability (or inability) of a material to undergo physical changes, such as melting, boiling, and bending, influence whether and how a material is used. In a **physical change**, the material remains the same, even though its form appears to have changed. A simple example of a physical change is the tearing of paper in Investigation 1 of Investigating Matter A.1. Although the paper has a different appearance after tearing, it is clearly the same material.

When a substance changes into one or more *new* substances, it has undergone a **chemical change**. A substance's **chemical properties**, which relate to any kind of chemical changes it undergoes, often determine its usefulness. Consider the common chemical change of iron rusting. The tendency of a metal to combine with oxygen, such as when iron rusts, is the chemical property that accounts for this chemical change. You can often detect a chemical change by observing one or more indications of a change, such as the formation of a gas or solid, a permanent color change, or a temperature change, which indicates that thermal energy has been absorbed or given off. For instance, you observed the formation of bubbles in Investigation 9 of Investigating Matter A.1. This is a good indication that mixing vinegar and baking soda results in a chemical change. Figure 1.7 illustrates some physical and chemical changes of copper.

Both chemical and physical properties are important when choosing materials for particular applications, such as coin-making. In the following activity, you will classify some characteristics of common materials as either physical or chemical properties.

> Be careful to avoid thinking that changes such as formation of a gas or solid, color change, or temperature change can ONLY indicate a chemical change. Some of these events can also be observed during a physical change. As an example, think of what you would observe as a container of water boils.

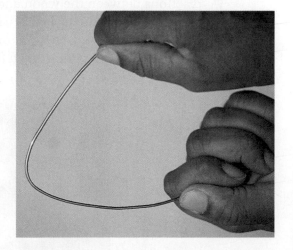

Figure 1.7 *Examples of physical and chemical changes involving copper. Bending does not alter the chemical identity of the copper (above), but allowing the copper to react with nitric acid (right) does. What evidence can you find in the photos to support these conclusions?*

DEVELOPING SKILLS

A.3 PHYSICAL AND CHEMICAL PROPERTIES

The blue crystals you investigated in Station C of Investigating Matter A.1 contain copper.

Sample Problem 1: Consider this statement: Substances containing copper are often blue in color. Does the statement describe a physical or chemical property?

To answer this, first ask yourself a question: Was the substance chemically changed for its color to be observed? If the answer is *no*, then the statement describes a physical property; if the answer is *yes*, then the statement describes a chemical property. You can observe this property—color—without changing the chemical makeup of the copper-containing substance. Color is a characteristic physical property of many substances.

Sample Problem 2: Consider this statement: Oxygen gas supports the burning of wood. Does the statement refer to a physical or chemical property of oxygen gas?

If you apply the same key question—is there a change in the identity of the wood and the oxygen?—you will arrive at the correct answer. As you have no doubt noted, the ash left over from a campfire looks nothing like the original wood that was burned. In fact, the burning—or **combustion**—of wood involves chemical reactions between the wood and oxygen that change both materials. The reaction products of ash, carbon dioxide, and water vapor are very different from wood and oxygen. Thus, the statement refers to a chemical property of oxygen (as well as of wood).

Wood burning

Now it's your turn. Classify each statement as describing either a physical property or a chemical property. (*Hint:* Decide whether the chemical identity of the material does or does not change when the property is observed.)

1. Pure metals have a high **luster** (are shiny and reflect light).

2. The surfaces of some metals become dull when exposed to air.

3. Nitrogen gas, a relatively nonreactive material at room temperature, can form nitrogen oxides at the high temperatures of an operating automobile engine.

Metallic luster

4. Milk turns sour if left too long at room temperature.

5. Diamonds are hard enough to be used as a coating for drill bits.

6. Metals are typically **ductile** (can be drawn into wires).

7. Leavened bread dough increases in volume if it is allowed to "rise" before baking.

8. Unreactive argon gas, rather than air, is used to fill many light bulbs to prevent the metal filament wire inside the bulb from being destroyed through chemical reactions.

9. Generally, metals are better conductors of heat and electricity than are nonmetals.

Bread rising

Sample Problem 3: *Consider the activities you completed in Investigating Matter A.1. Write a statement similar to those in Questions 1–9 based on one of your observations. Then identify whether your statement describes a physical or chemical property.*

It may be easiest to first identify a particular observation that you think illustrates either a chemical or physical property. Then write the statement. For instance, you might be relatively certain that burning paper in Investigation 2 resulted in a chemical change. You could then write "Paper turns black and curls up when burned" as your statement and identify it as a chemical change.

10. For each of the following settings or situations, write a statement similar to those in Questions 1–9 and then identify whether your statement describes a physical or chemical property.

    a. One of your observations in Investigating Matter A.1. (Do not refer to Investigation 2, which was used in Sample Problem 3.)

    b. Your everyday experiences.

    c. One desirable property of a dollar coin.

A.4 PROPERTIES MATTER: DESIGNING THE PENNY

As you might imagine, you need to consider many factors when selecting materials for a specific use. A material with properties well suited to a purpose may be either unavailable in sufficient quantity or too expensive. Alternatively, a material may have undesirable physical or chemical properties that limit its use. In these and other situations, you can often find another material with most of the sought-after properties and use it instead.

The cost of a material is an issue when manufacturing coins and paper money, for example. Just imagine what would happen if the declared value of a coin were less than the cost of its component metals. How would this affect the production and circulation of the coin? This situation nearly occurred in the United States about one-quarter century ago. In the early 1980s, copper became too expensive to be used as the primary metal in pennies. In other words, the cost of the copper composing a penny was becoming just about as great as the face value of the penny. Zinc, another metallic element, was chosen to replace most of the copper in all post-1982 pennies. Zinc is about as hard as copper and has a density (7.14 g/cm^3) quite close to the density of copper metal (8.94 g/cm^3). Zinc is also readily available and is less expensive than copper.

Unfortunately, zinc is more chemically reactive than copper. During World War II, copper metal was in short supply. To conserve that resource, zinc-plated steel pennies—known to coin collectors as "white cents" or "steel cents"—were created in 1943. The new pennies quickly corroded. As you can see in Figure 1.8, these pennies also looked considerably different from traditional copper pennies. Production of zinc-plated pennies ended within a year.

Figure 1.8 *Zinc-copper and copper pennies (top); "new" and corroded zinc-plated steel pennies (bottom).*

The problems associated with using zinc in pennies were solved in the early 1980s. In the new fabrication, the properties of copper were used where they were most needed—on the coin's surface—and the properties of zinc, where they were useful—within the coin's body. Current pennies are composed of a zinc core surrounded by a thin layer of copper metal, added to increase the coin's durability and maintain its familiar appearance. Figure 1.9 shows a cross-section of a post-1982 penny.

The story of the penny highlights some important physical and chemical properties that will be important for you to consider as you evaluate the use of dollar coins. To be able to identify the benefits and drawbacks of different forms of currency, you need to learn more about the building blocks of all materials—the atoms of the chemical elements. As you do so, you will also begin to learn and practice a new "language," using symbols and chemical formulas to represent the elements and compounds that make up all materials.

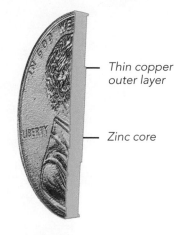

Thin copper outer layer

Zinc core

Figure 1.9 *Cross-section showing the structure and composition of a post-1982 penny.*

concept check 2

You determined the density of post-1982 pennies in Investigating Matter A.1.
1. Compare your experimentally determined coin density to that of pure copper (8.94 g/cm^3) and pure zinc (7.14 g/cm^3).
2. Based on this comparison, estimate the relative quantities of zinc and copper in post-1982 pennies. (For instance, do you think pennies are composed of equal quantities of zinc and copper?)
3. Briefly describe your reasoning, or show mathematically how you arrived at your estimation.

A.5 THE PARTICULATE VIEW OF MATTER

So far in this investigation of materials, you have focused on properties that are observable with your senses and simple instruments. Have you wondered why metals are shiny or why mixing vinegar and baking soda solutions produces bubbles? To understand why materials have specific properties, you must investigate them at the **particulate level**—that is, at the level of their atoms and molecules.

The solid materials and solutions you encountered in Investigating Matter A.1, and indeed all materials around you, are examples of matter. As you may recall from previous science courses, **matter** is anything that occupies space and has mass. All matter is composed of atoms. **Atoms** are often called the building blocks of matter. Matter that is made up of only one kind of atom is known as an **element**. For example, oxygen is considered an element because it is composed of only oxygen atoms. Because hydrogen gas contains no atoms other than hydrogen atoms, it too is an element. Approximately 90 different elements are found in nature, each having its own type of atom and identifying properties.

An *element* cannot be broken down into any simpler substances.

Consider one of the most common examples of matter—the water that you drink everyday. What type of matter is a sample of liquid water? Is it an element? As you probably know already, water contains atoms of two elements—oxygen and hydrogen. Thus, water is not an element. Instead, water is an example of a **compound**—a substance composed of atoms of two or more elements linked together chemically in certain fixed proportions. To date, chemists have identified more than 30 million compounds.

Compounds and elements are represented by chemical formulas. Although you will learn more about chemical symbols and formulas in the next section, you are probably familiar with some already. In addition to water (H_2O), some other compounds and formulas with which you may be familiar include table salt (NaCl), ammonia (NH_3), baking soda ($NaHCO_3$), chalk ($CaCO_3$), table sugar ($C_{12}H_{22}O_{11}$), and octane (C_8H_{18}).

Each element and compound is considered a **substance** because each has a uniform and definite composition as well as distinct properties. The smallest unit of a molecular compound is a **molecule**, a collection of atoms that move and act together as a single entity. Atoms of a molecule are held together by **chemical bonds**. You can think of chemical bonds as the "glue" that holds atoms of a molecule together. Oxygen is an example of an *element* typically found in molecular form—two oxygen atoms bonded to one another—whereas water is a molecular *compound*. One molecule of water is composed of two hydrogen atoms bonded to one oxygen atom, hence H_2O. An ammonia molecule (NH_3) contains three hydrogen atoms bonded to a nitrogen atom. Figure 1.10 shows representations of some atoms and molecules.

> You will learn more about how chemical bonds form in Unit 3.

Figure 1.10 *Top row: hydrogen (H), oxygen (O), and nitrogen (N) atoms. Bottom row: hydrogen (H_2), water (H_2O), and ammonia (NH_3) molecules. Models similar to these are used throughout this book to depict atoms and molecules.*

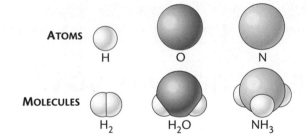

A.6 SYMBOLS, FORMULAS, AND EQUATIONS

An international "chemical language" for use in oral and written communication has been developed to represent atoms, elements, and compounds. The "letters" in this language's alphabet are **chemical symbols**, which are used and understood by scientists throughout the world. Each element is assigned a chemical symbol. Only the first letter of the symbol is capitalized; all other letters are lowercase. For example, C is the symbol for the element carbon and Ca is the symbol for the element calcium. Symbols for some common elements are listed in Table 1.1 (page 38).

"Words" in the language of chemistry are composed of "letters" (symbols representing elements). Each "word" is a **chemical formula**, which represents a different chemical compound. In the chemical formula of a compound, a chemical symbol represents each element present. A **subscript** (a small number written below the normal line of letters) indicates how many atoms of the element just to the left of the subscript are in one unit of the compound.

For example, the ammonia solution you used in Investigating Matter A.1 contains ammonia, whose chemical formula is NH_3. See Figure 1.11. The subscript 3 indicates that each ammonia molecule contains three hydrogen atoms. Each ammonia molecule also contains one nitrogen atom. However, the subscript 1 is understood in the absence of any subscript and is therefore not included in chemical formulas.

If formulas are words in the language of chemistry, then **chemical equations** can be regarded as chemical "sentences." Each chemical equation summarizes the details of a particular chemical reaction. **Chemical reactions** entail the breaking and forming of chemical bonds, causing atoms to become rearranged into new substances. These new substances have different properties from those of the original materials.

Figure 1.11 *Nitrogen is added to crops as ammonia fertilizer.*

In Investigating Matter A.1, you used a solution of household ammonia, which is made by dissolving gaseous ammonia, NH_3, in water. Gaseous ammonia is produced by the reaction of hydrogen gas with nitrogen gas. This reaction can be represented by the following chemical equation:

$$3\ H_2 \quad + \quad N_2 \quad \longrightarrow \quad 2\ NH_3$$

Hydrogen gas   Nitrogen gas   Ammonia

Reactants   Product

The equation shows that three hydrogen molecules (H_2) and one nitrogen molecule (N_2) react to produce ($\longrightarrow$) two molecules of ammonia (NH_3). The original (starting) substances in a chemical reaction are called **reactants**; their formulas are always written on the left side of the arrow. The new substance or substances formed from the rearrangement of the reactant atoms are called **products**; their formulas are always written on the right side of the arrow. Note that this equation, like all chemical equations, is **balanced**—the total number of each type of atom (six H atoms and two N atoms) is the same for both reactants and products.

■ DEVELOPING SKILLS

A.7 CHEMICAL SYMBOLS AND FORMULAS

> *Sample Problem:* *The chemical formula for propane, a compound commonly used as a fuel, is C_3H_8. What elements are present in propane, and how many atoms of each are there in one molecule of propane?*
>
> You are correct if you said each propane molecule consists of *three* carbon atoms and *eight* hydrogen atoms.

Table 1.1

Common Elements	
Name	**Symbol**
Aluminum	Al
Bromine	Br
Calcium	Ca
Carbon	C
Chlorine	Cl
Cobalt	Co
Copper	Cu
Gold	Au
Hydrogen	H
Iodine	I
Iron	Fe
Lead	Pb
Magnesium	Mg
Mercury	Hg
Nickel	Ni
Nitrogen	N
Oxygen	O
Phosphorus	P
Potassium	K
Silver	Ag
Sodium	Na
Sulfur	S
Tin	Sn

1. Using Table 1.1, name the element represented by each of these symbols.

   a. P   d. Co   g. Na

   b. Ni   e. Br   h. Fe

   c. Cu   f. K

2. Which elements in Question 1 have symbols corresponding to their English names?

3. Which is more likely to be the same throughout the world—an element's symbol or its name? Explain.

4. For each formula, name the elements and give the number of atoms of each element in each compound.

   a. H_2O_2 Hydrogen peroxide   Antiseptic

   b. $CaCl_2$ Calcium chloride   Winter deicer for sidewalks

   c. $NaHCO_3$ Sodium hydrogen carbonate   Baking soda

   d. H_2SO_4 Sulfuric acid   Battery acid

5. Translate this written description into a chemical equation: *Two molecules of hydrogen (H_2) react with one molecule of oxygen (O_2) to form two molecules of water (H_2O).*

■ MODELING MATTER

A.8 PICTURES IN THE MIND

You live in a **macroscopic** world—a world filled with large-scale ("macro"), readily observed things. As you experience the properties and behavior of bulk materials, you probably give little thought to the particulate world of atoms and molecules. If you wrap leftover cake in aluminum foil, it is unlikely that you think about how the individual aluminum atoms are arranged in the wrapping material (see Figure 1.12). It is also unlikely that you consider what the mixture of molecules making up air looks like as you breathe. And you probably seldom wonder about atomic and molecular behavior when you observe water boiling or a rusted iron nail.

Nevertheless, having a sense of how atoms and molecules might look in elements and compounds and behave in reactions is a useful tool for understanding the chemistry that is occurring. To develop this sense, it is useful to construct and evaluate **models**, or representations, of atoms and molecules. This activity will give you practice in observing, interpreting, evaluating, and creating visual models of matter, "pictures in the mind," at the particulate (atomic and molecular) level. You will also practice using and interpreting chemical symbols and formulas that make up the language of chemistry.

Figure 1.12 *One common use of aluminum foil is to wrap food.*

Use the following key to draw and interpret models of elements and compounds containing hydrogen, carbon, nitrogen, oxygen, and chlorine:

Sample Problem 1: *Draw a particulate level model of carbon dioxide (CO_2). (Hint: Carbon is in the center of the molecule.)*

The chemical formula CO_2 shows that carbon dioxide contains one carbon atom and two oxygen atoms. We're told that carbon is in the middle, so we'll draw a model that looks like this:

Sample Problem 2: *Write the chemical formula for the element represented by this particulate-level model:*

Using the key above, we identify the atoms as nitrogen. Two nitrogen atoms make up the molecule, so the chemical formula for elemental nitrogen must be N_2.

Now it's your turn to create and interpret particulate-level models of elements and compounds.

1. In Investigating Matter A.1, you mixed ammonia solution and vinegar solution.

    a. Draw a molecular-level model of ammonia, which has chemical formula NH_3.

    b. The substance in vinegar that reacted with the ammonia is called acetic acid. Below is a molecular-level model of acetic acid. Write its chemical formula. (*Hint:* Write the symbol for carbon first.)

2. Draw molecular-level models of these compounds:

    a. Methane (CH_4): The primary component of natural gas. (*Hint:* Carbon is in the center of the molecule.)

    b. Water (H_2O): The most abundant substance on Earth. (*Hint:* Oxygen is in the center of the molecule.)

    c. Elemental chlorine (Cl_2): Poisonous gas used to make plastics and disinfect water.

    d. Hydrogen chloride (HCl): Important industrial chemical also sometimes used in pure form to etch semiconductor crystals.

    e. Carbon monoxide (CO): Toxic gas used in many industrial processes, including purifying nickel.

3. You melted and burned paraffin wax in Investigating Matter A.1 (page 28). Write the chemical formula of paraffin wax given its model below. (*Note:* The carbon and hydrogen atoms are smaller than in the key so that this molecule can fit on the page.)

So far, none of the models you have drawn or interpreted are of metals. How can you visualize solid metals? Figure 1.12 (page 39) contains a photograph of aluminum foil. Although we use the chemical symbol "Al" to represent aluminum, what we visualize is actually a large collection of aluminum atoms, like this:

This model, like all models, has some limitations. It is really only showing one small segment of the aluminum foil; the atoms extend outward from what is shown. The main point, though, is that we still only use a single symbol, Al, without any subscripts, to represent an entire collection of aluminum atoms. Think about how that is different from the molecular compounds you worked with in Questions 1–3.

4. In Section A.4, you learned about the composition of current pennies. Re-examine Figure 1.9 (page 35), which shows a cross-section of a penny.

    a. What chemical symbols represent copper and zinc?

    b. Zero in on a small area of the cross-section in Figure 1.9. Draw a particulate-level model of what you think the copper and zinc atoms look like in that small area.

    c. Describe in words how you decided to draw your model.

5. In chemistry, we often try to relate the atomic and molecular structure of materials to their properties. One property of metals is that they can be pounded into thin sheets.

    a. Consider the particulate-level model of aluminum foil and other metals. Does this model help you to explain why metals can be pounded into thin sheets?

    b. Describe your reasoning.

Investigating Matter A.1 provided you with opportunities to observe matter and its changes on a large scale. In fact, you observe every phenomenon you encounter using your senses of sight, touch, smell, and hearing. One of the most important ways of thinking like a chemist is to take in what you observe and visualize what is happening at a level you cannot see—the level of atoms and molecules. Another skill is using chemical symbols and formulas to represent matter and its changes. As you continue to read, challenge yourself to visualize what the atoms and molecules might look like when you encounter a symbol or formula. Do the same when you look at a photograph. While these ways of thinking are often taken for granted by experienced chemists, they are difficult to develop. You will have opportunities to practice these skills throughout this course, but you will become even more skilled if you challenge yourself to think this way any time you are making an observation in everyday life.

> You may also experience some phenomena using your sense of taste—but never in the laboratory!

A.9 THE ELEMENTS

You know that all matter is composed of atoms. One element differs from another because its atoms have properties that differ from those of other elements (see Figure 1.13, page 42). More than 100 chemical elements are now known. Table 1.1 (page 38) lists some common elements and their symbols. An alphabetical list of all elements (names and symbols) can be found inside the back cover, or on pages 96–97.

Figure 1.13 *Each of these elements (clockwise from top right: sulfur, antimony, iodine, phosphorus, copper, and bismuth) is composed of chemically identical atoms. Which elements appear metallic? Which appear nonmetallic?*

Elements can be classified in several ways according to similarities and differences in their properties. Two major classes are metals and nonmetals. **Metals** include such elements as iron (Fe), tin (Sn), zinc (Zn), and copper (Cu). Carbon (C) and oxygen (O) are examples of **nonmetals**. Everyday experience has given you some awareness of metallic and nonmetallic properties. The upcoming investigation will let you further explore the properties of metals and nonmetals.

Several elements called **metalloids** have properties that are intermediate to those of metals and nonmetals. That is, they exhibit both metallic and nonmetallic properties. Examples of metalloids include silicon (Si) and germanium (Ge), both commonly used in the computer industry.

What properties of matter can we use to distinguish metals, nonmetals, and metalloids? The next activity will help you find out.

■ INVESTIGATING MATTER

A.10 METAL OR NONMETAL

Preparing to Investigate

In this investigation, you will explore several properties of seven elements and then decide whether each element is a metal, nonmetal, or metalloid. You will examine the color, luster, and form of each element, and you will also attempt to crush each sample with a hammer. In addition, you or your teacher (as a demonstration) will test each substance's ability to conduct electricity. Finally, you will determine the reactivity of each element with two solutions: hydrochloric acid, HCl*(aq)*, and copper(II) chloride, $CuCl_2(aq)$.

Before you begin, read *Gathering Evidence* to learn what you will need to do and note safety precautions. *Gathering Evidence* also provides guidance about when you should collect and record data. Construct a data table appropriate for recording the data you will collect. In your data table, create six columns: one column will be used to list the elements tested, and the other five columns will be used to record each result for appearance, conductivity, crushing, reactivity with copper(II) chloride solution, and reactivity with hydrochloric acid (HCl).

> The symbol *(aq)* means that the substance is dissolved in water—thus it indicates an aqueous solution.

Gathering Evidence

1. Before you begin, put on your goggles, and wear them properly throughout the investigation.

2. *Appearance:* Observe and record the appearance of each element, including physical properties such as color, luster, and form. You can record the form as nonmetallic (like table salt, NaCl, or baking soda, $NaHCO_3$), or metallic (like iron, Fe).

3. *Conductivity:* If an electrical conductivity apparatus is available, use it to test each sample. *(**Caution:** Avoid touching the bare electrode tips with your hands; some may deliver an uncomfortable electric shock.)* Touch both electrodes to the element sample, but do not allow the electrodes to touch each other. See Figure 1.14. If the bulb lights, even dimly, electricity is flowing through the sample. Such a material is called a **conductor**. If the bulb fails to light, the material is a **nonconductor**.

Figure 1.14 *Testing a sample for electrical conductivity.*

4. *Crushing:* Gently tap each element sample with a hammer as shown in Figure 1.15. Based on the results, decide whether the sample is **malleable**, which means it flattens without shattering when struck, or **brittle**, which means it shatters into pieces.

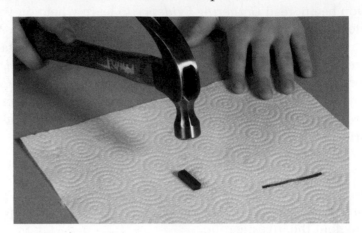

Figure 1.15 *Testing a sample for malleability.*

5. *Reactivity with copper(II) chloride.*
   a. Label seven wells of a clean 24-well plate *A* through *G*.
   b. Place a sample of each element in its well. The ribbon or solid wire samples provided by your teacher will be < 1 cm in length. Other samples should be no larger than the size of a match head.
   c. Add 15 to 20 drops of copper(II) chloride (CuCl₂) solution to each sample.
   d. Observe each well for three to five minutes—changes may be slow. Decide which elements reacted with copper(II) chloride solution and which did not. Record these results.
   e. Discard the well plate contents as instructed by your teacher.

6. *Reactivity with acid.*
   a. Repeat Steps 5a and 5b.

   b. Add 15 to 20 drops of hydrochloric acid (HCl) to each well that contains a sample. *(**Caution:** 0.5 M hydrochloric acid (HCl) can chemically attack skin if allowed to remain in contact for a long time. If any hydrochloric acid accidentally spills on you, ask a classmate to notify your teacher immediately. Wash the affected area immediately with tap water and continue rinsing for several minutes.)*
   c. Observe and record each result. Remember that the formation of gas bubbles or a change in a sample's appearance may indicate that a chemical reaction has occurred. Decide which elements reacted with hydrochloric acid and which did not. Record these results.
   d. Discard the well plate contents as instructed by your teacher.

7. Wash your hands thoroughly before leaving the laboratory.

Interpreting Evidence

1. Classify each property tested in this activity as either a physical property or a chemical property.

2. Sort the seven coded elements into two groups based on similarities in their physical and chemical properties.

3. Which element or elements could fit into either group? Why?

4. Using the following information, classify each tested element as a metal, a nonmetal, or a metalloid:

   - Metals have luster, are malleable (can be hammered into sheets), and conduct electricity.

   - Many metals react with acids; many metals also react with copper(II) chloride solution.

   - Nonmetals are usually dull in appearance, are brittle, and do not conduct electricity.

   - Metalloids have some properties of both metals and nonmetals.

You have been introduced to one classification scheme for elements: metals, nonmetals, and metalloids. However, the quantity of detailed information that is available about all the elements is enormous. When you are selecting or designing materials for particular uses, the more information you have about the materials (including similarities and differences among them), the better your decisions will be.

■ MAKING DECISIONS
A.11 IT'S ONLY MONEY

Based on what you have learned so far, you can begin to consider the questions posed in the unit-opening Web page. A good first step is to specify some properties that are *necessary* or *desirable* in currency. For example, a high-melting point material is *required* for coins; after all, who would want their money to melt in hot sunlight? However, a metallic luster is only a *desirable* coin property.

Part I:

Apply your knowledge of existing currency, as well as what you have learned about properties of materials, to answer the following questions.

1. For a suitable coin:

   a. What physical properties must the material have?

   b. What other physical properties are desirable?

   c. What chemical properties are required of the material?

   d. What other chemical properties are desirable?

2. For a suitable banknote (bill):

   a. What physical properties must the material have?

   b. What other physical properties are desirable?

   c. What chemical properties are required of the material?

   d. What other chemical properties are desirable?

3. Which would make the best primary material for a new coin: a metal, nonmetal, or metalloid? Explain.

4. What factors or desirable coin characteristics did you consider when you answered Question 3?

Part II:

To make knowledgeable decisions, it is useful to collect information from several reliable sources. As you consider the use and sustainability of U.S. currency, it will be helpful to find the composition of dollar coins and bills now in circulation.

5. How will you find the composition of current U.S. dollar bills and coins? (*Hint:* Think about the organizations that are responsible for producing money in the U.S.)

6. How will you decide whether the information you find is reliable?

7. What materials are found in the U.S. one-dollar coin? Write the name and symbol of each element in this coin.

8. What materials compose the U.S. dollar bill?

Save your answers to these questions—they will help guide your decision-making later in this unit.

SECTION A SUMMARY

Reviewing the Concepts

The physical properties of a material can be determined without altering the material's chemical makeup; physical changes alter a material's physical properties. Chemical properties describe how a material reacts chemically through its transformation into one or more different materials.

1. Classify each as a chemical or a physical property:
   a. Copper has a reddish brown color.
   b. Propane burns readily.
   c. Carbon dioxide gas extinguishes a candle flame.
   d. Honey pours more slowly than does water.

2. Classify each as a chemical or a physical property:
   a. Metal wire can be bent.
   b. Ice floats in water.
   c. Paper is flammable.
   d. Sugar is soluble in water.

3. Classify each as a chemical or a physical change:
   a. A candle burns.
   b. An opened carbonated beverage fizzes.
   c. Hair curls as a result of a "perm."
   d. As shoes wear out, holes appear in the soles.

4. Classify each as a chemical or a physical change:
   a. A cut apple left out in the air turns brown.
   b. Flashlight batteries lose their "charge" after extended use.
   c. Dry cleaning removes oils from clothing.
   d. Italian salad dressing separates into layers over time.

5. For each of your answers in Question 4, give evidence for your classification as a chemical or physical change.

6. a. List steps involved in making chocolate-chip cookies from scratch.
   b. Classify each step in Question 6a as involving either a chemical change or a physical change.

An element is composed of only one type of atom; compounds consist of two or more types of atoms. Both elements and compounds are considered substances.

7. Define substance and give two examples.

8. Classify each of these substances as an element or a compound.
   a. CO   c. HCl   e. $NaHCO_3$ g. I_2
   b. Co   d. Mg   f. NO

9. Look at these models.

   a. Which represent elements?
   b. Which represent compounds?

> A chemical formula indicates the composition of a substance.

10. What two pieces of information does a chemical formula provide?

11. Distinguish between chemical symbols and chemical formulas.

12. Name the elements and list the number of each atom in the following formulas for substances:

    a. phosphoric acid, H_3PO_4 (used in soft drinks and fertilizers)

    b. sodium hydroxide, NaOH (found in some drain cleaners)

    c. sulfur dioxide, SO_2 (a by-product of coal combustion)

> Elements can be classified as metals, nonmetals, or metalloids, according to their physical and chemical properties.

13. Classify each property as characteristic of metals or nonmetals:

    a. shiny in appearance

    b. does not react with acids

    c. shatters easily

    d. electrically conductive

14. List the names and symbols of two elements that are metalloids.

15. What would you expect to happen if you tapped a sample of nickel with a hammer?

16. List two properties that make nonmetals unsuitable for electric wiring.

17. List three properties that make metals suitable for coins.

> Scientific investigations often involve making accurate observations of phenomena and collecting these observations as organized data. These data can be used to draw inferences and as evidence in proposing and supporting scientific explanations.

18. What is an inference?

19. Classify each of the following as an observation or an inference:

    a. Bubbles are produced when baking soda solution and vinegar are mixed.

    b. Baking soda solution and vinegar react when mixed.

    c. The dog with muddy paws has been digging in the garden.

    d. Katie was late to chemistry class on Wednesday.

    e. Universal indicator turned red when added to hydrochloric acid.

    f. Universal indicator showed that hydrochloric acid is acidic.

20. Why is it important to organize laboratory observations?

21. Why is it necessary to read an entire investigation before beginning laboratory work?

> Models help in visualizing and understanding particles and phenomena that cannot be directly observed.

22. Draw a molecular-level model of oxygen (O_2).

23. Draw a molecular-level model of carbon tetrachloride (CCl_4), a toxic compound once used in the production of refrigerants.

24. Write the formula of the compound represented by this model:

25. Draw a particulate-level model to show how copper can be drawn into a wire.

26. What is meant by the statement "All models have some limitations"?

How do chemists describe matter?

Now that you have begun your study of chemistry, it is time to tackle the question that opened Section A. Think about what you have learned, then answer the question in your own words in organized paragraphs. Your answer should demonstrate (and perhaps clarify!) your understanding of the key ideas in this section.

Be sure to consider the following in your response: properties of substances, the particulate view of matter, the language of chemistry, models, and the role of observation and investigation.

Connecting the Concepts

27. Three kinds of observations that may indicate a chemical change are listed below. However, a physical change may also result in each observation. Describe a possible chemical cause and a possible physical cause for each observation:

    a. change in color

    b. change in temperature

    c. formation of a gas

28. Describe an inference that you have made—outside of chemistry class—in the last week.

29. Describe a model that is used outside of chemistry class to visualize or understand phenomena.

30. Which is more clearly represented by models, a molecular compound or a metal? Explain your reasoning.

Extending the Concepts

31. Some elements in Table 1.1 (page 38) have symbols that are not based on their modern names (such as K for potassium). Look up their historical names and explain the origin of their symbols.

32. The symbols of elements are accepted and used by chemists in all nations, regardless of their country's official language. However, the name of an element often depends on language. For example, the element N is *nitrogen* in English but *azote* in French. The element H is *hydrogen* in English but *Wasserstoff* in German. Investigate the names of some common elements in a foreign language of your choice. What are the meanings or origins of the foreign element names that you find? How do those meanings or origins compare with those for the corresponding English element names?

33. How is mercury different from most other metallic elements? Using outside resources, describe some applications that take advantage of the properties of metallic mercury.

34. Depending on how a sample of iron is heated and cooled, it can either be hard and brittle or malleable. Explain how the same metal can have both characteristics.

35. Classify the components of one or more pieces of jewelry you might possibly wear as being composed of metals, nonmetals, or metalloids.

SECTION **B** PERIODIC TRENDS

How can the periodic table be used to help explain and predict the properties of chemical elements?

More than 100 chemical elements have been discovered and explored. Of those, a small number are considered highly prized and used for coins, jewelry, and art. Why are some more valuable than others? How do scientists classify and organize the characteristics of the elements? In this section, you will examine properties of some elements and construct statements about groups of elements and their reactivity. You will also be introduced to the periodic table, one of a chemist's most important tools. As you study the ideas in this section, think about how this tool could help you to choose or explain the use of certain elements in forms of currency.

GOALS

- Use the periodic table to
    a. predict physical and chemical properties of an element,
    b. identify elements by their atomic numbers, and
    c. locate periods and groups (families) of elements.
- Recognize and distinguish characteristics of the basic subatomic particles: protons, neutrons, and electrons.
- Describe what constitutes an ion. Indicate the electrical charge of an ion containing a specified number of protons and electrons.
- Use the basic structure of the atom to explain the organization of the periodic table.
- Collect, organize, and represent data.
- Explore periodic trends of groups of elements.
- Write the formula and name of an ionic compound, given the compound's anion and cation names and electrical charges.

concept check 3

1. How can you distinguish between chemical and physical properties?
2. What information does a chemical formula contain about a compound?
3. Draw and label a diagram that illustrates the structure of an atom.

■ MAKING DECISIONS

B.1 GROUPING THE ELEMENTS

By the mid-1800s, chemists had identified about 60 elements. Five of these elements were nonmetals that are gases at room temperature: hydrogen (H), oxygen (O), nitrogen (N), fluorine (F), and chlorine (Cl). Two liquid elements were also known, the metal mercury (Hg) and the nonmetal bromine (Br). The rest of the known elements were solids with widely differing properties.

To organize information about the known elements, several scientists tried to place elements with similar properties near one another in a chart. Such an arrangement is called a **periodic table**. Dimitri Mendeleev, a Russian chemist, published a periodic table in 1869. We use such a table today. In some respects, the periodic table has a pattern that resembles a monthly calendar, in which weeks repeat in a regular (periodic) seven-day cycle. Figure 1.16 shows a moon chart, which is another type of periodic chart.

You will receive a set of 20 element data cards. Each card lists some properties of a particular element.

1. Arrange the cards in order of increasing atomic mass.

2. Try sorting the cards into several different groups. Each group should include elements with similar properties. You might need to try several methods of grouping according to properties before you find one that makes sense to you.

3. Examine the cards within each group for any patterns. Arrange the cards within each group in some logical sequence. Again, trial and error may be a useful method for accomplishing this task.

4. Observe how particular element properties vary from group to group. Record your observations as well as any inferences that you make.

5. Arrange all the card groups into some logical sequence.

6. Select the most reasonable and useful patterns within and among card groups. Then tape the cards onto a sheet of paper to preserve your pattern for later classroom discussion.

7. Write an explanation that summarizes how and why you (1) grouped the elements, (2) arranged the elements within groups, and (3) arranged the groups. Support your explanation with evidence from the data cards.

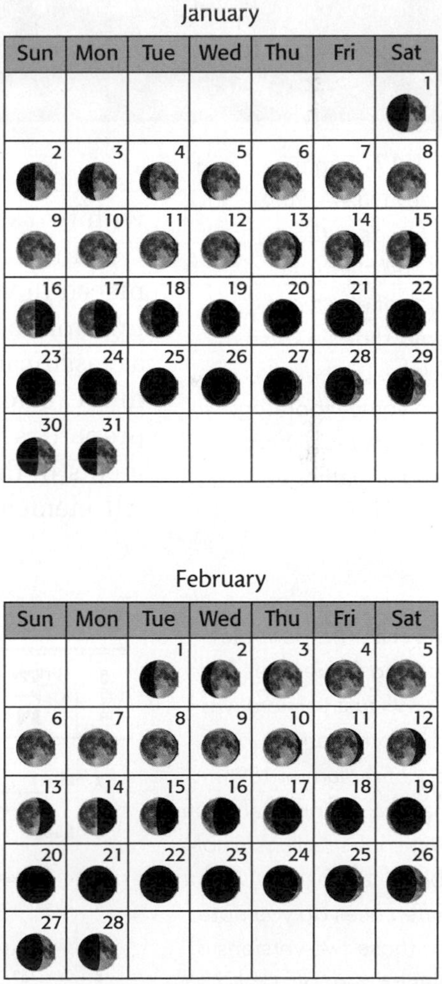

Figure 1.16 *This chart of the moon shows that the phases repeat at regular intervals or periods. Scientists use the term periodicity to describe trends such as this. Can you think of other common periodic examples?*

B.2 EARLY PERIODIC TABLES

The periodic tables of the 1800s were organized according to two characteristics of elements, one physical and one chemical. First, chemists knew that atoms of different elements have different masses. For example, hydrogen atoms have the lowest mass, oxygen atoms are about 16 times more massive than hydrogen atoms, and sulfur atoms are about twice as massive as oxygen atoms (making them about 32 times more massive than hydrogen atoms). Based on such comparisons, an average atomic mass was assigned to each element in Mendeleev's periodic table. This atomic mass then became one of the two criteria for arranging elements. See Figures 1.17 and 1.18.

The other criterion for organizing elements was their respective "combining capacity" with other elements, such as chlorine and oxygen. This is a chemical property. Atoms of various elements differ in the way that they combine with another element. For example, one atom of potassium (K) or cesium (Cs) combines with only one atom of chlorine (Cl) to produce the compound KCl or CsCl. One atom of magnesium (Mg) or strontium (Sr), however, combines with two atoms of chlorine to produce the compound $MgCl_2$ or $SrCl_2$.

In the first periodic tables, elements with similar chemical properties were placed in vertical columns. Horizontal arrangements were based on increasing atomic masses of the elements. In Making Decisions B.1, you developed a classification scheme for some elements in much the same way Mendeleev did. Creators of early periodic tables were unable to explain similarities in properties found among neighboring elements. We know now that all elements in the leftmost column of the periodic table are very reactive metals. All elements listed in the rightmost column are unreactive (noble) gases.

Figure 1.17 *Dimitri Mendeleev (1834–1907) studied trends in the physical and chemical properties of elements. This stamp shows a version of Mendeleev's periodic table and some laboratory equipment that may have been used to gather evidence to construct it.*

Figure 1.18 *This periodic table (rotated 90 degrees from Mendeleev's first version) was created before Marie Curie's discovery of radium in 1898. Note that the periodic table on the stamp shows radium in column II below Ba (barium). What other differences do you note between those two versions of Mendeleev's periodic table? What differences exist between those periodic tables and the modern version on page 60?*

T a b e l l e II.

Reihen	Gruppe I. — R^2O	Gruppe II. — RO	Gruppe III. — R^2O^3	Gruppe IV. RH^4 RO^2	Gruppe V. RH^3 R^2O^5	Gruppe VI. RH^2 RO^3	Gruppe VII. RH R^2O^7	Gruppe VIII. — RO^4
1	H=1							
2	Li=7	Be=9,4	B=11	C=12	N=14	O=16	F=19	
3	Na=23	Mg=24	Al=27,3	Si=28	P=31	S=32	Cl=35,5	
4	K=39	Ca=40	—=44	Ti=48	V=51	Cr=52	Mn=55	Fe=56, Co=59, Ni=59, Cu=63.
5	(Cu=63)	Zn=65	—=68	—=72	As=75	Se=78	Br=80	
6	Rb=85	Sr=87	?Yt=88	Zr=90	Nb=94	Mo=96	—=100	Ru=104, Rh=104, Pd=106, Ag=108.
7	(Ag=108)	Cd=112	In=113	Sn=118	Sb=122	Te=125	J=127	
8	Cs=133	Ba=137	?Di=138	?Ce=140	—	—	—	— — — —
9	(—)	—	—	—	—	—	—	
10	—	—	?Er=178	?La=180	Ta=182	W=184	—	Os=195, Ir=197, Pt=198, Au=199.
11	(Au=199)	Hg=200	Tl=204	Pb=207	Bi=208	—	—	
12	—	—	—	Th=231	—	U=240	—	— — — —

der chemischen Elemente.

The reason for such patterns, which was proposed more than 50 years after Mendeleev's work, serves as the basis for the modern periodic table.

B.3 THE ELECTRICAL NATURE OF MATTER

Section A introduced the concept of elements and compounds. How do the atoms in compounds "stick" together to form molecules? Are atoms made up of even smaller particles? The answers to these questions require understanding the electrical properties of matter.

You have already experienced the electrical nature of matter, most probably without realizing it. Clothes often display static cling when taken from the dryer. The pieces of fabric stick firmly together and can be separated only with effort. The shock that you sometimes receive after walking across a rug and touching a metal doorknob is another reminder of matter's electrical nature. And if you rub two inflated balloons against your hair, both balloons will attract your hair but repel each other, a phenomenon best observed when the humidity is low. See Figure 1.19.

Figure 1.19 *Rubbing a balloon against your hair results in static electricity. The balloon attracts your hair, even when you hold it away from your head.*

The electrical properties of matter can be summarized as follows:

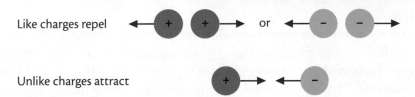

What are these positive and negative charges? How do they relate to the idea of atoms and molecules? Every electrically neutral (uncharged) atom contains equal numbers of positively charged particles called **protons** and negatively charged particles called **electrons**. For instance, an electrically neutral sodium atom contains 11 protons and 11 electrons. An electrically neutral chlorine atom contains 17 protons and 17 electrons. Knowing the total protons and electrons within an atom, you can determine its overall electrical charge. For example, a potassium atom with 19 protons and 18 electrons would be an ion with a 1+ charge, or K^+.

In addition to protons and electrons, most atoms contain one or more electrically neutral particles called **neutrons**. Positive–negative attractions between protons in one atom and electrons in another atom provide the attachment that holds atoms together. This "sticky force" is the chemical bond you read about on page 36 (A.5).

You will use these ideas in later sections and in future units to learn about and explain the properties of materials (and why they are useful in currency and other applications), the process of dissolving, and chemical bonding. Now you will combine these ideas with your knowledge of atoms to explain the organization of the modern periodic table.

B.4 THE MODERN PERIODIC TABLE

In the modern periodic table of the elements, elements are placed in sequence according to their increasing **atomic number** (number of protons). Because electrically neutral (uncharged) atoms contain equal numbers of protons and electrons, the periodic table is also sequenced by the number of electrons contained in neutral atoms of each element.

Early periodic tables, much like the one you just constructed, used average atomic masses to organize the elements. Although this method produces reasonable results for elements with relatively small atomic masses, it does not work well for more massive atoms. The reason for this is the existence of another small particle that also contributes to the atomic mass, the electrically uncharged neutron. The total mass of an atom is largely determined by the combined mass of protons and neutrons in its nucleus. The **nucleus** is a concentrated region of positive charge (due to protons) in the center of an atom. See Figure 1.20.

The total number of protons and neutrons in the nucleus of an atom is called the **mass number**. Electrons make up the rest of an atom, but because each electron is about 1/2000th the mass of a proton or neutron, the total mass of the electrons does not contribute significantly to the mass of an atom.

While all atoms of a particular element have the same number of protons, the number of neutrons can differ from atom to atom of an element. For example, carbon atoms always contain 6 protons, but they may contain 6, 7, or 8 neutrons. Thus, individual carbon atoms can have mass numbers of 12, 13, or 14. For example, 6 protons + 7 neutrons = mass number 13. Atoms with the same number of protons but different numbers of neutrons, such as these carbon atoms, are called **isotopes**. In other words, isotopes are atoms of the same element with different mass numbers.

Is there a connection between the atomic numbers used to organize the modern periodic table and the properties of elements used by 19th-century chemists to create their periodic tables? If there is, what is that connection? Continue reading to explore the relationship between atomic numbers and the properties of elements.

Protons, electrons, and neutrons are all referred to as *subatomic particles*, since they are smaller than atoms.

Mass number and atomic number are not the same.

Isotopes are the major reason for fractional atomic masses found on the periodic table. For example, the atomic mass of chlorine is listed as 35.45. This value represents the *average* mass of all of the atoms of chlorine.

Scientific American Conceptual Illustration

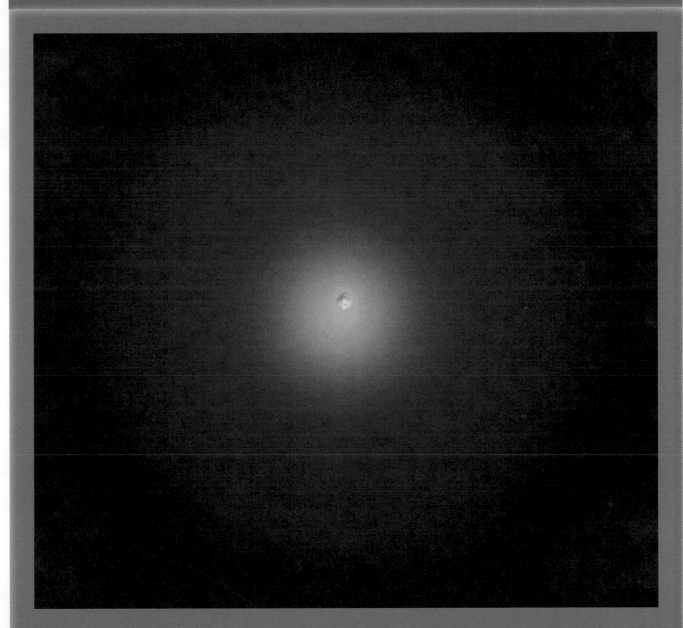

Figure 1.20 *Model of a lithium (Li) atom. Each electrically neutral lithium atom contains three protons, three electrons, and either three or four neutrons. In this model, the protons and three neutrons reside in the Li nucleus (center). Electrons are not depicted as individual particles; they occupy "electron clouds" around the nucleus. Two electrons (yellow cloud) remain close to the nucleus; the third electron (fainter cloud) is relatively farther away. Electron clouds occupy most of an atom's volume; the nucleus accounts for nearly all the atom's mass. Note that an atom does not have a distinctly defined "outer edge," but, instead, has a rather fuzzy outer region.*

DEVELOPING SKILLS

B.5 COUNTING SUBATOMIC PARTICLES

Sample Problem 1: Write the symbol, number of protons, number of neutrons, and number of electrons for the hydrogen isotope with mass number 1. See table below.

You can find element symbols on the periodic table (page 60). Hydrogen's symbol is H. An element's identity specifies its atomic number, which is equal to the number of protons in each atom of that element. Hydrogen's atomic number is one, so it contains one proton. If the atom is a neutral atom, then the number of electrons must equal the number of protons, so hydrogen atoms must each contain one electron. The mass number for the isotope is equal to the sum of the protons and neutrons, so the number of neutrons can be determined by subtracting the number of protons from the mass number. The mass number for this isotope of hydrogen is one, so there are no neutrons in this hydrogen isotope.

1. Copy and complete the table for each electrically neutral atom.

	Element Name	Element Symbol	Number of protons	Number of neutrons	Number of electrons
a.	Hydrogen	H	1	0	1
b.	Sodium			12	
c.	Boron			5	
d.		B		6	
e.				115	77
f.		Pa		140	

Sample Problem 2: Determine whether an atom of scandium with 21 protons and 18 electrons is electrically neutral or has a net charge. If it is charged, calculate the net charge.

With 21 positively charged protons and only 18 negatively charged electrons, the scandium atom would have an excess charge of 3+ and would not be electrically neutral.

Ex.   scandium: 21 protons   18 electrons   Charge: 3+

2. Decide whether each of these atoms is electrically neutral and, if not, calculate the net charge.

    a. chlorine: 17 protons   18 electrons   Neutral? __   Charge: __

    b. cobalt: 27 protons   24 electrons   Neutral? __   Charge: __

    c. gold: 79 protons   76 electrons   Neutral? __   Charge: __

    d. strontium: 38 protons   36 electrons   Neutral? __   Charge: __

    e. argon: 18 protons   18 electrons   Neutral? __   Charge: __

DEVELOPING SKILLS

B.6 PERIODIC VARIATION IN PROPERTIES

Your teacher will assist you in identifying the atomic numbers of the 20 elements you considered earlier in the unit. Use these atomic numbers and information about each element's properties from the element data cards used in Making Decisions B.1 to prepare the two graphs described below. Look for patterns between atomic numbers and element properties as you construct the graphs.

Graphing Guidelines

Creating and interpreting graphs are essential skills for scientists—and informed citizens! Whether you are looking at environmental data to determine trends or organizing data that will allow you to see a predictable relationship, you will be looking for regularities or patterns among the values. The following suggestions will help you prepare and interpret such graphs.

- Choose your scale so that the graph is large enough to fill most of the available space on the graph paper.

- Assign each regularly spaced division on the graph paper a convenient, constant value. The graph-paper line interval value should be easily "divided by eye", such as 1, 2, 5, or 10, rather than awkward values such as 6, 7, 9, or 14.

- An axis scale does not have to start at zero, particularly if the plotted values cluster in a range far from zero.

- Label each axis with the quantity and unit being plotted.

- Plot each point. Then draw a small symbol around each point, like this: ⊙. If you plot more than one set of data on the same graph, distinguish each by using a different color or small geometric shape to enclose the points, such as: ⊡, ▽, or △.

- Give your graph a title that will readily convey its meaning and purpose to readers.

> You have probably graphed equations in mathematics class in which the points always perfectly fit a line. Scientific data may not perfectly fit a line because the points represent observations of phenomena, not all of which are well represented by a line. Even data that show a linear trend include measurement error and usually do not form a perfect line.

- If you use technology—such as a graphing calculator or computer software—to prepare your graphs, ensure that you follow the guidelines just given. Different devices and software have different ways to process data. Choose the appropriate type of graph (e.g., scatter plot or bar graph) for your data.

Graph 1: Trends in a chemical property

1. On a sheet of graph paper, draw a set of axes and title the graph *Trends in a Chemical Property*.

2. Label the *x*-axis *Atomic Number*. Use your element data cards to determine the range of possible numbers that you will need to display on your graph. Mark your *x*-axis with these values in mind.

3. Label the *y*-axis *Number of Oxygen Atoms (per Atom) in Oxide*. What is the range of possible numbers that you will have to display on your graph? Mark your *y*-axis with these values in mind.

4. Construct a bar graph, as demonstrated in Figure 1.21, by plotting the oxide data from the element cards.

> A bar graph is used when graphing discontinuous data; that is, data that have only certain values. Since atomic numbers can only be positive integers, the data you are graphing in this activity are discontinuous.

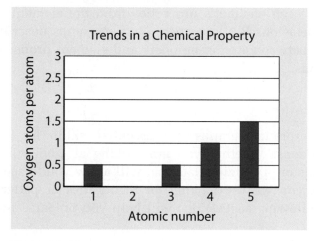

Figure 1.21 *Sample bar graph for oxide data.*

5. Label each bar with the actual symbol of the element involved in that compound.

Graph 2: Trends in a physical property

6. On a separate sheet of graph paper, draw a set of axes, and title the plot *Trends in a Physical Property*.

7. Label the *x*-axis *Atomic Number*. What range of possible numbers will you have to display on your graph? Mark your *x*-axis with these values in mind.

8. Label the *y*-axis *Boiling Point (K)*. What range of possible numbers will you have to display on your graph? Remember to consult your element cards. (*Hint:* Do not include data for the element with atomic number 6. Including the boiling point of this element (carbon) would

distort the graph.) Mark your *y*-axis with these values in mind, as shown in Figure 1.22. Use as much of the graph paper as possible to plot these kelvin temperatures. (*Hint:* See the margin note on this page to learn how to convert temperature values from degrees Celsius to kelvins.)

> Temperature in kelvins (K) is related to temperature in degrees Celsius (°C) by K = 273.15 + °C. You will learn more about the kelvin temperature scale in Unit 2.

9. Construct a bar graph as in Step 4, this time using the boiling point data from the element cards, as shown in Figure 1.22.

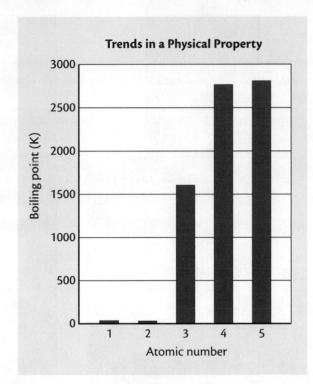

Figure 1.22 *Sample bar graph for boiling point data.*

10. Label each constructed bar with the actual symbol of the element it represents.

Questions

1. Does either bar graph reveal a repeating, or cyclic, pattern? (*Hint:* Focus on elements represented by very large or very small values.) Describe any patterns you observe.

2. Are these graphs consistent with patterns found in your earlier (Making Decisions B.1) grouping of the elements? Explain.

3. Based on these two bar graphs, why is the chemist's organization of elements called a periodic table? (*Hint:* Look up the meaning of *periodic* in the dictionary.)

4. Where are elements with the highest proportion of oxygen in their oxides located on the periodic table?

5. Where are elements with the highest boiling points located on the periodic table?

6. Describe any trends you noted in your answers to Questions 4 and 5.

PERIODIC TABLE OF THE ELEMENTS

Key

1	— Atomic number
Hydrogen	— Name
H	— Symbol
1.008	— Atomic weight

Metals
Metalloids
Nonmetals

Group																	
1 1A	**2** 2A	**3** 3B	**4** 4B	**5** 5B	**6** 6B	**7** 7B	**8** 8B	**9** 8B	**10** 8B	**11** 1B	**12** 2B	**13** 3A	**14** 4A	**15** 5A	**16** 6A	**17** 7A	**18** 8A
1 Hydrogen **H** 1.008																	2 Helium **He** 4.003
3 Lithium **Li** 6.94	4 Beryllium **Be** 9.01											5 Boron **B** 10.81	6 Carbon **C** 12.01	7 Nitrogen **N** 14.01	8 Oxygen **O** 16.00	9 Fluorine **F** 19.00	10 Neon **Ne** 20.18
11 Sodium **Na** 22.99	12 Magnesium **Mg** 24.31											13 Aluminum **Al** 26.98	14 Silicon **Si** 28.09	15 Phosphorus **P** 30.97	16 Sulfur **S** 32.07	17 Chlorine **Cl** 35.45	18 Argon **Ar** 39.95
19 Potassium **K** 39.10	20 Calcium **Ca** 40.08	21 Scandium **Sc** 44.96	22 Titanium **Ti** 47.87	23 Vanadium **V** 50.94	24 Chromium **Cr** 52.00	25 Manganese **Mn** 54.94	26 Iron **Fe** 55.85	27 Cobalt **Co** 58.93	28 Nickel **Ni** 58.69	29 Copper **Cu** 63.55	30 Zinc **Zn** 65.38	31 Gallium **Ga** 69.72	32 Germanium **Ge** 72.64	33 Arsenic **As** 74.92	34 Selenium **Se** 78.96	35 Bromine **Br** 79.90	36 Krypton **Kr** 83.80
37 Rubidium **Rb** 85.47	38 Strontium **Sr** 87.62	39 Yttrium **Y** 88.91	40 Zirconium **Zr** 91.22	41 Niobium **Nb** 92.91	42 Molybdenum **Mo** 95.96	43 Technetium **Tc** [98]	44 Ruthenium **Ru** 101.07	45 Rhodium **Rh** 102.91	46 Palladium **Pd** 106.42	47 Silver **Ag** 107.87	48 Cadmium **Cd** 112.41	49 Indium **In** 114.82	50 Tin **Sn** 118.71	51 Antimony **Sb** 121.76	52 Tellurium **Te** 127.60	53 Iodine **I** 126.90	54 Xenon **Xe** 131.29
55 Cesium **Cs** 132.91	56 Barium **Ba** 137.33	71 Lutetium **Lu** 174.97	72 Hafnium **Hf** 178.49	73 Tantalum **Ta** 180.95	74 Tungsten **W** 183.84	75 Rhenium **Re** 186.21	76 Osmium **Os** 190.23	77 Iridium **Ir** 192.22	78 Platinum **Pt** 195.08	79 Gold **Au** 196.97	80 Mercury **Hg** 200.59	81 Thallium **Tl** 204.38	82 Lead **Pb** 207.2	83 Bismuth **Bi** 208.98	84 Polonium **Po** [209]	85 Astatine **At** [210]	86 Radon **Rn** [222]
87 Francium **Fr** [223]	88 Radium **Ra** [226]	103 Lawrencium **Lr** [262]	104 Rutherfordium **Rf** [265]	105 Dubnium **Db** [268]	106 Seaborgium **Sg** [271]	107 Bohrium **Bh** [272]	108 Hassium **Hs** [277]	109 Meitnerium **Mt** [276]	110 Darmstadtium **Ds** [281]	111 Roentgenium **Rg** [280]	112 Copernicium **Cn** [285]						

Period 1–7

Lanthanides

57 Lanthanum **La** 138.91	58 Cerium **Ce** 140.12	59 Praseodymium **Pr** 140.91	60 Neodymium **Nd** 144.24	61 Promethium **Pm** [145]	62 Samarium **Sm** 150.36	63 Europium **Eu** 151.96	64 Gadolinium **Gd** 157.25	65 Terbium **Tb** 158.93	66 Dysprosium **Dy** 162.50	67 Holmium **Ho** 164.93	68 Erbium **Er** 167.26	69 Thulium **Tm** 168.93	70 Ytterbium **Yb** 173.05

Actinides

89 Actinium **Ac** [227]	90 Thorium **Th** 232.04	91 Protactinium **Pa** 231.04	92 Uranium **U** 238.03	93 Neptunium **Np** [237]	94 Plutonium **Pu** [244]	95 Americium **Am** [243]	96 Curium **Cm** [247]	97 Berkelium **Bk** [247]	98 Californium **Cf** [251]	99 Einsteinium **Es** [252]	100 Fermium **Fm** [257]	101 Mendelevium **Md** [258]	102 Nobelium **No** [259]

7. Predict which element should have the lowest boiling point: selenium (Se), bromine (Br), or krypton (Kr). Use evidence from your graphs to explain how you decided.

8. Using your graphs, roughly predict the pattern in boiling points and oxide numbers for the next five to eight elements, starting with gallium (Ga).

B.7 ORGANIZATION OF THE PERIODIC TABLE

When the first 20 elements are listed in order of increasing atomic numbers and grouped according to similar properties, they form horizontal rows called **periods**. This **periodic relationship** among elements is summarized in the modern periodic table, which you can see on page 60. To become more familiar with the periodic table, locate the 20 elements you grouped earlier. How do their relative positions compare with those shown in the arrangement you constructed in Making Decisions B.1?

Each vertical column in the periodic table contains elements with similar properties. Each column is called a **group** or **family** of elements. For example, the **alkali metal family** consists of the six elements (starting with lithium) in the first column at the left of the table. See Figure 1.23. Each element in this family is a highly reactive metal that forms a chloride compound with a 1:1 alkali metal atom to chlorine atom ratio and an oxide with a 2:1 (alkali metal atom to oxygen atom) ratio. By contrast, at the right of the table, the **noble gas family** consists of very unreactive (or even chemically inert) elements; only xenon (Xe) and krypton (Kr) are known to form compounds under normal conditions. The group containing fluorine, chlorine, bromine, and iodine—in the column just to the left of noble gases—is called the **halogen family**. Halogens are highly reactive and readily form binary compounds with hydrogen.

Figure 1.23 *Sodium is an alkali metal. All alkali metals have similar physical and chemical properties, so they are placed in the same column (family) in the periodic table.*

Binary compounds are composed of only two elements.

The arrangement of elements in the periodic table provides an orderly summary of key characteristics of each element. By knowing the major properties of a certain chemical family, you can predict some properties of any element in that family. This knowledge can be very helpful in evaluating elements for possible uses.

DEVELOPING SKILLS

B.8 PREDICTING PROPERTIES

Some chemical or physical properties, including those you graphed in Developing Skills B.6, are called **periodic properties**. Such properties vary among elements according to trends that repeat as atomic number increases. Because of these trends, some element properties can be estimated by averaging the respective properties of the elements located just above and just below an element on the periodic table. That is how Mendeleev predicted the properties of several elements unknown in his time. He was so convinced these elements existed that he purposely left gaps for them in his periodic table, along with a prediction of some of their properties. When those elements were discovered shortly thereafter, they fit into the gaps as expected. Mendeleev's fame rests largely on the accuracy of these predictions.

For example, germanium (Ge) was not known when Mendeleev proposed his periodic table. However, in 1871 he predicted the existence of germanium, calling it *ekasilicon*. Not only did Mendeleev predict germanium's existence, he also accurately predicted many of its chemical and physical properties, based on known properties of other elements in its same family.

Eka- means "standing next in order."

> **Sample Problem:** *Given that the density of silicon (Si) is 2.3 g/cm^3 and the density of tin (Sn) is 7.3 g/cm^3, estimate the density of germanium (Ge).*
>
> These three elements are in the same group in the periodic table. Germanium is below silicon and above tin. (You can verify this by locating these elements in the periodic table.) The predicted density of germanium can be estimated by averaging the densities of silicon and tin, arriving at a calculated value of 4.8 g/cm^3.
>
> When germanium was discovered in 1886, its density was found to be 5.3 g/cm^3, which is within ~10% of the earlier estimated density. The periodic table helped guide Mendeleev (and now you, too) to make a useful prediction.

Formulas for chemical compounds can also be predicted from relationships established in the periodic table. For example, carbon and oxygen form carbon dioxide (CO_2). What formula would be predicted for a compound of carbon and sulfur? The periodic table indicates that sulfur (S) and oxygen (O) are in the same family. Knowing that carbon and oxygen form CO_2, a logical—and quite correct—prediction would be CS_2 (carbon disulfide). Now it's your turn.

1. The element krypton (Kr) was not known in Mendeleev's time. Given that the boiling point of argon (Ar) is −186 °C and of xenon (Xe) is −107 °C, estimate the boiling point of krypton.

2. a. Estimate the melting point of rubidium (Rb). The melting points of potassium (K) and cesium (Cs) are 337 K and 302 K, respectively.

   b. Do you expect the melting point of sodium (Na) to be higher or lower than that of rubidium (Rb)? Explain. On what evidence did you base your answer?

3. Mendeleev knew that silicon tetrachloride ($SiCl_4$) existed. Using his periodic table, he correctly predicted the existence of *ekasilicon*, which we now know as germanium. Predict the formula for the compound formed by germanium and chlorine.

4. Here are formulas for several known compounds: NaI, $MgCl_2$, CaO, Al_2O_3, and CCl_4. Using that information, predict the formula for a compound formed from:

   a. C and F   b. Al and S   c. K and Cl   d. Ca and Br   e. Sr and O

5. Elemental nickel is white to light gray in color, highly ductile, lustrous, and resistant to tarnish. Which other element would you expect to share these properties: Pt, Cu, Rb, or Kr? Explain your choice.

6. Manganese, which has melting point 1519 K and is one of the metals used in the Presidential dollar coin, was the only known element in its family for 150 years. When rhenium, which has melting point 3459 K, was discovered in 1925, what could chemists predict about the melting point of element 43 (which was still unknown at the time)?

■ INVESTIGATING MATTER

B.9 PERIODIC TRENDS

Preparing to Investigate

As you learned earlier, elements were placed in columns on the periodic table because they had similar chemical and physical properties. In this investigation, you will examine some characteristics of elements from one family. Group VIIA, which consists of the elements fluorine, chlorine, bromine, iodine and astatine, is called the **halogen** group. When elements in this group form compounds, those compounds are known as *halide* compounds. In this investigation, you will determine which halogens and halide compounds react with one another and develop a rule to summarize those interactions.

Before you begin, read *Gathering Evidence* to learn what you will need to do and note safety precautions. *Gathering Evidence* provides a procedure for systematically combining the substances. As you read *Gathering Evidence*, make a chart or diagram of the solutions that will be added to each test tube.

> Group VIIA is also known as group 17.

> When a halogen gains an electron, giving it a 1− charge, it is called a *halide*.

Note that three liquids will be added to each test tube: a halogen, hexane, and a halide. After reading *Gathering Evidence*, construct a data table that clearly indicates the substances that will be added to each test tube and provides space to record observations after each addition.

Gathering Evidence

1. Before you begin, put on your goggles, and wear them properly throughout the investigation.

2. Place six clean, dry test tubes in a test tube rack. Mark the test tubes A–F.

3. Add 10 drops of chlorine water (Cl_2) to test tubes A and B. Record your observations. *(**Caution:** The liquids used in this investigation are toxic and irritating to skin. If you spill any liquid on yourself or others, wash it off thoroughly and ask a classmate to inform your teacher immediately. Avoid inhaling any fumes.)*

4. Add 10 drops of bromine water (Br_2) to test tubes C and D. Record your observations.

5. Add 10 drops of iodine water (I_2) to test tubes E and F. Record your observations.

6. Add 20 drops of hexane to each test tube A–F.

7. Place a rubber stopper on each test tube. Holding the stopper in place, carefully mix the liquids.

8. Record your observations, including the number of layers and the color of each layer. See Figure 1.24.

9. Add 10 drops of 0.1 M NaBr to test tube A. Mix the solutions as instructed in Step 7. Record detailed observations of your results.

10. Add 10 drops of 0.1 M NaI to tube B and mix.

11. Add 10 drops of 0.1 M NaCl to tubes C and E.

12. Add 10 drops of 0.1 M NaI to tube D.

13. Add 10 drops of 0.1 M NaBr to tube F.

14. Record all observations.

15. Dispose of solutions as instructed by your teacher.

Figure 1.24 *Which of these layers contains hexane? How do you know? What caused the color in the hexane layer?*

Interpreting Evidence

1. What did you observe about halogens by themselves?

2. Do halogens mix more easily with water or with hexane? How do you know?

3. Look at your results for tubes A and B.

    a. Did chlorine water react

        i. with NaBr?

        ii. with NaI?

b. For each answer to Question 3a, what evidence shows that a chemical reaction did or did not take place?

Making Claims

4. Which halogen is most reactive? Give evidence to support your claim.

5. Which halogen is least reactive? Give evidence to support your claim.

6. Construct a general statement to describe the periodic trend of halogen reactivity.

concept check 4

1. a. How are elements arranged in the periodic table?
   b. What is periodic about this arrangement?
2. Where are metals that are used to make coins generally found on the periodic table?
3. What changes in an atom when it becomes an ion?

B.10 IONS AND IONIC COMPOUNDS

In Section A.5 (page 35), you learned that some substances are made up of molecules. For instance, the elements in Investigating Matter B.9—the halogens—are found in the form of diatomic molecules, while hexane, C_6H_{14}, is a molecular compound. By contrast, the sodium halide compounds (sodium chloride, sodium bromide, and sodium iodide) in Investigating Matter B.9 are not composed of molecules. They are examples of another type of compound, an *ionic compound*. **Ionic compounds** are substances that are composed of positive and negative **ions**, which are electrically charged atoms or groups of atoms.

Atoms gain or lose electrons to form negative or positive ions, respectively. For instance, a sodium atom can easily lose one electron, resulting in an ion containing 11 protons and 10 electrons. A chlorine atom readily gains one electron. Summing protons and electrons reveals the resulting electrical charge for these two ions:

$$
\begin{array}{llll}
\text{Sodium ion:} & \text{11 protons} & \text{(11+ charge)} & \\
& +\ \underline{\text{10 electrons}} & \underline{\text{(10– charge)}} & \\
& \text{Sodium ion} & (\ 1+\ \text{charge}), & \text{thus Na}^+ \\
\\
\text{Chloride ion:} & \text{17 protons} & \text{(17+ charge)} & \\
& +\ \underline{\text{18 electrons}} & \underline{\text{(18– charge)}} & \\
& \text{Chloride ion} & (\ 1–\ \text{charge}), & \text{thus Cl}^- \\
\end{array}
$$

How readily an atom gains or loses electrons to form ions is another example of a periodic property. Take a moment to locate sodium and chlorine on the periodic table. What do you notice? Would you expect potassium to readily form ions? What about iodine?

An ionic compound has no net electrical charge; it is neutral because positive and negative electrical charges offset each other. The most familiar example of an ionic compound is table salt, sodium chloride (NaCl). Solid sodium chloride, NaCl*(s)*, consists of equal numbers of positive sodium ions (Na^+) and negative chloride ions (Cl^-) arranged in a 3-D network called a **crystal**. In solid ionic compounds, such as table salt, the ions are held together in crystals by attractions among the negative and positive charges. See Figure 1.25. A negatively charged ion is called an **anion**, and a positively charged ion is called a **cation**.

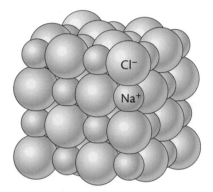

Figure 1.25 *You probably use sodium chloride, NaCl, to salt your food (far right). A scanning electron micrograph shows the cubic structure of NaCl crystals (center). A space-filling model of NaCl (left) provides information about how the individual chloride ions (Cl⁻) and sodium ions (Na⁺) are arranged within the salt crystal. What else does this model suggest about sodium and chloride ions?*

The ionic compound calcium chloride, $CaCl_2$, presents a picture similar to NaCl. However, unlike sodium ions, calcium ions (Ca^{2+}) each have a charge of 2+. Since ionic compounds are not found as individual molecules, chemists use the term **formula unit** when referring to the simplest unit of an ionic compound. A formula unit of NaCl contains one sodium cation and one chloride anion. How many ions make up a formula unit of $CaCl_2$?

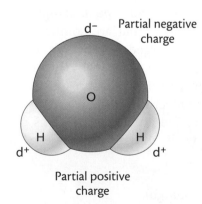

Figure 1.26 *Regions of charge on a water molecule. The δ⁺ and δ⁻ indicate partial electrical charges.*

Because ionic compounds consist of charged particles, they can often be dissolved in substances that have an uneven distribution of electrical charge. Water is such a compound. You will learn more about water in a later unit, but for now you should know that a water molecule has regions that tend to be positively charged around each of its hydrogen atoms and a region that tends to be negatively charged around the oxygen atom. See Figure 1.26.

Unlike water, hexane does not have positively and negatively charged regions. See Figure 1.27. Substances that do not contain ions (or uneven charge) tend to be more soluble in this kind of substance. You probably noticed that the hexane changed color when it was added to halogen water (Cl_2, Br_2, or I_2) in Step 6 of Investigating Matter B.9 (page 64). The halogen molecules, like hexane, have an even distribution of electrical charge. Since halogen molecules and hexane molecules have similar electrical properties, they tend to dissolve in one another.

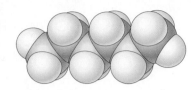

When you added a solution containing a halide compound to a hexane/halogen solution and shook it, a reaction took place only when a *more* reactive halogen was added to a *less* active halogen ion. The hexane layer changed color when a reaction took place because a different halogen was then dissolved in that layer. This type of reaction is sometimes referred to as a *single replacement reaction*. A sample equation is the following:

Figure 1.27 *Hexane molecule. Electrical charge is evenly distributed within this molecule.*

$$Cl_2(aq) + 2\,NaI(aq) \longrightarrow 2\,NaCl(aq) + I_2(aq)$$

In the next activity, you will use what you have learned about ions to construct models of several ionic compounds that you have used in investigations, as well as some other important ionic substances.

> In writing chemical formulas for substances, the symbols for solid *(s)*, liquid *(l)*, gas *(g)*, and aqueous solution *(aq)* are sometimes added. These symbols indicate the physical state of each substance under conditions of the reaction.

■ MODELING MATTER

B.11 IONIC COMPOUNDS

In Modeling Matter A.8 (page 38), you saw that it can be helpful to use models to translate between real-world (macroscopic) and particulate representations of matter. In this activity, you will use cards to represent ions as you practice writing ionic formulas. Keep in mind that each formula and set of cards represents a formula unit of that ionic compound, while the compound itself always exists in a much larger crystal containing many ions.

To begin, you can write formulas for ionic compounds by following two simple rules.

- Write the cation first, then the anion.
- The correct formula contains the fewest positive and negative ions needed to make the total electrical charge zero.

Why are the numbers of chloride ions different in sodium chloride (NaCl) and calcium chloride ($CaCl_2$)? In sodium chloride, the ion charges are 1+ and 1–. Because one cation and one anion results in a total charge of zero, the formula for sodium chloride must be NaCl. When cation and anion charges do not add up to zero, ions of either type must be added until the charges are the same. In calcium chloride, one calcium ion (Ca^{2+}) has a charge of 2+. Each chloride ion (Cl^-) has a charge of 1–; two Cl^- ions are needed to balance a charge of 2+. Thus, two chloride ions (2 Cl^-) are needed for each calcium ion (Ca^{2+}). The subscript 2 written after chlorine's symbol in the formula indicates this. The correct formula for calcium chloride is $CaCl_2$.

Some ions, called **polyatomic** *(many atom)* **ions**, consist of a group of bonded atoms with an electrical charge. Table 1.2 on page 70 includes several common polyatomic ions, as well as formulas and names of common **monatomic** *(one atom)* **ions**. Formulas for compounds containing polyatomic ions follow the same rules for writing formulas. However, if more than one polyatomic ion is needed to bring the total charge of the compound to zero, the formula for the polyatomic ion is enclosed in parentheses before the needed subscript is written. For instance, calcium nitrate is composed of Ca^{2+} ions and NO_3^- ions. Its correct formula, written to balance electrical charges, is $Ca(NO_3)_2$. Ammonium sulfate is composed of ammonium (NH_4^+) and sulfate (SO_4^{2-}) ions. *Two* ammonium cations with a total charge of 2+ are needed to match the 2– charge of *one* sulfate anion. Thus, the formula for ammonium sulfate is $(NH_4)_2SO_4$. Note that the entire ammonium ion symbol is enclosed in parentheses; the subscript 2 indicates that two cations are needed to complete the formula of ammonium sulfate.

Part I:

The written name of an ionic compound is composed of two parts. The cation is named first, then the anion. As Table 1.2 suggests, many cations have the same name as their original elements. Anions composed of a single atom, however, have the last few letters of the element's name changed to the suffix *-ide*. For example, the anion formed from fluorine (F) is fluor*ide* (F⁻). Thus KF is named potassium fluor*ide*. The following activity will provide practice in expressing names and writing formulas for ionic compounds, according to the universal language of chemistry.

The models (cards) your teacher will provide show the formula and charge of each ion. The size of the card varies depending upon the charge of the ion. Thus, when forming a model of calcium chloride, you will notice that calcium ion, with a 2+ charge, is twice the height of chloride ion, with a 1– charge, as shown in the margin.

| Ca^{2+} | Cl⁻ |
| | Cl⁻ |

Sample Problem: *Using models, determine the composition of potassium chloride, the primary ingredient in table-salt substitutes used by people on low-sodium diets. Place your models and answers in a table similar to the one shown below.*

Potassium chloride has been done as an example in the following sample data table.

	Cation	Anion	Model	Formula	Name
Sample Problem	K$^+$	Cl$^-$	K$^+$ | Cl$^-$	KCl	potassium chloride
1					
2					

Prepare a table similar to the one shown here that includes a model, and the name and composition of each ionic compound described. Use the model cards provided by your teacher and refer to Table 1.2 on page 70 as needed to complete this activity.

1. Sodium iodide was one of the sources of halide ions used Investigating Matter B.9.

2. The blue solution you used in Investigating Matter A.1 was probably copper(II) chloride.

3. Baking soda, also used in Investigating Matter A.1, is sodium hydrogen carbonate.

4. CaSO$_4$ is a component of plaster.

5. A substance composed of Ca^{2+} and PO$_4^{3-}$ ions is found in some brands of phosphorus-containing fertilizer.

6. Ammonium nitrate, a rich source of nitrogen, is often used in fertilizer mixtures.

7. Aluminum sulfate is a compound sometimes used to help purify water.

8. Magnesium hydroxide is called milk of magnesia when it is mixed with water.

9. Limestone and marble are two common forms of the compound calcium carbonate.

10. Silver nitrate, a compound used to make photographic film, will also be used in an upcoming investigation.

Table 1.2

Common Ions

Cations

1+ Charge		2+ Charge		3+ Charge	
Formula	Name	Formula	Name	Formula	Name
H^+	hydrogen	Mg^{2+}	magnesium	Al^{3+}	aluminum
Na^+	sodium	Ca^{2+}	calcium	Fe^{3+}	iron(III)*
K^+	potassium	Ba^{2+}	barium		
Cu^+	copper(I)*	Zn^{2+}	zinc		
Ag^+	silver	Cd^{2+}	cadmium		
NH_4^+	ammonium	Hg^{2+}	mercury(II)*		
		Cu^{2+}	copper(II)*		
		Pb^{2+}	lead(II)*		
		Fe^{2+}	iron(II)*		

Anions

1– Charge		2– Charge		3– Charge	
Formula	Name	Formula	Name	Formula	Name
F^-	fluoride	O^{2-}	oxide	PO_4^{3-}	phosphate
Cl^-	chloride	S^{2-}	sulfide		
Br^-	bromide	SO_4^{2-}	sulfate		
I^-	iodide	SO_3^{2-}	sulfite		
NO_3^-	nitrate	CO_3^{2-}	carbonate		
NO_2^-	nitrite				
OH^-	hydroxide				
HCO_3^-	hydrogen carbonate (bicarbonate)				
OCl^-	hypochlorite				
CH_3COO^-	acetate				

*Some metals form ions with one charge under certain conditions and another charge under different conditions. To specify the electrical charge for these ions, Roman numerals are used in parentheses after the metal's name.

Part II:

Recall that atoms of different elements have different numbers of protons and, thus, different atomic numbers. Similarly, neutral (uncharged) atoms of different elements have different numbers of electrons. Although any atom is uniquely identified by the number of protons contained in its nucleus, the physical and particularly the chemical properties of that atom (how it interacts with other atoms) are governed largely by the number and arrangement of the atom's electrons. How is arrangement of electrons related to the structure of the periodic table? Answer the following questions to discover this connection.

11. Sort the ion cards representing monatomic (one-atom) ions into groups according to their charges.

12. Locate the elements corresponding to your ions on the periodic table. What patterns do you notice?

13. Which ions do not fit a pattern?

14. Write a statement that makes a claim about arrangements of electrons and properties within a family of elements.

Understanding properties of atoms is the key to predicting and even manipulating the chemical behavior of materials. Combined with a bit of imagination, this information allows chemists to find new uses for materials and to create new compounds to meet specific needs. In the next few sections, you will look closely at a subset of elements—metals.

A major difference between atoms of metals and nonmetals is that metal atoms lose electrons much more easily than do nonmetal atoms. Under suitable conditions, one or more electrons may be removed from a metal atom. This results in metallic elements forming positive ions (cations), because the number of positively charged protons remains unchanged, while the total number of negatively charged electrons has been decreased. The tendency of a metal to lose electrons affects its reactivity. You will explore trends in this property in the next investigation.

> Chemists synthesize several thousand new compounds every year.

■ INVESTIGATING MATTER

B.12 RELATIVE REACTIVITIES OF METALS

> Two common compounds of copper and oxygen are CuO and Cu_2O. Because the name "copper oxide" could be applied to both, a Roman numeral is added to indicate copper's ionic charge. Copper(I) oxide is Cu_2O because it contains Cu^+ ions. Copper(II) oxide is CuO; it contains Cu^{2+} ions.

Preparing to Investigate

In this investigation, you will observe the reactions of the metals copper, magnesium, and zinc with four different solutions. Each solution contains a particular cation. The solutions you will use are copper(II) nitrate, $Cu(NO_3)_2$ (containing Cu^{2+}); magnesium nitrate, $Mg(NO_3)_2$ (containing Mg^{2+}); zinc nitrate, $Zn(NO_3)_2$ (containing Zn^{2+}); and silver nitrate, $AgNO_3$ (containing Ag^+).

Before you begin, read *Gathering Evidence* to learn what you will need to do and note safety precautions. Devise a systematic procedure that allows you to observe the reaction (if any) between each metal and each of the four ionic solutions. You will conduct each reaction in a separate well of your well plate using 10 drops of the specified solution and a small strip of metal (see Figure 1.28). How many different combinations of metals and solutions will you need to observe? How will you arrange things so you can complete your observations efficiently, yet know which metal and which solution are in each well?

Figure 1.28 *Metal strips and metal nitrate solutions should be arranged in an orderly manner in the 24-well plate.*

Remember that an organized, concise data table is preferable to several pages of notes about reactions.

Write a procedure to guide your investigation and construct a data table that clearly indicates which metals and solutions will be combined and provides a place to record results and observations. Your teacher will review and approve your procedure and data table before you begin.

Gathering Evidence

1. Before you begin, put on your goggles, and wear them properly throughout the investigation.

2. Obtain 1-cm strips of each of the metals to be tested. Clean the surface of each metal strip by rubbing it with sandpaper or emery paper. See Figure 1.29. Record observations of each metal's appearance.

3. Begin your planned procedure. If no reaction is observed, write "NR" in your data table. If a reaction occurs, record the changes you observe. (*Caution: Don't allow the AgNO$_3$ solution to come in contact with skin or clothing; it causes dark, unwashable stains.*)

Figure 1.29 *Clean each metal strip with sandpaper or emery paper.*

4. Dispose of all solid samples and solutions as directed by your teacher.

5. Wash your hands thoroughly before leaving the laboratory.

Interpreting Evidence

1. Which metal reacted with the most solutions?
2. Which metal reacted with the fewest solutions?

Making Claims

3. With which solutions (if any) would you expect silver metal to react? Explain your answer, citing evidence from your data and observations.

4. List the metals (including silver) in order, placing the most reactive metal first (the one reacting with the most solutions) and the least reactive metal last (the one reacting with the fewest solutions).

5. Refer to your "metal activity series" list from Question 4.

   a. Write a brief explanation of why the outside surface of a penny is made of copper instead of zinc.

   b. Which of the four metals mentioned in this investigation might be an even better choice than copper for the outside surface of a penny? What observational evidence supports your conclusion?

   c. Why do you think the metal you chose in Question 5b is not used for the outside surface of a penny?

6. Given your new knowledge about the relative chemical activities of these four metals,

   a. which metal is most likely to be found in an uncombined, or "free," (metallic) state in nature?

   b. which metal is most likely to be found combined with other elements?

Reflecting on the Investigation

7. Reconsider your experimental design for this activity.

   a. Would it have been possible to eliminate one or more of the metal–solution combinations and still obtain all information needed to complete chemical activity comparisons for the metals?

   b. If so, which combination(s) could have been eliminated? Why?

B.13 METAL REACTIVITY

When copper metal is heated, it gradually reacts with oxygen gas in the air to produce a black substance:

$$2\ Cu(s) \quad + \quad O_2(g) \quad \longrightarrow \quad 2\ CuO(s)$$
Copper   Oxygen gas   Copper(II) oxide

Although it reacts to form black copper(II) oxide when heated, at room temperature the metal remains relatively unreactive in air. You are probably familiar with this fact from observing that copper wire and the copper surface on pennies do not turn black under normal conditions.

Magnesium metal also reacts with oxygen gas. But unlike copper metal, magnesium heated in air quickly burns in a flash of light. See Figure 1.30. The equation for this reaction is:

$$2\ Mg(s) \quad + \quad O_2(g) \quad \longrightarrow \quad 2\ MgO(s)$$
Magnesium   Oxygen   Magnesium oxide

Figure 1.30 *Magnesium and oxygen react so spectacularly that small samples of magnesium are used in some fireworks.*

By contrast, gold (Au) does not react with any components in air, including oxygen gas. This is one reason gold is highly prized in long-lasting, decorative objects, such as jewelry. Gold-plated electrical contacts, such as those used in automobile air-bag circuitry and audio cable connectors, perform reliably because nonconducting oxides do not form on the gold-plated contact surfaces.

Observing how readily a certain metal reacts with oxygen provides information about the metal's chemical reactivity. In Investigating Matter B.12 (page 71), you ranked elements in relative order of their chemical reactivities; such a ranking is called an **activity series**. Based on what you have just read about gold and magnesium and what you already know about copper, how would you rank the three metals in terms of their relative chemical reactivity? What evidence supports your ranking?

CHEM**QUANDARY**

DISCOVERY OF METALS

Copper, gold, and silver are far from being the most abundant metals on Earth. Aluminum, iron, and calcium, for example, are all much more plentiful. Why, then, were copper, gold, and silver among the first metallic elements discovered?

A list of metals in order from most to least reactive is often called an activity series.

You have explored some chemistry of metals and know, for example, that copper metal is more reactive than silver but less reactive than magnesium. A more complete activity series is given in Table 1.3, which also includes brief descriptions of common methods for retrieving each metal from its ore.

You can use such an activity series to predict whether certain reactions may occur. For example, you observed in Investigating Matter B.12 that zinc metal, which is more reactive than copper, reacted with copper ions in solution. However, zinc metal did not react with magnesium ions in solution. Why? Zinc is less reactive than magnesium. In general, a more reactive metallic element (higher in the activity series) will cause ions of a less reactive metallic element (lower in the activity series) to change to neutral metal atoms.

Table 1.3

Metal Activity Series

Elements (in order of decreasing reactivity)	Metal ion(s) found in minerals	Process used to obtain the metal	Metal
Lithium (Li)	Li^+	Pass direct electric current through the molten mineral salt (electrometallurgy)	Li(s)
Potassium (K)	K^+		K(s)
Calcium (Ca)	Ca^{2+}		Ca(s)
Sodium (Na)	Na^+		Na(s)
Magnesium (Mg)	Mg^{2+}		Mg(s)
Aluminum (Al)	Al^{3+}		Al(s)
Manganese (Mn)	Mn^{2+}, Mn^{3+}	Heat mineral with coke (C) or carbon monoxide (CO) (pyrometallurgy)	Mn(s)
Zinc (Zn)	Zn^{2+}		Zn(s)
Chromium (Cr)	Cr^{3+}, Cr^{2+}		Cr(s)
Iron (Fe)	Fe^{3+}, Fe^{2+}		Fe(s)
Lead (Pb)	Pb^{2+}	Heat (roast) mineral in air (pyrometallurgy) or find the metal free (uncombined)	Pb(s)
Copper (Cu)	Cu^{2+}, Cu^+		Cu(s)
Mercury (Hg)	Hg^{2+}		Hg(l)
Silver (Ag)	Ag^+		Ag(s)
Platinum (Pt)	Pt^{2+}, Pt^{4+}		Pt(s)
Gold (Au)	Au^{3+}, Au^+		Au(s)

DEVELOPING SKILLS

B.14 TRENDS IN METAL REACTIVITY

Sample Problem: Will Pb metal react with Ag^+ ions?

Yes; according to Table 1.3, lead is a more reactive metal than silver, so lead metal will cause silver ions to change to silver metal.

Use Table 1.3 and the periodic table (page 60) to answer these questions.

1. a. What trend in metallic reactivity is found as you move from left to right across a horizontal row (period) of the periodic table? (*Hint:* Compare the reactivity of sodium with magnesium and aluminum.)

    b. In which part of the periodic table are the most-reactive metals found?

    c. Which part of the periodic table contains the least-reactive metals?

2. a. Will iron (Fe) metal react with a solution of lead(II) nitrate, $Pb(NO_3)_2$?

   b. Will platinum (Pt) metal react with a lead(II) nitrate solution?

   c. Explain your answers to Questions 2a and 2b.

3. Use specific examples from the activity series in your answers to these two questions:

   a. Are the least-reactive metals also the least expensive metals?

   b. If not, what other factor(s) might influence the market value of a metal?

■ MAKING DECISIONS

B.15 CHARACTERISTICS OF CURRENCY

In Making Decisions A.11, you created a list of some of the properties that are necessary or desirable in currency. Reflect upon that list as well as facts and ideas you learned in this section, including periodic trends and characteristics of metals, as you complete the following activities.

Part I:

1. Think about coins that are currently in circulation, including the dollar coin.

   a Which elements compose these coins?

   b. Where are these elements located on the periodic table?

2. Consider the metals you observed or learned about in this section.

   a. Which metals would be unsuitable for use in coins? Why?

   b. Would you eliminate the other metals in that family? Why or why not?

3. Review your list of characteristics from Making Decisions A.11 and make changes or additions as needed.

Part II:

The Web page that opened this unit noted that dollar coins have been introduced in the United States in the past without resulting widespread circulation and use. To make recommendations about currency, you will need to be familiar with arguments that groups and individuals make for use of particular forms of currency.

4. How will you find information about support for or resistance to dollar coins and bills?

5. List several objections to dollar coins, as well as the source for each objection.

6. What objections to, or support for, dollar bills were you able to find? List the source of each item.

7. After completing this task, how would you modify your information-gathering strategy?

SECTION B SUMMARY

Reviewing the Concepts

An atom is composed of smaller partcles (protons, neutrons, and electrons), each possessing a characteristic mass and electrical charge. The number of protons in an atom of a given element distinguishes it from atoms of all other elements. Atoms containing the same number of protons but different numbers of neutrons are considered isotopes.

1. For each of these elements, identify the number of protons or electrons needed for an electrically neutral atom.

   a. carbon: 6 protons   __ electrons

   b. aluminum: __ protons   13 electrons

   c. lead: 82 protons   __ electrons

   d. chlorine: __ protons   17 electrons

2. Decide whether each of these atoms is electrically neutral.

   a. sulfur: 16 protons   18 electrons

   b. iron: 26 protons   24 electrons

   c. silver: 47 protons   47 electrons

   d. iodine: 53 protons   54 electrons

3. Copy and complete the table located below for each electrically neutral atom.

Element Symbol	Number of Protons	Number of Neutrons	Number of Electrons
	6	6	6
	6	7	6
Ca		21	
		117	78
U		146	

4. A student is asked to explain the formation of a lead(II) ion (Pb^{2+}) from an electrically neutral lead atom (Pb). The student says that a lead atom must have gained two protons to make the ion. How would you correct this student's explanation?

5. Refer to the table provided for Question 3.

   a. Calculate the mass number for each element in the table.

   b. Which element has two isotopes in the table?

Isotopes of Magnesium			
Isotope Symbol	Mass Number	Number of Protons	Number of Neutrons
Mg-24	24	a	b
Mg-25	25	c	d
Mg-26	26	e	f

6. How many protons and neutrons are needed for each of the isotopes in the table above?

7. How does the mass of the electron compare to the masses of the proton and neutron?

8. A scientist announces the discovery of a new element. The only characteristic given in the report is the element's mass number of 266. Is this information sufficient, by itself, to justify the claim of the discovery of a new element? Explain.

> Elements are arranged in the periodic table based on their properties. Elements with similar chemical properties are placed in the same columns. Physical properties vary in predictable patterns across rows and down columns.

9. Give another term for these features of the periodic table:

   a. Row   b. Column

10. Give the names and symbols of two elements other than lithium in the alkali metal family.

11. Consider the noble gas family.

    a. Where are they located on the periodic table?

    b. Name one physical property that they share.

    c. Name one chemical property that they share.

12. Given a periodic table and the formulas $BeCl_2$ and AlN, predict the formula for a compound containing

    a. Mg and F   b. Ga and P

13. The melting points of sulfur (S) and tellurium (Te) are 115 °C and 450 °C, respectively. Estimate the melting point of selenium (Se).

14. Would you expect the boiling point of chlorine to be higher or lower than that of iodine? Explain.

> The properties of an element are determined largely by the number and arrangement of electrons in its atoms.

15. Are atoms of metallic or non-metallic elements more likely to lose one or more electrons?

16. Predict whether each element would be more likely to form an anion or a cation. (Note that anions are *negatively* charged; cations are *positively* charged.)

    a. sodium   d. copper   g. tin

    b. calcium   e. oxygen   h. iodine

    c. fluorine   f. lithium

17. Noble gas elements rarely lose or gain electrons. What does this indicate about their chemical reactivity?

> Ionic compounds are composed of positively and negatively charged ions (atoms that have lost or gained electrons), combined so that the compound has no net electrical charge.

18. Classify each of these as an electrically neutral atom, an anion, or a cation.

    a. O^{2-} b. Li   c. Cl   d. Ag^+ e. Hg^{2+}

19. For each particle in Question 18, indicate whether the electrical charge or lack of electrical charge resulted from an atom gaining electrons, losing electrons, or neither.

20. Write the symbol and show the electrical charge (if any) on the following atoms or ions:

    a. hydrogen with 1 proton and 1 electron

    b. sodium with 11 protons and 10 electrons

    c. chlorine with 17 protons and 18 electrons

    d. aluminum with 13 protons and 10 electrons

21. Write the name and formula for the ionic compound that can be formed from these cations and anions:

    a. K^+ and I^- d. Ba^{2+} and OH^-

    b. Ca^{2+} and S^{2-} e. NH_4^+ and PO_4^{3-}

    c. Fe^{3+} and Br^- f. Al^{3+} and O^{2-}

> Tables, graphs, and models are all used to represent scientific data and illustrate scientific ideas so that they are easier to analyze, interpret, and understand.

22. Why was the periodic table created?

23. What type of graph should be used to represent discontinuous data? Why?

24. How is a data table helpful

    a. before an investigation?

    b. during an investigation?

    c. after an investigation?

25. Think about the ion cards you used in Modeling Matter B.11.

   a. How are they helpful in learning about ions and ionic compounds?

   b. How are they imperfect models of ions?

> Metals react with one another in predictable patterns according to their reactivities.

26. Which of these reactions is more likely to occur? Why? (Refer to Table 1.3, page 75.)

   a. Calcium metal with chromium(III) chloride solution.

   b. Chromium metal with calcium chloride solution.

27. Consider these two equations. Which represents a reaction that is more likely to occur? Why?

   a. $Zn^{2+}(aq) + 2\ Ag(s) \longrightarrow Zn(s) + 2\ Ag^{+}(aq)$

   b. $2\ Ag^{+}(aq) + Zn(s) \longrightarrow 2\ Ag(s) + Zn^{2+}(aq)$

28. a. Why would it be a poor idea to stir a solution of lead(II) nitrate with an iron spoon? (See Table 1.3, page 75.)

   b. Write a chemical equation to support your answer.

How can the periodic table be used to help explain and predict the properties of chemical elements?

Throughout Section B, you have observed and studied elements and the periodic table. Think about what you have learned, then answer the question in your own words in organized paragraphs. Your answer should demonstrate your understanding of the key ideas in this section.

Be sure to consider the following in your response: properties of elements, subatomic particles, periodicity and trends, and reactivity.

Connecting the Concepts

29. Which pair is more similar chemically? Defend your choice.

   a. copper metal and copper(II) ions

      or

   b. oxygen with mass number 16 and oxygen with mass number 18

30. The diameter of a magnesium ion (Mg^{2+}) is 156 pm (picometers, where $1\ pm = 10^{-12}$ m); the diameter of a strontium ion (Sr^{2+}) is 254 pm. Estimate the diameter of a calcium ion (Ca^{2+}).

31. Identify each element in the periodic table described by the statements below:

   a. This element is a nonmetal. It forms anions with a 1– charge. It is in the same period as the metals used in a penny.

   b. This element is a metalloid. It is in the same period as the elements found in table salt.

32. Mendeleev arranged elements in his periodic table in order of their atomic masses. In the modern periodic table, however, elements are arranged in order of their atomic numbers. Cite two examples from the periodic table for which these two schemes would produce a different ordering of adjacent elements.

33. In building ships, common practice is to attach a piece of magnesium to the hull to act as a "sacrificial anode." In terms of metal activity, explain how this helps to prevent the corrosion of other metal parts of the ship.

Extending the Concepts

34. Construct a graph of the price per gram of an element versus its atomic number for each of the first 20 elements. Can the current cost of those elements be regarded as a periodic property? Explain. (*Hint:* Use a chemical supply catalog or the Web to locate the current price of each element.)

35. Although aluminum is a more reactive metal than is iron, it is often used for outdoor products. Investigate why this makes sense.

CHEMISTRY *AT WORK*
Q&A

Fireworks are one of the most striking parts of every Independence Day. But have you ever wondered how fireworks, well, work? Pyrotechnic chemists are experts on how these and other types of explosives function. They study how to make fireworks and other pyrotechnic devices safe to use and better for the environment. Have a blast reading about one pyrotechnic chemist and his work!

John Conkling, Pyrotechnic Chemist at Washington College in Chestertown, Maryland

Q. What is pyrotechnic chemistry?

A. Pyrotechnic chemistry is the chemistry of producing heat from chemical reactions and using that heat to produce color and light and audible effects. We use pyrotechnic chemistry not only for entertainment, but also for practical purposes like emergency signaling and military applications.

Q. How did you get into this field?

A. While I was teaching undergraduate chemistry, I was approached by a fireworks company that wanted to hire me for a side project developing chemical compositions for fireworks that made them safe to transport and store. I got really interested in the chemistry of fireworks, so I wrote some articles about it. One ended up being the cover story for a chemistry magazine! Once that published, the army called and wanted to work with me on some military pyrotechnic applications, and my pyrotechnic chemistry career really shot off. Nowadays, I do training seminars for people interested in anything that explodes: from people who design and manufacture fireworks to people who dispose of bombs.

Q. How do fireworks work?

A. Every fireworks mixture needs at least one chemical that's oxygen rich and one chemical that acts as a fuel. By choosing these compounds carefully, we can determine how much heat will be produced once a firework's fuse is lit, how fast a reaction will take place, and what solid and gas products will be produced. To make fireworks more attractive, we include compounds that color the flames produced when they explode. Different elements produce different colors when they burn. For example, strontium compounds burn bright red. Barium compounds produce a green light. Sodium compounds have a yellow-orange flame.

Q. How do you create a firework's looks, patterns, and sounds?

A. This is where chemistry and artistry intersect. To create beautiful pyrotechnic effects, we work with an artist. Based on the colors and effects we want to produce, we prepare a mixture of compounds, which is usually in a powder form. This powder is packed into pellets that we call "stars," which are placed around an explosive material that we call the "charge." The size of these stars determines the duration of the light you see once the charge ignites. For example, if the stars are small, like peppercorns, you get an explosion of short duration. If you have bigger stars, from the size of peas to sugar cubes, the burn lasts longer. If you paste these stars into a geometric figure onto stiff paper, you can make them explode into patterns like an expanding heart or circle. All fireworks are packaged into a cardboard casing that's wrapped in layers of papier mâché to make a hard exterior surface. When the firework is lit, the resulting combustion reaction produces hot gas that makes the casing expand, then shatter. A canister that's tightly wrapped will hold the pressure of a gas longer, then explode with a big bang. A canister that's more loosely wrapped will explode with less energy, producing a softer sound.

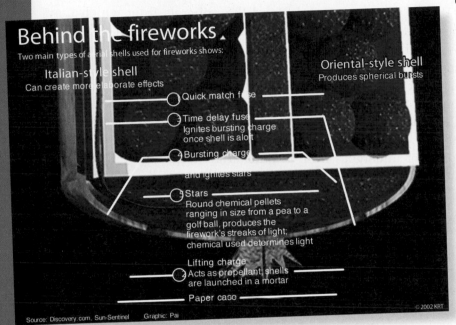

Behind the fireworks.
Two main types of aerial shells used for fireworks shows:

Italian-style shell
Can create more elaborate effects

Oriental-style shell
Produces spherical bursts

Quick match fuse

Time delay fuse
Ignites bursting charge
once shell is aloft

Bursting charge
and ignites stars

Stars
Round chemical pellets
ranging in size from a pea to a
golf ball, produces the
firework's streaks of light;
chemical used determines light

Lifting charge
Acts as propellant; shells
are launched in a mortar

Paper case

Source: Discovery.com, Sun-Sentinel   Graphic: Pai   © 2002 KRT

Q. How do you make fireworks safe?

A. Although the artistic effect is important, we also have to keep other considerations in mind when we design new fireworks. One very important consideration is safety: You don't want compositions that explode if you drop them on the floor. It's important to develop stable compositions that ignite only when desired. Fireworks were invented hundreds of years ago, and we've learned over the centuries to avoid certain chemicals and compositions that are too easy to ignite accidentally. There is also a big push now to make fireworks as "green" as possible—I'm not talking about those with a beautiful green flame, but those that don't pose a hazard to the environment.

Q. How do I become a pyrotechnic chemist?

A. It's an explosive field and one that is a lot of fun! You should take as many chemistry and physics classes as you can while you're in school. These classes will give you the background you need to understand the reactions that take place in pyrotechnic devices. Also, don't experiment on your own with pyrotechnics! There are a lot of formulas for explosives that are easy to obtain, but it's important to stay safe.

CELEBRATE JULY 4TH SAFELY

SECTION C MINERALS AND MOLES

What is the role of chemistry in the life cycle of metals?

Among Earth's resources, metals—and the minerals from which they are extracted—have long been used by humans. Those uses have ranged from tool making, energy transmission, and construction to works of art, decoration, and coin making. In this section, you will explore the properties and uses of minerals and metals. You will also learn about Earth's mineral resources and how some are converted to pure metals. This will allow you to begin to consider the life cycles of materials and products that you use. A **life cycle** is the sequence of steps that a material or product undergoes from raw materials to product to final disposal. In the opening Web page, Ms. Harris hints at the life cycles of various U.S. dollars when she discusses time of use and recycling options. Studying the topics in this section will help you to assess the statements in the Web page.

GOALS

- Describe or recognize factors that determine the feasibility of mining an ore at a specific site.

- Gather and evaluate information from a variety of sources.

- Describe Earth's atmosphere, hydrosphere, and lithosphere, including the distribution of resources among them.

- Apply the mole concept to calculations, including finding the molar mass of a compound given its formula and average atomic masses of its elements.

- Calculate the percent composition by mass of a specified element in a given compound.

- Define oxidation and reduction in terms of electron loss or gain. Identify and distinguish between oxidation and reduction processes.

- Represent oxidation and reduction processes using chemical equations and electron-dot structures.

- Explain why minerals of more reactive metals are more difficult to refine and process than are minerals of less reactive metals.

concept check 5

1. What is a substance?
2. Given a sample of a material, how could you determine whether or not it is a metal?
3. Are naturally occurring metals usually pure? Explain.

■ MAKING DECISIONS

C.1 METAL RESOURCES, PRODUCTION, AND USE

The extraction and processing of coinage metals (see Figure 1.31) can serve as a model for the life cycle of metals in general. Over the centuries, many metallic elements have been used in coins, including the following: aluminum, antimony, chromium, cobalt, copper, gold, iron, lead, magnesium, manganese, molybdenum, nickel, niobium, platinum, rhenium, selenium, silver, tin, titanium, tungsten, vanadium, zinc, and zirconium.

Figure 1.31 *Gold miners in Indonesia use a high-pressure water jet to blast through soil at a mine.*

Table 1.4 provides information on worldwide production of several metals. Note that the quantity of each metal produced varies. For example, nearly 100 times more aluminum is produced worldwide than tin. Why is this? Is aluminum more abundant than tin? Or are there more uses and demand for aluminum than for tin? How do the physical and chemical properties of these metals affect their production and eventual use?

With your teacher's assistance, decide which metal you will investigate. Each individual or group will share their results with the rest of the class. You will need these answers at the end of this section, so be sure to keep them for future reference.

Use library and Internet resources to answer the following questions about your selected metal.

Resources and Distribution

1. How abundant is it?
2. What is its origin; that is, on what continents and in which countries is your metal commonly found?
3. In what form(s) is it most likely to exist in nature?
4. What types of mining are used to remove it from the Earth?

Production

> For the purposes of this activity, important forms of energy include electrical, thermal (heat), mechanical, and chemical energy.

5. How is it processed?
6. What forms of energy are needed for this processing?
7. In what form is the metal transported for further processing or final use?

Properties and Uses

8. Investigate your metal's properties (both physical and chemical). Create a summary table listing its important properties.
9. List at least three important current uses for your metal.
10. Choose one use from your list in Question 9. Identify particular properties that make the metal an appropriate choice for this application.
11. Are the uses and demand for the metal adequately met by its production?

Table 1.4

Production of Selected Metals Worldwide, 2008

Metal	Country	Percent Production	Actual Production (1000 metric tons)	World Total Production (1000 metric tons)
Aluminum	China	34%	13 500	39 700
	Russia	11%	4200	
	Canada	8%	3100	
	United States	7%	2640	
	Australia	5%	1960	
Copper	Chile	36%	5600	15 700
	United States	8%	1310	
	Peru	8%	1220	
	China	6%	1000	
	Australia	5%	850	
Iron ores	China	35%	770 000	2 200 000
	Brazil	18%	390 000	
	Australia	15%	330 000	
	India	9%	200 000	
	Russia	5%	100 000	
Lead	China	41%	1540	3 800
	Australia	15%	576	
	United States	12%	440	
	Peru	9%	335	
	Mexico	4%	145	
Nickel	Russia	17%	276	1 610
	Canada	16%	250	
	Indonesia	13%	211	
	Australia	11%	180	
	New Caledonia	6%	92.6	
Silver	Peru	17%	3.6	21
	Mexico	14%	3.0	
	China	12%	2.6	
	Chile	10%	2.0	
	Australia	9%	1.8	
Tin	China	45%	150	333
	Indonesia	30%	100	
	Peru	11%	38	
	Bolivia	5%	16	
	Brazil	4%	12	
Zinc	China	28%	3200	11 300
	Australia	13%	1510	
	Peru	13%	1450	
	United States	7%	770	
	Canada	6%	660	

Source: http://minerals.usgs.gov/minerals/pubs/commodity/

C.2 SOURCES AND USES OF METALS

Human needs for resources—whether to create a new coin, make clothing, construct a space-vehicle rocket engine, or fertilize food crops—must all be met by chemical supplies currently present on Earth. These supplies of resources are often cataloged by where they are found. The table in Figure 1.32 indicates the chemical composition of Earth.

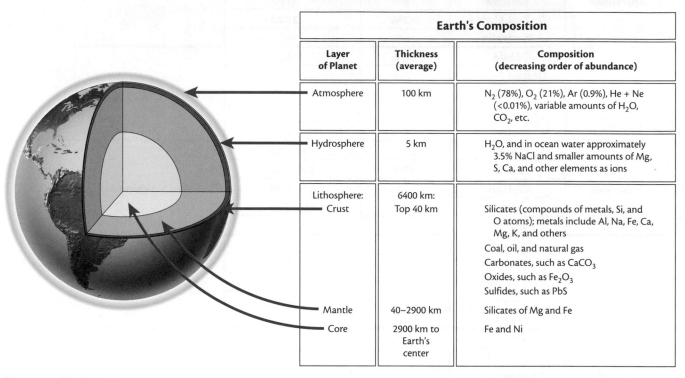

Earth's Composition		
Layer of Planet	**Thickness (average)**	**Composition (decreasing order of abundance)**
Atmosphere	100 km	N_2 (78%), O_2 (21%), Ar (0.9%), He + Ne (<0.01%), variable amounts of H_2O, CO_2, etc.
Hydrosphere	5 km	H_2O, and in ocean water approximately 3.5% NaCl and smaller amounts of Mg, S, Ca, and other elements as ions
Lithosphere: Crust	6400 km: Top 40 km	Silicates (compounds of metals, Si, and O atoms); metals include Al, Na, Fe, Ca, Mg, K, and others Coal, oil, and natural gas Carbonates, such as $CaCO_3$ Oxides, such as Fe_2O_3 Sulfides, such as PbS
Mantle	40–2900 km	Silicates of Mg and Fe
Core	2900 km to Earth's center	Fe and Ni

Figure 1.32 *Earth's composition. From what layer do most resources come that support human activities?*

Earth's atmosphere, hydrosphere, and outer layer of the lithosphere supply resources for all human activities. The **atmosphere** provides nitrogen, oxygen, neon, and argon. From the **hydrosphere** come water and some dissolved minerals. The **lithosphere**, which is the solid part of Earth, provides the greatest variety of chemical resources. For example, petroleum and metal-bearing ores are found in the lithosphere. An **ore** is a naturally occurring rock or mineral that can be mined, and from which it is profitable to extract a metal or other material. Naturally occurring collections of ores in the lithosphere are known as **deposits**. An ore deposit contains several components. Of these, **minerals** are the most important. They are naturally occurring solid compounds containing the element or group of elements of interest.

The deepest mines on Earth barely scratch the surface of its crust. If Earth were the size of an apple, all accessible resources of the lithosphere would be located within the apple's skin. From this thin band of soil and rock, we obtain the major raw materials needed to build homes, automobiles, appliances, computers, DVDs, and sports equipment—in fact, all manufactured objects.

As you can see from Table 1.4 (page 85), many of Earth's resources are not uniformly distributed. There is no relationship between a nation's supply of

Figure 1.33 *Mining metal ores has economic and environmental consequences. What impact might mining have on a local community?*

Figure 1.34 *Some minerals among these samples are mined and processed to yield useful metals.*

these resources and either its land area or its population. A particular region may be the predominant supplier of certain metals to industry. For example, South Africa supplies more manganese and platinum-group metals than any other nation, while Chile is the top supplier of copper worldwide.

The growth of the United States as a major industrial nation has been facilitated, in part, by the quantity and diversity of its mineral resources. Yet, in recent years, the United States has imported increasing quantities of some mineral resources. For example, about 79% of the nation's tin (Sn) is imported. See Figures 1.33 and 1.34 for examples of minerals and mining.

The greatest challenge regarding mineral resources is deciding on the wisest uses of available supplies. Making these decisions involves weighing alternatives and considering consequences at every step in a metal's life cycle. For example, is it worthwhile to mine a certain metallic ore at a particular site? The answer to this depends on several considerations:

- quantity of useful ore found at the site,
- percent of metal in the ore,
- type of mining and processing needed to extract the metal from its ore,
- distance of the mine from metal refining facilities and markets,
- metal's supply-versus-demand status, and
- environmental impact of the mining and processing.

Copper, a common metal in coins, provides a case study of a vital chemical resource. Copper is one of the most familiar and widely used metals in modern society. Among all the elements, it is second only to silver in electrical conductivity. This property and its relatively low cost, corrosion resistance, and ductility (ease of being drawn into thin wires) make copper the world's most common metal for electrical wiring. Copper is also used to produce brass, bronze, and other alloys; a variety of copper-based compounds; jewelry; and works of art. Table 1.5 summarizes copper's physical and chemical properties.

Table 1.5

Properties of Copper	
Malleability and ductility	High
Electrical conductivity	High
Thermal conductivity	High
Chemical reactivity	Relatively low
Resistance to corrosion	High
Useful alloys formed	Bronze, brass, etc.
Color and luster	Reddish, shiny

The first copper ores mined were relatively rich in copper—from 35% to 88% by mass. Such rich ores are no longer available, but ores containing less copper can be used. In fact, it is now possible to economically mine ores containing less than 1% copper. Copper ore is chemically processed to produce metallic copper, which is then transformed into a variety of useful products. Figure 1.35 summarizes the copper cycle from sources to common uses to waste products.

How do we determine the quantity of useful ore at a site? How does this relate to the profitability of mining at that site? What roles do processing and demand play in making resource use decisions? How do all of these considerations affect your choice of a coin material? The following activity will help you begin to address these questions.

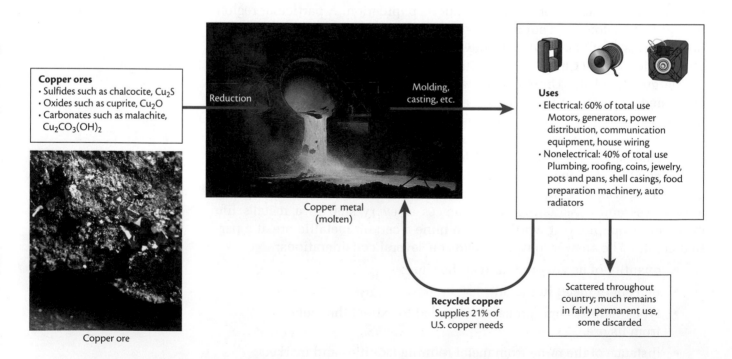

Figure 1.35 *The life cycle of copper includes mining copper ores, reducing the ore to obtain the metal, fashioning the metal for final use, and then either recycling the metal or discarding it. What effect does recycling have on the energy these processes require?*

INVESTIGATING MATTER
C.3 EXTRACTING ZINC

Preparing to Investigate

Early in this unit, you investigated some characteristic properties of matter. In that investigation, you learned that the inner core of modern pennies is made from a different substance than its exterior. In this investigation, you will extract that core material, zinc.

Before you begin, read *Gathering Evidence* to learn what you will need to do and note safety precautions. *Gathering Evidence* outlines a procedure to remove the zinc core from a penny. Construct a data table that provides space to record observations and measurements. You will record several masses (in grams) throughout this investigation; it will be important to distinguish among them. It may be helpful to use the left column of your data table to identify the objects for which mass is determined (e.g., *clean, nicked penny before reaction*) and the right column or columns for measured masses and observations.

Gathering Evidence

1. Before you begin, put on your goggles, and wear them properly throughout the investigation.

2. Obtain and clean, as directed by your teacher, a post-1982 U.S. penny.

3. Notch the edge of the penny using a metal file. Ensure that you nick the penny deep enough so that its inner zinc core is visible.

4. Measure and record the mass of your clean, nicked penny. ←——

5. Obtain 50 mL of 3.0 M HCl in a clean 100- or 250-mL beaker. (***Caution:*** *3.0 M hydrochloric acid (HCl) may irritate or damage your skin. If any hydrochloric acid accidentally spills on you, ask a classmate to notify your teacher immediately. Wash the affected area immediately with tap water and continue rinsing for several minutes.)*

6. Using tongs, gently place the clean, nicked penny into the beaker containing HCl. (***Caution:*** *Do not look directly down into the beaker or place your face directly above the beaker.)* ←

7. Observe the interactions between the penny and the solution. What happens to the penny? What happens to the solution? Does anything happen outside the penny/solution system? Record your observations.

> In science, a system refers to the part of the world that we want to study—in this case, everything in the beaker.

8. Store your beaker in the place indicated by your teacher so that the system can continue to react overnight.

9. Remove any remaining pieces of the penny and place on a pre-weighed paper towel or weighing boat. Allow to dry overnight.

10. Record the mass of the dry penny piece and weighing container. ←

Analyzing Evidence

1. Calculate the mass of "copper remaining from penny" by determining the mass of the pieces of penny that remained after reaction with HCl.

2. Calculate the mass of "zinc extracted" by subtracting mass of "copper remaining from penny" from mass of "clean, nicked penny before reaction."

Interpreting Evidence

1. What observations indicate that a chemical reaction took place?

2. Do you think that all of the zinc was extracted from the penny? Explain.

Reflecting on the Investigation

3. Think about a procedure that could recover the extracted zinc from solution. To do this, consider what you learned about metal reactivity in Section B (pages 71–76).

   a. Which metals can release zinc from its compounds?

   b. Create a table listing these metals in the left column and title the right column "Suitability for This Investigation," as shown below:

Metal	Suitability for This Investigation
Sodium	

   Complete the table, noting suitability considerations in the right column. For instance, sodium, as an alkali metal, is very reactive and not suitable for use in this investigation.

   c. After completing your table, draft a procedure that could recover zinc from your solution and determine the mass recovered. Consider not only the identity, but the form (see Figure 1.36) of the metal that you would choose.

4. Would it be easier or more difficult (compared to zinc) to recover silver from solution? Explain your reasoning.

5. Would it be feasible to remove the copper outer layer and recover it from solution instead of the zinc core? Why or why not?

Figure 1.36 *Zinc, shown here, and other metals are commercially available in many forms, including sheets, shot, granules, wire, and powder. Why are different forms needed for different applications? Which forms would work best for this investigation?*

C.4 COMPOSITION OF MATERIALS

One decision coin designers must make is whether to use only one material or a combination of materials. If the design uses more than one material, designers must decide how much of each material to use in the coin. The percent by mass of each component found in a sample such as a coin is called its **percent composition**.

In Section A, you learned that the composition of the U.S. penny has changed several times. During 1943, it was made of zinc-coated steel. After this date and up until 1982, the penny was made mostly of copper. Since 1982, U.S. pennies have been made primarily of zinc. A post-1982 penny with a mass of 2.500 g is composed of 2.4375 g zinc and 0.0625 g copper. The percent composition of the penny can be found by dividing the mass of each constituent metal by the mass of the penny and multiplying by 100%:

$$\frac{2.4375 \text{ g zinc}}{2.500 \text{ g total}} \times 100\% = 97.50\% \text{ zinc}$$

$$\frac{0.0625 \text{ g copper}}{2.500 \text{ g total}} \times 100\% = 2.50\% \text{ copper}$$

DEVELOPING SKILLS

C.5 APPLICATIONS OF PERCENT COMPOSITION

Percent composition helps geologists to describe how much metal or mineral is present in a particular ore and to evaluate whether the ore should be mined. Percent composition is also commonly used in describing contents of foods and other products, as well as the composition of the human body.

> *Sample Problem: A Roosevelt dime, the U.S. 10-cent coin since 1946, has a mass of 2.268 g. It is composed of 8.33% nickel and 91.67% copper. Calculate the mass of each metal present.*
>
> Given the percent composition, the mass of each metal can be calculated by multiplying the mass of the coin by the percent of that metal.
>
> 2.268 g $\times$ 8.33% nickel = 0.189 g nickel
>
> 2.268 g $\times$ 91.67% copper = 2.079 g copper

1. Calculate the percent composition of the penny you used in Investigating Matter C.3.

2. The bodies of elite distance runners are often composed of far less body fat than the average U.S. citizen. Calculate:
   a. The percent body fat of a 58-kg female distance runner with 9.1 kg of body fat.
   b. The lean body mass (body mass–body fat) of a 67-kg male marathoner with 4.8% body fat.

3. In the United States, whole milk contains 3.25% butterfat, whereas reduced-fat milk contains ~2% butterfat. If a gallon of whole milk has a mass of ~3.9 kg, about how much more butterfat does it contain compared to a gallon of reduced-fat milk?

4. Strict guidelines apply to the labeling of coffee produced from Hawaii-grown coffee beans. Consider a coffee labeled as follows:

50% MOLOKAI COFFEE ALL HAWAIIAN
Contains: 25% Kauai Coffee, 15% Kona Coffee, 10% Maui Coffee

   a. What mass of coffee beans from Kauai would be contained in 1 lb (454 g) of this coffee?
   b. What mass of Kona coffee beans would be contained in 62 kg of Molokai coffee?

5. Current U.S. \$1 coins (both the Presidential \$1 coin and the Sacagawea golden dollar) are composed of a manganese–brass consisting of 6.0% zinc, 3.5% manganese, 2.0% nickel, and the balance copper. Each dollar has a mass of 8.1 g. Calculate the mass of each of the four metal constituents in the dollar coin.

C.6 INTRODUCTION TO THE MOLE CONCEPT

As you know, metals are most often found in nature as minerals contained in ores. Common minerals in copper-bearing ores include sulfide and oxide compounds of copper. Chalcopyrite, $CuFeS_2$, is currently the most abundant mineral source of copper. See Figure 1.37. To determine the profitability of mining a certain deposit, geologists first need to be able to calculate the copper content of the mineral.

You learned earlier in this unit that the formula $CuFeS_2$ means that a formula unit of chalcopyrite contains one atom of copper, one atom of iron, and two atoms of sulfur. Although correct, this interpretation involves very small quantities of material. Also, we still cannot calculate the copper content of the mineral without more information.

Figure 1.37 *Chalcopyrite (pronounced "kal-ko-pie-rite") is found in many ore deposits, including the Olympic Dam deposit in Australia.*

The Mole

Chemists have devised a counting unit called the **mole** (symbolized *mol*) that is useful in calculating the metal content of ores and solving similar problems. You are familiar with other common counting units such as "pair" and "dozen." The mole can be thought of as the chemist's "dozen."

A pair of water molecules is two water molecules. One dozen water molecules is the same as 12 water molecules. One mole of water molecules is 602 000 000 000 000 000 000 000 water molecules. This special number—the number of particles (or "things") in one mole—is more conveniently written as 6.02×10^{23}. Either way, this is a very large number!

To help you get a better idea of the size of a mole, consider this analogy: Imagine stringing a mole of paper clips (6.02×10^{23} paper clips) together and wrapping the paper-clip string around the world. It would circle the world about 400 trillion (4×10^{14}) times! And even if you connected a million paper clips each second, it would take you 190 million centuries to finish stringing together one mole of paper clips.

6.02×10^{23} is called *Avogadro's number.*

The mole concept permits us literally to "count by weighing."

As large as one mole seems, however, drinking only one mole of water molecules would leave you quite thirsty on a hot day. One mole of water is

less than one-tenth of a cup of water—only 18 g (or 18 mL) of water. Because atoms and molecules are so very small, the number of those tiny building blocks needed to make up the "stuff" we see is nearly countless.

That is the reason the mole is so useful for work in chemistry. It represents a number of atoms, molecules, or formula units large enough to be conveniently weighed or measured in the laboratory. Furthermore, the average atomic masses of elements can be used to find the mass of one mole of any substance, a value known as the **molar mass** of a substance. Figure 1.38 shows a mole of several familiar substances.

Figure 1.38 *One mole each of copper, table salt, and water. If these samples each represent one mole of substance, why do they occupy different volumes?*

Molar Mass

Because mass is a readily measured quantity, the molar mass can be used to count particles by simply measuring the mass of material. Specific examples will help you to better understand this powerful idea. (See Figure 1.39.) Suppose you need to find the molar masses of carbon (C) and of copper (Cu). In other words, you want to know the mass of one mole of carbon atoms (6.02×10^{23} atoms) and the mass of one mole of copper atoms (6.02×10^{23} atoms). Rather than counting that collection of atoms onto a laboratory balance, you can quickly get the correct answers from average atomic mass data. The average atomic mass of each element is found on the periodic table. (Table 1.6, pages 96–97, organizes the elements alphabetically and also pro-

Figure 1.39 *One mole of carbon (left) and copper (right). What do the different samples have in common?*

vides average atomic mass data.) Carbon's average atomic mass is 12.01; copper's is 63.55. If the unit *grams* is attached to these numbers, the resulting values represent their molar masses:

$$1 \text{ mol C} = 12.01 \text{ g C} \quad 1 \text{ mol Cu} = 63.55 \text{ g Cu}$$

In summary, the mass (in grams) of one mole of an element's atoms equals the numerical value of that element's average atomic mass. Any element's molar mass can be easily found from the periodic table. The molar mass of a diatomic element, such as O_2, is twice the mass given in the periodic table.

The molar mass of a compound is simply the sum of the molar masses of its component atoms. For example, consider two copper-containing minerals—copper(I) oxide, Cu_2O, and malachite, $Cu_2CO_3(OH)_2$.

We begin by taking an atom inventory: one mole of Cu_2O contains two moles of Cu atoms and one mole of O atoms. So, to calculate the molar mass of Cu_2O, we add twice the molar mass of copper to the molar mass of oxygen giving

$$2 \text{ mol Cu} \times \frac{63.55 \text{ g Cu}}{1 \text{ mol Cu}} = 127.1 \text{ g Cu}$$

$$1 \text{ mol O} \times \frac{16.00 \text{ g O}}{1 \text{ mol O}} = 16.00 \text{ g O}$$

Molar mass of Cu_2O = (127.1 g + 16.00 g) = 143.1 g Cu_2O

The molar mass of malachite is found in a similar way. However, we must carefully count the total atoms of each element in more complex compounds such as this.

One mole of malachite, $Cu_2CO_3(OH)_2$, contains 2 mol Cu, 1 mol C, 5 mol O, and 2 mol H.

$$2 \text{ mol Cu} \times \frac{63.55 \text{ g Cu}}{1 \text{ mol Cu}} = 127.1 \text{ g Cu}$$

$$1 \text{ mol C} \times \frac{12.01 \text{ g C}}{1 \text{ mol C}} = 12.01 \text{ g C}$$

$$5 \text{ mol O} \times \frac{16.00 \text{ g O}}{1 \text{ mol O}} = 80.00 \text{ g O}$$

$$2 \text{ mol H} \times \frac{1.008 \text{ g H}}{1 \text{ mol H}} = 2.016 \text{ g H}$$

Molar mass of $Cu_2CO_3(OH)_2$ = 221.1 g $Cu_2CO_3(OH)_2$

Thus, the molar mass of a compound is found by first multiplying the moles of each element represented in the formula by the molar mass of that element. Then all of the element masses are added.

As you learned in Section B, average atomic masses are weighted averages of the masses of an element's isotopes.

Recall that *mol* is the symbol for the unit *mole*.

Copper(I) oxide is known as "cuprite" in the mining industry.

Table 1.6

Chart of the Elements							
Element	Symbol	Atomic Number	Average Atomic Mass	Element	Symbol	Atomic Number	Average Atomic Mass
Actinium	Ac	89	[227]	Erbium	Er	68	167.26
Aluminum	Al	13	26.98	Europium	Eu	63	151.96
Americium	Am	95	[243]	Fermium	Fm	100	[257]
Antimony	Sb	51	121.76	Fluorine	F	9	19.00
Argon	Ar	18	39.95	Francium	Fr	87	[223]
Arsenic	As	33	74.92	Gadolinium	Gd	64	157.25
Astatine	At	85	[210]	Gallium	Ga	31	69.72
Barium	Ba	56	137.33	Germanium	Ge	32	72.64
Berkelium	Bk	97	[247]	Gold	Au	79	196.97
Beryllium	Be	4	9.01	Hafnium	Hf	72	178.49
Bismuth	Bi	83	208.98	Hassium	Hs	108	[277]
Bohrium	Bh	107	[272]	Helium	He	2	4.003
Boron	B	5	10.81	Holmium	Ho	67	164.93
Bromine	Br	35	79.90	Hydrogen	H	1	1.008
Cadmium	Cd	48	112.41	Indium	In	49	114.82
Calcium	Ca	20	40.08	Iodine	I	53	126.90
Californium	Cf	98	[251]	Iridium	Ir	77	192.22
Carbon	C	6	12.01	Iron	Fe	26	55.85
Cerium	Ce	58	140.12	Krypton	Kr	36	83.80
Cesium	Cs	55	132.91	Lanthanum	La	57	138.91
Chlorine	Cl	17	35.45	Lawrencium	Lr	103	[262]
Chromium	Cr	24	52.00	Lead	Pb	82	207.2
Cobalt	Co	27	58.93	Lithium	Li	3	6.94
Copernicium	Cn	112	[285]	Lutetium	Lu	71	174.97
Copper	Cu	29	63.55	Magnesium	Mg	12	24.31
Curium	Cm	96	[247]	Manganese	Mn	25	54.94
Darmstadtium	Ds	110	[281]	Meitnerium	Mt	109	[276]
Dubnium	Db	105	[268]	Mendelevium	Md	101	[258]
Dysprosium	Dy	66	162.50	Mercury	Hg	80	200.59
Einsteinium	Es	99	[252]	Molybdenum	Mo	42	95.96

Chart of the Elements

Element	Symbol	Atomic Number	Average Atomic Mass	Element	Symbol	Atomic Number	Average Atomic Mass
Neodymium	Nd	60	144.24	Silicon	Si	14	28.09
Neon	Ne	10	20.18	Silver	Ag	47	107.87
Neptunium	Np	93	[237]	Sodium	Na	11	22.99
Nickel	Ni	28	58.69	Strontium	Sr	38	87.62
Niobium	Nb	41	92.91	Sulfur	S	16	32.07
Nitrogen	N	7	14.01	Tantalum	Ta	73	180.95
Nobelium	No	102	[259]	Technetium	Tc	43	[98]
Osmium	Os	76	190.23	Tellurium	Te	52	127.60
Oxygen	O	8	16.00	Terbium	Tb	65	158.93
Palladium	Pd	46	106.42	Thallium	Tl	81	204.38
Phosphorus	P	15	30.97	Thorium	Th	90	232.04
Platinum	Pt	78	195.08	Thulium	Tm	69	168.93
Plutonium	Pu	94	[244]	Tin	Sn	50	118.71
Polonium	Po	84	[209]	Titanium	Ti	22	47.87
Potassium	K	19	39.10	Tungsten	W	74	183.84
Praseodymium	Pr	59	140.91	Ununhexium	Uuh	116	[293]
Promethium	Pm	61	[145]	Ununoctium	Uuo	118	[294]
Protactinium	Pa	91	231.04	Ununpentium	Uup	115	[288]
Radium	Ra	88	[226]	Ununquadium	Uuq	114	[289]
Radon	Rn	86	[222]	Ununtrium	Uut	113	[284]
Rhenium	Re	75	186.21	Uranium	U	92	238.03
Rhodium	Rh	45	102.91	Vanadium	V	23	50.94
Roentgenium	Rg	111	[280]	Xenon	Xe	54	131.29
Rubidium	Rb	37	85.47	Ytterbium	Yb	70	173.05
Ruthenium	Ru	44	101.07	Yttrium	Y	39	88.91
Rutherfordium	Rf	104	[265]	Zinc	Zn	30	65.38
Samarium	Sm	62	150.36	Zirconium	Zr	40	91.22
Scandium	Sc	21	44.96				
Seaborgium	Sg	106	[271]				
Selenium	Se	34	78.96				

DEVELOPING SKILLS

C.7 MOLAR MASSES

Sample Problem: Draw a molecular-level model and find the molar mass of carbon dioxide, CO_2

The chemical formula and drawing show that one mole of carbon dioxide molecules contains one mole of carbon atoms and two moles of oxygen atoms. Adding the molar mass of carbon and twice the molar mass of oxygen gives

$$1 \text{ mol } C \times \frac{12.01 \text{ g C}}{1 \text{ mol } C} = 12.01 \text{ g C}$$

$$2 \text{ mol } O \times \frac{16.00 \text{ g O}}{1 \text{ mol } O} = 32.00 \text{ g O}$$

Molar mass of CO_2 = (12.01 g + 32.00 g) = 44.01 g CO_2

> The phrase "one mole of nitrogen" is quite unhelpful. It could refer to one mole of N atoms or to one mole of N_2 molecules. Thus, it is important to specify clearly which substance is involved.

Draw a molecular-level model and find the molar mass of each substance:

1. Nitrogen atoms: N
2. Nitrogen molecules: N_2
3. Methane: CH_4

Find the molar mass of each substance:

4. Sodium chloride (table salt): NaCl
5. Chalcopyrite: $CuFeS_2$
6. Sucrose (table sugar): $C_{12}H_{22}O_{11}$
7. Magnesium phosphate: $Mg_3(PO_4)_2$
8. Caffeine: $C_8H_{10}N_4O_2$
9. Calcium hydroxyapatite (a mineral found in teeth): $Ca_{10}(PO_4)_6(OH)_2$
10. Alunite (an aluminum mineral): $KAl_3(SO_4)_2(OH)_6$
    (*Hint:* Verify that this formula includes 14 oxygen atoms.)

concept check 6

1. Describe the relationship between a mineral and an ore.
2. How is a mole like a dozen? How is it different from a dozen?
3. Why is the molar mass of a diatomic element twice the average atomic mass value found on the periodic table?
4. How does the distributive property of mathematics apply to chemical formulas?

C.8 MOLES AND PERCENT COMPOSITION

A compound's formula indicates the relative number of atoms of each element present in the substance. For example, one common commercial source of copper metal is the mineral chalcocite—copper(I) sulfide (Cu_2S). Its formula indicates that the mineral contains twice as many copper atoms as sulfur atoms. The formula also reveals how much copper can be extracted from a certain mass of the mineral—an important consideration in copper mining and production.

Consider how the ideas of molar mass and percent composition are useful in determining how much copper can be obtained from copper-containing minerals and ores. Some copper-containing minerals are listed in Table 1.7 and shown in Figure 1.40. The percent of copper in chalcocite can be calculated, based on what you know about molar masses.

Table 1.7

Some Copper-Containing Minerals	
Common Name	**Formula**
Bornite	Cu_5FeS_4
Chalcocite	Cu_2S
Chalcopyrite	$CuFeS_2$
Malachite	$Cu_2CO_3(OH)_2$

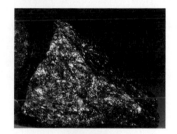

The formula for chalcocite indicates that one mole of Cu_2S contains two moles of Cu, or 127.10 g Cu, and one mole of S, or 32.07 g S. The molar mass of Cu_2S is $(2 \times 63.55 \text{ g}) + 32.07 \text{ g} = 159.17 \text{ g}$. Therefore,

$$\% \text{ Cu} = \frac{\text{Mass of Cu}}{\text{Mass of Cu}_2\text{S}} \times 100\%$$

$$\frac{127.10 \text{ g Cu}}{159.17 \text{ g Cu}_2\text{S}} \times 100\% = 79.85\% \text{ Cu}$$

A second calculation indicates that Cu_2S contains 20.15% sulfur. The sum of percent copper and percent sulfur equals 100.00%. Why?

Knowing the percent composition of metal in a particular mineral helps us decide whether a particular ore should be mined; however, it is not the sole criterion. What else do we need to consider?

Suppose an ore contains the mineral chalcocite, Cu_2S. Because nearly 80% of this mineral is Cu (see calculation), it seems likely that this ore is worth mining for copper. However, we must also consider the quantity of mineral actually contained in the ore. All other factors being equal, an ore that contains only 10% chalcocite would be a less desirable copper source than one containing 50% chalcocite. Thus, two factors must be taken into

Figure 1.40 *Ores of copper include (top to bottom) chalcocite, bornite, and malachite.*

account when deciding on the quality of a particular ore source: (1) the percent mineral in the ore and (2) the percent metal in the mineral.

Diagrams may be useful in understanding how these two percentage values relate to the total metal found in a particular ore. Consider an ore containing 10% chalcocite. Look at Figure 1.41. Suppose the larger box in Figure 1.41 represents a piece of this ore. According to this box, 10% of the ore is composed of the mineral chalcocite. (Ten squares are shaded to show this.). You know that chalcocite itself is approximately 80% copper by mass. To represent this, 80% of each chalcocite box is shaded in Figure 1.41 (one box is enlarged to show detail). This representation may help estimate visually how much copper is in this particular ore.

Figure 1.41 *The large square represents the total ore sample. (left) Each square within the large square represents 1% of the sample. The 10 colored squares indicate an ore that contains 10% chalcocite. (right) Eighty percent of each chalcocite square is colored to show the percent copper in chalcocite. Overall, only 8% of the total ore sample is copper. How would the diagram differ for an ore that is 50% chalcocite?*

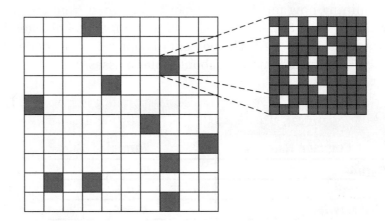

DEVELOPING SKILLS

C.9 PERCENT COMPOSITION

Sample Problem: One source of manganese is the mineral rhodochrosite, $MnCO_3$. Calculate the percent manganese in rhodochrosite.

The percent manganese can be calculated by dividing the molar mass of manganese by the molar mass of rhodochrosite:

$$\% \text{ Mn} = \frac{\text{Mass of Mn}}{\text{Mass of MnCO}_3} \times 100\%$$

One mole of $MnCO_3$ contains one mole of Mn, or 54.94 g Mn; one mole of C, or 12.01 g C; and three moles of O, or 48.00 g O. The molar mass of $MnCO_3$ is 54.94 g + 12.01 g + (3 × 16.00 g) = 114.95 g. So,

$$\% \text{ Mn} = \frac{54.94 \text{ g Mn}}{114.95 \text{ g MnCO}_3} \times 100\% = 47.79\% \text{ Mn}$$

1. The formula for azurite, a copper-containing mineral, is $Cu_3(CO_3)_2(OH)_2$.

   a. Complete an atom inventory (see page 95) for this compound.

   b. Calculate the molar mass of azurite. Use information provided in Developing Skills C.7 (page 98) as an example.

2. Use Table 1.7 on page 99 to answer the following. In your calculations, assume that each mineral is present at the same concentration in an ore. Also assume that copper metal can be extracted from a given mass of any ore at the same cost. (Note that the percent copper in chalcocite was calculated on page 99.)

   a. Calculate the percent copper in bornite.

   b. Calculate the percent copper in malachite.

   c. Which of these minerals—chalcocite, bornite, or malachite—could be mined most profitably?

3. Suppose you can either mine ore that is 20% chalcocite (Cu_2S) or ore that is 30% chalcopyrite ($CuFeS_2$). Assuming all other factors are equal, which would you choose? Draw diagrams similar to those in Figure 1.41 on page 100 to support your answer.

4. Two common iron-containing minerals are hematite (Fe_2O_3) and magnetite (Fe_3O_4). See Figure 1.42. If you had the same mass of each, which sample would contain the larger mass of iron? Support your answer with calculations.

Figure 1.42 *Hematite (Fe_2O_3, left) and magnetite (Fe_3O_4, right) are two iron-containing ores. If all other factors were equal, which ore would you mine for iron? Why?*

C.10 MINING AND REFINING

Knowing the percent composition of metal in a mineral and the percent composition of the mineral within an ore provides geologists and engineers valuable information as they determine the feasibility and profitability of mining a particular ore. However, once these ores are mined from the lithosphere, they must be **refined** in order to be used. Refining, in general, refers to the removal of impurities from a desired material. In metallurgy, refining means to use various methods to produce the free metal from an ore or ores. The process of converting a combined metal (usually a metal ion) in a mineral to a free metal involves a particular kind of chemical change.

Formation of Copper Metal (Reduction)

In general, for metallic cations to be converted to atoms of the pure metal, each cation must gain a particular number of electrons:

$$Cu^{2+} \quad + \quad 2e^- \quad \longrightarrow \quad Cu$$

Copper(II) ion   Copper metal

Chemists classify any chemical change in which a reactant can be considered to gain one or more electrons as a **reduction**. Thus the conversion of copper(II) cations to copper metal is a reduction reaction, and we say that copper(II) cations were **reduced**. You can convince yourself that this is a reduction reaction by examining the preceding equation.

Formation of Copper(II) Ions (Oxidation)

Chemists classify the reverse reaction, in which an ion or other reactant can be considered to lose one or more electrons, as an **oxidation**:

$$Cu \quad \longrightarrow \quad Cu^{2+} \quad + \quad 2e^-$$

Copper   Copper(II)
metal   ion

Any reactant that appears to lose one or more electrons is said to be **oxidized**. In this case, a copper atom is oxidized to a copper(II) ion by removal of two electrons.

Oxidation–Reduction Reactions

One easy way to remember this is OIL RIG: Oxidation Is Loss of electrons, Reduction Is Gain of electrons.

Whenever one reactant loses electrons, another reactant must simultaneously gain them. In other words, oxidation and reduction reactions never occur separately. Oxidation and reduction occur together in what chemists call **oxidation–reduction reactions** or, to use a common chemical nickname, **redox reactions**.

You have already observed redox reactions in the laboratory. During Investigating Matter B.12 (pages 71–73), copper metal reacted with silver ions. The following equation shows the oxidation–reduction reaction you observed:

$$Cu(s) \quad + \quad 2\,Ag^+(aq) \quad \longrightarrow \quad Cu^{2+}(aq) \quad + \quad 2\,Ag(s)$$

Copper   Silver ion   Copper(II) ion   Silver
metal   (colorless)   (blue)   metal

Each metallic copper atom (Cu) was oxidized (converted to a Cu^{2+} ion by losing two electrons), and each silver ion (Ag^+ from $AgNO_3$ solution) was reduced (converted to an Ag atom by gaining one electron). What you *actually* observed in Investigating Matter B.12 was the formation of a solid (silver) and the solution becoming blue because of formation of colored Cu^{2+} ions; these observable changes indicated that a redox reaction was occurring, even though you did not actually *see* the transfer of electrons.

In the same investigation, you found that copper(II) ions could be recovered from solution as metallic copper by allowing the copper(II) ions to react with magnesium metal, an element more active than copper. Magnesium atoms were *oxidized*; copper(II) ions were *reduced*. Do you see why?

$$Cu^{2+}(aq) \quad + \quad Mg(s) \quad \longrightarrow \quad Cu(s) \quad + \quad Mg^{2+}(aq)$$

| Copper(II) ion (blue) | Magnesium metal | Copper metal | Magnesium ion (colorless) |

Note that the total electrical charge on both sides of this equation is the same. Electrical charges—as well as atoms—must balance within a correctly written chemical equation. For instance, the net charge on both the reactant side and product side is 2+ in that equation.

In some circumstances, this reaction might be a useful way to obtain copper metal. However, as is often the situation, the desired copper metal would be obtained at the expense of "using up" another highly valuable material—in this case, magnesium metal. See Figure 1.43.

Many metallic elements are found in minerals in the form of cations because they combine readily with other elements to form ionic compounds. Obtaining a metal from its mineral requires energy and a source of electrons. A reactant that provides electrons is known as a **reducing agent**.

Figure 1.43 *Another application of redox reactions that "uses up" a more reactive metal is shown at right. Sacrificial anodes are installed on the hulls of ships, as well as on bridges and offshore drilling platforms. These anodes are constructed of more reactive metals than the structure or hull, and thus corrode in sea water first, protecting the structure or ship.*

Using Redox Reactions to Obtain Pure Metal

Look again at Table 1.3 (page 75). The table lists several techniques that metallurgists use to reduce metal cations or, in other words, to supply one or more electrons to each cation. The specific technique chosen depends on the metal's reactivity and the availability of inexpensive reducing agents and energy sources.

Two approaches summarized in the table are electrometallurgy and pyrometallurgy. As the table suggests, *electrometallurgy* involves using an electric current to supply electrons to metal ions, thus reducing them (see Figure 1.44). This process is used when no adequate chemical reducing agents are available or when very high-purity metal is sought. *Pyrometallurgy*, the oldest ore-processing method, involves treating metals and their ores with thermal energy (heat), as in a blast furnace (see Figure 1.45). Carbon (coke) and carbon monoxide are common reducing agents in pyrometallurgy: They provide electrons; metal ions are thus reduced to form metal atoms. A more reactive metal can be used if neither of these reducing agents will do the job.

A third approach to obtaining metals from their ions is the process called *hydrometallurgy*, which involves treating ores and other metal-containing materials with reactants in water solution. You used such a procedure when you investigated the reactivity of different metals in Investigating Matter B.12.

Figure 1.44 *Electrometallurgy is used to reduce copper ions to copper metal. Starter sheets (also made of copper) are hung from anodes and inserted into a solution containing copper ions.*

Figure 1.45 *Iron is extracted from iron ore in a huge container called a blast furnace. Minerals containing iron, such as hematite (iron(III) oxide), are reduced to iron metal in the blast furnace. Other impurities are burned away or converted into a material that can be easily removed, leaving the iron behind.*

Hydrometallurgy is used to recover silver and gold from old mine tailings (the mined rock left after most of the sought mineral is removed) by a process known as *leaching*. As higher-grade ores become scarcer, it will become economically feasible to use hydrometallurgy and other "wet processes" on metal-bearing minerals that can dissolve in water.

■ MODELING MATTER

C.11 ELECTRONS AND REDOX PROCESSES

The processes of oxidation (apparent or actual loss of one or more electrons) and reduction (gain of one or more electrons) can be clarified by visual representations of these events. To develop such representations, you will consider atoms of each of the metals you encountered during Investigating Matter B.12 (page 71).

First, however, here is a review of some key details about an atom's composition. Magnesium (Mg), a reactive metal, changed into Mg^{2+} ions in several reactions you observed during Investigating Matter B.12. The atomic number of Mg is 12, indicating that an electrically neutral atom of magnesium contains 12 protons and 12 electrons. Remember that numbers of protons and electrons must be equal for a neutral atom.

If magnesium is to form a Mg^{2+} ion, two negatively charged electrons must be removed from each magnesium atom. The bookkeeping involved can be summarized this way:

$$Mg \longrightarrow Mg^{2+} + 2e^-$$

Mg	Mg^{2+}	$2e^-$
12 protons (+)	12 protons (+)	
12 electrons (−)	10 electrons (−)	2 electrons (−)
Net charge: 0	Net charge: 2+	Net charge: 2−

To build a useful picture of this process in your mind, it is necessary to keep track of only the two electrons *released* by each magnesium atom, rather than monitoring all 12 of the atom's electrons. (In fact, in any normal chemical reaction, a magnesium atom is not observed to release any of its other 10 electrons.)

Thus, for bookkeeping purposes, an atom of Mg will be depicted this way: Mg: (the element symbol with two dots attached). Each dot represents one of magnesium's two readily removable electrons. The symbol Mg represents the remaining parts of a magnesium atom, including its remaining ten electrons. The resulting expression for Mg is called an **electron-dot structure**, or just a **dot structure**. The equation for oxidation of Mg atoms can be represented this way in electron-dot terms:

$$Mg: \longrightarrow Mg^{2+} + 2e^-$$

1. Construct a similar electron-dot expression for the change that occurred during Investigating Matter B.12 (page 71) when each of these events took place:

   a. An atom of zinc, Zn, was converted to a Zn^{2+} ion. (*Hint:* Zn has two readily removable electrons.)

   b. A silver ion, Ag^+, was converted to a metallic silver atom, Ag*(s)*.

2. Apply the definitions of *oxidation* and *reduction* to your two equations in Question 1, and label each reaction appropriately.

Now consider one of the complete reactions you observed during Investigating Matter B.12. When you immersed a sample of copper metal, Cu, in silver nitrate solution, $AgNO_3$, a blue solution containing Cu^{2+} formed, as well as crystals of solid Ag. The redox reaction that occurred is

$$Cu(s) + 2\,Ag^+(aq) \longrightarrow Cu^{2+}(aq) + 2\,Ag(s)$$

This redox reaction can be represented through the use of dot structures:

$$Cu\mathbin{:} + Ag^+ + Ag^+ \longrightarrow Cu^{2+} + Ag\mathbin{\cdot} + Ag\mathbin{\cdot}$$

3. Which reactant (Cu or Ag^+) is reduced?

4. Why is only one Cu atom needed for each two Ag^+ ions that react?

Each copper atom involved in this reaction loses two electrons. Thus, copper atoms must be oxidized in the transfer process. It is clear from the dot structures that the two electrons lost by copper are gained by two Ag^+ ions. So, Ag^+ is the *agent* that caused the removal of electrons from Cu (resulting in the oxidation of Cu). The reactant involved in removing electrons from the oxidized reactant is called the **oxidizing agent**—in this case, Ag^+ ions.

5. a. Given this explanation of an oxidizing agent, how would you define a *reducing agent*?

   b. What must be the reducing agent in the reaction between Cu*(s)* and Ag^+ ions?

6. Now consider another reaction you observed during Investigating Matter B.12:

$$Zn(s) + Cu^{2+}(aq) \longrightarrow Zn^{2+}(aq) + Cu(s)$$

   a. Draw an electron-dot representation of this reaction.

   b. Which reactant is oxidized?

   c. Which is reduced?

7. Identify the following in the reaction represented in Question 6:

   a. the oxidizing agent

   b. the reducing agent

8. Consider both of the oxidation–reduction reactions you analyzed in this exercise. What general features of an oxidation–reduction reaction would allow you to answer Questions 6 and 7 without drawing electron-dot representations?

9. Now consider a new oxidation–reduction reaction. Answer Questions 6 and 7 for this system:

$$Zn^{2+}(aq) + Mg(s) \longrightarrow Zn(s) + Mg^{2+}(aq)$$

10. Return to Investigating Matter B.12 and write an oxidation–reduction equation for the reaction of silver ions with magnesium atoms. Answer Questions 6 and 7 for this system.

We have now considered key factors influencing whether a particular ore is mined, as well as the chemical processes involved in releasing the metal from its ores and minerals through refining. We will next consider what happens to the metal after refining and to its products after their use and begin to develop the idea of material life cycles.

INVESTIGATING MATTER
C.12 COPPER PLATING

Preparing to Investigate

In Section A, you learned that accelerating copper prices in the early 1980s forced the United States to use zinc as the primary metal in pennies. Copper covers the coin's surface. This copper exterior not only preserves the traditional appearance of the coin, it protects the coin from corrosion. Using one material to protect the surface of another, less durable, material is by no means new or unusual. A layer of protective material can enhance the performance or appearance of a manufactured product while allowing a less expensive or more available material to be used for most of the item.

In this investigation, you will use electricity to plate copper onto graphite (carbon) rods. Before you begin, read *Gathering Evidence* to learn what you will need to do and note safety precautions.

Gathering Evidence

1. Before you begin, put on your goggles, and wear them properly throughout the investigation.

2. Obtain two graphite (carbon) rods in the form of pencil lead. (Depending upon the size of the electrolysis apparatus you will use, the graphite rods may still be encased inside two wooden pencils. If so, just be sure graphite protrudes from both ends so electrical contact can be made.) These rods will serve as the terminals or electrodes for the electrolysis process.

Figure 1.46 *Apparatus for plating copper onto graphite electrodes.*

3. Set up the apparatus shown in Figure 1.46.
4. Attach the 9-V battery connector to the battery, but *do not* connect the wire leads to the electrodes or allow the two wires to touch each other.
5. Pour enough copper(II) chloride ($CuCl_2$) solution into the U-tube so that the graphite electrodes can be partially immersed in the solution.
6. After your teacher approves your apparatus, attach wires to the two graphite terminals.
7. Observe the reaction for ~5 minutes. Record your observations.
8. Cautiously sniff each electrode.
9. Reverse the wire connections to the electrodes and repeat Step 7.
10. Wash your hands thoroughly before leaving the laboratory.

Interpreting Evidence and Making Claims

1. Describe changes observed during the electrolysis.
2. Were you successful in plating copper onto the graphite rod? What evidence supports your claim?
3. The **cathode** is the terminal (electrode) at which reduction occurs.
   a. To which terminal of the battery was the cathode connected in the electrolysis?
   b. What change did you observe at the cathode?
   c. Write a representation, similar to those you used in Modeling Matter C.11, for the reaction that occurred at the cathode.
4. The other electrode, the **anode**, is where oxidation occurs.
   a. Which electrode was the anode in the electrolysis?
   b. What change did you observe at the anode?
   c. Write a representation, similar to those you used in Modeling Matter C.11, for the reaction that occurred at the anode.

Reflecting on the Investigation

5. Draw and label a diagram of the system that you used to deposit copper.
6. Do you think that a metal could be substituted for one of the graphite rods? Explain.
7. Do you think this method would be useful for large-scale copper plating? Why?

C.13 ELECTROPLATING

The process you just used to cover graphite with copper involves using direct-current (DC) electricity, which causes redox reactions to occur. This general process is known as **electroplating**. As you have already learned, cations of most metallic elements can be reduced. This fact is exploited in electroplating. For example, metal bumpers on trucks are often made of steel. In wet or snowy climates, the exposed steel would quickly corrode, or rust. Manufacturers protect the steel by coating it with chromium and nickel. See Figure 1.47. Canadian 5-, 10-, and 25-cent coins are also plated with nickel to resist corrosion.

Writing **half-reactions**, which separately represent the reduction and oxidation parts of a redox process, can be helpful in understanding the plating process. The reduction half-reaction for plating nickel metal on steel can be written this way:

$$Ni^{2+}(aq) \ + \ 2 \ e^- \longrightarrow Ni(s)$$

Where do the electrons in this reduction reaction come from? Electroplating requires a power source, usually a battery or power supply when this reaction occurs in the laboratory. The power source transports electrons to the **cathode** (where reduction occurs). The ultimate source of electrons in any electrochemical process is the **anode**. In electroplating, the anode is usually made of the metal to be plated onto the object. As metal cations are reduced (removed from the solution) and deposited on the object attached to the cathode, metal atoms at the anode are oxidized and dissolve into the plating solution. Figure 1.48 illustrates this process.

Thus, the other half-reaction for the plating system is an oxidation, the reverse of the reaction shown above. Based on what you know about metal reactivities, would you expect this oxidation–reduction process to proceed on its own, without any external power source? Although systems that can plate metals without applied electric current have been developed, they must include a chemical reducing agent, which is sometimes more expensive than electric current.

Platings are usually bonded to the surface of the electroplated object. This is desirable because it keeps the metal finish firmly attached to the object. After several layers of atoms have been deposited, the plating has the properties of the plating metal; therefore, it can impart the properties for which it was selected. Electroplating is one of many ways in which chemists can modify materials to meet specific needs. Keep this idea in mind as you continue to consider the composition and use of currency.

Figure 1.47 *Chrome plating can protect exposed steel surfaces.*

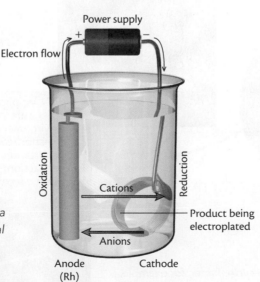

Figure 1.48 *Electroplating is caused by an oxidation–reduction process. The new coating is a layer of metal atoms deposited by reducing metal ions from the electroplating solution. Rhodium is sometimes plated onto jewelry to increase shine and protect against wear.*

C.14 THE LIFE CYCLE OF A MATERIAL

In designing a new product for human use, engineers and analysts consider the full life cycle of the materials involved. A **material's life cycle** has several distinct stages. Raw materials are first obtained and then refined and synthesized into the desired material (*materials acquisition*). That material is then used to make the product designed for a particular use (*manufacturing*). The product is transferred to the consumer, used, and, depending upon the product and circumstances, maintained or reused (*use/reuse/maintenance*). When the product is no longer useful, the materials may be recovered and recycled, or they may end up scattered in landfills (*recycle/waste management*). Figure 1.49 illustrates this general life cycle.

In every stage of the cycle, energy and resources are used. Wastes and emissions are also produced and must be included in assessing the overall impact of the process. Because energy, resource use, and waste management affect economics as well as the environment, each step in the life cycle of a material becomes a factor to consider when chemists and engineers design a new product.

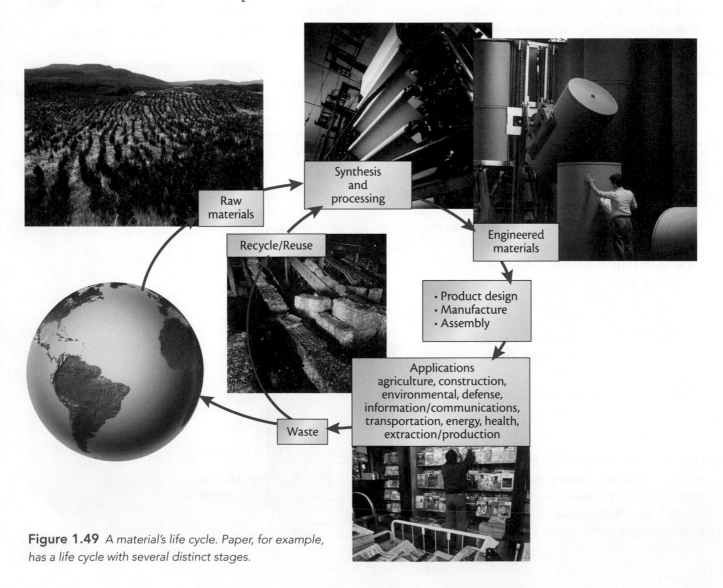

Figure 1.49 *A material's life cycle. Paper, for example, has a life cycle with several distinct stages.*

The next activity will allow you to model this process for a familiar product—a dollar coin.

▮ MAKING DECISIONS
C.15 LIFE CYCLE OF A COIN

So far in this unit you have considered properties that are necessary and desirable for coins and bank notes. You have also learned about the sources, uses, and properties of metals, and have begun to explore the life cycles of products and materials. Now you will use this knowledge to consider the life cycle of a particular metal product, the current U.S. dollar coin. Use Figure 1.35 (page 88) and Figure 1.49 to answer the following questions and continue to evaluate the statements in the unit-opening Web page.

1. In Making Decisions A.11, you found the composition of the U.S. one-dollar coin, and in Making Decisions C.1 you and your classmates gathered information about several metals. Construct a table similar to the one below to organize information about the metals that compose the dollar coin. For each metal, enter the following information into the table:

   - Major areas/countries of origin
   - Primary type of mining
   - Primary type of refining/processing
   - Cost of the refined metal
   - Competing uses of the metal

   (*Note*: With the exception of cost, this information was collected in Making Decisions C.1.)

	Metals			
	I	II	III	IV
Origin				
Mining type				
Processing type				
Cost				
Competing uses				

2. Referring to Figures 1.35 and 1.49, make a list of the steps involved in the life cycle of the U.S. one-dollar coin.

Using your table and the list you just created, answer the following questions:

3. Which steps in the life cycle of dollar coins consume energy? For each, explain the particular energy needs.

4. Which steps (such as the reduction of a mineral in an ore to produce the metal) require the use of other materials?

5. Which steps generate wastes or emissions? Explain.

6. How does the location or identity of the ore deposits affect the life cycle of the coins?

7. As you know, coins have a limited lifespan.
   a. How long do you expect each coin to stay in circulation?
   b. What will happen to the materials when used coins are removed from circulation?

8. Consider the cost of producing materials in the dollar coin. What might happen if the materials in the coin were worth more than the face value of the coin?

SECTION C SUMMARY

Reviewing the Concepts

> The resources for all human activities must be obtained from Earth's atmosphere, hydrosphere, and outer layer of its lithosphere. These resources are not uniformly distributed.

1. List two resources typically found in each of the three major "spheres" of Earth.

2. a. List and briefly describe three major parts of the lithosphere.

   b. Which layer serves as the main storehouse of chemical resources used in manufacturing consumer products?

3. Using Table 1.4 (page 85), identify the nation that produces the most

   a. silver.   b. copper.   c. tin.

4. According to information in Table 1.4 on page 85, which of these four nations—the United States, Australia, China, or Peru—produces the largest masses of the eight listed resources in the table?

> The feasibility of mining and extracting a mineral resource depends, in part, on how easily a particular metal can be processed and used, which largely depends on its chemical reactivity.

5. How do minerals differ from ores?

6. What factors determine the feasibility of mining a particular metallic ore at a certain site?

7. A 19th-century gold mine, inactive for over 100 years, has recently reopened for further mining. What factors may have influenced the decision to reopen the mine?

8. What is meant by the quantity of "useful ore" at a particular site?

9. Why are active metals more difficult to process and refine than are less active metals?

10. Based on your results from Investigating Matter B.12 (page 71), which metals involved in that investigation would be the easiest to process? Why?

11. Why do most metals exist in nature as minerals rather than as pure metallic elements?

> Gathering information is a form of data collection. Sources of information vary in their reliability.

12. Consider the information that you gathered in Making Decisions C.1 and Making Decisions C.15.

    a. What sources did you use?

    b. How did you determine whether a source provided accurate information?

    c. Did you note *bias* in any of the sources that you encountered? (A **bias** is a tendency toward a particular belief or perspective.) Did you choose to use these sources? Explain.

13. What sources of information do you consider to be consistently reliable? Why?

14. What are some potential consequences of collecting inaccurate data or information?

> One mole of substance contains 6.02×10^{23} particles. The molar mass of a substance can be calculated from average atomic masses of the component elements.

15. If you could spend a billion dollars (1×10^9 dollars) per second, how many years would it take to spend one mole of dollars?

16. Find the molar mass of each substance:
    a. oxygen gas, O_2
    b. ozone, O_3
    c. limestone, $CaCO_3$
    d. a typical antacid, $Mg(OH)_2$
    e. aspirin, $C_9H_8O_4$

17. How is it that samples of 63.6 g copper metal and 23.0 g sodium metal, with different masses, volumes, and densities, both correctly represent 1.00 mol of substance?

18. A major advantage of the mole concept is that it enables a chemist to "count by weighing." If one mole of potassium metal has a mass of 39.1 g, how many atoms are in
    a. 39.1 g potassium?
    b. 19.6 g potassium?
    c. 3.91 g potassium?

The percent composition of a material can be calculated by determining the proportion by mass of each constituent. The percent composition of a substance can be calculated from the relative number of atoms of each element in the substance's formula.

---

19. Calculate the percent composition of the U.S. quarter, which has a mass of 5.670 g and contains 5.198 g Cu and 0.472 g Ni.

20. Find the percent metal (by mass) in each compound:
    a. Ag_2S
    b. Al_2O_3
    c. $CaCO_3$

21. A 50.0-g sample of ore contains 5.00 g lead(II) sulfate, $PbSO_4$.
    a. What is the percent by mass:
        i. lead (Pb) in $PbSO_4$?
        ii. $PbSO_4$ in the ore sample?
        iii. Pb in the total ore sample?
    b. Use a diagram to represent the proportions of lead and lead(II) sulfate in the ore sample.

22. In carbon dioxide, two-thirds of the atoms are oxygen atoms; however, the percent oxygen by mass is not 67%. Explain.

The processes of oxidation and reduction occur together, resulting in oxidation–reduction (redox) reactions.

---

23. Define oxidation and reduction in terms of electron transfer.

24. Write an equation for each of the following processes:
    a. the reduction of gold(III) ions to gold metal
    b. the oxidation of elemental vanadium to vanadium(IV) ions
    c. the oxidation of Cu^+ to Cu^{2+} ions

25. Identify each of the following equations as representing either an oxidation reaction or a reduction reaction.
    a. $Fe^{2+} + 2e^- \longrightarrow Fe$
    b. $Cr \longrightarrow Cr^{3+} + 3e^-$
    c. $Al^{3+} + 3e^- \longrightarrow Al$

26. Consider the following equation:

    $Zn(s) + Ni^{2+}(aq) \longrightarrow Zn^{2+}(aq) + Ni(s)$

    a. Which reactant has been oxidized? Explain your choice.
    b. Which reactant has been reduced? Explain your choice.
    c. What is the reducing agent in this reaction?

27. Consider the following equation:

    $2 K(s) + Hg^{2+}(aq) \longrightarrow 2 K^+(aq) + Hg(s)$

    a. Which reactant has been oxidized? Explain your choice.
    b. Which reactant has been reduced? Explain your choice.
    c. What is the oxidizing agent in this reaction?

28. Write an equation for
    a. the oxidation of Al metal by Cr^{3+} ions.
    b. the reduction of Mn^{2+} ions by Mg metal.

Metal cations can be converted to metal atoms by electrometallurgy, pyrometallurgy, or hydrometallurgy, depending upon the metal's reactivity.

29. Explain how each of the following processes converts metal cations to metal atoms:

    a. electrometallurgy

    b. pyrometallurgy

    c. hydrometallurgy

30. What processes would be most useful in obtaining the following elements from their ores?

    a. magnesium

    b. lead

What is the role of chemistry in the life cycle of metals?

As you conclude your study of this section, look back on the essential ideas. You have encountered life cycles, moles, and metals, within the context of Earth's mineral resources. How are these ideas connected? Think about what you have learned, then answer the question in your own words in organized paragraphs. Your answer should demonstrate your understanding of the key ideas in this section.

Be sure to consider the following in your response: metal sources, metal processing, metal recycling, "counting by weighing," and percent composition.

Connecting the Concepts

31. Describe two ways in which electroplating is used in industry.

32. In Investigating Matter C.12 (page 107), electricity was used to convert Cu^{2+} ions to $Cu(s)$.

    a. Was this chemical change an oxidation or a reduction?

    b. Rather than obtaining electrons from electricity, you could use another metal to provide electrons. Based on the metal activity series (Table 1.3, page 75), name two metals that could be used for this purpose.

33. Can electroplating a rusty car bumper protect it from further rusting? Explain.

34. How does access to mineral resources contribute to a nation's economic success?

35. There are thousands of tons of gold in sea water. Explain why it is unlikely that ocean water will ever be "mined" for gold.

36. What information would you need to be able to "count" exactly 1000 nails of the same size by weighing them?

37. Chromium minerals are found at three different mine sites in these forms:

    Site 1: Chromite, $FeCr_2O_4$

    Site 2: Crocoite, $PbCrO_4$

    Site 3: Chrome ochre, Cr_2O_3

    Based only on percent composition of the minerals, at which site would chromium mining be most feasible?

38. Is there any connection between the process used to reduce a metal cation and the position of that element on the periodic table?

39. Both Investigating Matter B.9 and Investigating Matter B.12 dealt with reactivity series. In each case, the more reactive substance replaced the less reactive substance in a compound. Explain this behavior in terms of oxidation and reduction.

40. Suppose that you are serving on a committee seeking to make your school more "green." The worn cafeteria tables must be replaced and the committee has been asked to choose between new tables with steel (mostly iron) or aluminum frames. Assuming performance is similar, what life cycle questions would you ask to determine which choice is "greener"?

41. A 10.0-g sample of sports drink powder contains 7.0 g sucrose ($C_{12}H_{22}O_{11}$), 2.8 g glucose ($C_6H_{12}O_6$), and 0.20 g sodium chloride (NaCl). Calculate the percent composition (by mass) of

   a. sucrose

   b. glucose

   c. NaCl

Extending the Concepts

42. Using the composition of the sports drink in Question 41, calculate the percent composition (by mass) of

   a. sodium

   b. carbon

43. Find and describe a historical example of uneven distribution of mineral resources and its impact on relations among nations.

44. History documents that copper has been used by humans for 10 000 years, whereas aluminum has been used for only about 100 years. Suggest and explain some reasons for this difference.

45. The reactive metal aluminum is often used in containers for acidic beverages. Investigate and describe the technology that makes this possible.

46. What is a patina? Explain its value both aesthetically and chemically.

47. In your laboratory experiences, you may encounter some compounds called hydrates. Examples of hydrates include (i) $Na_2S_2O_3 \cdot 5H_2O$, (ii) $CaSO_4 \cdot 2H_2O$, and (iii) $Na_2CO_3 \cdot 10H_2O$.

   a. Why are these compounds called "hydrates"?

   b. Calculate the molar mass of each listed hydrate.

   c. Calculate the percent water in each hydrate.

   d. Although hydrates contain significant amounts of water, they are found as dry solid substances at room temperature. How is this possible?

SECTION D CONSERVATION AND CHEMICAL EQUATIONS

How is conservation of matter demonstrated in the use of resources?

In some chemical reactions, matter seems to be created—as a nail rusts, a new material appears. In other reactions, matter seems to disappear—a sheet of paper burns and apparently vanishes. In terms of fundamental particles, neither creation nor destruction of matter actually occurs. In this section, you will learn what happens to atoms in chemical reactions and how the atoms can be tracked through the use of chemical equations. This information will help you consider the fate of Earth's resources as well as the materials and products developed from them.

GOALS

- State and apply the law of conservation of matter.
- Relate balanced chemical equations to the law of conservation of matter.
- Write and explain balanced chemical equations.
- Distinguish between renewable and nonrenewable resources.
- Identify methods of conserving Earth's resources.
- Describe how alloys and their constituent elements differ in their chemical and physical properties.

concept check 7

1. What information about a compound is contained in its chemical formula?
2. What does it mean to "count by weighing"?
3. You have observed changes in matter as a result of chemical reactions. What do you think happens at the particulate level during these changes?

D.1 EXAMINING CHEMICAL EQUATIONS

Throughout this unit, you have seen and used several chemical equations. You learned in Section A (page 37) that chemical equations summarize the results of chemical reactions and can be regarded as chemical sentences in the language of chemistry. To review and extend what you know about chemical equations, we will now examine carefully a chemical reaction from one of this unit's investigations.

In Investigating Matter B.9, you observed the reaction of chlorine with sodium bromide solution. This reaction can be represented with the following chemical equation:

$$Cl_2 \quad + \quad 2\,NaBr \quad \longrightarrow \quad Br_2 \quad + \quad 2\,NaCl$$

Chlorine	Sodium bromide solution		Bromine		Sodium chloride solution

Reactants   Products

The reactants are chlorine and sodium bromide, written on the left side of the equation. Sodium chloride and bromine, the products of this reaction, are written on the right side of the equation. You already know that sodium, chlorine, and bromine readily form ions and that sodium bromide and sodium chloride are ionic substances. With that information, you can now interpret the equation in words: one chlorine molecule (Cl_2) and two sodium bromide formula units (2 NaBr) react to produce ($\longrightarrow$) one molecule of bromine (Br_2) and two formula units of sodium chloride (2 NaCl).

You may have noticed that the subscript 2 is included every time that the formula of a halogen (or H_2 and O_2) is written. Although most uncombined elements in chemical equations are represented as single atoms (e.g., Cu, Fe, Na, and Mg), a handful of elements are **diatomic molecules**; they exist as two bonded atoms of the same element.

"GEN-U-INE DIATOMICS" serves as a useful memory device for all common diatomic elements. The names of all diatomic elements either end in GEN or INE, and U should remember them!

Table 1.8

Elements That Normally Occur as Diatomic Molecules

Element	Formula
Hydrogen	H_2
Nitrogen	N_2
Oxygen	O_2
Fluorine	F_2
Chlorine	Cl_2
Bromine	Br_2
Iodine	I_2

Oxygen and chlorine are two examples of diatomic molecules. Table 1.8 (page 117) lists all the elements that exist as diatomic molecules at normal conditions. It will be helpful for you to remember these elements. Take a moment to find the diatomic elements in the periodic table. Where are they located?

In the next activity, you will practice representing diatomic molecules, as well as other reactants and products, as you increase your fluency in the language of chemistry.

■ MODELING MATTER

D.2 REPRESENTING REACTIONS

In Modeling Matter A.8 (page 38), you saw how chemical compounds can be represented with formulas and pictures. As you move through Section D, you will again be asked to model elements, compounds, and equations, and translate among symbolic, particulate, and macroscopic representations of substances.

Until now, we have focused on illustrating simple molecular compounds, such as water and carbon dioxide. As you learned in Section B (page 66), representing ionic compounds is more complex, since ionic compounds are found in large crystals. Although the formula for sodium iodide, which you used in Investigating Matter B.9, is written NaI, it is not made up of individual NaI units. See Figure 1.50. The formula NaI indicates the relative number of atoms in the larger crystal structure and represents a formula unit of sodium iodide.

Figure 1.50 *Sodium iodide, which is sometimes used to treat iodine deficiency, is shown here in a particulate representation.*

Sample Problem 1: *Consider one of the reactions you observed in Investigating Matter B.9. Sodium iodide and bromine react to form sodium bromide and iodine. The equation for this reaction can be written as follows:*

$$Br_2 \ + \ 2\,NaI \ \longrightarrow \ I_2 \ + \ 2\,NaBr$$

How can we represent this equation using particulate models?

Sodium iodide and sodium bromide are ionic compounds. We can use representations of formula units of each as long as we recognize the limitations of these representations. So, using a key similar to the one used in Modeling Matter A.8, we can write:

Model:

Atom: Br   Na   I

The complete equation is represented as:

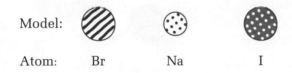

$$Br_2 \qquad + \qquad 2\,NaI \qquad \longrightarrow \qquad I_2 \qquad + \qquad 2\,NaBr$$

Consider the substances you have used in investigations throughout this unit. Many of these substances are more complex than the simple binary compounds above. For instance, silver nitrate (used in Investigating Matter B.12) contains a silver ion, Ag^+, and a polyatomic nitrate ion, NO_3^-. Baking soda, used in Investigating Matter A.1, contains sodium hydrogen carbonate, $NaHCO_3$. Although we must keep in mind the limitations of formula units and individual drawings—as with any ionic compound—we can construct simple drawings to represent compounds containing polyatomic ions.

Binary compounds contain two different elements.

Limestone is used in everything from buildings and statues to antacid tablets.

Sample Problem 2: Draw a particulate model for calcium carbonate, $CaCO_3$, the primary mineral in limestone.

Using a key similar to the one used in Modeling Matter A.8 (page 38), we have the following:

Model:

Atom: C   O   Ca

A formula unit for $CaCO_3$ can be represented as:

Now it is your turn to represent substances and reactions.

1. Represent an atom, molecule, or formula unit of each of the following using particulate models. Include a key.

   a. MgO, substance created when magnesium burns

   b. CO_2, product of combustion and respiration

   c. KCl, used in manufacturing fertilizer

   d. NaOH, also known as lye, one of the most important industrial chemicals

   e. iron(III) oxide, a component of rust

   f. silver nitrate, used in Investigating Matter B.12

   g. oxygen, gas essential to life

   h. $MnCO_3$, the main component in the mineral rhodochrosite (see Figure 1.51)

Figure 1.51 *Large red crystals of pure rhodochrosite are found in only a few places on Earth, including Colorado, where it is the official state mineral. Impure rhodochrosite ranges in color from pink to brown and is found worldwide.*

2. In Developing Skills C.9 (page 100) you determined the molar mass of azurite, a copper-containing mineral with the formula $Cu_3(CO_3)_2(OH)_2$.

   a. Name the polyatomic ions within azurite. (*Hint:* Refer to Table 1.2, page 70, if necessary.)

   b. How many of each of the ions you listed above are represented in one formula unit of azurite?

   c. How many total atoms would you need to create a particulate model of a formula unit of azurite? (*Note:* You do not need to create this model!)

3. As you observed in Investigating Matter A.10 (page 42), when an acid reacts with an active metal, hydrogen gas and an ionic compound are often formed. The following equation shows the reaction of hydrochloric acid (HCl) with magnesium metal:

$$2\ HCl\ +\ Mg\ \longrightarrow\ H_2\ +\ MgCl_2$$

   a. Write an interpretation of the statement in words.

   b. Identify each reactant and product as either a compound or an element.

   c. Draw a particulate model of the chemical reaction.

D.3 KEEPING TRACK OF ATOMS

Think for a moment about what happens to molecules of gasoline as they burn in an automobile engine (Figure 1.52). The carbon and hydrogen atoms that make up these molecules react with oxygen atoms in the air to form carbon monoxide (CO), carbon dioxide (CO_2), and water vapor (H_2O). These products are released as exhaust and disperse in the atmosphere. Thus, atoms of carbon, hydrogen, and oxygen originally present in the gasoline and air have not been destroyed; rather, they have become rearranged into different molecules.

Figure 1.52 *Where do the atoms in the gasoline go as a car's gas tank empties?*

Figure 1.53 *When a pile of coal burns, what becomes of the atoms that made up the coal?*

In short, "using things up" means materials are chemically changed, not destroyed. The **law of conservation of matter**, like all scientific laws, summarizes what has been learned by careful observation of nature: In a chemical reaction, matter is neither created nor destroyed. Molecules can be converted and decomposed by chemical processes, but atoms are forever. In a chemical reaction, matter—at the level of individual atoms—is always fully accounted for.

Because chemical reactions cannot create or destroy atoms, chemical equations representing such reactions must always be balanced. What does a "balanced equation" mean? Symbols or formulas for the reactants are placed on the left of the arrow; symbols or formulas for the products are placed on the right. In a **balanced chemical equation**, the number of atoms of each element is the same on the reactant and product sides.

Consider burning coal as an example (Figure 1.53). Coal is mostly carbon (C). If carbon burns completely, it combines with oxygen gas (O_2) to produce carbon dioxide (CO_2).

Here is a representation of the atoms and molecules involved in this reaction:

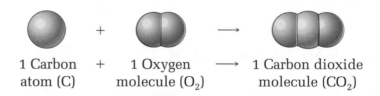

1 Carbon   +   1 Oxygen   ⟶   1 Carbon dioxide
atom (C)   molecule (O_2)   molecule (CO_2)

Note that the numbers of carbon and oxygen atoms on the reactant side equal the numbers of carbon and oxygen atoms on the product side. This indicates that the equation is balanced.

The representation of the coal-burning reaction shows that one carbon atom reacts with one oxygen molecule to form one carbon dioxide molecule:

$$C(s) \quad + \quad O_2(g) \quad \longrightarrow \quad CO_2(g)$$

Carbon   and   Oxygen   React to   Carbon
(in coal)   gas   produce   dioxide gas

Consider the reaction that occurs when copper is heated in air, discussed in B.13 (page 73). Copper metal (Cu) reacts with oxygen gas (O_2) to form copper(II) oxide (CuO). We can represent this process with the following:

2 Copper   **1 Oxygen**   **2 Copper(II) oxide**
atoms (Cu)   molecule (O_2)   formula units (CuO)

Again, note that the numbers of copper and oxygen atoms on the reactant side equal the respective numbers of copper and oxygen atoms on the product side. Written as a chemical equation, the reaction is

$$2\ Cu(s) \quad + \quad O_2(g) \quad \longrightarrow \quad 2\ CuO(s)$$

Copper   and   Oxygen   React to   Copper(II)
metal   gas   produce   oxide

You may have noticed that numbers have been placed in front of the copper and copper(II) oxide formulas. A "one" is also implied for the O_2. These numbers are called *coefficients*. **Coefficients** indicate the relative number of units of each substance involved in the chemical reaction. Reading this equation from left to right, you would say, *"Two copper atoms react with one oxygen molecule to produce two formula units of copper(II) oxide."*

Study Figure 1.54, which represents the copper–oxygen reaction in three ways. The photograph shows copper(II) oxide that you can observe, representing this reaction at the large-scale matter (or macroscopic) level. The particulate-level drawings model individual atoms and molecules for the reactants, and formula units within the ionic crystal of the product. The chemical equation represents this reaction symbolically. The ability to think about particles involved in chemical reactions and to represent reactions with symbols will help you link your observations to what you are unable to see (that is, what happens with atoms, ions, and molecules) and keep track of reactants and products.

In the following activity, you will practice recognizing and interpreting chemical equations.

> It is standard procedure to imply but not write the coefficient "1."

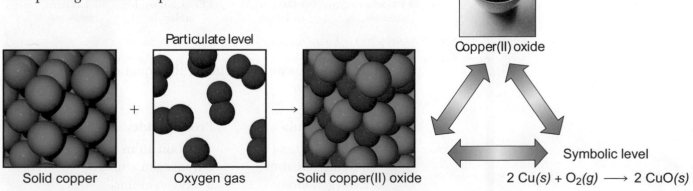

Figure 1.54 *Chemists describe the reaction between copper and oxygen in several ways: through* **symbolic** *equations, through models of what occurs at the* **particulate** *level, and through* **macroscopic** *descriptions of the observed properties of reactants and products.*

DEVELOPING SKILLS
D.4 ACCOUNTING FOR ATOMS

Sample Problem: *The combustion reaction between propane* *(C_3H_8) and oxygen gas (O_2) is a common source of thermal energy* *for campers, recreational-vehicle users, and others. See Figure 1.55.* *A chemical statement showing the reactants and products is*

$$C_3H_8(g) \; + \; O_2(g) \; \longrightarrow \; CO_2(g) \; + \; H_2O(g)$$

This chemical reaction produces thermal energy, which can also be considered a product.

Interpret this reaction in terms of (a) words, (b) molecular models, *and (c) an atom inventory. (d) Is this a balanced chemical equation?*

a. This statement can be interpreted in words as: "Propane gas reacts with oxygen gas to produce carbon dioxide gas and water vapor."

b. Using  to represent a propane

molecule, the chemical statement can be represented as:

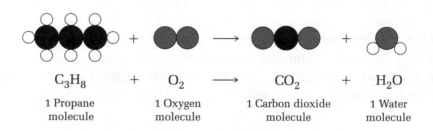

C₃H₈	+	O₂	⟶	CO₂	+	H₂O
1 Propane molecule		1 Oxygen molecule		1 Carbon dioxide molecule		1 Water molecule

c. Counting the atoms on each side of the equation gives this atom inventory:

Reactant side	**Product side**
3 carbon atoms	1 carbon atom
8 hydrogen atoms	2 hydrogen atoms
2 oxygen atoms	3 oxygen atoms

d. The respective numbers of carbon, hydrogen, and oxygen atoms are different in reactants and products. The original statement is not a balanced chemical equation.

For each of the following chemical statements,

    a. write an interpretation of the statement in words,

    b. draw a representation of the chemical statement (some structures are provided),

    c. complete an atom inventory of the reactants and products, and

    d. decide whether the chemical statement, as written, is balanced.

1. Many people use natural gas as a source of household heat. Natural gas contains methane (CH_4), which burns with oxygen gas (O_2) in air according to the equation

$$CH_4 \;+\; 2\,O_2 \;\longrightarrow\; CO_2 \;+\; 2\,H_2O$$

Use ⬤ to represent a methane molecule.

2. When an acid reacts with an active metal, hydrogen gas and an ionic compound are often formed. An expression for hydrobromic acid (HBr) reacting with magnesium metal to form hydrogen gas and magnesium bromide ($MgBr_2$) is:

$$HBr \;+\; Mg \;\longrightarrow\; H_2 \;+\; MgBr_2$$

    Let ⬤ represent a formula unit of magnesium bromide ($MgBr_2$). (Note that formula units are potentially misleading, because, under ordinary circumstances, ionic compounds are not found as individual units, such as molecules, but as large crystals made up of many ions. Thus this and similar representations of ionic compounds are technically inaccurate; however, for an atom inventory, they will suffice. Keep these limitations in mind.)

See page 66 in Section B.10 to view a representation of an ionic crystal structure.

Figure 1.55 *In what ways could you describe the reaction of propane with oxygen?*

3. Hydrogen sulfide (H_2S) and metallic silver react in air to form silver sulfide (Ag_2S), commonly known as silver tarnish, and water:

$$4\,Ag + 4\,H_2S + O_2 \longrightarrow 2\,Ag_2S + 4\,H_2O$$

You learned about the refining of metals using redox reactions on page 101 in Section C.

4. Iron can be extracted from the ore hematite (Fe_2O_3) by reduction with carbon monoxide in a blast furnace according to the equation.

$$Fe_2O_3 + 3\,CO \longrightarrow 2\,Fe + 3\,CO_2$$

5. Aluminum cookware must be treated to prevent it from reacting with acids found in foods. For example, aluminum can react with hydrochloric acid as follows:

$$2\,Al + 6\,HCl \longrightarrow 2\,AlCl_3 + 3\,H_2$$

Try Questions 6 and 7 without drawing representations. (Why might this be a good decision?)

6. Wood or paper, composed mainly of cellulose, $C_6H_{10}O_5$, can burn in air, forming carbon dioxide and water vapor.

$$C_6H_{10}O_5 + 6\,O_2 \longrightarrow 6\,CO_2 + 5\,H_2O$$

There are nine oxygen atoms in one $C_3H_5(NO_3)_3$ molecule. Can you see why?

7. Nitroglycerin, $C_3H_5(NO_3)_3$, the active constituent of dynamite, decomposes explosively, forming N_2, O_2, CO_2, and water.

$$2\,C_3H_5(NO_3)_3 \longrightarrow 3\,N_2 + O_2 + 6\,CO_2 + 5\,H_2O$$

D.5 NATURE'S CONSERVATION: BALANCED CHEMICAL EQUATIONS

The law of conservation of matter holds under normal circumstances. You may be familiar with the equation $E = mc^2$, which indicates that small quantities of matter (mass) can be converted to large quantities of energy within events such as nuclear reactions.

The law of conservation of matter is based on the notion that atoms are indestructible. All changes observed in matter can be interpreted as rearrangements among atoms. Correctly written (balanced) chemical equations represent such changes. In the preceding activity, you practiced recognizing balanced chemical equations. Now you will learn how to write them.

As an example, consider the reaction of hydrogen gas with oxygen gas to produce gaseous water. First, correctly write the reactant formula or formulas to the left of the arrow and the product formula or formulas to the right, keeping in mind that hydrogen and oxygen gas are diatomic (two-atom) molecules.

For a complete listing of elements found as diatomic molecules, see page 117.

$$H_2(g) + O_2(g) \longrightarrow H_2O(g)$$

Check the preceding expression by completing an atom inventory: Two hydrogen atoms appear on the left side and two on the right. Thus, hydrogen atoms are balanced. However, two oxygen atoms appear on the left and only one on the right. Because oxygen is not balanced, the expression requires additional work.

Here is an *incorrect* way to complete the equation:

$$H_2(g) \quad + \quad O_2(g) \quad \longrightarrow \quad H_2O_2(g) \quad \textbf{Incorrect!}$$

| 1 Hydrogen molecule | 1 Oxygen molecule | 1 Hydrogen peroxide molecule |

Although this chemical statement meets an atom-inventory check (there are two hydrogen and two oxygen atoms on both sides), this chemical expression is *quite wrong*. By changing the subscript of O from 1 to 2 in the product, the identity of the product has been changed from water (H_2O) to hydrogen peroxide (H_2O_2). Because hydrogen peroxide is not produced by this reaction, the chemical expression is incorrect. When balancing chemical equations, *subscripts* remain unchanged once correct formulas have been written for reactants and products. *Coefficients* (the numbers placed in front of formulas) must be adjusted to balance equations instead.

Additional hydrogen, oxygen, and water molecules must be added to the appropriate sides of the equation to balance the numbers of oxygen and hydrogen atoms. Another oxygen atom is needed on the product side to bring the number of oxygen atoms on both sides to two. Therefore, a water molecule is added:

$$H_2 \quad + \quad O_2 \quad \longrightarrow \quad H_2O \quad + \quad H_2O$$

| 1 Hydrogen molecule | 1 Oxygen molecule | 2 Water molecules |

As a result, two oxygen atoms appear on each side of the equation. Unfortunately, hydrogen atoms are no longer balanced: Now there are two hydrogen atoms on the left side and four atoms on the right. How can two hydrogen atoms be added to the reactant side? You are correct if you said by adding one hydrogen molecule to the left side:

$$H_2 \quad + \quad H_2 \quad + \quad O_2 \quad \longrightarrow \quad H_2O \quad + \quad H_2O$$

| 2 Hydrogen molecules | 1 Oxygen molecule | 2 Water molecules |

The atoms are now balanced. Count them to verify that!

Under certain conditions, H_2 reacts with O_2 in a violent explosion. However, the reaction can be controlled and, in fact, powers some types of rockets (Figure 1.56). Used in fuel cells, this reaction can generate electricity.

Figure 1.56 *Some rockets are propelled by the thrust generated when liquid hydrogen and liquid oxygen react to form gaseous water and considerable thermal energy.*

Chemical expressions are symbolic statements describing chemical reactions that may or may not be balanced.

It is neither convenient nor efficient to draw representations for every reactant and product in every chemical reaction. Thus this information is usually summarized in a chemical equation:

$$2 \text{ H}_2(g) \quad + \quad \text{O}_2(g) \quad \longrightarrow \quad 2 \text{ H}_2\text{O}(g)$$

This equation can be "read" as indicating, *"Two molecules of hydrogen gas react with one molecule of oxygen gas to produce two molecules of water vapor."* See Figure 1.57.

The following additional suggestions may help you as you write correctly balanced chemical equations:

- If polyatomic ions, such as NO_3^- and CO_3^{2-}, appear as both reactants and products, treat them as units, rather than balancing their atoms individually.

- If water is involved in the reaction, balance hydrogen and oxygen atoms last.

- Re-count all atoms after you think an equation is balanced—just to be sure!

Figure 1.57 *The primary reaction powering this hydrogen fuel cell vehicle is $2 \text{ H}_2 + \text{O}_2$. You will learn more about hydrogen fuel cells in Unit 3.*

DEVELOPING SKILLS

D.6 WRITING CHEMICAL EQUATIONS

Sample Problem: *The reaction of methane gas (CH₄) with chlorine gas (Cl₂) occurs in sewage treatment plants and often in chlorinated water supplies (Figure 1.58). Common products are liquid chloroform (CHCl₃) and hydrogen chloride gas (HCl). Write a balanced chemical equation for this reaction.*

A chemical statement describing this reaction follows. As you can see from a quick glance, the equation is not balanced.

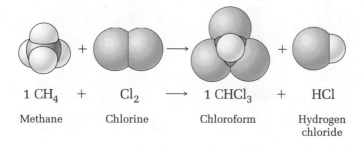

$$CH_4(g) \;+\; Cl_2(g) \;\longrightarrow\; CHCl_3(l) \;+\; HCl(g)$$

| Methane | Chlorine | Chloroform | Hydrogen chloride |

Figure 1.58 *A water-treatment plant, a portion of which is shown here, often uses chlorine to purify municipal water.*

To complete this chemical equation, you can follow this line of reasoning: One carbon atom appears on each side of the arrow, so carbon atoms balance. Thus, the coefficients in front of CH_4 and $CHCl_3$ are regarded (at least for the moment) as "1." We will write it in explicitly, just as a reminder, but the coefficient 1 is removed from the final equation.

$$1\,CH_4 \;+\; Cl_2 \;\longrightarrow\; 1\,CHCl_3 \;+\; HCl$$

| Methane | Chlorine | Chloroform | Hydrogen chloride |

For convenience, the symbols *(g)* and *(l)* are removed. They will reappear in the final equation.

Four hydrogen atoms are on the left, but only two hydrogen atoms are on the right (one in $CHCl_3$, a second in HCl). To increase the number of hydrogen atoms on the product side, the coefficient of HCl can be adjusted. Because two more hydrogens are needed on the right, the number of HCl molecules must be changed from 1 to 3. This gives four hydrogen atoms on the right:

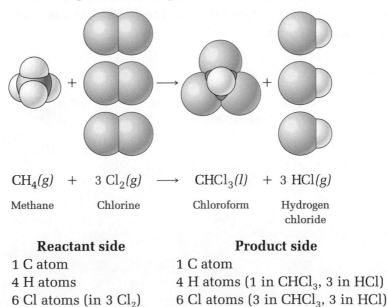

$$1\ CH_4\ +\ Cl_2\ \longrightarrow\ 1\ CHCl_3\ +\ 3\ HCl$$

Methane   Chlorine   Chloroform   Hydrogen
                                                                    chloride

Now both carbon and hydrogen atoms are balanced. What about chlorine? Two chlorine atoms appear on the left and six on the right side. These six chlorine atoms (three in $CHCl_3$, three in 3 HCl) must have come from three chlorine (Cl_2) molecules. Thus 3 must be the coefficient of Cl_2.

The chemical equation appears to be balanced. An atom inventory verifies that the equation is complete and correct as written.

$$CH_4(g)\ +\ 3\ Cl_2(g)\ \longrightarrow\ CHCl_3(l)\ +\ 3\ HCl(g)$$

Methane   Chlorine   Chloroform   Hydrogen
                                                                    chloride

Reactant side	**Product side**
1 C atom	1 C atom
4 H atoms	4 H atoms (1 in $CHCl_3$, 3 in HCl)
6 Cl atoms (in 3 Cl_2)	6 Cl atoms (3 in $CHCl_3$, 3 in HCl)

Copy the following chemical expressions onto a separate sheet of paper, and balance each if needed. For Questions 1–4, draw a representation of your final equation to verify that it is balanced. Structures unfamiliar to you will be provided.

1. Two blast furnace reactions are used to obtain iron from its ore:
   a. $C(s) + O_2(g) \rightarrow 2\ CO(g)$
   b. $Fe_2O_3(s) + CO(g) \longrightarrow Fe(l) + 3\ CO_2(g)$

   Let represent a formula unit of Fe_2O_3.

   Let  represent a molecule of CO_2.

2. The final step in the refining of a copper ore is:

$$CuO(s) + C(s) \longrightarrow Cu(s) + CO_2(g)$$

   Let ⬤⬤ represent a formula unit of CuO.

3. Ozone (O_3) can decompose to form oxygen gas (O_2):

$$O_3(g) \longrightarrow O_2(g)$$

   Let ⬤⬤⬤ represent an ozone molecule.

4. Ammonia (NH_3) in the soil reacts continuously with oxygen gas (O_2):

$$NH_3(g) + O_2(g) \longrightarrow NO_2(g) + H_2O(l)$$

5. As you observed in Investigating Matter B.12, copper metal reacts with silver nitrate solution to form copper(II) nitrate solution and silver metal:

$$Cu(s) + AgNO_3(aq) \longrightarrow Cu(NO_3)_2(aq) + Ag(s)$$

   (*Hint:* Look at the formula for copper(II) nitrate, $Cu(NO_3)_2$. The subscript of two outside the parentheses indicates that this formula contains two nitrate (NO_3^-) anions. So one formula unit of $Cu(NO_3)_2$ contains one copper ion, two nitrogen atoms, and six oxygen atoms.)

6. Combustion of gasoline in an automobile engine can be represented by the burning of octane (C_8H_{18}):

$$C_8H_{18}(l) + O_2(g) \longrightarrow CO_2(g) + H_2O(g)$$

concept check 8

1. Why must chemical equations always be balanced?
2. Is it possible to balance a chemical equation by changing subscripts? Explain.
3. What does it mean for a resource to be renewable?

INVESTIGATING MATTER
D.7 STRIKING IT RICH

Preparing to Investigate

Seeing is believing—or so it is said. In this investigation, you will study the effects of heat and chemical treatment on pennies. As you do so, keep in mind what you have learned about chemical and physical changes and conservation of matter, as well as the choices you will make in the unit project.

Before you begin, read *Gathering Evidence* to learn what you will need to do and note safety precautions. *Gathering Evidence* also provides guidance about when you should collect and record data. Construct a data table appropriate for recording the data you will collect.

Gathering Evidence

A *control* is an untreated sample that can be compared to treated samples.

1. Before you begin, put on your goggles, and wear them properly throughout the investigation.

2. Obtain three pennies. Use steel wool to clean each penny until it is shiny. Record the appearance of the pennies.

3. Set aside one of the clean pennies to serve as a **control**.

4. Weigh a 2.0-g to 2.2-g sample of granulated zinc (Zn) and place it in a 250-mL beaker.

5. Use a graduated cylinder to measure 25 mL of 1 M zinc chloride ($ZnCl_2$) solution. Add the solution to the beaker containing the granulated zinc. *(**Caution:** 1 M zinc chloride solution can damage skin. If any accidentally spills on you, ask a classmate to notify your teacher immediately, and wash the affected area with tap water immediately.)*

6. Cover the beaker with a watch glass and place it on a hot plate. Gently heat the solution until it just begins to bubble, then lower the temperature to sustain gentle bubbling. Do not allow the solution to boil vigorously or heat to dryness. *(**Caution:** Note the warning in Step 5 about zinc chloride solution.)*

7. Carefully remove the watch glass, then use forceps or tongs to carefully lower two clean pennies into the solution in the beaker. To avoid causing a splash, do not drop the coins into the solution. Put the watch glass on the beaker and boil gently for two to three minutes.

8. Carefully remove the watch glass, then use forceps or tongs to remove the two coins from the beaker. Rinse under running tap water, then gently dry with a paper towel. Set one treated coin aside for later comparisons and use the other treated coin in Step 9.

9. Set a hot plate to medium-high heat. Using the tongs, place the second treated penny on the hotplate and count to five. Use the tongs to flip the coin and count to five again. Repeat until you observe a color change (see Figure 1.59).

10. Pick up the coin from the hot plate surface with the tongs, rinse the heated coin under running tap water, and gently dry it with a paper towel. Record your observations.

11. Observe and compare the appearances of all three pennies. Record your observations.

12. Discard the used zinc chloride solution and solid zinc as directed by your teacher.

13. Wash your hands thoroughly before leaving the laboratory.

Figure 1.59 *Heating the treated penny on a hot plate.*

Interpreting Evidence

1. Compare the color of the three coins—untreated (the control), heated in zinc chloride solution only, and heated in zinc chloride solution and then on a hot plate.

2. Do the treated coins appear to be composed of metals other than copper? If so, explain.

Making Claims

3. If someone claimed that a precious metal was produced in this investigation, how would you decide whether the claim was correct?

4. What happened to the copper atoms originally present on the surface of the treated pennies? Provide evidence to support your answer.

Reflecting on the Investigation

5. Identify at least two practical uses for metallic changes similar to those you observed in this investigation.

6. Do you think the treated pennies could be converted back to ordinary coins? If so, what procedures would you use to accomplish this? (**Caution:** *Do not perform any laboratory work without your teacher's approval and direct supervision.*)

D.8 COMBINING ATOMS: ALLOYS

The investigation you just completed demonstrated how metallic properties can be modified by creating an **alloy**, a solid combination of atoms of two or more metals. The immersion of a penny in hot zinc chloride solution produced a silvery alloy of zinc and copper called *brass*. When you heated the penny on the hot plate, the zinc and copper atoms mixed more completely. The overall mixing of zinc and copper atoms to form brass is depicted in Figure 1.60. The resulting solid solution has a different concentration of zinc and copper and is known as one form of brass. Most brass materials have a golden color, unlike either copper or zinc. Brass is also harder than copper metal. Some other common alloys with familiar names are listed in Table 1.9.

Figure 1.60 *The steps to forming a brass layer onto a penny (a-c). Unlike a metal or a chemical compound with a specific formula, an alloy's composition can vary. Brass samples contain from 10 to 40% zinc (by mass). What is the remaining material?*

(a)
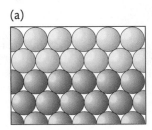

(b)
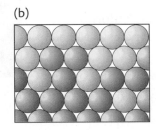

(c)
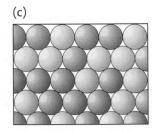

It is clear that one way to modify the properties of a particular metal is to form it into an alloy, just as you did when you produced a gold-colored penny. Often the results of alloying metals are unexpected, as you are about to discover.

A *mixture* is a combination of materials in which each material retains its separate identity. You will learn more about mixtures in upcoming units.

CHEM**QUANDARY**

FIVE CENTS' WORTH

A U.S. nickel is composed of an alloy of nickel and copper. Based on your familiarity with the appearance of that common five-cent coin, you might be surprised to learn it is composed of more copper than nickel! Specifically, each U.S. five-cent coin contains only 25% nickel and 75% copper by mass. What does this suggest about the difference between an alloy and a simple **mixture** of powdered copper and powdered nickel?

Table 1.9

Common Alloy Compositions and Uses

Alloy and Composition	Examples (composition in mass percent)	Comments
Brass *Copper and zinc*	**Red brass** **(90% Cu, 10% Zn)** **Yellow brass** **(67% Cu, 33% Zn)** **Naval brass** **(60% Cu, 39% Zn, 1% Sn)**	**Properties of brass vary with the proportion of copper and zinc used and with addition of small amounts of other elements.** **Brass is used in plumbing and lighting fixtures, rivets, screws, and ships.**
Bronze *Primarily copper with tin, zinc, and other elements*	**Coinage bronze** **(95% Cu, 4% Sn, 1% Zn)** **Aluminum bronze** **(90% Cu, 10% Al)** **Hardware bronze** **(89% Cu, 9% Zn, 2% Pb)**	**Bronze is harder than brass. Its properties depend on the proportions of its components.** **Bronze is used in bearings, machine parts, telegraph wires, gunmetal, coins, medals, artwork, and bells.**
Steel *Primarily iron with carbon and small amounts of other elements*	**Steel** **(99% Fe, 1% C)** **Nickel steel** **(96.5% Fe, 3.5% Ni)** **Stainless steel** **(90–92% Fe, 0.4% Mn, <0.12% C, Cr (trace))**	**Properties of steel are often determined by carbon content. High-carbon steel is hard and brittle; low- or medium-carbon steel can be welded and shaped.** **Steel is used in automobile and airplane parts, kitchen utensils, plumbing fixtures, and architectural decoration.**
Other common alloys	**Pewter** **(85% Sn, 6.8% Cu, 6% Bi, 1.7% Sb)**	**Pewter is often used for figurines and other decorative objects.**
	Mercury amalgams **(50% Hg, 20% Ag, 16% Sn, 12% Cu, 2% Zn)**	**Mercury amalgams have often been used for dental fillings.**
	14 Carat gold **(58% Au, 14–28% Cu, 4–28% Ag)**	**14-carat gold is popular in jewelry.**
	White gold (90% Au, 10% Pd)	**White gold is also principally used for jewelry.**

Some other useful alloys have a constant, definite ratio of metallic atoms. One example is Ni_3Al, a low-density, strong metallic alloy of nickel and aluminum used as a component of jet aircraft engines. A very hard chromium-platinum alloy, Cr_3Pt, forms the basis of some commercial razor blade edges. And a special group of alloys, including the niobium-tin compound Nb_3Sn, displays **superconductivity**—the ability to conduct an electric current without any electrical resistance—if cooled to a sufficiently low temperature.

D.9 CONSERVATION IN THE COMMUNITY

Now that you have seen how conservation works at the particulate level, and have learned how to express that conservation using chemical equations, it is time to think about conservation at the macroscopic level. The law of conservation of matter applies at all levels, a fact you will need to take into account as you consider the implications for resources in your choices about currency.

As you read more about resources, consider how chemistry is related to an idea you may have heard about—**sustainability**. Sustainable practices and processes are those present-day activities that preserve the ability of future generations to thrive and meet their resource needs on a habitable Earth.

Depleting Resources

In some ways, Earth is like a spaceship. The resources "on board" are all that are available to the ship's inhabitants. Some resources—such as fresh water, air, fertile soil, plants, and animals—can eventually be replenished by natural processes. These are called **renewable resources**. As long as natural cycles are not disturbed too much, supplies of renewable resources can be maintained indefinitely. Other materials—such as metals, natural gas, coal, and petroleum—are considered **nonrenewable resources** because they cannot be readily replenished. If atoms are always conserved, why do some people say that a resource may be "running out"? Can a resource actually "run out"?

The answer can be found by first remembering that atoms are conserved in chemical processes, but molecules might not be. For example, the current production of new petroleum molecules in nature is much, much slower than the current rate at which petroleum molecules are being burned to produce thermal energy, carbon dioxide, water, and other molecules. Thus, the total inventory of petroleum on Earth is declining, but the total number of carbon and hydrogen atoms on Earth remains constant.

Resources such as petroleum are regarded as *nonrenewable resources*, because they take millions of years to form.

A resource—particularly a metal—can be depleted in another way. As you have learned, profitable mining depends on finding an ore with at least some minimum metal content. This minimum level depends on the metal and its ore: from as low as 1% for copper or 0.001% for gold to as high as 30% for aluminum.

Once ores with high metal content are depleted, lower-grade ores with less metal content are processed. Meanwhile, atoms of the metal that were previously concentrated in rich deposits in limited parts of the world gradually become spread out (dispersed) over wider areas of Earth. This dispersion makes the used metal far less useful for new applications. Eventually, the mining and extraction of certain metals may become prohibitively expensive, thus making the metal too costly for general use. At that time, for practical purposes, the supply of that resource can be considered depleted.

Conserving Resources

Can such depletion scenarios be avoided? Can Earth's mineral resources be conserved? One strategy for conservation is to slow the rate at which resources are used. This can be best addressed by considering the four Rs—*rethinking, reusing, reducing,* and *recycling.*

Part of this strategy includes *rethinking* personal and societal habits and practices involving resource use. Rethinking can take the form of re-examining old assumptions, identifying resource-saving strategies, and, perhaps, uncovering new solutions to old problems. Rethinking can involve decisions such as whether it is better to choose paper or plastic bags in grocery stores, or, as is becoming more common, if it is even better to consider *reusing* grocery bags or bringing your own cloth bags.

Another approach is *reducing* the use of a resource by finding substitute materials with similar properties, preferably from renewable resources, or simply using less of the resource. In addition, some manufactured items can be refurbished or repaired for reuse rather than sent to a landfill. Common examples include used car parts and printer cartridges, both of which can often be reused after rebuilding or refilling. Possibly the fundamental part of resource conservation and management is to consider the most direct option—*source reduction.* That simply means decreasing the amount of resources used. The fewer resources used now, the more remain available for future generations.

Finally, certain items can be recycled, or gathered, for reprocessing. Such *recycling* allows the resources present in the items to be used again. Figure 1.61 (page 138) illustrates the steps involved in recycling aluminum cans.

Recycling is one option for the final step in the life cycle of a material—a topic you will now consider as you continue to evaluate how "green" the forms of U.S. currency actually are.

Minimum profitability levels for mining depend not only on the metal's abundance in the ore, but also on the metal's chemical activity. Less reactive metals (such as gold) are often found in their native, or uncombined, state and are more readily released from their compounds to form pure metal than are more reactive metals (such as aluminum). Thus, more reactive metals must usually be mined from ores with high metal content to be profitable.

What other Rs have you heard of in the context of ecology?

Green Chemistry is an approach to chemistry that aims to eliminate pollution in industrial settings by preventing it from happening in the first place through enactment of one or more of the four Rs.

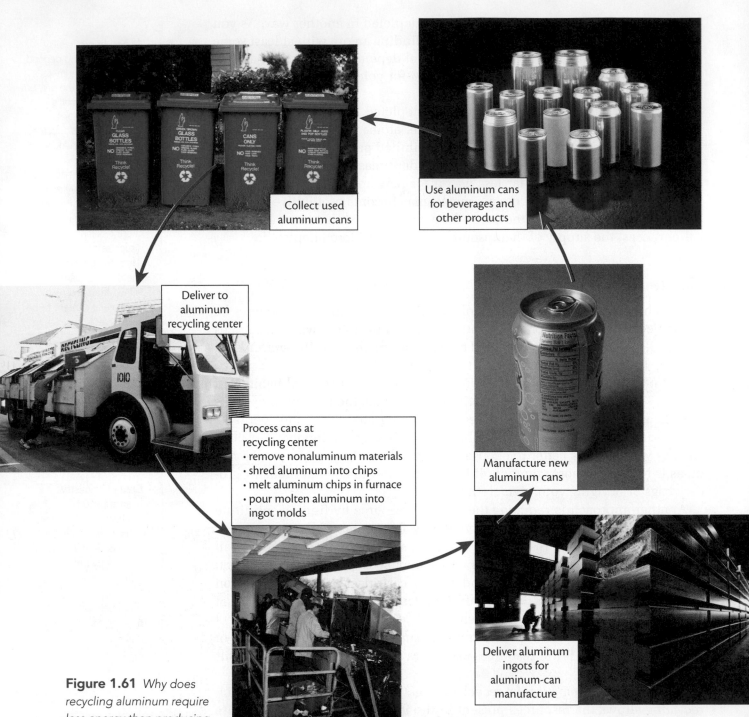

Use aluminum cans for beverages and other products

Collect used aluminum cans

Deliver to aluminum recycling center

Process cans at recycling center
• remove nonaluminum materials
• shred aluminum into chips
• melt aluminum chips in furnace
• pour molten aluminum into ingot molds

Manufacture new aluminum cans

Deliver aluminum ingots for aluminum-can manufacture

Figure 1.61 *Why does recycling aluminum require less energy than producing aluminum from its ore?*

■ MAKING DECISIONS

D.10 RETHINKING, REUSING, REDUCING, AND RECYCLING

In Section C (Making Decisions C.15), you considered the life cycle of the U.S. one-dollar coin. In this activity, you will apply what you have learned about conservation and product life cycles to the U.S. one-dollar bill. Use your responses to Making Decisions A.11 and C.15 as you answer the following questions.

1. First, consider the 4 Rs.

    a. In what ways are dollar bills reused?

    b. How could the resources necessary for dollar bill production be reduced?

    c. Are dollar bills recycled? If yes,

        i. What products are made from recycled banknotes?

        ii. About what percent of bills are recycled?

    d. How could rethinking be applied to the life cycle of dollar bills?

2. Think about all of the materials and energy that are required to produce dollar bills.

    a. Which materials are renewable?

    b. Which materials are not renewable?

    c. Which energy sources are renewable?

    d. Which energy sources are not renewable?

    e. Do the production and use of dollar bills meet criteria for sustainability? Explain.

3. Finally, use the answers to all of your previous Making Decisions activities and the questions in this activity to sketch and label a life cycle diagram for dollar bills. Be sure to include all raw materials and their sources, production, use, and final disposition. (*Hint:* Remember that recycled products also have a finite life span.)

Figure 1.62 *Recycled-content shingles can be made from materials such as milk jugs, used carpet, recycled tires and reclaimed wood. They are fire and impact resistant, look like traditional shingles, and last as long as or longer than shingles made from new materials.*

SECTION D SUMMARY

Reviewing the Concepts

> A chemical equation represents a reaction of one or more substances to form one or more new substances.

> The atoms that compose matter are neither created nor destroyed in a chemical reaction.

1. Represent each chemical equation by drawing particulate-level models of the reactants and products. Use circles of different sizes, colors, or shading for each element.

   a. $H_2(g) + Cl_2(g) \longrightarrow 2\ HCl(g)$

   b. $2\ H_2O_2(aq) \longrightarrow 2\ H_2O(l) + O_2(g)$

   Let  represent a hydrogen peroxide molecule, H_2O_2.

   c. Using complete sentences, write word equations for the chemical equations given in Questions 1a and 1b. Include the numbers of molecules involved.

2. Write chemical equations that represent these word equations:

   a. Baking soda ($NaHCO_3$) reacts with hydrochloric acid (HCl) to produce sodium chloride, water, and carbon dioxide.

   b. During respiration, one molecule of glucose ($C_6H_{12}O_6$) reacts with six molecules of oxygen gas to produce six molecules of carbon dioxide and six molecules of water.

3. State the law of conservation of matter.

4. What is a scientific law?

5. Complete atom inventories to decide whether each of the following equations is balanced:

   a. The preparation of tin(II) fluoride, a component of some toothpastes (called *stannous fluoride* in some ingredient lists):

   $$Sn(s)\ +\ HF(aq)\ \longrightarrow\ SnF_2(aq)\ +\ H_2(g)$$
   Tin   Hydrofluoric   Tin(II)   Hydrogen
   metal   acid   fluoride   gas

   b. The synthesis of carborundum for sandpaper:

   $$SiO_2(s)\ +\ C(s)\ \longrightarrow\ SiC(s)\ +\ CO(g)$$
   Silicon   Carbon   Silicon   Carbon
   dioxide   carbide   monoxide
   (sand)   (carborundum)

   c. The reaction of an antacid with stomach acid (hydrochloric acid):

   $$Al(OH)_3(s) + 3\ HCl(aq) \longrightarrow AlCl_3(aq) + 3\ H_2O(l)$$
   Aluminum   Hydrochloric   Aluminum   Water
   hydroxide   acid   chloride

6. Why are expressions such as "using up" and "throwing away" misleading, considering the law of conservation of matter?

> Coefficients in a chemical equation indicate the relative number of units of each reactant and product involved. Subscripts in a substance's formula may not be changed to balance an equation.

7. Consider this equation:

$$N_2(g) + 3\,H_2(g) \longrightarrow 2\,NH_3(g)$$

a. What is the coefficient for hydrogen gas?

b. What is the coefficient for NH_3 gas?

c. What is the coefficient for nitrogen gas?

8. For each process below, draw a representation of the chemical statement, balance the representation, and verify your answer.

a. Preparing tungsten from one of its minerals:

$$\_WO_3 + \_H_2 \longrightarrow \_W + \_H_2O$$

b. Heating lead sulfide in air:

$$\_PbS + \_O_2 \longrightarrow \_PbO + \_SO_2$$

c. Rusting (oxidation) of iron metal:

$$\_Fe + \_O_2 \longrightarrow \_Fe_2O_3$$

9. Balance each of these chemical expressions:

a. Preparing phosphoric acid (used in making soft drinks, detergents, and other products) from calcium phosphate and sulfuric acid:

$$\_Ca_3(PO_4)_2 + \_H_2SO_4 \longrightarrow \_H_3PO_4 + \_CaSO_4$$

b. Completely burning octane, C_8H_{18}, a component of gasoline:

$$\_C_8H_{18} + \_O_2 \longrightarrow \_CO_2 + \_H_2O$$

10. A student is asked to balance this expression:

$$Na_2SO_4 + KCl \longrightarrow NaCl + K_2SO_4$$

The student decides to balance it this way:

$$Na_2SO_4 + K_2Cl \longrightarrow Na_2Cl + K_2SO_4$$

a. Complete an atom inventory of the student's answer. Are the atoms conserved?

b. Did the student create a correctly balanced chemical equation? Explain.

c. If your answer to Question 10b is "no," write a correctly balanced equation.

> An alloy possesses properties that differ, sometimes significantly, from the properties of its constituent elements.

11. What is an alloy?

12. Give examples of two alloys you use regularly. (*Hint:* See Table 1.9 on page 135.)

13. What nonmetal is a component of both steel and stainless steel? (*Hint:* See Table 1.9 on page 135.)

14. Give the formula, application, and important physical property of an alloy that is also a well-defined compound.

> Resources are either renewable or nonrenewable. Resources can be conserved by recycling, reducing (controlling rate of use), or replacing them with substitute resources.

15. What is sustainability?

16. a. What is the difference between reusing and recycling?

b. Give two examples of each, other than those used in the textbook.

17. In addition to those found in the textbook, list four examples of

a. renewable resources.

b. nonrenewable resources.

18. Classify each use as either *recycling* or *reusing*:

    a. storing water in used juice bottles for an emergency.

    b. converting plastic milk containers into fibers used to weave clothing fabric.

    c. packing breakable items with shredded newspapers.

19. How would the life cycle of a light bulb compare to that of a newspaper? Consider material sources and disposal/recycling.

How is conservation of matter demonstrated in the use of resources?

In this section, you balanced chemical statements and then, once again, considered Earth's resources. Think about how these ideas are connected, then answer the question in your own words in organized paragraphs. Your answer should demonstrate your understanding of the key ideas in this section.

Be sure to consider the following in your response: conservation of matter, what it means to have a balanced chemical equation, sustainability, renewability, and resource depletion.

Connecting the Concepts

20. How is a scientific law, such as the law of conservation of matter, different from a law enacted by a government?

21. Earth has been compared to a space station.

    a. In what ways is this analogy useful?

    b. In what ways is it misleading?

22. Describe a real-world example of conservation of matter.

23. Describe at least two benefits of discarding less waste material and recycling more of it.

Extending the Concepts

24. Atoms that presently make up your body may have once been part of the body of a Tyrannosaurus rex, or even Alexander the Great or Cleopatra. Explain how this is possible.

25. Investigate recent advances and potential applications in the field of superconductivity.

26. What conclusions about materials can be drawn from a study of substances used for currency in ancient civilizations? Explain your ideas by citing examples.

27. Why is aluminum metal more easily produced from recycled aluminum cans than from aluminum contained in clay, bauxite, or aluminum oxide ore?

PUTTING IT ALL TOGETHER

MAKING THE CASE FOR CURRENCY

You probably realize that new products are not easily created and that even the design process requires multiple stages. One common step in designing a new product is soliciting proposals from several individuals or teams. Once the proposals are submitted, a panel of experts or consumers evaluates them to choose the best design. You will follow a similar process in selecting the best "dollar recommendation" from the proposals submitted by your classmates. Each team will present its recommendations to the class, which will also serve as the selection committee. Here is a summary of essential features expected of your proposal:

PRESENTATION

Recommendation

Clearly outline your recommendations for printing and minting dollars in the United States.

- Should both the dollar bill and dollar coin continue to be produced?
- Should one or the other be discontinued?
- Should the bill or the coin (or both) be made from different materials or otherwise redesigned?

Rationale

Support your recommendations with evidence, including:

- The factors that most influenced your recommendations and why they are important.
- Descriptions of the materials that make up the forms of currency that you recommend.
- Details about the raw materials needed to produce the currency, including
  a. for metals:
     - Where major deposits are located.
     - Important ores of the chosen raw materials.
  b. for all materials:
     - How the materials are mined, collected, or harvested.
     - How the materials are processed for use or production.
     - Your rationale for choosing the selected materials.
     - An analysis of both necessary and desirable properties of the selected materials.

Environmental Implications

Present an analysis of the life cycle of each form of currency to be retained in your recommendation, including:

- How obtaining the raw materials will impact local and global environments.
- What forms and relative quantities of energy will be required and what wastes will be generated in production of the currency.
- How long each form of recommended currency is expected to last in general circulation.
- How material(s) making up the recommended currency will be treated after the useful life of the bill or coin (i.e., How will they be recycled, reused, or disposed?)
- How the life cycle of the bill or coin will be affected by your suggested material modifications.

Advocacy Plan

Include a discussion that addresses how the general public and Congress would be encouraged to support your recommendations. Be sure to address:

- Identified concerns with the form or forms of currency that you recommend and how they will be addressed. (Indicate whether your suggested material modifications address any of these concerns.)
- Benefits of, and possible incentives for, adopting your recommendations.
- Strategies to ease a transition to new or lesser used forms of currency, if applicable.

Publication

Develop a web page or brochure that will inform others about your recommendations and persuade them to adopt your plan. Include portions of your presentation that are appropriate for the general public and that support your conclusions.

LOOKING BACK AND LOOKING AHEAD

As you come to the end of this unit, pause and reflect on what you have learned so far. You have learned some of the working language of chemistry, such as symbols, formulas, and equations; laboratory techniques; and major ideas in chemistry, such as the organization of the periodic table, the law of conservation of matter, and the structure of atoms.

This knowledge can help you better understand some societal issues. Central among these are the use and management of Earth's chemical resources, including water, metals, petroleum, food, and air.

Chemistry-related societal problems are far too complex for a simple technological fix. Issues of policy are not usually "either/or" situations, but often involve weighing many considerations. As a voting citizen, you will deal with issues that require some scientific understanding. Tough decisions may be needed. The remaining chemistry units that you study will continue to prepare you for that responsibility.

Unit 2 addresses gases, focusing on air in the atmosphere that surrounds and sustains us. As you learn about the properties of the substances in Earth's atmosphere, you will also have the opportunity to design a scientific investigation to study an air-quality issue of your choice.

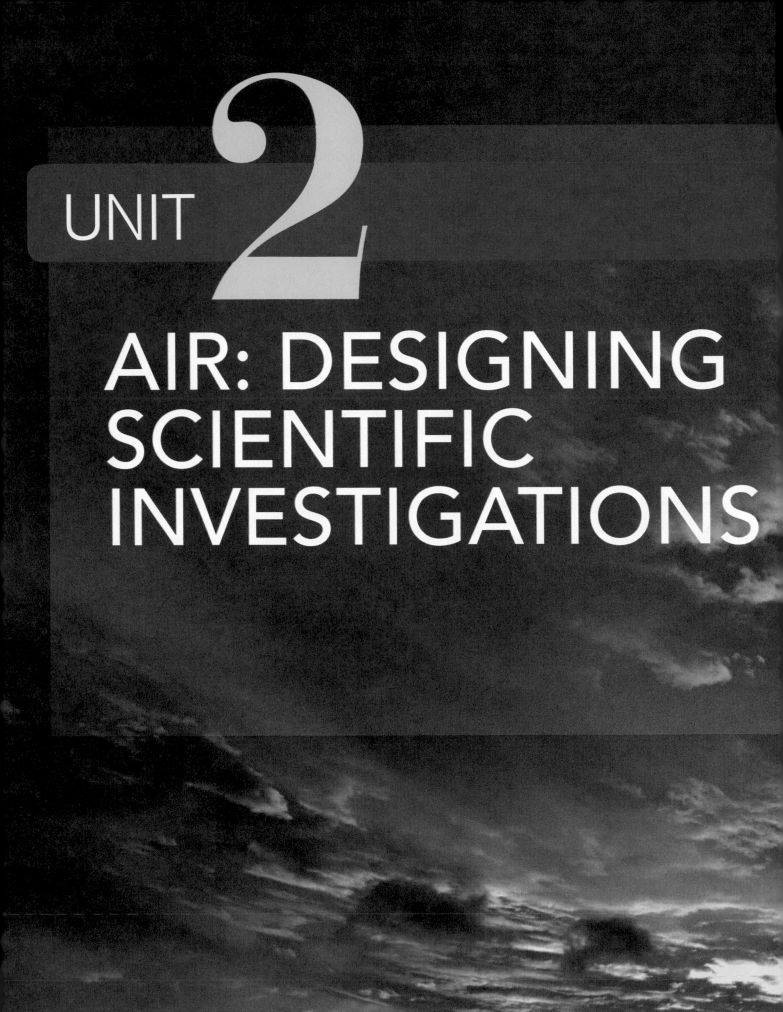

UNIT 2

AIR: DESIGNING SCIENTIFIC INVESTIGATIONS

?

As you read news online or in the local newspaper, you probably notice that there are many stories about the environment. One common theme that you have almost certainly observed is Earth's air quality. Do air-quality articles trigger your concerns? How does local and global air quality affect you? *Turn the page to begin exploring and investigating air-quality issues.*

What questions came to mind as you looked at the images and read the captions on the previous pages? Did you wonder whether your air is safe to breathe or if you should exercise outdoors? Did you think about effects on wildlife? Perhaps you wondered whether emissions tests are useful or what you can really do to help prevent air pollution.

Did you find any of the ideas unlikely, or did you ask yourself, "How do they know?" If you did, you are already starting to think like a scientist. As you learn about Earth's atmosphere and the gases that make it up, you will also be developing the skills of scientific inquiry.

At the end of this unit, you will report on an investigation that you have designed to test an air-quality claim or product. As you progress through this unit, you will develop a scientific question to guide your investigation, outline a procedure, and give and receive feedback on proposals within your class. Your final report will include both a poster outlining your planned investigation and a letter to those who would be interested in or affected by the results of your investigation. As you begin studying gases, keep in mind the thoughts you had as you viewed the opening photo essay and consider which air-qualilty questions and issues make a difference to you.

SECTION  PROPERTIES OF GASES

What information can investigations provide about gas behavior?

The air that surrounds and sustains us—Earth's atmosphere—is composed of several gases. Any air-quality investigation, including the one you will develop in this unit, must account for the behavior of gases in the atmosphere. This section addresses the properties and behavior of gases in general; other sections in this unit will focus on particular gases, interactions among gases, and the structure of Earth's atmosphere.

In addition to considering the properties of gases, an air-quality investigation must ask an answerable scientific question and gather meaningful and reliable information. In this section, you will practice developing good scientific questions and identifying the information that investigations can provide. As you learn about gases and investigations, continue to think about an air-quality question that affects or intrigues you. How might you address that question through scientific investigation? In this section, you will apply knowledge of gas behavior to help develop a question to guide your investigation. In later sections, you will design procedures to gather evidence that could support an answer to your question, and then refine these procedures before preparing a scientific poster to describe your proposed investigation in Putting It All Together.

GOALS

- Define and apply the concept of pressure, using various units.
- Describe and apply the relationship between pressure and volume of a gas sample at constant temperature.
- Describe and apply the relationship between kelvin temperature and volume of a gas sample at constant pressure.
- Describe and apply the relationship between pressure and kelvin temperature of a gas sample at constant volume.
- Identify and write good scientific questions.
- Explain processes of and reasons for designing scientific investigations.

concept check 1

In this unit, you will be designing an investigation to test an air-quality claim or product. As a first step, take a few minutes to observe the atmosphere that surrounds you. Can you see, smell, or taste the gases that make up the atmosphere? Is there anything that makes you wonder whether gases are forms of matter?

1. What is "matter"?

2. What observations have you made in your daily life that suggest that atmospheric gases are a form of matter? Explain.

■ INVESTIGATING MATTER
A.1 PROPERTIES OF GASES

Preparing to Investigate

Before you begin, read *Gathering Evidence* to learn what you will need to do and note safety precautions. Predict what you think will happen in each of the nine investigations; write down your predictions. Plan for data collection and construct an appropriate data table.

Gathering Evidence

Five stations, A through E, have been set up around the laboratory. At each station, you will complete the investigations indicated for that station. These investigations can be done in any order; that is, work at Station D can be completed before Station B investigations, and so on.

   Before you begin, put on your goggles, and wear them properly throughout the investigation.

Follow this general procedure:

- Reread the instructions for the investigation.
- Review your prediction.
- Complete the investigation.
- Record your observations.
- Restore the station to its original condition.

Station A

Investigation 1

1. Draw some air into a syringe.
2. Seal the tip by placing the cap on the *open* end.

3. Holding the cap in place, gently push the plunger down with your thumb, as shown in Figure 2.1.

4. Release the plunger.

5. Record your observations.

Investigation 2

1. Inflate and tie off two new balloons so that they are approximately the same size, about the size of a grapefruit.

2. Use tongs to submerge one inflated balloon in an ice–salt water bath.

3. Use tongs to submerge the other inflated balloon in a container of hot tap water.

4. Record your observations.

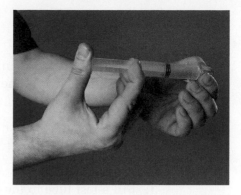

Figure 2.1 *Manipulating a sealed syringe (Investigation 1).*

Station B

Investigation 3

1. Inflate a balloon to approximately the size of a grapefruit and tie off the end.

2. Place the inflated balloon on a balance, using a piece of tape to hold it in place.

3. Record the mass of the balloon (and attached tape).

4. Remove the balloon from the balance. Use a pin to gently puncture the balloon near the neck and release most of the gas within the balloon.

5. Place the deflated balloon on the balance (with the tape still attached) and record its mass.

6. Record your observations.

Investigation 4

1. As shown in Figure 2.2, insert the rounded end of a new, uninflated balloon part way into an empty soft-drink bottle, stretching the balloon's neck over the mouth of the bottle.

2. Try to blow up the balloon so that it fills the bottle.

3. Remove and discard the used balloon.

4. Record your observations.

Figure 2.2 *Placement of a balloon in a bottle (Investigation 4).*

Station C

Investigation 5

1. Lower an empty drinking glass, with its open end facing downward, into a larger container of water.
2. With the open end still under water, slowly tilt the glass.
3. Record your observations.

Investigation 6

1. Fill a test tube to the rim with water.
2. Cover the test tube opening with a piece of stiff plastic.
3. Press down the plastic to make a tight seal with the mouth of the test tube.
4. While continuing to press the plastic to the test tube, invert the test tube above a sink or a pan.
5. Without causing any jarring, gently remove your hand from the piece of plastic.
6. Repeat the process with the test tube half-full of water.
7. Record your observations.

Station D

Investigation 7

1. Fill a test tube to the rim with water.
2. Cover the test tube opening with a piece of plastic wrap.
3. While continuing to press the plastic wrap to the test tube opening, invert the test tube and partially immerse it in a container of water.
4. Remove the piece of plastic wrap.
5. Move the test tube up and down, keeping its lower (open) end under water.
6. Repeat the process with the test tube half-full of water.
7. Record your observations.

Investigation 8

1. Locate the plastic bottle with a small hole in its side.
2. Cover the hole in the side of the bottle with your finger.
3. Fill the bottle with water.
4. Replace the cap tightly.
5. Holding the bottle over a sink, remove your finger from the hole.
6. Still holding the bottle over a sink, remove the cap.
7. Record your observations.

Station E

Investigation 9

1. Place ~10 mL of water in a clean, empty aluminum soft-drink can.

2. Place the can on a hot plate and bring the water to a rapid boil.

3. Using tongs to handle the soft-drink can, quickly remove the can from the heat and immediately invert it into a container of ice-cold water, as shown in Figure 2.3.

4. Record your observations.

Interpreting Evidence

1. Which of the previous investigations suggest that air is composed of matter? Explain your choices, using observational evidence to support your answer.

2. For each investigation, briefly describe how well your results corresponded with your predictions and propose explanations for any differences between your predictions and the observed results of the investigations.

Figure 2.3 *Use tongs to quickly invert a can containing boiling water into a container of ice water (Investigation 9).*

Making Claims

3. For any three investigations,

   a. describe your observations in detail.

   b. explain the role of air in the investigation.

4. Describe another investigation or experience you have had that suggests that

   a. air is matter.

   b. air exerts pressure.

5. Complete this investigation at home (or in the school's lunchroom): Put one end of a straw in a glass of water. Hold another straw outside the glass. Place the ends of both straws in your mouth and try to drink the water through the straw in the glass. See Figure 2.4.

   a. Describe what you observed when you tried this activity.

   b. Based on your observations, what makes it possible to drink liquid through a straw?

Figure 2.4 *What will happen when you try to drink water through the straw inside the glass?*

A.2 PRESSURE

In everyday language, the word *pressure* can have many meanings. Perhaps you use it to mean that you feel too busy or that you feel forced to behave in certain ways. The greater the pressure, the more "boxed in" you feel. To scientists, pressure also refers to force and space, but in quite different ways.

In science, **pressure** refers to the force applied divided by the surface area:

$$\text{Pressure} = \frac{\text{Force}}{\text{Area}}$$

The **force** (or weight) of any object depends on its mass and gravitational attraction. **Area** refers to the total surface of an object. For a square or a rectangular surface, area is calculated by multiplying the lengths of its two adjacent sides. The unit of length is the **meter (m)**. Area is expressed as m × m, or square meters (m²). You can see from the equation that as force increases, pressure also increases. Thus we can say that pressure is *directly* proportional to the force applied. By contrast, as area increases, the pressure decreases. This means that pressure is *inversely* proportional to the area upon which the force acts.

An inverse proportion is a mathematical relationship where an *increase* in one variable results in a *decrease* of the other variable.

To illustrate these relationships, think back to a common experience: Someone steps on your foot. The pressure (or maybe even pain) that you feel depends on the two variables involved in the equation above—*force* and *area*. You surely can feel the difference between a 120-pound person or a 180-pound person stepping on your foot, assuming that both people have the same size foot and wear similar shoes. The larger the force, in this case weight, the more pressure that is exerted. To illustrate the inverse relationship between area and pressure, suppose the same person stepped on your foot two different times. The first time, she wears sneakers; the second time, she is wearing high heels. In each case, the applied weight (force) is the same, but the area over which the force is applied differs. As the area of the shoe decreases, the pressure exerted on your foot increases. Consider how the "shoes" in Figure 2.5 relate to the concepts of force, area, and pressure.

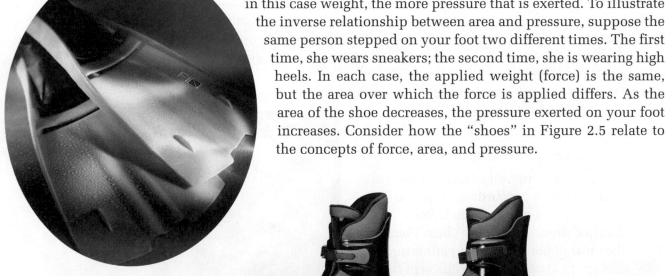

Figure 2.5 *If someone were to accidentally step on your foot, which "shoes" would you prefer that person wore? Explain your choice in terms of the concept of pressure.*

In the same way that someone exerts pressure on your foot, gas molecules exert pressure on the walls of their container and on other objects. You observed this in Investigating Matter A.1. For instance, in Investigation 6 (page 154), you may have concluded that air pressure holds the piece of plastic against the water in an inverted test tube. How much pressure did the air exert on the piece of plastic, and how could you measure this pressure?

You probably are familiar with several units used to report pressure. For example, U.S. weather forecasters report barometric air pressure in *inches of mercury*. When you check air pressure in car or bicycle tires, most likely your tire gauge reads *pounds per square inch*, or *psi*. Other pressure units you may have already heard of are *millimeters of mercury* (mmHg), and *atmospheres* (atm). With so many different units available for measuring and reporting pressure, how do scientists communicate with one another?

Scientists have agreed to use certain units when communicating results. This system of units, called the **International System of Units (SI)**, allows scientists from around the world to communicate clearly with each other. Table 2.1 (page 158) lists some SI units. Note that some, such as *mass* and *length*, are **base units**; they express the fundamental physical quantities of the modernized metric system. Others, such as *area, force, volume,* and *pressure*, are **derived units**; they are formed by mathematically combining two or more base units.

Recall that pressure can be expressed as *force per area*. It is useful to understand the SI units of force and area before learning about the SI derived unit for pressure. The SI derived unit for force is the **newton (N)**. To visualize one newton of force, imagine holding a personal-sized bar of soap (with a mass slightly greater than 100 g) in your open hand. That bar of soap "pushes" downward on your open hand with a force of about one newton (1 N). Figure 2.6 provides other examples of objects that exert about one newton of force.

Scientists have established standards upon which all measurements are based. These include units of *temperature, length, time,* and *mass*. For example, the standard for length (the meter) is currently based on the distance traveled by light in a vacuum during 1/299 792 458 of a second.

The abbreviation *SI* is based on *Système International d'Unités,* its official French name.

Is the SI unit for density a base unit or a derived unit?

Figure 2.6 *A wireless mouse, a medium-sized apple, and a stick of butter each exert about 1 N of force.*

Pascal is pronounced so it rhymes with *rascal*.

When a one-newton (1-N) force is exerted upon an area of one square meter (1 m²), the pressure equals one **pascal (Pa)**. The pascal (Pa) is the SI derived unit of pressure. One pascal (1 Pa) is a relatively small pressure; it is roughly the pressure exerted downward on a slice of bread by a thin layer of soft butter spread on its top surface. Because a pascal is so small, the *kilopascal (kPa),* a unit 1000 times larger than the pascal, is commonly used.

Table 2.1

Some automobile tires specify inflation pressure in kilopascals (kPa) in addition to pounds per square inch (psi).

SI Units for Selected Physical Quantities		
The Seven Base Quantities of SI	**Name**	**Symbol**
Length	meter	m
Mass	kilogram	kg
Time	second	s
Electric current	ampere	A
Thermodynamic temperature	kelvin	K
Amount of substance	mole	mol
Luminous intensity	candela	Cd
Some Derived Quantities		
Area	square meter	m²
	square centimeter	cm²
Volume	cubic meter	m³
	cubic centimeter	cm³
Force	newton	N (equal to kg·m/s²)
Pressure	pascal	Pa (equal to N/m²)

Another volume unit with which you are familiar— the milliliter (mL)— is equal to 1 cm³.

concept check 2

Figure 2.7 shows two bricks of the same mass lying on the ground.
1. Is each brick exerting the same total *force* on the ground? Explain.
2. Is each brick exerting the same *pressure* on the ground? Explain.

DEVELOPING SKILLS

A.3 APPLICATIONS OF PRESSURE

Sample Problem: As a preliminary step in calculating the pressure (in pascals) exerted on the ground by the brick on the right side of the photo in Figure 2.7, determine the area of the brick that is in contact with the ground. The dimensions of the brick surface touching the ground are 9.3 cm × 5.5 cm.

To answer this, first note that one pascal has units of one N/m^2. (Why is this true? If you are not sure, review Section A.2.) Thus, area must be expressed in square meters. First, convert each length from cm to m:

9.3 cm × 1 m/100 cm = 0.093 m

5.5 cm × 1 m/100 cm = 0.055 m

Multiply these lengths together to get area in square meters:

0.093 m × 0.055 m = 0.0051 m^2 = 5.1 × 10^{-3} m^2

> Recall that 0.0051 and 5.1 × 10^{-3} are equivalent expressions. If you need to review scientific notation, see Appendix B.

Using what you have learned about pressure, answer the following questions.

1. The brick on the right side of the photo in Figure 2.7 is exerting a force of 18 N. Use the results of the *Sample Problem* to calculate the pressure, expressed in pascals, that this brick exerts on the ground.

2. The brick on the left side of the photo weighs the same as the brick on the right side. Calculate the pressure, expressed in pascals, that the brick on the left side of the photo exerts on the ground. The dimensions of the brick surface touching the ground are 9.3 cm × 21.3 cm.

3. Compare the pressure values you calculated for each brick. Do your results make sense in light of your reasoning in Concept Check 2? Explain.

4. You need to chop some wood to build a fire. You—quite obviously—reach for an axe rather than a hammer to complete this task. Explain your choice in terms of the concept of pressure.

Figure 2.7 *Two similar bricks lie on the ground. Are they exerting the same pressure upon the ground?*

A.4 ATMOSPHERIC PRESSURE

You have just learned what pressure means to a scientist, and how it can be expressed in quantitative units. On a typical day at sea level, air in the atmosphere exerts a force of about 100 000 N on each square meter of your body; resulting in air pressure of about 100 kPa. This pressure is also roughly equal to *one atmosphere* (1 atm), which is another unit commonly used by scientists and one with which you will become familiar. One atmosphere is equal to 101.3 kPa. You observed the effects of atmospheric pressure (see Figure 2.8) several times in Investigating Matter A.1. For instance, atmospheric pressure caused water to remain in the sealed 2-L bottle despite the hole in its side.

Recall that in Investigating Matter A.1, you covered a test tube filled with water and inverted it into a container of water. You then uncovered the test tube. What did you notice about the level of water in the tube? What force supported the weight of the column of water in the test tube? Now imagine repeating the laboratory activity with taller and taller test tubes. If the test tube were tall enough, water would no longer remain at its initial level when inverted in a container of water. This puzzling effect was first observed in the mid-1600s. Scientists discovered that 1 atm of air pressure could support a column of water only as tall as about 10.3 m (33.9 ft). If the activity were tried with even taller inverted tubes, the water column still would reach only the 10.3-m level.

Figure 2.8 *Tornados are powerful reminders of atmospheric pressure.*

A *barometer* (a device that measures atmospheric pressure) based on a tube filled with water would be much too tall to be useful (see Figure 2.9a). Therefore, scientists replaced the water with mercury, a liquid 13.6 times denser than water. The resulting mercury column, illustrated by the barometer depicted in Figure 2.9b, is shorter than the water column by a factor of 13.6.

At one atmosphere of pressure, the mercury column has a height of 760 mm: 1 atm = 760 mmHg. This means that pressure can also be expressed in units of *mmHg*, or *inches of mercury*, corresponding to the height of a column of mercury supported by that pressure.

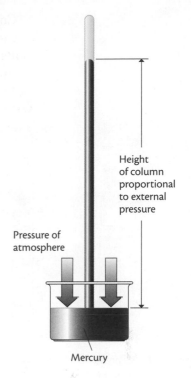

Figure 2.9a *Visualize the height of a column of water that can be supported by 1 atm of air pressure. For instance, think of the average height of your friends. If they stood on each others' shoulders, how many people would be needed in order to hold the top of the tube? Would you need to stand on a two-story house or a three-story house to hold the top of the tube? Describe or draw your mental picture. Pictured here is the world's largest barometer, the Bert Bolle Barometer in the Denmark Visitor Center in Denmark, Western Australia.*

Figure 2.9b *A mercury barometer. On a typical day at sea level, the atmosphere will support a column of mercury 760 mm high. The pressure unit of atmospheres is related to pressure expressed in millimeters of mercury: 1 atm = 760 mmHg. What common object could represent the height of a column of mercury that is supported by 1 atm of air?*

A.5 PRESSURE–VOLUME BEHAVIOR OF GASES

When you pushed down on a sealed syringe filled with air in Investigating Matter A.1, you observed that a gas sample can easily be compressed. A gas can be compressed much more easily than a liquid or solid; that is, the volume of a gas sample easily changes when an external pressure is applied to it. See Figure 2.10.

Figure 2.10 *A student pushes a plunger into a closed syringe. What's keeping him from pushing it in all the way?*

Consider your experience with the syringe. The more pressure you applied to the plunger, the smaller the volume of gas held inside the syringe became. Consequently, the pressure of the gas sample trapped below the plunger increased. Think of it this way: as the syringe volume within which the particles are confined continues to become smaller, the particles of gas collide with each other and with the syringe walls more frequently. The net force exerted by all those collisions across the interior walls of the syringe, or pressure, increases.

Assume that the pressure of gas inside the syringe was 1 atm before you started pushing on the plunger (at this point, the atmosphere was exerting the only pressure on the plunger). If, by pushing on the plunger, the volume of gas in the syringe were then lowered to *one-half* of its original volume, the pressure of the gas sample would be *doubled*. If the gas volume in the syringe were reduced to *one-fourth* of its original value, the gas pressure would become *four times larger*. In each case, the gas pressure inside the syringe increases as the gas volume decreases. Figure 2.11 illustrates this relationship between gas pressure and its volume.

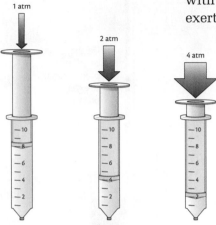

Figure 2.11 As pressure increases, the volume of gas trapped in the syringe decreases proportionally.

You will have more opportunities to visualize gas molecules in Section B.

The Pressure–Volume Relationship

How can we visualize this relationship at the molecular level? Gas molecules are in constant, random motion; gas pressure is caused by molecules colliding with the walls of their container.

The relationship between the volume and pressure of a gas sample at constant temperature can be described in several ways. The changes in volume and pressure for a particular gas sample at constant temperature can be described with words or pictures, such as those in Figures 2.11 and 2.12. These changes can also be described graphically, as shown in Figure 2.13. Note the shape of the curve. This shape indicates that gas pressure is *inversely proportional* to volume.

This relationship is useful for predicting the new gas pressure or volume resulting from a change in one of these two variables.

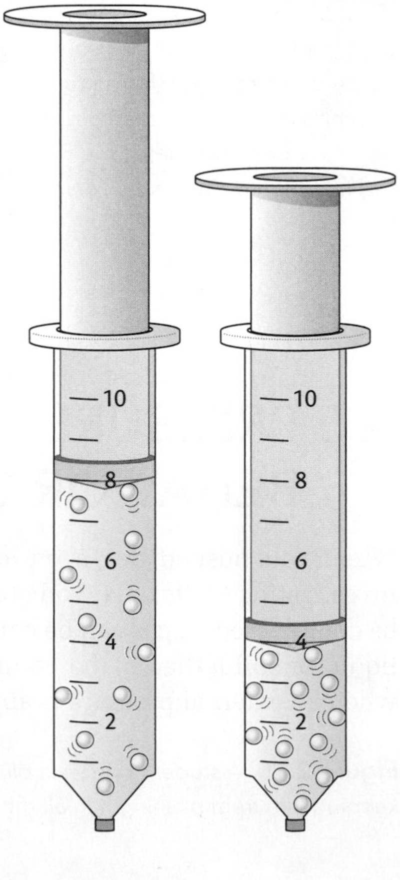

Figure 2.12 *As the volume of a sample of gas is reduced, the total number of molecular collisions with the container wall—and thus the gas pressure—increases proportionally. The gas molecules in each syringe have been greatly magnified to depict the relative distances among them, and thus to account for the more frequent collisions with the container wall (and, thus, higher gas pressure) in the syringe on the right.*

Again, consider the syringe in Figure 2.11. If initially a gas sample occupies a volume of 8.0 mL and exerts a pressure of 1.0 atm, how would the pressure of the gas sample change if its volume were increased to 10.0 mL? Take a few moments to think about this question, using your own experiences with the syringe and the illustrations in Figure 2.11 to guide your thinking. If you increase the volume of the syringe by pulling out on the plunger, what will happen to the gas pressure within the syringe? Will the new pressure be less than or greater than the initial pressure?

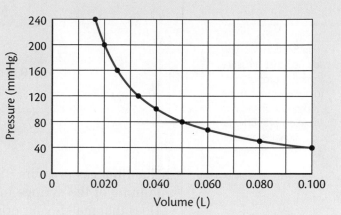

Did you reason that the new pressure would be *lower* than 1.0 atm? That is the correct conclusion. By now, you probably have gained an understanding of the relationship between gas volume and pressure that can guide you in predicting the general results of such changes. Now you're ready to learn how to make *quantitative* predictions of this type of gas behavior.

Figure 2.13 *The volume (V) of a gas sample, maintained at constant temperature and amount, is inversely proportional to its pressure (P). Therefore P x V is constant. A plot of pressure versus volume for any gas sample at constant temperature and amount will be similar to this one.*

Boyle's Law

Robert Boyle, an English scientist who studied gases, was the first to propose a quantitative law based upon the volume–pressure relationship for gases. He found that, for a sample of gas at constant temperature, the product of its measured pressure (P) and volume (V) remains the same, or $P \times V = k$. This relationship is known as **Boyle's law**. When either the pressure or volume of that same sample of gas is changed, the product of pressure multiplied by volume will still be equal to that same value (k) if the temperature and amount of gas are kept constant:

$$P_2 \times V_2 = k$$

Boyle's law expresses an inverse relationship. When changes in volume and pressure are made to a gas sample at constant temperature, a relationship can be expressed between the initial and final conditions of the gas sample. If P_1 and V_1 represent the initial pressure and volume of a sample of gas, and P_2 and V_2 represent its final values, then this equation can be written:

$$P_1 \times V_1 = P_2 \times V_2$$

Or, simply,

$$P_1 V_1 = P_2 V_2$$

This equation, which is an expanded expression of Boyle's law, can be used to analyze the gas sample problem you considered earlier.

Look closely at the mathematical relationship $P \times V = k$. Does it make sense to you that P and V are inversely proportional? What happens to P as V increases? As V decreases?

Note that $P \times V = k$ holds true only when temperature and amount of gas are kept constant.

$$P_1 = 1.0 \text{ atm} \longrightarrow P_2 = ? \text{ atm}$$
$$V_1 = 8.0 \text{ mL} \longrightarrow V_2 = 10.0 \text{ mL}$$

$$P_1 V_1 = P_2 V_2$$

$$(1.0 \text{ atm}) \times (8.0 \text{ mL}) = (P_2) \times (10.0 \text{ mL})$$

$$P_2 = \frac{(1.0 \text{ atm})(8.0 \text{ mL})}{10.0 \text{ mL}}$$

$$P_2 = 0.80 \text{ atm}$$

> In Section A.2, you learned that pressure is inversely proportional to area. Note that the product of the two inversely proportional variables remains the same.

The final gas pressure (0.80 atm) is less than the initial gas pressure: As the volume of the syringe in Figure 2.11 *increases*, the gas pressure *decreases*. Does that match your original prediction?

In the following activity, you will explain several common observations based on what you know about the relationship between gas pressure and volume. You will also have the opportunity to apply Boyle's law to solve several pressure–volume problems.

DEVELOPING SKILLS

A.6 PREDICTING GAS BEHAVIOR: PRESSURE–VOLUME

Sample Problem: A weather balloon with a volume of 4200 L at 1.0 atm is tested by placing it in a chamber and decreasing external pressure to 0.72 atm. What will be the final volume of the balloon?

In this case, we know the initial volume, V_1, and pressure, P_1, as well as the final pressure, P_2. Applying Boyle's law:

$$P_1 = 1.0 \text{ atm} \longrightarrow P_2 = 0.72 \text{ atm}$$
$$V_1 = 4200 \text{ L} \longrightarrow V_2 = ??$$

$$P_1 V_1 = P_2 V_2$$

$$(1.0 \text{ atm}) \times (4200 \text{ L}) = (0.72 \text{ atm}) \times (V_2)$$

$$V_2 = \frac{(1.0 \text{ atm})(4200 \text{L})}{0.72 \text{ atm}}$$

$$V_2 = 5800 \text{ L}$$

1. Explain each of the following observations:

   a. Even if they have ample supplies of oxygen gas, airplane passengers experience discomfort when the cabin undergoes a drop in air pressure.

   b. New tennis balls are sold in pressurized containers.

c. After descending from a high mountain, the capped, half-filled plastic water bottle from which you drank while standing at the summit now appears dented or slightly crushed.

2. You buy helium gas in small pressurized cans to inflate party balloons. The can label indicates that the container delivers 7100 mL of helium gas at 100.0-kPa pressure. The volume of the gas container is 492 mL.

   a. Do you think that the initial pressure of helium gas inside the can before use is greater or less than 100.0 kPa? Explain.

   b. Calculate the initial pressure of helium gas inside the container.

   c. Was your prediction in Question 2a correct?

3. Two glass bulbs are separated by a closed valve (see Figure 2.14). The 0.50-L bulb on the left contains a gas sample at a pressure of 6.0 atm. The 1.7-L bulb on the right is evacuated; it contains no gas:

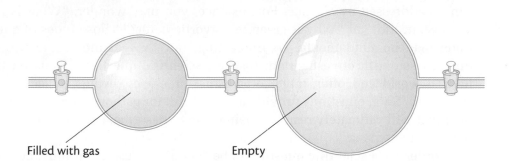

Filled with gas   Empty

Figure 2.14 *What will happen to the gas pressure if someone opens the middle valve?*

   a. Predict, in general, what will happen to the total volume of the gas sample if you open the middle valve. Explain.

   b. Predict, in general, what will happen to the total pressure of the gas sample if you open the middle valve. Explain.

   c. Calculate the actual pressure of the gas sample after the valve is opened.

A.7 DESIGNING SCIENTIFIC INVESTIGATIONS: GETTING STARTED

The predictions you just made about pressures and volumes of gases are possible because scientists have investigated these relationships. Why do scientists conduct investigations? How do they design and carry out investigations? These questions will be addressed throughout this unit, as you develop and practice the skills necessary for designing scientific investigations.

Asking Good Scientific Questions

Scientists conduct investigations for many reasons. They may wish to make new discoveries about the natural world, such as new life forms on this and other planets. They may hope to explain a wide range of phenomena or to develop a model that accounts for a set of observations. Scientific investigations are also conducted to test ideas, hypotheses, or claims, such as those featured in the unit-opening photo essay.

Often, an investigation begins with a question that is meaningful to a scientist or group of scientists. Questions can arise from analyzing results of prior or current investigations, making an unexpected observation, or simply by thinking, "What if we tried this? How would changing that variable affect our results?" While writing a scientific question may seem like a trivial task, writing a "good scientific question" can be quite challenging. To begin to pose "good scientific questions," we need to focus on what it means for a question to be "scientific" and then what it means to be "good."

We ask a lot of important questions during our lifetime, but not all of them can be addressed by science. For instance, you may wonder, "What is my purpose in life?" or "Why is green my favorite color?" Those types of questions may be fundamental or interesting, but they cannot be answered through scientific investigation. Instead, **scientific questions** focus on the natural world and often target organisms, processes, events, objects, and structures. Such questions provide a framework for gathering and analyzing data that will ultimately result in being able to describe, explain, or predict natural phenomena.

In order for a scientific question to be "good" (or useful), it must be both testable and answerable using current scientific knowledge and evidence from investigations. Some questions are too broad to be answered with only one or even several investigations. An example in chemistry might be, "How do gases behave?" A more focused question, and thus a better question for framing an individual investigation, would be, "What relationship exists between volume and pressure of a gas?" Scientific questions help the researcher to focus on the variables involved, and suggest a design for collecting and analyzing data.

> Testable questions are those that can be *disproved* through experimentation.

Gathering Evidence: Collecting Data and Making Observations

Developing a good scientific question often coincides with designing an investigation to address that question. The goal of the investigation design is to produce the best possible *evidence* to support claims based upon the investigation. **Evidence** includes qualitative observations or quantitative data, and its quality depends on how carefully and systematically observations are made, the tools and instruments available for making measurements, and whether the investigation can be repeated with similar results.

Investigations in chemistry often use an **experimental design**; that is, they rely on making measurements on one variable while changing another variable. An example of a question that implies an experimental design is "How does change in pressure of a gas affect its volume?" This question suggests that pressure will be the **independent variable**, which is manipulated by the investigator, and that volume will be the *response* or *dependent variable*. The **dependent variable** in an experiment is measured or observed in order to draw conclusions about effects of changes to the independent variable. Through analysis and interpretation of collected evidence, good scientific questions can be answered and scientific claims can be supported.

DEVELOPING SKILLS

A.8 DESIGNING AN EXPERIMENT TO INVESTIGATE TEMPERATURE–VOLUME RELATIONSHIPS

Sample Problem: What are the variables in Investigating Matter A.9?

Step 5 of *Gathering Evidence* indicates that temperature and air-column length will be measured. Thus, temperature and air-column length are the variables in this investigation.

Read *Preparing to Investigate* and *Gathering Evidence* in Investigating Matter A.9 Exploring Temperature–Volume Relationships. Then return here to apply what you have learned about experimental design to this upcoming investigation by answering the following questions:

1. Write a good scientific question that frames the investigation in Investigating Matter A.9.
2. Compare your scientific question to the question or questions written by your laboratory partner or group.
   a. How are your questions similar?
   b. How do your questions differ?
   c. Write a consensus scientific question for your laboratory group.
3. Which variables must be held constant? How do you know?
4. Describe how each variable will be measured. For each:
   a. Identify the variable.
   b. Describe whether the variable is a dependent or independent variable.
   c. Describe how the variable will be measured or manipulated, and in what units it will be recorded and reported.
5. Why do you think that a hot oil bath will be used instead of a hot water bath?

6. The procedure calls for measurement of the length of the air bubble within the tube, not the volume of the bubble.

   a. Why do you think that length will be measured instead of volume?

   b. Will measuring length instead of volume provide adequate data to address the investigation question? Explain.

7. Do you think the tests you will perform will provide enough evidence to answer the question you posed in #1? Explain.

■ INVESTIGATING MATTER

A.9 EXPLORING TEMPERATURE–VOLUME RELATIONSHIPS

Preparing to Investigate

Most matter is observed to expand when heated and contract when cooled. As you observed in Investigating Matter A.1, gas samples expand and shrink to a much greater extent than either solids or liquids. In this investigation, you will study how temperature changes influence the volume of a gas sample, assuming pressure and amount of gas remain unchanged. To do this, you will heat a thin glass tube containing a trapped air sample and record changes in volume as the air sample cools.

Before you begin, read *Gathering Evidence* again to review what you will need to do and note safety precautions.

Gathering Evidence

1. Before you begin, put on your goggles, and wear them properly throughout the investigation.

2. Using two small rubber bands, fasten a capillary tube to the lower end of a thermometer. (See Figure 2.15.) Place the open end of the tube close to the thermometer bulb and 5 to 7 mm from the bulb's tip.

3. Immerse the tube and thermometer in a hot oil bath that has been prepared by your teacher. Be sure the entire capillary tube is immersed in oil. Wait for your tube and thermometer to reach the temperature of the oil (~100 °C). Record the temperature of the bath.

4. After your tube and thermometer have reached a steady temperature, lift them up until only about one quarter of the capillary tube is still in the oil bath. Pause for ~3 seconds to allow some oil to rise into the

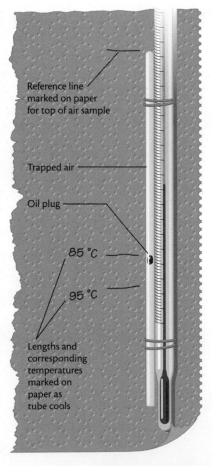

Reference line marked on paper for top of air sample

Trapped air

Oil plug

85 °C

95 °C

Lengths and corresponding temperatures marked on paper as tube cools

Figure 2.15 *Apparatus for studying how gas volume changes with temperature.*

tube. Then quickly place the tube and thermometer on a paper towel
(to avoid dripping) and carry them back to your desk. (**Caution:** *Be
careful not to touch the hot end of the thermometer or the drips of
hot oil.*)

5. Lay the tube and thermometer on a clean piece of paper towel on the
    desk. Make a reference line on the paper at the sealed end of the
    capillary tube. Also mark the upper end of the oil plug, as shown in
    Figure 2.15. Alongside this mark, write the temperature at which the
    mark was made corresponding to that air-column length.

6. As the temperature of the gas sample drops, make at least six marks
    to represent the length of the air column trapped above the oil plug at
    various temperatures. Write the corresponding temperature next to
    each mark. Allow enough time so the temperature drops by 50 to
    60 °C.

7. When the thermometer shows a steady temperature (near room
    temperature), make a final observation of length and temperature.
    Discard the tube and the rubber bands according to your teacher's
    instructions. Wipe the thermometer clean.

8. Measure the length (in millimeters) from each marked line to the
    mark for the sealed end of the tube. Record each length of the gas
    sample. Ask your teacher to check your data before you discard your
    paper towel. Your teacher may ask you to submit your paper towel
    along with your report, so be sure to get specific instructions before
    disposing of the towel.

9. Wash your hands thoroughly before leaving the laboratory.

Analyzing Evidence

1. Plot the length–temperature data for your air sample with length on
    the vertical axis and temperature on the horizontal axis. The *y*-axis
    (length) should range from 0 to 10 cm; the *x*-axis (temperature)
    should include values from −350 to 150 °C. Label your axes, and
    arrange the scales so the graph nearly fills the space available.

2. Draw the best straight line through your plotted points with a
    ruler. Using a dashed line, extend this straight graph line so that it
    intersects the *x*-axis. Use your completed graph to help you answer
    the following questions.

3. Renumber the temperature scale on your graph, assigning the value
    "zero" to the temperature at which your plotted graph line intersects
    the *x*-axis. The new scale now expresses temperature in *kelvins* (K)—
    the **Kelvin temperature scale**. One kelvin is the same size as one
    degree Celsius. However, unlike zero degrees Celsius, zero kelvins
    is the lowest temperature theoretically possible. It is called **absolute
    zero.**

Interpreting Evidence

1. Locate the intersection of your extended graph line with the *x*-axis.

   a. At what temperature does the line intersect the *x*-axis?

   b. What would be the volume of your gas sample at that temperature?

   c. Could the gas sample actually reach the predicted volume? Explain.

2. Based on your renumbered (K) graph,

   a. what temperature in kelvins (K) would correspond to 0 °C, the freezing point of water?

   b. what kelvin temperature would correspond to 100 °C, the normal boiling point of water?

Making Claims

3. Describe the relationship between gas volume and temperature at constant pressure. Support your answer with evidence from this investigation.

4. Revisit your consensus scientific question developed in Section A.8.

   a. What was your question?

   b. Can the question be answered based upon the evidence you gathered in this investigation?

   c. If yes, provide an answer to the question and support it with evidence. If no, what additional evidence is required to answer the question?

A.10 TEMPERATURE–VOLUME BEHAVIOR OF GASES

Your plotted data from Investigating Matter A.9 should indicate a relationship between volume and temperature for a sample of gas at constant pressure. The volume–temperature relationship for gases was also demonstrated in Investigation 2 of Investigating Matter A.1 (page 153), when you placed two balloons of equal size into water baths of differing temperatures. As you observed, increasing or decreasing the temperature of a gas sample results in changes in its volume (Figure 2.16).

In the 1780s, French chemists (and hot-air balloonists) Jacques Charles and Joseph Gay-Lussac studied the changes in gas volume caused by temperature changes at constant pressure. Data for oxygen gas and nitrogen gas are shown in Figure 2.17. The plots for different gases and different sample sizes have different appearances. However, if all the graph lines are

Figure 2.16 *The gas held in hot-air balloons expands when heated.*

smoothly extended down to the *x*-axis (an extrapolation), the lines meet at the same low temperature. Lord Kelvin (an English scientist) used the work of Charles and Gay-Lussac to establish a simple mathematical temperature–volume relationship for gases, known as Charles' law. Lord Kelvin based his own new temperature scale on this relationship.

Doubling the kelvin temperature of a gas sample doubles its volume at constant pressure and amount of gas. Reducing the kelvin temperature by one half causes the gas volume to decrease by one half, and so on. These relationships are summarized in **Charles' law**. At constant pressure and amount of gas, the volume (*V*) of a gas sample divided by its temperature in kelvins (*T*) is always a constant value:

$$\frac{V}{T} = k$$

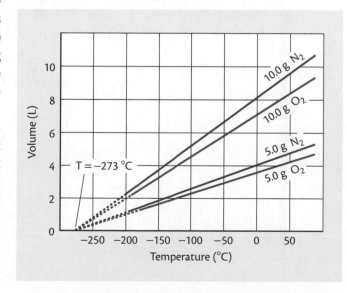

Figure 2.17

Temperature–volume measurements of various gas samples at 1.0 atm pressure. Extrapolation has been made for temperatures below liquefaction.

where k is a constant. This is true even when the temperature and volume of that same sample of gas are changed, as long as pressure and amount of gas are kept constant:

$$\frac{V_2}{T_2} = k$$

Because both expressions are equal to the same value (k), a temperature–volume relationship can be expressed between the initial and final state of the gas sample:

$$\frac{V_1}{T_1} = \frac{V_2}{T_2}$$

This expression of Charles' law is useful in predicting changes in gas volume and temperature at constant pressure and amount of gas.

DEVELOPING SKILLS

A.11 PREDICTING GAS BEHAVIOR: TEMPERATURE–VOLUME

Sample Problem: Since temperature will affect weather balloons, consider the same weather balloon you investigated in Developing Skills A.6 (page 164). At constant pressure, this weather balloon, with an initial volume of 4200 L, is tested by placing it in a chamber and decreasing temperature from 20 to –60 °C. What will be the final volume of the balloon?

In this case, we know the initial volume, V_1, and temperature, T_1, as well as the final temperature, T_2. First, convert both temperatures to kelvins by adding 273 to the temperatures measured in degrees Celsius: $T_1 = 20 + 273 = 293$ K and $T_2 = -60 + 273 = 213$ K. Then, applying Charles' law:

$$K = °C + 273$$

$$T_1 = 293 \text{ K} \longrightarrow T_2 = 213 \text{ K}$$

$$V_1 = 4200 \text{ L} \longrightarrow V_2 = \text{ ??}$$

$$\frac{V_1}{T_1} = \frac{V_2}{T_2}$$

$$\frac{4200 \text{ L}}{293 \text{ K}} = \frac{V_2}{213 \text{ K}}$$

$$V_2 = \frac{(4200 \text{ L})(213 \text{ K})}{293 \text{ K}}$$

$$V_2 = 3100 \text{ L}$$

Apply Charles' law to answer the following questions. For some questions, it may be necessary to convert temperatures to kelvins.

1. What will happen to the volume of an inflated balloon, originally at 20 °C, if you take it outdoors where the temperature is 40 °C? Assume that pressure is constant and no gas escapes from the balloon.

2. In planning to administer a gaseous anesthetic to a patient,
   a. why must the anesthesiologist take into account the fact that during surgery the gaseous anesthetic is used both at room temperature (18 °C) and at the patient's body temperature (37 °C)?
   b. what problems might arise if the anesthesiologist did not allow for the patient's higher body temperature?

3. An air bubble trapped in bread dough at room temperature (291 K) has a volume of 1.0 mL. The bread bakes in the oven at 623 K (350 °C).
   a. Predict whether the air-bubble volume will increase or decrease as the bread bakes.
   b. Calculate the new volume of the air bubble, using Charles' law. (*Hint:* Remember that Charles' law only applies to temperatures expressed in kelvins.)

4. You buy a 3.0-L helium balloon in a mall and place it in a car sitting in hot summer sunlight. The temperature in the air-conditioned mall is 22 °C, and the temperature inside the closed car is 45 °C.
   a. What will you observe happening to the balloon as it sits in the warm car?
   b. What will be the new volume of the balloon?
   c. What do you think is happening to gas atoms within the balloon as it warms?

concept check 3

Use your own words to describe each of the following:
1. The relationship between temperature and volume of a sample of gas at constant pressure.
2. The relationship between pressure and volume of a sample of gas at constant temperature.
3. How the relationships described in your answers to Questions 1 and 2 differ.

A.12 TEMPERATURE–PRESSURE BEHAVIOR OF GASES

Picture a closed cylinder of gas, such as a deep-sea scuba tank where volume is constant. See Figure 2.18. What would happen to the motion of gas molecules in the cylinder if you were to heat the tank? How would this affect the gas pressure?

If you concluded that raising the temperature of the gas at constant volume should cause an increase in gas pressure, you are correct. Increasing temperature increases the kinetic energy (and thus the average velocity) of the gas molecules. Because the molecules, on average, are traveling faster, the number of molecular collisions with the container walls increases, and the energy involved with each collision also increases. These effects cause a corresponding increase in gas pressure. The mathematical expression for this relationship is:

$$\frac{P}{T} = k$$

where k is a constant. Particular changes in gas pressure and temperature at constant volume can be found, using this equation:

$$\frac{P_1}{T_1} = \frac{P_2}{T_2}$$

■ DEVELOPING SKILLS

A.13 USING GAS RELATIONSHIPS

Figure 2.18 *Scuba divers in the Arctic ocean (top) and in the tropics (bottom). Assuming the cylinders are of equal volume and contain the same amount of gas, which of the cylinder pressure-gauge readings will be higher?*

Sample Problem: As you drive up a mountain pass, you hear a loud "pop" from the rear of the car. Later, you discover that your previously unopened bag of potato chips is now open. You have maintained a constant, comfortable temperature in the vehicle, so what happened?

As you ascend, the atmospheric pressure within and outside of the vehicle decreases, while the pressure within the potato-chip bag remains constant. As a result, the volume of the bag increases until the strength of the seal is exceeded and the bag "pops" open.

Solve the following problems using appropriate gas relationships:

1. If the kelvin temperature of a gas sample held in a steel tank increases to three times its original value, predict what will happen to the pressure of the gas. Will it increase or decrease? By what factor do you expect the gas pressure to change?

2. A gas sample at a constant pressure shrinks to one-fourth its initial volume. What must have happened to its temperature? Did it increase or decrease? By what factor did the kelvin temperature of the gas change?

3. Explain why automobile owners in severe northern climates often add air to their tires in wintertime and release some air from the same tires in the summertime.

4. Incandescent light bulbs are filled with inert (unreactive) gas. When the light is turned on, what will happen to the gas pressure inside the bulb? Why?

5. Use gas laws to explain why a weather balloon expands in size as it rises from Earth's surface.

6. Why does the label on an aerosol container caution you not to dispose of the container in a fire?

■ MAKING DECISIONS

A.14 ASKING QUESTIONS AND GATHERING EVIDENCE

Part I:

Suppose that you wished to design an investigation to test the relationship between the temperature of a gas and its pressure.

1. Write a scientific question that frames the investigation.

2. Identify the variables of interest.

3. a. What conditions must be held constant?

   b. How would you design your investigation so that these conditions remain constant?

4. Consider how you might measure each variable. For each:

   a. Identify the variable.

   b. State whether the variable is a dependent or independent variable.

   c. Describe how the variable will be measured or manipulated, and what units will be reported.

Part II:

Now that you have a little experience analyzing and designing scientific investigations, it is time to begin developing your own air-quality investigation. In this activity, you will think about scientific questions that could guide your investigation, sometimes called "beginning questions" or "guiding questions." Remember that your ultimate goal is to design and refine an investigation to test an air-quality claim or product. Although you probably will not actually conduct experiments and gather evidence, you will create a scientific poster to report on the investigation and its design at the end of the unit.

5. Reread the captions in the unit opener. Which air-quality issues or claims highlighted there interest you or are important in your community?

6. What other air-quality issues or claims (not represented in the unit opener) interest you or are important in your community?

7. Choose two issues or claims that you identified in Questions 5 and 6. For each issue or claim, write a good, scientific, testable question.

8. Share your questions with a partner or group as directed by your teacher.

   a. What common features do these questions have?

   b. On the basis of constructive feedback from your peers, refine and rewrite each question.

SECTION A SUMMARY

Reviewing the Concepts

> Pressure involves a force applied over a particular area. Air pressure is often expressed in units of atmospheres (atm), kilopascals (kPa), or millimeters of mercury (mmHg).

1. It is much easier to slice a piece of pie with the edge of a sharp knife than with the edge of a pencil. Explain this in terms of applied pressure.

2. Heavy vehicles that must move easily over loose sand are often equipped with special tires.
   a. Would you expect these tires to be wide or narrow?
   b. Explain your answer using the concept of pressure.

3. U.S. weather reports generally express air pressure in units of inches of mercury. During a severe storm, the barometric pressure can drop as low as 27.2 inches of mercury. Convert this air-pressure value to:
   a. millimeters of mercury (mmHg).
      (*Hint:* 1 inch = 25.4 mm)
   b. atmospheres (atm).
   c. kilopascals (kPa).

4. Which is more likely to cause damage to a wooden floor: A 1068-N (240-pound) basketball player in athletic shoes (floor-contact area = 420 cm^2) or a 534-N (120-pound) sports reporter standing in high heels (floor-contact area = 24 cm^2). Support your answer with suitable pressure calculations.

> The volume of a sample of gas in a flexible, impermeable container will increase if external pressure is reduced, and decrease if external pressure is increased.

5. Explain why a sealed bag of potato chips "puffs out" (increases in volume) if it is carried from sea level up to a high mountain pass.

6. If the same bag of chips (see Question 5) were carried under water by a scuba diver, what would happen to its volume as the sealed bag descended into deep water?

7. A small quantity of the inert gas argon (Ar) is added to incandescent light bulbs to reduce vaporization of tungsten (W) atoms from the solid filament. What volume of argon gas measured at 760 mmHg is needed to fill a 0.21-L light bulb at a pressure of 1.30 mmHg? Assume the argon gas temperature remains constant.

8. A party-supply store sells a helium-gas cylinder with a volume of 1.55 x 10^{-2} m^3. If the cylinder provides 1.81 m^3 of helium for balloon inflation (at 0 °C and 1 atmosphere), what must be the pressure inside the cylinder?

Scientific investigations are conducted for many reasons, including explaining natural phenomena, developing models, and testing hypotheses or claims. Most scientific investigations are based upon a testable scientific question.

9. List at least three sources of questions for beginning a scientific investigation.

10. Identify each of the following questions as scientific or not scientific.

   a. Why is my favorite color blue?

   b. Why is the sky blue?

   c. How should this new organism be classified?

   d. Why did this new organism evolve?

   e. Are gases matter?

   f. Are there gases on Mars?

   g. Which gas is most important?

11. List two "good scientific questions" regarding physical properties of gases.

Scientific investigations are designed to produce evidence that will support claims or conclusions, using observations, data, or both.

12. What is evidence?

13. List three factors that affect the quality of evidence obtained in a scientific investigation.

14. Explain the role of each of the following in an experimental investigation:

   a. independent variable

   b. dependent variable

   c. testable question

The volume of a sample of gas in a flexible, impermeable container at constant pressure will increase if its temperature is increased and decrease if its temperature is decreased.

15. A 2.50-L balloon at 25 °C is inflated with helium. At constant pressure, the temperature of the balloon is then lowered to −5 °C. What will be the balloon's volume at this new temperature?

16. What do you think a marshmallow, composed mainly of small pockets of air, would look like at temperatures close to absolute zero? Explain your answer.

The pressure exerted by a sample of gas in a rigid, impermeable container will increase if its temperature is increased, and decrease if its temperature is decreased.

17. Experts recommend measuring automobile tire pressure when the tires are cold rather than just after driving for several hours. Why?

18. Soccer players often notice that kicking the ball feels different on cold days compared to warm days. Explain why this might be so.

19. A steel tank at 299 K contains 0.285 L of nitrogen gas at 1.92-atm pressure. The tank is capable of withstanding a maximum pressure of 7.34 atm. At what temperature will the tank burst?

20. An oxygen gas sample in a rigid container is heated from 15 to 30 °C. By what factor will the oxygen gas pressure increase?

What information can investigations provide about gas behavior?

In this section, you have been introduced to experimental design, as well as the behavior of gases in several contexts. Think about what you have learned and how these ideas are related, then answer the question in your own words in organized paragraphs. Your answer should demonstrate your understanding of the key ideas in this section.

Be sure to consider the following in your response: properties of gases, pressure (including atmospheric pressure), scientific questions, and relationships among temperature, pressure, and volume of a gas.

Extending the Concepts

23. Some people suggest that they could dive to the bottom of a swimming pool and stay there, breathing through an empty garden hose that extends above the water surface. ⚠ (*Caution! Do not try this!*) How would you argue against this plan?

24. The gas behavior described by Boyle's law is a matter of life and death to scuba divers. On the water surface, the diver's lungs, tank, and body are at atmospheric pressure. However, under water, a diver's body experiences the combined pressures of the atmosphere and the water.

    a. Why do scuba divers need to use pressurized tanks?

    b. What would happen to the tank volume if it were not strong enough to withstand the pressure of the outside water?

    c. How do the problems of a diver compare with those of a pilot climbing in an unpressurized airplane to a higher altitude?

Connecting the Concepts

21. If the temperature of one mole of gas increases while its volume decreases, would you expect the gas pressure to increase, decrease, or remain the same? Explain your answer in terms of the gas laws.

22. The volume of a hot-air balloon remains constant, while the temperature of the gas in the balloon changes. How is this possible?

SECTION B PHYSICAL BEHAVIOR OF GASES

How are models and theories useful in explaining and predicting behavior of gases?

Throughout the 17th and 18th centuries, scientists conducted investigations that provided important information about the properties and behavior of gases. This behavior was summarized in scientific laws that mathematically described the results of changing gas pressure, temperature, and volume. Laws and observations, however, provide only some of the answers that scientists seek. To predict novel behavior and design new investigations, scientific explanations are needed. This section introduces some of these explanations, including the kinetic molecular theory.

As you explore these theories and models, think about how they help to explain the behavior of gases in Earth's atmosphere. Stay focused as well on the air-quality investigation that you began to design in Section A. The models and theories you will encounter in this section can help you refine your guiding question and start to construct a procedure for collecting evidence to support an answer to that question.

GOALS

- State and apply postulates of the kinetic molecular theory.
- State and explain Avogadro's law for gases.
- Use the ideal gas law in problems involving gases.
- Describe conditions under which the assumptions of the ideal gas law are not valid.
- Distinguish between temperature and heat.
- Describe changes matter undergoes as energy is added or removed.
- Describe similarities and differences among scientific laws, theories, and models.

concept check 4

1. Using what you learned about pressure in Section A, explain how you are able to draw soda up a straw.
2. Why does every tea kettle have a vent?
3. What happens to the molecules in a sample of gas as it is heated?

B.1 ATOMS AND MOLECULES IN MOTION

As you get up on a cold winter morning, you look out your window to see clouds of exhaust coming from cars, trucks, and buses driving past your home. Your hand recoils from hot spray coming from the shower head; it is too uncomfortable for you to enjoy your morning shower. Turning on too much cold water, however, leads to a chilling experience.

Your experiences with gases, liquids, and solids are all at a sensory level; that is, you observe matter and its changes through your senses of sight, smell, touch, taste, and hearing. In so doing, you arrive at some generalizations about the world around you. Making observations is the first step in "doing science." A more difficult task is crafting a theory that explains all observations and that also can be used to predict outcomes of experiments yet to be completed. **Scientific theories** and models attempt to offer "how"-type explanations of phenomena in the natural world.

One such theory concerns the motion of atoms and molecules within all states of matter. What do you already know about the three states of matter from observing them? You know that a solid has a definite shape and that its structure is rigid. Experience also tells you that a liquid flows and, unlike a solid, depends on a container to define its shape. A sample of gas, likewise, does not have a definite shape. If you have blown up a balloon and then let go of it without tying it off, you also know that air rapidly leaves the balloon and the balloon changes shape.

The atoms, molecules, or ions that make up solids are held tightly to one another. Although the particles in solids vibrate about a position, they are tightly packed and cannot move past one another. The molecules and ions within liquids are more mobile. They can move past each other, but intermolecular attractive forces prevent them from moving too far apart. The arrangement and motion of particles in solids and liquids are depicted in Figure 2.19.

> You will learn more about intermolecular forces later in this section. For now, just think of these as the forces that cause molecules to stick to each other.

Figure 2.19 *You can observe the properties of solids (zinc, left) and liquids (mercury, right) using your senses. Models help explain how atoms interact with each other and their surroundings.*

Most molecules in the gaseous state are not strongly attracted to each other. In contrast to particles in solids and liquids, particles in gases are very far apart. In fact, their size is negligible compared to the great distances that separate them. Gas molecules move in straight-line paths at very high speeds and change direction only when they collide with each other or with another object. Collisions of gaseous particles with container walls produce gas pressure in the container. Figure 2.20 illustrates the motion of particles in a gas sample. Compare this model of gases to the models of solids and liquids in Figure 2.19.

Gas molecules move at average speeds that depend on how much kinetic energy they possess. **Kinetic energy**, sometimes called *energy of motion*, is the energy associated with any moving object. Its value depends on both the mass of the moving object and its velocity. Traveling at the same velocity, a more massive object has greater kinetic energy than does a less massive object. Thus, a baseball has much greater kinetic energy than a ping-pong ball when both are traveling at the same speed. If two moving objects have equal mass, the object moving faster has greater kinetic energy. This helps explain the difference in damage to a car bumper when it taps the wall of a parking garage at 10 km/h compared to colliding with the same wall at 30 km/h.

> 10 km/h = 6.3 miles per hour.

The behavior of moving gas molecules can be explained by the idea of kinetic energy. This explanation is based on several *postulates*, some of which you have just considered. A **postulate**, also called an axiom, is an accepted statement used as the basis for developing an argument or explanation. Many scientific and mathematical theories begin with postulates. Since postulates may be approximations or simplifications, their validity is sometimes tested through experiments.

> Einstein's special theory of relativity is based upon postulates.

The following list summarizes these postulates:

- Gases consist of tiny particles (molecules) whose size is negligible compared with the great distances that separate them from each other. On average, the separation between gaseous particles at 1-atm pressure and 0 °C is about 11 times greater than the diameter (size) of the particles.

- Gas molecules are in constant, random motion. They collide with each other and with the walls of their container or surrounding objects. Gas pressure is caused by molecular collisions with container walls.

- Molecular collisions are **elastic**. This means that although individual molecules in a gas sample may gain or lose kinetic energy, there is no gain or loss in total kinetic energy from all of these collisions.

> Two of these postulates are really approximations used to simplify discussions of gas behavior. You will encounter situations later in this section for which these approximations are not valid.

- At a given temperature, gas molecules have a range of kinetic energies. However, the average kinetic energy of the molecules is constant and depends only on the temperature of the gas sample. Therefore, molecules of different gases at the same temperature have equal average kinetic energies. As gas temperature increases, the average velocity and kinetic energy of the gas molecules also increase.

- Gas molecules only interact during collisions. Attractive or repulsive forces among gas particles are negligible.

These postulates serve as the basis of **kinetic molecular theory** (KMT) of gases, which can be used to explain the gas behaviors you investigated in Section A. It is useful to describe a gas sample that follows all of these postulates as an **ideal gas**. In the next few sections, you will learn more about both ideal gas behavior and conditions under which these postulates are no longer valid.

Scientific American Conceptual Illustration

Figure 2.20 *Particulate-level model of gases. A helium-filled balloon soars above the landscape. In magnification, helium atoms are depicted within the balloon, He(g). Nitrogen molecules (N₂, blue) and oxygen molecules, (O₂, red), representing 98% of air, are shown outside the balloon. What does this model suggest about spacing and movement of gas particles compared to those in liquids and solids*

MODELING MATTER

B.2 UNDERSTANDING KINETIC MOLECULAR THEORY

Previously, you modeled the structure of matter and changes in matter using pictures and symbols. Another way that matter can be modeled is through the use of analogies. An analogy can help you relate certain features of an abstract idea or theory to a situation that is familiar to you. Read the analogy provided in the next paragraph and answer the questions that follow concerning the kinetic molecular theory of gases.

Imagine that you cause several highly elastic, small "super-bounce balls" (see Figure 2.21) to bounce randomly around inside a box that you steadily shake; this serves as an analogy for gas molecules randomly bouncing around inside a sealed container. See Figure 2.22.

Figure 2.21 *Super-bounce balls can model gas-molecule behavior.*

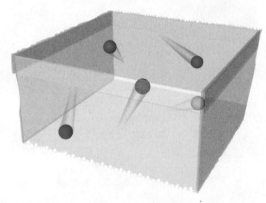

Figure 2.22 *Super-bounce balls moving about randomly in a box.*

1. Decide which of these four gas variables—*volume, temperature, pressure,* and *number of molecules*—best matches each of the following, and explain each choice.

    a. the number of super-bounce balls in the box

    b. the size of the box

    c. the vigor with which you shake the box

    d. the number and force of collisions with the box walls of the randomly moving super-bounce balls

2. How does each of the following changes relate to what you have learned about gas behavior?

    a. The vigor of shaking and the number of super-bounce balls remain the same, but the size of the box is decreased.

   b. The size of the box and the number of super-bounce balls remain the same, but the shaking becomes more vigorous.

   c. The size of the box and the vigor of shaking are kept the same, but the number of super-bounce balls is increased.

3. Imagine that you make the box smaller, and smaller, and smaller. Which postulate or postulates of KMT might no longer be represented by this analogy? Explain.

4. Suggest another situation similar to those in Question 2 that can serve as an analogy for the behavior of gases. Explain how the situation is analogous.

5. All analogies have limitations. For example, the super-bounce ball analogy fails to represent some characteristics of gases. Gas molecules travel at very high velocities (on the order of 6000 km/h), whereas the super-bounce balls move much more slowly (on the order of 1 km/h). Suggest two other characteristics of actual gases that are not properly represented by this super-bounce ball analogy.

6. Describe your own analogy that might be useful for modeling gas behavior.

   a. Identify features of your analogy that relate to features of KMT and *T-V-P* relationships (Section A) for gases.

   b. Point out some key limitations of your analogy.

B.3 SCIENTIFIC THEORIES, MODELS, AND LAWS

In Section A, you learned about scientific questions and some of the reasons that scientists conduct investigations. A primary goal of science is to understand how the natural world works. When a coherent set of ideas explains many related observations or events within the natural world, it is known as a **scientific theory**. KMT is an example of a scientific theory. Theories explain observations and data and can be used to predict results of investigations. Explanations are regarded as theories once they are widely accepted among scientists as a result of their predictive ability, agreement with experimental data, and consistency with other accepted theories and knowledge.

The opening question for this section asks you to consider the role of scientific models and theories in predicting and explaining gas behavior. Scientists and philosophers of science do not all agree on the distinctions between scientific models and scientific theories. For our purposes, a **scientific model** is a representation of either a part of the natural world or of a scientific theory. Scientific models are developed either from observations and data or from the formal statements and postulates of a scientific theory. In a sense, models provide a bridge between theory or data and the observable, testable world.

Scientific theories and models not only explain the results of past investigations, they are important guides for designing future investigations. Models may influence the types of questions investigators ask or the forms of evidence they seek. All theories and models are dynamic or fluid; that is, they are subject to revision or rejection based upon further investigation. Later in this unit, you will think about how the theories and models you have encountered in this section will influence your air-quality investigation.

You might wonder how the visualizations you have constructed in Modeling Matter activities relate to scientific models as discussed here. Although physical replicas and drawings are types of models, they cannot themselves explain phenomena. Such simple models are, however, powerful tools for learning and constructing your own understanding of natural phenomena.

Unlike models and theories, scientific laws do not provide explanations for observed behavior. Laws, like Boyle's law, describe the behavior of matter in nature, sometimes expressed as a single mathematical expression. Statements are usually considered scientific laws only after they are confirmed by many investigations. Boyle's law holds only under the conditions of constant temperature within a closed system; such restrictions are common for scientific laws.

B.4 THE IDEAL GAS LAW

So far, you have considered gas behavior under differing conditions of volume, pressure, and temperature. Another variable that must be included in a comprehensive model of gas behavior is the actual amount of the gas sample; that is, the number of gas molecules contained in a particular gas sample. If you have the same volumes of oxygen gas, nitrogen gas, and carbon dioxide gas in three different balloons at the same temperature and pressure, how do the numbers of gas molecules compare? See Figure 2.23.

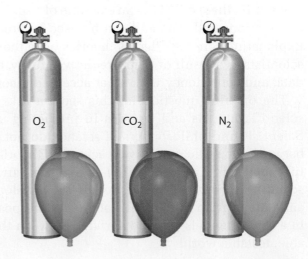

Figure 2.23 *These three equally sized balloons contain three different gases at the same temperature and pressure. Are the total gas molecules in each balloon the same or different?*

That same question was investigated in the early 1800s by the Italian lawyer and mathematical physics professor Amedeo Avogadro. By making careful observations of gas samples like those you have worked with, he proposed: *Equal volumes of all gases at the same temperature and pressure contain the same number of molecules.* This important statement is commonly known as **Avogadro's law.**

Starting on page 161, you learned about a relationship between the pressure and volume of an ideal gas sample, if its temperature is held constant (Boyle's law). Similarly, you are now familiar with the relationship between the temperature and volume of an ideal gas sample (Charles' law) and the relationship between the amount of gas (in moles) within a sample and the volume of the gas sample (Avogadro's law).

In 1834, a more general equation—the **ideal gas law**—was derived from the experimentally based laws of researchers such as Boyle and Charles. All of the variables represented in the laws of Boyle, Charles, and Avogadro are included in the ideal gas law:

$$P \times V = n \times R \times T$$

or

$$PV = nRT$$

where P = pressure, V = volume, n = moles (amount) of gas, R = a constant, and T = temperature.

As you probably noticed, the ideal gas law contains a constant (R). In order to apply this equation, the value of the constant, R, must be known. This value depends on the units in which pressure, volume, temperature, and amount of gas are expressed. The value of R given below is used when pressure is expressed in atmospheres, volume is measured in liters, temperature is given in kelvins, and amount of gas is expressed in moles. Several values of R are listed in Table 2.2 (page 188).

$$R = 0.0821 \frac{\text{atm} \times \text{L}}{\text{mol} \times \text{K}}$$

Under specific circumstances, the ideal gas law can be reduced to each of the gas laws upon which it is based. Boyle discovered that for a gas sample at constant temperature, the product of its measured pressure (P) and volume (V) remains the same, or:

$$P \times V = k \quad \text{(Boyle's law)}$$

In this case, a relationship between pressure and volume is apparent when both the temperature and the amount of gas are kept constant. Consider these variables as you reexamine the ideal gas law:

$$P \times V = n \times R \times T$$

If n and T are both held constant, then the product $n \times R \times T$ must also be equal to a constant value. In other words, the equation could be rewritten as Boyle's law:

$$P \times V = k$$

Avogadro's complete name was Lorenzo Romano Amedeo Carlo Avogadro, Conte di Quaregna e di Cerreto.

Recall that an *ideal gas* is one that follows all of the postulates of the kinetic molecular theory as described in Section B.1

Reporting data as if it were collected at standard conditions (STP) makes it easier for scientists to compare data taken at different locations under varying conditions.

The ideal gas law can be similarly simplified to Charles' law when P and n are held constant and to Avogadro's law when P and T are held constant (see Figure 2.24). This simplification to Avogadro's law leads to an important consequence—*all* ideal gases have equal molar volumes if they are measured at the same temperature and pressure. The **molar volume** is the volume occupied by one mole of a substance. To see this, calculate the volume of 1 mole of gas at 0 °C and 1 atmosphere; these conditions are commonly called standard temperature and pressure (STP).

$$V = \frac{n \times R \times T}{P}$$

$$V = \frac{1.00 \text{ mol} \times 0.0821 \frac{\text{L} \times \text{atm}}{\text{mol} \times \text{K}} \times 273 \text{ K}}{1.00 \text{ atm}}$$

$$V = 22.4 \text{ L}$$

Thus, at conditions of 0 °C and 1 atm, the ideal gas law predicts that the molar volume of any ideal gas sample is 22.4 L (see Figure 2.25). There is no corresponding simple relationship between moles of various solids or liquids and their corresponding volumes.

Figure 2.24 *What would you need to know to estimate the number of molecules of gas inside this basketball?*

Figure 2.25 *If these 11 2-L bottles were filled with air, they would contain approximately 1 mole of gas.*

Table 2.2

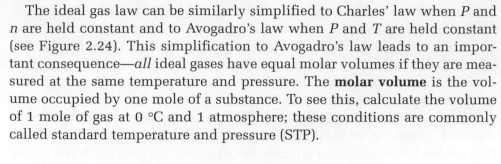

Values of the Ideal Gas Constant, R, in Common Units	
Pressure Unit	**R Value and Units**
mmHg (torr)	$62.4 \frac{\text{mmHg} \times \text{L}}{\text{mol} \times \text{K}}$
atmosphere	$0.0821 \frac{\text{atm} \times \text{L}}{\text{mol} \times \text{K}}$
inHg	$2.455 \frac{\text{inHg} \times \text{L}}{\text{mol} \times \text{K}}$
pounds per square inch	$1.206 \frac{\text{psi} \times \text{L}}{\text{mol} \times \text{K}}$
bar	$0.0831 \frac{\text{bar} \times \text{L}}{\text{mol} \times \text{K}}$
pascal	$8311.8 \frac{\text{Pa} \times \text{L}}{\text{mol} \times \text{K}}$

DEVELOPING SKILLS

B.5 USING THE IDEAL GAS LAW

Sample Problem: *What pressure would be exerted by 2.5 moles of nitrogen gas in a 3.5-L tank at 25 °C?*

Because this problem is not about changes in conditions, it is appropriate to use the ideal gas law.

Rearranging the ideal gas law to solve for pressure gives:

$$P = \frac{n \times R \times T}{V}$$

Substituting the given values we obtain:

$$P = \frac{(2.5 \text{ mol}) \times (0.0821 \frac{L \times atm}{mol \times K}) \times (298 \text{ K})}{3.5 \text{ L}}$$

$$P = 17 \text{ atm}$$

Remember that when gas laws are applied, temperature must always be expressed in kelvins.

Why is

$R = 0.0821 \dfrac{L \times atm}{mol \times K}$

used in this problem?

Solve the following problems using the ideal gas law.

1. A 0.50-L canister of hazardous sarin gas was discovered at an old, abandoned military installation. To properly dispose of this gas, technicians must know how much gas is held in the canister. The gas pressure is measured as 10.0 atm at room temperature (25 °C). How many moles of sarin are held inside the canister?

2. What volume will 2.0 mol $H_2(g)$ occupy at 40.0 °C and 0.50-atm pressure?

3. When the volume of a gas sample is measured, its pressure and temperature must also be specified.
   a. Why?
   b. That practice is normally not necessary for measuring the volumes of liquids or solids. Why?

4. What volume would be occupied by
   a. 1.0 mol of $CO_2(g)$ at STP?
   b. 3.5 mol of $CH_4(g)$ at STP?

concept check 5

1. What makes Charles' law a scientific law?
2. Why does the pressure of a gas sample in a closed container increase as its temperature is increased?
3. What would happen to the pressure exerted by a gas sample if both the temperature of the gas and the volume occupied by the gas were doubled, while holding the amount of gas constant?
4. What can cause a gas to transform into a liquid?

B.6 NON-IDEAL GAS BEHAVIOR

All gas relationships considered up to now have related to ideal gases. As you know, a gas sample that behaves under all conditions as the kinetic molecular theory predicts is called an ideal gas. Most gas behavior approximates that of an ideal gas and is satisfactorily explained by the kinetic molecular theory. However, at very high gas pressures or very low gas temperatures, real gases do not behave ideally. That is, the gas laws you have considered do not accurately describe gas behavior under all conditions.

Why do some gases behave in ways that seem to "violate" the ideal gas law? You learned earlier that the following two postulates of the kinetic molecular theory are assumptions that are not always valid:

- Gases consist of tiny particles (molecules), whose size is negligible compared with the great distances that separate them from each other.

- Gas molecules only interact during collisions. Attractive or repulsive forces among gas particles are negligible.

Can you think of conditions that could cause these assumptions to be invalid? Consider a situation of very high gas pressure. At very high pressures, gas particles are very close together. Thus, the volume of a gas particle may no longer be negligible compared to the empty space around it. Also, the weak attractive or repulsive forces between gas molecules may no longer be negligible.

Now think about decreasing the temperature of a gas sample. As the gas cools, the average kinetic energy of the gas molecules decreases. If gas molecules move very slowly, the weak attractive forces between the particles may again become important. These attractions cause non-ideal gas behavior.

Finally, for most substances that are gases at room temperature and pressure, the assumptions of the kinetic molecular theory are valid and we can apply the ideal gas law. However, some gases, such as carbon tetrachloride (CCl_4), have relatively strong intermolecular forces, resulting in inaccurate predictions when applying the ideal gas law.

B.7 TEMPERATURE, HEAT, AND PHASE CHANGES

In everyday conversation, the words *heat* and *temperature* are commonly interchanged, even though they mean different things. A similar mistake is to use the terms *weight* and *mass* interchangeably. Many students—and others—do not understand the difference between heat and temperature. Think about these two terms. What differences can you identify between temperature and heat?

One difference you may have noted is that temperature can be easily measured. As you know, atoms and molecules in all states of matter are constantly moving. Temperature is a measure of the average kinetic energy associated with this molecular motion. So more motion at the molecular level— more internal energy—results in a higher temperature. Temperature is often measured with a thermometer, which relies on the change of a physical property in response to temperatures. Mercury and alcohol-filled thermometers (see Figure 2.26) use the variation in liquid density with temperature. By constrast, electronic thermometers usually rely on a change in electrical resistance with temperature.

The mercury thermometer, no longer commonly used in schools, was invented by German physicist Daniel Gabriel Fahrenheit.

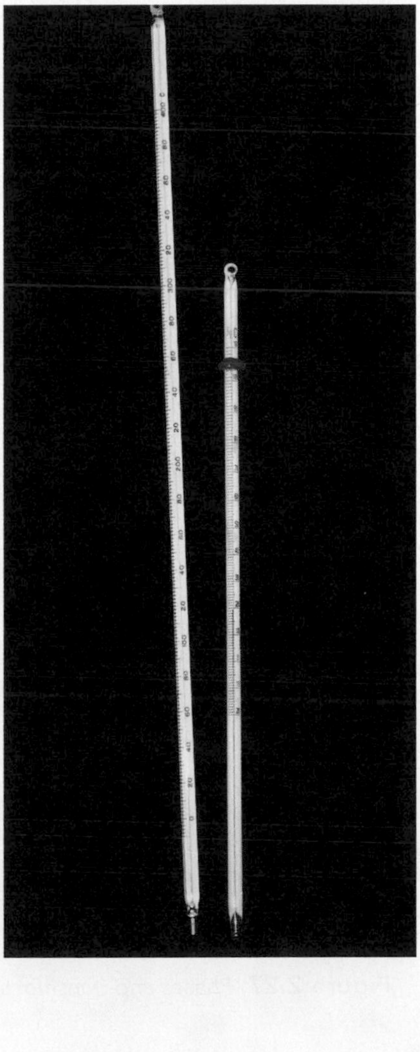

Figure 2.26 *Bulb thermometers filled with mercury (left) and colored alcohol (right). The liquid in a bulb thermometer expands as external temperature increases, causing it to move up the tube within the thermometer. The height of the liquid column can then be marked at known temperatures (such as the freezing and boiling points of water) to calibrate the thermometer.*

Energy of molecular motion within an object is known as thermal energy when it is not being transferred.

How is temperature different from heat? To answer this, we first need to develop some ideas about heat. Consider what happens if you let two objects touch that are at different temperatures—perhaps your hand and cold water in a swimming pool. The water feels cold on your hand. At the molecular level, the average kinetic energy of the molecules in the cold water is lower than the average kinetic energy of the molecules in your skin. When the two meet, some energy from the molecules in your hand is transferred and molecules in the water then move faster. As a result, the molecules in your hand have less energy than before the interaction, and the molecules in the water have a little more energy. This transferred energy from one object to another due to temperature difference is called **heat**. The definition of heat implies the flow of energy from a warmer object to a cooler object.

Think about what happens when you hold something hot. Which way is the energy moving? The difference between heat and temperature may seem subtle or not all that important. However, it is always important to be as clear as possible and to use accurate language when describing the natural world or scientific investigations.

Now that you can distinguish heat from temperature, consider one of the potential outcomes of non-ideal gas behavior. Think about what happens as a gas is cooled. As energy is transferred away from the gas, the gas particles move more slowly. The temperature of the gas decreases as energy (in the form of heat) flows away from it. At some point, the gas particles do not have enough kinetic energy to overcome the attractive forces among themselves and begin to "stick together." Once the molecules are no longer moving rapidly and independently, they cannot be considered to be in the gas phase. Cooling a gas such that it turns into a liquid in this way is called **condensation**. The reverse of this process, heating a liquid until it turns into a gas, is called **vaporization**. During these changes, the chemical identity of the substance is not altered, so they are physical changes. Physical changes that alter the state of a substance are known as phase changes. Common phase changes are illustrated in Figure 2.27.

How would you explain why adding energy to a liquid can cause it to turn into a gas?

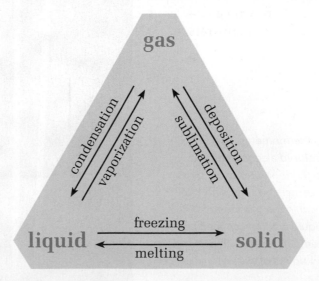

Figure 2.27 *Phases and transitions between them are identified in this illustration.*

INVESTIGATING MATTER

B.8 PHASE CHANGES

Preparing to Investigate

In this investigation, you will observe the effects of adding energy, in the form of heat, to a sample of ice. Before reading *Gathering Evidence*, think about how you might carry out this investigation. What would you need to measure and record? Record your thinking before continuing.

Now read *Gathering Evidence* carefully. Also, note safety precautions and plan necessary data collecting and observations.

Making Predictions

Predict what you think will happen as you gather evidence and write down your predictions.

Gathering Evidence

1. Before you begin, put on your goggles, and wear them properly throughout the investigation.
2. Fill a 100-mL beaker about 2/3 full with crushed ice. Put as much ice in the beaker as possible without breaking the beaker (it should be packed tightly).
3. Place the beaker on a hot plate.
4. Insert a thermometer or temperature probe into the ice and clamp it in place as shown in Figure 2.28. Make sure the thermometer is not touching the sides or bottom of the beaker.

Figure 2.28 *Secure the beaker of ice on a hot plate using a ring stand.*

5. Turn the hot plate on and record the temperature.

6. Record the temperature of the sample and the time as well as a description of the contents of the beaker at regular time intervals.

7. Continue to make measurements until the water has been boiling for four measurements. Be careful to not boil all the water from the beaker.

8. Remove the beaker from the hot plate and turn off the hot plate.

9. Wash your hands thoroughly before leaving the laboratory.

Analyzing Evidence

1. Plot the time–temperature data for your investigation with time on the horizontal axis and temperature on the vertical axis. Label your axes, and arrange the scales so the graph nearly fills the space available.

2. Label the points on your graph at which ice (1) began to melt and (2) had completely melted.

3. Indicate the point on your graph at which water began to boil.

4. Draw a curve through your plotted points.

Interpreting Evidence

Refer to your completed graph as you answer the following questions.

1. What do you notice about the graph?

2. How well did your actual results correspond with your predictions? Propose explanations for any differences between your predictions and observed results.

Making Claims

3. What scientific claims can you make as a result of this investigation?

4. For each claim listed in your answer to Question 3, cite evidence from the investigation to support your claim.

5. Consider the phase change from ice to water:

   a. What is the name of this phase change?

   b. What do the data, that is, temperature measurements, tell you about energy required or released by this phase change?

   c. What do your observations tell you about this phase change?

6. **Specific heat capacity** is a characteristic property of a material. It is defined as the quantity of heat needed to raise the temperature of one gram of the substance by one degree Celsius.

   a. What feature of your graph would provide information about the specific heat capacity of water?

  b. The specific heat capacity of liquid water is ~4.2 J/(g•°C) (joules per gram per °C). What additional information would you need to calculate this value from your graph?

  c. Is the specific heat capacity for ice the same as the specific heat capacity for liquid water? How do you know?

Reflecting on the Investigation

7. Why is it important that the temperature probe or thermometer not touch the sides or bottom of the beaker?

8. Why is it important to make both measurements and observations during an investigation? (*Hint:* What could happen if you made only observations without measurements or vice versa?)

■ MAKING DECISIONS

B.9 USING KMT TO REFINE YOUR AIR-QUALITY INVESTIGATION

In Section A, you began developing an air-quality investigation by writing beginning questions. Since then you have learned more about gases and their behavior, and about a model used to explain gas behavior. You also have some experience making measurements with gases. Using your new knowledge, revisit the beginning questions you developed—and start to think about the procedures that you will describe in upcoming sections—as you answer the following questions:

1. Do you still find your questions relevant, interesting, and engaging enough to serve as the basis for your unit project? If not, draft new questions and get feedback from your group members to improve them as you did in Making Decisions A.14.

2. a. How could knowledge of kinetic molecular theory improve your questions?

    b. Revise your questions to reflect your knowledge of KMT.

3. Think about the gases that you would study in your investigation. Are those gases well-represented by an ideal gas model or would you need to use a real/non-ideal gas model to represent them? Explain.

4. Would you be able to make measurements on the actual system or phenomenon of interest in your investigation? If not, could a model be used to represent the system of interest? Explain.

5. Could your investigation be helpful in constructing a scientific model of an air-quality phenomenon? Explain.

SECTION B SUMMARY

Reviewing the Concepts

> The kinetic molecular theory states that gases are composed of particles of negligible size that are in constant motion and that undergo elastic collisions. The average kinetic energy of a gas sample is directly related to its kelvin temperature.

1. Use KMT to explain each of these observations:

   a. Decreasing the volume of a gas sample at constant temperature causes the gas pressure to increase.

   b. At constant volume and constant amount of gas, the pressure of a gas sample changes if its temperature is changed.

2. Think of a pool table as a model (analogy) for gas behavior.

   a. How would you use pool balls to demonstrate the effect of an increase in the temperature of a gas sample?

   b. How would you use this model to demonstrate the effect of adding more gas molecules to the sample?

   c. In what ways is the pool-table model unsuccessful and misleading in modeling gas behavior?

3. Sketch the gas particles in

   a. a balloon submerged in hot water.

   b. a balloon submerged in ice water.

> For a given gas sample, the ideal gas law expresses the relationship among its volume, pressure, temperature, and amount of gas present.

4. Determine the pressure exerted by 0.122 mol oxygen gas in a 1.50-L container at room temperature (25 °C).

5. At what temperature will 2.5 mol nitrogen gas exert a pressure of 10.0 atm in a 2.0-L container?

6. How many moles of gas are contained in a sample that exerts a pressure of 5.5 atm in a 4.6-L container at 300.0 K?

7. What volume will be occupied by a balloon containing 0.65 mol H_2 at 25 °C and 1.0-atm pressure?

> Equal numbers of gas molecules at the same temperature and pressure occupy the same total volume. One mole of a gas sample at 0 °C and 1 atm occupies a volume of 22.4 L.

8. What volume would 8.0 g helium (He) gas occupy at 0.0 °C and 1.0 atm?

9. How many moles of gas molecules would be in a 2.0-L bottle of air at 0.0 °C and 1.0 atm?

10. a. How many moles of oxygen gas would be in a 1.0-mL gas sample at 0.0 °C and 1.0 atm?

    b. Express your answer to Question 10a in molecules.

11. What would be the mass of gas inside a 3.0-L balloon at 0 °C and 1.0 atm filled with

    a. He(g)?   b. CO_2(g)?   c. CH_4(g)?

12. If the three filled, tied-off balloons from Question 11 were released, would each balloon rise or fall?

Scientific theories and models explain observations and data. Scientific laws describe, but do not explain, natural phenomena.

13. Recall Investigating Matter A.9 (page 168).

    a. Give an example of a scientific law that describes this phenomenon.

    b. Describe how KMT helps to explain this phenomenon.

14. You may have heard someone say, "It's just a theory" in an everyday conversation.

    a. What does this phrase imply about the person's understanding or definition of a theory?

    b. How does this everyday use of the term *theory* differ from the definition of a scientific theory?

15. Some people perceive scientific theories as "immature laws," meaning that theories will eventually develop into laws if adequate evidence is provided.

    a. Comment on the validity of this view.

    b. Why might this perception perpetuate the "it's only a theory" sentiment in everyday conversation?

16. In Modeling Matter B.2, you used an analogy—a type of model—to better understand kinetic molecular theory. Why is it useful to construct models to represent theories?

17. In your own words, explain why the ideal gas law is an example of a scientific law rather than a scientific theory.

Under certain conditions, real gases do not behave as predicted by the kinetic molecular theory.

18. Referring to postulates of the kinetic molecular theory, explain why gases do not behave ideally under each of the following conditions:

    a. very high gas pressure

    b. very low gas temperature

19. Describe changes in motion of gas particles as a gas is cooled.

20. Why is the ideal gas law so widely used when it is not always valid?

Temperature is a measure of the average kinetic energy of a sample of matter, while heat is a form of energy that transfers from one "object" to another due to a difference in temperature.

21. Explain how temperature is measured.

22. Will heat transfer between two objects with the same temperature? Explain.

23. Consider the statement, "The hot rock contains a lot of heat."

    a. Is this an accurate statement?

    b. What would be a better way to describe the hot rock?

Transitions among the physical states of a substance, called phase changes, require the addition or release of energy.

24. Indicate whether energy is required for or released by each of the following phase changes:

    a. vaporization   d. melting

    b. sublimation   e. deposition

    c. condensation   f. freezing

25. Why are phase changes considered physical changes?

26. What happens to the temperature of a substance as it undergoes a phase change? Explain.

How are models and theories useful in explaining and predicting behavior of gases?

Section A introduced several gas laws that were discovered by scientific experimentation. This section introduced tools for explaining and extending these laws. Think about what you have learned and how these ideas are related, then answer the question in your own words in organized paragraphs. Your answer should demonstrate your understanding of the key ideas in this section.

Be sure to consider the following in your response: the kinetic molecular theory of gases; scientific models, theories, and laws; non-ideal gas behavior; and phase changes.

Connecting the Concepts

27. Locate three Modeling Matter activities you have completed in Unit 1 or Unit 2. For each, describe what you represented (data, theory, observations). Explain how these activities fit with your understanding of the use of models in science.

28. A sample of a pure gas has a mass of 6.7 g and exerts a pressure of 1.5 atm on a 2.5-L container at 298 K. What is the gas?

29. Based on your knowledge of gas properties, predict what will happen to the density of a 100.0-g helium gas sample if you heat it from room temperature to 62 °C at constant pressure.

30. If air at 25 °C has a density of 1.28 g/L and your classroom has a volume of 2.0×10^5 L,

   a. what mass of air is contained in your classroom?

   b. assume that the room temperature is increased. How would that affect the total mass of air contained in the room,

       i. if the room were tightly sealed?

       ii. if the room were not sealed?

31. Heat is often described as "lost energy." For instance, automobile engines release about 75% of the energy from gasoline combustion as heat. Explain this idea of energy loss in terms of your understanding of heat.

32. Consider Investigating Matter B.8, Phase Changes.

   a. What would be an appropriate beginning question for this investigation?

   b. What are the variables of interest?

33. Given what you now know about the ideal gas law, consider an investigation to determine the identity of an unknown pure gas.

   a. What would be an appropriate beginning question for this investigation?

   b. What variables would you need to measure?

   c. What variables, if any, would you need to control?

   d. How could you measure each of the variables you identified in Question 33b?

Extending the Concepts

34. Consider the behavior of an ideal gas as described by the postulates of the kinetic molecular theory. For each postulate, predict how gas behavior might change if the postulate did not hold.

35. The Kelvin temperature scale is used extensively in theoretical and applied chemistry work.

a. Find out how closely scientists have approached a temperature of zero kelvins (0 K) in a laboratory setting.

b. Results from research in cryogenics have suggested intriguing applications of low-temperature chemistry and physics concepts. Report on activities in this field.

CHEMISTRY *AT WORK*
Q&A

History isn't just a class you take in school. It's all around us! Almost every city in the United States is loaded with places and objects much older than you and your classmates—combined. Every day, these historical artifacts and buildings are under pressure from chemical forces that threaten to age them. Read on to see how one chemist is working to save history by stopping time in its tracks.

Carol Chin, Preservation Scientist at the National Center for Preservation Technology and Training in Natchitoches, Louisiana

Elfreth's Alley in Philadelphia is one of the oldest residential streets in the country, dating back to the early 1700s.

Q. What is historic preservation, and how does chemistry fit in?

A. Historic preservation is the process of preserving, conserving, or protecting historic sites or buildings, statues, or monuments. Materials constantly undergo chemical reactions with surrounding air and precipitation that cause them to deteriorate, stain, or age. Using chemistry, we can slow these changes.

Q. How did you get interested in chemistry?

A. Both my parents were scientists. My mom was a chemist, and my dad was a biologist, so science runs in the family. I started off thinking I'd be a chemist, but I also wanted to spend time outdoors doing field work. In college, I did an independent study project with a geochemist, a scientist who combines the study of chemistry with geology. I learned that I could study chemistry at sea. I worked for a year at an oil field research company, then got my master's degree in marine chemistry. This field studies how ocean water is affected by natural processes, such as what rivers add, or human activities, such as agricultural and industrial runoff. I got my doctoral degree in marine geology.

Q. How did you get involved in historic preservation?

A. While in graduate school, I started making personal choices about how to minimize my impact on the environment. I decided that if I lived in an old house instead of a new house made of new materials, I could greatly reduce the effect I have on the environment. Later, in an effort to save a neighborhood landmark from demolition, I got to know other community members who were involved in historic preservation. I began looking for ways that my scientific degree could make a difference in the historic preservation field.

Q. What does your job entail?

A. I work for the National Center for Preservation Technology and Training (NCPTT), a part of the National Park Service. NCPTT does research to develop preservation technologies. My group is mostly involved with finding new ways to preserve monuments and structures made of marble or limestone. These materials are especially susceptible to acid rain.

Q. What is acid rain, and how does it affect these stones?

A. Carbon dioxide and sulfur dioxide, which mostly come from burning fossil fuels, produce acid when they combine with water. Marble and limestone are made of calcium carbonate. When acid rain interacts with calcium carbonate, an acid-base reaction takes place that deteriorates the stone. Want to see how acid affects calcium carbonate? Try putting an antacid tablet in a glass of vinegar. The vinegar breaks down the antacid. We want to protect monuments from the same fate.

Q. How do you protect these monuments?

A. One way is by using consolidants, such as synthetic sealants, to strengthen the surfaces of stones. There's a fine line between sealing a surface and protecting it, though. You don't want to create an impermeable surface that traps moisture that can cause further degradation. Some consolidants work by chemically binding with the surface and helping to strengthen it, the way fluoride in toothpaste protects your teeth. Others provide a protective coating on the surface, which limits the effects of the environment on the surface of the stone. Since every historic object is different, we evaluate different consolidants to see which works best in each situation.

Q. How do I become a preservation chemist?

A. It is important to have a strong background in chemistry. Some people in this field have degrees in historic preservation, materials science, or art conservation. But there are lots of unconventional paths to this kind of job, like mine! I got here because I was looking for a way to use my scientific background to make a more direct difference in environmental issues. If we learn how to preserve and continue to use sites, objects, and structures, then we won't have to use as many new resources and materials and ultimately we will have less of an effect on the environment. I think you have to do what you enjoy—otherwise, why do it?

SECTION **C** INTERACTIONS OF MATTER AND ENERGY IN ATMOSPHERES

What does evidence reveal about properties of Earth's atmosphere?

As you posed and refined the guiding question for your air-quality investigation, you may have noted that the study of Earth's atmosphere is quite complex. You have begun to understand some of the variables in this complex system by collecting evidence about gas behavior in Sections A and B. To be able to make specific claims about air-quality issues, however, one must often study and collect evidence about inter-actions among variables within the system. In Earth's atmosphere, this means considering mixtures and reactions of atmospheric gases, as well as the role of solar radiation.

As you explore the complexity of Earth's atmosphere in this section, start to decide which components and interactions you will need to include in your air-quality investigation. Remember that you will design procedures to gather evidence that could support an answer to your question, and then refine these procedures before preparing a scientific poster to describe your proposed investigation in Putting It All Together.

GOALS

- Describe the major components of Earth's atmosphere.
- Describe the relationships among electromagnetic radiation's energy, frequency, and wavelength and identify types of electromagnetic radiation.
- Explain how Earth's atmosphere interacts with solar radiation, including how the greenhouse effect works.
- Apply Avogadro's law in calculations, including stoichiometric problems.
- Explain the effect of a catalyst on a reaction using the collision theory.
- Describe how evidence gathered from scientific investigations is used to make claims about the natural world.

concept check 6

1. According to the ideal gas law, which of the properties of gases—pressure, volume, temperature, and number of moles—are directly proportional?
2. What does Avogadro's law state regarding equal volumes of gases?
3. Use KMT to explain why
   a. gases expand to fill the space available.
   b. gas pressure is proportional to temperature at constant volume.
4. What are the key components of Earth's atmosphere?

C.1 GASES IN OUR ATMOSPHERE

Throughout this unit, you have studied gases and their behavior, often within closed or controlled systems. In fact, most of Earth's gaseous substances are found within its atmosphere. Derived from the Greek words for vapor and sphere, an **atmosphere** is the collection of gases held by a planet or other astronomical body through gravitational attraction. Although many planets and some moons are known to have atmospheres, far more is known about Earth's atmosphere than any others. Scientists study atmospheres for varied reasons: geologists can learn about the geological history of a planet from its atmosphere; meteorologists can understand its climate; and biologists can evaluate the likelihood of past, present, or future life. Chemists study atmospheres primarily to understand interactions of substances within atmospheres. On Earth, these interactions are affected significantly by the actions of living organisms, particularly human beings.

Recall that a *system* refers to the part of the world that we want to study. In terms of Earth and atmospheric systems, it also refers to a group of interrelated materials and processes. Both definitions are helpful in considering gas behavior.

CHEM**QUANDARY**

TRAVELING THE ATMOSPHERE

Suppose it were possible for you to fly from Earth's surface up to the farthest regions of the atmosphere (see Figure 2.29 on page 204). What do you think you would encounter as you traveled 5, 10, or 50 km away from Earth? Would the air temperature change? Would you be surrounded by the same mixture of gas molecules as your altitude increased?

Figure 2.29 *Just how thin is the atmosphere that surrounds us? Looking up from Earth, the skies can seem endless. This view highlights the narrow strip of atmosphere that separates Earth from space.*

The only components of the atmosphere visible from space are condensed water vapor in the form of clouds and, at times, traces of colored nitrogen oxide (NO_x) pollutants near urban areas.

Although you cannot take a trip such as the one described in the ChemQuandary right now, you can find answers to many questions about the atmosphere. Technology allows scientists to measure a range of air characteristics at different altitudes. In the following activity, you will analyze atmospheric data collected at altitudes up to 80 km. As you do so, think about how these data are used to describe the properties of Earth's atmosphere.

DEVELOPING SKILLS
C.2 GRAPHING ATMOSPHERIC DATA

Sample Problem: Predict the shape of the graph for a y *versus* x *plot of air temperature versus altitude.*

With increasing altitude, air temperature decreases, then increases, then decreases again. The graph will not be linear or regular; it will go down, then up, then down again.

Table 2.3 summarizes atmospheric data gathered at various altitudes. Use the information to answer the questions that follow:

1. Predict the shape of the graph for a *y* versus *x* plot of air pressure versus altitude.

Table 2.3

	Atmospheric Data			
Altitude (km)	**Air Temperature (°C)**	**Air Pressure (mmHg)**	**Mass (g) of 1-L Air Sample**	**Molecules in 1-L Air Sample**
0	20	760	1.20	250×10^{20}
5	−12	407	0.73	150×10^{20}
10	−45	218	0.41	90×10^{20}
12	−60	170	0.37	77×10^{20}
20	−53	62	0.13	27×10^{20}
30	−38	18	0.035	7×10^{20}
40	−18	5.1	0.009	2×10^{20}
50	2	1.5	0.003	0.5×10^{20}
60	−26	0.42	0.0007	0.2×10^{20}
80	−87	0.03	0.00007	0.02×10^{20}

The unit mmHg, millimeters of mercury, is a common unit of gas pressure (see page 160).

2. Prepare graphs according to the following instructions:
   a. Plot air temperature versus altitude data.
   b. Plot pressure versus altitude data.
   c. Draw a best-fit line through the plotted points for each graph. (*Note:* The graph line may be straight or curved.)

3. Consider your graphs.
   a. Does the shape of either graph differ from your prediction? If so, how?
   b. Which graph follows a more regular pattern? Why?
   c. Explain the trend of your air pressure versus altitude graph using KMT.

4. Based on the data in Table 2.3, would you expect air pressure to rise or fall if you traveled from sea level (0 km) to
   a. Pike's Peak (4301 m above sea level)?
   b. Death Valley (86 m below sea level)?

5. Suppose you gathered 1-L samples of air at several altitudes.
   a. How would the following values change with altitude?
      i. mass of the air sample
      ii. total molecules in the air sample
   b. If you were to plot air-sample mass versus the number of molecules contained in the sample, what would the graph look like? Why?
   c. If you were to plot air pressure versus the number of molecules contained in the sample, what would the graph look like? Why?
   d. Would each sample have the same composition? How do you know?

6. Scientists often characterize Earth's atmosphere as being composed of four general layers.

   a. Mark the graphs you prepared in Question 2 with lines at appropriate altitudes to indicate where you think the general transitions between atmospheric layers might be.

   b. How confident are you in marking these transitions on your graph? Explain.

   c. Which of these layers will your planned experiment investigate?

C.3 STRUCTURE OF EARTH'S ATMOSPHERE

As you learned in Developing Skills C.2, scientists have identified four distinct layers of Earth's atmosphere. In order, the four layers are the *troposphere* (nearest Earth's surface), the *stratosphere*, the *mesosphere*, and the *thermosphere* (outermost layer). Trends in measurements of atmospheric properties such as temperature change with altitude (as in Developing Skills C.2), chemical composition, density, and patterns of movement allow the layers to be distinguished from one another. Most of the atmosphere's mass and all of its weather are located within 10 to 15 km of Earth's surface in the layer known as the **troposphere**. This is a very small distance compared to the size of Earth. If Earth were roughly the size of a softball, the troposphere would be about the thickness of a coat of varnish or paint applied to the ball. Put another way, the highest point on Earth, Mt. Everest, is 8.8 km high. If it were only twice as tall, it would reach the top of the troposphere.

> The word *troposphere* is based on the Greek words *tropos*, which means "turn," and *sphaira*, which means "sphere."

Gases mix continuously in the troposphere, so its composition is reasonably uniform around the world. This mixture is what we commonly refer to as *air*. Table 2.4 lists the main components of the troposphere. The relative abundance of a component in a mixture is described by its **concentration**, which expresses the quantity of a substance in a specified quantity of mixture or other substance. In gaseous mixtures, concentration is often measured in terms of percent by volume. For example, if a sample of natural gas (a common fuel) contains 700 cubic meters (m^3) of methane (CH_4) in 1000 m^3 of gas, it has a methane concentration of 70%. Smaller concentrations may be expressed in parts per million (ppm), which for gases means 1 cm^3 (1 mL) of substance in 1 m^3 (1000 L) of gaseous mixture.

> The amount of water vapor in air determines its humidity.

In the mixture of gases we call air, nitrogen is the most abundant component, followed by oxygen and substantially lower amounts of argon and carbon dioxide. In addition to the gases listed in Table 2.4, air samples can contain up to 5% water vapor; in most locations the water vapor range is from 1 to 3%. Other gases naturally present in air at concentrations below 0.0001% (1 ppm) include hydrogen gas (H_2), xenon (Xe), ozone (O_3), nitrogen oxides (NO and NO_2), carbon monoxide (CO), and sulfur dioxide (SO_2). Chemical analysis reveals that ancient air trapped in glacial ice has about the same chemical makeup as does our current atmosphere (Figure 2.30).

Figure 2.30 *Analysis of gases trapped in glacial ice shows that the composition of the atmosphere has not changed much since ancient times.*

This finding suggests that there has been little change, except for carbon dioxide concentrations, in tropospheric air over a very long time.

Human activity and natural phenomena such as volcanic eruptions can alter the concentrations of some trace gases and add other substances to air. This may lead to decreased air quality, as you will learn later in this unit. In the next activity, you will construct models of gaseous mixtures in order to make a macroscopic substance (air) that is usually invisible a little easier to imagine.

Table 2.4

Gaseous Components of Dry Tropospheric Air		
Substance	**Formula**	**Percent of All Gas Molecules**
Major components		
Nitrogen	N_2	78.08
Oxygen	O_2	20.95
Minor components		
Argon	Ar	0.93
Carbon dioxide	CO_2	0.033
Trace components		
Neon	Ne	0.0018
Ammonia	NH_3	0.0010
Helium	He	0.0005
Methane	CH_4	0.0002
Krypton	Kr	0.0001

■ MODELING MATTER

C.4 MIXTURES OF GASES

Think back to the models that you have constructed so far in this course. You have drawn atoms and molecules, modeled the movement of electrons, represented reactions, and illustrated movements of gas molecules. In this activity, you will combine the skills and concepts you have learned in earlier activities with some ideas about mixtures to draw models of gaseous mixtures that might be found in the laboratory or in the atmosphere.

To begin, consider this example: Suppose you want to draw a model of two gaseous compounds in a **homogenous mixture**. We first need to look closely at this description. You know that compounds are composed of atoms of two or more different elements linked together by chemical bonds. The fact that the compounds are in the gaseous state tells us that the molecules are relatively far apart and moving rapidly. A **mixture** is created when two or more substances combine, yet each retains its individual properties. A **homogeneous** mixture is uniform throughout, so the two compounds in this example should be intermingled and evenly distributed.

> A *heterogeneous mixture* is a combination of two or more materials that are not uniform in composition.

Suppose a molecule of one of these gaseous compounds contains two different atoms. To represent this molecule, you could draw two differently shaded or labeled circles to denote atoms of the two elements and a line connecting the atoms to indicate a bond, like this:

Suppose the other compound is composed of molecules made up of three atoms, and that two atoms are of the same element. You now need to decide on the order that the atoms should be connected: the unique atom (Y) could be in the middle, X–Y–X, or on the end, X–X–Y. As long as you draw this imaginary compound in the same way every time, it doesn't matter which way you do it for this activity. However, the way in which atoms are connected in real compounds does, in fact, make a difference; X–Y–X would be quite a different molecule from X–X–Y.

a

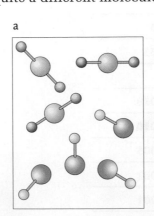

b

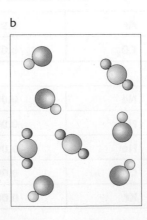

c

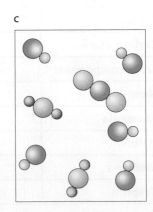

Examine the three models (*a, b,* and *c*) in the illustration on page 208. Which best represents a homogeneous mixture of the two compounds just described? You are correct if you said that *b* is the best visual model. The two types of molecules are uniformly mixed, and the atoms are shaded to indicate that they represent different elements. In *a* the mixture is not homogeneous because the molecules are not uniformly mixed. Model *c* contains three different compounds instead of two. Notice that in *a*, bonded atoms in each molecule are connected by lines. In *b* and *c*, bonded atoms just touch each other. Both representations are used by chemists; either one is acceptable in this activity. Note that it was the key features of each model provided—the number of distinct substances and their distribution in the container—that allowed you to choose the best representation.

Now it is your turn to create and evaluate visual models of gaseous matter.

1. Draw a model of a homogeneous mixture composed of three different gaseous elements.

2. What kind of matter does the following model represent? Explain your answer.

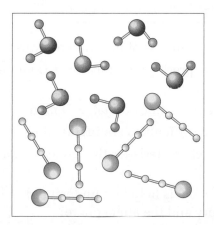

3. Draw a model of a container full of each of these samples of matter.

    a. a mixture of gaseous elements X and Z

    b. a two-atom gaseous compound of X and Z

    c. a four-atom gaseous compound of X and Z

4. Compare each visual representation that you created in Question 3 with those of your classmates.

    a. What are the key features of the models?

    b. Although the models may look a little different, does each set depict the same type of sample? Comment on any similarities and differences.

    c. Do the differences help or hinder your ability to visualize the type of matter being depicted? Explain.

5. Draw a model of a sample of tropospheric air containing at least 20 molecules. Refer to Table 2.4 (page 207) for appropriate proportions of substances. Remember that you cannot draw partial molecules, so some trace components may have to be omitted from your drawing. Also, recall that some elements in air are diatomic, meaning that they normally exist as a two-atom molecule.

6. Another student was asked to draw a model of a mixture composed of an element and a compound:

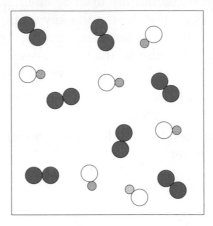

Comment on the usefulness of the student's drawing.

7. You have been interpreting and creating 2-D models of 3-D molecules.

   a. What are some limitations of flat, 2-D models?

   b. What particular difficulties are involved in representing gases?

   c. What are some characteristics that good models of matter should have?

   d. How do these models help you understand the behavior of gases in the laboratory and the atmosphere?

   e. How could models be helpful as you design your investigation?

   f. How could you use models to help others understand your investigation?

As you continue studying chemistry, you will encounter visual models of matter similar to those in this activity. When you see them, think about their usefulness as well as their possible limitations.

concept check 7

1. How do scientists distinguish the layers of the atmosphere?
2. Would you consider the atmosphere a heterogeneous or homogeneous mixture? Explain.
3. Do gases in the atmosphere ever react with one another? How do you know?

C.5 COLLISION THEORY

In Section B, you were introduced to the kinetic molecular theory (KMT) as a way to explain gas behavior. One postulate of KMT includes the idea that atoms and molecules collide with one another and with the container in which they are stored. When a gas behaves ideally, all such collisions are elastic; that is, a molecule bounces away from the container wall or other molecule with no loss of energy. Experience tells us, however, that under some conditions gas molecules do not just "bounce off," but react to form different substances. For instance, methane gas burns in oxygen to produce carbon dioxide and water, and nitrogen gas can combine with hydrogen gas to produce ammonia. In each of these processes, gas molecules collide with enough energy and at an angle that results in bond-breaking in the starting substances. The "released" atoms can form new bonds, making different compounds. Energy is required to break the bonds between two atoms, whereas energy is released when bonds are formed. This exchange of energy can be illustrated using a potential energy diagram, as shown in Figure 2.31.

Potential energy can be considered "stored energy." In this case, the energy is stored within molecules. You will learn more about energy and chemical bonds in Unit 3.

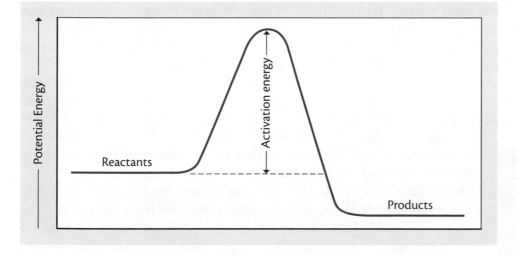

Figure 2.31 *Potential energy diagram for a typical chemical reaction. The activation energy represents the minimum energy needed to start the reaction.*

Because gas molecules must have a certain energy and orientation to react upon collision, reactive collisions tend to be rare under standard conditions. This results in a very slow rate of reaction or product formation. Industrial chemists are concerned about controlling the rate of chemical reactions—often in order to make more product in less time—and thus may introduce a *catalyst* into a reaction between gases. A **catalyst** is a material that speeds up a chemical reaction that would otherwise proceed far more slowly. Enzymes that aid in digesting food and in promoting other bodily functions are organic catalysts. Although a catalyst participates in the chemical reaction, it is not considered a reactant because it emerges unchanged when the reaction is complete.

How can a catalyst speed up a reaction and remain unchanged? **Collision theory** states that reactions can occur only if molecules collide with sufficient energy and with suitable orientation to disrupt chemical bonds.

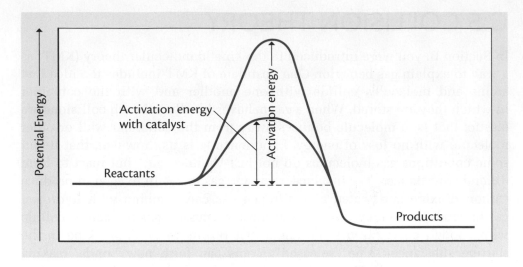

Figure 2.32 *The catalyst provides an alternative pathway with lower activation energy.*

The minimum energy required for such effective collisions is called the **activation energy**. You can think of the activation energy as an energy barrier that stands between the reactants and the products. See Figure 2.32. Reactants must have enough energy to get over the barrier before a reaction can occur. That is, the reactants must have sufficient kinetic energy to participate in effective molecular collisions. When the energy barrier is high, few molecules will have sufficient energy to get over it and the overall reaction will proceed slowly. A catalyst works by providing a different reaction pathway, one with lower activation energy. In effect, the catalyst lowers the energy barrier. The result is that more molecules have sufficient energy to react and form products within a given period of time.

Since the Clean Air Act of 1970 was passed, the Environmental Protection Agency (EPA) has set emissions standards for automobiles. The EPA sets allowable limits for hydrocarbons, nitrogen oxides, and carbon monoxide emissions. One major contribution toward meeting those standards was the development of the **catalytic converter**, which is a reaction chamber built into the exhaust system of motor vehicles, as shown in Figure 2.33. In automotive catalytic converters, a few grams of platinum, palladium, or rhodium embedded in a noncatalytic material such as aluminum oxide (Al_2O_3) act as the catalyst.

Catalytic converter

Figure 2.33 *A catalytic converter (top) is installed in an automobile's exhaust system (bottom).*

In the catalytic converter, exhaust gases and outside air pass over several solid catalysts that help speed the conversion of potentially harmful gases to harmless products—nitrogen oxides to nitrogen gas, carbon monoxide to carbon dioxide, and hydrocarbons to carbon dioxide and water. Exhaust gases enter the catalytic converter, where, for example, nitrogen oxides and carbon monoxide are removed as shown below:

$$2\,NO(g) \;+\; 2\,CO(g) \xrightarrow{\text{Catalyst}} \; N_2(g) \;+\; 2\,CO_2(g)$$

Despite improvements made by catalytic converters, transportation is still a major air-pollution source in many states (see Figure 2.34).

Figure 2.34 *Many cities and towns, and some states, require annual testing of exhaust emissions.*

C.6 REACTIONS OF GASES

In order for automobile companies to evaluate the success of a catalytic converter, they must be able to measure the amount of gases produced from the reactions within the engine of the car. You may also need to measure the amount of gas produced by a gaseous reaction in your air-quality investigation. Recall from your study of Avogadro's law (page 187) that all gases have the same molar volume under the same conditions. This realization greatly simplifies our thinking about chemical reactions involving gases and allows industrial chemists to measure the emissions from the tailpipe. For example, consider the following chemical equations that involve gaseous reactions:

$$N_2(g) \;+\; O_2(g) \longrightarrow 2\,NO(g)$$
$$2\,H_2(g) \;+\; O_2(g) \longrightarrow 2\,H_2O(g)$$

You have learned that coefficients in chemical equations indicate the relative numbers of molecules or moles of reactants and products. Based on Avogadro's law, similar calculations also involve volumes of gaseous reactants and products.

The reaction of $N_2(g)$ and $O_2(g)$ to form $NO(g)$ occurs within the cylinders of internal combustion engines. Catalytic converters later transform $NO(g)$ back to $N_2(g)$.

Although nitric oxide (NO) in the environment is a precursor to both smog and acid rain (which you will learn more about in Section D), in the human body it is important in transmitting information between cells.

Hence, the equation representing the reaction between nitrogen and oxygen gases can now be interpreted: 1 volume of $N_2(g)$ and 1 volume of $O_2(g)$ combine to form 2 volumes of $NO(g)$, assuming all gases were measured at the same conditions of temperature and pressure.

Similar relationships hold for the volumes of all gaseous reactants and products: Using Avogadro's law, you can interpret the coefficients in terms of gas volumes.

$$1 \text{ volume } N_2(g) + 1 \text{ volume } O_2(g) \longrightarrow 2 \text{ volumes } NO(g)$$

$$2 \text{ volumes } H_2(g) + 1 \text{ volume } O_2(g) \longrightarrow 2 \text{ volumes } H_2O(g)$$

The measured volumes could be expressed in any convenient units, such as liters (L) or cubic centimeters (cm^3). In the second equation, which represents the formation of water, you could combine 200.0 L $H_2(g)$ and 100.0 L $O_2(g)$ and expect to produce 200.0 L $H_2O(g)$, if all gases were measured at the same conditions. Note that, unlike mass, gas volumes are *not* necessarily conserved in a chemical reaction—using the second reaction, note that 200 L + 100 L of reactants generates 200 L of product and *not* 300 L. See Figure 2.35.

Figure 2.35 *Two volumes of $H_2(g)$ combine with one volume $O_2(g)$ to form two volumes $H_2O(g)$.*

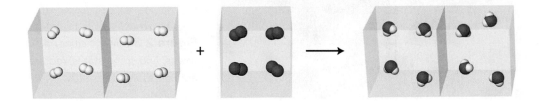

DEVELOPING SKILLS

C.7 STOICHIOMETRY IN REACTIONS OF GASES

Stoichiometry is the quantitative interpretation of chemical equations.

Sample Problem 1: In the following equation, what volume of $C_2H_6(g)$ is needed to produce 12 L $CO_2(g)$? (All measurements are made at the same temperature.)

$$2 \text{ } C_2H_6(g) + 7 \text{ } O_2(g) \longrightarrow 4 \text{ } CO_2(g) + 6 \text{ } H_2O(g)$$

From the chemical equation, we know that 2 L $C_2H_6(g)$ will produce 4 L $CO_2(g)$. If we want to produce 12 L of $CO_2(g)$, which is three times more than is represented in the equation, then we must start with *three times* as much $C_2H_6(g)$, or 6 L $C_2H_6(g)$.

Sample Problem 2: *Using the same scenario as Sample Problem 1, how many liters of $O_2(g)$ would be needed to produce 12 L $CO_2(g)$? Think through this question and make a prediction.*

Problems such as this can also be solved using ratios made from coefficients from the balanced equation. If we know from the previous equation that

$$7 \text{ L } O_2(g) \longrightarrow 4 \text{ L } CO_2(g)$$

and we want to know how many liters $O_2(g)$ are needed to produce 12 L $CO_2(g)$, we can set up a ratio to solve for the answer:

$$? \text{ L } O_2 = 12 \text{ L } CO_2 \times \frac{7 \text{ L } O_2}{4 \text{ L } CO_2} = 21 \text{ L } O_2$$

According to this calculation, 21 L $O_2(g)$ are needed to produce 12 L $CO_2(g)$. Was your prediction correct?

Now it is your turn to check your understanding of this concept.

1. In a gaseous reaction, 2 mol NO react with 1 mol O_2:

$$2 \text{ NO}(g) + O_2(g) \longrightarrow 2 \text{ NO}_2(g)$$

   a. Given the same conditions of temperature and pressure, what volume of $O_2(g)$ would react with 4 L NO gas?

   b. How might chemists use their knowledge of molar volumes to monitor the progress of this reaction?

2. Toxic carbon monoxide (CO) gas is produced when fossil fuels, such as gasoline, burn without sufficient oxygen gas. The CO can eventually be converted to CO_2 in the atmosphere. Automobile catalytic converters are designed to speed up this conversion:

   Carbon monoxide gas + Oxygen gas $\longrightarrow$ Carbon dioxide gas

   a. Write the balanced equation for this conversion.

   b. How many moles of oxygen gas would be needed to convert 50.0 mol carbon monoxide to carbon dioxide?

   c. What volume of oxygen gas would be needed to react with 968 L carbon monoxide? (Assume both gases are at the same temperature and pressure.)

3. A common way to produce ammonia (NH_3) is by the catalyzed reaction:

   Hydrogen gas + Nitrogen gas $\xrightarrow{\text{Catalyst}}$ Ammonia gas

   a. Write the balanced equation for this conversion.

   b. How many moles of hydrogen gas would be needed to convert 20.0 mol nitrogen to ammonia?

   c. What volume of hydrogen gas would be needed to react with 182 L nitrogen gas? (Assume both gases are at the same temperature and pressure.)

■ INVESTIGATING MATTER
C.8 GENERATING AND ANALYZING CO$_2$

Preparing to Investigate

As you have learned, air is a mixture of nitrogen gas (N$_2$) and oxygen gas (O$_2$) together with much smaller amounts of carbon dioxide and other gases (see Figure 2.36). Each of the gases within air has distinct physical and chemical properties.

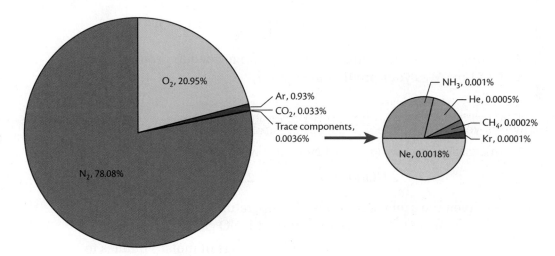

Figure 2.36 *This chart shows the composition of the gases in the atmosphere. Notice that carbon dioxide is a very small component of the air we breathe.*

In this investigation, you will generate carbon dioxide and explore some of its properties. To generate carbon dioxide, you will use the same reaction that you observed in Investigating Matter A.1 in Unit 1:

$$NaHCO_3(s) + HC_2H_3O_2(aq) \longrightarrow NaC_2H_3O_2(aq) + H_2O(l) + CO_2(g)$$

Before you begin, read *Gathering Evidence* to learn what you will need to do and note safety precautions. Plan for data collection, and construct an appropriate data table.

Making Predictions

Predict what you think will happen in each of the tests in Part II and write down your predictions.

Gathering Evidence

Part I: Generating Carbon Dioxide

1. Before you begin, put on your goggles, and wear them properly throughout the investigation.

2. Use a balance to measure a 0.22-g sample of sodium hydrogen carbonate (baking soda), NaHCO₃. Place the sample into a vial cap.

3. Obtain a syringe and remove the plunger and syringe cap. Hold your finger over the tip of the syringe and fill the syringe with water.

4. Float the vial cap containing the NaHCO₃ on top of the water in the syringe. See Figure 2.37.

5. While holding the syringe over a sink or a beaker, remove your finger from the syringe tip, allowing the vial cap to lower into the syringe. See Figure 2.38.

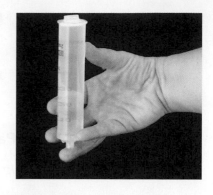

Figure 2.37 *A vial cap floating on top of water in a syringe barrel.*

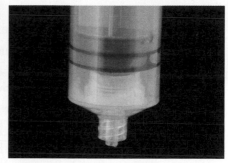

Figure 2.38 *The vial cap (left) is lowered by allowing the water to drain out of the syringe (center) leaving it seated at the base of the syringe barrel (right).*

6. Replace the plunger in the syringe and lower until the vial cap sits firmly in the depression at the base of the syringe.

7. Dispense some dilute acetic acid solution (vinegar), $HC_2H_3O_2(aq)$, into a small beaker or other container.

8. Draw 5 mL of vinegar into the syringe. Be careful not to tip the vial cap, as that may cause the reaction to begin before you are ready. See Figure 2.39.

9. Securely place the syringe cap on the tip of the syringe.

10. Shake the syringe vigorously. The reactants will mix, causing the reaction to proceed.

11. Observe the reaction until no more bubbles are formed.

12. Tip the syringe up, remove the syringe cap, then tip the syringe down and discharge the liquid into the sink or a beaker by pressing gently on the plunger. (*Note:* Do NOT push the plunger completely down, as this will discharge the CO_2 along with the liquid.)

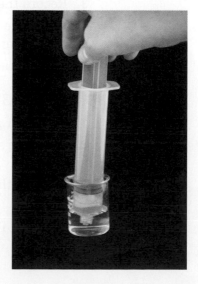

Figure 2.39
Draw 5 mL acetic acid into the syringe and then attach the syringe cap.

13. Wash the CO_2 by drawing 5 mL of water into the syringe, replacing the syringe cap, and shaking the syringe vigorously. Discharge the liquid as in Step 12. Repeat the washing process.

14. Replace the syringe cap. Label the syringe as "CO_2 gas" and store it until you are ready to perform investigations in Part II.

> Washing removes excess acetic acid from the CO_2.

Part II: Analyzing Carbon Dioxide

Investigation 1

1. Obtain or generate a syringe of CO_2 as described in Part I.

2. Dispense a few milliliters of calcium hydroxide solution (limewater), $Ca(OH)_2(aq)$, into a small beaker.

3. Remove the syringe cap from your CO_2 gas sample and attach a 5 to 10 cm piece of rubber tubing to the syringe.

4. Bubble a small amount (<5 mL) of gas from the syringe into the limewater. See Figure 2.40.

5. Record your observations.

6. Bubble an additional 5 mL of gas from the syringe and observe the results.

7. Record your observations.

Figure 2.40 *After attaching the tubing, bubble CO_2 through the $Ca(OH)_2$ solution.*

Investigation 2

1. Obtain or generate a syringe of CO_2 as described in Part I.

2. Place ~30 mL of pH indicator solution (prepared by your teacher) into a small beaker.

3. Record the color and the pH of the solution.

4. Discharge 5 to 10 mL of CO_2 gas onto the surface of the solution in the beaker.

5. Record the color and the pH of the solution.

Investigation 3

1. Obtain or generate a syringe of CO_2 as described in Part I.

2. Dispense some sodium hydroxide solution, NaOH(aq), into a small beaker.

3. Remove the syringe cap and draw 5 mL of NaOH solution into the syringe.

4. Securely place the syringe cap on the tip of the syringe.

5. Shake the syringe.

6. Record your observations.

Investigation 4

1. Obtain or generate a syringe of CO_2 as described in Part I.

2. Tape a candle to a glass stirring rod.

3. Hold the syringe upright with cap in place and plunger up.

4. Ignite the candle.

5. Remove the plunger from the syringe and carefully lower the lit candle into the syringe. See Figure 2.41.

6. Record your observations.

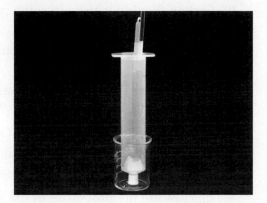

Figure 2.41 *A candle is attached to the base of a stirring rod and lowered into the syringe. (An unlit candle is shown here for illustration only.)*

Interpreting Evidence and Making Claims

1. A precipitate is an insoluble solid that separates from a liquid mixture. Was a precipitate formed in any of these investigations? How do you know?

2. Consider your results in Investigation 1.

   a. What do you think caused the changes that you observed?

   b. If the products formed in this investigation are calcium carbonate, $CaCO_3(s)$, and water, write a balanced equation for the reaction that takes place.

   c. The procedure in Investigation 1 is commonly used as a test for the presence of CO_2 gas. If a gas is bubbled through limewater and no cloudiness is apparent, would it be reasonable to claim that the gas is not CO_2? Explain.

3. Consider your observations during Investigation 2.

   a. Does CO_2 dissolve in the indicator solution? How do you know?

   b. How does CO_2 affect the pH of the solution?

4. Think about your results in Investigation 3.

   a. What claim could you make about the interaction of CO_2 and NaOH?

   b. What evidence supports your claim?

5. Compare your observations to your predictions for each investigation. For each prediction, explain why your reasoning was correct or incorrect. Be sure to address the property responsible for the experimental results.

Reflecting on the Investigation

6. Why was it important to remove excess acetic acid from the CO_2 you generated?

7. If Earth's atmosphere contained a higher concentration of CO_2, what might be some consequences?

8. Carbon dioxide is used in some types of fire extinguishers. Explain.

9. You may have learned that human lungs expel carbon dioxide. Describe two procedures that could provide evidence to support this claim.

10. What do you think would happen if you repeated the investigations in Part II using O_2 instead of CO_2? Explain.

11. Now consider the design of your air-quality experiment.

   a. Will it be necessary to generate a gas in your investigation?

   b. Which, if any, gases will you study in your investigation?

   c. How are the properties of the gases you identified in b similar to the properties of CO_2? How are they different?

   d. How will the properties of the gases you plan to study affect the design of your experiment?

   e. Will any of the procedures used in this investigation be necessary or applicable to your experimental design? Explain.

concept check 8

1. How does Avogadro's law help to interpret chemical equations involving gases?
2. What two requirements for the reaction of gas molecules are identified by collision theory?
3. What is a greenhouse gas?

C.9 THE ELECTROMAGNETIC SPECTRUM AND SOLAR RADIATION

In Developing Skills C.2, you constructed a graph of atmospheric temperature versus altitude. As you know, the temperature of a substance or mixture depends upon the average kinetic energy of its molecules. What is the source of energy for air molecules? Why does this energy vary with altitude? How do energy and matter—specifically air pollutants—interact within Earth's atmosphere? To answer these questions, and gain additional tools for posing your own questions, read on.

Figure 2.42 *Because the Moon lacks an atmosphere, it has very different temperatures on its Sun-facing side than it does on its other side.*

The Sun's radiant energy and the gases that make up Earth's atmosphere combine to maintain a hospitable climate for life on this planet. Without Earth's atmosphere, midday sunshine would heat a rock hot enough to fry an egg and the Sun's ultraviolet rays would burn exposed skin quickly; nights would be so cold that carbon dioxide gas would become solid.

Along with its vital role in storing the gases necessary for life, Earth's atmosphere provides protection from some of the Sun's ultraviolet radiation. Additionally, the atmosphere helps moderate Earth's temperature by controlling how much solar radiation is trapped close to Earth's surface (see Figure 2.42).

The enormous quantity of energy produced by the Sun, which is Earth's main external source of energy, is a result of the fusion of hydrogen nuclei into helium. **Nuclear fusion** (discussed in Unit 6) is the combining of two nuclei to form a new, heavier nucleus, with the accompanying release of energy. Nuclear fusion occurs at high temperature and pressure, powering the Sun and other stars (see Figure 2.43).

Solid carbon dioxide, $CO_2(s)$, is often called, somewhat misleadingly, dry ice.

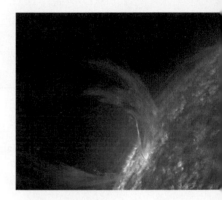

Figure 2.43 *The Sun is powered by nuclear fusion— a process that releases huge quantities of energy. This energy is used by growing plants, drives the hydrologic cycle, and warms Earth's atmosphere.*

Electromagnetic Radiation

Some energy liberated through nuclear fusion in the Sun travels to Earth as **electromagnetic radiation**. Electromagnetic radiation is composed of **photons**, or bundles of energy. Photons travel as waves and, as you might expect, move at the speed of light. Unlike sound waves or ocean waves, electromagnetic waves do not require a medium, or substance, to support their movement. Electromagnetic radiation can move through a vacuum as well as through air and other media (Figure 2.44).

The speed of light is about 3.0×10^8 m/s.

Figure 2.44 *Electromagnetic radiation is useful in a wide variety of applications including operation of household appliances and computers, collection of satellite data using radio telescopes, and chemiluminescent detection of forensic evidence.*

Several types of electromagnetic radiation make up the **electromagnetic spectrum**, with each type representing a particular energy range. Figure 2.45 shows major types of electromagnetic radiation. Radio waves and microwaves are at the low-energy end of the spectrum; X-rays and gamma rays are at the high-energy end of the spectrum. Between these extremes are infrared and ultraviolet radiation and the familiar visible spectrum, which is the only type of electromagnetic radiation that the unaided human eye can see.

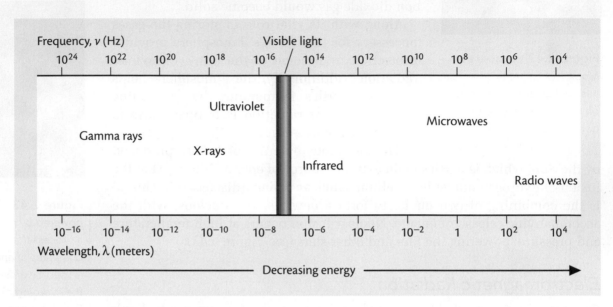

Figure 2.45 *The electromagnetic spectrum.*

Photons in each of these types of electromagnetic radiation have characteristic energy ranges related to their wave properties. All waves, including photon waves, involve oscillation. The rate of oscillation, or the number of waves that pass a given reference point per second, is called **frequency**. The frequency of a wave is directly proportional to its energy; high-frequency radiation is also high-energy radiation.

Another characteristic of waves is **wavelength**, shown in Figure 2.46. The distance between the tops (or any corresponding parts) of successive waves

> Wavelength and frequency are inversely proportional.

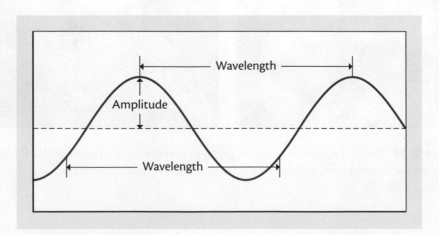

Figure 2.46 *Parts of a wave.*

is the wavelength. The wavelength and energy of a photon are inversely proportional; radiation with longer wavelengths is less energetic than is radiation with shorter wavelengths.

Photons can transfer their energy as they collide and interact with matter. The photon's energy, and thus its wavelength or frequency, largely determines its effect on living things and other types of matter.

The complete *solar spectrum* is shown in Figure 2.47. Most of the radiant energy emitted by the Sun is spread over a large portion of the electromagnetic spectrum: About 45% is in the infrared (IR) region, 46% is in the visible region, and 9% of this radiation is in the ultraviolet (UV) region. In the following paragraphs, you will consider these three regions of the solar spectrum in more detail. Less than 1% of solar radiation falls outside of these three regions.

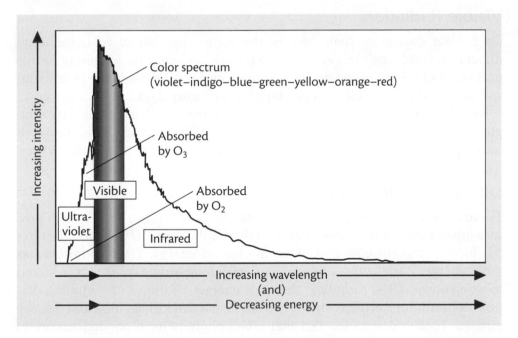

Figure 2.47 *The solar spectrum. Intensity, plotted on the y-axis, is an expression of the quantity of radiation at a given wavelength.*

CHEM**QUANDARY**

ALWAYS HARMFUL

How accurate is the following statement: "All radiation is harmful and should be avoided"?

Infrared Radiation

Electromagnetic radiation with a frequency slightly lower than that of red visible light is called **infrared (IR)** radiation. Infrared radiation causes some bonded atoms to vibrate more energetically, causing an increase in the material's temperature. For this reason, infrared heat lamps are sometimes used to keep cooked food warm before serving. Most of the infrared radiation from the Sun cannot reach Earth's surface because it is absorbed by CO_2 and gaseous H_2O molecules in Earth's atmosphere. However, some of the shorter-wavelength (higher-energy) solar radiation absorbed by Earth is transformed and reradiated as infrared energy. This re-radiated energy is reflected back and retained by the atmosphere, a fact that will be important later when we consider the role of CO_2 in the atmosphere.

> Microwave radiation has a frequency similar to that of infrared radiation, but it causes some molecules to vibrate more energetically, also increasing temperature.

> In the 1860s, John Tyndall demonstrated the ability of CO_2 and of gaseous H_2O to absorb radiant heat.

Visible Radiation

On a clear day, more than 90% of the visible portion of solar radiation directed toward Earth travels down to Earth's surface. The scattering of this portion of the Sun's radiation by water, air, and dust is the cause of red sunsets and blue skies, such as those depicted in Figure 2.48.

Visible radiation can energize electrons in some chemical bonds. An example of this is the interaction of visible light with electrons in chlorophyll molecules, which provides the energy needed for photosynthesis reactions.

> Photon–electron interactions also occur in molecules within your eyes, making it possible for you to see.

Ultraviolet Radiation

There are three subcategories of ultraviolet (UV) radiation, all of which possess greater energy than does visible light. Of the three, *UV-A radiation* has the longest wavelengths and, thus, the lowest energy. *UV-B radiation* has more energy; it can cause sunburn, and with long-term exposure it is linked to skin cancer. *UV-C radiation*, the most energetic form of ultraviolet radiation, is useful for sterilizing materials because it can kill bacteria and destroy viruses. This is due to the fact that UV-C photons have enough energy to break chemical bonds. As a result, chemical changes, including damage to living tissue, can occur in materials exposed to ultraviolet radiation.

Figure 2.48 *Some visible radiation from the Sun is scattered by Earth's atmosphere. This scattering creates the colors we see in the sky.*

UV-C consists of ultraviolet radiation with wavelengths shorter than 280 nm, UV-B wavelengths range from 280 to 320 nm, and UV-A radiation has wavelengths longer than 320 nm.

UV-C radiation is absorbed in the stratosphere before reaching Earth's surface. Most UV-B radiation, and much UV-A radiation, does not reach Earth's surface; it is absorbed by the ozone layer in the stratosphere. If all the UV radiation reaching the atmosphere actually reached Earth's surface, it is likely that most life on Earth would be destroyed. Ultraviolet radiation, however, is not all bad. Humans and animals must have some exposure to it because vitamin D is produced when the skin receives moderate doses of ultraviolet radiation included in sunlight.

> Specially designed light bulbs are available that produce particular forms of UV radiation suitable for specific applications such as tanning booths.

> Vitamin D is discussed further in Unit 7.

C.10 EARTH'S ENERGY BALANCE

The mild average temperature (15 °C, or 59 °F) at Earth's surface is determined partly by a balance between the inward flow to Earth of the Sun's energy and the outward flow into space of solar energy following its interaction with Earth and its atmosphere. Properties of Earth's surface and atmosphere help determine how much thermal energy our planet can hold near its surface—where, of course, terrestrial life resides—and how much energy Earth radiates back into space. The combination of these two factors helps establish a balanced energy flow, leading to a hospitable climate here on Earth.

About 30% of incoming solar radiation never reaches Earth's surface, but is reflected directly back into space by clouds and atmospheric particles. Solar radiation is also reflected when it strikes materials such as snow, sand, or concrete on Earth's surface. In fact, visible light reflected in this way allows Earth's illuminated surface to be seen from space.

> You will learn more about the hydrologic cycle in Unit 4.

Of the remaining 70% of incoming solar radiation that actually reaches Earth's surface, about two-thirds is absorbed, warming the atmosphere, oceans, and continents. The other one-third of this energy powers the hydrologic cycle, which is the continuous cycling of water into and out of the atmosphere by evaporation and condensation. Solar energy causes water to evaporate from the oceans and land masses. The water condenses to form clouds, which then release water back to Earth as precipitation (see Figure 2.49).

Figure 2.49 *One-third of the solar energy that reaches Earth powers the hydrologic cycle. Imagine how much thermal energy was generated as water vapor condensed in this developing thunderhead.*

Greenhouse Gases

All objects with temperatures above zero kelvins (0 K) radiate energy. The quantity of this radiated energy is directly related to an object's kelvin temperature. Specifically, Earth's surface re-radiates most absorbed solar radiation, but usually at longer wavelengths (lower energy) than that of the original incoming radiation. This re-radiated energy plays a major role in Earth's energy balance. Some types of molecules in the air do not absorb the Sun's UV and visible radiation, allowing it to reach the surface of Earth, but *do* absorb any infrared radiation that is re-radiated from Earth's surface, thus holding warmth in the atmosphere.

Carbon dioxide and water readily absorb infrared radiation, as do methane (CH_4), nitrous oxide (N_2O), and halogenated hydrocarbons such as CF_3Cl and other chlorofluorocarbons (CFCs). Because clouds are composed of droplets of water or ice, they absorb infrared radiation. Energy absorbed by these molecules in the atmosphere is re-radiated in all directions. Thus, energy can pass back and forth between Earth's surface and molecules in the atmosphere many times before it finally escapes to outer space.

This trapping and returning of infrared radiation by carbon dioxide, water, and other atmospheric substances is known as the **greenhouse effect**, because this process resembles, to some extent, the way thermal energy is held in a greenhouse (or in a closed car) on a sunny day. (See Figure 2.50.) Atmospheric gas molecules that effectively absorb infrared radiation are classified as **greenhouse gases**.

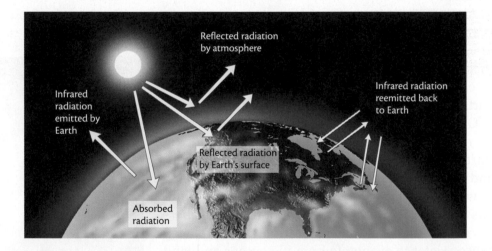

Figure 2.50 *When solar energy strikes Earth, clouds reflect about 30% back into space, and 70% moves through the atmosphere to reach Earth's surface. Energy that reaches the surface can be reflected or absorbed. Objects that absorb the Sun's UV and visible light reradiate it at lower-energy wavelengths such as infrared (IR) radiation, which is absorbed by atmospheric gases such as water, carbon dioxide, and methane (CH_4). IR radiation can be absorbed and re-radiated by atmospheric gases and Earth's surface several times before escaping into space. This "greenhouse effect" helps maintain warm temperatures on Earth.*

Without water and carbon dioxide molecules in the atmosphere to absorb and re-radiate thermal energy back to Earth, scientists estimate that our planet would have an average temperature of a frigid –18 °C (0 °F). At the other thermal extreme is the planet Venus, which demonstrates a runaway greenhouse effect. The Venusian atmosphere is composed of 96% carbon dioxide (and clouds made of sulfuric acid), which prevents the escape of most infrared radiation. As a result, the average surface temperature on Venus (450 °C) is much higher than that on Earth (15 °C). Although some of this difference is due to planetary positions relative to the Sun, Venus (the second planet from the Sun) is actually hotter than Mercury (the planet nearest the Sun).

Climate Connections

The interaction of solar radiation with Earth's atmosphere is a major factor in determining climates and weather. Radiant energy from the Sun warms Earth's land and water surfaces. Earth's warm surfaces, in turn, warm the air above them. As warmer air expands, its density decreases. This warmer air becomes displaced by colder, denser air, causing the warmer air to rise. These movements of warm and cold air masses help create continuous air currents that drive the world's weather (see Figure 2.51).

Climate, which refers to the average or prevailing weather conditions in a region, is influenced by other factors. One factor is Earth's rotation on its axis, which causes day and night and influences wind patterns. Other significant factors are Earth's revolution around the Sun and tilt on its axis. The combination of these factors causes uneven distribution of solar radiation, which results in four distinct seasons in Earth's mid-latitudes, a climatic pattern.

The differing thermal properties of materials on Earth's surface represent another factor influencing climate. Thermal properties include reflectivity and specific heat capacity.

Figure 2.51 *This hang glider depends on strong air currents as it is launched.*

DEVELOPING SKILLS

C.11 SOLAR RADIATION

Sample Problem: What type of electromagnetic radiation— ultraviolet, visible, or infrared radiation—is likely to be used in a sterilizing cabinet for safety goggles? Why?

Specially designed ultraviolet light bulbs are used in goggle sterilizing cabinets. The high-energy photons in ultraviolet radiation can kill bacteria and viruses.

Check your understanding of the interaction of radiation with matter and of radiation's role in controlling Earth's surface temperature by answering the following questions:

1. Why is human exposure to ultraviolet radiation potentially more harmful than exposure to infrared radiation?

2. Describe two essential roles played by visible solar radiation.

3. Explain why arid states, such as New Mexico and Arizona, experience wider air-temperature fluctuations from night to day than do states with more humid conditions, such as Florida.

4. Suppose Earth had a less dense atmosphere (fewer gas molecules) than it does now.

   a. How would average daytime temperatures be affected? Why?

   b. How would average nighttime temperatures be affected? Why?

5. Think about the air-quality issue that you plan to investigate in your unit project.

   a. What role does radiation play in this issue?

   b. Will radiation be controlled or variable in your experiment? Explain.

■ MAKING DECISIONS

C.12 MAKING CLAIMS AND PROVIDING EVIDENCE

You have now read about greenhouse gases and global warming and graphed the pressure and temperature of the atmosphere. You have also begun to plan a scientific study that would help advance atmospheric science. How do scientists draw conclusions from data gathered in investigations? How do they choose to organize and share those conclusions in a way that is compelling to the audience? Often, answering these questions requires re-examining the initial question, then deciding what claims can be made based upon the evidence collected in the investigation. A **claim** is a one- or two-sentence statement summarizing an important result of your investigation. Recall Investigating Matter A.9, in which you measured the change in volume as a result of the change in temperature. A claim made from the results of this investigation could be: If the temperature of a gas increases, then the volume of that gas increases. An unacceptable or inappropriate claim would be: The volume of gas was 4.5 cm^3.

To support your claims, it is critical that you provide **evidence**. Evidence is experimental support for claims. Evidence should be used to answer the questions, "How do I know what I know?" and "Why am I making this claim?" It should be presented in a usable format along with an explanation. Again recalling Investigating Matter A.9, your evidence would include a graph of length versus temperature and an explanation that the length of the air column represents (and is directly proportional to) the volume of gas trapped in the tube. The accompanying explanation should consist not only of balanced equations or mathematical calculations, but should also include interpretations of equations, data, and calculations. As you answer the following questions, focus on supporting, explaining, and interpreting claims, rather than listing or restating results.

Part I: Thinking about Claims

1. Identify a claim that you can make from Developing Skills C.2.

2. What evidence supports your claim in Question 1?

3. Do you think that another student would make the same claim based on the evidence you provided? Explain.

4. Look at your results from Investigating Matter C.8.

   a. What are some claims that you could make?

   b. What evidence would you use to support those claims?

Part II: Claims and Your Air-Quality Investigation

You have now developed and refined your initial question and considered the roles of KMT, models, and radiation in your air-quality investigation. At this point, it is important to think about possible answers to your initial questions, the evidence you would need to answer the initial question, and the methods that you would use to collect this evidence. Although you probably will not collect this evidence yourself, your procedure should be designed so that someone with access to the materials and equipment required could do so.

Remember that your final report will include both a poster outlining your planned investigation and a letter to those who would be interested in or affected by the results of your investigation. Keep these expectations in mind as you answer the following based upon your guiding question:

5. What claims do you anticipate being able to make?

6. What evidence would you need to be able to make those claims?

7. If you have not already done so, begin thinking about how you would gather evidence to answer your guiding questions.

   a. Write out a preliminary procedure for your investigation.

   b. Note any questions you have about developing the procedure or about planned steps. Save your procedures, notes, and questions.

   c. Will your planned investigation collect adequate and appropriate data to serve as evidence to make the claims you identified in Question 5? Explain.

SECTION C SUMMARY

Reviewing the Concepts

> Earth's atmosphere is composed of a mixture of gases.

1. List the percent values of the four most plentiful gases found in the atmosphere.

2. List three changes in the atmosphere as the altitude increases from sea level to high altitude.

3. a. Has the atmosphere's composition changed significantly throughout human history?

   b. What evidence can you cite to support your claim?

4. Sketch and label the four layers of the atmosphere.

> Molecules can break bonds and form new substances when they collide with sufficient energy and suitable orientation.

5. Does bond-breaking require or release energy?

6. Examine Figure 2.31 on page 211.

   a. Which have more potential energy, the reactants or the products?

   b. What does the activation energy barrier represent?

7. What is a catalyst?

8. Describe how a catalyst can speed up a chemical reaction.

> The coefficients in a balanced chemical equation that involves gases indicate the relative volumes of gaseous reactants or products.

9. This equation represents the production of ammonia (NH_3) by the reaction of nitrogen gas with hydrogen gas

$$N_2(g) + 3 H_2(g) \rightleftharpoons 2 NH_3(g)$$

   a. If 1 mol $N_2(g)$ reacts with 3 mol $H_2(g)$ in a flexible container at constant temperature and pressure, would you expect the total gas volume to increase or decrease? Why?

   b. How many moles of NH_3 would form if 2.0 mol N_2 react completely with hydrogen gas?

   c. How many grams of hydrogen gas would be needed to react completely with 2.0 mol N_2?

10. In a chemical reaction, 1 L hydrogen gas (H_2) reacts with 1 L chlorine gas (Cl_2) to produce 2 L hydrogen chloride gas (HCl). All volumes are measured at the same temperature and pressure. Create a depiction of that chemical reaction, using

    a. a sketch involving molecular models.

    b. a balanced chemical equation.

Electromagnetic radiation includes X-rays; gamma rays; ultraviolet (UV), visible, and infrared (IR) radiation; radio waves; and microwaves. The energy transmitted by radiation varies according to its wavelength—the shorter the wavelength, the higher the energy.

11. List the main types of electromagnetic radiation in order of increasing energy.

12. Describe how each type of radiation listed in your answer to Question 11 affects living things.

13. Write an equation or a sentence describing the relationship between the frequency of electromagnetic radiation and its energy.

14. Why is the word *spectrum* a good descriptor of the types of energy found in electromagnetic radiation?

15. Why is visible light useful in plant photosynthesis, while other forms of electromagnetic radiation are not?

16. Ultraviolet light is often used to sterilize chemistry-laboratory protective goggles. Why is ultraviolet light effective for this use, while visible light is not effective?

Some atmospheric gases, such as carbon dioxide, methane, and water vapor, absorb infrared radiation.

17. Describe how atmospheric CO_2 and water help maintain moderate temperatures at Earth's surface.

18. Identify both a natural process and a human activity that can increase the amount of

a. CO_2 in the atmosphere.

b. CH_4 in the atmosphere.

19. What changes in the composition of the atmosphere would cause the average surface temperature of the Earth to

a. increase?

b. decrease?

20. Explain why, on a sunny winter day, a greenhouse with transparent glass walls is much warmer than is a structure with opaque wooden walls.

21. Draw sketches to show how

a. a greenhouse works.

b. the global greenhouse effect works.

Earth's atmosphere protects living organisms by absorbing and distributing solar energy.

22. Compare infrared, visible, and ultraviolet radiation in terms of how well they are absorbed by the atmosphere.

23. Describe the two main effects of the solar radiation that reaches Earth's surface.

Scientific investigations are designed to collect evidence to make and support claims about the natural world.

24. What is a scientific claim?

25. Which of the following are stated as appropriate scientific claims?

a. CO_2 is formed in the reaction of sodium hydrogen carbonate and acetic acid.

b. 5.5 mL of CO_2 are formed.

c. CO_2 is the heaviest gas.

d. CO_2 dissolves in water.

e. Spilled baking soda caused the experiment to fail.

26. Why is simply displaying or restating data inadequate to support a scientific claim?

27. List at least three types of evidence that can be used to support a scientific claim.

What does evidence reveal about the properties of Earth's atmosphere?

In this section, you have encountered data collected in Earth's atmosphere and studied conclusions based upon these data. You have also considered processes and components of atmospheres, including mixtures and reactions of gases and the role of radiation. Think about what you have learned and how these ideas are related, then answer the question in your own words in organized paragraphs. Your answer should demonstrate your understanding of the key ideas in this section.

Be sure to consider the following in your response: structure (layers and composition) of the atmosphere, collision theory, Avogadro's law, properties of CO_2 and other gases, solar radiation, and energy processes/balance.

Connecting the Concepts

28. What would be the effect on the average global temperature if significant atmospheric increases occurred in
    a. carbon dioxide?
    b. methane?
    c. water vapor?
    d. How would such global temperature changes, in turn, affect Earth's atmosphere? Explain.

29. Many inexpensive sunglasses block visible light, but not ultraviolet radiation. Explain the hazards of wearing such sunglasses at the beach or on a ski slope during a bright, sunny day.

30. Consider the data you graphed in Developing Skills C.2.
    a. List two scientific claims you could make based on the data or the graph.
    b. For each claim, cite evidence for making that claim.

31. Collision theory states that molecules can break bonds and form new substances when they collide with sufficient energy and suitable orientation. What types of evidence could support each of the following aspects of collision theory?
    a. Formation of new substances.
    b. Requirement for sufficient energy.
    c. Requirement for suitable orientation.

Extending the Concepts

32. What evidence would be needed to support a claim that human activity has *not* influenced global climate?

33. In what ways is an analogy involving an "ocean of gases"
    a. useful in thinking about the atmosphere?
    b. not an accurate representation of the atmosphere's observed behavior?

34. Scientists use deep-ice samples from the Antarctic to estimate the atmospheric carbon dioxide concentrations many thousands of years ago. Research and prepare a report on how such ice samples are analyzed to obtain this information.

35. Oxygen gas is essential to life as we know it. Earth's atmosphere contains approximately 21% oxygen gas. Would a higher concentration of atmospheric oxygen gas be desirable? Explain.

36. Investigate and describe how night-vision goggles allow objects to become visible in low-light conditions.

37. Some animals "see" a portion of the electromagnetic spectrum other than the wavelengths to which human eyes respond. Investigate some examples and speculate on the benefit of these abilities to the animal(s) you research.

38. In terms of the kinetic molecular theory, describe changes in the behavior of gas molecules as they absorb electromagnetic radiation.

39. Investigate and describe characteristics of molecules that allow them to absorb various types of radiation.

SECTION D HUMAN IMPACT ON AIR QUALITY

How are claims about air quality supported by experimental evidence and chemistry concepts?

In the introduction to this unit, you read many claims about air quality. You may have seen similar claims in your local news or advertisements. How do you know whether an air-quality claim is accurate? What types of evidence and explanation are needed to support such claims?

So far in this unit, you have learned about the behavior of gases and interactions in Earth's atmosphere. In this section, you will analyze the effects of human activity on Earth's atmosphere. You will also consider how air-quality data are collected and decisions are made, particularly in the context of current problems such as smog and acid rain.

As you gain knowledge of air pollution and its consequences, critically examine the evidence and explanations you have proposed collecting and constructing for your own air-quality investigation. Are they adequate to answer your guiding questions? Do they support the production of an exemplary poster presentation? Would they be compelling to an audience outside of your class or school?

GOALS

- Identify primary and secondary air pollutants and their sources.
- Identify factors and contaminants that contribute to photochemical smog.
- Describe personal and global strategies that may help reduce air pollution.
- Explain why precipitation is naturally acidic and can become more acidic due to atmospheric contaminants.
- Describe effects of acidic precipitation (acid rain) and pH changes on natural systems.
- Describe the implications of experimental results.

concept check 9

1. What is air?
2. What are some common sources of air pollution near where you live?
3. What causes acid rain?

D.1 SOURCES OF AIR POLLUTANTS

As you learned in Section C (page 206), the overall makeup of Earth's tropospheric air has not changed significantly in thousands of years. The concentrations of nitrogen, oxygen, and trace gases remain essentially the same as they were during the last ice age, about 20 000 years ago. Despite this fact, local changes to air do take place on a regular basis. Such changes can impact the environment and human health. Some of these changes are natural, such as the emission of volcanic gases, while others result from human activity, often the burning of fuels. In either case, substances found in air that are not normal components of the atmosphere or that are present at elevated concentrations are considered **pollutants**. In the United States, polluted air is so common that weather reports for some cities include the levels of some contaminants in the air (see Figure 2.52).

Figure 2.52 *The downtown Denver skyline on a clear morning (top) and, two days later, shrouded in smog (bottom).*

Figure 2.53 *Small-scale, individual activities (such as mowing a lawn) can lead to buildup of air pollution in densely populated cities and towns.*

For safety reasons, an odoriferous substance is added to household natural gas, which is mainly methane; the methane itself is odorless.

Most human-generated air pollutants result from fossil-fuel-based energy production, including gasoline-powered vehicles and coal-based electricity generation. Depending on where you live, motor vehicles, power plants, and/or industries may contribute to local air pollution. See Figure 2.53.

Contaminated air often smells unpleasant and looks unsightly. Even when it cannot be smelled or seen, air pollution can be harmful. Air pollution causes billions of dollars of damage every year. It can corrode buildings and machines, and it can stunt the growth of agricultural crops and weaken livestock. Air pollution has been associated with diseases such as bronchitis, asthma (Figure 2.54), emphysema, and lung cancer, adding to human suffering and hospital costs worldwide.

Natural sources produce more of almost every major air contaminant (except SO_2) than do human sources. In most cases, natural air contamination occurs over wide regions and is seldom noticed. Furthermore, the environment may dilute, transform, or disperse such naturally emitted substances before they accumulate to harmful levels in the air.

By contrast, air pollutants from human activities are usually generated within localized areas, such as near smokestacks, exhaust pipes, or large population centers. When the amount of an air contaminant overwhelms the ability of natural processes to disperse or

Figure 2.54 *Air pollution can exacerbate asthma symptoms.*

dispose of it, air quality can become a serious problem. Many large cities are prone to high concentrations of air pollutants. If air pollutants generated by activity in large U.S. cities were evenly spread over the entire nation, they would have substantially lower concentrations and be much less noticeable. Unfortunately, weather-related air movement does not produce such even spreading.

Contaminants in air can be categorized as either *primary* or *secondary air pollutants*. **Primary air pollutants** directly enter the atmosphere, either as a result of a natural or human-initiated process. For example, some methane (CH_4) that enters the atmosphere is a by-product of fossil-fuel use and a component of natural gas. Methane is also produced naturally in large quantities by anaerobic bacteria as they break down organic matter, a process that occurs in wetlands, landfills, and within the digestive systems of animals such as cows and sheep.

Another primary air pollutant is actually a class of compounds known as **volatile organic compounds** (VOCs), a group of hydrocarbons that easily evaporate or are gaseous at room temperature. As their name indicates, hydrocarbons are molecules made up of C and H. VOCs may also contain smaller amounts of O, N, and other elements. Many VOCs released to the atmosphere are unburned fuel molecules, but release of these compounds has also been linked to natural processes (several types of trees are known to emit large amounts of VOCs—see Figure 2.55) and activities such as painting and dry cleaning. As you will later learn, VOCs are involved in smog formation.

Secondary air pollutants are formed in the atmosphere when primary air components react either with each other or with natural air components. For example, atmospheric sulfur dioxide (SO_2) and oxygen gas react to form sulfur trioxide (SO_3), a secondary air pollutant. Reactions with water in the atmosphere can convert sulfur trioxide to sulfates (SO_4^{2-}) or sulfuric acid (H_2SO_4), a secondary contaminant partly responsible for acid rain.

Particulate pollutants, as shown in Figure 2.56, include microscopic particles that enter the air from either human activities (such as power plants, waste burning, diesel combustion, road building, and mining) or natural processes (such as forest fires, wind erosion, and volcanic eruptions). Common particulate pollution includes emissions from smokestacks and vehicle tailpipes, which occasionally are observed as "smoke" but may not always be visible. In fact, virtually undetectable smaller particles that commonly are emitted from diesel-burning vehicles are more dangerous to human health because they can penetrate deep into the lungs.

Figure 2.55 *Urban planners need to balance the aesthetic and environmental benefits of planting trees with the quantity of VOCs released by trees, which varies widely from species to species.*

For the purposes of regulation, the U.S. Environmental Protection Agency (EPA) classifies some air pollutants as *hazardous air pollutants* or *criteria pollutants*. **Criteria pollutants** (see Table 2.5) are both commonly found throughout the United States and detrimental to human health or the environment. **Hazardous air pollutants**, sometimes called *air toxics*, are known or suspected to cause cancer or other serious health effects or adverse environmental effects. Although some are far more dangerous than criteria pollutants, they are generally much less common.

Nitrogen monoxide (NO)—also called nitric oxide—and nitrogen dioxide (NO_2) are sometimes referred to collectively as NO_x (pronounced "nocks").

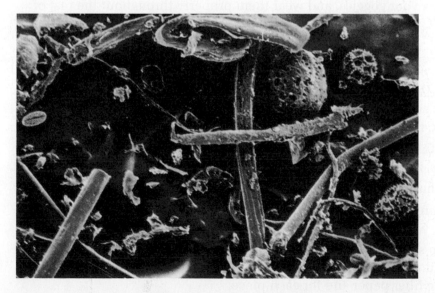

Figure 2.56 *Particulate contaminants found in air may include dust, pollen, and soot, all shown here under magnification.*

Table 2.5

U.S. EPA Criteria Pollutants
Carbon monoxide (CO)
Sulfur oxides (SO_x)
Nitrogen oxides (NO_x)
Ozone (O_3)
Lead (Pb)
Particulate matter (PM)

■ INVESTIGATING MATTER

D.2 DETECTING POLLUTANTS
IN AIR

Asking Questions

As you just learned, the EPA regulates allowable levels of pollutants in air. In order to enforce these regulations, the presence and concentration of air pollutants must be detected and measured. In this investigation, you will detect pollutants using chemical and physical measurements with tools prepared in the laboratory.

As you read *Gathering Evidence*, develop one or more scientific questions that you can answer in this investigation.

Preparing to Investigate

Before you begin, read *Gathering Evidence* to learn what you will need to do and note safety precautions. Plan for data collection based upon the question or questions you developed and construct an appropriate data table.

Gathering Evidence

Part I: Detecting Ozone

1. Identify at least three outdoor locations and a control location to measure ozone pollution. For this investigation, test sites must not be in direct sunlight. What is your independent variable? What other variables must you control? What variables are essentially the same for all sites?

2. Put on your goggles, and wear them properly throughout the rest of *Part I*.

3. Place 50 mL distilled water in a 150-mL beaker.

4. Add ~4 g cornstarch to the distilled water in the beaker.

5. Heat the beaker containing the mixture on a hot plate. Stir with a glass stirring rod and heat until the mixture gels; that is, it becomes thick and translucent.

6. Carefully remove the beaker from the hot plate and add ~1.5 g potassium iodide. Stir well.

7. Let the gelled solution cool for ~2 minutes.

8. Obtain four strips or pieces of filter paper. Place the filter paper on a large watch glass.

9. Use a small paint brush to uniformly apply the gelled solution to the filter paper as shown in Figure 2.57. Repeat on the other side of the filter paper and for each piece.

Figure 2.57 *Painting gel onto filter paper for ozone test strips.*

10. Devise methods for labeling and mounting your test strips. (*Note:* Strips must hang freely.)

11. Spray each strip with distilled water and install at ozone test sites and control site.

12. Leave each strip at its test site for ~30 minutes.

13. Collect the papers and store in a sealed plastic bag or glass jar out of direct light if results are not to be immediately observed.

14. Observe results by spraying each test strip with distilled water and comparing the strips to one another and the control strip. Record your results.

15. Wash your hands thoroughly before leaving the laboratory.

Part II: Detecting Particulates

16. Identify at least three outdoor locations (test sites) to measure particulate pollution. Make sure that the sites are chosen to design an investigation with an independent variable. What is that independent variable? What other variables must you control? What variables are essentially the same for all sites? What other factors determine whether a location is a good test site?

> Recall from Section A that an independent variable is manipulated by the researcher and the response of the dependent variable is then measured.

17. Obtain an index card for each of your test sites and an additional card to serve as a control. Cut a square hole, 3 × 3 cm, in the middle of each card.

18. Determine how you will mount your cards for sample collection. Again, consider the variables that you must control. Make any necessary alterations to the card for mounting without altering the square hole. Label each card with site information.

19. Place a piece of transparent packaging tape over each square hole as shown in Figure 2.58. Avoid touching the sticky side of the tape.

20. Mount the cards in the test sites with the sticky side facing outward.

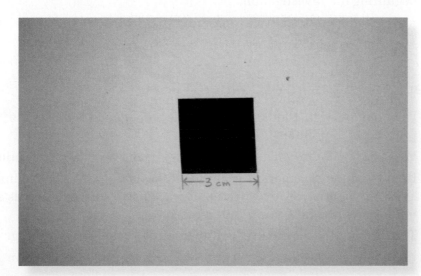

Figure 2.58 *Apply tape to the underside of an index card with a 3 x 3 cm square opening.*

21. Leave the cards in place at the test sites for at least one day. (*Note:* Cards can stay out several days, but should not be exposed to rain or snow. Temporarily retrieve the cards and store them inside in plastic bags if precipitation is likely.(

22. Collect the cards, taking care not to touch the sticky side of the tape. Cover the sticky portion with transparent packaging tape to seal particles between the two pieces of tape.

23. Observe the particle collectors under a microscope. Devise methods for quantifying and qualitatively describing the particles collected. Record your data.

Analyzing and Interpreting Evidence

1. Consider your ozone test strips:

    a. Which test site had the greatest concentration of ozone?

    b. Were there large differences in ozone concentration among the strips? Describe any differences you observed.

2. Explain your ozone test results in terms of possible sources of pollution.

3. Consider your particulate test cards:

    a. Which test site produced the card with the largest number of particulates?

    b. Were there large differences in particle density among the cards? Describe any differences you observed.

4. Explain your particulate test results in terms of possible sources of pollution.

Making Claims

5. Look again at the question or questions you developed before beginning the investigation.

    a. For each question, write one claim that addresses that question.

    b. What evidence supports each claim you made in part a?

Reflecting on the Investigation

6. Which of these methods used a chemical change to indicate the presence of a pollutant? Explain.

7. What factors, other than pollution, may have affected your results?

8. Based upon your experiences in this investigation, write a beginning question for another investigation using some of these materials and briefly describe how you would proceed.

9. Think about the air-quality investigation that you are designing.

    a. Would you need to *detect* or *measure* air pollutants? Explain.

    b. What procedures will you propose for detection or measurement?

D.3 MONITORING AND IDENTIFYING AIR POLLUTANTS

The methods you used to detect pollutants in Investigating Matter D.2 are similar to methods scientists still use to monitor air quality. The treated filter paper that you prepared is known as Schoenbein paper. The potassium iodide applied to the paper reacts with ozone in the troposphere according to the following equation:

$$2\,KI + O_3 + H_2O \longrightarrow 2\,KOH + I_2 + O_2$$

potassium iodide, ozone, water, potassium hydroxide, iodine, oxygen

When you add distilled water after exposure, iodine reacts with starch in the paper to form a blue starch–iodine complex. Thus, a darker blue color indicates that a larger amount of ozone was present.

Scientists use both simple chemical and physical methods, like the ones you used, and more sophisticated techniques. Methods of measuring air pollution can be divided into four categories: passive sampling, active sampling, automatic point monitoring, and remote open path monitoring, as described in Table 2.6.

You will use a starch–iodine complex again as an indicator in Unit 7.

Ambient means surrounding or encompassing; ambient air is a term used by the EPA and scientists to refer to outside air.

Table 2.6

Methods of Measuring Air Pollution				
Method	**Analysis site**	**Description**	**Advantages**	**Common materials measured**
Passive sampling	Laboratory	Collects pollutants through diffusion onto a filter or membrane; samples are sealed in the field	Reliable and cost-effective	Nitrogen dioxide, benzene
Active sampling	Laboratory	Collects samples by pumping air through a collector in the field for a known period of time	Allows for frequent sample collection; less expensive than automatic methods	Many particulate and gaseous pollutants
Automatic monitoring	Test site	Uses analytical methods to measure hourly pollutant concentrations at a single point	Continuous monitoring (real-time measurements); high-resolution	Ozone, nitrogen oxides, sulfur dioxide, carbon monoxide, particulates
Remote open path monitoring	Test site	Uses light waves to make concentration measurements of several pollutants at once	Continuous monitoring (real-time measurements); measures multiple pollutants	Sulfur dioxide, methane, ozone, nitrogen oxides

The current air-quality monitoring networks in the United States focus on the criteria pollutants, as well as visibility measurements and detection of acid rain precursors. There is currently no national air toxic monitoring network; however, each of approximately 300 monitoring sites provides data on some of 188 hazardous air pollutants (HAPs). Developing a national air toxic monitoring network will require cooperation among the EPA and state, tribal, and local agencies responsible for air-quality monitoring. See Figure 2.59.

Figure 2.59 *The National Air Monitoring Stations network emphasizes urban and multisource areas in placements of its 1080 stations.*

Ground-level sampling, as described above and practiced in Investigating Matter D.2, is crucial, but only tells part of the story. Sampling at higher altitudes is necessary to model atmospheric processes and determine the fate of pollutants after emission. To accomplish this, collection devices and monitoring systems must be transported to high altitude and must be designed to work under harsh conditions found there. Transportation options include specially designed balloons, retrofitted aircraft (see Figure 2.60), and even rockets. Atmospheres of other planets may also help scientists to understand changes in Earth's atmosphere, so spacecraft and satellites (see Figure 2.61) have been developed to study planetary atmospheres.

Figure 2.60 *Aircraft designed to sample air quality have sampling and data analysis instruments on board.*

All sampling and analysis procedures, including air-quality monitoring, seek to provide useful and valid data. In many cases, quantitative data are needed to track pollutant levels over time, compare sites, or provide inputs for computer modeling. The validity of the data depends on the *accuracy* and *precision* of the methods used to collect and analyze the samples and generate the data. **Accuracy** is the extent to which a measurement represents its corresponding actual values. That is, a measurement is accurate if a reading of 3.1 ppm is taken when the actual concentration of the pollutant is 3.1 ppm. **Precision**, on the other hand, describes how closely repeated measurements cluster around the same value. Measurements are valid only if they are both accurate and precise.

> Recall from Section C (page 206) that 3.1 ppm means that there are 3.1 particles of pollutant in every million particles air.

Accurate and precise   Precise but not accurate   Not accurate or precise   Accurate but not precise

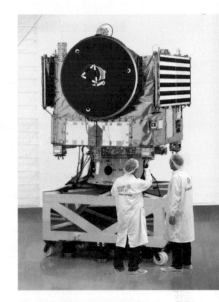

Figure 2.61 *Venus Express, launched in November 2005, is among the spacecraft designed to study extraterrestrial atmospheres. The data collected by such missions help scientists to understand the evolution of planets, atmospheres, and weather.*

DEVELOPING SKILLS

D.4 DRAWING CONCLUSIONS FROM AIR POLLUTION DATA

What proportion of total air contamination does human activity produce? Automobile use alone contributes about half the total mass of human-generated air contaminants. However, air is also contaminated each time we heat or cool a building, when we use fossil-fuel-based electricity, or when food and other products are delivered to retail stores.

Now that you have some ideas about the origins of air-quality data, this activity provides an opportunity to evaluate some of those data as well as human impacts on air quality. Table 2.7 (page 244) gives a detailed picture of sources of some major U.S. air contaminants. Use information provided in Table 2.7 to answer the following questions:

1. Consider the pollutants listed in Table 2.7 on the next page.
   a. What do all of these substances have in common?
   b. Which criteria pollutant is omitted?
   c. Why do you think it was omitted?

2. Overall, what is the main source of the air contaminants in Table 2.7 in the United States?

Selected U.S. Air Pollutants, 2005 (in 10^3 metric tons per year)

Source	CO	NO_x	VOCs	PM_{10}	SO_2	$PM_{2.5}$	Pb*	Totals**
Transportation	62 620	9807	6292	450	601	380	0.513	80 150
Fuel combustion								
Electric utilities	594	3498	44	567	9 497	461	0.065	14 661
Industrial	1150	1853	138	308	1 619	161	0.015	5229
Other (residential, commercial, institutional)	3035	665	1 248	420	525	382	0.374	6275
Industrial processes	3475	960	6800	1342	986	683	2.869	14 246
Miscellaneous (fires, agriculture and forestry, fugitive dust)	16 777	191	3602	16 223	122	2954	–	39 869
Totals	87 651	16 975	18 124	19 309	13 350	5021	3.559	

Key	CO	Carbon monoxide		$PM_{2.5}$	Particulate matter (<2.5 µm diam)
	NO_x	Nitrogen oxides		VOCs	Volatile organic compounds
	PM_{10}	Particulate matter (<10 µm diam)		SO	Sulfur dioxide

Source: U.S. EPA, National Emissions Inventory Data, 2005.
*Pb data from U.S. EPA, National Air Quality and Emissions Trends Report, 2001.
**Note: Totals for sources and pollutants in table. Additional sources and pollutants exist.

Table 2.7

3. For which contaminants is one-third or more contributed by
   a. industry?   b. transportation?   c. fuel combustion?

4. Based on data in Table 2.7, would the replacement of gasoline-fueled vehicles by gasoline–electric hybrid vehicles eliminate transportation as a source of sulfur dioxide (SO_2) emissions? Explain.

If air-quality decisions are to be made based upon measurements, it is important to have confidence in the validity of those data.

5. Are you confident that the data in Table 2.7 and the EPA air-quality data are valid? Explain.

6. Think about the data you gathered in Investigating Matter D.2.
   a. How confident are you in the validity of the data you collected? Why?

b. Were the data you collected accurate? How do you know?

c. Were the data you collected precise? Explain.

d. Would you feel comfortable using your data to make a decision about regulating pollutants? Why or why not?

7. How will you ensure that any data collected in the air-quality investigation you are designing will be valid?

✓ concept check 10

1. Why is the EPA more concerned about some air pollutants than about others?
2. Why are both accuracy and precision necessary for valid measurements?
3. How is air pollution influenced by weather and climate?

D.5 SMOG: HAZARDOUS TO YOUR HEALTH

In 1952, a deadly air pollution incident over London, England, lasted five days and contributed to the deaths of nearly 4000 people. Four years earlier in the coal-mining town of Donora, Pennsylvania (see Figure 2.62), pollutant-laden air killed 20 people and hospitalized hundreds of others. Similar, though less deadly, episodes have occurred historically in other cities. In each case, the culprit was a potentially hazardous condition called **smog**. Smog forms when large quantities of air pollutants are released into a concentrated area, usually a city, and meteorological conditions do not allow the pollutants to disperse. Indicators of smog include reduced visibility, irritation of eyes and lungs, and damage to materials. Fatality rates in severe smog episodes, such as those in London and Donora, have been higher than predicted from known hazards of sulfur oxides or particulates alone. According to some researchers, this increase may be due to **synergistic interactions**, where the combined effect of several substances is greater than the sum of their separate effects alone.

A brownish haze that irritated the eyes, nose, and throat and also damaged crops first appeared in the air above and around Los Angeles, California, in the 1940s. Researchers were puzzled for some time because Los Angeles has no significant industrial or heating activities, which were the primary pollutant sources in prior smog episodes. However, the city did have many motor vehicles and abundant sunshine. These factors contributed to the formation of *photochemical smog*. The geography of Los Angeles, bordered by mountains on three sides, was another factor. Although its valley location makes Los Angeles smog-prone, the smog experienced there was much worse than seemed reasonable. There had to be more to this story.

The word *smog* is a combination of *smoke* and *fog*.

Figure 2.62 *Donora, Pennsylvania, at midday in late October, 1948.*

Normally, air at Earth's surface is warmed by solar radiation and by re-radiation from surface materials. This warmer, less dense air rises, carrying pollutants away with it. Cooler, less polluted air then moves in below. In a **temperature inversion**, a cool air mass is trapped beneath a less dense warm air mass, often in a valley or over a city (see Figure 2.63).

(a) **NORMAL CONDITIONS**

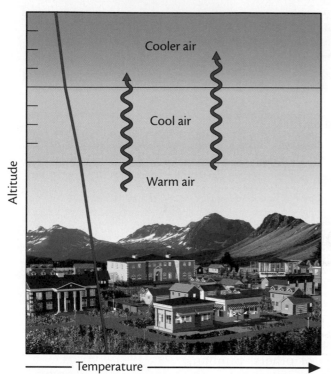

(b) **TEMPERATURE INVERSION**

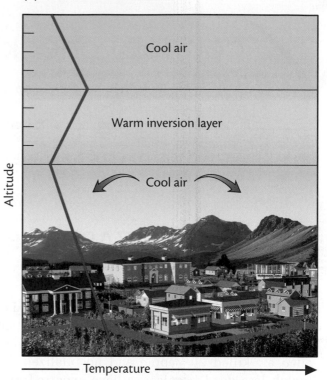

Figure 2.63 *A temperature inversion. Unlike normal conditions (a), in a temperature inversion (b), a layer of cool air is trapped close to Earth's surface below a warm layer, preventing air contaminants from escaping.*

Any reaction initiated by light is a photochemical reaction.

In Los Angeles, the combination of sunny weather and mountains can produce temperature inversions about 320 days annually. During a temperature inversion, air pollutants cannot escape, so their concentrations may rise to dangerous levels. The production and severity of photochemical smog, which can occur even without a temperature inversion (particularly in regions congested with automobiles, such as Los Angeles), are amplified by that accumulation of pollutants.

The simplified equation that follows represents the key ingredients and products of **photochemical smog**. Hydrocarbons, carbon monoxide, and nitrogen oxides from vehicle exhausts are irradiated by sunlight in the presence of oxygen gas. The resulting reactions produce a potentially dangerous mixture, including other nitrogen oxides, ozone, and irritating organic compounds, as well as carbon dioxide and water vapor:

$$\text{Hydrocarbons} + \text{Sunlight} + O_2(g) + CO(g) + NO_x(g) \longrightarrow O_3(g) + NO_x(g) + \text{Organic compounds} + CO_2(g) + H_2O(g)$$

(auto exhaust)   (oxidizing agents and irritants)

Nitrogen dioxide has a pungent, irritating odor at detectable concentrations. Even at relatively low concentrations (0.5 ppm), nitrogen dioxide can inhibit plant growth. At 3 to 5 ppm, this pollutant can cause human respiratory distress after an hour of exposure.

Ozone is a very powerful oxidizing agent (oxidant). At concentrations as low as 0.1 ppm, ozone can crack rubber, corrode metals, and damage plant and animal tissues. Ozone may also cause chest pain, shortness of breath, throat irritation, and coughing; worsen chronic respiratory diseases such as asthma; and compromise the body's ability to fight respiratory infections.

Because substances in smog can endanger health, their levels in the air are of major public interest; many weather forecasters report an *air-quality rating*, along with humidity and temperature data. The EPA has devised the Air-Quality Index (AQI) based on concentrations of pollutants that are major contributors to smog in metropolitan areas. The concentrations in air of substances associated with photochemical smog customarily vary over a 24-hour period. In the following activity, you will consider factors that may influence such concentration changes.

> Recall that the concentration unit "ppm" means "parts per million."

> You may have smelled ozone after a lightning storm or near an electrical motor. It is also generated, either intentionally or as a by-product, by some "air cleaners."

DEVELOPING SKILLS

D.6 VEHICLES AND SMOG

Use the data in Figure 2.64 to answer the following questions:

1. Consider pollutant level patterns.

   a. Between what hours do the concentrations of nitrogen oxides and hydrocarbons peak?

   b. Account for this fact in terms of automobile use.

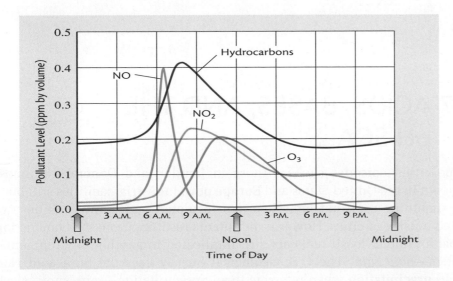

Figure 2.64 *Photochemical smog formation.*

2. List two reasons why a given pollutant may decrease in concentration over a period of several hours.

3. Although ozone is necessary in the stratosphere to protect Earth from excessive ultraviolet light, at Earth's surface it is a major component of photochemical smog.

   a. Using Figure 2.64, determine which substances are at their minimum concentrations when O_3 is at maximum concentration.

   b. What does this suggest about the production of O_3 in polluted tropospheric air?

4. Some state and local agencies declare Air-Quality Action Days when the levels of ozone are expected to be unhealthy. Explain why each of the following recommendations is made for such days:

   a. Use public transportation.

   b. Refuel vehicles only after dusk.

   c. Avoid prolonged or strenuous exertion outdoors.

   d. Do not burn leaves or other yard waste.

> In some communities, public transportation is free on Air-Quality Action Days.

CHEM**QUANDARY**

CONTROLLING AIR POLLUTION

From a practical viewpoint, why do you think that controlling air pollution from all U.S. automobiles, recreational vehicles, buses, and trucks would be more difficult than controlling air pollution from U.S. power plants and industries.

D.7 ACIDS, BASES, AND THE pH SCALE

> Smog is still an important health and environmental issue in many developing countries.

Air pollution episodes such as those in London and Donora are almost unknown in the United States and Europe now. Industrial facilities still burn fuel, including coal, but emissions are regulated and many facilities are located outside of cities. However, pollutants resulting from the combustion of coal and other fossil fuels are still implicated in another environmental problem—*acid rain*. The U.S. Geological Survey uses the term **acid rain** or **acid precipitation** when referring to any precipitation—rain, snow, sleet, or hail—that has a pH lower than what is natural for the area. The average pH of natural, unpolluted precipitation is 5.6, so precipitation with a pH value lower than 5.6 is considered acid rain.

You have probably heard the term *pH* used before, perhaps in connection with diets or cosmetics. The term was also used in Investigating Matter C.8 (page 218) when you tested the acid-base properties of carbon dioxide. What is pH, and how is it related to air quality? The **pH scale** is a convenient way to measure and report the acidic, basic, or chemically neutral character of a solution. Nearly all pH values are in the range from 0 to 14, although some extremely acidic or basic solutions may be outside this range. At room temperature, any pH value less than 7.0 indicates an acidic condition; the lower the pH, the more acidic the solution. Solutions with pH values greater than 7.0 are basic; the higher the pH, the more basic the solution. Basic solutions are also called **alkaline** solutions.

Quantitatively, a change of one pH unit indicates a tenfold difference in acidity or alkalinity. For example, lemon juice, with a pH of about 2, is nearly 10 times as acidic as soft drinks, which have a pH of about 3. This also means a change of 2 pH units results in a hundredfold (10×10) difference in acidity or alkalinity.

Acids and bases, some examples of which are shown in Figures 2.65 and 2.66, and listed in Table 2.8 (page 250), exhibit characteristic properties. For example, basic solutions react with the vegetable dye litmus to turn it blue, whereas acidic solutions turn litmus red. Both acidic and basic solutions conduct electricity. Each type of solution has a distinctive taste and a distinctive feel on your skin. (**Caution:** *Never test these sensory properties in the laboratory.*) In addition, acids and bases are able to react chemically with many other substances. You are probably familiar with the ability of acids and bases to corrode materials, which is a type of chemical behavior.

Most **acids** are made up of molecules, including one or more hydrogen atoms that can be released rather easily in water solution. These "acidic" hydrogen atoms are usually written first in the formula for an acid. See Table 2.8 on the next page.

> Vinegar is an acid you probably have tasted; that common kitchen ingredient is considered a dilute solution of acetic acid.

Figure 2.65 *Use Table 2.8 (page 250) to identify the bases contained in this oven cleaner and other common household items.*

Figure 2.66 *Vinegar contains acetic acid, and some soft drinks contain both carbonic acid and phosphoric acid. Many fruits and vegetables also contain acids.*

Table 2.8

Names, Formulas, and Common Uses of Some Acids and Bases		
Name	**Formula**	**Common Uses**
Acids		
Acetic acid	$HC_2H_3O_2$	In vinegar (typically a 5% solution of acetic acid)
Carbonic acid	H_2CO_3	In carbonated soft drinks
Hydrochloric acid	HCl	Used in removing scale buildup from boilers and for cleaning materials
Nitric acid	HNO_3	Used in making fertilizers, dyes, and explosives
Phosphoric acid	H_3PO_4	Added to some soft drinks to give a tart flavor; also used in making fertilizers and detergents
Sulfuric acid	H_2SO_4	Largest-volume substance produced by the chemical industry; used in automobile battery fluid
Bases		
Calcium hydroxide	$Ca(OH)_2$	Present in mortar, plaster, and cement; used in paper pulping and removing hair from animal hides
Magnesium hydroxide	$Mg(OH)_2$	Active ingredient in milk of magnesia
Potassium hydroxide	KOH	Used in making some liquid soaps
Sodium hydroxide	NaOH	A major industrial product; active ingredient in some drain and oven cleaners; used to convert animal fats into soap

Many **bases** are ionic substances that include hydroxide ions (OH^-). Sodium hydroxide, NaOH, and barium hydroxide, $Ba(OH)_2$, are two examples. Some bases, such as ammonia (NH_3) and baking soda (sodium hydrogen carbonate, $NaHCO_3$), contain no OH^- ions but still produce basic solutions because they react with water to generate OH^- ions, as illustrated by the following equation.

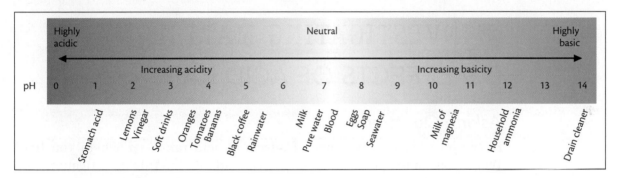

$$NH_3 \quad + \quad H_2O \quad \longrightarrow \quad NH_4^+ \quad + \quad OH^-$$

What about substances that display neither acidic nor basic characteristics? Chemists classify these substances as chemically neutral. Water, sodium chloride (NaCl), and table sugar (sucrose, $C_{12}H_{22}O_{11}$) are all examples of chemically neutral compounds. At 25 °C, a pH of 7.0 indicates a chemically neutral solution. The pH values of some common materials are shown in Figure 2.67.

Highly acidic							Neutral						Highly basic

Increasing acidity   Increasing basicity

pH   0   1   2   3   4   5   6   7   8   9   10   11   12   13   14

Stomach acid · Lemons Vinegar · Soft drinks · Oranges Tomatoes Bananas · Black coffee · Rainwater · Milk · Pure water Blood · Eggs Soap Seawater · Milk of magnesia · Household ammonia · Drain cleaner

Figure 2.67 *The pH values of some common materials.*

D.8 ACID RAIN

As you can see in Figure 2.67, rainwater is naturally slightly acidic. Is this consistent with data you collected when you tested carbon dioxide (CO_2) in Investigating Matter C.8 (page 216)? The pH of rainwater is normally slightly acidic, about 5.6, because the atmosphere contains some substances that produce acidic solutions when dissolved in water. The acidic pH of natural precipitation is mainly due to reaction of carbon dioxide with water in the atmosphere, forming carbonic acid, H_2CO_3:

$$CO_2(g) \quad + \quad H_2O(l) \quad \rightleftharpoons \quad H_2CO_3(aq)$$

This dilute solution of carbonic acid then falls to Earth as rain, snow, or sleet.

Other natural events can contribute to the acidity of precipitation. Volcanic eruptions, forest fires, and lightning bolts produce sulfur dioxide (SO_2), sulfur trioxide (SO_3), and nitrogen dioxide (NO_2). These gases can then react with atmospheric water in much the same way that carbon dioxide does:

$$SO_2(g) \quad + \quad H_2O(l) \quad \rightleftharpoons \quad H_2SO_3(aq)$$
Sulfurous acid

$$SO_3(g) \quad + \quad H_2O(l) \quad \rightleftharpoons \quad H_2SO_4(aq)$$
Sulfuric acid

The expression *acidic precipitation* is sometimes used instead of *acid rain* to indicate that any form of precipitation—fog, sleet, snow, or rain— can be acidic.

Double arrows indicate that both forward and reverse reactions occur simultaneously. You can learn more about *reversible reactions* in Unit 4.

The actual processes that generate acid rain are more complex than depicted here.

$$2\,NO_2(g) \;+\; H_2O(l) \;\rightleftharpoons\; HNO_3(aq) \;+\; HNO_2(aq)$$
Nitric acid   Nitrous acid

The result, as the previous equations suggest, is the formation of acidic precipitation.

If rain customarily is acidic, why does acid rain pose a problem? The fact that this precipitation is *more* acidic than normal leads to the problems associated with acid rain. In the following investigation, you will model the formation of acid rain in order to consider its effects on a variety of materials.

■ INVESTIGATING MATTER
D.9 EFFECTS OF ACID RAIN

Asking Questions

Have you seen any evidence of effects of acid rain near where you live? What types of materials are damaged by acid rain? In this investigation, you will model the effects of acid rain on plants, metals, and stones. To do this, you will mix sodium sulfite (Na_2SO_3) and hydrochloric acid (HCl) to form sulfur dioxide and water according to the following equation:

$$Na_2SO_3 \;+\; 2\,HCl \;\longrightarrow\; SO_2 \;+\; H_2O \;+\; 2\,NaCl$$

By trapping the SO_2 gas in a closed environment, you can test its effects—and thus model the effects of acid rain—on materials within that environment.

What materials can you use in the laboratory to model plants, metals, or stones? What characteristics must these materials have in order to be both useful in the investigation and adequate models of the materials that they represent? As you read *Gathering Evidence*, answer these questions and develop one or more scientific questions about the effects of acid rain that you can answer based upon this model.

Preparing to Investigate

Before you begin, read *Gathering Evidence* to learn what you will need to do and note safety precautions. Plan for data collection based upon the question or questions you developed and construct an appropriate data table. In your data table, be sure to clearly identify the materials you have chosen to represent plants, metal, and stones.

Gathering Evidence

1. Before you begin, put on your goggles, and wear them properly throughout the investigation.

2. Set up a Petri dish with materials as shown in Figure 2.68. Use a few drops of universal indicator and place a small quantity, about the size of a match head, of sodium sulfite (Na_2SO_3) inside a shallow weighing boat or vial cap in the center of the Petri dish.

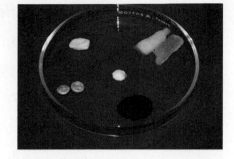

3. Place two drops of phenol red on top of your stone sample.

4. Place one drop of distilled water on top of your metal sample.

5. With the Petri dish lid ready, place a few drops of HCl onto the sodium sulfite in the weighing boat or vial cap. Replace the Petri dish lid.

Figure 2.68 *Typical Petri dish setup.*

6. Make observations of anything that happens within the Petri dish and record those observations.

7. Set the Petri dish aside to make observations later.

8. After 20 minutes, record your additional observations.

9. Dispose of the contents and clean the Petri dish as directed by your teacher.

10. Wash your hands thoroughly before leaving the laboratory.

Interpreting Evidence

1. SO_2 gas was formed in the Petri dish.

   a. What effect did this gas have on the universal indicator?

   b. What evidence supports your answer to part a?

2. As you have learned, precipitation with a pH < 5.6 is defined as acid rain. How does the pH inside your Petri dish compare to this value?

Making Claims

3. What effect might an atmosphere similar to that within your Petri dish have if it were to surround

   a. steel girders?

   b. a forest or field?

   c. What evidence can you give for your answers to parts a and b?

4. Consider the reaction between your "acid rain" (H_2SO_3) and a substance you may have chosen for the stone sample, marble chips ($CaCO_3$).

   a. Write an equation for the reaction. (*Hint:* Carbon dioxide and calcium sulfite solution are two of three products formed.)

   b. What conclusions can you draw about the effects of acid rain on marble statues and building materials from the reaction in part a?

5. Review the question or questions you developed before beginning the investigation.

   a. For each question, write a claim that addresses that question.

   b. What evidence supports each claim you made in part a?

> Recall from Unit 1 that properly written chemical equations are always balanced.

Reflecting on the Investigation

6. This investigation uses a model of acid rain production to investigate its effects. Describe, with both an equation and an explanation, how the production of acid rain is modeled.

7. How would you design an experiment to determine whether the changes you noted within the Petri dish were actually caused by reactions of SO_2 gas?

8. Is the short-term, intense production of acid in this investigation a reasonable model for long-term damage caused acid precipitation? Explain.

9. Would it be useful to develop and use a small-scale model, similar to the Petri dish model of acid rain, for your planned air-quality investigation? Explain.

D.10 CONSEQUENCES AND CONTROL OF ACID RAIN

In Investigating Matter D.9, you observed the effects of acid rain on materials in a closed system. What are the consequences of acid rain in Earth's environment? When sulfur oxides and nitrogen oxides are emitted from power plants, various industries, and fossil-fuel-burning vehicles, they can react with water vapor to form acids. These acids lower the pH of rainwater—at times to 4.5 or lower in the northeastern United States. See Figure 2.69. Similar changes in pH have been observed for precipitation in Scandinavia, Central Europe, and other areas downwind from large industrial centers and power plants.

You have observed that acid rain can directly damage plant material. Tree leaves and needles can be harmed when acid rain falls and when they are surrounded by acid clouds and fog that are even more acidic than rainfall. Acid rain also damages plants by leaching needed minerals from the soil and releasing toxic substances into the soil. See Figure 2.70. Excessively acidic rain can also lower the pH of lakes and streams, enough to kill fish eggs and other aquatic life.

As you discovered in Investigating Matter D.9, damage caused by acid rain is not limited to effects on natural ecosystems. Most buildings and other structures contain metal, limestone, or concrete, which are all materials susceptible to damage by acids. Statues and monuments (such as the Parthenon in Greece) that have stood for centuries show signs of significant surface damage, due, in part, to acid rain. Acid attacks calcium carbonate in limestone, marble, and cement according to this equation:

> Aluminum (Al) is one substance released in soil by acid rain that is toxic to plants and trees.

$$H_2SO_4(aq) \ + \ CaCO_3(s) \ \longrightarrow \ CaSO_4(s) \ + \ H_2O(l) \ + \ CO_2(g)$$

| Sulfuric acid (in acid rain) | Calcium carbonate | Calcium sulfate | Water | Carbon dioxide |

Hydrogen ion concentration as pH from measurements made at the Central Analytical Laboratory, 2008

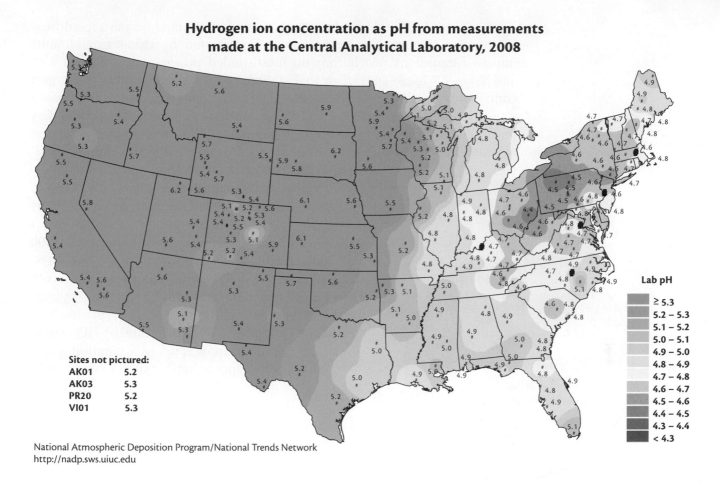

Sites not pictured:

AK01	5.2
AK03	5.3
PR20	5.2
VI01	5.3

Lab pH

- ≥ 5.3
- 5.2 – 5.3
- 5.1 – 5.2
- 5.0 – 5.1
- 4.9 – 5.0
- 4.8 – 4.9
- 4.7 – 4.8
- 4.6 – 4.7
- 4.5 – 4.6
- 4.4 – 4.5
- 4.3 – 4.4
- < 4.3

National Atmospheric Deposition Program/National Trends Network
http://nadp.sws.uiuc.edu

Figure 2.69 *pH of precipitation throughout the continental United States (National Atmospheric Deposition Program).*

Calcium sulfate is much more soluble in water than is calcium carbonate. Thus, as calcium sulfate forms, it washes away, uncovering fresh solid calcium carbonate that reacts further with acid rain. See Figure 2.71.

Given these consequences, many nations, states, and communities have acted to limit emissions of gases that cause acid rain. Control of acid rain is difficult, however, because air pollution does not respect political boundaries. For example, sulfur oxides can be carried by air currents for several days, so acid rain often shows up hundreds of kilometers from the original sources of air contamination.

Figure 2.70 *This forest in the Czech Republic has been badly damaged by acid rain. Several forests within the United States show similar damage.*

Figure 2.71 *Acid rain has disfigured these limestone gargoyles on the Cathedral of Notre Dame in Paris, France.*

The Clean Air Act Amendments of 1990 were enacted, in part, to address issues of acid rain. The compliance plan created by these amendments imposed emissions restrictions on fossil-fueled power plants. Figure 2.72 and Table 2.7 (page 244) show that stationary fuel combustion is the largest contributor to SO_2 emissions. National SO_2 emissions have decreased steadily since the early 1980s. Unlike SO_2 emissions, which are produced by burning fuels containing sulfur, nitrogen oxide (NO_x) emissions are formed by reaction of atmospheric nitrogen gas with oxygen gas at the high temperatures associated with most forms of combustion—including internal combustion engines.

Figure 2.73 indicates that fuel combustion by power plants is joined by transportation as major sources of nitrogen oxide emissions. Between 1990 and 2007, national SO_2 emissions decreased 44%, while total national NO_x emissions were 33% lower. The EPA estimates that reducing NO_x emissions from fixed-location sources, such as power plants, is a cost-effective strategy. But the issue of NO_x emissions from mobile sources, such as automobiles and other modes of transportation, is much more complicated.

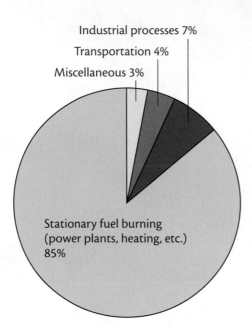

Figure 2.72 *SO_2 emissions by source, 2005. EPA National Emissions Inventory data.*

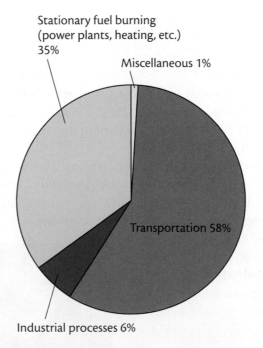

Figure 2.73 *NO_x emissions by source, 2005. EPA National Emissions Inventory data.*

CHEM**QUANDARY**

THE RAIN IN MAINE...

The state of Maine has no coal-fired power plants and relatively few residents. (The average population density of Maine is approximately 41 people per square mile; Maine's largest city, Portland, has a population of about 65 000 residents.) However, Figure 2.69 (page 255) indicates that the average pH of precipitation in the state is lower than that of rainfall free of air contaminants (pH ~5.6). Why might this be?

■ MAKING DECISIONS

D.11 IMPROVING EARTH'S AIR QUALITY

As you have learned, human activity can create large-scale impact on gases surrounding our planet. Such impact may influence the world's average temperature or increase potential exposure of living systems to ultraviolet radiation by thinning the protective stratospheric ozone layer. Air pollution can also lead to regional effects, such as acid rain and smog, or even to localized problems, such as the potential health hazards associated with elevated levels of air pollutants.

Atmospheric chemistry is quite complex and difficult to study. However, increased knowledge of chemical and physical processes involved in the atmosphere, all gained by scientific research, already has helped guide decision making to protect and enhance the vital envelope of gases encircling Earth.

After preparing a poster that describes the investigation you designed to investigate an air-quality issue or claim, you will write a letter to members of your local or national community summarizing your investigation. Think about this letter and your air-quality investigation as you answer the following questions:

1. What organizations or groups of people would be interested in the results of your investigation?

2. How could you communicate the results to these people or groups?

3. How would the results of your investigation be useful in improving Earth's air quality?

SECTION D SUMMARY

Reviewing the Concepts

> Air pollution results from contributions by both primary and secondary pollutants.

1. What distinguishes a primary air pollutant from a secondary air pollutant?

2. Characterize each of the following pollutants as primary or secondary and identify its major source.

   a. CO_2 c. NO_x

   b. CH_4 d. H_2S

3. Write and balance the equation for formation of SO_3 from SO_2 and oxygen.

4. What are criteria pollutants?

> Pollutant concentrations in ambient air are measured using various techniques in order to provide data for air-quality compliance and modeling.

5. Compare passive and active air-sampling methods.

6. How are air-quality measurements made at high altitude?

7. What features are required for valid measurements?

> Photochemical smog can intensify due to temperature inversions and adverse wind patterns.

8. What ingredients are needed to create photochemical smog?

9. What conditions are shared by cities that suffer from photochemical smog?

10. Are all cities subject to temperature inversions? Explain your answer.

11. What are some ways that photochemical smog can be reduced?

> Solutions can be characterized as acidic, basic, or chemically neutral on the basis of their observed properties. Water solutions with pH 7.0 at room temperature are neutral; those with lower pH are acidic and those with higher pH are basic.

12. What ion is found in many bases?

13. What element is found in most acids?

14. Classify each sample as acidic, basic, or chemically neutral:

   a. seawater (pH = 8.6)

   b. drain cleaner (pH = 13.0)

   c. vinegar (pH = 2.7)

   d. pure water (pH = 7.0)

15. Using Figure 2.67 on page 251, decide which is more acidic:

   a. a soft drink or a tomato.

   b. coffee or seawater.

   c. milk of magnesia or ammonia.

16. Coffee has a pH of 4.0 and pure water a pH of 7.0. How many more times acidic is coffee than pure water?

> Rainwater is naturally acidic, due to dissolved CO_2, but contaminants in the atmosphere can produce precipitation that is even more acidic than normal.

17. What is the pH range of

   a. typical rainwater?   b. acid rain?

18. Provide an explanation (including a chemical equation) for why rain is naturally acidic.

19. Cite evidence from Investigating Matter D.9: Effects of Acid Rain, page 252, that low pH has a harmful effect on plant material.

Gaseous sulfur oxides and nitrogen oxides generated from natural and human sources contribute to acid rain.

20. List two natural sources of sulfur oxides and nitrogen oxides and two human-generated sources of these two pollutants.

21. What strategies can reduce emissions of sulfur oxides and nitrogen oxides to the atmosphere?

22. Identify the major acidic components in acid rain.

23. Write a balanced equation representing what happens when SO_3 gas dissolves in water.

Air pollution can be affected by everyday activities and changes in chemical technologies.

24. List five ways that air pollutants from motor vehicles and industries can be reduced.

25. Listed below are some examples of decisions that one might make to reduce personal contributions to air pollution. For each of these examples, identify the particular air pollutant(s) involved and explain how their release will potentially be reduced.

    a. Ordering a fast-food lunch inside a restaurant, rather than waiting in a line of cars at the drive-up window.

    b. Hanging wet laundry outside to dry.

    c. Avoiding "topping off" a car's gas tank by adding more gasoline once the pump has stopped automatically.

    d. Buying locally grown groceries and other locally produced items.

    e. Turning off lights, the TV, and other electrical devices (computer, music, and DVD players) when exiting rooms.

26. Green Chemistry seeks to prevent pollution before it starts. What are advantages of this strategy over combating existing air pollution?

How are claims about air quality supported by experimental conclusions and chemistry concepts?

In this section, you have examined air pollution data from a variety of investigations and sources. You have learned about some chemical reactions that take place in Earth's atmosphere and the conditions that lead to these reactions. Think about what you have learned, then answer the question in your own words in organized paragraphs. Your answer should demonstrate your understanding of the key ideas in this section.

Be sure to consider the following in your response: sources of air pollutants, validity of data, air monitoring and collection methods, smog, and acid rain.

Connecting the Concepts

27. Explain why a substance that could be called "hydrogen hydroxide" has neither acidic nor basic properties. What is a more familiar name of this compound?

28. How might monitoring practices be different for primary and secondary air pollutants?

29. What are advantages and disadvantages of scrubbing as a strategy for air-pollution reduction?

30. Why is the presence of ozone regarded as a problem at ground level, while the absence of ozone is regarded as a problem in the stratosphere?

31. What pattern would you expect to find among pH readings taken of precipitation in the immediate vicinity of an acid-rain producing smokestack and at regular intervals downwind and upwind from the site? Sketch a graph of what you would expect to find.

32. Considering the identity of the criteria pollutants and the sources of air pollutants in the United States (Table 2.7, page 244), where should air-quality monitoring devices be located? Explain.

33. Think about the investigations you have completed throughout this unit. Why is it important to:

    a. Compare your results to those of your classmates?

    b. Compare your results to published data, when available?

    c. Discuss your ideas with your classmates?

    d. Reflect upon how your ideas have changed?

Extending the Concepts

34. Consider these three air-pollutant control technologies: scrubbers, catalytic converters, and electrostatic precipitators. Which technique is most effective in combating each of these types of air pollutants?

    a. sulfur oxides

    b. particulates

    c. hydrocarbons

35. Use the Internet and library to research and report on serious acid rain effects over the past decade.

36. The pH of rainwater is ~5.6. Rainwater flows into the ocean. The pH of ocean water, however, is ~8.7. Investigate reasons for this difference in pH.

37. Why would scrubbing probably not be a good strategy for removing NO_x from exhaust gases?

38. What advances have been recently accomplished in designing internal combustion engines to prevent pollution?

39. Investigate the topic of diesel engines. Are catalytic converters used with diesel engines? If so, do they differ from catalytic converters used with gasoline engines?

PUTTING IT ALL TOGETHER

INVESTIGATING AIR-QUALITY CLAIMS

As you learned in this unit, Earth's atmosphere is composed of various gases and particulates, some naturally occurring and others produced by human activities. Health and environmental concerns sometimes arise when these atmospheric components interact with one another and with solar radiation.

Many suggestions and products have been put forth to improve air quality. How do we know if an air-cleaning product works or if a change in our behavior will, in fact, decrease air pollution, acid rain, or global warming? How can we find out? At this point, you have designed an experiment to do just that.

As you conclude this unit, you will present your design to your peers, much as scientists present research to their peers, using a poster session. Your classmates will evaluate your investigation and help find improvements. You will then write a letter summarizing your proposed investigation to an organization or group of people who would be interested in or affected by its results.

As you prepare your poster and get ready to be a constructive contributor in your community, keep the following guidelines in mind.

Guidelines

Poster

- Choose a title that reflects your investigation and grabs the reader's attention.

    At scientific meetings and science fairs, there are often hundreds of posters, and your title determines whether someone stops to look at yours.

- Give your guiding question prominent placement.

    After all, this is why you planned the investigation. Make sure to present your question in clear language and an engaging format.

- Briefly, but clearly, describe your experimental approach.

    Your classmates should be able to understand how one would perform the investigation and what materials would be needed.

- Describe (or display) the evidence to be collected.

  Be sure the reader knows what data would be collected in your investigation to answer the guiding question. Point out how validity of the data will be ensured.

- Discuss the expected results.

  Explain how the planned investigation provides evidence that addresses the initial air-quality claim.

- Create visual interest.

  Again, draw readers in by using a graphic organizer, a chart, a data table, or a relevant photo or drawing. Also, be careful not to put too much text on your poster.

Providing and Using Feedback

- Be constructive.

  Aim for suggestions that will help your classmates improve their investigations or posters.

- Ask questions.

  If something is unclear to you, write a question for the poster's author.

- Consider the impact.

  Would this investigation make a difference to you in a way the author has not considered? Let the author know (constructively).

- View feedback as help.

  When receiving feedback on your own poster, remember that your evaluators are not trying to attack your work; they are trying to help you create the best possible investigation and poster.

Letter

- Target your audience.

  Use the appropriate level of language and formality for the individuals who will be reading your letter.

- Get to the point.

  Explain why your proposed investigation is important and why you have chosen to communicate with this group.

- Be convincing.

  Support your claims with reasoning and evidence.

- Ask for support, encouragement, or information.

  Give your reader a reason to be involved with your letter and your project.

LOOKING BACK AND LOOKING AHEAD

Throughout this unit, you have studied gases and their behavior, increased your knowledge of Earth's atmosphere, and learned to design a scientific investigation. You have also begun to study energy, and continued to develop your fluency in the language of chemistry.

The knowledge you have gained in the first two units can help you better understand some societal issues. Central among these concerns are the use and management of Earth's chemical resources, which include water, metals, petroleum, food, and air.

However, many problems are far too complex for a simple technological fix. Issues of policy are not usually "either/or" situations, but they often involve weighing many considerations. As a voting citizen, you will deal with issues that require some scientific understanding. Tough decisions may be needed. The remaining chemistry units that you study will continue to prepare you for those responsibilities.

Unit 3 deals with petroleum, a nonrenewable chemical resource, and the consequences of human use of this resource.

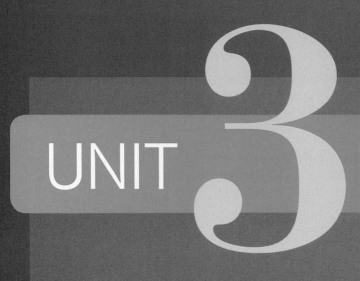

UNIT 3

PETROLEUM: BREAKING AND MAKING BONDS

How can the physical properties of petroleum be explained by its molecules and their interactions?

SECTION A
Petroleum—What Is It?
(page 268)

Why are carbon-based molecules so versatile as chemical building blocks?

SECTION B
Petroleum: A Building-Material Source (page 297)

What are the benefits and consequences of burning hydrocarbons?

SECTION C
Petroleum: An Energy Source
(page 330)

What alternatives to petroleum are available for burning and building?

SECTION D
Alternatives to Petroleum
(page 358)

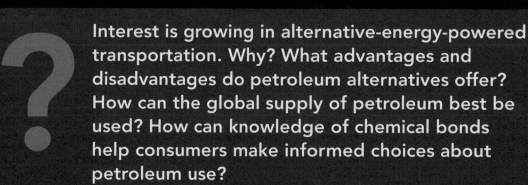

Interest is growing in alternative-energy-powered transportation. Why? What advantages and disadvantages do petroleum alternatives offer? How can the global supply of petroleum best be used? How can knowledge of chemical bonds help consumers make informed choices about petroleum use?
Turn the page to learn more about this energy-rich resource.

TLC PLUG-IN HYBRID VEHICLE TV COMMERCIAL— WORKING SCRIPT

Text/Script	Image/Sound

Text/Script

You already love the *TLC*, the most luxurious hybrid car on the market. Wait until you meet *TLC-p*, the first luxury plug-in hybrid.

What does the "p" stand for? Lots of things—plug-in, peppy, petroleum-free . . .

Unlike older petroleum-fueled vehicles, the *TLC* plug-in hybrid automobile is virtually emission-free. It's as close as you can get to "carbon-less transportation." In fact, it is considered a Super Ultra Low Emission Vehicle by the U.S. Environmental Protection Agency. Why? Because the *TLC-p* uses an energetic electric motor combined with a gasoline engine for back-up to deliver maximum power and efficiency. The electric motor is powered by batteries that are recharged simply by plugging in the *TLC-p* when you go to sleep. In addition, any excess energy generated by the gasoline engine is stored to later power the electric motor. You can drive knowing that you're helping, not hurting, the environment.

While you are decreasing your carbon footprint with the *TLC-p*, you'll enjoy the comfort of organic leather seats and triple-filtered air. And—thanks to the electric motor that takes over in city driving—the ride can be so quiet that driving the *TLC-p* seems more like a walk in the park.

There will be fewer stops on your road to fun with the *TLC-p*. Because it is powered by your home outlets, the *TLC-p* uses almost no fuel and can extract nearly all the energy in each drop of gasoline. With averages of

Image/Sound

(Video image of luxurious setting, such as a "green" spa hotel room with big glass windows, a comfy couch and a bubble bath.)

(Show an open road, perhaps along a coast.)

(Show a peaceful forest scene with nature sounds.)

(Show an open meadow with blue skies and fluffy, white clouds blowing by.)

(Show a scenic, but quiet park on a sunny day.)

(Show a gasoline station with a beach in the background.)

(Show beach-goers.)

Text/Script	Image/Sound
up to 100 miles per gallon, driving your plug-in hybrid automobile is easy on your pocketbook—as little as two cents per mile to operate. Compare that to more than 11 cents per mile for conventional gasoline-burning vehicles. Not only will you transform your own life, you will transform the planet by slashing your fuel requirements and automobile emissions. Help conserve petroleum resources! Visit your *TLC* dealer today.	*(View of a TLC-p, rear quarter panel with plug-in and logo.*  (*"TLC-p: Still the same great-looking hybrid, now available with plug-in power."* appears on the screen above.)

Are the claims made in this advertisement for the new *TLC-p* accurate? Is it truly a "transformative" vehicle? Will it help reduce owners' reliance on petroleum and other carbon-based fuels? Do better transportation options exist? Could you produce an advertisement for a "greener" automobile?

To answer these questions, you will learn about petroleum and how it fuels our modern lives. You'll consider the challenges of providing energy and building materials to an ever-increasing population, while protecting the planet we all share.

At the end of the unit, you will be challenged to create your own video advertisement for an alternative-fuel vehicle. Your ad will reflect what you have learned and make evidence-based claims, while revealing your vision for the future of automobile transportation.

Developing a vision of the future requires knowing something about the past, the present, and the possibilities. As you create your commercial message, you will learn about the origin and structure of the hydrocarbon compounds presently used to power most automobiles and why the molecules of these compounds are so useful as fuels and raw materials for the manufacture of important products. You will explore the consequences of using hydrocarbon fuels and investigate alternative sources of energy and matter. As you move toward understanding the role of petroleum, keep in mind the impact of decision-making at the personal, community, and global levels.

SECTION A PETROLEUM— WHAT IS IT?

How can the physical properties of petroleum be explained by its molecules and their interactions?

Figure 3.1 *Much time and energy will be expended to refine this mixture of petroleum before it is viable as a fuel or as a source of "builder molecules." What about petroleum makes it such an attractive resource?*

You live in a world of newly developed products and materials. Whether on billboards, on radio, on television, on the Internet, or in magazines, each advertisement attempts to sell a product by educating its audience about the product's unique features. The product may be faster, lighter, easier to use, newer, or better tasting—to sample a few of many common product claims. The advertisement for the *TLC-p* is no exception; it highlights several energy- and fuel-related features of the new vehicle that will (or so it is claimed) "help conserve petroleum resources." In Unit 2, you started thinking about how scientists—and consumers— evaluate claims. In this unit, you will continue to evaluate such claims, as well as make product claims of your own.

The word *petroleum* is quite familiar to you. But do you know what petroleum *is* or what it is *made of*? See Figure 3.1. Can you explain what properties make petroleum useful for both burning and building? In this section, you will explore characteristics of some key compounds found in petroleum. Specifically, you will focus on their physical properties, their molecular structures, and how these molecules interact with one another.

GOALS

- Describe the chemical makeup of petroleum and how it is refined.
- Describe the use of distillation as a separation technique and the application of fractional distillation to petroleum refining.
- Identify and write formulas for alkanes.
- Define *isomer* and draw structural formulas for possible isomers of a particular hydrocarbon.
- Predict and explain relative boiling points of hydrocarbons in terms of their intermolecular forces.

concept check 1

1. List three claims made in the *TLC-p* advertisement that opens this unit.
2. Name some physical properties that can be used to identify or distinguish substances.
3. Petroleum is composed of hydrocarbons. What elements make up hydrocarbons?
4. List some items that you use every day that are made from petroleum.

■ INVESTIGATING MATTER
A.1 PROPERTIES OF PETROLEUM

Asking Questions

In this investigation, you will explore some of the physical properties of several petroleum-based materials. In particular, you will measure the densities and relative viscosities of several materials separated from petroleum, as well as the same properties of water.

Viscosity is the term for resistance to flow. A material with high viscosity flows slowly and with difficulty, like honey. A material with low viscosity flows readily, like water. You will determine relative viscosities, which means ranking materials on a scale from most viscous to least viscous.

As you look at Table 3.1 (page 270) and *Gathering Evidence*, develop one or more scientific questions that you can answer in this investigation.

> *Viscous* materials are typically thick, syrupy liquids that have relatively high resistance to flow.

Preparing to Investigate

Before you begin, read *Gathering Evidence* to learn what you will need to do and note safety precautions. Develop procedures to find the density of solid and liquid samples. Ask your teacher to approve your procedures before continuing. Finally, think about the data and observations you will need to record in each part of the investigation and prepare appropriate data tables.

Gathering Evidence
Appearance and State

1. Before you begin, put on your goggles, and wear them properly throughout the investigation.
2. Obtain the six materials listed in Table 3.1 (page 270).
3. Record the physical state of each sample.
4. Describe the appearance of each sample.

Table 3.1

Physical Properties of Petroleum Samples

Material	Carbon Atoms per Molecule	Physical State at Room Temperature	Appearance
Mineral oil	12–20		
Asphalt	More than 34		
Kerosene	12–16		
Paraffin wax	More than 19		
Motor oil	15–18		
Household lubricating oil	14–18		

Figure 3.2 *Capped viscosity tube with bead.*

Relative Viscosity

1. Hold the capped tube containing water upside down until the bead inside is in the cap of the tube.

2. Gently turn the tube horizontally (the bead will stay at one end). See Figure 3.2.

3. Quickly turn the tube upright so that the bead is at the top and start the timer or stopwatch.

4. Stop the timer when the bead reaches the bottom of the tube.

5. Record the time needed for the bead to reach the bottom of the tube.

6. Repeat three times and calculate the average time required for the bead to fall.

7. Repeat Steps 1–6 for each liquid, petroleum-based sample.

8. Rank your samples in order of relative viscosity, assigning number 1 to the least-viscous material.

Density

1. Develop procedures to find the density of solid and liquid samples. Get your teacher's approval before conducting your investigation.

2. Without opening any sealed tube, find the density of each liquid sample. Your teacher will provide a sealed tube containing only a bead, as well as a tube that you may open that contains only a bead.

3. Record your data and calculations.

4. Determine the density of each solid sample.

5. Record your data and calculations.

6. Wash your hands thoroughly before leaving the laboratory.

Interpreting Evidence

1. Rank the materials (including water) from least to most viscous.

2. Create a graphic organizer that illustrates the differences in viscosity among the samples.

3. Propose a rule, based on your observations in this investigation, about the connection between number of carbon atoms in a substance's molecules and the viscosity of that substance.

4. In oil spills, the oil's density plays a major role. Explain.

5. You should never use water to extinguish a gasoline or oil fire. Why not?

Reflecting on the Investigation

6. Think about your experimental design for the density portion of the investigation. Did you need to make any modifications so that your data produced accurate results? Explain.

7. Suppose a classmate suggests that the petroleum materials studied in this investigation can be separated at room temperature on the basis of their viscosities.

    a. Do you agree? Explain your answer.

    b. What would be some advantages of such a separation procedure?

8. What conditions might change the viscosity of a substance? In one or two sentences, describe an experiment you could design to test this.

9. In your own words and using evidence from the investigation, answer the scientific question or questions you posed at the beginning of the investigation.

A.2 PETROLEUM: A SOURCE OF ENERGY AND HYDROCARBONS

Petroleum is a vital chemical world resource. Petroleum pumped from underground is called **crude oil**. This liquid varies from colorless to greenish-brown to black, and can be as fluid as water or as resistant to flow as soft tar.

We cannot use crude oil commercially in its natural state. Instead, we transport it by pipeline, ocean tanker, train, or barge to oil refineries, where it is separated into simpler mixtures. See Figure 3.3. Some mixtures are ready to use, while others require further refinement. Refined petroleum is chiefly a mixture of various **hydrocarbons**—molecular compounds that contain only atoms of the elements hydrogen and carbon. You can readily see how this class of compounds got its name.

> The word *petroleum* comes from the Latin words *petr-* (rock) and *oleum* (oil).

> Crude oil was used as lamp oil in ancient times.

Figure 3.3 *The largest oil tanker in the world can carry up to 4.1 million barrels (over 150 billion gallons) of crude oil. Such ships are so big, onboard crew members often need a bicycle to travel from one point to another.*

Burning petroleum provides nearly half of the total annual U.S. energy needs. Most petroleum is used as a fuel. Converted to gasoline, petroleum powers millions of U.S. automobiles, each traveling an average of 14 000 miles annually. Other petroleum-based fuels provide heat to homes and office buildings, deliver energy to generate electricity, and power diesel engines and jet aircraft.

Petroleum's other major use is as a raw material from which a stunning array of familiar and useful products is manufactured—from CDs, sports equipment, clothing, automobile parts, plastic charge cards, and carpeting to prescription drugs and artificial limbs. Figure 3.4 (at right) shows some petroleum-based products.

Based on your experiences with petroleum fuels and products, what percent of petroleum would you estimate is burned for energy? What percent of petroleum would you estimate is used for producing other useful substances? Can you identify other uses of petroleum? The answers in the next paragraph may surprise you.

What did you predict for the percent of petroleum burned for energy? Did you predict 50%? Was your prediction 60%? Astonishingly, 89% of all petroleum is burned as fuel. Only about 7% is used for producing new substances, such as medications and plastics. The remaining 4% is used as lubricants, road-paving materials, and an assortment of miscellaneous products. In fact, for every gallon of petroleum used to produce useful products, more than five gallons are burned for energy.

What happens to molecules contained in petroleum when they are burned or used in manufacturing? As in all chemical reactions, the atoms become rearranged to form new molecules.

When hydrocarbons burn, they react with oxygen gas in the air to form carbon dioxide (CO_2) gas and water vapor. (See equations below representing the burning of two typical hydrocarbons.) These gases then disperse in the air. The hydrocarbon fuel is used up; it will take millions of years for natural processes to replace it. Thus petroleum is a *nonrenewable resource*, as were the minerals you studied in Unit 1.

$$CH_4(g) \; + \; 2\,O_2(g) \; \longrightarrow \; CO_2(g) \; + \; 2\,H_2O(g) \; + \; energy$$

$$2\,C_2H_6(g) \; + \; 7\,O_2(g) \; \longrightarrow \; 4\,CO_2(g) \; + \; 6\,H_2O(g) \; + \; energy$$

Figure 3.4 *It is likely you use petroleum-based products in almost every area of your life. How many can you identify?*

A.3 ORIGIN AND DISTRIBUTION OF PETROLEUM

Fossil fuels are the products of geological heat and pressure acting on biomolecules of prehistoric plants and animals (Figure 3.5). The energy released by burning these fuels represents energy originally captured from sunlight by prehistoric green plants during photosynthesis. Thus **fossil fuels**—petroleum, natural gas, and coal—can be thought of as forms of buried sunshine.

Most evidence indicates that fossil fuels originated from living matter in ancient seas some 500 million years ago. These organisms died and eventually became covered with sediments. Pressure, heat, and microbes converted what was once living matter into petroleum, which became trapped in porous rocks. It is likely that some petroleum is still being formed from sediments of dead matter. However, such a process is far too slow to consider petroleum a renewable resource.

To use petroleum, it must be extracted from the ground in the form of crude oil. A drilling rig, as in Figure 3.6, bores into Earth's crust to create a well that is used to access the oil stored in the rock formations below ground. Once the well has been drilled, the drilling rig is replaced with a pumping station. As of 2008, more than 363 000 oil wells were actively pumping oil in the United States alone.

> Energy is a measure of the ability of a system to do work.

Figure 3.5 (above) *A few botanical gardens, including the Zilker Botanical Gardens shown here, have created "prehistoric gardens" containing plants similar to those that may have been decomposed by microorganisms to form petroleum.*

Figure 3.6 (left) *Oil rig drilling platform in Texas.*

Figure 3.7 *This large oil platform is located about 150 miles southeast of New Orleans. It is extracting oil from 6000 feet below the surface in the Gulf of Mexico. The drill ship shown in the background is working about five miles away from the platform.*

Figure 3.8 *Oil from the leaking BP Deep Horizon oil rig is seen swirling through the currents in the Gulf of Mexico in May 2010. Large-scale spills like this one are among the risks posed by extraction of offshore oil.*

Oil exploration is not limited to drilling on land. More than 25% of the oil extracted in the United States in 2008 came from 4000 offshore oil platforms, mostly in the Gulf of Mexico. Modern offshore oil platforms (see Figures 3.7 and 3.8) can operate in water up to 8000 feet deep and draw from wells that may extend another 6000 feet below the surface of the sea floor. Each oil platform collects crude oil from several wells, then pumps the oil to shore via underwater pipeline or transfers it to tankers for transportation and processing.

Like other resources, petroleum is not evenly distributed around the world. The worldwide distributions are summarized in Figure 3.9. The North American petroleum reserves are only about 5% of the world's known supply. Central Asia, the Far East, and Oceania account for 56% of the

The composition of petroleum also varies from place to place.

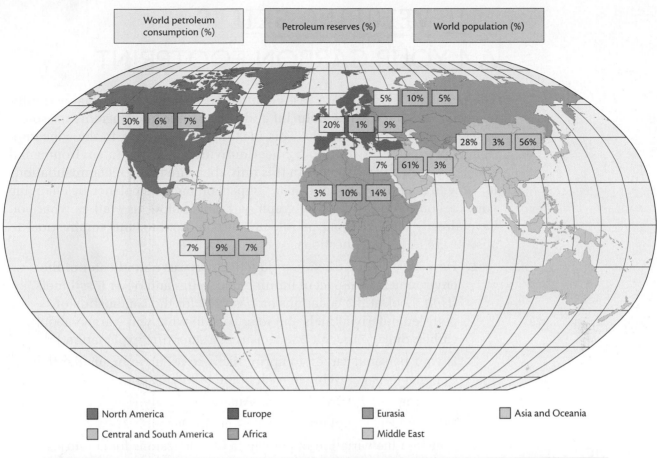

| North America | Europe | Eurasia | Asia and Oceania |
| Central and South America | Africa | Middle East | |

| | **Consumption 2004** | | **Reserves 2007** | | **Population 2004** | |
Regions	10^3 **Barrels/day**	**Percent**	10^9 **Barrels**	**Percent**	**Millions**	**Percent**
North America	25 046.0	30.4	69.295	5.6	430.25	6.8
Central and South America	5349.1	6.5	111.211	9.0	442.77	7.0
Europe	16 259.8	19.7	15.570	1.3	587.57	9.2
Eurasia	4040.8	4.9	128.146	10.4	286.10	4.5
Africa	2819.5	3.4	117.482	9.5	874.75	13.7
Middle East	5539.4	6.7	755.325	61.0	180.15	2.8
Asia and Oceania	23 353.2	28.3	40.847	3.3	3565.50	56.0
Total	82 407.7	100.0	1237.876	100.0	6367.10	100.0

Figure 3.9 *Distribution of the world's petroleum reserves, population, and consumption of petroleum.*

world's population, but this region has only about 4% of the world's petroleum reserves.

By contrast, approximately 66% of the world's known crude oil reserves are located in Middle Eastern nations. The consumption of petroleum is also not evenly distributed around the world, as you see in Figure 3.9. In the next activity, you will consider patterns of petroleum consumption and associated environmental impact.

Reserves are resources that can be tapped by available technology at costs consistent with current market prices.

DEVELOPING SKILLS

A.4 YOUR CARBON FOOTPRINT

The quantity of greenhouse gases emitted based upon individual activities has been called a **carbon footprint** and is measured in kilograms (or sometimes tons) of carbon dioxide. This idea, which is related to the consumption of petroleum, is used to connect individual choices to environmental effects. As you will learn later in this unit, there are many recommendations for how to measure and reduce your impact on nature's balance by changing your carbon footprint. Use the graph and table below, as well as what you have already learned about petroleum, carbon dioxide, and combustion, to answer the following questions.

1. How does carbon dioxide production serve as an indicator of the environmental impact of burning petroleum and other fossil fuels? (*Hint:* Look at the equations on page 272 for the combustion of petroleum-derived fuels and refer back to what you learned about carbon dioxide in Unit 2 to construct a scientific explanation.)

2. Is your personal carbon footprint likely to be larger or smaller than that of a teenager living in each of the following countries?

   a. China   b. Brazil   c. Australia   d. Kenya

3. What evidence supports your answers to Question 2?

4. Consider the variation in per capita CO_2 emissions for countries around the world, as shown in Figure 3.10. How can you explain the differences that you observe?

Figure 3.10

Population and CO_2 emissions per capita.

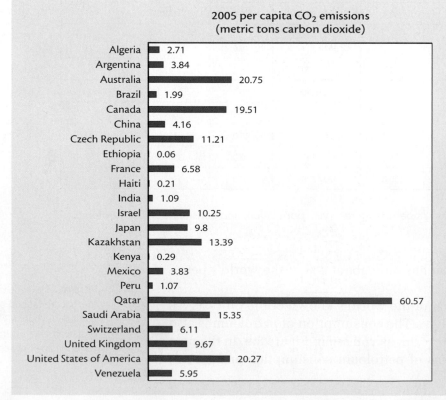

Country	Population (millions)
Algeria	32.56
Argentina	39.54
Australia	20.09
Brazil	186.11
Canada	32.39
China	1306.31
Czech Republic	10.24
Ethiopia	73.05
France	62.94
Haiti	8.3
India	1093.56
Israel	6.74
Japan	127.54
Kazakhstan	15.19
Kenya	34.91
Mexico	106.2
Peru	28.05
Qatar	0.86
Saudi Arabia	26.42
Switzerland	7.49
United Kingdom	60.44
United States of America	295.73
Venezuela	25.27

2005 per capita CO_2 emissions (metric tons carbon dioxide)

Algeria	2.71
Argentina	3.84
Australia	20.75
Brazil	1.99
Canada	19.51
China	4.16
Czech Republic	11.21
Ethiopia	0.06
France	6.58
Haiti	0.21
India	1.09
Israel	10.25
Japan	9.8
Kazakhstan	13.39
Kenya	0.29
Mexico	3.83
Peru	1.07
Qatar	60.57
Saudi Arabia	15.35
Switzerland	6.11
United Kingdom	9.67
United States of America	20.27
Venezuela	5.95

5. What roles do population and availability of fossil fuels play in a country's average carbon footprint?

You have now learned what petroleum is, how we use it, and where we find it. Petroleum is a complex mixture of hydrocarbons that must be refined or separated into simpler mixtures to become useful. In the following investigation, you will learn about this basic separation process as you study a mixture of two liquids.

■ INVESTIGATING MATTER
A.5 SEPARATION BY DISTILLATION

Preparing to Investigate

You know that you can often separate substances by taking advantage of their different physical properties. One physical property commonly used to separate liquids is their density. However, density differences will work only if substances are insoluble (do not dissolve) in each other, which is not the case with petroleum; its components are soluble (can dissolve) in each other. Another physical property chemists often use to separate substances is *boiling point*. The separation of liquid substances according to their differing boiling points is called **distillation**.

As you heat a liquid mixture containing two components, the component with the lower boiling point will vaporize first and leave the distillation flask. That component will then condense back to a liquid as it passes through a condenser—all before the second component begins to boil. See Figure 3.11. Each condensed liquid component, called the **distillate**, can thus be collected separately.

In this investigation, you will use distillation to separate a mixture of two liquids. Then you will identify the two substances in the mixture by comparing the observed distillation temperatures with the boiling points of several possible compounds listed in Table 3.2.

Figure 3.11
A simple distillation apparatus.

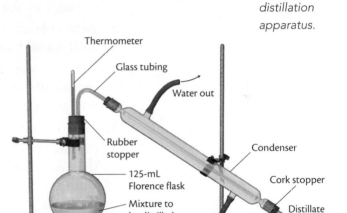

The boiling points listed in Table 3.2 are based on normal sea-level atmospheric pressure.

Possible Components of Distillation Mixture			
Substance	**Formula**	**Boiling Pt. (°C)**	**Appearance with I$_2$**
2-Propanol (rubbing alcohol)	C_3H_8O	82.4	**Bright yellow**
Acetone	C_3H_6O	56.5	**Yellow to brown**
Water	H_2O	100.0	**Colorless to light yellow**
Cyclohexane	C_6H_{12}	80.7	**Magenta**

Table 3.2

Before you begin, read *Gathering Evidence* to learn what you will need to do and note safety precautions. *Gathering Evidence* also provides guidance about when you should collect and record data (see Step 8). Think about the data and observations you will need to record and create data tables to record your observations and measurements. Remember to include proper units for each data column.

Gathering Evidence

*(**Caution:** This distillation should only be completed with a hot plate or other electric heating source. The presence of open flames near the distillation apparatus represents a fire hazard.)*

1. Before you begin, put on your goggles, and wear them properly throughout the investigation.

2. Assemble an apparatus similar to that shown in Figure 3.11 (page 277). Label two beakers *Distillate 1* and *Distillate 2*.

3. Using a clean, dry, graduated cylinder, measure a 50-mL sample of the distillation mixture. Pour the mixture into the distillation flask and add a boiling chip.

4. Record your observations of the starting mixture.

5. Connect the flask to a condenser, as shown in Figure 3.11. Ensure that the hoses are attached to the condenser and to the water supply as shown. Position the Distillate 1 beaker at the outlet of the condenser so that it will catch the distillate, as shown in Figure 3.12.

6. Ensure that all connections are tight and will not leak.

7. Turn on the water to the condenser, and then turn on the hot plate to start gently heating the flask. *(**Caution:** The substances, other than water, are volatile and highly flammable. Be sure that no flames or sparks are in the area.)*

8. Record the temperature every minute until the first drop of distillate enters the beaker. Then continue to record the temperature every 30 seconds. Continue to heat the flask and collect distillate until the temperature begins to rise again. At this point, replace the Distillate 1 beaker with the Distillate 2 beaker.

9. Continue heating and recording the temperature every 30 seconds until the second substance just begins to distill. Record the temperature at which the first drop of second distillate enters the beaker. Collect 1 to 2 mL of the second distillate. *(**Caution:** Do not allow all of the liquid to boil from the flask.)*

Figure 3.12 *This distillation apparatus is used to separate a mixture of two liquids. The component with the lowest boiling point vaporizes first, converts back to a liquid in the condenser, and collects (here) in a beaker.*

10. Turn off the hot plate and allow the distillation apparatus to cool. While the apparatus is cooling, test to what extent solid iodine dissolves in each distillate by adding a few crystals of iodine to each beaker and stirring. Record your observations. (**Caution:** *Iodine is corrosive on contact. It will stain skin and clothing.*)

11. Disassemble and clean the distillation apparatus and dispose of your distillates as directed by your teacher.

12. Wash your hands thoroughly before leaving the laboratory.

Analyzing Evidence

1. Plot your data on a graph of time versus temperature. Which variable is the independent variable? That variable should be plotted on the x-axis. (*Hint:* Recall the graph that you constructed in Investigating Matter B.8 in Unit 2 on page 194.)

> Recall that the *independent variable* is manipulated by the investigator.

2. a. Using your graph, identify the temperatures at which Distillate 1 and Distillate 2 were collected.

   b. How well do the horizontal plateaus in your graph match the temperatures at which you collected the first drops of each distillate?

3. Combine your data with the data of other students who distilled the same mixture. Examine the combined data, and find

   a. the average temperature (*mean*) for each of the two plateaus you observed.

   b. the most frequently observed temperature (*mode*) for each of the two plateaus you observed.

> The *mean* and *mode* are useful values for describing "middle" characteristics for a set of related data.

Interpreting Evidence

1. In which of the distillates did solid iodine dissolve to a greater extent? What observational evidence leads you to that conclusion?

2. Use data in Table 3.2 (page 277) to identify each distillate sample. Support your identification claims with evidence.

Reflecting on the Investigation

3. All laboratory teams investigating the same mixture did not observe the same distillation temperatures. Describe some factors that may contribute to this inconsistency.

4. What laboratory tests could you perform to decide whether the liquid left behind in the flask is a mixture or a pure substance?

5. Of the substances listed in Table 3.2 (page 277), which two would be most difficult to separate from each other by distillation? Why?

6. How would a graph of time versus temperature look for the distillation of a mixture of all four substances listed in Table 3.2 (page 277)? Sketch the predicted graph and describe its features.

A.6 PETROLEUM REFINING

In Investigating Matter A.1 (page 269), you examined a variety of components of crude oil, while in Investigating Matter A.5 (page 277), you separated a simple laboratory mixture using distillation. Crude oil, in its natural state, is a mixture of many compounds. Separating such a complex mixture requires applying distillation techniques to large-scale oil refining. The refining process does not separate each compound in crude oil. Rather, it produces several distinctive mixtures called **fractions**. This process is known as **fractional distillation**. See Figure 3.13. Compounds in each fraction have a particular range of boiling points and specific uses.

Figure 3.13 *These fractionating towers contain many different levels of condensers to cool the oil vapor as it rises. Temperatures range from ~400 °C (at the base) to 40 °C (at the top).*

Figure 3.14 illustrates fractional distillation (fractionation) of crude oil. First, the crude oil is heated to about 400 °C in a furnace and then pumped into the base of a distilling column (fractionating tower), which is usually more than 30 m (100 ft) tall. Many components of the heated crude oil vaporize. The temperature of the distilling column is highest at the bottom and decreases toward the top. Trays arranged at appropriate heights inside the column collect the various condensed fractions.

During distillation, the vaporized molecules move upward in the distilling column. The smaller, lighter molecules have the lowest boiling points and either condense high in the column or are drawn off the top of the tower as gases. Fractions with larger molecules have higher boiling points and are more difficult to separate from one another and thus require more thermal energy (heat energy) to vaporize. These larger molecules condense back to liquid in trays lower in the column. Substances with the highest boiling points never do vaporize. Recall from Investigating Matter A.1 (page 269) that petroleum components vary in viscosity. Which components would you expect to find in the lower trays? These components are called **bottoms** and drain from the column's base. Each arrow in Figure 3.14 indicates the name of a particular fraction and its boiling-point range.

As you learn more about the characteristics of the fractions obtained from petroleum, think about how people use petroleum-based products in both traditional and hybrid vehicles.

Although the names given to various fractions and their boiling ranges may vary somewhat, crude oil refining always has the same general features.

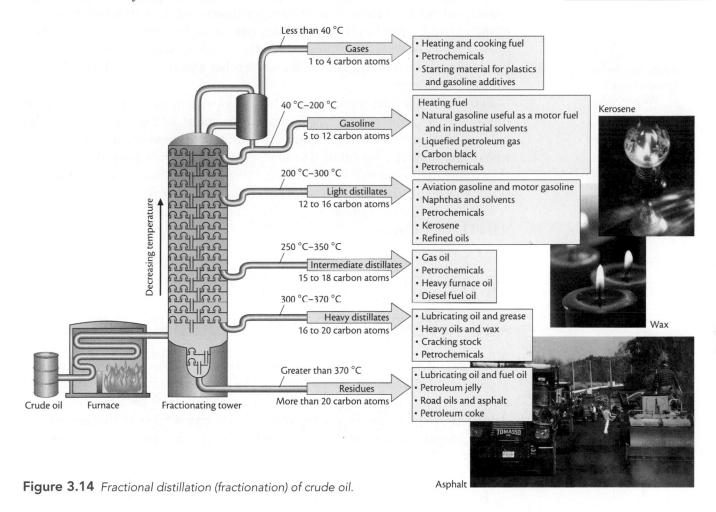

Figure 3.14 *Fractional distillation (fractionation) of crude oil.*

concept check 2

1. Where did petroleum's energy originate?
2. How are the components of petroleum
   a. similar to one another?
   b. different from one another?
3. For a given component of petroleum, how does the number of carbon atoms within its molecules relate to its viscosity and boiling point?

A.7 EXAMINING PETROLEUM'S MOLECULES

Petroleum's gaseous fraction contains compounds with low boiling points (less than 40 °C). These small hydrocarbon molecules, which contain from one to four carbon atoms, have low boiling points because they are only slightly attracted to each other or to other molecules in petroleum. Forces of attraction and repulsion between molecules are called **intermolecular forces**.

As a result of weak intermolecular attractive forces, these small hydrocarbon molecules readily separate from each other and rise through the distillation column as gases.

Petroleum's liquid fractions—including gasoline, kerosene, and heavier oils—consist of molecules having from 5 to about 20 carbon atoms. Molecules with even more carbon atoms are found in the greasy fraction (the bottoms) that does not vaporize. These viscous compounds have the strongest intermolecular attractive forces among all substances found in petroleum. It is not surprising that they are solids at room temperature.

Now complete the following activity to learn more about physical properties of hydrocarbons.

> Just as interstate highway refers to a road that runs between states, intermolecular forces act *between* molecules.

DEVELOPING SKILLS

A.8 HYDROCARBON BOILING POINTS

Chemists often gather and analyze data about the physical and chemical properties of substances. These data can be organized in many ways, but the most useful techniques are those that uncover trends or patterns among the data. The development of the periodic table is an example of this approach.

You can examine patterns among boiling points of some hydrocarbons to make useful predictions.

Use the data found in Table 3.3 to answer these questions.

> Examine the names of the hydrocarbons in Table 3.3. What do they all have in common?

1. a. In what pattern or order are the data in Table 3.3 organized?

   b. Is this a useful way to present the information? Explain.

2. Suppose that you were searching for a trend or pattern among these boiling points.

   a. Propose a more useful way to arrange these data.

   b. Reorganize the data table based on your idea.

Now use your reorganized data table to answer the following questions:

3. Which substance or substances are gases at room temperature (22 °C)?

4. If a substance is a gas at 22 °C, what can you infer about the boiling point of that substance?

> *Infer* means to reach a conclusion based on evidence and reasoning.

Table 3.3

Boiling Points of Selected Hydrocarbons		
Hydrocarbon	**Boiling Point (°C)**	**Molecular Formula**
Butane	−0.5	C_4H_{10}
Decane	174.0	$C_{10}H_{22}$
Ethane	−88.6	C_2H_6
Heptane	98.4	C_7H_{16}
Hexane	68.7	C_6H_{14}
Methane	−161.7	CH_4
Nonane	150.8	C_9H_{20}
Octane	125.7	C_8H_{18}
Pentane	36.1	C_5H_{12}
Propane	−42.1	C_3H_8

5. Which substance or substances boil between 22 °C (room temperature) and 37 °C (body temperature)?

6. What can you infer about intermolecular forces among decane molecules compared to intermolecular forces among butane molecules?

▮ MODELING MATTER
A.9 MODELING ALKANES

In this activity, you will assemble models of several simple hydrocarbons. See Figure 3.15. Your goal is to associate the 3-D shapes of these molecules with the names, formulas, and pictures used to represent them on paper.

Figure 3.15 *Due to the extremely small size of atoms and molecules, people often use model kits to visualize the 3-D geometry of hydrocarbons and other molecular structures.*

Figure 3.16 *Scientists use many types of physical and mental models to better understand concepts related to atoms and molecules.*

There are several useful ways to construct models of molecules; there are advantages and disadvantages to each method of representing molecules (Figure 3.16). Two common types of molecular models are shown in Figure 3.17. Most likely, you will use **ball-and-stick models**. Each ball represents an atom, and each stick represents a pair of shared electrons (a single covalent bond) connecting two atoms.

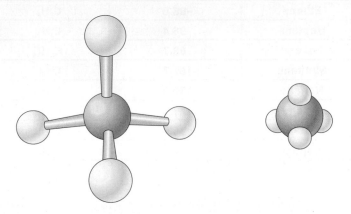

Figure 3.17 *Three-dimensional CH_4 models: ball-and-stick (left) and space-filling (right).*

Of course, molecules are not composed of ball-like atoms located at the ends of stick-like bonds. Experimental evidence shows that atoms are in contact with each other. This is more closely represented by **space-filling models**. However, ball-and-stick models are still useful because they can clearly represent the bonding and geometry of molecules.

Methane, the simplest hydrocarbon, is the first member of a series of hydrocarbons known as **alkanes**. Each carbon atom in an alkane shares electrons with four other atoms. You will explore alkanes in this activity.

1. Assemble a 3-D model of a methane (CH_4) molecule. Note that the angles defined by the bonds between atoms are not 90°, as you might have expected by looking at the structural formula. If you were to build a close-fitting box to surround a CH_4 molecule, the box would be shaped like a pyramid with a triangle as its base. This 3-D shape is called a **tetrahedron**.

The tetrahedral geometry that you observe results from the fact that similar electrical charges repel. The four pairs of electrons in the bonds surrounding the carbon atom repel each other and arrange themselves to be as far away from one another as possible. In this spatial arrangement, they point to the corners of a tetrahedron. The angle formed between each pair of C—H bonds is 109.5°, a value that has been verified experimentally.

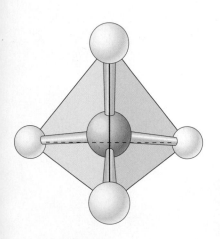

Figure 3.18 *The tetrahedral shape of a methane molecule.*

2. Convert your 3-D model into a 2-D drawing (similar to the one shown in Figure 3.18) that conveys the tetrahedral structure of methane. Shade the carbon atom to distinguish it from the hydrogen atoms.

3. Assemble models of a two-carbon alkane molecule and a three-carbon alkane molecule. Recall that each carbon atom in an alkane is bonded to four other atoms.

   a. How many hydrogen atoms are in a two-carbon alkane?

   b. How many hydrogen atoms are in a three-carbon alkane?

   c. Draw a ball-and-stick model, similar to the one shown in Figure 3.17, of the three-carbon alkane. Shade carbon atoms to distinguish them from hydrogen atoms.

4. **Molecular formulas** specify the number of each atom type within a molecule. The molecular formulas of the first two alkanes are CH_4 and C_2H_6. Identify the molecular formula for the third alkane.

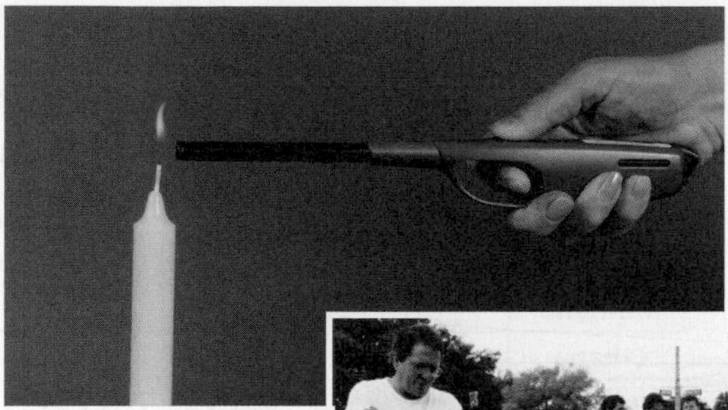

Examine your three-carbon alkane model. Note that the middle carbon atom is attached to *two* hydrogen atoms, but the carbon atom at each end is attached to *three* hydrogen atoms. This molecule can be represented as $CH_3—CH_2—CH_3$, or $CH_3CH_2CH_3$. Formulas such as these provide convenient information about how atoms are arranged in molecules. For many purposes, such **condensed formulas** are more useful than molecular formulas such as C_3H_8.

Consider the molecular formulas of the first few alkanes: CH_4, C_2H_6, and C_3H_8. Given the pattern represented by that series, predict the formula of the four-carbon alkane.

The general molecular formula of all open-chain alkane molecules can be written as C_nH_{2n+2}, where n is the number of carbon atoms in the molecule. Therefore, even without assembling a model, you can predict the formula of a five-carbon alkane: If $n = 5$, then $2n + 2 = 12$, and the resulting formula thus is C_5H_{12}.

The names of the first 10 alkanes are given in Table 3.4 (page 286) and some examples of uses of these compounds are shown in Figure 3.19 and Figure 3.20. As you can see, each name is composed of a prefix, followed by *-ane* (designating an alk*ane*). The prefix indicates the number of carbon atoms in the backbone carbon chain. *Meth-* means one carbon atom, *eth-* means two, *prop-* means three, and *but-* means four. For alkanes with 5 to 10 carbon atoms, the prefix is derived from Greek—*pent-* for five, *hex-* for six, and so on.

Figure 3.19
Butane is a liquid commonly used as fuel in lighters.

Figure 3.20 *Propane often serves as an emergency energy source during power outages. Here, workers weigh filled tanks to determine how much propane they contain.*

n can be any positive integer.

Table 3.4

			Alkane Formulas	
Name	**Total Carbon Atoms**	**Boiling Point (°C)**	**Molecular Formula**	**Condensed Formula**
Methane	1	−161.7	CH_4	CH_4
Ethane	2	−88.6	C_2H_6	CH_3CH_3
Propane	3	−42.1	C_3H_8	$CH_3CH_2CH_3$
Butane	4	−0.5	C_4H_{10}	$CH_3CH_2CH_2CH_3$
Pentane	5	36.1	C_5H_{12}	$CH_3CH_2CH_2CH_2CH_3$
Hexane	6	68.7	C_6H_{14}	$CH_3CH_2CH_2CH_2CH_2CH_3$
Heptane	7	98.4	C_7H_{16}	$CH_3CH_2CH_2CH_2CH_2CH_2CH_3$
Octane	8	125.7	C_8H_{18}	$CH_3CH_2CH_2CH_2CH_2CH_2CH_2CH_3$
Nonane	9	150.8	C_9H_{20}	$CH_3CH_2CH_2CH_2CH_2CH_2CH_2CH_2CH_3$
Decane	10	174.0	$C_{10}H_{22}$	$CH_3CH_2CH_2CH_2CH_2CH_2CH_2CH_2CH_2CH_3$

Some Members of the Alkane Series

5. Disassemble your molecular models and replace all parts in their container.

6. Write structural formulas for butane and pentane.

7. a. Name the alkanes with the following condensed formulas:

    i. $CH_3CH_2CH_2CH_2CH_2CH_2CH_3$

    ii. $CH_3CH_2CH_2CH_2CH_2CH_2CH_2CH_2CH_3$

   b. Write molecular formulas for the two alkanes in Question 7a.

8. a. Write the formula of an alkane containing 25 carbon atoms.

   b. Did you decide to write the molecular formula or the condensed formula for this compound? Why?

9. Find the molar mass of pentane. (*Hint:* Remember to take subscripts into account, as you first learned to do in Unit 1, page 95.)

10. Name the alkane molecule that has a molar mass of

    a. 30 g/mol.   b. 58 g/mol.   c. 114 g/mol.

DEVELOPING SKILLS
A.10 TRENDS IN ALKANE BOILING POINTS

In Investigating Matter A.5, you used the technique known as *distillation* to separate liquid mixtures according to the boiling points of the components. You also learned that fractions of petroleum are separated based on their boiling points. Why do fractions with the largest molecules have the highest boiling points? Why are the smallest molecules found in fractions with the lowest boiling points? In this activity, you will explore this trend in alkane boiling points.

Using data for the alkanes found in Table 3.4, prepare a graph of boiling points. The x-axis scale should range from 1 to 13 carbon atoms (even though you will initially plot data for 1 to 10 carbon atoms). The y-axis scale should extend from −200 °C to +250 °C.

1. Label the axes appropriately and plot the data. Draw a best-fit line through your data points according to the following guidelines:

   • The line should follow the trend of your data points.

   • The data points should be roughly equally distributed above and below the line.

   • The line should not extend past your data points.

   Figure 3.21 shows an example of a best-fit line.

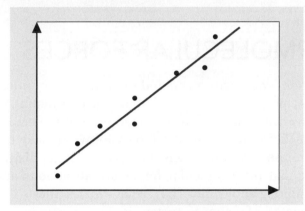

Figure 3.21 *An example of a best-fit line drawn through several data points.*

2. Estimate the average change in boiling point (in °C) when one carbon atom and two hydrogen atoms (—CH_2—) are added to a particular alkane chain.

3. The pattern of boiling points among the first 10 alkanes allows you to predict boiling points for other alkanes.

   a. Using your graph, estimate the boiling points of undecane ($C_{11}H_{24}$), dodecane ($C_{12}H_{26}$), and tridecane ($C_{13}H_{28}$). To do this, follow the trend of your best-fit line by extending a dashed line from the best-fit line you drew for the first 10 alkanes. This procedure is called **extrapolation**. Then read your predicted boiling points for C_{11}, C_{12}, and C_{13} alkanes on the y-axis.

   b. Compare your predicted boiling points to actual values provided by your teacher.

   c. Is your extrapolated line a good model for the relationship between number of carbon atoms in an alkane molecule and the substance's boiling point? Why or why not?

4. You learned that a substance's boiling point depends in part on its intermolecular forces, which are the attractions among its molecules. For the alkanes you have studied, what is the relationship between the strength of these attractions and the number of carbon atoms in each molecule?

concept check 3

1. In a hydrocarbon molecule, how many bonds will be formed by a
   a. carbon atom?
   b. hydrogen atom?
2. a. List several ways that scientists represent molecules.
   b. Identify at least one advantage and one disadvantage of each method.
3. Describe the relationship between the number of carbon atoms in each molecule of an alkane and intermolecular attractive forces among those molecules.
4. What is an isomer?

A.11 INTERMOLECULAR FORCES

Throughout this section, you have considered the influence of intermolecular forces on physical properties of hydrocarbons in petroleum. Viscosity, boiling point, and physical state are all impacted by intermolecular forces. In fact, differences in these intermolecular forces are what make it possible to separate petroleum into useful fractions through fractional distillation.

You already know that intermolecular forces are attractions and repulsions between and among molecules, but what causes these forces and determines their strength? The simple answer is that intermolecular forces are the result of uneven distribution of electrical charge within a molecule. Regions of charge in one molecule can then be attracted to (or repulsed by) regions of charge in other molecules. You may have noticed that so far we have only considered intermolecular forces of *attraction*. Intermolecular *repulsions* become important only when atoms and molecules are very close together. These intermolecular repulsive forces affect how much a substance can be compressed, but do not noticeably affect boiling points.

There are two major ways that electrical charge can become unevenly distributed within a molecule. First, as you know, electrons are in constant motion in electron clouds surrounding atoms and molecules. Sometimes the electrons "gather" or "pile up" in certain regions of the molecule. Areas of instantaneous negative charge result near the electrons, while areas of instantaneous positive charge are created where atomic nuclei are partially exposed. This motion of electrons and the resulting intermolecular forces are present to some extent in all molecules.

In addition to uneven charge distribution due to electron motion, some molecules also have permanent partial charges resulting from differences in the ability of their atomic nuclei to attract electrons. Molecules that exhibit permanent partial charges are known as **polar molecules**. Opposite partial charges in neighboring polar molecules attract one another, resulting in intermolecular forces between the molecules.

You will learn more about polar molecules in Unit 4.

The trend towards stronger intermolecular forces, and thus higher boiling points, as alkane molecules increase in size (number of carbons atoms) can be explained by two factors. First, as the number of atoms in a molecule increases, so also does the number of electrons in that molecule. More electrons result in greater fluctuations in electron movement and distribution of electrical charge, giving rise to stronger intermolecular forces. Second, larger molecules have a larger surface area along which they can interact with other molecules. Thus, the increase in boiling point with number of carbon atoms among straight-chain alkanes is consistent with evidence-based explanations of intermolecular forces caused by electron motion.

In the following activities, you will explore shapes of different alkane molecules and consider the influence of molecular shape on intermolecular forces. Keep the underlying causes of intermolecular forces in mind as you use modeling and data to further investigate the boiling points of alkanes.

▌ MODELING MATTER
A.12 ALKANE ISOMERS

The alkane molecules you have considered so far are **straight-chain alkanes**, where each carbon atom is only linked to one or two other carbon atoms. In alkanes with four or more carbon atoms, other arrangements of carbon atoms are possible. In **branched-chain alkanes**, one carbon atom can be linked to three or four other carbon atoms. An alkane composed of four or more carbon atoms can have either a straight-chain or a branched-chain structure. In this activity, you will use ball-and-stick molecular models (see Figure 3.22) to investigate such variations in alkane structures—variations that can lead to different properties.

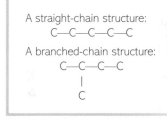

A straight-chain structure:
C—C—C—C—C

A branched-chain structure:
C—C—C—C
       |
      C

Figure 3.22 *Students working together to construct molecular models.*

1. Assemble a ball-and-stick model of a molecule with the formula C_4H_{10}. Compare your model with those built by others. How many different arrangements of atoms in the C_4H_{10} molecule did your class construct?

Molecules that have identical molecular formulas but different arrangements of atoms are called **structural isomers**. By comparing models, convince yourself that there are only two structural isomers of C_4H_{10}. The formation of isomers helps to explain the very large number of compounds composed of carbon chains or rings.

2. a. Draw a sketch of each C_4H_{10} isomer.

   b. Write a structural formula for each C_4H_{10} isomer.

3. As you might expect, alkanes containing larger numbers of carbon atoms also have larger numbers of structural isomers. In fact, the number of different isomers increases rapidly as the number of carbon atoms increases. For example, chemists have identified three pentane (C_5H_{12}) isomers, as shown in Table 3.5. Try building these structural isomers. How might you convince a classmate that there exist only these three C_5H_{12} isomers?

Table 3.5

Pentane Isomers and Their Boiling Points

Structural Formula	Boiling Point (°C)
$CH_3-CH_2-CH_2-CH_2-CH_3$	36.1
$CH_3-CH-CH_2-CH_3$ $\quad\quad\quad\mid$ $\quad\quad CH_3$	27.8
$\quad\quad CH_3$ $\quad\quad\quad\mid$ CH_3-C-CH_3 $\quad\quad\quad\mid$ $\quad\quad CH_3$	9.5

4. Now consider possible structural isomers of C_6H_{14}, hexane.

   a. Working with a partner, draw structural formulas for as many different C_6H_{14} isomers as possible. Compare your structures with those drawn by other groups.

   b. How many different C_6H_{14} isomers were found by your class?

5. Build models of one or more C_6H_{14} isomers, as assigned by your teacher.

   a. Compare the 3-D models built by your class with corresponding structures drawn on paper.

   b. Based on your careful examination of the 3-D models, how many different C_6H_{14} isomers are possible?

DEVELOPING SKILLS

A.13 BOILING POINTS OF ALKANE ISOMERS

You have already observed that boiling points of straight-chain alkanes are related to the number of carbon atoms in their molecules. Increased intermolecular forces are associated with the greater molecule-to-molecule contact possible for larger alkanes. Now read on to analyze how the different shapes of alkane isomers affect boiling point.

1. Boiling points for several isomers of pentane and octane are listed in Table 3.5 and Table 3.6. For each set of isomers, how does the boiling point change as the extent of carbon-chain branching increases? Based on your conclusions, assign each of these boiling points to one of the following C_7H_{16} isomers: 98.4 °C, 92.0 °C, 79.2 °C.

   a. CH_3—CH_2—CH_2—CH_2—CH_2—CH_2—CH_3

   b. CH_3—CH_2—CH—CH_2—CH_2—CH_3
       |
       CH_3

   c.
                        CH_3
                        |
       CH_3—CH_2—CH_2—C—CH_3
                        |
                        CH_3

Table 3.6

Some Octane Isomers and Their Boiling Points	
Structural Formula	**Boiling Point (°C)**
CH_3—CH_2—CH_2—CH_2—CH_2—CH_2—CH_2—CH_3	125.7
CH_3—CH_2—CH_2—CH_2—CH_2—CH—CH_3 | CH_3	117.7
CH_3—CH—CH_2—C—CH_3 with CH_3 branches	99.2

2. Here is the structural formula of a C_8H_{18} isomer:

$$CH_3-CH_2-CH_2-\overset{\displaystyle CH_3}{\underset{\displaystyle CH_3}{\overset{|}{\underset{|}{C}}}}-CH_2-CH_3$$

a. Compare this isomer to each C_8H_{18} isomer listed in Table 3.6 (page 291). Predict whether this isomer has a higher or lower boiling point than each of the other listed C_8H_{18} isomers.

b. Would the C_8H_{18} isomer shown here have a higher or lower boiling point than each of the three C_5H_{12} isomers depicted in Table 3.5 (page 290)? Why?

3. How do you explain the boiling point trends that you investigated in this activity? Use pictures and words to support your claim.

CHEM**QUANDARY**

FUELS AND CLIMATE

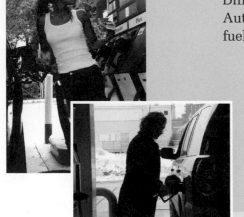

Different parts of the United States experience very different climates. Automobile fuel used in Maine during its cold winter months differs from fuel used, say, in Arizona during hot summer months (Figure 3.23).

- Why do differing climates require different automobile fuels? What practical reasons are there (in terms of transport, storage, and transfer of fuel) for tailoring fuel to a specific climate?

- What aspects of the components of fuel used in a Maine winter would differ from the components of fuel used during a summer in Arizona? (*Hint:* Think in terms of molecular structure.)

Figure 3.23 *How would the properties of the gasoline being pumped in these two situations vary?*

■ MAKING DECISIONS

A.14 WHO HAS THE OIL?

This unit opened with an advertisement for a plug-in hybrid automobile. In the past few years, hybrids and other alternative-fuel vehicles have become more common in the United States. Why do you think this is? Did you note that oil or petroleum has become more expensive or that the U.S. has limited supplies of petroleum? These answers highlight issues involved in dealing with resource distribution across the global community.

Like many other chemical resources, Earth's petroleum is unevenly distributed. Large amounts are concentrated in small areas. In addition, nations with large reserves do not necessarily consume large amounts of the resource. Regions with a high demand for petroleum may be far away from regions with high petroleum reserves. As a result, nations often make exporting and importing arrangements, known as *trade agreements*. Figure 3.9 (page 275) shows worldwide distribution of petroleum reserves, along with regional distribution of the world's population, and regional shares of petroleum consumption. Use Figure 3.9 to help you think about and answer the following questions:

1. Look at identified regional petroleum reserves.
   a. Which region has the most petroleum reserves relative to its population?
   b. Which region has the least petroleum reserves relative to its population?

2. Consider the consumption of petroleum in each region.
   a. Which regions consume a greater percent of the world's supply of petroleum than they possess?
   b. Which regions consume a smaller percent of the world's supply of petroleum than they possess?
   c. Given your answers to Questions 2a and 2b,
      i. Which regions are likely to export petroleum?
      ii. Which regions are likely to import petroleum?

3. List several pairs of regions that might make petroleum trade agreements with each other.

4. Not all possible trade agreements are actually made. What factors might prevent trade relationships between regions?

5. What current or recent world conflicts involve disagreements over or attempts to secure control of oil reserves?

6. What two regions are the largest petroleum consumers?

7. Why do these two regions consume such large quantities of petroleum?

8. Some countries and regions have large populations, but relatively low petroleum consumption. As these areas strive to improve their standard of living, how could each of the following be impacted?
   a. Petroleum demand
   b. Petroleum prices
   c. Trade agreements involving regions with significant reserves
   d. Conflict over oil
   e. Carbon dioxide emissions
   f. Need for alternative-fuel technologies

Thus far, we have recognized how greatly we depend on petroleum. Society's increasing reliance on this nonrenewable, limited resource must be analyzed carefully. However, before we can address such issues, we need a better understanding of petroleum itself. What features make petroleum so valuable for both "building and burning"? How is this related to choices about alternative-fuel vehicles? In Section B, you will learn how chemical bonding helps explain the use of petroleum as a raw material for the manufacture of many useful products.

SECTION A SUMMARY

Reviewing the Concepts

Petroleum (crude oil), a nonrenewable resource that must be refined prior to use, consists of a complex mixture of hydrocarbon molecules.

---

1. What is a hydrocarbon?
2. What does it mean to *refine* a natural resource?
3. What characteristics of petroleum make it a valuable resource?
4. What is the likelihood of discovering a pure form of petroleum that can be used directly as it is pumped from the ground? Explain your answer.
5. What is meant by saying that oil is *crude*?

Petroleum is a source of fuels that provide thermal energy. It is also a source of raw materials for the manufacture of many familiar and useful products.

---

6. On average, the United States uses ~20 million barrels of petroleum daily:
   a. What is the average number of barrels of petroleum used daily in the United States for building (nonfuel) purposes?
   b. How many barrels of petroleum, on average, are burned as fuel daily in the United States?
7. Name several fuels obtained from crude petroleum.
8. a. List four household items made from petroleum.
   b. What materials could be substituted for each of these four items if petroleum were not available to make them?

Liquid substances can often be separated according to their differing boiling points in a process called distillation.

---

9. Under what conditions could density be used to separate two different liquids?
10. Refer to Table 3.2 (page 277). Which possible mixture of two listed substances would be the easiest to separate by distillation? Explain your reasoning.
11. Sketch the basic setup for a laboratory distillation. Label the key features of your sketch.
12. Referring to Table 3.2 (page 277), sketch a graph of the distillation of a mixture of acetone and water. Label its key features.

Fractional distillation of crude oil produces several distinctive and usable mixtures (fractions). Each fraction contains molecules of similar sizes, boiling points, and intermolecular forces.

---

13. How does fractional distillation differ from a simple distillation?
14. Petroleum fractions include light, intermediate, and heavy distillates and residues. List three useful products derived from each of these three fractions.
15. Where in a distillation tower—top, middle, or bottom—would you expect the fraction with the highest boiling point range to be removed? Why?
16. After fractional distillation, each fraction is still a mixture. Suggest a way to further separate the components of each fraction.

> Alkane molecules, hydrocarbons linked by single covalent bonds, are represented by the general formula C_nH_{2n+2}.

> The structures of hydrocarbons affect their intermolecular attractions and, thus, their boiling points.

17. Use the general molecular formula to write the molecular formula for an alkane containing

    a. 9 carbons.   c. 16 carbons.

    b. 10 carbons.   d. 18 carbons.

18. Calculate the molar mass of each alkane listed in Question 17.

19. Name and give the molecular formula for the alkane with a molar mass of

    a. 44 g/mol.

    b. 72 g/mol.

> Isomers are molecules with identical molecular formulas, but different arrangements of atoms. Each isomer is a distinct substance with its own characteristic properties.

20. Are the following three molecules isomers of one another? Explain your answer.

$$
\begin{array}{ccc}
CH_3 & CH_3 & CH_3 \\
| & | & | \\
CH_2-CH_2-CH_2 & & CH_2 \\
& & | \\
& & CH_2-CH_2-CH_3
\end{array}
$$

$$CH_3-CH_2-CH_2-CH_2-CH_3$$

21. Draw structural formulas for at least three structural isomers of C_9H_{20}.

22. What is the shortest-chain alkane that can demonstrate isomerism?

23. An unbranched hydrocarbon molecule can be represented as a linear chain or as a zigzag chain. Explain in what way both representations are correct.

24. Rank the following straight-chain hydrocarbons from lowest boiling point to highest: hexane (C_6H_{14}), methane (CH_4), pentane (C_5H_{12}), and octane (C_8H_{18}). Explain your rankings in terms of intermolecular forces.

25. Consider hexane isomers.

    a. Draw two hexane structural isomers, one a straight-chain molecule and the other a branched-chain molecule.

    b. Which of the two isomers you drew would have the lower boiling point? Explain your choice.

26. Which molecule in each of the following pairs of hydrocarbon molecules would have the lower boiling point? In each case, describe your reasoning.

    a. a short, straight chain or a long, straight chain

    b. a short, branched chain or a long, branched chain

    c. a short, branched chain or a long, straight chain

How can the physical properties of petroleum be explained by its molecules and their interactions?

In this section, you have learned about petroleum, an important natural resource. You have explored its composition and physical properties and considered the interactions among its molecules. Think about what you have learned, then answer the question in your own words in organized paragraphs. Your answer should demonstrate your understanding of the key ideas in this section.

Be sure to consider the following in your response: intermolecular forces, boiling points, viscosity, isomers, and fractions.

Connecting the Concepts

27. Why is petroleum considered a non-renewable resource?

28. Using a Venn diagram, distinguish fractional distillation from simple distillation.

29. Simple distillation is never sufficient to separate two liquids completely. Explain.

30. In a fractionating tower, petroleum is generally heated to 400 °C. What would happen if it were heated to only 300 °C?

31. The molar masses of methane (16 g/mol) and water (18 g/mol) are similar. At room temperature, methane is a gas and water is a liquid. Explain this difference in terms of intermolecular forces.

32. Which mixture would be easier to separate by distillation—a mixture of pentane and straight-chain octane or a mixture of pentane and a branched-chain octane isomer? Explain the reasoning behind your choice.

33. The traditional unit of volume for petroleum is the barrel, which contains 42 gallons. Assume that those 42 gallons provide 21 gallons of gasoline. How many barrels of petroleum does it take to operate an automobile for one year, assuming the vehicle travels 10 000 miles and gets 27 miles per gallon of gasoline?

34. Explain why thermal energy is added at one point and removed at another point in the process of distillation.

Extending the Concepts

35. Is it likely that the composition of crude oil found in Texas is the same as that of crude oil found in Kuwait? Explain your answer.

36. Gasoline's composition is varied by oil companies for use in different parts of the nation and for use in different seasons. What factors help determine the composition of gasoline blended for different seasons?

37. What kind of petroleum trade relationship would be expected between North America and the Middle East? If other world regions become more industrialized and global petroleum supplies decrease, how might the North America–Middle East trade relationship change?

38. How would the hydrocarbon boiling points listed in Table 3.3 (page 283) change if they were measured under increased atmospheric pressure? (*Hint:* Although butane is stored as a liquid inside of a butane lighter, it escapes through the lighter's nozzle as a gas.)

39. The two isomers of butane have different physical properties, as illustrated by their different boiling points. They also have different chemical properties. Explain how isomerism may contribute to their differences in chemical behavior.

40. What properties of petroleum make it an effective lubricant?

41. One problem with diesel fuel is the fact that it solidifies at cold temperatures. How can its formulation be altered to prevent this problem?

SECTION B PETROLEUM: A BUILDING-MATERIAL SOURCE

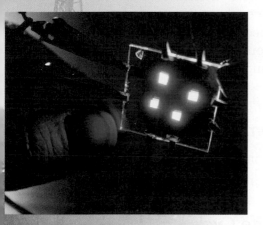

Figure 3.24 *This blue phosphorescent OLED (organic light emitting diode) is made from multiple layers of "designer molecules."*

Why are carbon-based molecules so versatile as chemical building blocks?

Just as an architect uses knowledge of available construction materials in designing a building, a chemist—a "molecular architect"—uses knowledge of available molecules in designing new molecules (Figure 3.24). Architects must know about the structures and properties of common building materials. Likewise, chemists must understand the structures and properties of their raw materials. The "builder molecules"—that small portion of petroleum not used for energy, lubricants, or road tar—are available for building by molecular architects. In this section, you will explore the structures of common hydrocarbon builder molecules and some materials made from them. As you do so, consider the properties that allow these carbon-based molecules to serve as the foundation of so many different materials and products.

GOALS

- Describe the life cycle of a polymer and identify everyday items that are made from petroleum-based polymers.
- Define and distinguish *monomer* and *polymer*.
- Describe and represent electron arrangements between covalently bonded atoms.
- Explain how molecular structure affects physical properties of a polymer.
- Give examples of how functional groups, including rings and multiple bonds, impart characteristic properties to organic compounds.
- Identify and describe the roles of addition and condensation reactions in polymerization.

concept check 4

1. How many bonds does each carbon atom form in an alkane?
2. Describe how you would determine whether two structural formulas are isomers or simply different representations of the same molecule.
3. What are some useful physical properties of plastics?

■ MAKING DECISIONS
B.1 LIFE CYCLE OF A POLYMER

Polymers are an important category of "designed molecules" in the 21st century. You have undoubtedly heard the term polymer and may even know that it refers to a large molecule typically composed of 500 to 20 000 or more repeating units of simpler molecules known as **monomers**. But where do polymers originate, and what do they have to do with our study of petroleum? The following activity will help connect the dots—or the monomers.

Consider the household materials that your class has gathered. Look for the symbol (see Figure 3.25) on each item and group the items according to these symbols.

> The symbols found on many plastics are actually known as resin identification codes. This code system was developed by The Society of the Plastics Industry to facilitate recycling.

Part I: Examining Household Polymers

1. What similarities can you identify among items within a single group?
2. What differences exist between items in different groups?

Part II: Considering a Polymer's Life Cycle

Your teacher will assign each group to a plastics code. Use the Internet or other resources to answer the following questions about the polymer resin referred to by your code. Be prepared to share your results with the class.

3. List your assigned code and the name of the polymer to which it refers.
4. Draw the chemical structure of the monomer that forms your polymer.

Uses

5. List at least five common applications of your polymer.
6. List the materials collected by the class that are made from your polymer.

Figure 3.25 *Each recyclable symbol represents a type of polymer resin. Here, the symbol for polyvinylchloride is shown, along with examples of items made from this polymer resin.*

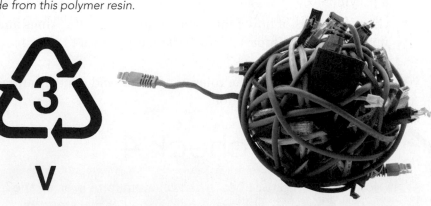

Recycling and Disposal

7. Is this material recycled in your community? If so, how?

8. What products are made from your polymer after it is recycled?

9. If it is not recycled, how is this material disposed?

Carbon Footprint

10. How might use of this material impact your personal carbon footprint? Our national carbon footprint?

As you present your research and listen to your classmates, you may note that these polymers have many similarities. They are composed primarily of carbon and hydrogen atoms with a few other atoms. If their chemical composition is similar, how can their uses and properties be so different? To answer this question, we need to delve deeply into the structure of these hydrocarbon-based compounds to learn how they are "put together." In the next section, you will start this process at the most fundamental level, considering the bond between carbon and hydrogen.

B.2 CHEMICAL BONDING

Hydrocarbons and their derivatives, including the polymers you investigated in Making Decisions B.1, are the focus of the branch of chemistry known as **organic chemistry**. These substances are called *organic compounds* because early chemists thought that living organisms—plants or animals—were needed to produce them. However, chemists have known for more than 150 years how to make most organic compounds without any assistance from living systems.

In hydrocarbon molecules, carbon atoms are joined to form a backbone called a **carbon chain**. Hydrogen atoms are attached to the carbon chain. Carbon's versatility in forming bonds helps explain the abundance of different hydrocarbon compounds. To illustrate this versatility, recall the large number of alkanes that could be made from six carbons in Modeling Matter A.12 (page 290). Hydrocarbons can be regarded as the starting point of an even larger number of compounds that contain atoms of other elements attached to a carbon chain.

Before learning more about hydrocarbon compounds, it will be helpful to understand how atoms are joined together to form molecules through chemical bonding. You probably already learned in other science courses that electrons are involved in chemical bonding. In Unit 1, you learned about chemical bonds that are formed when atoms gain or lose electrons—ionic bonds. Electrons are also central to bonding in hydrocarbons, so let us begin our study of chemical bonds by considering the role of electrons.

The *carbon chain* forms a framework to which a wide variety of other atoms can be attached.

Electron Shells

How are atoms of carbon or other elements held together in compounds? The answer is closely related to the arrangement of electrons in atoms. You already know that atoms are made up of neutrons, protons, and electrons. In addition, you know that neutrons and protons are located in the small, dense, central region of the atom, the *nucleus*. So where are electrons found within the atom and with respect to one another?

In Unit 1 (page 63), you investigated the periodic properties of some elements. You also learned that the tendency of metals to lose electrons is an example of a periodic property. Experimental investigations have revealed other periodic properties of the elements. The analysis of these experimental results has led scientists to suggest that electrons reside in separate energy levels in space surrounding the nucleus. Similar energy levels are grouped into **shells**, each of which can hold only a certain maximum number of electrons. For example, the first shell surrounding the nucleus of an atom has a capacity of two electrons. The second shell can hold a maximum of eight electrons.

Consider an atom of helium (He), the first member of the noble-gas family. See Figure 3.26 (left image). A helium atom has two protons (and two neutrons) in its nucleus, and two electrons occupying the first, or innermost, shell. Because two is the maximum this shell can hold, the shell is completely filled.

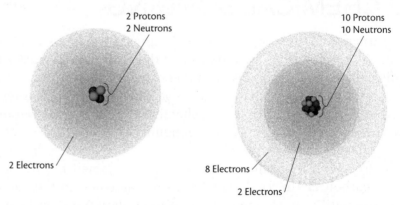

2 Protons
2 Neutrons

10 Protons
10 Neutrons

2 Electrons

8 Electrons

2 Electrons

Figure 3.26 *Atomic structure of helium (left). Note that the nucleus is composed of two protons and two neutrons. Two electrons are located within—and fill—the first electron shell, depicted here by an electron cloud.*

Atomic structure of neon (right). Note that the nucleus is composed of 10 protons and 10 neutrons. Two electrons are located—and fill—the first electron shell, and eight electrons are located within—and fill—the second electron shell.

The next noble gas, neon (Ne), has atomic number 10. See Figure 3.26 (right image). Each electrically neutral neon atom contains 10 protons and 10 electrons. Two electrons occupy and completely fill the first shell. The remaining eight electrons fill the second shell. In neon, each shell has reached its electron capacity.

Helium (He, atomic number 2) and neon (Ne, atomic number 10) are not chemically reactive; their atoms do not combine with each other or with atoms of other elements to form compounds. By contrast, sodium (Na) atoms—with atomic number 11 and one more electron than for neon atoms—are extremely reactive (Figure 3.27). Chemists explain that sodium's reactivity is due to its tendency to lose that one electron in its third, unfilled shell. Doing so leaves 10 electrons, giving the atom the same electron occupancy as a neon atom. In Unit 1 Modeling Matter B.11, you sorted common monatomic ions into groups that corresponded to the location of each element on the periodic table. Recall that lithium and potassium, found along with sodium in group 1 (the leftmost column of the periodic table), also form ions with a 1+ charge. You similarly observed that elements in group 2, such as magnesium and calcium, tend to form ions with a 2+ charge, or lose two electrons.

Fluorine (F) atoms each have nine electrons—one fewer than neon atoms—and are also extremely reactive. Their reactivity is due to their tendency to gain an additional electron, achieving a filled outer shell of electrons. Other halogen (group 17) family elements, including chlorine and bromine, also tend to gain a single electron to form an ion with a 1− charge.

Noble-gas elements are essentially unreactive because their separate atoms have filled electron shells. All but helium have eight electrons in their outer shell; helium needs only two electrons to attain its first-shell maximum. A useful key to understanding the chemical behavior of many elements is to recognize that atoms and ions with *filled electron shells* are particularly stable; that is, they tend to be chemically unreactive.

Figure 3.27 *One product of the reaction between sodium (Na) and water is hydrogen gas. So much heat is produced in this violent reaction that the hydrogen gas often undergoes combustion accompanied by flames.*

Covalent Bonds

In most substances composed of nonmetals, atoms achieve filled electron shells by *sharing* electrons. Typically, only **valence electrons**, or those electrons within an atom's unfilled, outer shell, participate in bonding. As you will see, the sharing of two or more valence electrons between two atoms—called a **covalent bond**—allows both atoms to fill their outer shells completely.

A hydrogen molecule (H_2) provides a simple example of electron sharing. Each hydrogen atom contains only one electron, so one more electron is needed to fill the first shell. Two hydrogen atoms can accomplish this if they each share their single electron. If an electron is represented by a dot (•), then formation of a hydrogen molecule can be depicted this way:

> Remember that ionic bonds are the attractive forces between positively charged cations and negatively charged anions. Ionic charges result from gain or loss of electrons.

$$H\cdot \; + \; \cdot H \longrightarrow H{:}H$$

The chemical bond formed between two atoms that share a pair of electrons is called a **single covalent bond** (see Figure 3.28, page 302). Through such sharing, both atoms achieve the stability associated with completely filled electron shells. A carbon atom, atomic number 6, has six electrons—two in its first shell and four more in the second shell. To fill the second shell to its capacity of eight, four more electrons are needed. These electrons can be obtained through covalent bonding.

The number of valence electrons in an atom can often be predicted from the periodic table. For instance, sodium and potassium are found in group 1,

> In Modeling Matter A.9 (page 283), you built models of alkanes. Recall that each carbon atom had four other atoms attached to it.

and each has one electron in the outermost shell. Atoms of nitrogen family elements have five electrons in their outermost shells and are found in group 15. As you know, halogen atoms need one electron to complete their outermost electron shells. This means that each halogen (group 17) atom has seven valence electrons. Can you identify a pattern that links an element's group number to the number of valence electrons in its outer shell? Locate each family on the periodic table to confirm your pattern, then practice using this pattern to predict the number of valence electrons in sulfur and calcium.

Scientific American Conceptual Illustration

Figure 3.28 *Model of a hydrogen (H₂) molecule.* *A hydrogen molecule is held together by a single covalent bond. Each hydrogen atom in the molecule shares two electrons, thereby filling each atom's outer electron shell. Electrical attractions between shared electrons and each atom's positively charged nucleus (depicted by the high cloud density between the two nuclei) hold the H₂ molecule together.*

Consider the simplest hydrocarbon molecule, methane (CH_4). The formula for this molecule can be determined by evaluating the number of valence electrons in carbon and hydrogen. Each carbon atom has four valence electrons and thus needs four more to attain a completely filled second electron shell. Each hydrogen atom has one valence electron that can be shared, thus four hydrogen atoms are required to provide filled electron shells for one carbon atom. A representation of this arrangement is:

$$4\,H\cdot \; + \; \cdot\overset{\displaystyle\cdot}{\underset{\displaystyle\cdot}{C}}\cdot \; \longrightarrow \; H:\overset{\displaystyle H}{\underset{\displaystyle H}{C}}:H$$

Other familiar molecules can be similarly explained. For instance, an oxygen atom has eight electrons. Two of these electrons are in the first shell, and six are in the second, or valence, shell. Oxygen thus needs two more electrons to fill its valence shell. Therefore one oxygen atom will combine with two hydrogen atoms to make a stable compound, H_2O.

DEVELOPING SKILLS

B.3 PREDICTING AND REPRESENTING CHEMICAL BONDS

As in the formula for a hydrogen molecule, dots surrounding each element's symbol represent the valence electrons for that atom. Structures such as these are called **electron-dot formulas**, also known as **Lewis dot structures** or, simply, **Lewis structures**. The two electrons in each covalent bond "belong" to both bonded atoms. Dots placed between the symbols of two atoms represent electrons that are shared by those atoms.

When the electrons associated with each atom are determined, each shared electron in a covalent bond is "counted" twice, once for each element. For example, count the dots surrounding each atom in methane (see above). Note that each hydrogen atom has a filled outer electron shell with two electrons. The carbon atom also has a filled outer electron shell with eight electrons. Each hydrogen atom is associated with one pair of electrons; the carbon atom has four pairs of electrons, or eight electrons.

For convenience, each pair of electrons in a covalent bond can be represented by a line drawn between the symbols of each atom. This yields another common representation of a covalently bonded substance called a **structural formula**.

> The American chemist G. N. Lewis is credited with laying the foundation for our current understanding of bond formation.

$$H:\overset{\cdot\cdot}{\underset{\cdot\cdot}{C}}:H$$

Electron-dot formula
of methane, CH_4

$$H-\overset{H}{\underset{H}{C}}-H$$

Structural formula
of methane, CH_4

Although you can draw flat, 2-D pictures of molecules on paper, assembling 3-D models gives a far more accurate representation. You used both structural formulas and molecular models to represent simple alkanes in Section A. Such models can help predict a molecule's physical and chemical behavior. You will have an opportunity to build more of these models later in this section.

Sample Problem: *The two-carbon alkane ethane has formula C_2H_6. To what combination of atoms is each carbon atom bonded? Draw a Lewis structure for this molecule.*

To answer, consider the formula for ethane, C_2H_6. Each carbon must form four bonds; one to the other carbon atom and three to hydrogen atoms. The Lewis structure represents these bonds as shown:

$$H:\overset{H\ H}{\underset{H\ H}{C:C}}:H$$

1. Nitrogen forms a very stable compound with hydrogen.
   a. How many hydrogen atoms would be needed to combine with one nitrogen atom to form a stable compound? Explain your answer.
   b. Draw a Lewis structure for the stable compound you identified in Question 1a.
2. A common type of organic compound is an alkyl halide. These are hydrocarbon compounds in which a halogen atom replaces a hydrogen atom in the structure, such as C_3H_7Cl.
   a. What similarity of halogen and hydrogen atoms allows the halogen to substitute for hydrogen in organic compounds?
   b. Draw all possible Lewis structures for C_3H_7Cl.
3. Consider an alkane with seven carbon atoms.
   a. Write the formula of this hydrocarbon.
   b. Draw Lewis structures for at least three isomers of this compound.

concept check 5

1. In your own words, explain how the periodic table can be used to determine the number of valence electrons in an atom.
2. How is the number of valence electrons in an atom related to the number of chemical bonds that atom will form?
3. Why are polymers so versatile?

B.4 CREATING NEW OPTIONS: PETROCHEMICALS

In Making Decisions B.1, you investigated several polymers that together account for the basis of countless modern products. These synthetic polymers are used in paints, fabrics, rubber, insulating materials, foams, adhesives, molding, and structural materials. But how long have these polymers been part of people's daily lives, and what led to their preponderance?

Until the early 1800s, all objects and materials used by humans were either created directly from wood or stone or crafted from metals, glass, and clays. Available fibers included cotton, wool, linen, and silk. All medicines and food additives came from natural sources. Celluloid (from wood) and shellac (from animal materials) were the only sources for commercially produced polymers.

Today many common objects and materials created by the chemical industry are unlike anything seen or used by citizens of the 1800s or even the mid-1900s. See Figure 3.29. Many of these differences are due to the use of compounds produced from oil or natural gas, called **petrochemicals**. Some petrochemicals, such as detergents, pesticides, pharmaceuticals, and cosmetics, are used directly. Most petrochemicals, however, serve as raw materials in producing other synthetic substances, particularly a wide range of polymers. Worldwide production of these petroleum-based polymers is more than four times that of aluminum products.

The astonishing fact is that it takes relatively few *builder molecules* (small-molecule compounds) to make thousands of new substances, including many polymers. One builder molecule is *ethene*, C_2H_4, a hydrocarbon compound commonly called *ethylene*. The structural formula for ethene is shown in the equation on the next page. The two carbon atoms in an ethene molecule share two pairs of electrons. This produces an arrangement of electrons called a **double covalent bond**. Because of the high reactivity of its double bond, ethene is readily transformed into many useful products.

Figure 3.29 *Two street scenes—one from the early 1900s (top), and one from the early 2000s (bottom). What similarities and differences do you notice in the use of materials?*

Double bonds will be discussed in more detail on page 310.

A simple example—the formation of ethanol (ethyl alcohol) from water and ethene—illustrates how ethene reacts:

| Ethene | Water | Ethanol |

In this reaction, the water molecule (HOH) "adds" to the double-bonded carbon atoms by adding an H to one carbon atom and an OH group to the other carbon atom. This type of chemical change is called an **addition reaction**.

Ethene can also undergo an addition reaction with itself. Because the added ethene molecule contains a double bond, another ethene molecule can be added, and so on. This creates a long-chain substance called *poly-ethene*, commonly known as *polyethylene*, a polymer consisting of 500 to 20 000 or more repeating units of ethene (ethylene) monomer. The chemical reaction that produces polyethene can be represented this way:

Ethene (ethylene) The growing polymer chain Polyethene (polyethylene)

Polymers formed in reactions such as this are called—sensibly enough—**addition polymers**. Polyethene is commonly used in grocery bags and packaging. The United States produces millions of kilograms of polyethene annually. Recall that in Making Decisions B.1, two classes of recyclable materials were made from polyethene, high-density polyethylene (HDPE) and low-density polyethylene (LDPE).

We can make various addition polymers from monomers that closely resemble ethene. The most common is to replace one or more hydrogen atoms in ethene with an atom or atoms of another element. In the following examples, note the replacements for hydrogen atoms in each monomer and polymer:

Vinyl chloride Polyvinyl chloride (PVC)

Acrylonitrile Polyacrylonitrile (PAN)

Styrene Polystyrene

Perhaps you wonder why two names are shown for the same substance— *ethene/ethylene* and *polyethene/polyethylene*. Chemists have developed a system for naming substances that makes it easy to communicate clearly with each other. However, some substances were given names long before this system was universally accepted. Some "common names" of substances, such as *ethylene* and *polyethylene*, continue in wide use.

The hexagon-shaped ring depicted in styrene and polystyrene represents benzene, a substance discussed later on page 315.

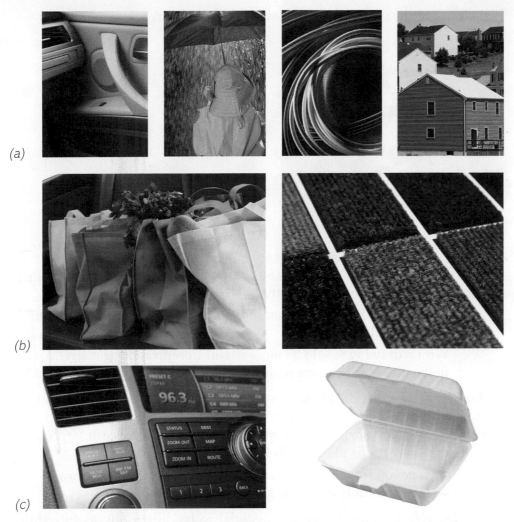

Figure 3.30 *Products commonly made from (a) polyvinyl chloride, (b) polypropylene, and (c) polystyrene.*

The atoms that compose the monomers determine the properties of the resulting polymer. Thus polyethene and polyvinyl chloride are fundamentally different materials. However, as with many modern materials, the properties of a polymer are often altered to meet a variety of needs and to produce many products. See Figure 3.30.

■ MODELING MATTER

B.5 POLYMER STRUCTURE AND PROPERTIES

Unmodified, the arrangement of covalent bonds in long, stringlike polymer molecules causes the molecules to coil loosely. A collection of polymer molecules (such as those in a sample of molten polymer) can intertwine, much like strands of cooked spaghetti. In this form, the polymer is flexible and soft.

1. Look at the two drawings below. Each represents a polymer strand.

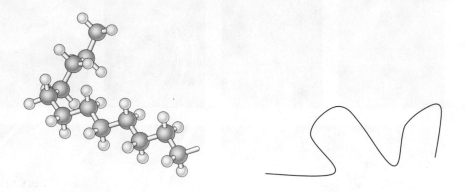

    a. What information does the structure on the left provide that the structure on the right does not?

    b. Which drawing would be more helpful in explaining the bonding within a polymer?

    c. Which drawing would be more useful in representing many molecules and showing their general relationship to one another?

Throughout the rest of this activity, use simple drawings, similar to the drawing on the right above, to represent polymers. Ball-and-stick models are provided in the text so that you continue to gain experience interpreting multiple representations of molecules.

2. Using a pencil line on paper to represent a linear polymer, draw a collection of loosely coiled polymer molecules.

Ductility refers to the ability of a material to be drawn out, as into thin strands. You used this term when learning about metals in Unit 1. We can use the same term when referring to polymers. For most polymers like the ones you just drew, flexibility and ductility depend on temperature. When the material is warm, the polymer chains can slide past one another easily. The polymer becomes more rigid when it cools. We use such polymers, classified as *thermoplastics*, in many everyday products, such as soft drink bottles, milk bottles, and plastic grocery bags.

The flexibility of a polymer can also be enhanced by adding molecules that act as internal lubricants among the polymer chains. For example, untreated polyvinyl chloride (PVC) is used in rigid pipes and house siding. With added lubricant molecules, polyvinyl chloride becomes flexible enough for use in such consumer goods as raincoats and inflatable pool toys.

The reactions that form polymer chains can also take place perpendicular to the main chain, forming side chains. These polymers are called **branched polymers**. The branched form of polyethylene is shown in the following illustration. The extent of branching can be controlled by adjusting reaction conditions.

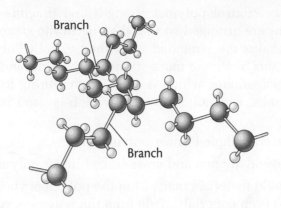

Branch

Branch

3. Draw at least two different models of branched-chain polymers. Try to vary the representations—the forms of branched polymers can differ greatly.

Branching changes the properties of a polymer by affecting the ability of chains to slide past one another and by altering intermolecular forces. Recall from Section A that the boiling points of branched alkanes were also affected by differences in intermolecular forces.

4. Which of the molecules in Step 3 would have the higher boiling point? Support your claim with reasoning.

Another way to alter the properties of polymers is through **cross-linking**. Polymer rigidity can be increased if the polymer chains are cross-linked so that they can no longer move or slide readily. You can see this for yourself if you compare the flexibility of a plastic soda bottle with that of its screw-on cap. Polymer cross-linking is much greater in the cap. The following illustration shows a cross-linked form of polyethene.

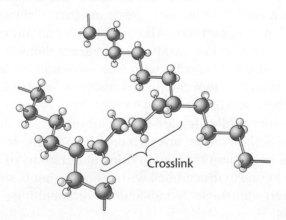

Crosslink

5. Draw several linear polymer chains that have been cross-linked.
6. Draw several cross-linked, branched polymers.
7. Explain how cross-linked, branched polymers differ from cross-linked, linear polymers.

Chemists can control polymer strength and toughness. To do this, the polymer chains are arranged so that they lie in the same general direction. The aligned chains are stretched until they uncoil. Polymers that remain uncoiled after this treatment make strong, tough films and fibers. Such materials include polyethene, which is used in everything from garbage bags to artificial ice rinks, and polypropylene, which is used in bottles and some carpeting.

8. Draw several aligned linear polymer chains.

9. Draw several aligned and cross-linked linear polymer chains.

10. How would materials made from the polymers you drew in Step 8 be different from materials made from the polymers you drew in Step 9?

11. For what types of products could the materials made from the polymer you drew in Step 8 be used?

12. Based on this activity and what you learned in Section B.4, how do you think the structures of LDPE and HDPE differ?

B.6 BEYOND ALKANES

Carbon atoms are versatile building blocks. Carbon can form bonds with other atoms in several different ways. As you learned in Section A, each carbon atom in an alkane molecule is bonded to four other atoms. Compounds such as alkanes are called **saturated hydrocarbons** because each carbon atom forms as many single covalent bonds as it can. In some hydrocarbon molecules, carbon atoms bond to three other atoms, not four. Members of this series of hydrocarbons are called **alkenes**. The simplest alkene was briefly introduced earlier in this section—ethene, C_2H_4.

The carbon–carbon bonding that characterizes alkenes is a double covalent bond. In a double bond, four electrons (two electron pairs) are shared between the bonding partners. Alkenes, which contain carbon–carbon double bonds, are described as **unsaturated hydrocarbons**. Not all carbon atoms are bonded to their full capacity with four other atoms. Because of their double bonds, alkenes are more chemically reactive—and therefore better builder molecules—than are alkanes.

Not all builder molecules are hydrocarbons. In addition to carbon and hydrogen, some also contain one or more other atoms, such as oxygen, nitrogen, chlorine, or sulfur. One way to think about many of these substances is as carbon-backbone hydrocarbons with other elements substituted for one or more hydrogen atoms. Such molecules are sometimes called **substituted hydrocarbons**.

Adding atoms of other elements to hydrocarbon structures significantly changes their chemical reactivity. Even molecules composed of the same elements can have quite different properties. The molecules that make up a permanent-press shirt and permanent antifreeze may each contain the same elements—carbon, hydrogen, and oxygen. The dramatic differences in their properties and uses result from the arrangements of atoms in these two molecules.

Some of these differences can be explained by intermolecular forces. As you know, strength of intermolecular forces is related to molecular structure. Compounds in which a hydrogen atom is bonded directly to an atom of oxygen, nitrogen, or fluorine exhibit particularly strong intermolecular forces, evidenced by exceptionally high boiling points. Intermolecular forces among these molecules are known as **hydrogen bonds** and are about 10 times stronger than other types of intermolecular forces. Hydrogen bonding can also explain much of the unique behavior of water, which you will investigate in depth in Unit 4.

■ MODELING MATTER

B.7 BUILDER MOLECULES

This activity, in which you will use models to represent various arrangements of atoms, will help you to become more familiar with alkenes and builder molecules containing oxygen, and connect your knowledge of polymers to the structure of these molecules. Be prepared to record your observations and answers to questions in your notebook.

Part 1: Alkenes

1. Examine the Lewis dot formula and structural formula for ethene, C_2H_4. Confirm that each atom has attained its filled outer shell of electrons:

$$
\begin{array}{ccc}
\overset{\displaystyle H\ \ H}{H\!:\!\overset{..}{C}\!:\!:\!\overset{..}{C}\!:\!H} & \overset{\displaystyle H\ \ \ H}{H\!-\!C\!=\!C\!-\!H} & CH_2CH_2 \text{ or } C_2H_4 \\
\text{Lewis dot formula} & \text{Structural formula} & \text{Molecular formula}
\end{array}
$$

The names of alkenes follow a pattern much like that of the alkanes. The first three alkenes are *ethene, propene,* and *butene.* The same prefixes that you learned for alkanes are used to indicate the number of carbon atoms in the molecule's longest carbon chain. However, each alkene name ends in *-ene* instead of *-ane.*

2. Recall that the alkane general formula is C_nH_{2n+2}. Examine the molecular formulas of ethene (C_2H_4) and butene (C_4H_8). What general formula for alkenes do these molecular formulas suggest?

3. Assemble models of an ethene (C_2H_4) molecule and an ethane (C_2H_6) molecule. See Figure 3.31. Compare the arrangements of atoms in the two models. Rotate the two carbon atoms in ethane about the single bond. Then try a similar rotation with ethene.

   a. What do you observe?

   b. Can you build a molecule in which you can complete a rotation about a C=C double bond?

   c. Write a general rule to summarize your findings.

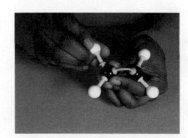

Figure 3.31 *You can use the flexible connector pieces to model the double bonds in molecules such as ethene.*

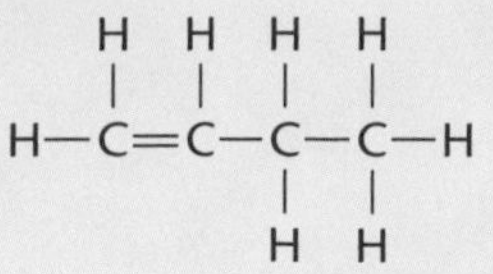

H—C=C—C—C—H (with H atoms above and below as shown)

1-Butene, or simply butene

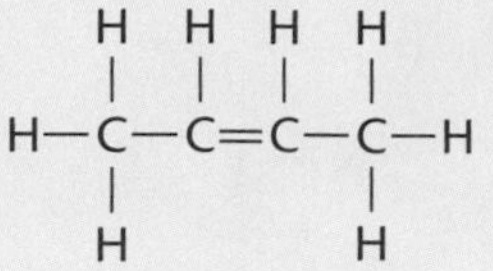

H—C—C=C—C—H (with H atoms as shown)

2-Butene

H—C—C=C—H with H—C—H below

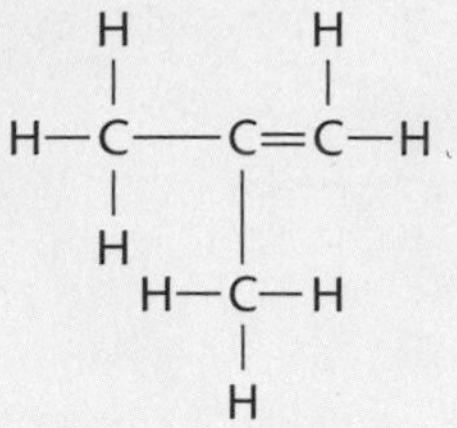

Methylpropene

Figure 3.32 *Three isomers of butene, C_4H_8.*

4. Build a model of butene (C_4H_8). Remember that alkenes must contain a double bond. Compare your model to those made by others.
   a. How many different arrangements of atoms in a C_4H_8 chain appear possible? Each arrangement represents a different substance—another example of isomers.
   b. Which structural formulas in Figure 3.32 correspond to models built by you or your classmates?

As with alkanes, we name alkenes according to the length of their longest carbon chain. The carbon atoms are numbered, beginning at the end of the chain closest to the double bond. The name of each isomer starts with the number assigned to the first double-bonded carbon atom. Look again at the butene isomer structures in Figure 3.32 and confirm these naming rules with 1-butene and 2-butene.

5. Does each pair of formulas represent isomers or the same substance?
   a. $CH_2 = CH— CH_2—CH_3$ or $CH_3—CH_2—CH=CH_2$
   b. 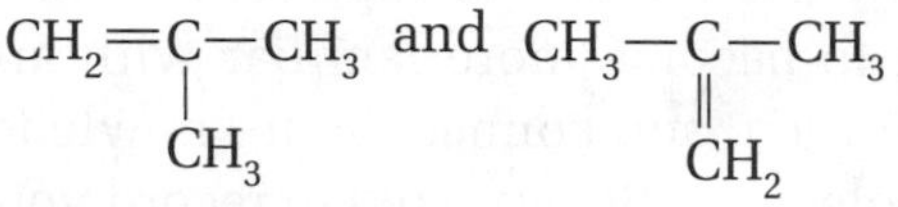 $CH_2=C—CH_3$ and $CH_3—C—CH_3$ (with CH_3 below first, CH_2 double-bonded below second)

6. How many isomers of propene (C_3H_6) are there? Support your answer with the appropriate structures.

7. Do the following two structures represent isomers or the same substance? Explain.

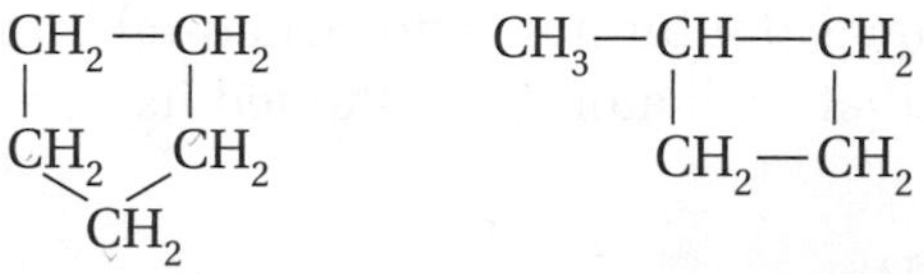

   $CH_2—CH_2$ / CH_2 CH_2 / CH_2 $CH_3—CH—CH_2$ / $CH_2—CH_2$

8. Assemble a model of a hydrocarbon molecule containing a *triple* covalent bond. Base this on your knowledge of hydrocarbon molecules with either single or double bonds between carbon atoms. Your completed model represents a member of the hydrocarbon series known as **alkynes**. Given your understanding of the naming of alkanes and alkenes, write structural formulas for

   a. ethyne, commonly called acetylene (Figure 3.33).
   b. 2-butyne.

9. Are alkynes saturated or unsaturated hydrocarbons? Explain.

10. Would you expect alkynes to be more or less reactive than alkenes? Explain.

Figure 3.33 *Ethyne (commonly called acetylene) is the primary fuel in blowtorches. When burned, it generates the high temperatures needed to weld and to cut through metal.*

Part 2: Compounds of Carbon, Hydrogen, and Singly Bonded Oxygen

11. Assemble as many different molecular models as possible, using all nine of these atoms:

    - 2 carbon atoms (each forming four single bonds)
    - 6 hydrogen atoms (each forming a single bond)
    - 1 oxygen atom (forming two single bonds)

12. On paper, draw a structural formula for each compound you have assembled in Step 11, indicating how the nine atoms are connected. Compare your structures with those built by other classmates. Be sure that you are satisfied that all possible structures have been produced, then answer these questions:

    a. How many distinct structures did you identify?

    b. Are all of these structures isomers? Explain.

13. Each structure you have identified possesses distinctly different physical and chemical properties. Which compound should have the highest boiling point? Why?

> Remember that boiling points are related to strengths of intermolecular forces.

Part 3: Alkene-Based Polymers

In this part of the activity, you and your classmates will use models to simulate the formation of several addition polymers.

14. Build models of two ethene molecules.

15. Using the information on page 306 as a guide, combine your two models into a **dimer**, a two-monomer structure. How did you have to modify the monomer structure to accomplish this?

16. Combine your dimer with that of another lab team. Continue this process until your class has created one long-chain structure. Although your resulting molecular chain is not yet long enough to be regarded as a model of a polymer, you have simulated the processes involved in creating a typical addition polymer.

17. Repeat Steps 14 and 15, first by replacing ethene with vinyl chloride, and then by replacing ethene with propene. Then find out whether you can follow the same steps for styrene. (See page 306 for molecular structures of these substances.)

18. Assume the structures you built in Steps 16 and 17 became significantly larger. Give the name of each resulting polymer.

19. Are the chains you built linear or branched? How would this affect the properties of the polymer?

20. Build a polyethene chain that includes some cross-linking.

    a. How does cross-linking change the behavior of the model, if at all?

    b. How would the cross-linking change the observed properties of the polymer?

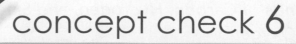

concept check 6

1. If you were going to design a new material for a car tire, would you want to use a polymer that was or was not cross-linked? Explain.
2. Consider the unsaturated hydrocarbons you have learned about and modeled.
   a. Why are alkenes useful as monomers?
   b. Could an alkyne be used as a monomer? Explain.
3. Why might an organic compound containing the arrangement of atoms shown here behave as an acid?

B.8 CARBON RINGS AS BUILDER MOLECULES

So far, you have examined only a small part of the inventory of builder molecules available to chemical architects. Now you will explore two classes of compounds in which carbon atoms are joined in *rings* rather than in chain structures. Many important naturally occurring compounds are composed of carbon atoms in rings. These types of compounds often have unique chemical and physical properties that make them useful as starting materials for other compounds. Like many of the builder molecules you have already studied, these carbon-ring compounds are commonly found in—or produced from—petroleum.

As a first step to doing this, picture a straight-chain hexane molecule, $CH_3-CH_2-CH_2-CH_2-CH_2-CH_3$. Next, remove one hydrogen atom from the carbon atom at each end. Then imagine those two carbon atoms bonding to each other. The result is the molecule known as cyclohexane.

$$CH_3CH_2CH_2CH_2CH_2CH_3$$

Hexane

Cyclohexane

Cyclohexane, obtained by petroleum refining, is a starting material for making nylon, a familiar petrochemical polymer. Cyclohexane is representative of the **cycloalkanes**, which are saturated hydrocarbons made up of carbon atoms joined in rings.

Another class of hydrocarbon builder molecules is **aromatic compounds**. Aromatic compounds have chemical properties distinctly different from those of the cycloalkanes and their derivatives. The structural formula of benzene (C_6H_6), which is the simplest aromatic compound, is shown in the following illustration. In the representation on the right, each "corner" of the six-carbon (hexagonal) ring represents a carbon atom with its hydrogen atom.

Although chemists who first investigated benzene proposed these structures, the chemical properties of the compound did not support their model. Recall that carbon–carbon double bonds (C=C) are usually very reactive. But, in terms of its reactivity, benzene behaves as though it does not contain any such double bonds. A deeper understanding of chemical bonding was needed to explain benzene's puzzling structure.

Substantial experimental evidence indicates that all carbon–carbon bonds in benzene are identical. Thus, its structure is not well represented by alternating single and double bonds. Instead, chemists often represent a benzene molecule in the following way:

The inner circle represents the equal sharing of bonding electrons among all six carbon atoms. The hexagonal ring represents the bonding of six carbon atoms to each other. Each "corner" in the hexagon ring is the location of one carbon atom and one hydrogen atom, thus accounting for benzene's formula, C_6H_6.

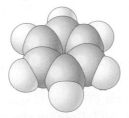

Benzene, C_6H_6

Although only small amounts of aromatic compounds are found in petroleum, fractionation and cracking produce large amounts of these aromatic compounds. Gasoline contains benzene and other aromatic compounds as octane-number enhancers, but they are used primarily as builder molecules. Entire chemical industries (dye and drug manufacturing, in particular) are based on the unique chemistry of aromatic compounds (Figure 3.34).

Figure 3.34 *Aromatic compounds are the primary builder molecules for dyes such as these.*

B.9 BUILDER MOLECULES CONTAINING OXYGEN

Some petroleum-based builder molecules are not components of petroleum but are instead synthesized from molecules found in petroleum. For example, the builder molecule ethanol can be produced by reacting ethene found in petroleum with water, as you saw in Section B.4 (page 306). In assembling molecular models with C, H, and O atoms, you probably discovered one of the following compounds:

$$CH_3{-}OH \qquad\qquad CH_3{-}CH_2{-}OH$$

Methanol (methyl alcohol)   Ethanol (ethyl alcohol)

As you can see, each molecule has an —OH group attached to a carbon atom. This general structure is characteristic of a class of compounds known as **alcohols**. The —OH group is recognized by chemists as one type of **functional group**—an atom or group of atoms that imparts characteristic properties to organic compounds. If the letter R is used to represent all the rest of the molecule other than the functional group, then the general formula of an alcohol can be written as

$$R{-}OH$$

Any alcohol

In this formula, the line indicates a covalent bond linking the oxygen of the —OH group with the adjacent carbon atom in the molecule. In methanol (CH_3OH), the letter R would represent CH_3—; in ethanol (CH_3CH_2OH), R would represent CH_3CH_2—.

The formulas and structures of two common alcohols are shown below. What does R represent in each compound?

$$CH_3CH_2CH_2OH$$

$$\begin{array}{ccc} & CH_2 & \\ CH_2 & & CH_2 \\ | & & | \\ CH_2 & & CH{-}OH \\ & CH_2 & \end{array}$$

1-Propanol   Cyclohexanol

Two other classes of oxygen-containing compounds, **carboxylic acids** and **esters**, are versatile builder molecules. Both can be synthesized from alcohols. The functional group in each of these classes of compounds contains two oxygen atoms, as shown in the following structures:

$$\overset{\displaystyle O}{\underset{\displaystyle R'{-}C{-}OR}{\|}} \qquad\qquad \overset{\displaystyle O}{\underset{\displaystyle R{-}C{-}OH}{\|}}$$

Ester   Carboxylic acid

Note that both classes of compounds have one oxygen atom double-bonded to a carbon atom and a second oxygen atom single-bonded to the same carbon atom. *Ethanoic acid* (a carboxylic acid that is better known as *acetic acid*) and *methyl ethanoate* (an ester more commonly called *methyl acetate*) are examples of these two classes of compounds:

Ethanoic acid   Methyl ethanoate

Some familiar examples that include or involve the use of alcohols, carboxylic acids, and esters are shown in Figure 3.35. These builder molecules can be modified by adding other functional groups that include nitrogen, sulfur, or chlorine atoms. The rich variety of functional groups greatly expands the types of molecules that chemists can ultimately build.

(a) Carboxylic acids

(b) Esters and alcohols

Figure 3.35 *Examples that include or involve (a) carboxylic acids and (b) esters and alcohols.*

■ INVESTIGATING MATTER

B.10 CONDENSATION

Preparing to Investigate

In this investigation, you will synthesize esters. As you learned in Section B.9, esters are often synthesized from alcohols and carboxylic acids. The esters you will produce have familiar, pleasing fragrances. Many perfumes and artificial flavorings contain esters. In fact, the characteristic aromas of many herbs and fruits arise from esters found naturally in those plants.

The compounds you will be making are also naturally occurring compounds; the chemical structure of these compounds is already known from other investigations. The three esters you will synthesize are listed in Table 3.7.

Table 3.7

Table of Chemical Structures of Three Esters

Name	Structure
Methyl 2-hydroxybenzoate (methyl salicylate)	
Pentyl ethanoate (pentyl acetate)	$H_3C-C-O-C_5H_{11}$ with $\|$ O
Octyl ethanoate (octyl acetate)	$C_8H_{17}-O-\overset{O}{\overset{\|}{C}}-CH_3$

One example of the formation of an ester is the production of methyl ethanoate (methyl acetate) from ethanoic acid (acetic acid) and methanol in the presence of sulfuric acid as a catalyst:

$$CH_3-\overset{O}{\overset{\|}{C}}-OH \ + \ H-O-CH_3 \ \xrightarrow{H_2SO_4} \ CH_3-\overset{O}{\overset{\|}{C}}-O-CH_3 \ + \ H-OH$$

Ethanoic acid   Methanol   Methyl acetate   Water

Identify the carbon atoms in the ester that came from the carboxylic acid. Which carbon atoms originated in the alcohol?

To emphasize the roles of functional groups in the formation of an ester, a general equation can be written using R notation.

$$R-\overset{O}{\overset{\|}{C}}-OH \ + \ H-O-R' \ \xrightarrow{H_2SO_4} \ R-\overset{O}{\overset{\|}{C}}-O-R' \ + \ H-OH$$

Carboxylic acid   Alcohol   Ester   Water

Note how the functional groups of the acid and alcohol combine to form a water molecule, while the remaining atoms join to form an ester molecule.

Examine the three esters you will synthesize in this investigation (Table 3.7). Identify the alcohol and the carboxylic acid needed to make each ester. (*Hint:* Use Table 3.8 to find the structures of some common alcohols and carboxylic acids.)

Construct a data table that lists each ester to be synthesized, along with the alcohol and carboxylic acid you will use in the synthesis reaction. Provide space to note observations about the properties of each reactant and product. Your teacher will review and approve your data table before you begin.

Table 3.8

Some Common Acids and Alcohols Used in Ester Synthesis

Carboxylic Acids		Alcohols	
Ethanoic (acetic) acid	CH₃COOH	Methanol	CH₃OH
Benzoic acid		Ethanol	CH₃CH₂OH
Butanoic (butyric) acid	CH₃CH₂CH₂COOH	Propanol	CH₃CH₂CH₂OH
Methanoic (formic) acid	HCOOH	Pentanol	CH₃CH₂CH₂CH₂CH₂OH
2-Hydroxybenzoic (salicylic) acid		Octanol	CH₃CH₂CH₂CH₂CH₂CH₂CH₂CH₂OH

Gathering Evidence

You will be using a procedure that has been developed to optimize the ester synthesis reaction. When chemists synthesize a compound, they must often conduct many investigations to find the best conditions for the reaction. A good synthesis not only produces the maximum quantity of the desired product, it also minimizes waste and use of hazardous compounds.

As you synthesize these esters, you will be using very small quantities of materials. Why is it important to do synthesis in the classroom with small amounts of materials?

Read the procedure that follows to learn what you will need to do. Be sure to note safety precautions.

> Preventing waste is the first principle of green chemistry.

1. Before you begin, put on your goggles, and wear them properly throughout the investigation.

2. Prepare a hot-water bath by adding ~50 mL tap water to a 100-mL beaker. Place the beaker on a hot plate, and heat the water until it is near boiling. Be sure that no flames or sparks are in the area. (*Note:* Use a hot plate, *not* a Bunsen burner, to heat the water bath.)

3. Obtain a small, clean test tube. Place 20 drops of the appropriate alcohol into the tube. Next add 0.1 g of the carboxylic acid (as identified in your table). Then add 2 drops of concentrated sulfuric acid to the tube. (***Caution:*** *Concentrated sulfuric acid is corrosive and will cause burns to skin or fabric. Add the acid slowly and very carefully. If any sulfuric acid accidentally spills on you, ask a classmate to notify your teacher immediately. Wash the affected area immediately with tap water and continue rinsing for several minutes.*

Figure 3.36 *Proper setup for heating substances with a water bath.*

Other than water and sulfuric acid, the substances used in this investigation are volatile and flammable.)

 4. As you dispense these compounds, note their odors. (***Caution:** Do not directly sniff any reagent fumes—some may irritate or burn nasal passages.)* Record any odors you happen to note.

5. Place the test tube in the hot-water bath you prepared in Step 2, as shown in Figure 3.36.

6. Keep the tube in the hot water, and do not spill the contents. Note any color changes. Continue heating for five minutes.

7. If you have not noticed an odor from the test tube after five minutes, remove the test tube from the water bath, hold the test tube away from you with the tongs, and wave your hand across the top of the test tube to waft any vapors toward your nose (Figure 3.37). Record your observations about the odor of the product. Compare your observations with those of other class members.

Figure 3.37 *To avoid burning or irritating your nasal passages, gently "waft" vapors toward your nose with your hand.*

8. Repeat Steps 3 through 7 using 20 drops of alcohol and 20 drops or 0.1 g (depending upon physical state) of carboxylic acid, along with 2 drops of concentrated sulfuric acid for each of the esters in your table. (***Caution:** Avoid inhaling any ethanoic (acetic) acid fumes.)*

9. Dispose of your products as directed by your teacher.

10. Wash your hands thoroughly before leaving the laboratory.

Interpreting Evidence

1. Describe the odors of the three esters produced in this laboratory activity.

2. Write a chemical equation for the formation of methyl 2-hydroxybenzoate (methyl salicylate).

3. Repeat Question 2 for the second and third esters you produced.

4. Besides the ester, what other substance or substances are formed in each reaction?

5. Use the Internet or other resources to find information about the odors of the esters in Table 3.7 (page 318). Based on this information, answer the following questions.

    a. What claims can you make about each synthesis reaction you conducted?

    b. What evidence supports your claims?

Reflecting on the Investigation

6. You used sulfuric acid in each reaction. What is its purpose in the synthesis?

7. In Question 5b, you provided evidence to support claims about your synthesis of esters. What other types of evidence might chemists use to support claims that they have synthesized a particular substance?

8. Why do you think that the type of reaction in this synthesis is called *condensation*? (*Hint:* Look at the products of the reaction.)

B.11 CONDENSATION POLYMERS

Earlier in this section, you learned that many polymers can be formed from alkene-based monomers through addition reactions. Not all polymers are formed in this way, however. Natural polymers, such as proteins, starch, cellulose (in wood and paper), and synthetic polymers, including the familiar nylon and polyester, are also formed from monomers. However, unlike addition polymers, these polymers form with the loss of simple molecules when monomer units join. You just conducted a similar reaction, called a **condensation reaction**, in Investigating Matter B.10. As each ester was synthesized from the combination of an alcohol and a carboxylic acid, a small molecule was produced. Examine the carboxylic acid and alcohol represented below.

$$\underset{\text{Succinic acid}}{\text{H}-\text{O}-\overset{\overset{\text{O}}{\|}}{\text{C}}-\text{CH}_2-\text{CH}_2-\overset{\overset{\text{O}}{\|}}{\text{C}}-\text{O}-\text{H}} \qquad \underset{\text{Ethylene glycol}}{\text{H}-\text{O}-\text{CH}_2-\text{CH}_2-\text{O}-\text{H}}$$

How do these reactants differ from the carboxylic acids and alcohols you used in Investigating Matter B.10? How would the resulting ester be different from the esters that you produced?

Condensation reactions can be used to make a second type of polymer, and the resulting product is called a **condensation polymer**. Monomers for condensation polymers typically contain an –OH group. Recall that monomers for addition polymers contained a carbon–carbon double bond. Here is a simple representation of this process:

$$\underset{\text{Monomer 1}}{-\text{R}_1-\text{O}-\text{H}} + \underset{\text{Monomer 1}}{\text{H}-\text{O}-\text{R}_2-} \longrightarrow \underset{\substack{\text{Condensation} \\ \text{polymer}}}{-\text{R}_1-\text{O}-\text{R}_2-} + \underset{\text{Water}}{\text{H}-\text{OH}}$$

As you probably noted, starting materials for condensation polymers (as opposed to simple esters) have functional groups on each end, so the result of each condensation reaction is another (larger) molecule that can now undergo additional condensation reactions.

One common condensation polymer is polyethylene terephthalate (PET). It is often used in bottles for soft drinks and other beverages. It has many other applications, including as polyester in clothing and fiberfill and as Dacron for surgical tubing and other high-strength fabrics. More than 2 million kilograms of PET are produced each year in the United States. Some examples of products made from PET are shown in Figure 3.38 (page 322).

The structure of the PET polymer

Figure 3.38 *Some common applications of PET.*

While some PET may be recycled, material containing high quantities of additives, such as dyes or other polymers, is either disposed of in a landfill or incinerated. Another option is a process that breaks down PET into its monomers, which are then repolymerized into high-quality PET products. This technology decreases the need for new petroleum-based builder molecules as well as the quantity of PET discarded in landfills.

■ MAKING DECISIONS

B.12 BUILDER MOLECULES IN
TRANSPORTATION

You have surveyed how some petroleum components are used as building blocks for countless everyday objects. You can easily find examples of products built from petroleum-based materials in your surroundings, products that were unknown even a few decades ago. See Figure 3.39. Synthetic polymers have many advantages over traditional materials, including favorable strength-to-weight ratios and recyclability. How are these advantages of synthetic polymers useful to automobile designers when they are trying to improve the fuel efficiency of new cars? On the other hand, what are some potential downfalls of the use of synthetic polymers in automobile manufacturing?

In the TV ad that opened this unit (see page 266), a new automobile was described as helping to conserve petroleum resources. Is this possible? Is it plausible? In this activity, you will consider this claim and evaluate how builder molecules are used in automobile construction.

Figure 3.39 *How many petroleum-based components do you see?*

1. List several automobile components commonly made of polymers.

2. The TV advertisement claimed that driving the *TLC-p* will "help conserve petroleum resources."

    a. Does that claim refer to petroleum-based builder molecules used in the manufacture of the *TLC-p*?

    b. Would the components on your list from Question 1 possibly differ between a *TLC-p* and a traditional automobile? Why?

3. Try to think of a replacement material (not another polymer or petrochemical) for each component on your list.

4. Decide whether any disadvantages are associated with each replacement material you have suggested. List the two biggest disadvantages for each component, or write "none" if the replacement material provides similar properties at a comparable cost. A table such as the one shown here will help you organize your ideas and answers.

Summary for Question 4			
Automobile Component	Petroleum Based? (yes or no)	Potential Replacement Material	Disadvantages of Replacement Material

5. Advocates of petroleum conservation typically emphasize increasing an automobile's energy efficiency, rather than replacing auto components made from petroleum's polymer-based builder molecules. Why might that be?

SECTION B SUMMARY

Reviewing the Concepts

> The atoms in hydrocarbons and in other molecules are held together by covalent bonds.

1. What is a covalent bond?

2. Why do atoms with filled outer electron shells not form covalent bonds?

3. It has been suggested that a covalent bond linking two atoms is like two dogs tugging at the same sock. Explain how this analogy describes the way that shared electrons hold together atoms in a covalent bond.

> Molecules can be represented by Lewis dot structures and structural formulas.

4. What does each dot in a Lewis dot structure represent?

5. What is the advantage of using a dash instead of electron dots to represent a covalent bond?

6. a. What information does a structural formula convey that a molecular formula does not?

   b. In what ways is a structural formula an inadequate representation of an actual molecule?

7. Choose a branched six-carbon hydrocarbon molecule.

   a. Draw a Lewis dot structure to represent its structure.

   b. Draw a structural formula for the same molecule.

8. Each carbon atom has six total electrons. Why, then, does the electron dot representation of a carbon atom show only four dots?

> The chemical combination of small, repeating molecular units (monomers) forms large molecules (polymers). The molecular structure of polymers can be designed or altered to produce materials with desired flexibility, strength, and durability.

9. How many repeating units are found in each of these structures?

   a. a monomer   c. a trimer

   b. a dimer   d. a polymer

10. a. Draw pictures of polymers with and without cross-links.

    b. How does cross-linking alter the properties of a polymer?

11. List four examples of natural polymers and four examples of synthetic polymers.

12. What structural features make the properties of one polymer different from those of another?

13. List and explain two methods for altering the characteristics of a polymer.

> Carbon atoms can be joined by double bonds, in alkenes, or triple bonds, in alkynes.

14. Why is the term unsaturated used to describe the structures of alkenes and alkynes?

15. Rank the following in order of decreasing chemical reactivity: alkyne, alkane, and alkene. Explain your answer.

16. Why is an unsaturated hydrocarbon likely to be more reactive than a saturated hydrocarbon?

17. Predict the molecular formula for a compound named

   a. octane.

   b. pentyne.

   c. decene.

18. How does the presence of a double or triple bond affect the ability of a hydrocarbon molecule to rotate about its carbon chain?

19. Draw a Lewis dot structure and a structural formula for each of the following molecules:

   a. propane.

   b. propene.

   c. propyne.

20. Why is it impossible to have methene, a one-carbon alkene?

> Addition reactions involve the chemical combination of molecules. Some polymers can be synthesized by combining monomers in an addition reaction.

21. Use structural formulas to illustrate how propylene (propene), shown below, can polymerize to form polypropylene.

$$
\begin{array}{c}
\quad\ \ H\ \ \ H\ \ H \\
\quad\ \ | \quad\ | \quad | \\
H-C-C=C-H \\
\quad\ \ | \\
\quad\ \ H
\end{array}
$$

22. Explain why alkanes cannot be used to make polymers.

23. What change occurs in the monomer structure during polymerization?

> Ring compounds include cycloalkanes and aromatic compounds.

24. In what ways are cycloalkanes different from aromatic compounds?

25. Draw a structural formula and write the molecular formula for cyclopentane.

26. How did the term aromatic compounds originate?

27. Why is the circle-within-a-hexagon representation of a benzene molecule a better model than the hexagon with alternating double bonds?

28. What do the six "corners" in the hexagon-shaped benzene representation signify?

> Functional groups—such as alcohols, carboxylic acids, and esters—impart characteristic properties to organic compounds.

29. a. Write the structural formula for a molecule containing three carbon atoms that represents

     i. an alcohol.

     ii. an organic acid.

     iii. an ester.

  b. Circle the functional group in each structural formula.

  c. Name each compound.

30. a. What does the R stand for in ROH?

  b. What type of compound would this be?

31. Classify each of the following compounds as a carboxylic acid, an alcohol, or an ester:

  a. $CH_3-CH_2-CH_2-CH_2-OH$

  b. $CH_3-\overset{\overset{\textstyle O}{\|}}{C}-O-CH_2-CH_3$

  c.
$$
\begin{array}{c}
CH_3-CH-CH_2-CH_3 \\
\quad\quad | \quad\quad\ O \\
\quad\quad\quad\quad\quad\ \| \\
CH_2-C-OH
\end{array}
$$

  d. $CH_3-\overset{\overset{\textstyle O}{\|}}{C}-OH$

32. What functional group is responsible for the pleasing odors of many herbs, fruits, and perfumes?

> Condensation reactions involve the chemical combination of two larger molecules with the loss of a small molecule.

33. Ethanoic (acetic) acid, CH_3COOH, and butanoic (butyric) acid, $CH_3(CH_2)_2COOH$, are common reactants in condensation reactions. Their structural formulas are

and

$$CH_3 - \overset{\overset{\displaystyle O}{\|}}{C} - OH$$

$$CH_3 - CH_2 - CH_2 - \overset{\overset{\displaystyle O}{\|}}{C} - OH$$

Predict the products of a condensation reaction between *each of these two acids and each of the alcohols* in the list below. Provide the name and structural formula for the product. For example, the product of a condensation reaction between ethanoic acid and ethanol is ethyl ethanoate. (*Hint:* To organize your response, construct a table with two columns—one for each acid—and three rows—one for each alcohol.)

a. methanol (CH_3OH)

b. ethanol (C_2H_5OH)

c. propanol (C_3H_7OH)

34. Why is the word *condensation* used to describe the reaction that forms esters?

35. Name three examples of polymers formed by condensation reactions.

> Many everyday objects are constructed from petroleum-based polymers.

36. For each of the resin codes listed below, identify the monomer used and one possible use of the polymer.

a. LDPE

b. PS

c. V

37. What type of polymer material would you use if you were going to make a bag for bread? Explain your choice.

38. What are some issues associated with the recycling of materials made from polymers?

39. Nylon is a very important synthetic polymer. It is composed of the two monomers shown below.

$$HO - \overset{\overset{\displaystyle O}{\|}}{C} - CH_2CH_2CH_2CH_2 - \overset{\overset{\displaystyle O}{\|}}{C} - OH$$

$$\underset{H}{\overset{H}{\diagdown}} N - CH_2CH_2CH_2CH_2CH_2CH_2 - N \underset{H}{\overset{H}{\diagup}}$$

a. Draw the structure of the polymer formed by the reaction of these two monomers.

b. Is this polymer formed by a condensation or addition process?

Why are carbon-based molecules so versatile as chemical building blocks?

In this section, you have learned about chemical bonding and its application to organic chemistry. You have explored the structure of hydrocarbons, other "builder" molecules from petroleum, and polymers. You have also considered (and in one case, investigated) two important methods of synthesizing polymers. Think about what you have learned, then answer the question in your own words in organized paragraphs. Your answer should demonstrate your understanding of the key ideas in this section.

Be sure to consider the following in your response: covalent bonding, electron shells, Lewis dot structures, builder molecules, functional groups, and polymers.

Connecting the Concepts

40. Identify the missing molecule (represented by a question mark) in each of the following equations. (*Hint:* If you are uncertain about an answer, start by completing an atom inventory. Remember that the final equation must be balanced.)

a. $CH_2{=}CH_2 + ? \longrightarrow CH_3{-}CH_2Br$

b.
$$CH_3{-}\overset{\overset{\displaystyle H}{|}}{C}{=}CH_2 + ? \longrightarrow CH_3{-}\underset{\underset{\displaystyle OH}{|}}{CH}{-}CH_3$$

c.
$$C_6H_{13}{-}\overset{\overset{\displaystyle O}{\|}}{C}{-}OH + ? \longrightarrow$$
$$C_6H_{13}{-}\overset{\overset{\displaystyle O}{\|}}{C}{-}O^-Na^+ + HOH$$

41. Assemble models of benzene and cyclohexane. Describe differences between the two molecular structures.

42. Architects have a saying: "Form follows function." Does this saying apply to chemistry, or is the opposite—"function follows form"—more correct? Explain your answer.

43. List five products manufactured from cyclic compounds.

44. How does adding a double or a triple bond change the shape of the carbon backbone in a hydrocarbon?

45. Draw structural formulas for both a saturated form of C_4H_8 and an unsaturated form of C_4H_8.

Extending the Concepts

46. Considering that petroleum is a natural resource, why aren't all petroleum-based polymers classified as natural?

47. Research and write about one example in which substituting a different functional group in a drug's molecular structure has had a dramatic affect on its medicinal properties.

48. Organic molecules that contain alcohol or carboxylic acid functional groups are often more soluble in water than similarly sized hydrocarbon molecules. Explain.

49. Investigate several different polyesters. How are they formed? What explains their differences in properties? Give two structural formulas for polyesters. List five everyday products made of polyester.

50. C_2H_6O is the formula for two isomers: one has a boiling point of 78 °C, and the other has a boiling point of –22 °C.

   a. Draw the structural formula for each isomer.

   b. Which of the two isomers has the higher boiling point? Explain your choice.

CHEMISTRY *AT WORK*
Q&A

You can do your part to protect Earth by recycling trash, turning off lights, or riding your bike instead of driving. But how do you save the planet after a major environmental disaster, such as an oil spill? Environmental chemists are a pivotal part of teams that help evaluate the severity of oil spills and assist decision-makers in developing plans to speed recovery. Read on to see what one chemist is learning about how the environment behaves after an oil spill—even if it took place decades ago.

Christopher Reddy, Associate Scientist in the Department of Marine Chemistry and Geochemistry at Woods Hole Oceanographic Institution in Woods Hole, Massachusetts

Bird covered in oil at Fort Baker Cove in Sausalito, CA.

Q. What is environmental chemistry, and what do environmental chemists do?

A. Environmental chemistry is the study of how elements and compounds behave and interact in the environment, as well as their fate over time. Most environmental chemists are interested in gathering data and using data to weave a story. When a chemical spills in a river, what happens over an hour, a week, a decade?

Q. How did you end up in this career?

A. After I got my chemistry degree in college, I wanted to study something with real-world, practical value that people can apply right away. Eventually, I decided to study marine pollution. My research in graduate school set an important standard that I still keep in my job at Woods Hole: I won't do research unless I can explain it to my Uncle Bob or anyone else I meet. I want to be able to say what I'm doing, why it's important, and how it will help people make decisions.

Q. What does your lab study?

A. Part of what we do is track oil spills that happened a long time ago. Our big research question is, how long does oil last in the environment, and why? When you think of an oil spill, you imagine big black puddles coating sandy beaches and rocks. Some people think that when those puddles disappear, the oil is gone. Our studies show that oil from spills can stick for decades. Oil spills are a tragedy, but they also give us a tremendous opportunity to understand how nature handles unwanted chemical guests. Studying old spills helps people who will be in charge of cleaning up after a spill to develop a plan.

Q. Why would oil from a spill linger so long?

A. We know that microbes like to eat oil. Oil is made up of many different types of molecules, and each has a different structure that affects whether bacteria will devour it quickly or pass it by. You'd suspect that something long and straight—like a Popsicle—would be easier to eat than something with lots of branches. We found that long, straight hydrocarbon molecules called *n*-alkanes are usually the first to get eaten, but branched alkanes and cyclic molecules can stay in the environment. If these molecules are tucked away from wind and rain, you might still find them long after a spill.

Helen White takes a core of sediment in Wild Harbor to analyze how bacteria decomposed oil from the 1969 spill.

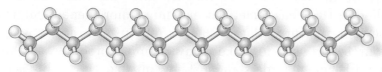

A long *n*-alkane called *n*-hexadecane.

Q. Describe one of the spills you have studied.

A. We recently tested an area on a beach in Cape Cod, Massachusetts, where a tanker spilled oil in 1969—more than 40 years ago! We grabbed samples from places where oil might remain by hammering tubes into the sand and pulling out chunks of soil 10 inches deep. We ran extracts from these samples through a gas chromatograph, an instrument that analyzes what types of molecules were in the samples. Most of the oil was gone, but we found compounds left from the oil in one spot, a little marsh tucked behind the bay where the spill happened.

Wild Harbor appears pristine, but oil from the 1969 spill still lies buried in marsh sediments.

Q. Is this oil still affecting the environment?

A. Oil spills, even old ones, can devastate marine creatures. Our team worked with a student who studied fiddler crabs. She gathered crabs from oily areas and from places with little or no leftover oil. Back in the lab, we compared how they reacted to a pendulum in front of them, which is scary for crabs. Crabs from less oily places ran away much faster than crabs from heavily oiled places. The slower crabs are basically intoxicated from oil—which can make them easier for predators to nab.

Q. How do I become an environmental chemist?

A. Keep in mind that before you become an environmental chemist, you have to become a chemist first. You'll need a strong background in chemistry and math. Get a head start by volunteering in your community. If you live near a body of water, volunteer with organizations that monitor water quality. These organizations give volunteers kits to measure water's oxygen content and to test for pollutants. It's a great way to get out in the field, get your hands dirty, and get data to make decisions about how to protect the environment.

SECTION C PETROLEUM: AN ENERGY SOURCE

What are the benefits and consequences of burning hydrocarbons?

For almost 5000 years, humans have used petroleum in small amounts. Only in the last 150 years, roughly since the first U.S. oil well was drilled in 1859, has society become much more dependent on this non-renewable resource (Figure 3.40). You just learned how petroleum resources are used to make many common products. In this section, you will begin to understand just how much everyday life is influenced by petroleum's role as an energy source—not just in powering automobiles and other vehicles, but in energizing modern society itself. In addition, you will have the opportunity to consider the potential costs of widespread hydrocarbon combustion.

Figure 3.40 *In 1859, Edwin Drake struck oil at the world's first successful oil well, in Titusville, PA. Oil was already refined and used for many purposes at that time. Drake developed an efficient way to retrieve petroleum, leading to an oil boom in that region.*

GOALS

- Identify and give examples of kinetic energy, potential energy, and the law of conservation of energy.
- Describe endothermic and exothermic reactions in terms of total energy involved in bond breaking and bond making using words, equations, and potential energy diagrams.
- Use experimental data to compare fuels and calculate molar heats of combustion.
- Write balanced equations for combustion of hydrocarbon fuels, including energy involved.
- Discuss the carbon cycle and how natural or human factors can affect it.
- Consider the impacts of personal and community behavior on global climate.

✓ concept check 7

Consider what occurs when you burn a candle.
1. After the candle burns for a while, you notice it getting smaller. Where has the mass gone?
2. You also notice energy is released in the form of heat and light. What is the source of that energy?
3. Candle burning involves a chemical change. Describe the reactants and products in this reaction.

■ INVESTIGATING MATTER

C.1 COMPARING FUELS

Asking Questions

You turn on your gas stove to boil some water in a pot. Perhaps you grill burgers on a propane grill. You burn candles, watch an explosive car crash on the television, or burn gasoline in your car's engine to make it go. You encounter and harness energy produced by burning petroleum-based fuels every day. Have you ever wondered how much energy is given off in these reactions? Is one fuel "better" or "more efficient" than others? How would you compare fuels? You will begin to explore these ideas in this investigation.

Preparing to Investigate

Several different sources of energy are available for this investigation. Use the steps outlined in *Gathering Evidence* to guide your investigation of each fuel or energy source. Before you begin, read *Gathering Evidence* to learn what you will need to do and note safety precautions. Think about data and observations you will need to record and prepare an appropriate data table.

Gathering Evidence

1. Before you begin, put on your goggles and wear them properly throughout the investigation.

2. Carefully measure ~100 mL of chilled water. Record the volume to the nearest milliliter. Pour the water into a clean, empty soft-drink can.

3. Set up the apparatus as shown in Figure 3.41. Do not ignite or turn on the energy source yet. Adjust the can so that the top of the flame or energy source will be ~2 cm from the bottom of the can.

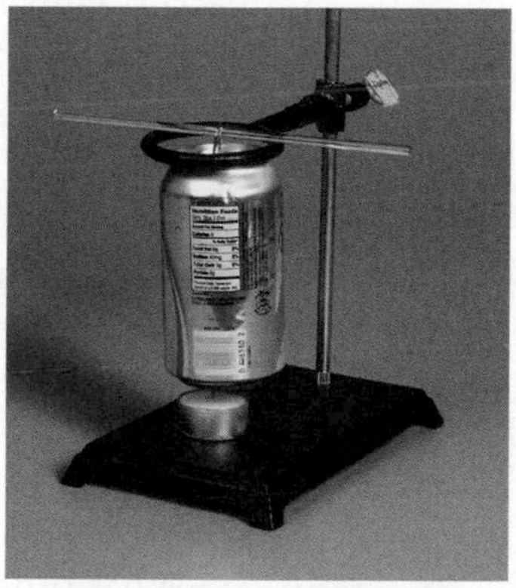

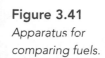

Figure 3.41
Apparatus for comparing fuels.

4. Measure the water temperature to the nearest 0.1 °C. Record this value.

5. Place the energy source under the can of water. Ignite or turn on the energy source and begin heating. As the water heats, stir it gently. (*Caution: Do not stir with the thermometer—use a stirring rod.*)

6. After three minutes, discontinue heating and extinguish or turn off the energy source.

7. Continue stirring the water until its temperature stops rising. Record the highest temperature reached by the warmed water.

8. Discard the warmed water from the can. Repeat Steps 2–7 for each fuel or energy source specified by your teacher.

9. Wash your hands thoroughly before leaving the laboratory.

Interpreting Evidence

Use your own group's data to answer the following questions.

1. Rank the energy sources from "best" to "worst." Explain how you made these decisions.

2. When comparing energy sources:

   a. How would you or did you account for different initial water temperatures?

   b. How would you or did you account for different volumes of water?

3. Look at the data for your "best" energy source. If you repeated this investigation using half the volume of water, how would the temperature change of the water differ? Explain your prediction.

It would be helpful to be able to use compiled class data to compare the different tested energy sources.

4. As a class, determine a unit that can be used to "standardize" data for comparison.

   a. Record and explain that unit.

   b. Describe in your own words what it means to standardize data.

5. Once class data have been collected:

   a. Rank the energy sources from "best" to "worst" based upon class data.

   b. Did your ranking process differ from the process that you used for your group's data?

   c. Did your rankings differ? Explain.

Reflecting on the Investigation

6. In this investigation, you varied the energy source and standardized for volume and initial temperature of water. Consider other variables in this investigation.

   a. Which other variable was controlled in this investigation?

   b. Do you think that this was the most useful variable to control if your goal was to compare energy sources? Why or why not?

   c. What other variables could be controlled or standardized to more effectively compare fuels and energy sources?

7. Think about the units and quantities you observed and used in this investigation.

   a. Did you measure anything using "energy units"?

   b. What are some units of energy with which you are familiar?

   c. Describe what you would need to know in order to calculate the energy involved in heating the water in each of your trials.

C.2 ENERGY FROM FOSSIL FUELS

Each fuel source you tested in Investigating Matter C.1 released energy in the form of heat and light as it burned. Why does burning such fuels release energy? What is happening at the level of atoms and molecules that can account for this?

Fossil-fuel energy is comparable in some ways to the energy stored in a wind-up toy race car. The "winding-up" energy tightens a spring in the toy. Most of that energy, stored within the coiled spring, is a form of **potential energy**, which is energy of position (or condition). As the spring unwinds, it provides energy to the moving parts of the car, which moves the car forward. Energy related to motion is called **kinetic energy**. Thus the movement of the car is based on converting potential energy into kinetic energy. Eventually, the toy "winds down" to a lower-energy, more-stable state and stops.

In a similar manner, **chemical energy**, which is another form of potential energy, is stored in chemical compounds. When an energy-releasing chemical reaction takes place (such as when fuel burns), bonds break and reactant atoms reorganize to form new bonds. The process yields products with different and more stable bonding arrangements of their atoms. That is, the products have less potential energy (chemical energy) than did the original reactants. Some of the energy stored in the reactants has been released in the form of heat and light.

The combustion, or burning, of methane (CH_4) gas illustrates such an energy-releasing reaction. It can be summarized this way:

$$CH_4 \quad + \quad 2\,O_2 \quad \longrightarrow \quad CO_2 \quad + \quad 2\,H_2O \quad + \quad Energy$$

Methane   Oxygen   Carbon   Water
gas   gas   dioxide gas

> A rock poised at the top of a hill has more potential energy— energy of position— than the same rock resting at the bottom of the hill.

> Methane is a major component of natural gas.

The reaction releases considerable **thermal energy**. In fact, laboratory burner flames are based primarily on that reaction—conducted, of course, under very controlled conditions. To gain a better understanding of the energy involved, imagine that the reaction takes place in two simple steps: One step involves bond breaking and one step involves bond making.

The actual reaction is more complicated than this, but simplifying it into two steps allows us to more easily analyze the energy involved in the overall chemical reaction.

In the first step, suppose that all the chemical bonds in one CH_4 molecule and two O_2 molecules are broken. The result of this bond-breaking step is that separated atoms of carbon, hydrogen, and oxygen are produced. All such bond-breaking steps are energy-requiring processes, or **endothermic** changes.

> An endothermic reaction can proceed only if energy is continuously supplied.

In an endothermic change, energy must be added to "pull apart" the atoms in each molecule. Thus energy appears as a *reactant* in Step 1, as shown in the following chemical equation:

$$\text{Energy} + CH_4 + 2\,O_2 \longrightarrow C + 4\,H + 4\,O$$

To complete the methane-burning reaction, suppose that the separated atoms now join to form the new bonds needed to make the product molecules: one CO_2 molecule and two H_2O molecules. The formation of chemical bonds is an energy-releasing process, or an **exothermic** change. Because energy is given off, it appears as a *product* in Step 2, as shown in the following chemical equation:

$$C + 4\,H + 4\,O \longrightarrow CO_2 + 2\,H_2O + \text{Energy}$$

> Once an exothermic reaction begins, it releases energy until the reaction stops

When methane burns, the energy released in forming carbon–oxygen bonds in CO_2 and hydrogen–oxygen bonds in H_2O is greater than the energy used to break the carbon–hydrogen bonds in CH_4 and the oxygen–oxygen bond in O_2. That is why this overall chemical change is exothermic. The complete energy "accounting summary" for burning methane is shown in Figure 3.42.

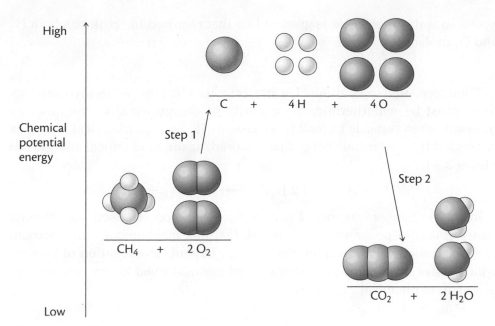

High

Chemical potential energy

C   +   4 H   +   4 O

Step 1

Step 2

CH$_4$ +   2 O$_2$

CO$_2$ +   2 H$_2$O

Low

Figure 3.42 *A potential energy diagram for the formation of carbon dioxide and water from methane and oxygen gases. In Step 1, bonds break. This is an endothermic process. Step 2, bond making, releases energy, so it is an exothermic process. Because more energy is released in Step 2 than is required in Step 1, the overall reaction is exothermic.*

Whether an overall chemical reaction is exothermic or endothermic depends on how much energy is added (endothermic process) in bond breaking and how much energy is given off (exothermic process) in bond making. If more energy is given off than is added, the overall change is exothermic. However, if more energy is added than is given off, the overall change is endothermic. For chemical reactions, this energy is generally measured in joules (J), the SI unit for energy. A joule is a derived unit that is defined as the energy exerted by a force of one newton moving an object through a distance of one meter; for example, it would require approximately one joule to lift a stick of butter one meter into the air. The exothermic combustion of methane releases 55 600 J for every gram of CH$_4$ (or 890 kJ for each mole of CH$_4$) that is burned.

In general, if a process converts potential energy into kinetic energy, then the reverse process converts kinetic energy back to potential energy. For example, if you wind the spring of a model race car, you are converting your energy of motion (winding) into energy stored in the spring. A biker or skateboarder uses the transformation of potential energy into kinetic energy in order to complete jumps and tricks (Figure 3.43).

Likewise, if a particular chemical reaction is exothermic (releasing thermal energy), then the reverse reaction is endothermic (converting thermal into potential energy). For example, burning hydrogen gas—involving the formation of water—is exothermic. The energy released by formation of H—O

If a rock rolling downhill is an exothermic process, then pushing the same rock back uphill must be an endothermic process.

Figure 3.43 *Bicycling is an example of continuous conversions between potential and kinetic energy.*

bonds in water molecules is greater than that required to break bonds in H_2 and O_2 molecules.

$$2\,H_2 \;+\; O_2 \;\longrightarrow\; 2\,H_2O \;+\; \text{Energy}$$

Therefore, the separation of water into its elements—the reverse reaction—must be endothermic, the quantity of energy equal to the quantity released when water is formed from gaseous H_2 and O_2. Electrical energy is converted into potential energy that is stored, again, as chemical energy. See Figure 3.44.

$$\text{Energy} \;+\; 2\,H_2O \;\longrightarrow\; 2\,H_2 \;+\; O_2$$

In these changes, no energy is actually consumed or "used up," though transformations occur among *chemical, thermal, mechanical,* and *electrical energy.* This concept is summarized by the **Law of Conservation of Energy,** which states that energy is neither created nor destroyed in any mechanical, physical, or chemical processes.

Figure 3.44 *An electric current provides energy to decompose water, producing hydrogen (right test tube) and oxygen (left test tube) gas. This process is referred to as the "electrolysis" of water. As follows from the formula of water (H_2O), hydrogen and oxygen gas are produced in a 2 to 1 ratio.*

DEVELOPING SKILLS
C.3 ENERGY IN CHEMICAL REACTIONS

1. One way to understand energy involved in breaking chemical bonds is to use an analogy, such as the process of pulling apart two magnets.
   a. How is pulling apart magnets similar to breaking chemical bonds?
   b. Does pulling apart magnets require energy or release energy?
   c. How is the energy involved in pulling apart magnets analogous to the energy involved in breaking chemical bonds?
   d. Think about holding two magnets close to each other. You let go, and they snap together. How is this similar to energy involved in making chemical bonds?

2. Consider the chemical reaction in which water is decomposed into its elements.
   a. Write a chemical equation for this reaction.
   b. Draw Lewis structures for each reactant and product.
   c. Draw a potential energy diagram for this reaction.
   d. Describe your energy diagram in words.

3. Consider the reaction in which CO_2 and H_2O form CH_4 and O_2.
   a. Write a balanced chemical equation for this reaction.
   b. Include the appropriate quantity of energy (in joules) as either a reactant or product in the chemical equation. (*Hint:* See Section C.2, page 335.)
   c. Explain how you knew whether to consider energy as a reactant or product.

d. Is the reaction exothermic or endothermic?

e. Draw a potential energy diagram for this reaction.

4. Evidence indicates that fossil fuels originate from biomolecules of prehistoric plants and animals.

   a. Burning fossil fuels releases energy. Does it make sense to say that, overall, the chemical reactions that produced the fossil fuels are endothermic? Why or why not?

   b. What is the origin of the energy required to make fossil fuel molecules?

5. Think about an ice cube melting in a glass. (Recall Investigating Matter B.8 in Unit 2.)

   a Is the physical process of melting an ice cube exothermic or endothermic?

   b. Explain your reasoning.

   c. Which is more important in this process: making and breaking chemical bonds *within* the water molecules or making and breaking intermolecular forces *among* the water molecules? Explain.

6. Think about one of the fuel sources you used in Investigating Matter C.1.

   a. Was energy released or absorbed by the combustion of the fuel?

   b. Was energy released or absorbed by the water in the can?

   c. How is the quantity of energy released by combustion of the fuel related to the energy involved in heating the water?

■ INVESTIGATING MATTER

C.4 COMBUSTION

Asking Questions

You strike a match and a hot, yellow flame appears. If you bring the flame close to a candlewick, the candle ignites and burns (Figure 3.45). Candle burning involves chemical reactions of wax, which is composed of long-chain alkanes, with oxygen gas at elevated temperatures. Although many chemical reactions are involved in combustion, chemists simplify the process by usually focusing on the overall changes. For example, the complete burning of one component of candle wax, $C_{25}H_{52}$, can be summarized this way:

Figure 3.45 *As with all combustion reactions, thermal energy is released when candles burn. What other form(s) of energy result from this process?*

$$C_{25}H_{52}(s) \quad + \quad 38\ O_2(g) \longrightarrow 25\ CO_2(g) \quad + \quad 26\ H_2O(g) \quad + \quad \text{Energy}$$

| Paraffin wax (Alkane) | Oxygen gas | Carbon dioxide gas | Water vapor | |

You already know that fuels provide thermal energy as they burn. But how much energy is released? How can we measure the quantity of released energy? You began to develop ideas about how to quantify energy released during combustion in Investigating Matter C.1. You will follow a very similar procedure in this investigation, except that you will need to make a few additional measurements and record your measurements with more precision. As a result, you will be able to experimentally determine the *heat of combustion* of a candle (paraffin wax) and compare this quantity with known values for other hydrocarbons.

Write a scientific question that summarizes the goal of this investigation.

Preparing to Investigate

Before you begin, review *Gathering Evidence* in Investigating Matter C.1. As a class, you will design a similar procedure for this investigation. Review *Reflecting on the Investigation* Questions 6 and 7 (page 332) in Investigating Matter C.1 as you develop your procedure. Once your class procedure is completed, think about the data and observations you will need to record and prepare an appropriate data table. Remember that you will need to record data both before and after the candle is burned. Be sure to include a column that can be used to report the difference between the *before* and *after* values that you noted.

Gathering Evidence

Before you begin, put on your goggles, and wear them properly throughout the investigation. Follow the procedure developed by your class, including appropriate safety considerations. Collect data as required. When you are finished, clean up your materials as indicated by your teacher and wash your hands thoroughly before leaving the laboratory.

Analyzing Evidence

Use your laboratory data to complete calculations that will allow you to answer your beginning question. The following diagram and steps will serve as a guide. Steps 1–3 guide you to calculate the thermal energy absorbed by the water. This is equal to the thermal energy released by the paraffin that was combusted. Steps 4–6 guide calculations that will allow you to answer your beginning question.

Specific heat capacity is sometimes shortened to the term *specific heat*.

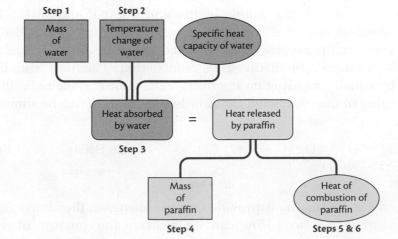

1. Calculate the mass of water heated. (*Hint:* The density of liquid water is 1.0 g/mL. Thus each milliliter of water has a mass of 1.0 g.)

2. Calculate the total increase in the temperature of the water.

3. Using values from the two preceding steps, calculate how much thermal energy was used to heat the water sample. (*Hint:* Use the method illustrated in the sample problem.)

Sample Problem: *Suppose a 10.0-g water sample is heated from 25.0 to 30.0 °C, an increase of 5.0 °C. How much thermal energy must have been added to the water?*

The answer can be reasoned this way. As you learned in Unit 2 Investigating Matter B.8, the specific heat capacity of a substance describes the quantity of energy required to cause a change in temperature. The specific heat capacity of water is 4.2 J/g•°C. That is, it takes 4.2 J to raise the temperature of 1 g water by 1 °C.

In this example, however, there is 10 times more water and a temperature increase that is 5 times greater. Thus 10.0 × 5.0, or 50 times more thermal energy is needed. Therefore, the specific heat must be multiplied by 50 to obtain the answer: 50 × 4.2 J = 210 J. It takes 210 J to increase the temperature of 10.0 g water by 5.0 °C.

> The exact specific heat capacity of water is 4.184 J/g • °C. However, a rounded-off value of 4.2 J/g • °C is adequate.

---

4. Calculate the total mass of paraffin wax burned.

The quantity of thermal energy given off when a certain amount of a substance burns is called the **heat of combustion**. See Table 3.9 (page 340). The heat of combustion can be expressed as the thermal energy released when either one gram or one mole of substance burns.

5. Calculate the heat of combustion of paraffin, expressed in units of

    a. joules per gram (J/g) of paraffin.

    b. kJ/g.
    (*Hint:* Assume that all the energy released by the burning paraffin wax is absorbed by the water.)

> 1 kJ = 1000 J

If the amount of substance burned is one mole, the quantity of thermal energy involved is called the **molar heat of combustion**.

6. Calculate the molar heat of combustion of paraffin, expressed in units of kJ/mol. (*Hint:* You just calculated the thermal energy released when one gram of paraffin burns. Because one mole of paraffin ($C_{25}H_{52}$) has a mass of 352 g, the molar heat of combustion will be 352 times greater than the heat of combustion expressed as kJ/g.)

Interpreting Evidence

1. Quantitatively describe how the temperature of a sample of water changes when energy is added. (*Hint:* Use the value and units of water's specific heat capacity to help prepare your answer.)

Your teacher will direct the collection of class heat of combustion data, expressed in units of kilojoules per gram (kJ/g) of paraffin. Use the collected data to make interpretations based upon the investigation.

2. Use the combined results of your class to determine a best estimate for the heat of combustion of paraffin in kJ/g. Use the class estimate as you answer the following questions.

3. How does your experimental heat of combustion (in kJ/g) for paraffin wax, $C_{25}H_{52}$, compare to the accepted heat of combustion for propane, C_3H_8? See Table 3.9.

4. How do the molar heats of combustion (in kJ/mol) for paraffin and propane compare?

5. Explain the differences in your answers to Questions 3 and 4 for paraffin and propane.

6. In your view, which hydrocarbon—paraffin or propane—is the better fuel? Explain your claim, citing evidence from your observations and data.

Table 3.9

Heats of Combustion for Selected Hydrocarbons

Hydrocarbon	Formula	Heat of Combustion	Molar Heat of Combustion
Methane	CH_4	55.6	890
Ethane	C_2H_6	52.0	1560
Propane	C_3H_8	50.0	2200
Butane	C_4H_{10}	49.3	2859
Pentane	C_5H_{12}	48.8	3510
Hexane	C_6H_{14}	48.2	4141
Heptane	C_7H_{16}	48.2	4817
Octane	C_8H_{18}	47.8	5450
Nonane	C_9H_{20}	44.3	5685
Decane	$C_{10}H_{22}$	44.2	6294

Reflecting on the Investigation

7. In calculating the heat of combustion of paraffin, you assumed that all thermal energy from the burning wax went to heating the water. Was this a good assumption? Explain.

8. What other laboratory conditions or assumptions might cause errors in your calculated values?

9. The combustion of wax is exothermic; that is, a burning candle releases energy.

   a. Explain this statement in terms of the energy stored in the reactant and product molecules.

   b. Draw a potential energy diagram for this reaction.

10. Consider the complete combustion of paraffin.

   a. Write an equation for this reaction.

   b. Calculate the mass of CO_2 produced by your candle as it burned.

   c. Use your data and the number of students enrolled in chemistry at your school to estimate the CO_2 produced by this investigation at your school.

   d. Do you think candle burning is a significant contributor to CO_2 production? Explain your answer.

11. Restate and answer your beginning question, using evidence to support your claims.

concept check 8

1. What does it mean to describe a reaction as exothermic?
2. What claims can you make about the energy required to break bonds versus energy released by making bonds during fuel burning?
3. Aluminum has a specific heat capacity of 0.900 J/(g·°C). You have a 50-g block of aluminum and a glass containing 50 g of water, both at room temperature. You heat both at the same rate on a hot plate. Which will reach a higher temperature after three minutes of heating? Explain.
4. Identify two processes that affect the global carbon cycle.

C.5 USING HEATS OF COMBUSTION

With abundant oxygen gas and complete combustion, the burning of a hydrocarbon can be described by the equation

Hydrocarbon + Oxygen gas $\longrightarrow$ Carbon dioxide + Water + Thermal energy

"Thermal energy" is written as a product of the reaction because energy is released when a hydrocarbon burns (Figure 3.46). The combustion of a hydrocarbon is a highly exothermic reaction.

The equation for burning ethane (C_2H_6) is

$$2\,C_2H_6(g) + 7\,O_2(g) \longrightarrow 4\,CO_2(g) + 6\,H_2O(g) + \underline{?}\ \text{kJ thermal energy}$$

To complete this equation, the correct quantity of thermal energy involved must be included. Table 3.9 indicates that ethane's molar heat of combustion is 1560 kJ/mol. That is, burning one mole of ethane releases 1560 kilojoules of energy. According to the chemical equation above, two moles of ethane ($2\,C_2H_6$) are burned. Of course, the total thermal energy must correspond to

Figure 3.46 *Natural gas (often used as a fuel to heat homes and for cooking) is a mixture of hydrocarbons. Its primary component is methane, the smallest of the hydrocarbon molecules, consisting of only one carbon and four hydrogen atoms.*

the amounts of all other reactants and products involved. Thus, the total thermal energy released will be twice that quantity released when one mole of ethane burns. Thus, the complete combustion equation for ethane is:

$$2 \, C_2H_6(g) + 7 \, O_2(g) \longrightarrow 4 \, CO_2(g) + 6 \, H_2O(g) + 3120 \text{ kJ}$$

As you found in Investigating Matter C.4, and as suggested in Table 3.9 (page 340), heats of combustion can also be expressed as the energy involved when one gram of hydrocarbon burns (kJ/g). That information is useful in finding out how much energy is released when a certain mass of fuel is burned.

DEVELOPING SKILLS
C.6 HEATS OF COMBUSTION

This activity will help you develop a better understanding of the energy involved in burning hydrocarbon fuels. Use Table 3.9 (page 340) to answer these questions.

Sample Problem 1: *How much thermal energy would be produced by burning 12.0 g octane, C_8H_{18}?*

Table 3.9 indicates that burning 1.00 g octane releases 47.8 kJ. Burning 12.0 times more octane produces 12.0 times more thermal energy, or

$$12.0 \times 47.8 \text{ kJ} = 574 \text{ kJ}$$

The calculation can also be written this way:

$$12.0 \text{ g octane} \times \frac{47.8 \text{ kJ}}{1 \text{ g octane}} = 574 \text{ kJ}$$

Sample Problem 2: *How much energy (in kilojoules) is released by completely burning 25.0 mol hexane, C_6H_{14}?*

According to Table 3.9, the molar heat of combustion of hexane is 4141 kJ. This means that 4141 kJ of energy is released as 1.00 mol hexane burns. So, burning 25.0 times more hexane will liberate 25.0 times more energy.

Burning 25.0 mol hexane would thus release 104 000 kJ of thermal energy:

$$25 \text{ mol } C_6H_{14} \times \frac{4141 \text{ kJ}}{1 \text{ mol } C_6H_{14}} = 104\,000 \text{ kJ}$$

Since the amount of hexane is specified to three significant figures, the answer in Sample Problem 2 is also rounded to three significant figures. See Appendix B if you need more practice using significant figures.

1. Write a chemical equation, including thermal energy, for the complete combustion of each of the following alkanes:

   a. propane   c. octane

   b. butane   d. decane

2. a. How much thermal energy is produced by burning two moles of octane?

   b. How much thermal energy is produced by burning one gallon of octane? (A gallon of octane occupies a volume of ~3.8 L. The density of octane is 0.70 g/mL.)

   c. Suppose a car operates so inefficiently that only 16% of the thermal energy from burning fuel is converted to useful "wheel-turning" (mechanical) energy. How many kilojoules of useful energy would be stored in a 20.0-gallon tank of gasoline? (Assume that octane burning and gasoline burning produce the same results.)

3. The molar heat of combustion of carbon contained in coal is 394 kJ/mol.

   a. Write a chemical equation for burning the carbon contained in coal. Include the thermal energy produced.

   b. Gram for gram, which is the better fuel—carbon or octane? Explain your answer using calculations.

   c. In what applications might coal replace petroleum-based fuel?

   d. Describe one application in which coal would be a poor substitute for petroleum-based fuel.

■ MODELING MATTER

C.7 INCOMPLETE COMBUSTION

You have now seen and written chemical equations representing combustion of several different hydrocarbon fuels. In each case, products were shown only as CO_2 and H_2O. However, this is based on the assumption that enough oxygen is available to completely burn the hydrocarbon, converting all of the carbon atoms into CO_2. You have likely encountered evidence that this is not always a valid assumption.

When not enough oxygen gas is available to convert all of the carbon atoms from the hydrocarbon fuel into CO_2, other products such as CO or C may be formed. You may have a carbon monoxide (CO) detector in your home, or you may recall seeing soot (C) in a fireplace. These products form when hydrocarbons undergo incomplete combustion. In such a case, oxygen would be called the **limiting reactant** in the combustion process; it *limits* the amount of hydrocarbon that can be completely burned.

CO is an air pollutant you studied in Unit 2.

This is somewhat analogous to a situation you might encounter if you bought one package of hot-dog buns and one package of hot dogs for a picnic. Hot dogs are normally sold in packages of ten, but hot-dog buns are eight to a package. Thus, hot-dog buns would become the "limiting reactant" if you had just one package of each. They limit the total number of complete hot-dog sandwiches that can be assembled—a total of eight sandwiches with two "excess" hot dogs.

1. Consider the combustion of propane, C_3H_8.
   a. Draw a molecular model of one propane molecule.
   b. Imagine allowing that one propane molecule to react with as much oxygen gas (O_2) as needed for all the carbon atoms to form CO_2 and all the hydrogen atoms to form H_2O. Draw all the CO_2 and H_2O molecules that can be formed by complete combustion of one propane molecule.
   c. How many oxygen molecules are needed for complete combustion of one molecule of propane?

2. Now consider the combustion of butane, C_4H_{10}.
   a. Draw models of one butane molecule and four molecules of oxygen gas.
   b. Form (and draw) as many molecules of CO_2 and H_2O as possible from the models you drew in Question 2a.
   c. Do the four molecules of oxygen gas provide sufficient oxygen atoms to support the complete combustion of one butane molecule?
   d. Which is the limiting reactant in this process—butane or oxygen gas?

3. Assume that 1.0 mol C_4H_{10} is completely burned in excess oxygen to form carbon dioxide and water.
   a. How many moles of CO_2 would be produced?
   b. How many moles of H_2O would be produced?
   c. What is the minimum number of moles of oxygen gas required to support that reaction?
   d. If it were possible to form CO instead of CO_2, how many moles of oxygen gas would be required?
   e. Consider your answers to Questions 3c and 3d. Under what conditions would CO formation be favored over CO_2 formation? Why?

4. Recall your investigations of fuels in Investigating Matter C.1 and C.4. In particular, think about the apparatus you used to heat water.
   a What evidence of incomplete combustion did you note?
   b. Did the degree of incomplete combustion vary among the different fuels? Explain.

5. You learned that complete combustion of two moles of ethane releases 3120 kJ of energy. Would you expect incomplete combustion of this fuel to release more or less than 3120 kJ of energy? Explain.

C.8 THE CARBON CYCLE

By now you are very familiar with combustion reactions that convert carbon in the form of hydrocarbon fuels (which may be solid, liquid, or gas) into gaseous carbon dioxide, along with water and energy. Without the influence of human activity, the distribution of carbon within its various forms and locations on Earth would remain relatively unchanged over time. However, as you will see in Developing Skills C.9, atmospheric CO_2 levels have increased by about 30% since 1800. How do scientists explain this?

The increase in atmospheric CO_2 levels is the result of several processes. For example, clearing forests removes vegetation that would ordinarily consume CO_2 through photosynthesis. As cuttings and scrap timber are burned, they release CO_2 into the atmosphere. Most significantly, burning fossil fuels releases CO_2 into the air, as these equations illustrate:

Burning coal: $C(s) \ + \ O_2(g) \longrightarrow CO_2(g)$

Burning natural gas: $CH_4(g) \ + \ 2\,O_2(g) \longrightarrow CO_2(g) \ + \ 2\,H_2O(g)$

Burning gasoline: $2\,C_8H_{18}(g) \ + \ 25\,O_2(g) \longrightarrow 16\,CO_2(g) \ + \ 18\,H_2O(g)$

> Do these chemical equations represent complete or incomplete combustion?

More than a century ago, it was suggested that a significant increase in burning fossil fuels might release enough carbon dioxide into the atmosphere to affect Earth's surface temperature. This suggestion was based on the idea that human activity can affect processes in natural ecosystems, producing changes that might not always be beneficial. In particular, burning fossil fuels might perturb the natural movement of carbon within Earth's systems—the global **carbon cycle**.

> Recall what you learned about the greenhouse effect in Unit 2. See pages 226–227.

Remember that the Law of Conservation of Matter implies that the total atoms of all stable elements on Earth are essentially unchanged. However, these atoms are found in many physical states and combined in many different compounds.

In the carbon cycle, illustrated in Figures 3.47 and 3.48 (page 346), the different forms and compounds in which carbon atoms are found can be

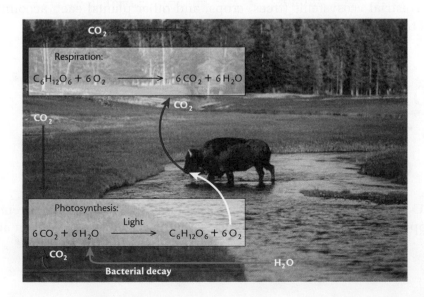

Figure 3.47 *The carbon cycle—major pathways within the biosphere.*

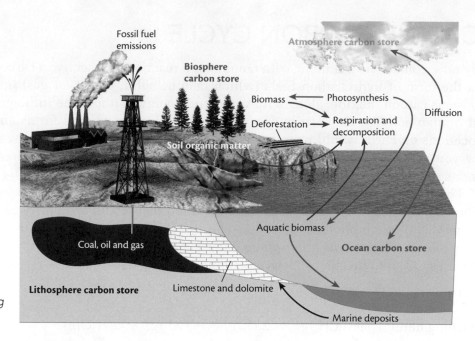

Figure 3.48 *The carbon cycle—relationships among major carbon reservoirs.*

considered *chemical reservoirs* of carbon atoms. These reservoirs include atmospheric CO_2 gas, solid calcium carbonate ($CaCO_3$) in limestone, natural gas (methane, CH_4), and organic molecules, to name a few.

All movements within the carbon cycle—and thus among these reservoirs—either require energy or release energy. For instance, plants use CO_2 and energy from the Sun to form carbohydrates by means of photosynthesis. The carbohydrates are consumed by other organisms (or by the plant itself), and are eventually broken down or oxidized, releasing energy for use by the organisms that consumed them.

The carbon atoms used and circulated in photosynthesis represent only a tiny portion of available global carbon. Gaseous CO_2 continually moves between the atmosphere and the oceans. In fact, 71% of Earth's carbon atoms (in the form of CO_2) are dissolved in its oceans. Another 22% are trapped in fossil fuels and in carbonate rocks formed when dissolved CO_2 reacted with water, producing carbonates, then sediments, then rocks. Dead organisms and terrestrial ecosystems (trees, crops, and other plants) each account for half of the remaining global carbon-atom inventory.

DEVELOPING SKILLS

C.9 TRENDS IN ATMOSPHERIC CO₂ LEVELS

The CO_2-level data reported in Table 3.10 summarize average measurements of trapped gas bubbles in Antarctic ice or, more recently, of air taken at the

Mauna Loa Observatory in Hawaii. These data are also available on the Internet. If possible, locate the most recent annual data before you start this activity. Graph the data summarized in Table 3.10. Prepare the x-axis scale to include the years 1800 to 2050, and the y-axis to represent CO_2 levels from 280 to 600 ppm. Plot the data and draw a smooth curve to indicate the trend among plotted points.

1. Assuming the trend in your smooth curve will continue, extrapolate your curve with a dashed line from the last year for which you have data to the year 2050. You can now make and evaluate some predictions, using the graph you have just completed.

2. What does your graph indicate about the general change in CO_2 levels since 1800?

3. Based on your extrapolation, predict CO_2 levels for

   a. next year.

   b. the year 2030.

   c. the year 2080.

4. Consider your predictions in Question 3.

   a. Which of your predictions from Question 3 is likely to be the most accurate?

   b. Why?

5. One of the scenarios considered in many global warming models is a doubling of atmospheric CO_2.

   a. Does your graph predict a doubling of the 1900 CO_2 level?

   b. If yes, in what year will this doubling presumably occur?

6. In the last few questions, you have been asked to draw conclusions outside the range of collected data.

   a. What assumptions must you make when extrapolating from known data trends?

   b. How do these assumptions affect the accuracy of your predictions?

7. Why might CO_2 data for air samples collected at Mauna Loa Observatory be different from data collected at other locations around the planet?

Table 3.10

Approximate Carbon Dioxide Levels in the Atmosphere, 1800–2009	
Year	Approximate CO₂ Level (ppm by volume)
1800	283
1820	284
1840	285
1860	286
1880	291
1900	297
1920	303
1940	309
1960	317
1965	320
1970	326
1975	331
1980	339
1985	346
1990	354
1995	361
2000	369
2001	371
2002	373
2003	376
2004	378
2005	380
2006	382
2007	384
2008	386
2009	387

Note: Data before 1960 are based on air bubbles trapped in ice-core samples, Law Dome, East Antarctica. Data since 1960 are based on air samples collected at Mauna Loa Observatory, Hawaii.

C.10 GREENHOUSE GASES AND GLOBAL CLIMATE CHANGE

Now that you have graphed the levels of carbon dioxide in the atmosphere and examined some physical and chemical properties of CO_2, we turn to its importance in Earth's atmosphere. What roles, both positive and negative, does CO_2 play, and what evidence exists for the claims made about it?

You learned about the greenhouse effect in Unit 2. Recall that as long as concentrations of carbon dioxide and other greenhouse gases in the atmosphere remain relatively constant, the greenhouse effect will comfortably maintain Earth's average temperature. Both the water cycle and the carbon cycle work well to maintain stable concentrations of water and carbon dioxide in their respective reservoirs, including the atmosphere. However, as you already realize, human activity changes things. Burning fossil fuels, for instance, releases more carbon dioxide than Earth's system may be capable of absorbing.

> You will learn more about the water cycle in Unit 4.

Spiraling Up

If more CO_2 is added to the atmosphere than can be removed by natural processes, its concentration will increase. Eventually, if sufficient CO_2 were added, the atmosphere could retain enough additional infrared radiation from the Sun to increase Earth's average surface temperature. With an increased surface temperature, CO_2 stored in ice, water, and the frozen floors of northern forests could also be released. A spiraling-up effect could occur, where warmer temperatures produce more carbon dioxide, which produces warmer temperatures, and so on.

> Note that both products of hydrocarbon combustion reactions—CO_2 and H_2O— are greenhouse gases.

The role of water must also be considered, as it is also a greenhouse gas. The atmosphere contains about 12 trillion metric tons of water vapor, a quantity so large that it might seem impossible that human activity could significantly affect it. However, if global temperatures increase, oceans and other bodies of water will also become warmer. The amount of water vapor released from Earth's oceans increases as temperature increases, so more of this slightly warmer water will evaporate, increasing the atmospheric concentration of water vapor. As a greenhouse gas, increased water vapor may cause an even greater increase in global temperatures due to absorption and release of infrared radiation, which causes another upward spiral.

This spiraling-up effect is also commonly called a *runaway greenhouse effect*. Increased water vapor concentration would also lead to increased cloud cover, which would reflect more solar radiation, thus counteracting some of the predicted temperature increase.

Two other naturally occurring greenhouse gases are also produced by human activity: nitrous oxide (N_2O) and methane (CH_4). Many agricultural and industrial activities, as well as burning solid waste and fossil fuels, increase the concentration of N_2O in the atmosphere. CH_4 occurs naturally as a decomposition product of plant and animal wastes, but also is produced from refining fossil fuels and raising livestock. See Figure 3.49. Finally, some gases that do not occur naturally can also contribute to the greenhouse effect.

Figure 3.49 *In addition to carbon dioxide, atmospheric gases such as methane (CH_4) and nitrous oxide (N_2O) act as greenhouse gases. Burning solid waste (left) increases nitrous oxide, while termite activity (center) and raising cattle (right) increase methane concentrations.*

Of particular significance are fluorocarbons used in refrigeration and air conditioning, several of which are no longer permitted to be sold in many countries. Some of these gases are much more effective than CO_2 in retaining heat at Earth's surface.

Temperature Increases

What are some implications of increased atmospheric concentrations of greenhouse gases? Examine Figure 3.50. The 20th century's 14 warmest years all occurred since 1990, with eight occurring since 1998. Of these, 2005 was the warmest year on record and 1998 and 2007 were tied for second.

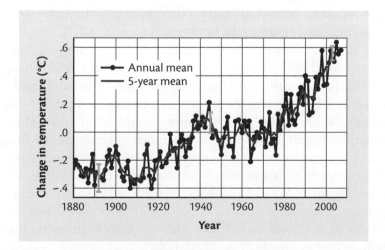

Figure 3.50 *Global annual mean surface air temperature trends since 1880. The 1951–1980 mean surface air temperature has been used as the zero-point base for comparison.*

An international panel of climate scientists has predicted that under a business-as-usual scenario, in which no steps are taken to control the release of human-generated CO_2 and other greenhouse gases, the average global surface temperature could rise 0.8 to 2.9 °C (1.4 to 5.8 °F) in the next 50 years and 1.4 to 5.8 °C (2.2 to 10 °F) in the next 100 years, with significant regional variation. This prediction can be compared to evidence that Earth has warmed only 3 to 5 °C since the depths of the last Ice Age, some 20 000 years ago. These observed and predicted increases in average global surface temperatures are often referred to as **global warming**. The 2007 report of the United Nations Intergovernmental Panel on Climate Change (IPCC) reported that Earth's average global surface air temperature increased by 0.4 to 0.8 °C over the past 100 years. The 2000-page report was prepared by 500 climate scientists and reviewed by another 500 climate experts.

Although most climate specialists agree that Earth's average temperature will increase, there is still some disagreement about causes of this predicted warming. New data indicate that the lower atmosphere may not be warming at the same rate as Earth's surface, which suggests that additional factors contribute to the buildup of greenhouse gases and to the warming trend. Among these other factors is circulation of ocean water. Ocean currents, such as El Niño and La Niña, transport thermal energy within the oceans and can affect climates when this energy is transferred to land and to the atmosphere.

The Nobel Peace Prize committee recognized the IPCC "for their efforts to build up and disseminate greater knowledge about man-made climate change, and to lay the foundations for the measures that are needed to counteract such change" by awarding IPCC the 2007 Nobel Peace Prize.

Possible Impact

What are the projected effects of global warming? The IPCC reports a sea level rise of 3 mm per year in the decade from 1993 to 2003. This rise is due to expansion of ocean water with increasing temperature and melting of glaciers and polar ice caps. There is also a small contribution from the calving and melting of the Greenland and Antarctic Ice Sheets. Based upon these trends and current models, global average sea levels are expected to rise 0.2 to 0.6 m by 2099. These increased levels could cause flooding in coastal cities and islands around the world.

The first report of projected climate increases was made in 1990 and predicted a temperature increase of between 0.15 and 0.3 °C per decade. The models can now be assessed, and data confirm their accuracy. Current models predict a global surface temperature rise of 2.4 to 6.4°C during the 21st century. Warming would be greater over land, and the models suggest that there would be a greater impact in northern latitudes. Possible regional climate changes, including reduced summer precipitation and soil moisture in North America, might be expected. Northern regions might benefit from lengthened growing seasons, while growers in southern regions would probably shift to crops that can benefit from warmer winters. Snow cover would contract (see Figure 3.51) and polar ice would continue to shrink.

The predicted changes also increase the likelihood of intense weather events. Tropical cyclones would become more intense, with higher wind speeds and greater amounts of precipitation.

Figure 3.51 *A pair of northeast-looking photographs, both taken from the same location on the west shoreline of Muir Inlet, Glacier Bay National Park and Preserve, Alaska. In the 105 years or so between photographs, Muir Glacier has retreated more than 50 kilometers (31 miles) and is completely out of the field of view.*

C.11 RESPONSE TO GLOBAL CLIMATE CHANGE

Scientists today understand the influence of CO_2 and other greenhouse gases on world climate better than they did even a decade ago. Sophisticated computer modeling has enhanced their research efforts. This new understanding supports the idea of a surface global warming trend. In 1992, representatives of more than 150 nations developed the Framework Convention on Climate Change, where they agreed to develop policies and procedures to reduce greenhouse gas emissions. In 1997, the third meeting of parties to this agreement, held in Kyoto, Japan, resulted in a protocol to address climate change. The agreement became effective in 2005, and to date 183 Parties to the Convention have ratified the Protocol.

The *Kyoto Protocol* sets ambitious goals for reducing greenhouse gas emissions. The major feature is that it sets binding targets for the reduction of greenhouse gas emissions. These amount to an average five percent reduction over the five-year period from 2008–2012. For industrialized nations, this involves developing energy-efficient technologies, relying more heavily on renewable energy, and applying alternative processes that do not release greenhouse gases into the atmosphere. In 2007, more than 10 000 representatives

As of 2010, the United States was the only industrialized nation that had not ratified the Kyoto Protocol.

from 180 countries met in Bali, Indonesia, to begin developing new policies and negotiating processes to achieve greater reductions in emissions necessary to tackle climate change beyond 2012. This work continued with the development of the Copenhagen Accord in late 2009.

One example of an alternative process to reduce CO_2 emissions is the production of polystyrene foam used to make such items as plastic egg cartons and meat trays. This process uses CO_2 that would normally be released during the production of ammonia. The captured CO_2 replaces chlorofluorocarbons (CFCs) that had been used, thus reducing greenhouse gases in two ways. First, CO_2 that would normally be released by another industry is instead consumed. Second, CO_2 replaces CFCs, which have a much greater global-warming potential than CO_2 does.

As important as it is for hundreds of countries to work together to fight global climate change, there are actions that individuals may take to reduce their own carbon footprints. You can think of your carbon footprint as a combination of those activities that directly put CO_2 into the atmosphere (burning fossil fuels through transportation and home energy use) and the life cycle of products you buy and use (whose manufacturing and eventual breakdown may release CO_2 into the atmosphere). See Figure 3.52.

Numerous Web sites exist that allow you to calculate your own carbon footprint by entering data such as the number of miles you drive each day, whether you use public transportation, what type of energy is used to heat your home, what you eat, how and where you buy clothing, and whether you recycle used containers and goods. These Web sites also provide suggestions for decreasing your personal carbon footprint from relatively easy lifestyle changes (like turning off lights when not in use) to more drastic and higher-impact changes (like trading in your car for a bicycle).

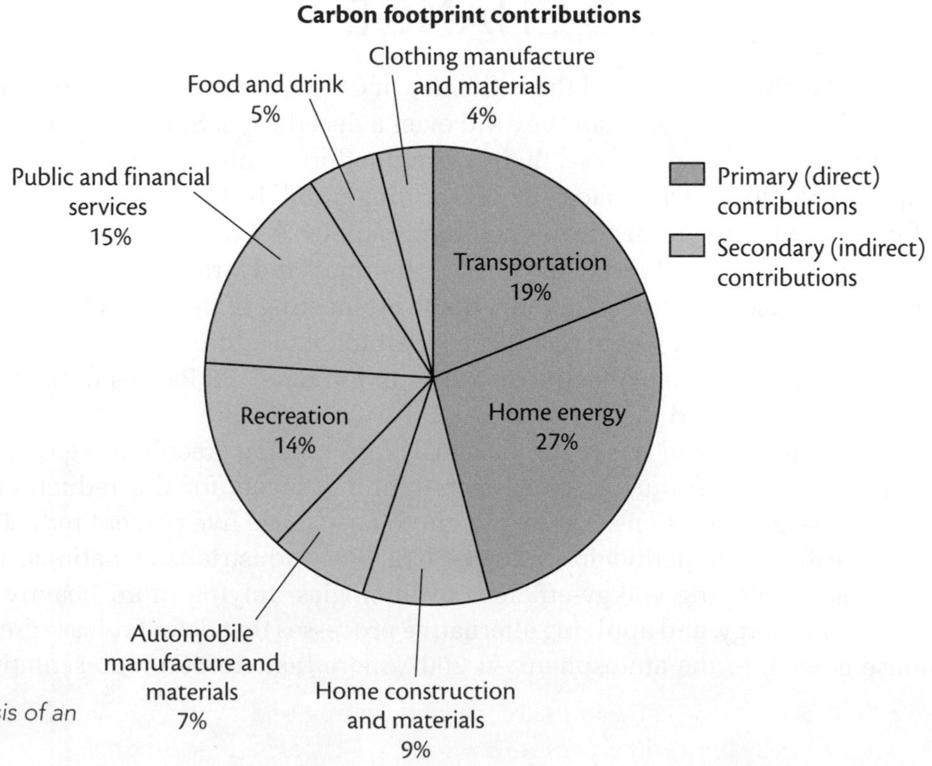

Carbon footprint contributions

Legend:
- Primary (direct) contributions
- Secondary (indirect) contributions

Pie chart segments:
- Food and drink 5%
- Clothing manufacture and materials 4%
- Public and financial services 15%
- Transportation 19%
- Recreation 14%
- Home energy 27%
- Automobile manufacture and materials 7%
- Home construction and materials 9%

Figure 3.52 *A sample analysis of an individual's carbon footprint.*

■ MAKING DECISIONS

C.12 FUEL FOR TRANSPORTATION

Now that you have examined the importance of petroleum as a fuel, particularly for transportation (Figure 3.53), it is time to revisit the television ad for the *TLC-p* that opened this unit. Later, you will have the opportunity to design an original advertisement for an alternative-powered vehicle.

Use what you have learned about petroleum to answer the following questions:

1. Consider this claim made in the television ad: Unlike older petroleum-fueled vehicles, the *TLC-p* plug-in hybrid automobile is virtually emission free. It's as close as you can get to "*carbon-less transportation.*"

    a. Evaluate this claim in terms of CO_2 emissions that may directly result from operating the vehicle.

    b. Evaluate this claim in terms of CO_2 emissions that may be created through the manufacturing and transportation of the vehicle to your local dealership.

2. Identify a claim made within the television ad concerning the energy source(s) for operating the *TLC-p*.

    a. State the claim.

    b. Evaluate this claim based on what you know about the stated energy source(s).

3. In terms of overall energy use, do you think the *TLC-p* is an improvement over conventional gasoline-powered vehicles? Support your claim with evidence.

4. Can you imagine a scenario for which daily operation of the *TLC-p* would be virtually free of burning fossil fuels? If so, describe the scenario. If not, explain why no such scenario exists.

5. Many energy conversions are required to propel a traditional gasoline-powered vehicle.

    a. Identify several of these conversions.

    b. Identify additional energy conversions involved in operating and charging the *TLC-p* that are not involved in propelling a gasoline-powered vehicle.

    c. With these additional energy conversions, how likely is it that the *TLC-p* can "extract nearly all the energy in each drop of gasoline"?

6. The television ad claims that the *TLC-p* will help you to decrease your carbon footprint.

    a. Describe a scenario in which purchasing and driving this vehicle would result in a person *decreasing* her carbon footprint.

    b. Describe a scenario in which purchasing and driving this vehicle would result in a person *increasing* his carbon footprint.

Figure 3.53 *Petroleum, precursor to the fuels used by these vehicles, is a nonrenewable resource. What alternative fuels can be devised to power vehicles such as these?*

SECTION C SUMMARY

Reviewing the Concepts

> Chemical energy, a form of potential energy, is stored within chemical compounds.

1. From a chemical viewpoint, why is petroleum sometimes considered "buried sunshine"?

2. Define and give one example of

a potential energy.

b. kinetic energy.

3. In terms of chemical bonds, what happens during a chemical reaction?

4. Based on its structural formula, which has more potential energy, a molecule of methane or a molecule of butane? Explain your answer.

5. Classify each as primarily a demonstration of kinetic energy or potential energy.

a. a skateboard positioned at the top of a hill

b. a charged battery in a flashlight that's turned off

c. a rolling soccer ball

d. gasoline in a parked car

e. water flowing over a waterfall

6. State the law of conservation of energy.

> The energy change in a chemical reaction equals the difference between the energy required to break reactant bonds and the energy released in forming product bonds. Reactions are classified as either exothermic or endothermic.

7. Why is energy required to break chemical bonds?

8. For each of the following events, determine whether the reaction is exothermic or endothermic. Explain your answers in terms of bond breaking and bond making.

a. burning wood in a campfire

b. cracking large hydrocarbon molecules

c. digesting a candy bar

9. Burning a candle is an exothermic reaction. Explain this fact in terms of the quantity of energy stored in the reactants compared with the quantity of energy stored in the products.

10. Using Figure 3.42 on page 335 as a model, draw a potential energy diagram that illustrates the energy change when hydrogen gas reacts with oxygen gas to produce water and thermal energy.

> When a hydrocarbon burns completely, it reacts with oxygen gas from the air to liberate thermal energy and produce carbon dioxide gas and water vapor.

11. Write a balanced chemical equation, including the quantity of thermal energy (refer to Table 3.9, page 340), for the complete combustion of

a. pentane.   b. propane.   c. hexane.

12. The combustion of acetylene, C_2H_2 (used in a welder's torch), can be represented as:

$$2\,C_2H_2 + 5\,O_2 \longrightarrow 4\,CO_2 + 2\,H_2O + 2512\text{ kJ}$$

   a. What is the molar heat of combustion of acetylene in kilojoules per mole?

   b. If 12 mol acetylene burns fully, how much thermal energy will be produced?

13. a. List two factors that would help you decide which hydrocarbon fuel to use in a particular application.

   b. Explain how you would use each factor in making your decision.

14. When candle wax (a mixture of hydrocarbons) burns, it seems to disappear. What actually happens to the wax?

15. Water gas (a 50–50 mixture of CO and H_2) is made by the reaction of coal with steam. Because the United States has substantial coal reserves, water gas might serve as a substitute fuel for natural gas (composed mainly of methane, CH_4). Water gas burns according to this equation:

$$CO + H_2 + O_2 \longrightarrow CO_2 + H_2O + 525\text{ kJ}$$

   a. How does water gas compare to methane in terms of thermal energy produced when fully burned?

   b. If a water gas mixture containing 10.0 mol CO and 10.0 mol H_2 were completely burned in O_2, how much thermal energy would be produced?

16. In a laboratory activity, a student team measures the heat released by burning heptane (C_7H_{16}). Using the following data, calculate the molar heat of combustion of heptane in kJ/mol. The specific heat capacity of water is 4.18 J/(g·°C).

   • Mass of water   179.2 g
   • Initial water temperature   11.6 °C
   • Final water temperature   46.1 °C
   • Mass of heptane burned   0.585 g

Cycles, such as the carbon and water cycles, help explain movement of matter on Earth.

17. Identify two examples of each of the following:

   a Natural sources of atmospheric CO_2.

   b Natural reservoirs of CO_2.

   c. Human activities that lead to a net increase in atmospheric CO_2 levels.

18. Explain how, over time, a particular carbon atom can be part of the atmosphere, biosphere, lithosphere, and hydrosphere.

19. Write a chemical equation depicting the transfer of a carbon atom between any two "spheres" listed in Question 18.

20. With global temperature increase, more water vapor would enter the atmosphere. How would this

   a. further increase global temperature?

   b. counteract predicted increases in global temperature?

Scientific evidence indicates that Earth's climate is affected by human activities.

21. What is the Kyoto Protocol?

22. List two strategies that industrialized nations can apply to meet Kyoto Protocol requirements.

23. How does each of the following activities impact your personal "carbon footprint"?

   a. Using a bicycle instead of a car.

   b. Attending a concert.

   c. Taking a trip by airplane.

   d. Purchasing locally grown food.

   e. Buying a new backpack made from recycled materials.

What are the benefits and consequences of burning hydrocarbons?

In this section, you have learned about chemical energy in hydrocarbon molecules. You have considered the origin and application of this energy and compared hydrocarbon fuels. You have also examined the role of hydrocarbon combustion in global change. Think about what you have learned, then answer the question in your own words in organized paragraphs. Your answer should demonstrate your understanding of the key ideas in this section.

Be sure to consider the following in your response: kinetic and potential energy, conservation of energy, heat of combustion, the carbon cycle, global climate change, and your carbon footprint.

Connecting the Concepts

24. In a laboratory activity, a student completely burns 4.2 g ethanol (C_2H_5OH). The molar heat of combustion of ethanol is 1366 kJ/mol.

    a. How much thermal energy is released?

    b. The thermal energy from this reaction is used to warm a 468-g sample of water in a calorimeter. The water temperature changes from 21 to 89 °C. How much thermal energy is absorbed by the water? The specific heat capacity of water is 4.18 J/(g·°C).

    c. Compare the quantity of heat released by the reaction to the quantity of heat absorbed by the calorimeter. What accounts for the difference?

25. What modifications would have to be made to the laboratory procedure in Section C.4 if a liquid or gaseous fuel were used?

26. Why is the energy stored in petroleum more useful than solar energy for most applications? Give at least three reasons in your explanation.

27. Many organizations are touting possible methods to avoid greenhouse gas emissions and decrease your personal carbon footprint. How do you know if their claims are reasonable? Let's consider two cases.

    a. Public transit: Individual commuters switching from single occupancy driving on a 20-mile round-trip commute to public transport can reduce their daily CO_2 emissions by 20 pounds (9.08 kg). For 240 commuting days, this works out to 4800 pounds (2180 kg) less CO_2 emitted in a year.

       i. In Developing Skills C.6, you calculated the quantity of thermal energy produced by burning one gallon of octane. (Recall that a gallon of octane occupies a volume of ~3.8 L and has a density of 0.70 g/mL.) How much CO_2 is produced by the complete combustion of a gallon of octane?

       ii. If the average U.S. worker commutes 20 miles round trip per day in a vehicle that can travel 17 miles on one gallon of gasoline (octane), how much CO_2 is produced each day by this commute?

       iii. How much CO_2 is produced annually per worker, if the average worker commutes 240 days per year? Express your answer in both grams and metric tons CO_2.

       iv. Can the claim be supported?

       v. What assumptions were made in the calculations?

b. Light bulbs: It is claimed that replacing one conventional light bulb with a compact fluorescent light bulb can save 300 lb (136 kg) per year of CO_2 emissions.

   i. The average rate of CO_2 produced by electrical generation from all sources in the U.S. is ~0.7 kg CO_2/kWh. How much CO_2 is emitted annually in powering a 60-W incandescent bulb for 8 hours each day?

   ii. How much CO_2 would be saved by replacing the incandescent bulb in part i with a compact fluorescent bulb that requires 14 W?

   iii. Can the claim be supported?

Extending the Concepts

28. Write a word equation for photosynthesis. Include energy in your equation.

29. Using Figure 3.42 on page 335 as a model, draw a potential energy diagram for photosynthesis.

30. Compare the equation for burning hydrocarbons to the equation for photosynthesis. How are these two reactions related?

SECTION D ALTERNATIVES TO PETROLEUM

What alternatives to petroleum are available for burning and building?

Petroleum is a nonrenewable resource; its total available inventory on Earth is finite. Thus, other sources of energy and builder molecules must eventually be found to meet the needs of modern society. In this section, you will explore several alternatives to petroleum as a fuel and a raw material. You will also learn about emerging ways to power personal and commercial vehicles—information you can use as you prepare your own alternative-fuel vehicle advertisement.

GOALS

- Describe major sources of energy used throughout history.
- Describe octane rating in terms of chemical composition and burning characteristics and identify methods used to increase a fuel's octane rating.
- Define catalyst and describe the role of catalysts in modifying fuels and in green chemistry.
- Evaluate biodiesel as an alternative to petroleum-based diesel fuel.
- Describe and evaluate potential alternative sources of fuels and builder molecules.
- Describe alternative-fuel vehicles in use or in development.

concept check 9

1. What is petroleum? Describe it as best you can in your own words.
2. Why is petroleum considered a nonrenewable resource?
3. About what percent of the energy in gasoline is used to propel a car: 100%, 80%, 40%, or 20%? Explain and justify your choice.

D.1 ENERGY: PAST AND PRESENT

The Sun is our planet's primary energy source. As you learned in Section C (page 346), the Sun provides power to the global carbon cycle. How can we interpret this in terms of chemical bonds and energy conversions? Consider photosynthesis, in which green plants use the Sun's radiant energy to convert carbon dioxide and water into carbohydrates and oxygen gas. In photosynthesis, green plants store radiant energy from the Sun as chemical potential energy by converting molecules of carbon dioxide and water to carbohydrates and oxygen (Figure 3.54). Stored chemical energy is later transformed and used by plant and animal cells in a complex stepwise process called **cellular respiration**. The energy released by cellular respiration is used by organisms to power internal energy-consuming processes, including forming other organic molecules. Such organic molecules found in plants and animals are called **biomolecules**.

Solar energy and energy stored in biomolecules are the key energy sources for life on Earth. To understand the magnitude of the quantity of energy coming from the Sun, think about the energy used in one week in the United States, approximately 2×10^{18} J. That quantity is equivalent to the energy released by a hydrogen-bomb explosion, a severe earthquake, or by burning 2×10^{10} gallons of gasoline. By contrast, the radiant energy received by Earth daily from the Sun is approximately 1×10^{26} J—nearly 100 million times more. Since the discovery of fire, human use of stored solar energy held in the molecules that make up wood, coal, and petroleum has had a major influence on civilization's development. In fact, the forms, availability, and cost of energy greatly influence how—and even where—people live.

In the past, abundant supplies of inexpensive energy were available. Until about 1850, wood, water, wind, and animal power satisfied the United States' slowly growing energy needs. Wood, then the predominant energy source, was readily available to most people to use for heating, cooking, and lighting. People used water, wind, and animal power, all forms of energy whose primary source was the Sun, for transportation and to power machinery and industrial processes. With industrialization and population growth, the demand for energy increased and the primary fuel sources changed.

> Even though energy is required in digestion to break the chemical bonds in carbohydrates and oxygen, the digestion of plants is, overall, exothermic. More energy is released by forming bonds in the products than is required to break bonds in the reactants. These ideas were introduced in Section C.2 (page 333).

> In 1850, the population of the United States was 23 million. By 2009, the U.S. population had increased to 307 million.

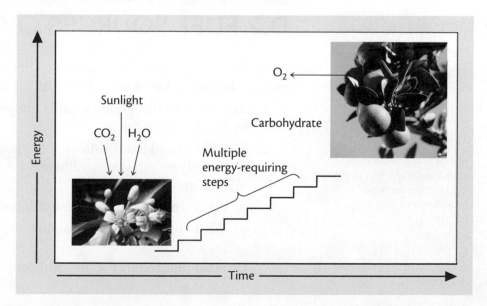

Figure 3.54 *Through photosynthesis, solar energy is converted into chemical energy stored in biomolecules of these plants.*

Figure 3.55 illustrates how U.S. energy sources have changed since 1860. In the next activity, you will explore how energy supplies and fuel use have shifted in the United States during the past 150 years.

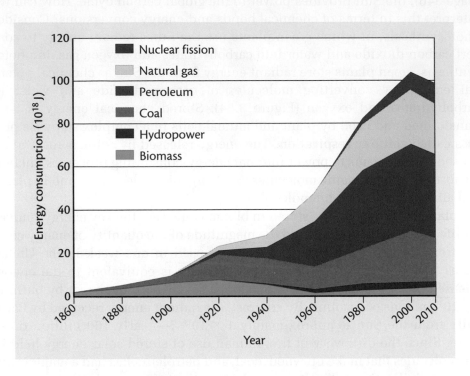

Figure 3.55 *Annual U.S. consumption of energy from various sources (1860–2010).*

DEVELOPING SKILLS
D.2 FUEL SOURCES OVER THE YEARS

As you can see from Figure 3.55, there have been shifts in the types of fuels used in the United States since 1860. Use Figure 3.55 to answer these questions.

1. According to this graph, what is the primary trend in the use of fuels in the U.S. since 1860?

2. a. Identify and list the dates of the period during which biomass (mainly wood) was used to meet at least half of the United States' total energy needs.

   b. What were the chief modes of travel during that period?

   c. What factors might explain the declining use of biomass after that period?

   d. What main energy source did people use to replace biomass?

> The renewable energy source known as *biomass* includes material from plant matter and other living (or recently living) organisms.

3. Compared with other energy sources, only a small quantity of petroleum was used as fuel before the 1920s. What do you think petroleum's main uses might have been before that time?

4. Petroleum use increased and coal use began to decline at about the same time.
   a. When did this occur?
   b. What could explain the growing use of petroleum after this date?
   c. Which energy sources have been used comparatively more since 1980?
   d. What are major uses of these energy sources?

5. a. Describe the trends in petroleum and coal use since 1980.
   b. What factors could account for these trends?

As you know, data can be represented in multiple ways. Consider the representation used in Figure 3.55 as you answer the following questions:

6. What features of this graph make it
   a. easy to interpret?
   b. difficult to interpret?

7. If the data were displayed in a table,
   a. Would you be able to identify the trends and answer the questions above?
   b. Would it be easier or more difficult? Explain.

8. Describe some other methods that could be used to display these data.

D.3 ENERGY EFFICIENCY

As you analyzed the trends in fuel use in the United States, you probably noted that some fuels are better suited to certain applications than are others. For instance, it would be difficult to power automobiles with coal or wood, so that significant energy demand is met by refined petroleum. Petroleum, however, is neither limitless nor inexpensive. In response to concerns about limitations and costs of petroleum, as well as impact on the environment, scientists and engineers have been asked to develop *energy-efficient* products and processes. **Energy efficiency** refers to the use of smaller quantities of energy to achieve the same effect.

One way to maximize the benefits from available supplies of petroleum-based fuels is to reduce the total number of energy conversions the fuel undergoes. We can also seek ways to increase the efficiency of energy-conversion devices. Although energy-converting devices definitely have increased the usefulness of petroleum and other fuels, some useful energy is always "lost" whenever energy is converted from one form to another. That is, no energy conversion is totally efficient; some energy, usually liberated as

Increasing energy efficiency—by designing processes to use smaller quantities of energy, decreasing the number of energy conversions, or switching to more efficient conversions—is one of the principles of green chemistry.

heat energy, always becomes unavailable to do useful work. Unfortunately, devices that convert chemical energy, such as that stored in fuels, to thermal energy and then to mechanical energy are typically less than 50% efficient. Solar cells, which convert solar energy to electrical energy, and fuel cells, which convert chemical energy to electrical energy, may be able to replace petroleum or increase the efficiency of its use.

In one sense, an automobile—whether powered by electricity, gasoline, or even solar energy—can be considered a collection of energy-converting and energy-powered devices. Consider an automobile with 100 units of chemical energy stored in the mixture of molecules that make up gasoline in its fuel tank. See Figure 3.56. Even a well-tuned automobile converts only about 25% of that chemical energy (potential energy) to useful mechanical energy (kinetic energy). The remaining 75% of gasoline's chemical energy is lost to the surroundings as heat (thermal energy). The following activity will allow you to realize what this means in terms of gasoline consumption and expense.

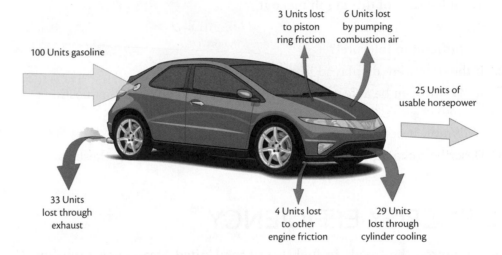

Figure 3.56 *Energy use in a moving automobile. Note that only about 25% of the potential energy in the gasoline is used to power the vehicle.*

DEVELOPING SKILLS

D.4 ENERGY CONVERSION EFFICIENCY

You now know that much of the chemical energy in gasoline is not transformed into useful mechanical energy. This activity will help you to evaluate the costs of this inefficiency and the potential benefits of more efficient vehicles.

Sample Problem: *Assume that a family drives 225 miles each week in a car that can travel 23.0 miles on one gallon of gasoline. How much gasoline does the car use in one year?*

Questions such as this can be answered by attaching proper units to all values, then multiplying and dividing them as though they were arithmetic expressions.

For example, the relevant information can be expressed as:

$$\frac{225 \text{ miles}}{1 \text{ week}} \quad \text{and} \quad \frac{23.0 \text{ miles}}{1 \text{ gal}}$$

or, if needed, as inverted expressions:

$$\frac{1 \text{ week}}{225 \text{ miles}} \quad \text{and} \quad \frac{1 \text{ gal}}{23.0 \text{ miles}}$$

Calculating the desired answer also involves using information you already know—there are 52 weeks, for example, in one year. You also know that the desired answer must have units of "gallons per year" (gal/year). When care is taken to ensure that units are multiplied and divided to produce gal/year, the following expression is formed:

$$\frac{225 \text{ miles}}{1 \text{ week}} \times \frac{1 \text{ gal}}{23.0 \text{ miles}} \times \frac{52 \text{ weeks}}{1 \text{ year}} = 509 \text{ gal/year}$$

Now answer these questions.

1. Assume that a conventional automobile averages 23.0 miles per gallon of gasoline and travels 11 000 miles annually.

    a. How much fuel will be burned in one year?

    b. If gasoline costs $3.00 per gallon, how much would be spent in one year?

2. Assume that a hybrid-powered automobile averages 55.0 miles per gallon of gasoline and travels 11 000 miles annually.

    a. How much gasoline will the hybrid-powered automobile burn in one year?

    b. If gasoline costs $3.00 per gallon, how much would be spent on fuel annually?

3. Assume the automobile in Question 1 uses only 25.0% of the energy released by burning gasoline.

    a. How many gallons of gasoline are wasted each year due to energy conversion inefficiency?

    b. How much does this wasted gasoline cost at $3.00 per gallon?

4. Suppose a new car travels 70.0 miles on one gallon of gasoline with a 40.0% efficient engine.

   a. How much fuel is saved annually compared to cars in Questions 1 and 2?

   b. How much is wasted annually due to energy inefficiency?

concept check 10

1. Only a small portion of the stored chemical energy in fuels is actually converted into motion in an automobile. What happens to the rest of the energy? Explain.
2. How would a transition to electric cars affect fuel use distributions over the next 50 years?
3. Think about the gasoline and kerosene fractions of petroleum that you learned about in Section A.
   a. How do the molecules in these fractions differ?
   b. How might these differences affect the uses of gasoline and kerosene?

D.5 ALTERING FUELS

As automobile use grows throughout the world, the demand for gasoline continues to increase rapidly. Because the gasoline fraction in a barrel of crude oil normally represents only about 18% of the total, researchers have been anxious to find a way to increase this yield. One promising method has been based on the discovery that it is possible to alter the structures of some of petroleum's hydrocarbon molecules so that 47% of a barrel of crude oil can be converted to gasoline.

Chemists and chemical engineers are adept at modifying or altering available chemical resources to meet new needs. Such alterations sometimes involve converting less-useful materials to more-useful products, or converting a low-demand material into high-demand materials. Read on to learn about several applications of chemical technology to the production of automobile fuels.

Cracking

By 1913, chemists had devised a process for converting larger molecules in kerosene into smaller, gasoline-sized molecules by heating the kerosene to 600 to 700 °C. The process of converting large hydrocarbon molecules into smaller ones through the application of heat and a catalyst is known as **cracking**.

Today, more than a third of all crude oil undergoes cracking. The process has been improved by adding catalysts. As you know, a *catalyst* increases the speed of a chemical reaction by participating in it; however, the catalyst is not used up. Catalytic cracking is more energy efficient because it occurs

> Recall that the gasoline fraction obtained from crude oil refining includes hydrocarbons with 5 to 12 carbons per molecule. (See Figure 3.14, page 281).

at lower temperature, 500 °C, rather than the 700 °C required for non-catalytic cracking. The catalyst acts to lower the energy requirements of reacting molecules, thus allowing the reaction to proceed at lower temperature. Through cracking, a 16-carbon molecule, for example, might be changed into two 8-carbon molecules:

$$C_{16}H_{34} \longrightarrow C_8H_{18} + C_8H_{16}$$

Use of catalysts is an important aspect of green chemistry. In this case, the catalyst decreases the quantity of energy that must be added to the system. In other cases, use of a catalyst decreases the amounts of starting materials that are required.

In practice, molecules with up to about 14 carbon atoms can be produced through cracking. Molecules with 5 to 12 carbon atoms are particularly useful in gasoline, which remains the most important commercial product of refining. Some C_1, C_2, C_3, and C_4 molecules produced in cracking are immediately burned, keeping the temperature high enough for more cracking to occur.

Octane Rating

Gasoline is composed mainly of straight-chain alkanes, such as hexane (C_6H_{14}), heptane (C_7H_{16}), and octane (C_8H_{18}). In most automobiles, the gasoline–air mixture is compressed in cylinders just prior to being ignited by spark plugs. This compression can be enough to heat the alkanes to the point where they burn before the spark plug ignites them. The premature burning causes engine "pinging" or "knocking," as the piston bangs backwards against the crankshaft at the wrong time and may contribute to engine problems. A "pinging" engine is less efficient than an engine with proper combustion, since a smaller portion of the fuel's energy acts to propel the car.

Modern, well-tuned engines rarely knock under normal conditions.

Branched-chain alkanes burn more satisfactorily, being less likely to undergo combustion due to compression in engine cylinders; they do not ping as much. The structural isomer of octane shown here has excellent combustion properties in automobile engines. This octane isomer is known chemically as 2,2,4-trimethylpentane. Can you see how the name of the molecule relates to its structure? For convenience, this substance is frequently referred to by its common name, isooctane.

$$CH_3-\underset{\underset{CH_3}{|}}{\overset{\overset{CH_3}{|}}{C}}-CH_2-\underset{\underset{CH_3}{|}}{CH}-CH_3$$

Isooctane, C_8H_{18}

As you probably know, gasoline is sold in a variety of grades—and at corresponding prices. A common reference standard for gasoline quality is the octane scale. On this scale, isooctane, the branched-chain hydrocarbon you just learned about, is assigned an octane number of 100. Straight-chain heptane (C_7H_{16}), a fuel with very poor engine performance, is assigned an octane

Assigning an octane number of 100 to isooctane is arbitrary and does not imply that it has the highest possible octane number. In fact, several fuels burn more efficiently in engines than isooctane does; they are assigned octane numbers higher than 100.

number of zero. Gasoline samples can be rated in comparison with isooctane and heptane.

The **octane rating** for a particular fuel is determined by testing the fuel's burning efficiency (Figure 3.57). Gasoline with a higher octane number is less likely to knock than is gasoline with a lower octane number. High-performance engines are more prone to knocking due to the design of their cylinders, and thus may require higher-octane fuel. Octane ratings in the high 80s and low 90s (87, 89, 92) are quite common, as a survey of nearby gasoline pumps, like those pictured in Figure 3.58, will reveal.

From the 1920s until the 1970s, the octane rating of gasoline was increased at low cost by adding a substance such as tetraethyl lead, $(C_2H_4)_4Pb$, to the fuel. Unfortunately, lead from the treated gasoline was discharged into the atmosphere along with other vehicle exhaust products. Lead also causes the catalytic converters used in automobiles since the 1970s to be ineffective by coating the surface of the catalyst. Both of these issues result in harm to the environment; therefore, lead-based gasoline additives are no longer used in the United States.

Figure 3.57 *Researchers use test engines to determine the octane rating of fuel.*

Figure 3.58 *Octane ratings posted on gasoline pumps.*

You learned about automobile catalytic converters in Unit 2 (page 212).

Other octane-boosting strategies involve altering the structures of hydrocarbon molecules in petroleum. This works because branched-chain hydrocarbons burn more satisfactorily than straight-chain hydrocarbons. (Recall isooctane's octane number compared with that of heptane.) Straight-chain hydrocarbons are converted to branched-chain hydrocarbons by a process called **isomerization**. During isomerization, hydrocarbon vapor is heated with a catalyst:

$$CH_3-CH_2-CH_2-CH_2-CH_2-CH_3 \xrightarrow[\text{Catalyst}]{\text{Heat}} CH_3-\overset{\displaystyle CH_3}{\underset{\displaystyle CH_3}{\underset{|}{\overset{|}{CH}}}}-CH-CH_3$$

$C_6H_{14}(g)$ $C_6H_{14}(g)$

Straight-chain isomer   Branched-chain isomer

The branched-chain alkanes produced by isomerization are blended with C_5 to C_{12} molecules obtained from cracking and distillation, producing a high-quality gasoline. Although cracked and isomerized molecules improve how gasoline burns, they also increase its cost. One reason for this increase is the extra fuel needed to produce such gasoline.

Oxygenated Fuels

The phaseout of lead-based gasoline additives in the United States meant that alternative octane-boosting supplements were required. This led to the blending of a group of additives called **oxygenated fuels** with gasoline. The molecules of these additives contain oxygen in addition to carbon and hydrogen. Although oxygenated fuels actually deliver less energy per gallon than regular gasoline hydrocarbons do, their economic appeal stems from their ability to increase the octane number of gasoline while reducing exhaust-gas pollutants. In most conditions, oxygenated fuels encourage more complete combustion, producing lower emissions of air pollutants such as carbon monoxide (CO). Addition of oxygenated compounds to gasoline was further encouraged by Clean Air Act requirements designed to reduce formation of ozone in certain areas of the United States.

Refer to Section C.7, page 343, to review how CO is formed by incomplete combustion of hydrocarbons.

Methyl tertiary-butyl ether, MTBE, with an octane rating of 116, was initially introduced in the late 1970s as an octane-boosting fuel additive.

$$CH_3-\overset{\overset{\displaystyle CH_3}{|}}{\underset{\underset{\displaystyle CH_3}{|}}{C}}-O-CH_3$$

In the 1990s, MTBE became the most common oxygenated fuel additive in gasoline. The EPA credited MTBE with substantial reductions in emissions of air pollutants from gasoline-powered vehicles in the 1990s. By the late 1990s, however, evidence began to mount that groundwater and drinking water supplies had been contaminated due to MTBE seeping from defective underground gasoline storage systems. MTBE dissolves readily in water and is difficult to remove in water-treatment processes. The unpleasant taste and odor that MTBE imparts to water, even at concentrations below those regarded as a public health concern, triggered consumer complaints. In light of these concerns, many states passed legislation restricting or banning the use of MTBE as an oxygenated fuel. The Energy Policy Act of 2005 (EPACT 2005) removed the federal requirement for oxygenated fuels in high-pollution areas, which resulted in further decreases in the use of MTBE.

EPACT 2005 affected the formulation of gasoline in another way by setting standards for the inclusion of renewable fuels in U.S. fuel supplies. The use of fuels such as alcohols and biodiesel, which you will synthesize in an upcoming investigation, is encouraged by this renewable fuel standard. Alcohols such as *methanol* (methyl alcohol, CH_3OH) and *ethanol* (ethyl alcohol, CH_3CH_2OH) are added to gasoline at distribution locations. Methanol boosts octane and can be made from natural gas, coal, corn, or wood—a contribution toward conserving nonrenewable petroleum resources. Ethanol is made primarily from corn in the U.S., but can also be produced from sugar cane or switchgrass.

An advantage of ethanol is that a blend of 10% ethanol and 90% gasoline can be used as an oxygenated fuel in nearly all modern automobiles without engine adjustments or problems. Several automakers now manufacture vehicles that have been modified to run on a blend of 85% ethanol and 15% gasoline, known as E85.

The blend of 10% ethanol and 90% gasoline was once called *gasohol*, but is now more commonly referred to as *E10*.

■ MAKING DECISIONS

D.6 OXYGENATED FUELS

Most areas of the United States that implemented winter oxygenated fuel programs in order to meet federal carbon monoxide (CO) requirements have since achieved compliance and dropped their oxygenated fuel programs. However, newer federal regulations require the use of reformulated gasoline to decrease ground-level ozone (O_3) formation and set standards for inclusion of renewable fuels in the overall fuel supply. Both of these policies result in the inclusion of oxygenated fuels in gasoline. Methanol (CH_3OH) and ethanol (CH_3CH_2OH), sometimes produced from corn, are used as gasoline additives or substitutes. Think about the choices that you and your family make as consumers while you answer the following questions.

1. Since the phase-out of methyl tertiary-butyl ether (MTBE), the primary oxygenates added to gasoline have been methanol and ethanol.

   a. What are the sources of alcohols used as fuel additives or replacements?

   b. What are some benefits, other than air-quality improvement, of using alcohols in fuel?

   c. What are some drawbacks of using alcohols in fuel?

2. Gram for gram, the heats of combustion of alcohols are considerably lower than those of any hydrocarbon fuels considered so far. For example, the heats of combustion for methanol and ethanol are 23 kJ/g and 30 kJ/g, respectively.

   a. Explain this fact in terms of bond breaking and bond formation.

   b. What does this mean in terms of automobile fuel requirements?

   c. What characteristics of alcohol combustion would tend to offset the impact of lower heats of combustion? Explain.

3. Considering what you learned about ground-level ozone formation in Unit 2 (page 246):

   a. Why are oxygenates useful in preventing ozone formation?

   b. Some evidence suggests that oxygenated fuels produce higher emissions of NO_x than do hydrocarbon-only fuels. How would this impact ground-level ozone formation?

4. Ethanol is viewed by some advocates as a "carbon-neutral" fuel because the CO_2 emitted as it burns is roughly equivalent to the CO_2 taken in by the plants as they grow.

   a. What other CO_2 emissions should be considered when evaluating ethanol as a fuel?

   b. How would you compare ethanol and gasoline in terms of total CO_2 emissions?

5. One of the decisions you will likely face (if you have not already done so) is choosing a personal vehicle. List at least two benefits and two drawbacks to choosing each of the following vehicles based upon your knowledge of gasoline and oxygenated fuels.

   a. A gasoline-only vehicle.

   b. A flexible-fuel vehicle that can use either gasoline or E85.

D.7 ALTERNATIVE FUELS AND ENERGY SOURCES

Everyday life in the United States requires considerable quantities of energy. As you learned, the energy sources used in this country have changed over time. As energy demands have accelerated, the nation increasingly has relied on nonrenewable fossil fuels—coal, petroleum, and natural gas. What is the future for fossil fuels, particularly petroleum?

The United States is a mobile society. Some 70% of petroleum consumed in the United States is used for transportation. Although efforts to revitalize and improve public transportation systems merit attention, most experts predict our nation's citizens will continue to rely on personal vehicles well into the foreseeable future. And remember, even energy-conserving mass transit systems must have a fuel source. What options, then, does chemistry offer to extend, supplement, or even replace petroleum as an energy source?

Oil Shale and Oil Sands

Petroleum from oil sands and oil shale rock is an option with some promise. See Figure 3.59. Major deposits of oil shale are located west of the Rocky Mountains, while significant deposits of oil sands are found in Utah and in Alberta, Canada. **Oil shale** contains *kerogen*, which is partially formed oil. When the rocks are heated, kerogen decomposes into a material quite similar to crude oil. **Oil sands** contain *bitumen*, a viscous, heavy crude oil. This oil must be separated from the clay, sand, and water that surround it and then upgraded before it can be refined in facilities designed for petroleum.

Unfortunately, vast quantities of sand and rock must be processed to recover these fuels. Moreover, enormous volumes of water are also needed for processing, which poses a problem where water is scarce. Finally, present extraction methods use the equivalent

> A metric ton of oil shale typically contains the equivalent of 80 to 330 L of oil.

> Oil sands are sometimes called *tar sands*.

Figure 3.59 *Large-scale commercial oil sands surface mining activity north of Alberta, Canada. The shovel bucket holds approximately 100 tons of oil sands ore. An oil sands processing plant is visible in the background. Image from Suncor Energy, Inc.*

of one-sixth to one-half a barrel of petroleum to produce every barrel of oil from shale or sands. Currently, Canada is the only country that is producing oil from oil sands or shale on a large scale.

Coal Liquefaction

Because known coal reserves (see Figure 3.60) in the United States are much larger than known reserves of petroleum, another possible alternative to petroleum is a liquid fuel produced from coal. The technology to convert coal to liquid fuel (and also to convert coal to builder molecules) has been available for decades, having been used in Germany since the mid-1900s. Current coal-to-liquid-fuel technology is quite well developed here in the United States. However, the present cost of mining and converting coal to liquid fuel (commonly known as *coal liquefaction*) is considerably greater than that of producing the same quantity of fuel from petroleum.

In addition, converting coal to liquid fuel releases significantly more CO_2 than processing an equivalent quantity of crude oil. But if petroleum costs increase sufficiently, obtaining liquid fuel from coal—itself a nonrenewable resource—may become a more attractive option.

Renewable Petroleum Substitutes

Petroleum replacement candidates are not limited to other fossil fuels; currently there is rising interest in transitioning to renewable fuel sources. In addition to the corn-based alcohols—ethanol and methanol—discussed in Section D.5, advanced biofuels are under development. These newer petroleum substitutes include fuels made from plant-based cellulose and carbohydrates. Although there are challenges associated with breaking down plant cellulose or converting plant sugars into liquid transportation fuels, these technologies would allow the use of waste plant material. Using waste biomass avoids problems associated with upsetting supplies of food for humans and livestock.

Another such petroleum substitute under consideration is **biodiesel**, a fuel that can be burned in diesel engines. See Figure 3.61. One major benefit of biodiesel is that any source of plant or animal fat can be converted into

As you learned in Unit 2, one disadvantage of coal as an energy source is the air pollution generated when it burns.

Figure 3.60 *Trains often transport coal, an important source of energy.*

A mixture of 20% biodiesel and 80% petroleum diesel is referred to as B20, much like ethanol–gasoline blends.

Figure 3.61 *This consumer is filling a diesel-powered car with biodiesel. Biodiesel is clean, renewable, and can be easily integrated into the current petroleum infrastructure.*

biodiesel. Although biodiesel must be refined for use in automobiles, it offers a "green" strategy for producing a renewable petroleum substitute. Currently, biodiesel is generally sold blended with petroleum-based diesel and is seldom used as a pure fuel on its own.

In the coming Investigating Matter D.8, you will have the opportunity to synthesize crude biodiesel from common vegetable cooking oil.

Other Energy Strategies

To meet some U.S. energy requirements, it is possible to move away from petroleum and petroleum substitutes altogether. Alternative energy sources currently in use or under investigation include hydropower (water power), nuclear fission and fusion, solar energy, wind energy (Figure 3.62), burning biomass, and geothermal energy. Other approaches include constructing more energy-efficient buildings, vehicles, and machines, as well as using alternative fuels. In addition, reducing energy use—through such actions as carpooling, using public transportation, and turning off unneeded appliances—helps to conserve fossil fuels and to reduce energy needs overall. All of these steps are intended to further reduce our need to burn petroleum.

Figure 3.62 *Wind moves these turbines—generating electricity without burning fossil fuels.*

▮ INVESTIGATING MATTER

D.8 SYNTHESIZING AND EVALUATING BIODIESEL FUEL

Asking Questions

As you have just learned, biodiesel is a renewable fuel that can be made from any source of plant oil or animal fat. Chemically, biodiesel is a mixture of methyl esters of fatty acids. Recall that you have synthesized esters once before in Investigating Matter B.10. In this investigation, you will use ordinary vegetable cooking oil to synthesize and then evaluate crude biodiesel.

Look at this structure of a typical fat found in canola oil.

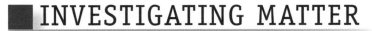

The red portion of the structure represents the glycerol backbone of the fat, while the long portions are the fatty acids that will serve as the source of carboxylic acids in this synthesis.

Think about the following questions and record your answers in your laboratory notebook before reading further.

> You can learn more about fatty acids, glycerol, and fats in Unit 7.

- Given your experience with ester synthesis, what substances do you think you will add to the cooking oil to produce methyl esters of these fatty acids?
- Again considering investigations that you have completed in this unit, how can you test your biodiesel to evaluate its usefulness as a fuel?

Preparing to Investigate

Before you begin, read *Gathering Evidence* to learn what you will need to do and note safety precautions. Develop a procedure to measure the heat of combustion of the biodiesel that you synthesize. Ask your teacher to approve your procedure before continuing.

> Recall that you measured the heat of combustion of paraffin in Investigating Matter C.4.

Think about the data and observations you will need to record during each part of the investigation and prepare appropriate data tables. Refer back to Investigating Matter B.10 and C.4 in your textbook and laboratory notebook for guidance. Note that Step 13 calls for completing a second trial.

Gathering Evidence

Synthesis

1. Before you begin, put on your goggles, and wear them properly throughout the investigation.
2. Using a graduated cylinder, measure 25 mL of canola oil and pour it into a clean, plastic, screw-top bottle. See Figure 3.63.
3. Carefully add 4 mL of methanol to the oil in the bottle. (**Caution:** *Methanol is flammable and toxic.*)
4. Slowly add 5 to 6 drops of 9 M potassium hydroxide (KOH) to the liquid in the bottle. (**Caution:** *Potassium hydroxide is corrosive.*)
5. Tightly cap the bottle. Taking turns with your partner, shake the bottle vigorously for 10 minutes.
6. Add 0.5 gram pure NaCl, then re-cap and shake vigorously for several seconds.
7. Allow the mixture to sit for 30 minutes or overnight so that it separates into two layers. (*Note:* The bottom layer may be small and difficult to see in the plastic bottle.)

Figure 3.63
Ordinary canola oil is used for this biodiesel synthesis.

Combustion

8. Set up the apparatus you specified in your procedure.
9. To prepare the biodiesel for combustion, carefully decant ~5 mL of it

into a metallic sample cup. Make a "wick system" by removing the wick and metal support from the candle and placing it in the empty metal cup. See Figure 3.64.

10. Collect initial data according to your procedure before igniting the biodiesel.

11. Ignite the biodiesel. When it reaches the endpoint you have specified, extinguish the flame by placing a watch glass or ceramic tile atop the sample cup.

12. Allow the cup to cool before collecting final data.

13. Repeat the combustion procedure a second time.

Figure 3.64 *Sample cup and wick setup for burning biodiesel.*

Analyzing Evidence

1. Calculate the heat of combustion for your sample of biodiesel.

2. Petroleum diesel produces 43 kJ/g of thermal energy when burned. How does your biodiesel compare to petroleum diesel?

3. Compare your heat of combustion with that of other laboratory teams.

   a. What is the class mean?

   b. Graph the class data (Total Thermal Energy Absorbed vs. Mass of Fuel Burned). What does the slope of the best fit line tell you?

Interpreting Evidence

1. In synthesizing the biodiesel sample, what changes did you observe in the characteristics of the starting materials compared to those of the final products?

2. Look again at the structure of the typical canola oil fat.

   a. Draw a structure representing a methyl ester of one of the fatty acid chains.

   b. What else (besides the methyl esters) would be produced in the reaction? Explain.

   c. What do you think is the purpose of shaking the reaction mixture?

Reflecting on the Investigation

3. What was the purpose of the 9 M KOH? Why did you only need to add a few drops?

4. Is biodiesel a "better" fuel than petroleum diesel? Support your claim with both evidence and reasoning.

5. Suppose that you wanted to make biodiesel using this method for your personal vehicle.

   a. What changes would you need to make to the procedure?

   b. What challenges might be encountered in "scaling up" the process?

■ MAKING DECISIONS

D.9 BIODIESEL AS A PETROLEUM SUBSTITUTE

In the United States and worldwide, a variety of raw materials are used to produce biodiesel. These include vegetable oil grown for biodiesel production, waste vegetable oil, animal fats, and algae. However, soybean oil grown specifically for this purpose is by far the most common beginning material.

In 2008, soybean farms in the United States (see Figure 3.65) produced, on average, 2.65 metric tons of soybeans per hectare. About 4.8 kg of soybeans must be crushed into oil to produce a liter of biodiesel. There are about 460 million hectares in use for agriculture in the United States and another 15 million surplus hectares. Of the active agricultural land, about 70% is used for livestock grazing; the remainder is used for crops.

When it is combusted, biodiesel releases 32 960 kJ of energy per liter, while combustion of an equivalent volume of petroleum diesel releases 34 790 kJ of energy. The production of a liter of biodiesel or petroleum-based fuel from raw materials requires about 6 600 kJ. Biodiesel and petroleum diesel have similar CO_2 emissions when combusted, but studies have shown that biodiesel burning produces less CO, fewer unburned hydrocarbons, and more nitrogen oxides than the burning of petroleum diesel.

Use the information presented here, as well as reliable Internet sources if needed, to answer the following questions.

1 metric ton = 1000 kg

Figure 3.65 *Soybeans are one possible source of biodiesel. What benefits of biodiesel can you identify?*

1. Given that ~200 billion liters of diesel are used in the United States each year:
   a. How many kilograms of soybeans would be required to produce enough biodiesel to meet this need?
   b. How many hectares would be required to grow the quantity of soybeans needed to produce enough biodiesel to meet all U.S. diesel needs?
2. Could the United States expect to grow enough soybeans to replace all fossil diesel fuel with soybean-based biodiesel? Explain.
3. Is petroleum, or a replacement fuel, required to produce biodiesel? Explain.
4. Based upon your experience in Investigating Matter D.8:
   a. What other substances would be required for the production of biodiesel from soybeans?
   b. What are the possible sources of these substances?
   c. Would byproducts or waste products be produced, as well as biodiesel? Explain.
5. If biodiesel and petroleum diesel produce similar quantities of CO_2 during combustion, why is biodiesel considered a *carbon-neutral* fuel?

6. Think about the sources of information that you used to answer these questions. How did you know whether they were reliable?

7. Create a list of pros and cons of using biodiesel as a petroleum replacement.

concept check 11

1. How do cracking and additives improve the octane rating of gasoline?
2. Why are oxygenated fuels controversial in some communities and regions?
3. List types of alternative-fuel vehicles that are familiar to you.

D.10 ALTERNATIVE-FUEL VEHICLES

As you now know, personal vehicles consume a significant portion—about 50%—of petroleum burned for fuel in the U.S. Because there is a limited supply of petroleum as a resource and emissions are produced by petroleum-burning engines, alternative-fuel vehicles are being developed, tested, and used. What are some of these fuels, and how are they used to propel vehicles? What are advantages and disadvantages of various alternative fuels? The overview that follows will help you prepare your own automobile advertisement.

Compressed Natural Gas (CNG)

Most passenger vehicles and buses can be converted to dual-fuel vehicles that run on either natural gas or gasoline. Natural gas, mainly methane (CH_4), is produced either from gas wells or during petroleum processing. Compressed and stored in high-pressure tanks, this product is commonly known as **compressed natural gas (CNG)**.

A refillable CNG tank, capable of powering an automobile up to 300 miles, can be comfortably installed in a car's trunk. Many CNG-powered vehicles are operating worldwide, particularly in government and mass transit fleets. See Figure 3.66.

Many homes are heated using compressed natural gas.

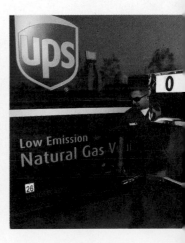

Figure 3.66 *About 12–15% of public transit buses (left) in the United States are powered by natural gas, and some package-delivery companies (right) also include CNG vehicles in their fleets.*

Among the advantages of CNG are wide availability and an 80% decrease (compared to gasoline) in carbon monoxide (CO) and nitrogen oxide (NO_x) emissions. However, refueling systems require a compressor, which increases the vehicle cost by $2000 to $4000. Also, there is a higher fire risk resulting from a collision of a CNG-powered vehicle.

Liquefied Petroleum Gas (LPG)

Another petroleum-based gaseous fuel is propane (C_3H_8), also known as **liquefied petroleum gas (LPG)**. Very similar to CNG, it is often used in dual-fuel vehicles. LPG is used mainly in fleet vehicles, such as taxis, delivery trucks, and buses. Advantages include longer engine life and lower maintenance costs due to cleaner fuel combustion. Also, since the fuel–air mixture is completely gaseous, LPG vehicles tend to start more readily in extremely cold conditions than do gasoline-powered vehicles. A newer technology, liquid propane injection, introduces the propane to the engine cylinder while still in the liquid state, increasing engine efficiency.

Since LPG is stored as a liquid, compressors are not required for refueling. Facilities for production, storage, and bulk distribution of LPG exist throughout much of the United States; expanded use in personal vehicles would require development of distribution sites.

Flexible Fuel and Diesel

As you learned in Section D.5, some vehicles are currently being manufactured to run on fuels containing up to 85% ethanol (C_2H_5OH). These vehicles are known as flexible-fuel vehicles. Unlike CNG and LPG vehicles, they can use a single fuel system for E85, gasoline, and mixtures of the two fuels. Their operation is similar regardless of fuel; however, they can travel farther on a gallon of gasoline than on a gallon of E85. Ethanol-based automobile fuel is distributed in the same manner as gasoline-based fuel, thus the existing distribution network can easily be modified to dispense ethanol.

Similarly, drivers of vehicles designed to operate using diesel fuel obtain their fuel through the established fueling station network. Such vehicles can often use either ultra-low sulfur petroleum-based diesel (ULSD) fuel or biodiesel blends. ULSD vehicles are currently among the most fuel-efficient vehicles available. Some newer diesel engines use both a particulate filter and a selective catalytic reduction (SCR) system to reduce emissions of air pollutants. Exhaust and a reducing agent known as "diesel exhaust fluid" react in the catalytic chamber, with the aid of a catalyst, to reduce nitrogen oxides (NO_x) to nitrogen and water.

Fuel Cell

One developing option for generating electricity to power vehicles is the **fuel cell**. It did not become a practical energy source until the 1960s, when fuel cells were used in the U.S. space program. Any fuel containing hydrogen (such as methanol or natural gas) can be used in a fuel cell.

As shown in Figure 3.67, one common form of fuel cell converts oxygen gas and hydrogen fuel into electrical energy and water, which is its only emission. From a chemical viewpoint, such an operating fuel cell represents another way to release and harness the chemical potential energy stored in hydrogen and oxygen molecules:

$$2\,H_2 \;+\; O_2 \longrightarrow 2\,H_2O \;+\; \text{Electrical energy} \;\;(\text{and some thermal energy})$$

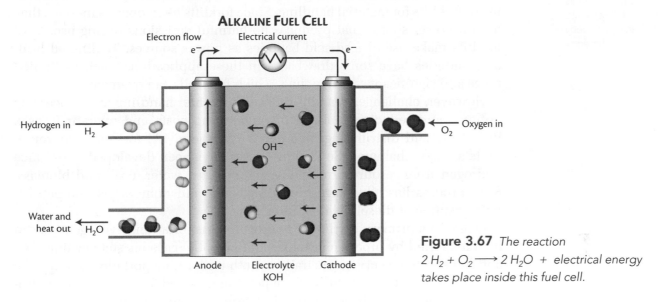

Figure 3.67 *The reaction* $2\,H_2 + O_2 \longrightarrow 2\,H_2O$ + *electrical energy takes place inside this fuel cell.*

The fuel cell involves platinum electrodes that catalyze the removal of electrons from hydrogen atoms. In one common type of fuel cell called a *proton exchange membrane* (PEM) fuel cell (see Figure 3.67), electrodes are in contact with a polymer that allows hydrogen ions to pass through, but not hydrogen molecules. The electrons flow from the fuel cell through an external circuit (where they do useful work), returning to the fuel cell at the other electrode. See Figure 3.68. That second electrode catalyzes the reaction of oxygen gas (the oxidant) with hydrogen ions and electrons to produce water molecules, thus completing the electrical circuit. In the PEM hydrogen-oxygen fuel cell, the net products are water molecules and electrical energy.

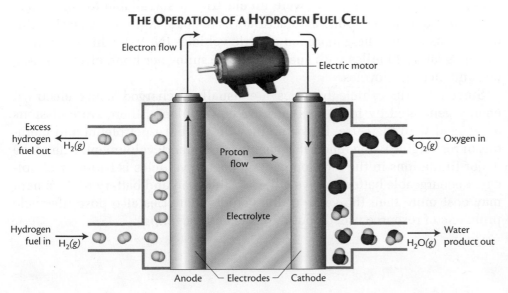

Figure 3.68 *Schematic diagram of a proton exchange membrane (PEM) fuel cell. Hydrogen ions form when electrons are removed from H_2 at the anode. The electrons flow through the circuit to the cathode, doing useful work along the way (note the electric motor in the circuit). The H^+ ions (protons) flow through the polymer to the cathode, where they combine with oxygen gas and electrons to produce water.*

The chemical energy stored in hydrogen—considered an energy carrier rather than an energy source—is converted directly into electrical energy, thereby increasing efficiency.

Fuel cells require no electrical recharging, and eliminate or substantially reduce release of air pollutants. They are more efficient than are internal-combustion engines; they can obtain more useful power from a given quantity of fuel than can be obtained by burning it.

Fuel cells are currently being used in some specialized applications, including in forklifts for material handling. Since forklifts often operate indoors, they need a power source that produces no harmful emissions during use. Most forklifts make use of lead-acid batteries as power sources. Traditional lead-acid batteries have some drawbacks in these applications, such as limited range and significant recharge time, which fuel cells can overcome.

However, challenges remain in developing fuel handling and processing options, and reducing fuel-cell manufacturing costs and operating costs. Producing and distributing high-purity hydrogen (H_2) for fuel cells represents a major challenge. Several processes have been developed to produce hydrogen from resources including natural gas, water, coal, and biomass. Each process has advantages and drawbacks and none is yet competitive with gasoline or diesel on a per-mile basis.

One of the primary barriers is hydrogen transportation. Hydrogen gas can be transported by pipeline or tanker trailer in the gaseous state or liquefied and transported by cryogenic trailers. Both trailer transportation options use significant quantities of energy from petroleum-based fuels, while the hydrogen pipeline system is very limited. Alternatives include producing hydrogen at local refueling stations, which would add to the cost of production.

Electricity

You can learn more about batteries in Unit 5.

Another option for powering vehicles with electricity uses an electric motor to propel the vehicle. Electrical energy to power the motor must be stored onboard in an energy storage device, usually a battery. Since storage capacity is limited, batteries must be regularly recharged by connecting to a source of electrical energy.

Although interest in electric vehicles surged when gasoline prices rose above $3.00 per gallon, there were no electric cars available to consumers from the major automobile manufacturers as of 2010. Available electric vehicle options include neighborhood electric vehicles (NEVs), which can often travel as far as 30 miles with speeds up to 25 miles per hour, electric scooters, and electric bicycles.

In the United States, 48% of electricity is generated by coal combustion and 21% by the burning of natural gas. Nuclear and hydroelectric power account for 19% and 6%, respectively, of the U.S. total.

Since electric vehicle batteries are usually recharged using electrical energy generated by fossil fuel-burning power plants, there are emissions associated with their use. However, they do not burn petroleum and methods of recharging using renewable energy are under development. One of the major limitations in the development of electric vehicles is battery technology. Rechargeable batteries have a limited life span and battery replacement may cost more than the value of the vehicle. Batteries also pose life-cycle problems of resource use, disposal, reuse, and recycling.

Hybrid and Plug-In Hybrid

Some car designers believe that a *gasoline–electric vehicle* will best meet consumer needs while reducing emissions and fuel costs. See Figure 3.69. *Hybrid gasoline–electric vehicles*, or **hybrid vehicles**, like the fictitious original *TLC* featured in this unit, come equipped with a gasoline-burning engine as well as a battery-powered electric motor. The batteries are recharged while driving, partially through conversion of the car's kinetic energy into stored chemical potential energy whenever the vehicle's brakes are applied.

Figure 3.69 *Hybrid cars are becoming more common. Some states give tax incentives or carpool lane advantages for purchasing and driving hybrid or other alternative-fuel vehicles.*

As you already learned, traditional gasoline-powered automobiles are unable to convert all the chemical energy stored in gasoline and oxygen into useful mechanical energy. Figure 3.56 (page 362) documented that only 25% of the original chemical energy is converted into usable horsepower, while the other 75% is lost to the surroundings as thermal energy.

A hybrid vehicle, however, is able to store some of the energy that would otherwise be "lost" and save it for later use. In essence, chemical potential energy—originally stored in gasoline and oxygen—is converted into kinetic energy as the car accelerates. When the car possesses more kinetic energy than is required—for example, as it goes downhill or when brakes are applied—excess kinetic energy is converted back into chemical potential energy stored in the car's battery.

By comparison, when brakes are applied in traditional gasoline-powered vehicles, the car's kinetic energy is converted into wasted thermal energy through friction as the brake's pads slow the car. In addition, hybrid cars are engineered to switch over from the gasoline engine to the electric motor whenever possible, such as when idling at a red light. This increases the car's overall energy efficiency; nearly 40% of the gasoline's chemical energy is converted into useful kinetic energy.

Hybrid vehicles typically achieve over 40 miles per gallon and can travel more than 650 miles between fueling stops. Although these vehicles produce the same kinds of emissions as fossil-fuel-burning vehicles, the quantity produced over a given distance is smaller.

Scientific American Working Knowledge Illustration

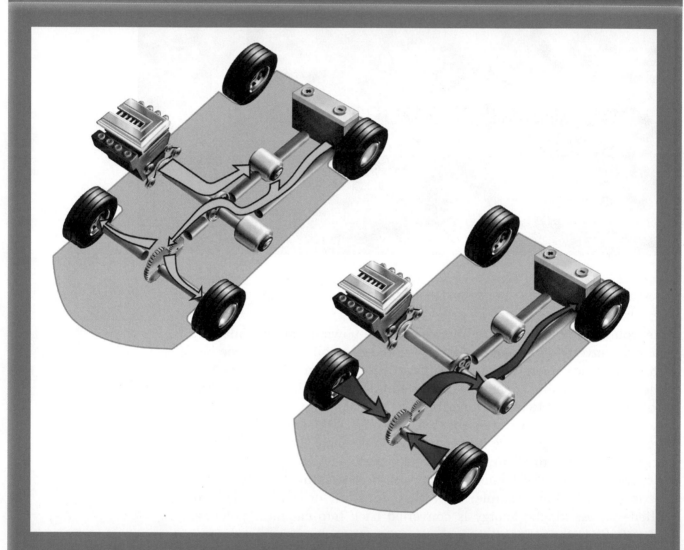

Figure 3.70 *Hybrid car. Hybrid gasoline–electric vehicles represent alternatives to traditional gasoline-powered vehicles. Left: As the hybrid car accelerates (green arrows), it draws power from both the gasoline engine and electric storage battery. In some models, the electric battery provides all the needed power at low speeds, which also lowers gasoline consumption. Right: When brakes are applied (red arrows), some kinetic energy is transformed into chemical potential energy stored in the car's electric battery. This energy is used when the car again accelerates. In traditional gasoline-powered vehicles, all of this energy would be lost as heat through friction by the brakes.*

Plug-in hybrid vehicles (like the *TLC-p*) add to the versatility of hybrid vehicles by allowing the vehicle batteries to be charged by plugging them into a source of electrical energy. This adaptation allows plug-in hybrid vehicles to run longer on electricity alone, without sacrificing the convenience of a vehicle that can be refueled along the road. Put another way, current hybrid vehicles have a gasoline engine assisted by an electric motor, while plug-in hybrids have an electric motor assisted by a gasoline engine.

Several major automakers have announced plans to make plug-in hybrid vehicles available to commercial fleets and consumers in the next few years. As is the case for electric vehicles, battery technology remains the primary challenge for plug-in hybrid vehicle development.

> A plug-in hybrid went on sale in China in late 2008.

D.11 ALTERNATIVE SOURCES FOR BUILDER MOLECULES

As you have just learned, there are many alternatives being considered to using petroleum as a transportation fuel. What about the other uses of petroleum? What alternatives exist to petroleum as a source of builder molecules, lubricants, and other specialty products? Given that only 13% of petroleum is used for non-fuel purposes, do we need to be concerned about these uses? Could alternative fuels and conservation make replacements for builders unnecessary?

Scientists in the Northwest United States are investigating the use of biomass, including wood pulp and wheat straw, to synthesize builder molecules. The process requires breaking down the biomass into starch, cellulose, and glucose. This step is among the most challenging, because it requires rupturing tough plant cell walls. The initial products could then be transformed into starting materials for polymers and common consumer products. Most of these starting materials currently are petroleum products.

Some plant-based polymer builder materials are already being used in floor coverings and refrigerators. Castor oil, a vegetable oil, is a promising builder material source already used to make the polymer polyurethane for use in car seats and mattresses. Both 1,3-propanediol, obtained as a byproduct of biodiesel production, and succinic acid, an organic acid produced by bacteria from plant matter, show promise as builder molecule sources.

> 1,3-propanediol and succinic acid are polyols, or compounds with more than one –OH functional group per molecule.

Finally, renewable alternatives are also being developed for other applications of petroleum products, such as lubrication and hydraulics. Vegetable oils (from soybeans, corn, sunflowers, and other plants) can replace petroleum-based substances in many of these uses. Challenges to the substitution of bio-based (vegetable) oils include a tendency to oxidize, which has been largely overcome through selective breeding, and poor performance in cold conditions. The U.S. federal government sponsors a program called "Bio-Preferred" that seeks to increase the use of bio-based products and has encouraged the development of bio-based lubricants and other products.

■ MAKING DECISIONS

D.12 EVALUATING ALTERNATIVE-FUEL OPTIONS

Based on what you have just learned about alternative-fuel vehicles, create a table that will easily enable you to compare alternative-fuel options in terms of fuel type, availability, efficiency, cost, environmental impact, renewability, and ease of refueling. Work as a group to enter appropriate information in the table. Refer to the information in your completed table to answer the following questions:

1. Under what circumstances does each type of alternative-fuel vehicle still rely on petroleum as a fuel? Explain.

2. Which of the alternative-fuel vehicle options involve, or could involve, completely renewable fuel sources? Explain.

3. Which of the options has the greatest potential for a completely renewable fuel source in the future? For each option you name, propose an approach that might support the transition from nonrenewable to renewable fuel.

4. The *TLC-p* advertised at the start of this unit is a hybrid vehicle. Why is this type of vehicle currently popular with buyers?

5. Which of the options do you consider the most promising overall for future use? Be prepared to list and explain the factors upon which you based your decision.

SECTION D SUMMARY

Reviewing the Concepts

> Throughout human history, society has relied on a variety of biomolecules as energy sources.

1. What is a biomolecule?
2. Refer to Figure 3.55, page 360. Identify the primary source of energy in the United States
   a. in 1900.
   b. in 1940.
   c. since 1985.
3. List three fuels that contain biomolecules used for energy.
4. What is the source of the energy stored in biomolecules?
5. Although petroleum has been used for thousands of years, only in relatively recent history has it become a major energy source. List three technological factors that help to explain this.

> Thermal energy can be converted into other forms of energy. Although energy can neither be created nor destroyed, useful energy is lost each time energy is converted from one form to another.

6. In powering an automobile, 25% of the energy is said to be useful. What happens to the other 75% of the energy?
7. Explain what energy-conversion efficiency means.

8. One gallon of gasoline produces ~132 000 kJ of energy when burned. Assume that an automobile is 25% efficient in converting this energy into useful work.
   a. How much energy (in kJ) is "wasted" when a gallon of gasoline burns in an automobile?
   b. What happens to this wasted energy?

> Larger molecules in petroleum's kerosene fraction (C_{12} to C_{16}) can be converted to smaller molecules useful in the gasoline fraction (C_5 to C_{12}). The speed of the process, called cracking, is increased by using a catalyst.

9. Write one possible balanced equation showing how each of the following molecules could be cracked into molecules for the gasoline fraction.
   a. $C_{17}H_{36}$ b. $C_{18}H_{38}$
10. In terms of energy efficiency, why are catalysts used during the cracking process?
11. Explain why only small amounts of catalysts are needed to crack large amounts of petroleum.

> The octane rating is an expression of a fuel's burning performance in an engine. Additives may enhance gasoline's burning performance.

12. Why do we no longer use tetraethyl lead as a gasoline additive?
13. Draw and compare the molecular structures of octane and isooctane.

14. How does the addition of oxygenated compounds affect a fuel's octane rating?

15. List two ways to increase a fuel's octane rating.

> Diesel fuel substitutes can be produced from renewable resources.

16. What is the primary functional group in biodiesel molecules?

17. Why was potassium hydroxide used in making biodiesel in Investigating Matter D.8?

18. What are three advantages of biodiesel over petroleum-based diesel fuel?

19. Could renewable diesel replace all diesel use in the United States? Support your claim with evidence.

> Since petroleum is a nonrenewable resource, it will eventually be necessary to develop alternative sources of energy and builder molecules.

20. List one advantage and one disadvantage of using each of the following alternative fuel sources:
    a. oil shale
    b. coal
    c. biodiesel

21. List three alternative energy sources that do not involve biomolecules.

22. What percent of petroleum consumed in the United States is used for transportation?

23. List and describe two emerging non-petroleum sources of builder molecules.

> Several types of alternative-fuel vehicles are being developed, tested, and used.

24. List an advantage and a disadvantage of each of the following alternative power sources for vehicles.
    a. compressed natural gas
    b. liquefied petroleum gas
    c. flexible fuel or biodiesel
    d. diesel
    e. hydrogen fuel cell
    f. electric
    g. hybrid gasoline–electric
    h. plug-in hybrid gasoline–electric

25. What are the emission products of fuel-cell powered vehicles?

26. What is the most common source of the hydrogen gas used in fuel cells?

What alternatives to petroleum are available for burning and building?

In this section, you have considered several alternatives to petroleum for both burning and building applications. Think about what you have learned, then answer the question in your own words in organized paragraphs. Your answer should demonstrate your understanding of the key ideas in this section.

Be sure to consider the following in your response: energy efficiency, oxygenated fuels, oil shale, coal, biomass, biodiesel, fuel cell, and hybrid vehicles.

27. Name the energy transformation that occurs in a hybrid vehicle when
    a. the vehicle is accelerated.
    b. the vehicle's brakes are applied.

Connecting the Concepts

28. Why is the use of catalysts encouraged by green chemistry?

29. Some energy authorities recommend exploring ways to use more renewable energy sources such as hydroelectric, solar, and wind power as replacements for nonrenewable fossil fuels.
    a. Why might this be a useful policy?
    b. Which of these renewable sources are least likely to replace fossil fuels at this time? Why?

30. If there is a dramatic increase in petroleum prices on the world market, explain which alternative source of energy you would expect to serve as the most practical substitute
    a. immediately.
    b. in 10 years.

31. Suggest three ways individuals could reduce their total consumption of petroleum products.

32. We often consider alternative energy sources to be pollution-free. Choose two alternative energy sources and evaluate them in terms of their possible impact on the environment.

33. Using Figure 3.55 on page 360 as a guide, make a pie chart showing the percent of all fuel sources used in
    a. 1900.   b. 2000.

34. Some people who do not own a car believe that they do not consume any fossil fuel. Evaluate their claim.

35. When biodiesel burns, it emits CO_2, like any other hydrocarbon-based fuel. Explain why it is not considered to be a *net* contributor to atmospheric CO_2.

36. Choose two alternative-fuel vehicles and discuss how choosing each one (instead of a gasoline-powered car) would affect your overall carbon footprint.

Extending the Concepts

37. Referring to Figure 3.56 (page 362), suggest specific engine-design modifications that might increase the proportion of energy available to propel the car.

38. MTBE was once widely used as an oxygen source for automobile fuels. Investigate the chemical properties of this additive and the controversy surrounding its use. Report on its current status.

39. Predict how your community would respond to a proposal for installing wind- or solar-power devices on a large scale to provide electrical power. Consider both advantages and disadvantages that might be considered in a ballot proposal or legislative action.

40. Analyze how the relatively low cost and abundant availability of petroleum have affected the search for and development of alternative energy sources.

41. Identify and evaluate some state or federal programs that encourage development or use of alternative energy and builder molecule sources.

42. Experts disagree about how long fossil-fuel supplies will last. Research and evaluate some of their current opinions.

PUTTING IT ALL TOGETHER

GETTING MOBILE

If you are an average U.S. high school student, you have already viewed about a half-million television commercials, many of them for automobiles. As alternative-fuel vehicles become more available, advertisements similar to the one that opened this unit will become more common.

To make informed, intelligent consumer decisions, it is important to analyze the information conveyed in such advertisements, much as you did when you investigated claims in Unit 2. You can further develop your analytical skills by devising and defending your own product claims.

You will now draft and produce your own automobile advertising message. After presenting your commercial message, you will answer questions posed by other students.

Designing and Evaluating Vehicle Ads

Each student team will create and present an advertisement featuring an imaginary but plausible vehicle that uses a particular type of fuel. Your teacher will indicate which power options you may choose. You will present the commercial message to the class for analysis and comment, based on ideas introduced and discussed in your chemistry class. You will also prepare a technical brief to explain and support the claims made in your advertisement.

Each advertising message must meet certain specifications. Use these specifications to guide the development of your advertisement and written technical brief, as well as to evaluate the presentations of your classmates.

Scientific Claims

- All scientific claims must be accurate and supported in the technical brief.
- Because the type of fuel (energy source) represents the unique feature of your vehicle, you should feature and explain it briefly in the advertisement.
- Compare your vehicle to those that depend solely on gasoline or other petroleum products for fuel.

Presentation

- Your message should take no more than one minute of "air time."
- Your message should be:
  - organized
  - visually stimulating
  - motivating to a potential buyer

Technical Brief

- Your written technical summary must:
  - Explain how your vehicle's power source works.
  - Discuss the origin of the energy in your vehicle's fuel and the energy cost involved in producing that fuel.
  - Include implications of emissions released by your vehicle, fuel efficiency, and how, other than in its use as a fuel, petroleum is involved in the production or use of your vehicle.

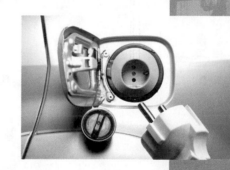

LOOKING BACK AND LOOKING AHEAD

This unit has illustrated once again that chemical knowledge can help inform personal and community decisions related to resource use and possible replacement options. Thus far in your *ChemCom* studies, you have focused on three types of resources: air, minerals, and petroleum. The next unit explores another kind of resource—water—the most abundant substance on Earth. As you will learn, use of water and responsibility for providing and maintaining clean water are important, and sometimes contentious, issues in every community.

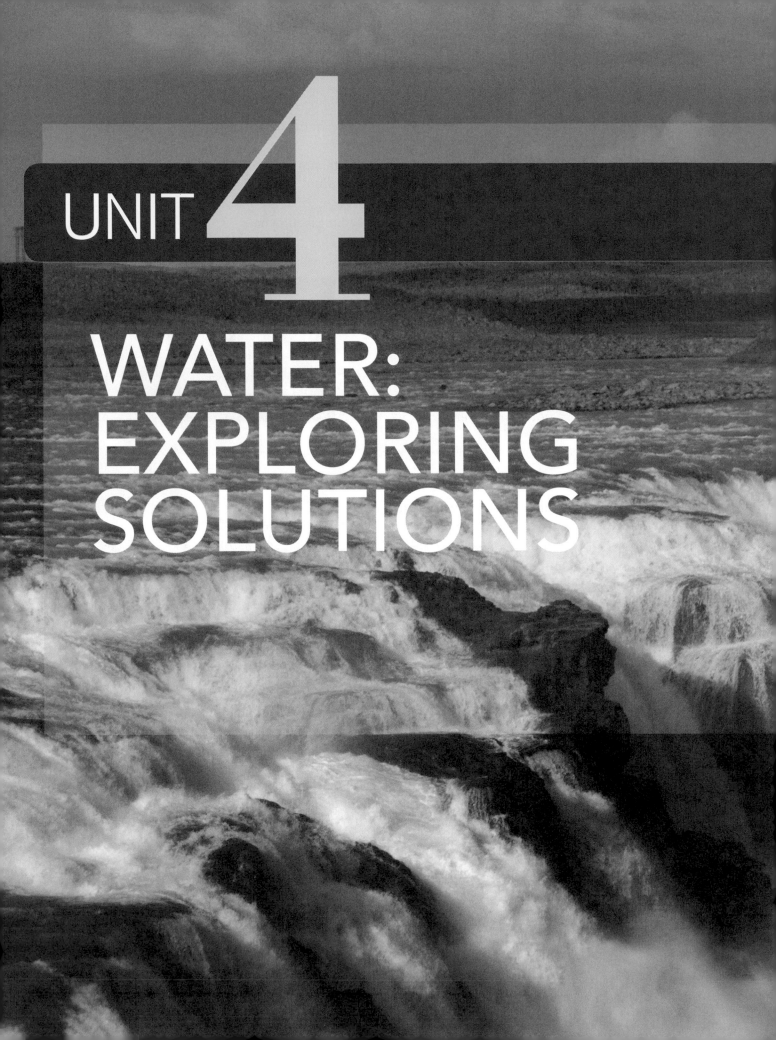

UNIT 4

WATER: EXPLORING SOLUTIONS

What makes water unique?

Why do some substances readily dissolve while others do not?

How do we describe chemical behavior in aqueous solutions?

How is chemistry applied to produce safe drinking water?

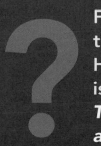

Fish are dying in the Snake River. Why? What are the consequences for the Riverwood community? How might this situation be related to water issues in your own community?
Turn the page to find out more about this crisis and the role of water in modern life.

RIVERWOOD NEWS

Breaking news at RiverwoodNewsLive.com

Fish Kill Triggers Riverwood Water Emergency

Severe Water Rationing in Effect

By Lori Katz
RIVERWOOD NEWS STAFF REPORTER

Citing possible health hazards, Mayor Edward Cisko announced today that Riverwood will stop withdrawing water from the Snake River and will temporarily shut down the town's water-treatment plant over water-quality concerns provoked by a massive fish kill. Starting at 6 p.m., river water will not be pumped to the plant for at least three days. If the cause of the fish kill has not been determined and corrected by that time, the shutdown will continue indefinitely.

During the plant shutdown, water engineers and chemists from the county sanitation commission (which operates Riverwood's water-treatment plant) and the U.S. Environmental Protection Agency (EPA) will investigate the cause of the major fish kill discovered yesterday. The fish kill extended from the base of the Snake River Dam, located upstream from Riverwood, to the town's water-pumping station. The dam is owned and operated by Riverwood Power. The reservoir behind the dam provides water storage for the town, as well as a source of cooling water for Riverwood Power's electric-generating station.

The initial alarm was sounded when Jane Abelson, 15, and Chad Wong, 16—both students at Riverwood High School—found many dead fish floating in a favorite fishing spot. "We thought maybe someone had poured poison into the reservoir," explained Wong.

Soon after discovering the fish kill, Riverwood High School students returned to the river to investigate.

Mary Steiner, a Riverwood High School biology teacher, accompanied the students back to the river. "We hiked downstream along the farms and past the old mine for almost a mile. Dead fish of all kinds were washed up on the banks and caught in the rocks," Abelson reported.

Ms. Steiner contacted county sanitation commission officers, who immediately collected Snake River water samples for analysis. Chief sanitation engineer Hal Cooper reported at last night's emergency meeting that the water samples appeared clear, colorless, and odorless. However, he indicated some concern. "We can't say for certain that the water supply is safe until the cause of the fish kill is determined. It's far better that we take no chances until then," Cooper advised.

See FISHKILL, page 5

FISHKILL, from page 1

Mayor Cisko canceled the community's "Riverwood Fish-In," which was scheduled to start next week. No plans to reschedule Riverwood's annual fishing tournament were announced. "The decision was made at last night's emergency town council meeting to start investigating the situation immediately," he said.

> **" We can't say for certain that the water supply is safe until the cause of the fish kill is determined . . . "**

After five hours of often-heated debate yesterday, the Riverwood town council finally reached agreement to stop drawing water from the Snake River. Council member Henry McLatchen (also a chamber of commerce member) commented that the decision was highly emotional and unnecessary. He cited financial losses that motels and restaurants will suffer because of the Fish-In cancellation, as well as potential loss of future tourism dollars due to adverse publicity. However, McLatchen and other council members sharing that view were out-voted by those holding the position that the fish kill, the only one within Riverwood's recorded history, may indicate a public health emergency.

Mayor Cisko assured residents that essential services will not be affected by the crisis. For example, he promised to maintain fire department access to adequate supplies of water to meet firefighting needs.

Arrangements have been made to transport emergency drinking water from Mapleton. The first water shipments are due to arrive in Riverwood by midmorning tomorrow. Distribution points are listed in Section 2 of today's Riverwood News, along with guidelines on conserving water during this emergency. This information is also available on the *Riverwood News* Web site.

All Riverwood schools will be closed Monday and possibly through Wednesday. No other closings or cancellations have been announced. Local TV and radio will report any schedule changes as they become available.

A public meeting tonight at 8 p.m. at the town hall features Dr. Margaret Brooke, a water expert at State University. She will answer questions concerning water safety

Dead fish washed up along the banks of the Snake River yesterday afternoon.

and use. Brooke was invited by the county sanitation commission to help clarify the situation for concerned citizens. The meeting will be streamed live on the *Riverwood News* Web site.

Asked how long the water emergency would last, Brooke refused to speculate, saying that she first needed to talk to other scientists conducting the investigation. EPA investigators, in addition to collecting and analyzing water samples, will examine the dead fish in an effort to determine what was responsible for the fish kill. Brooke reported that trends or irregularities in water-quality data from Snake River monitoring during the past two years also will play a part in the investigation.

—Lori Katz can be reached at lkatz@riverwoodnewslive.com

Townspeople React to Fish Kill and Riverwood Water Crisis

By Juan Hernandez
RIVERWOOD NEWS STAFF REPORTER

In a series of on-the-street interviews, Riverwood citizens expressed a variety of opinions earlier today about the crisis. "It doesn't bother me," said nine-year-old Jimmy Hendricks. "I'm just going to drink bottled water and juice."

"I knew that eventually they'd pollute the river and kill the fish," complained Harmon Lewis, a lifelong resident of Fieldstone Acres, located east of Riverwood. Lewis, who traces his ancestry to original county settlers, still gets his water from a well and will be unaffected by the water crisis. He said that he plans to pump enough well water to supply the children's ward at Community Hospital if the emergency extends more than a few days.

Bob and Ruth Hardy, co-owners of Hardy's Ice Cream Store, expressed annoyance at the inconvenience but were reassured by council actions. They were eager to learn the reason for the fish kill and its possible effects on water supplies. Their daughter Toni, who loves to fish, was worried that late-season fishing would be ruined. Toni and her father won first prize in last year's angling competition.

❝ *It doesn't bother me, I'm just going to drink bottled water and juice. . . .* **❞**

Riverwood Motel owner Don Harris expressed concern for both the health of town residents and the loss of business due to the tournament cancellation. "I always earn reasonable income from this event and, without the revenue from the Fish-In, I may need a loan to pay bills in the next six months."

David Price was excited that school was canceled.

The unexpected school vacation was "great," according to 12-year-old David Price. Asked why he thought schools would be closed Monday, Price said that all he could think of was that "the drinking fountains won't work."

— *Juan Hernandez* can be reached at jhern@riverwoodnewslive.com

The fish kill sparked diverse reactions among Riverwood residents. Don Harris voiced concerns about health and economic impact of the fish kill.

SECTION

SOURCES, USES, AND PROPERTIES OF WATER

What makes water unique?

Riverwood confronts at least a three-day water shortage. The water emergency has aroused understandable concern among Riverwood citizens, town officials, and business owners. What caused the fish kill? Does the fish kill mean that Riverwood's water supply poses hazards to humans? In the following pages, you will monitor the town's progress in answering these questions and develop a hypothesis as to the cause of the fish kill. As you follow the unfolding story, you will deepen your understanding of matter and investigate water's properties and uses.

GOALS

- Explain the relationship between the chemical structure of water and its unique properties.
- Distinguish several types of mixtures (solutions, colloids, and suspensions).
- Interpret and create models that represent mixtures at the particulate level.
- Analyze personal and community uses of water, including direct and indirect uses.

✓ concept check 1

1. a. How do hydrocarbon molecules stick together?
b. How do water molecules stick together?
2. Consider an oil-and-vinegar-based salad dressing.
a. Propose an explanation for the observation that the contents of a bottle of this dressing separate into two layers.
b. How does the density of the oil compare with that of the water? What evidence can you cite to support your answer?
3. What is a solution?

A.1 TOWN IN CRISIS

Although Riverwood is imaginary, its problems are not. Residents of many communities have faced these and similar problems. In fact, two water-related challenges confront each of us every day. Can we get enough water to supply our needs? Can we get sufficiently pure water? These two questions serve as major themes of this unit, and their answers require an understanding of water's chemistry and uses.

The notion of water purity must be given careful consideration. See Figure 4.1. You will soon learn that the cost of producing a supply of water that is 100% pure is prohibitively high. Is that level of purity needed—or even desirable? Communities and regulatory agencies are responsible for ensuring the availability of water of sufficiently high quality for its intended uses at reasonable cost. How do they accomplish this task?

Even the apparently simple idea of water use presents some fascinating puzzles, as the *ChemQuandary* below illustrates.

During the 20th century, the population of Earth increased threefold, and water use increased by a factor of six. Access to water has the potential to replace fossil fuel use as the most pressing global-economic issue in this century. In the following activities, you will investigate your household's water use and explore and model the properties of water and aqueous solutions.

One liter (1 L) is approximately one quart.

Figure 4.1 *This water is obviously clear, but is it pure? What is pure water?*

CHEM**QUANDARY**

WATER, WATER EVERYWHERE

It takes approximately 850 L of water to produce one 1.0-L plastic bottle of orange juice. It takes about 450 L of water to place one fried egg on your breakfast plate.

Think of possible explanations for these two facts by listing the steps involved in producing and delivering to your home the fruit juice and the egg. Then review each step and consider where water use would occur.

■ MAKING DECISIONS

A.2 USES OF WATER

Keep a diary of water use in your home for three days. On a data table like the one shown here, record how often various water-use activities happen. Ask each household member to cooperate and help you.

Data Table			
Per Household	**Day 1**	**Day 2**	**Day 3**
Number of persons			
Number of baths			
Number of showers			
Average duration of a shower (min.)			
Number of toilet flushes			
Number of hand-washed loads of dishes			
Number of machine-washed loads of dishes			
Number of washing-machine loads of laundry			
Number of lawn or garden waterings			
Average duration of a watering (min.)			
Number of car washes			
Number of cups of water (estimated) for cooking and drinking			
Number of times water runs in sink			
Average duration of water running (min.)			
Other uses and frequency			

Check the activities listed on the chart. See Figure 4.2. If family members use water in other ways within the three-day period, add those uses to your diary. Estimate the quantities of water used by each activity.

Figure 4.2 *Will you wash a car in the next three days? If so, record it in your water-use log.*

■ INVESTIGATING MATTER

A.3 PROPERTIES OF WATER

Preparing to Investigate

Determining the cause of the Riverwood fish kill will require knowledge of water's properties and the chemistry of water-based solutions, commonly called **aqueous solutions**. In addition to explaining the fish kill, an understanding of water and the chemistry that occurs in water is vital to understanding the chemistry in your body. In this investigation, you will explore some properties of water and aqueous solutions. You will refer back to these investigations as you progress through the unit.

Before you begin, read *Gathering Evidence* to learn what you will need to do, and note safety precautions. *Gathering Evidence* also provides guidance about when you should collect and record data.

Making Predictions

After reading *Gathering Evidence*, prepare an appropriate data table. Predict what you think will happen in Investigations 1, 2, 5, and 7 and write down your predictions.

Gathering Evidence

Before you begin, put on your goggles, and wear them properly throughout the investigation.

Six stations, A through F, have been set up around the laboratory. At each station, you will complete the investigation for that station. The stations can be completed in any order. When working at a particular station, though, you must complete the investigations at the station in the order indicated. Follow these general instructions:

- Take note of how the station is set up before you begin. You will be expected to reset the station after you are finished.
- Re-read the procedure and safety reminders.
- Review your predictions.
- Complete the investigation.
- Record your data and observations. (*Note:* Do not try to explain your observations. The purpose of these activities is for you to make observations. As you progress through the unit, you will develop explanations for many of the phenomena you observe in this investigation.)

Station A:

Investigation 1

1. Pour water into a Petri dish until it is half-full.
2. Using the tongs provided, gently place a *dry* paperclip on the surface of the water. See Figure 4.3

Figure 4.3 *Placing a paper clip on water in a Petri dish.*

3. Record your observations.

4. Repeat Steps 1 and 2 using a clean, dry Petri dish and 2-propanol instead of water.

5. Record your observations.

6. Remove the paper clips, dispose of the water and 2-propanol as directed by your teacher, and reset the station.

Station B:

Investigation 2

1. Rub the comb with the paper towel.

2. Open the buret so a stream of water is flowing into the container.

3. Bring the comb near (but not touching) the stream of water. See Figure 4.4.

4. Record your observations.

5. Refill the buret with water.

Station C:

Investigation 3

1. Using a dropper, form a small puddle of water (about the size of a dime) on the surface of a piece of wax paper.

2. Place the tip of the dropper in the center of the puddle and slowly drag the tip of the dropper around the wax paper.

3. Record your observations.

4. Using a different dropper, repeat the investigation using olive or corn oil.

5. Wipe off the wax paper for the next group.

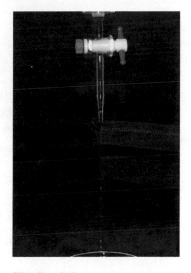

Figure 4.4 *Bring the comb near the stream of water from the buret.*

Investigation 4

1. You have two beakers labeled A and B. Using tongs, place an ice cube in each beaker.

2. Record your observations, then remove the ice cubes from the beakers.

3. Reset the station according to your teacher's instructions.

Station D:

Investigation 5

1. Pour 10 mL of cold water into the beaker labeled C.

2. Pour 10 mL of hot water into the beaker labeled H.

3. Begin adding cubes or packets of sugar to each beaker. After each addition, stir until dissolved. Keep adding cubes or packets until no more will dissolve.

4. Record the number of cubes or packets that dissolved in each beaker.

5. Dispose of each solution as directed by your teacher.

Station E:

At this station, you will find three solutions and three filtration setups. Your teacher has already poured some of each solution through the filter paper.

Investigation 6

1. Describe the appearance of the three original solutions.
2. For each solution, A, B, and C, shine a flashlight through the solution. See Figure 4.5.
3. Record your observations.
4. Describe what has been caught by the filter paper for each solution.

Figure 4.5 *Left: Particles are suspended in the sample in the beaker on the left. The particles are too small to see, but large enough to reflect light coming from a beam to the left of the beakers. This is called the **Tyndall effect**. Particles in the solution in the beaker on the right are too small to reflect light. Right: The Tyndall effect is also observable in nature.*

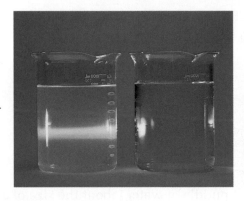

Station F:

Investigation 7

1. Half-fill the three labeled beakers with distilled water.
2. Test the pH and conductivity of the water (as directed by your teacher) in each beaker and record your value.
3. Add a small amount of Salt A to beaker A and stir until dissolved. Add a small amount of Salt B to beaker B and stir. Add a small amount of Salt C to beaker C and stir until dissolved.
4. Retest the pH and conductivity of each solution and record your values.
5. Dispose of your solutions as directed by your teacher.

Interpreting Evidence

1. For each investigation in which you made predictions,
    a. describe how well your observations agreed with your predictions.
    b. propose explanations for any differences between your predictions and observed results.
2. Did any of these investigations examine the tendency of water to "stick together"? Explain your answer using experimental evidence.
3. How do particles filtered from the three mixtures in Investigation 6 differ?

4. In Unit 2 Investigating Matter C.8 (page 216), you monitored the pH of water before and after adding CO_2. Which salt in Investigation 7 had an effect similar to that of CO_2 on the pH of the water?

Making Claims

5. In Investigation 4, what claim can you make about the relative densities of the two liquids?
6. Use your observations to either support or refute the claim that water has electrical properties.
7. Based on your observations, is the following a justifiable claim: "More of a solid will always dissolve in hot water than in cold water." Explain your reasoning.

Reflecting on the Investigation

8. You have explored many properties of water and how it differs from other liquids. Using your answers to the questions above, write a paragraph that sums up what you know about water from this investigation.
9. Write two questions about water that are still unanswered after this investigation.

A.4 PHYSICAL PROPERTIES OF WATER

Water is a common substance—so common that we usually take it for granted. We drink it, wash with it, swim in it, and sometimes grumble when it falls from the sky. But are you aware that water is one of the rarest and most unusual substances in the universe? As planetary space probes have gathered data, scientists have learned that the great abundance of water on Earth's surface is unmatched by any planet or moon in our solar system. Earth is usually half-enveloped by water-laden clouds, as you can see in Figure 4.6. In addition, more than 70% of Earth's surface is covered by oceans that have an average depth of more than three kilometers (two miles).

Kilo- (k) is a metric prefix meaning 1000. One kilometer (km) = 1000 meters (m).

Figure 4.6 *Earth as seen from space. Our planet's abundant supply of water makes it unlike any other in our solar system.*

The density of liquid water depends on its temperature—for example, the density of liquid water is 0.992 g/mL at 40 °C and 0.972 g/mL at 80 °C.

Figure 4.8 *One cubic centimeter (shown actual size). 1 cm³ = 1 mL*

0 °C = 32 °F

Water is a form of matter. See Figure 4.7. As you recall, matter is anything that occupies space and has mass. Water has many physical properties—properties that can be observed and measured without changing the chemical makeup of the substance. One physical property of a sample of matter is its *density*, the mass of material within a given volume. The density of water as a liquid is easy to remember. Because one milliliter (mL) of liquid water at room temperature (25 °C) has a mass of about 1.00 g, the density of this water is 1.00 g/mL.

Figure 4.7 *What states of water can be observed in this winter scene?*

One milliliter of volume is exactly equal to one cubic centimeter (1 cm³), which is pictured in Figure 4.8. Thus, the density of water at 25 ˚C can also be reported as 1.00 g/cm³.

The *density* of solid water—ice—is 0.917 g/mL at 0 °C. Unlike most substances, the density of the solid form of water is less than the density of the liquid form. This is why solid water (ice) floats on liquid water. For all other nonmetallic substances, the solid would sink. Imagine the consequences if ice were more dense than liquid water. For example, bodies of water would freeze from the bottom up. This would have a drastic impact on the aquatic life in the water!

The melting point of ice, another physical property, is 0 °C at normal atmospheric pressure. The boiling point of water is 100 °C at normal atmospheric pressure. Table 4.1 shows boiling and melting points of three molecular compounds, including water, with similar molecular masses. What characteristics of water molecules could explain the differences evident in Table 4.1?

Table 4.1

Some Properties of Water, Methane, and Ammonia			
Substance	**Molar Mass**	**Melting Point**	**Boiling Point**
H_2O	18 g/mol	0 °C	100 °C
NH_3	17 g/mol	−78 °C	−33 °C
CH_4	16 g/mol	−183 °C	−162 °C

Recall that in Unit 3 (page 250), you learned about intermolecular forces and how they influence the properties of a compound. To understand the unique properties of water, first consider the bond between oxygen and hydrogen in a water molecule. This is a covalent bond, formed by the sharing of an electron pair. However, the electrons in this bond are not shared equally. One atom in a covalent bond often has a greater attraction for the shared pair of electrons than does the other bonded atom. This tendency for an atom to attract the shared pair of electrons in a covalent bond is the atom's **electronegativity**. Oxygen is more electronegative than hydrogen, so it has a stronger pull on the pair of electrons it shares with hydrogen. The result of this unequal sharing is a polar bond—the hydrogen atom at one end of the bond has a slight positive charge, while the oxygen atom at the other end has a slight negative charge.

Just as individual bonds in molecules can be polar, a molecule can be polar if it has at least one polar bond and an appropriate molecular shape. A polar molecule has an uneven distribution of electrical charge, which means that it has regions of both partial positive and partial negative charge arranged asymmetrically. Evidence shows that a water molecule has a bent (asymmetric) or V-shape, as illustrated in Figure 4.9, rather than a linear, sticklike shape as in H–O–H. The oxygen end is an electrically negative region that has a greater concentration of electrons (shown as δ^-) compared with the two hydrogen ends, which are electrically positive (shown as δ^+). The large difference in electronegativity between hydrogen and oxygen, along with the bent shape of the water molecule, results in a quite polar molecule.

> The Greek symbol δ (delta) means "partial"—thus, partial positive and partial negative electrical charges are indicated. Because these charges are equal and opposite, the molecule as a whole is electrically neutral.

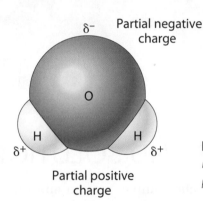

Figure 4.9 *Polarity of a water molecule. The δ^+ and δ^- indicate partial electrical charges.*

In Unit 3, you learned that the forces between and among molecules are called intermolecular forces. Because water molecules are very polar, they display unusually strong intermolecular forces known as hydrogen bonds. Note that these forces are not really bonds—they do not link atoms or ions in molecules or ionic substances. Hydrogen bonds are attractive forces between molecules. The word *bond* is simply used to indicate that hydrogen bonds are very strong compared to other intermolecular forces.

> *Intermolecular* means between molecules.

The ability of water molecules to form very strong intermolecular attractions with each other is responsible for many of its unique properties. Because of these strong attractions among molecules, a large quantity of energy is required to pull the molecules within the solid apart to form a liquid, resulting in a high melting point compared to similar molecular substances.

Another consequence of the strong intermolecular forces in water is that it has very strong **cohesive forces**, the attractive forces that molecules in a liquid have for each other. These strong cohesive forces are responsible for water's high boiling point compared to other substances of similar molar mass. The very strong cohesive forces in water are also responsible for some of the observations you made in Investigating Matter A.3. For instance, you were able to

observe a paperclip floating on the surface of water in Investigation 1. Recall how your results differed when you used 2-propanol. How could you explain this difference? Also, in Investigation 3, water formed a spherical drop on wax paper, but oil did not. What do your results suggest about the relative strength of the cohesive forces in water compared to oil? Consider these properties of water as you examine Figure 4.10.

Hydrogen bonding is also responsible for the low density of ice compared to liquid water. When water freezes to form a solid, it takes on a hexagonal structure that maximizes the number of hydrogen bonds that can form. This results in an increase in the space between water molecules as ice forms, as shown in Figure 4.11.

Figure 4.10 *What properties of water might account for its sheeting action under the swimmer's right arm?*

Figure 4.11
Three-dimensional structures of ice (left) and liquid water (right).

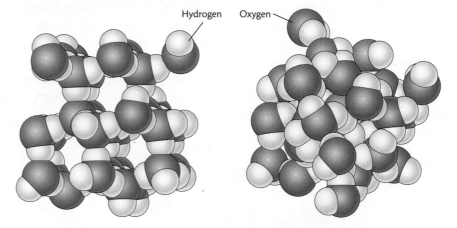

Hydrogen   Oxygen

A.5 MIXTURES AND SOLUTIONS

How can you decide if a water sample is clean enough to drink? How can substances in water be separated and identified? Answers to these questions will be helpful in understanding, and possibly solving, the fish-kill mystery.

When two or more substances combine yet retain their individual properties, the result is called a **mixture**. As you saw in Investigating Matter A.3, the components of a mixture can be separated by physical means, such as filtration. Some of the mixtures you encountered in Investigating Matter A.3 were examples of **heterogeneous mixtures**. That is, their composition is not the same, or uniform, throughout.

One type of heterogeneous mixture is called a **suspension** if the solid particles are large enough to settle out or can be separated by using filtration. Muddy water and water plus flour or chalk particles are examples of suspensions.

If the particles are smaller than those in a suspension, they may not settle out and thus may cause the water to appear cloudy. The scattering of the light, known as the **Tyndall effect** (see Figure 4.5, page 398), indicated that small, solid particles were still in the water. This type of mixture is called a **colloid**.

A more familiar example of a colloid is whole or low-fat milk (see Figure 4.12), which contains small butterfat particles dispersed in water. These colloidal butterfat particles are not visible to the unaided eye. The mixture appears **homogeneous**—uniform throughout. Under high magnification, however, individual butterfat globules can be observed suspended in the water. Milk no longer appears homogeneous.

A heterogeneous mixture's composition varies from sample to sample.

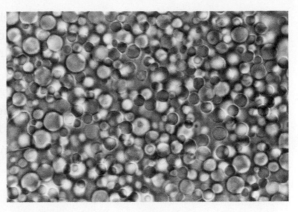

Figure 4.12 *Milk (left) is an example of a colloid. Small butterfat particles are visible under magnification (right).*

Particles far smaller than colloidal particles also may be present in a mixture. When small amounts of table salt are mixed with water, the salt dissolves in the water. That is, the salt crystals separate into particles so small that they cannot be seen even at high magnification, nor do the particles exhibit the Tyndall effect when a light beam is passed through the mixture. These particles become uniformly mingled with the particles of water, producing a homogeneous mixture. All **solutions** are homogeneous mixtures. In a salt solution, the salt is the **solute** (the dissolved substance) and the water is the **solvent** (the dissolving agent). All solutions consist of one or more solutes and a solvent. See Figure 4.13.

One type of evidence that something is dissolved in a water sample (as in Investigating Matter A.3) is a conductivity test. A positive result (the bulb lights up) indicates that electrically charged particles are dissolved in the mixture. Note that a conductivity test cannot always tell you if something is dissolved in a water sample or not. As you will soon learn, many molecular substances dissolve in water without producing electrically charged particles.

Figure 4.13 *Sugar cubes dissolve in tea. In this case, tea is the solvent and sugar is the solute.*

■ MODELING MATTER
A.6 REPRESENTING MIXTURES

Throughout this course, you have represented the macroscopic world with drawings of matter at the microscopic level. As you study aqueous mixtures, try to visualize what is in the mixture at the molecular level. This activity will give you some practice drawing representations of mixtures.

1. Draw a model of a container of a solution composed of a solvent that is a two-atom compound of L and R, and a solute that is a compound composed of two atoms of element D and one atom of element T.

2. Draw a model of the solution made when you dissolved sugar in Investigation 5 of Investigating Matter A.3,

   a. in cold water.

   b. in hot water.

   c. Identify differences between the representations in 2a and 2b.

3. The element iodine (I) has a greater density in its solid state than in its gaseous state. Draw models that depict and account for this difference at the particulate level. Iodine normally exists as a two-atom molecule.

4. Draw a model you could use to help explain to a friend what happened when you shined the flashlight through the solutions in Investigation 6 of Investigating Matter A.3.

5. Compare your drawings from Questions 1–4 with a classmate.

   a. How are your drawings similar?

   b. How are your drawings different?

   c. Do the differences in your drawings indicate different ideas or simply different representations? Explain.

6. Consider some limitations of the representations you have drawn.

   a. What ideas and events are difficult to represent in these drawings?

   b. What advantages would there be to using computer animations instead of static drawings?

concept check 2

1. Suppose that you placed water, olive oil (density = 0.92 g/mL), and corn syrup (density = 1.4 g/mL) in a graduated cylinder and separate layers formed. Draw a diagram showing which layer would be in each position, then explain your reasoning.
2. Some people claim that it is easier to "float" in the ocean than in a freshwater lake. What does this comparison reveal about the densities of seawater and freshwater?
3. How do heterogeneous and homogenous mixtures differ?
4. Identify the source of your drinking water.

A.7 WATER SUPPLY AND DEMAND

> A trillion liters is 1 000 000 000 000 liters, or 10^{12} liters.

The problems caused by the water shut-off in Riverwood raise a larger question. Is the United States in danger of running out of water? The answer is both no and yes. The total water available is far more than enough. Each day, some 15 trillion liters (4 trillion gallons) of rain or snow falls in the United States. Only 10% is used by humans. The rest flows back into large bodies of water, evaporates into the air, and falls again as part of Earth's perpetual **water cycle**, or **hydrologic cycle**, which contains a fixed quantity of water. So that is the *no* part of the answer. However, the distribution of rain and snow in the United States (and around the world) does not necessarily correspond to regions of high water use. Figure 4.14 summarizes how available water is used in various regions of the country, organized by five major water-use categories.

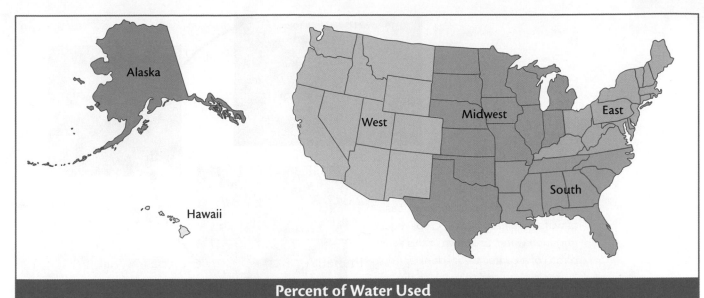

| Percent of Water Used | | | | | | |
| Purpose | Region | | | | | |
	Nation	East	South	Midwest	West	Alaska	Hawaii
Domestic	12	14	10	12	11	8	14
Irrigation/Livestock/Aquaculture	34	3	20	23	76	68	6
Industrial	4	5	7	5	1	1	2
Mining	1	1	1	2	1	19	0
Thermoelectric-power	49	78	63	58	11	3	78

Figure 4.14 *Water use in the United States. How does water use differ across the regions?*

A U.S. family of four (two adults, two children) uses an average of 1480 liters (390 gallons) of water daily. That approximate volume represents **direct water use**, which can be directly measured. There is also **indirect water use**, hidden uses of water that you may never have considered. Each time you eat a slice of pizza, some potato chips, or an egg, you are "using" water. Why? Because water was needed to grow and process the various ingredients of each food.

Consider again the ChemQuandary on page 394. At first glance, you probably thought that the volumes of water mentioned were absurdly large. How could so much water be needed to produce one fried egg or one plastic bottle of fruit juice? These two examples illustrate typical indirect (hidden) uses of water. The chicken that laid the egg needed drinking water. Water was used to grow the chicken's feed. Water was also used for various steps in the process that eventually brought the egg to your home. Even small quantities of water used for these and other purposes quickly add up when billions of eggs are involved!

In the Riverwood newspaper article you read earlier, Jimmy Hendricks was quoted as saying that he would drink bottled water and fruit juice until the water supply was turned on again. However, drinking fruit juice from a container involves the use of much more water than drinking a glass of tap water. Why? Because the quantity of liquid in the container is insignificant when compared with the quantity of water used to make the container and

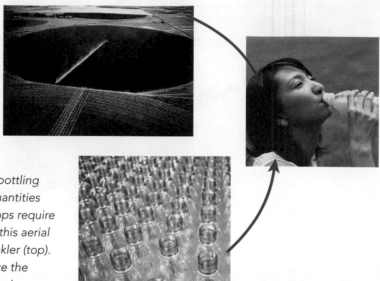

Figure 4.15 *Making and bottling juice (right) requires large quantities of water. Fruits and other crops require irrigation water, as shown in this aerial photo of a center pivot sprinkler (top). Water is also used to produce the plastic juice container (bottom).*

possibly to irrigate the fruit trees and process the fruit juice. See Figure 4.15. Growing the oranges, for example, is the primary source of the surprising 850 L of water mentioned on page 394. What examples of hidden water use do you encounter in daily life?

Although we depend on large quantities of water, most people are not aware of how much they actually use. This lack of awareness is understandable because water normally flows freely when taps are turned on—in Riverwood or in your home. Where does all this water come from? Check what you already know about the distribution of this seemingly plentiful resource.

■ MAKING DECISIONS
A.8 WATER-USE ANALYSIS

Use Table 4.2 and the data you collected for Making Decisions A.2 to answer the following questions about your household's water use, as well as those of your classmates.

1. Calculate the total water volume (in liters) used by your household during the three days.

2. How much water (in liters) did one member of your household use, on average, in one day?

3. Construct a **histogram** showing the average individual water use for the households represented by your class. To do so, organize the data into equal subdivisions, such as 500–599 L, 400–499 L, and so forth. Count the number of data points in each subdivision. Then use this number to represent the height of the appropriate bar on your histogram, as illustrated in Figure 4.16. In this sample histogram, for example, you can see that three data points fell between 500 and 599 L. (*Hint:* The range in values for each histogram bar must be equal and should be selected so that there are ~10 total bars.)

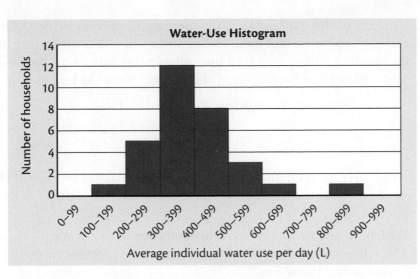

Figure 4.16 *Sample histogram of water-use data. The height of each bar represents the number of households with average daily individual water use in that range.*

4. What was the largest average daily use calculated by someone in your class? What was the smallest? The difference between the largest and smallest values in a data set is the **range** of those data points. What is the range of average daily personal water use within your class?

5. Compute a **mean** value for the class data set by adding all values together and dividing by the total number of values. The mean is a mathematical expression for the most "typical" or "representative" value for a data set.

Table 4.2

Water Required for Typical Activities	
Activity	**Water Volume (L)**
Bathing (per bath)	130
Showering (per minute)	
- Regular showerhead	19
- Water-efficient showerhead	9
Cooking and drinking (per 10 cups of water)	2
Flushing toilet (per flush)	
- Conventional toilet	19
- Water-saving toilet	13
- Low-flow toilet	6
Watering lawn (per hour)	1130
Washing clothes (per load)	170
Washing dishes (per load)	
- By hand (with water running)	114
- By hand (washing and rinsing in dishpans)	19
- By machine (full cycle)	61
- By machine (short cycle)	26
Washing car (running hose)	680
Running water in sink (per minute)	
- Conventional faucet	19
- Water-saving faucet	9

Table 4.2 indicates that a regular showerhead delivers 19 L each minute. In U.S. customary units, that's 5 gallons per minute, or 25 gallons of water during a five-minute shower! To visualize that volume of water, think of 47 two-liter beverage bottles or five filled 5-gallon buckets.

6. Another useful expression is the **median**, or middle value. To find the median for a set of data, list all values in either ascending or descending order. Then find the value in the middle of the list—the point where there are as many data points above as below.

Consider this data set: 1   2   3   4   5   6   7

three values lower   ↑   three values higher

median

If you have an even number of data points, take the average of the two values nearest the middle. Calculate the median value for the class data.

7. Which do you think is more representative of the data set—the mean or median value? That is, which is a better expression of the "average" for these data?

8. Compare your answer to Question 2 with the estimated average volume of water, 370 L, used daily by each person in the United States. What reasons can you give to explain any difference between your value and the national average value?

9. Which is closer to the national average (mean) for daily water use by each person, your answer to Question 2 or the class mean in Question 5? What reasons can you give to explain why that value is closer?

Recall that the Riverwood town council arranged to truck water from Mapleton to Riverwood (see Figure 4.17) for three days to meet the needs of Riverwood residents for drinking and cooking. The current population of Riverwood is ~19 500.

10. Based on water-use data collected and analyzed by your class (pages 395 and 406–408), explain how you would estimate the total volume of water that would need to be hauled to Riverwood during the three days.

Figure 4.17 *Emergency workers prepare to distribute water*

11. What additional information would help you improve your estimate in Question 10? Why?

12. What assumptions must you make to complete your estimate?

You are now quite aware of the volume of water you use daily. Suppose you had to live with much less. How would you ration your water for survival and comfort? This is exactly the question that now confronts Riverwood residents.

A.9 WHERE IS THE WORLD'S WATER?

You are probably not surprised to learn that most of Earth's total water (97% of it, in fact) is in the oceans. However, the next largest global water-storage place is not so obvious. Do you know what it is? If you said rivers and lakes, you and many others agree, but that answer is incorrect. The second-largest quantity of water is stored in Earth's ice caps and glaciers. See Figure 4.18. Figure 4.19 shows how the world's supply of water is distributed.

Figure 4.18 *Most of Earth's fresh water is stored in glaciers like this one.*

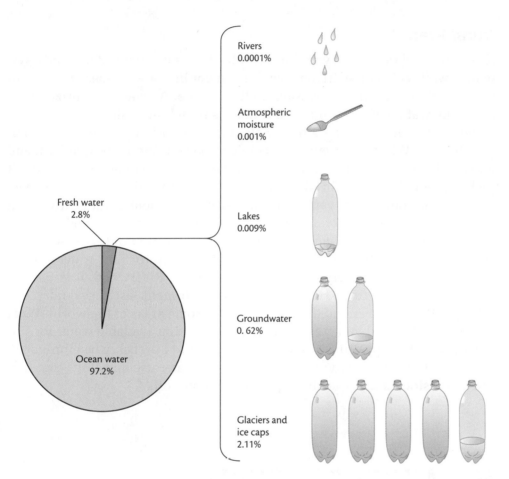

Rivers
0.0001%

Atmospheric
moisture
0.001%

Lakes
0.009%

Fresh water
2.8%

Groundwater
0. 62%

Ocean water
97.2%

Glaciers and
ice caps
2.11%

Figure 4.19 *Distribution of the world's water supply.*

As you know, water can be found in three different physical states at ordinary temperatures. Water vapor in the air is in the gaseous state. Water is most easily identified in the liquid state—in lakes, rivers, oceans, clouds, and rain. Ice is a common example of water in the solid state. What other forms of "solid water" can you identify?

At present, most of the United States is fortunate to have abundant supplies of high-quality water. You turn on the tap, use what you need, and go about your daily routine—giving little thought to how that seemingly unlimited water supply manages to reach you.

In some U.S. regions, water in the gaseous state is experienced as high humidity that contributes to summer discomfort.

City Water

About one-fifth of the U.S. water supply is held as groundwater.

If you live in a city or town, the water pipes in your home are linked to underground water pipes. These pipes bring water that was previously cleaned and purified at a water-treatment plant to all the faucets in the area. This water may have been pumped to the treatment plant from a reservoir, lake, or river. If your home's water supply originated in a lake, river, or other body of water, you are using **surface water**. If it originated in a well, you are using **groundwater**. Groundwater must be pumped to the surface.

Rural Water

If you live in a rural area, your home probably has its own water-supply system. A well with a pipe driven deep into an **aquifer** (a water-bearing layer of rock, sand, or gravel) pumps water to the surface. A small pressurized tank holds the water before it enters your home's plumbing system.

Not surprisingly, neither groundwater nor surface-water samples are completely pure. When water falls as precipitation (rain, snow, sleet, or hail) and joins a stream or when it seeps far into the soil to become groundwater, it picks up small amounts of dissolved gases, soil, and rock. These dissolved materials are rarely removed from water at the treatment plant or from well water. In the amounts normally found in water, they are harmless. In fact, some minerals found in water (such as iron, zinc, or calcium) are essential in small quantities to human health, or may actually improve the water's taste.

When the water supply is shut off, as it was in Riverwood, it is usually shut off for a short time. However, suppose a drought lasted several years. Suppose there is no end to the shortage, as in some areas of the world. What uses of water would you give up first? Clearly, using available water for survival would have priority. Nonessential uses would probably be eliminated. In the next activity, you will consider similar water-use questions that Riverwood residents will face during their water shutoff.

▮ MAKING DECISIONS
A.10 RIVERWOOD WATER USE

What is the difference between *clean* water and *pure* water?

Water supply systems collect, purify, and distribute water. Many water supply systems are operated by cities, towns, or counties.

Now that you have learned about water's properties and some water purification techniques, you are ready to return to the problem of Riverwood's fish kill. Recall that one response to the fish kill was to shut off the water. In this activity, you will evaluate the impact of this decision on Riverwood residents.

Families in most U.S. cities and towns receive an abundant supply of clean, but not absolutely pure, water at an extremely low cost. You can check the water cost in your own area. If you use water from a water supply system, your family's water bill will contain the current water cost.

It is useless to insist on absolutely pure water. The cost of processing water to make it completely pure would be extremely high. And, even if costs were not a problem, it would still be impossible to have absolutely pure water. The atmospheric gases nitrogen (N_2), oxygen (O_2), and carbon dioxide (CO_2) will always dissolve in the water to some extent. In Unit 2, you learned about how gaseous pollutants can dissolve in rainwater to produce acid rain. See Figure 4.20.

Now, however, Riverwood authorities have severely rationed home water supplies for three days while they investigate possible fish-kill causes. The County Sanitation Commission recommends cleaning and rinsing your bathtub, adding a tight stopper, and filling the tub with water. That water will be your family's total water supply for all uses other than drinking and cooking for up to three days. (Recall that water for drinking and cooking will be trucked in from Mapleton.)

Assuming that your household has just one tub of water (150 L) to use and considering the typical water uses (see Figure 4.21) listed here, answer the questions that follow the list.

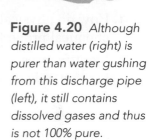

Figure 4.20 *Although distilled water (right) is purer than water gushing from this discharge pipe (left), it still contains dissolved gases and thus is not 100% pure.*

Figure 4.21 *Faced with water rationing, how essential do you consider these water uses to be? Could water from one activity be reused for the other?*

- washing cars, floors, windows, pets
- bathing, showering, washing hair, washing hands
- washing clothes, dishes
- watering indoor plants, outdoor plants, lawn
- flushing toilets

1. List three water uses that you could do without.
2. Identify one activity that you could *not* do without.
3. For which tasks could you reduce your water use? How?
4. Impurities added by using water for one particular use may not prevent its reuse for other purposes. For example, you might decide to save handwashing water and use it later to bathe your dog.
   a. For which activities could you use such impure water?
   b. From which prior uses could this water be taken?

Clean water is a valuable resource that must not be taken for granted. Unfortunately, water is easily contaminated. In the next section, as Riverwood deals with its water emergency, you will examine some of the causes of water contamination.

Potable water is pure enough for use in drinking and cooking. *Grey* water is left over from home use (such as showers, sinks, laundry, and dishwashers) and may be reused for other purposes, especially landscape watering. *Black* water has contacted toilet wastes and must be chemically treated and disinfected before reuse.

SECTION A SUMMARY

Reviewing the Concepts

> Strong hydrogen bonds between water molecules are responsible for many of the unique physical properties of water.

1. How do physical properties differ from chemical properties?

2. Identify three physical properties of water that are unique for a substance with such a small molar mass.

3. How does the density of solid water compare to the density of liquid water? Explain the cause of this difference.

4. If two atoms share electrons, how does knowledge of the electronegativity of each atom help determine if the bond will be polar?

> Mixtures are physical combinations of two or more substances. Mixtures can be classified as homogenous or heterogeneous.

5. Give an example of a homogeneous mixture and a heterogeneous mixture and describe differences you would see when observing a sample of each mixture.

6. When gasoline and water are poured into in the same container, they form two distinct layers.

   a. Describe how you could experimentally determine which liquid is on top.

   b. How could you determine which liquid would be on top *without* an experiment?

7. You notice beams of light passing into a darkened room through blinds on a window. Does this demonstrate that air in the room is a solution, a suspension, or a colloid? Explain.

8. Identify each of the following materials as a solution, a suspension, or a colloid. Explain your choice.

   a. a medicine accompanied by instructions to "shake before using"

   b. Italian salad dressing

   c. mayonnaise

   d. a cola soft drink

   e. an oil-based paint

   f. milk

9. Sketch a visual model on the molecular level, like those you drew in Modeling Matter A.6 (page 403), that shows each of the following types of mixtures. Label and explain the features of each sketch.

   a. a solution   b. a suspension

10. Suppose you have a clear, red liquid mixture. A beam of light is observed as it passes through the mixture. Over a period of time, no particles settle to the bottom of the container. Classify this mixture as a solution, a colloid, or a suspension, and provide evidence to justify your choice.

> Making informed decisions about water use requires consideration of direct and indirect uses as well as global distribution of water supplies.

11. Assume that Jimmy Hendricks drank just packaged fruit juice during the water shortage. Does that mean he did not use any water? Explain.

12. List at least three indirect uses of water associated with producing a loaf of bread.

13. Look at Figure 4.19 on page 409. Fresh water makes up 2.8% of Earth's water supply. Calculate the percent of fresh water found in

   a. glaciers and ice caps.   b. lakes.

14. Rank the following locations in order of greatest to least total water abundance on Earth: rivers, oceans, glaciers, water vapor.

15. Is water a renewable or nonrenewable resource? Explain.

What makes water unique?

In this section, you explored the physical properties of water and learned how these properties are influenced by the structure of the water molecule. You have also considered uses of water and properties of water mixtures. Think about what you have learned, then answer the question in your own words in organized paragraphs. Your answer should demonstrate your understanding of the key ideas in this section.

Be sure to consider the following in your response: physical properties, electronegativity, intermolecular forces, hydrogen bonds, mixtures, direct and indirect water uses, and water purity.

Connecting the Concepts

16. Explain why it might be possible that a molecule of water that you drank today was once swallowed by a dinosaur.

17. A carbonated beverage plant has separate water supplies for various uses, such as cleaning, producing the beverage, and employee uses.

    a. How do the water purity standards differ for the various uses?

    b. Why wouldn't it be more efficient to have one water supply for the entire plant?

18. One source reports that each person in the United States uses an average of 370 L of water daily. Other sources, however, report that U.S. per capita water use is 4960 L. If both values are correct, explain this apparent discrepancy.

19. Given an unknown mixture:

    a. what steps would you follow to classify it as a solution, a suspension, or a colloid?

    b. describe how each step would help you to distinguish among the three types of mixtures.

20. Explain the possible risks in failing to follow the direction "shake before using" on the label of a medicine bottle.

21. Is it possible for water to be 100% "chemical free"? Explain.

Extending the Concepts

22. Compare the physical properties of water (H_2O) with the physical properties of the elements from which it is composed.

23. Investigate and report on why "100% pure water" would be unsuitable for long-term human consumption—even if taste were not a consideration.

24. Using an encyclopedia or an Internet search engine, compare the maximum and minimum temperatures naturally found on the surfaces of Earth, the Moon, and Venus. The large amount of water on Earth is one of the factors limiting the natural temperature range on the planet. Suggest ways that water accomplishes this. As a start, find out (or review) what *heat of fusion, heat capacity*, and *heat of vaporization* mean.

25. Look up the normal freezing point, boiling point, heat of fusion, and heat of vaporization of ammonia (NH_3). If a planet's life forms were made up mostly of ammonia rather than water, what special survival problems might those life forms face? What temperature range would an ammonia-based planet need to support those life forms?

RIVERWOOD NEWS

TODAY'S WEATHER: abundant sunshine

Breaking news at RiverwoodNewsLive.com

MORNING EDITION

Meeting Raises Fish-Kill Concerns

By Carol Simmons
RIVERWOOD NEWS STAFF REPORTER

More than 300 concerned citizens, many prepared with questions, attended a Riverwood Town Hall public meeting last evening to hear from scientists investigating the Snake River fish kill.

Dr. Steven Schmidt, a chemist with the U.S. Environmental Protection Agency (EPA), expressed regret that the fishing tournament was canceled but strongly supported the town council's decision, saying it was the safest course in the long run. He reported that his laboratory is still conducting tests on the river water.

> **❝ I thought scientists would have the answers . . . ❞**

Dr. Margaret Brooke, a State University water specialist, helped interpret information and answered questions. Local physician Dr. Jason Martinez and Riverwood High School family and consumer sciences teacher Alicia Green joined the speakers for a question-and-answer session.

Brooke confirmed that preliminary water-sample analyses showed no obvious cause for the fish kill. She reported that EPA scientists have been unable to identify any microorganisms in the fish that could have been responsible for their death. Concerning possible fish-kill causes, Brooke concluded that "it must have been something dissolved or temporarily suspended in the water."

Even substances that do not dissolve much in water are being investigated, since even low concentrations of heavy metal ions or some organic molecules can cause harm to fish. In addition, EPA chemists will collect water samples hourly today to look for any unusual fluctuations in dissolved-oxygen (DO) levels. Fish require an adequate amount of oxygen gas dissolved in water for their survival. DO levels that are either too low or too high sometimes kill fish.

Brooke expressed confidence that further studies would shed more light on possible causes of the problem.

Dr. Martinez reassured citizens that "thus far, no illness reported by either physicians or the hospital can be linked to drinking water." Ms. Green offered water-conservation tips for housekeeping and cooking to make life easier for inconvenienced citizens. The information sheet she distributed is available on the Riverwood town Web site.

Mayor Cisko confirmed that water supplies will again be trucked in from Mapleton today, and he expressed hope that the crisis will last no longer than three days. Cisko also reported that the town has hired consulting engineers to examine Snake River flow data and investigate the structure and operations of both the Snake River Dam and tailing ponds at the abandoned Aurgent mining site.

> **❝ I'll never take tap water for granted again . . . ❞**

Those attending the meeting appeared to accept the emergency situation with good spirits. "I'll never take tap water for granted again," said Trudy Anderson, a Riverwood resident. "I thought scientists would have the answers," wondered Robert Morgan, head of Morgan Enterprises. "They don't know either! There's certainly more involved in all this than I ever imagined."

SECTION B LOOKING AT WATER AND ITS CONTAMINANTS

Why do some substances readily dissolve while others do not?

As this article indicates, water experts attribute the cause of the fish kill to something dissolved or suspended in the Snake River. You are already familiar with many properties of water, including that it is a polar substance. What might be dissolved in the Snake River water that could cause a fish kill? To narrow down the possibilities, you will have to consider three main questions. First, what types of substances dissolve in water? Second, how much of each substance can dissolve in water? Finally, at what concentrations (either too high or too low) are these substances lethal to fish? And, just like scientists and public policy decision makers, you will gather and use available evidence to make decisions regarding Riverwood's situation.

GOALS

- Use the concepts of polarity and intermolecular forces to account for water's ability to dissolve many ionic and molecular substances.
- Use words, pictures, and chemical equations to describe the process of dissolving ionic and molecular substances in water.
- Quantitatively describe and predict solution variables, including concentration, volume, temperature, mass and moles of solute, and solubility.
- Interpret and use solubility curves.
- Describe the effect of temperature on the solubility of gaseous substances.

concept check 3

1. Use intermolecular forces to explain one of water's unique and important properties.
2. How does a solution differ from a suspension?
3. You know from experience that both sugar and table salt dissolve in water.
   a. Describe in a few sentences (or with a drawing) what you visualize happening at the molecular level when salt dissolves in water.
   b. Do you have the same mental picture of sugar dissolving in water? If not, describe any differences.

■ INVESTIGATING MATTER

B.1 WHAT SUBSTANCES DISSOLVE IN WATER?

The *Riverwood News* reported that Dr. Brooke believes a substance dissolved in the Snake River is one likely cause of the fish kill. She based her judgment on her chemical knowledge and experience. Dr. Brooke also has a general idea about which contaminating solutes to focus on in the early stages of the investigation: those that dissolve easily in water and are harmful to fish at elevated concentrations. Such background knowledge helps Dr. Brooke (and other chemists) reduce the number of water tests required in the laboratory.

However, can Dr. Brooke and the EPA scientists completely rule out substances that appear not to dissolve in water? What does it mean for a substance to be *soluble* versus *insoluble*? Is anything truly *insoluble* in water? It is likely that at least a few molecules or ions of any substance will dissolve in water. Thus, the term *insoluble* actually refers to substances that dissolve only very, very slightly in water. For instance, lead(II) sulfide (PbS), which is considered insoluble, actually dissolves very slightly in water—10^{-14} g (0.000 000 000 000 01 g) PbS per liter of water at room temperature. See Figure 4.22. Could such low concentrations of lead(II) ions lead to a fish kill? Such questions are answered by examining toxicity data. (You will examine toxicity data later in this unit.)

Part I of this investigation will provide an opportunity to begin classifying various molecular and ionic solutes as *soluble* or *insoluble* (very, very slightly soluble) in water. You will extend your investigation in Part II.

Figure 4.22 *Galena is a naturally occurring mineral that contains lead(II) sulfide. Although lead compounds are toxic to humans and other organisms, lead sulfide is insoluble in water.*

Preparing to Investigate

Part I: What Solutes Dissolve in Water?

In Part I, you will design a procedure for testing whether several ionic and molecular substances dissolve in water. (Your teacher will tell you which solutes to investigate. Write these in your laboratory notebook.)

As a class, discuss how you could test whether substances like those listed in your laboratory notebook dissolve in room-temperature water. With your partner, design a step-by-step procedure to determine whether each solute is soluble (S) or insoluble (I) in room-temperature water (see Figure 4.23).

The following questions will help you design your procedure.

> In this activity, *insoluble* means that you could observe no evidence of solute dissolving.

1. What particular observations will allow you to judge how well each solute dissolves in water? That is, how will you decide whether to classify a given solute as soluble or insoluble in water?

2. Which variables will you have to control? Why?

3. How should the solute and solvent be mixed—all at once or a little at a time? Why?

Figure 4.23 *Planning a procedure for investigating whether various solutes dissolve in water.*

In designing your procedure, keep these concerns in mind:

- Do not let any solutes make contact with your skin.
- Follow your teacher's directions for waste disposal.

Construct a data table for recording observations and data during Part I of the investigation. When you and your partner have agreed upon a written procedure and an accompanying data table, ask your teacher to check and approve your written plans.

Part II: What Factors Affect Whether a Solute Will Dissolve?

In Part II, you will broaden your investigation. Your teacher may assign your group to investigate one of the questions listed below, or you may investigate a different question. Here are some possible questions:

- How does temperature affect whether a solid dissolves in water?
- Do particular groups of ionic compounds (e.g., ionic compounds containing sodium ions, carbonate ions, or chloride ions) exhibit similar dissolving behavior in water?
- For substances that do not dissolve noticeably in water, are there other solvents in which they will dissolve?

It is best to gather evidence for Part I before planning your procedure for Part II. However, keep your Part II question in mind as you carry out your procedure for Part I.

Gathering Evidence

Part I: What Solutes Dissolve in Water?

Before you begin, put on your goggles, and wear them properly throughout the investigation. Follow your teacher-approved procedure to investigate whether each specified solute dissolves in water. Record all data in your data table. Dispose of all materials as directed by your teacher.

Part II: What Factors Affect Whether a Solute Will Dissolve?

Once you know the question you will address in this investigation, design a procedure that will allow you to collect the data necessary to answer your question. Can you use the same procedure you designed for Part I? If not, what parts of that procedure should be revised? In considering your Part II procedure, keep the questions listed in Part I in mind. Write down the procedure and construct the data table or tables that you plan to use for this investigation.

Have your written procedure and data table or tables approved by your teacher. Then collect and record data for your investigation. Dispose of all materials as directed by your teacher.

Wash your hands thoroughly before leaving the laboratory.

Interpreting Evidence

1. Based on your observations, classify each solute investigated in Part I as either soluble or insoluble in water.

2. State in a few sentences how you decided whether each solute in Part I was soluble or insoluble.

3. Compare your classifications in Part I with those of the rest of the class. Are there any differences? If so, how can you explain those differences?

Making Claims

4. State the question you were addressing in your Part II investigation.
   a. Write two or three sentences that describe your conclusions (answering your question).
   b. On what evidence are your conclusions based?
   c. How confident are you in your conclusions?
   d. What additional tests would you like to perform to be more confident in your claims?

Reflecting on the Investigation

5. You investigated the dissolving behavior of both ionic and molecular substances. Many of the substances you studied in Unit 1 were ionic, while the majority of substances you encountered in Units 2 and 3 were molecular.
   a. At this point, what is your definition of an ionic compound?
   b. What is your definition of a molecular compound?
   c. Based on your current definitions, classify each substance you encountered in Part I and Part II (including water) as either ionic or molecular.
   d. Based on your current classifications, is it possible to make the claim that all ionic substances are soluble in water? Why or why not?

e. Based on your current classifications, is it possible to make the claim that all molecular substances are soluble in water? Why or why not?

6. According to your data, which solutes tested are least likely to be dissolved in the Snake River? Explain, and support your explanation with evidence from your completed investigation.

You now know how some solids and liquids behave in water. In the next sections, you will learn more about the process by which ionic and molecular substances dissolve in water. Applying what you have learned about intermolecular forces will help you understand and predict why some ionic and molecular substances dissolve in water and why others do not.

B.2 DISSOLVING IONIC COMPOUNDS IN WATER

Your classifications of ionic and covalent compounds as soluble or insoluble in the previous investigation were based on macroscopic observations. How can your observations be explained at the *particulate* level? To answer this question, first for ionic compounds, we will focus on interactions among ions and interactions between ions and water molecules. How do interactions among the ions in the solute and interactions between ions and water molecules explain why many ionic compounds dissolve? Can these same interactions explain why some ionic compounds dissolve only very slightly in water?

First, consider attractive forces within the ionic compound. Recall that ionic compounds are composed of positively charged cations (usually metal ions) and negatively charged anions. See Figure 4.24. Within the ionic crystal, cations are surrounded by anions, with the anions likewise surrounded by cations. Attractive forces between the charged particles are so strong that all ionic compounds are solids at room temperature. Think of table salt (sodium chloride, $NaCl$), baking soda (sodium hydrogen carbonate, $NaHCO_3$), and chalk (calcium carbonate, $CaCO_3$), which are all common ionic compounds.

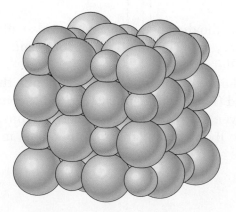

Figure 4.24 *Ionic compounds consist of positively charged cations (represented here by small gray spheres) and negatively charged anions (large green spheres) arranged in a crystal.*

Next, recall that water is a polar molecule. You learned in Section A that water molecules have particularly strong interactions due to intermolecular forces called hydrogen bonds. When an ionic compound is placed in water, then, what determines whether or not it will dissolve? A third set of interactions must be considered—those between the charged ions and the polar water molecules. The substance will dissolve only if its ions are so strongly attracted to water molecules that the water molecule can "tug" the ions away from the crystal.

Water molecules are attracted to ions located on the surface of an ionic solid, as shown by the models in Figure 4.25. The water molecule's negative (oxygen) end is attracted to the crystal's cations. The positive (hydrogen) ends of other water molecules are attracted to the anions of the crystal. When the attractive forces between the water molecules and the surface ions are strong enough, the bonds between the crystal and its surface ions become strained, and the ions may be pulled away from the crystal. Figure 4.25 uses models of water molecules and solute ions to illustrate the results of water tugging on solute ions. The detached ions become surrounded by water molecules, producing an ionic solution.

Using the description and illustrations of this process, can you decide which factors influence whether an ionic solid will dissolve in water? Because dissolving involves competition among three types of attractions—those between solvent and solute particles, between solvent particles themselves, and between particles within the solute crystals—properties of both solute and solvent determine whether two substances will form a solution. Water is highly polar, so it is effective at dissolving many ionic substances. However, if positive–negative attractions between cations and anions within the crystal are strong enough, a particular ionic compound may be only slightly soluble (or insoluble) in water.

Consider again two common ionic compounds, table salt and chalk. Table salt dissolves readily in water, but a piece of chalk does not. Why might the positive–negative attractions between cations and anions within chalk be stronger than in salt, preventing chalk from dissolving in water? One answer lies in examining the charges on the ions in each substance. Sodium chloride, which makes up table salt, consists of Na^+ and Cl^- ions—each has a charge of one. In calcium carbonate, which makes up chalk, each ion has a charge of two (Ca^{2+} and CO_3^{2-}). Higher charges lead to greater attractive forces among the ions, and the attractions between water molecules and ions cannot overcome them.

Besides charges on the ions, the distance between ions also affects attractive forces within the ionic crystal. The smaller the ions, and thus closer together they are, the greater the attractions.

These factors help to explain why many ionic compounds that contain heavy metal ions such as lead (Pb^{2+}) and mercury (Hg^{2+}) dissolve only very slightly in water. However, even at extremely low concentrations, such ions can cause damage or death to living organisms and, therefore, must still be considered as possible causes of the Riverwood fish kill.

Scientific American Conceptual Illustration

Figure 4.25 *Particulate-level model of sodium chloride (NaCl) dissolving in water (H₂O).* Sodium chloride crystals (common table salt) consist of positively charged sodium ions (Na⁺, gray spheres) and negatively charged chloride ions (Cl⁻, green spheres). Every water molecule is polar; the oxygen (red) side has partial negative electrical charge and the hydrogen (white) side has partial positive electrical charge. As water molecules approach the sodium chloride crystal, hydrogen attracts chloride ions and oxygen attracts sodium ions. Several water molecules surround each ion and carry it away from the crystal.

As you study this 2-D representation, try to visualize it in three dimensions and put the dissolving process "in motion." Think about these ions and molecules interacting at different times, such as right after the crystal drops into the water, as the crystal seems to disappear completely, or when the water has fully evaporated.

B.3 DISSOLVING MOLECULAR COMPOUNDS IN WATER

Now that you have a mental image of what happens when ionic compounds dissolve (or do not dissolve) in water, consider what might be similar or different about dissolving molecular compounds. As you know from everyday life and from the investigation in Section B.1, molecular substances such as sugar, ethanol, and ammonia dissolve in water. However, they do not form ions in solution. Atoms within these molecules are covalently bonded to each other, and the molecules themselves remain intact as they are dissolved.

> You learned about covalent bonding and molecular substances in Unit 3.

Unlike ionic substances, which are crystalline solids at normal conditions, molecular substances may be found as solids, liquids, or gases at room temperature. Some molecular substances, such as oxygen (O_2) and carbon dioxide (CO_2), have little attraction among their molecules and are, thus, gases at normal conditions. Molecular substances such as ethanol (ethyl alcohol, C_2H_5OH) and water (H_2O) have larger intermolecular attractions, which cause these "stickier" molecules to form liquids at normal conditions. Other molecular substances with even greater between-molecule attractions—table sugar ($C_{12}H_{22}O_{11}$), for example—are solids at normal conditions. As you learned earlier, stronger attractive forces hold molecules together more tightly, determining in which state the substance will be found under normal conditions.

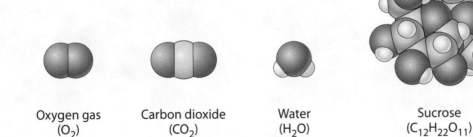

Oxygen gas (O_2)   Carbon dioxide (CO_2)   Water (H_2O)   Sucrose ($C_{12}H_{22}O_{11}$)   Ethanol (C_2H_5OH)

What determines whether a molecular substance will dissolve in water? Just as in explaining why only some ionic compounds dissolve, the strength of three sets of attractive forces must be considered—the attractive (intermolecular) forces among molecules in the solute, the relatively strong intermolecular attractions among water molecules, and the interactions between solute and water molecules. If the attractive forces between solute molecules and water molecules are strong enough to overcome the attractive forces within the molecular substance, the substance will dissolve in water. But what causes these attractions? One important factor is the distribution of electrical charge within the molecules.

Most molecular compounds contain atoms of nonmetallic elements. These atoms share electrons and thus form covalent bonds. Atoms of different elements do not share electrons equally within the bond. As you learned in Section A, the ability of an element's atoms to attract shared electrons when bonding is known as the element's *electronegativity*. In molecular substances, these differences in electron attraction are not large enough to cause ions to form, but they can cause electrons to be distributed unevenly among the atoms.

You already know that the oxygen region of a water molecule is electrically negative compared with the regions around the hydrogen atoms, which are electrically positive. This happens because oxygen is more electronegative than hydrogen. Therefore, electrons are partially pulled away from the two hydrogen atoms toward the oxygen atom within each H—O covalent bond. Because water is "bent," or shaped like a "V," there is an overall net negative charge near the oxygen and an overall net positive charge on the hydrogen end. The result is a *polar molecule*. If water molecules were linear, the pull of electrons from hydrogen to oxygen in one direction would cancel out the pull from hydrogen to oxygen in the opposite direction and there would be no net imbalance of charge. This is the reason, in fact, that carbon dioxide is nonpolar. In CO_2, the oxygen atoms are more electronegative than carbon, but each is pulling on the electrons equally and in opposite directions, so the molecule itself is nonpolar. (Picture two people with equal strength playing tug-of-war.) See Figure 4.26.

> Polar molecules were introduced in Unit 3 (page 288).

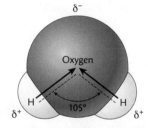
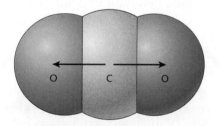

Figure 4.26 *Oxygen is much more electronegative than either carbon or hydrogen, so electrons in each of the illustrated molecules are pulled toward the oxygen atom. In H_2O, the bent shape of the molecule results in a net imbalance of electrical charge, whereas the linear shape of CO_2 results in equal pull from both sides and no net imbalance of charge.*

Water's polarity results in strong attractions between water molecules and solutes that are polar or ionic, because both have regions (or particles) with opposite electrical charges. Water is a common example of a polar solvent. Polar molecules tend to dissolve readily in polar solvents. For example, water is a good solvent for sugar and ethanol, which are both composed of polar molecules. On the other hand, nonpolar liquids are good solvents for nonpolar molecules, since both have symmetrical electron distributions. For instance, nonpolar cleaning fluids are used to dry-clean clothes because they readily dissolve nonpolar body oils found in the fabric of used clothing.

> Unfortunately, many nonpolar dry-cleaning solvents damage both human health and the environment. However, recent development of new technologies allows environmentally benign nonpolar solvents such as liquefied carbon dioxide to be used in dry cleaning.

By contrast, nonpolar molecules (such as those of oil and gasoline) do not dissolve well in polar solvents (such as water or ethanol) or vice versa. This pattern of dissolving behavior—polar substances dissolve in polar solvents, while nonpolar substances dissolve in nonpolar solvents—is often summed up as "like dissolves like." See Figure 4.27.

Were dissolved molecular substances present in the Snake River water where the fish died? Most likely yes, at least in small amounts. Were they responsible for the fish kill? That depends on which molecular substances were present and at what concentrations. And that, in turn, depends in part on how each solute interacts with water's polar molecules.

Figure 4.27 *A painter must use paint thinner—a nonpolar solvent—to clean oil-based paint (also nonpolar) from this paintbrush. Water, a polar solvent, will not dissolve the paint. This is an example of "like dissolves like."*

> Various molecular substances may normally be present at very low concentrations—so low that we observe no harm to living things.

In the upcoming Modeling Matter activity, you will model and compare the dissolving behavior of some typical molecular and ionic substances.

concept check 4

1. Based on "like dissolves like," classify each molecular substance from Investigating Matter B.1 as either polar or nonpolar. What is your reasoning?
2. Does "like dissolves like" apply to dissolving ionic compounds in water? Why or why not?
3. Oxygen is not very soluble in water, but fish depend on dissolved oxygen to live. Would you classify O_2 as polar or nonpolar? Explain your reasoning.

■ MODELING MATTER

B.4 THE DISSOLVING PROCESS

So far in this section, you have investigated and learned about the process of dissolving both ionic and molecular compounds in water. Figure 4.25 (page 421) uses a single image to illustrate what happens when salt crystals dissolve in water. However, it is hard to capture in one "snapshot" the dynamic process of dissolving, so you might visualize a short "movie" of the dissolving process in your mind. In this activity, you will represent these mental movies as a series of drawings illustrating the dissolving process. You will also represent this process using chemical language by writing chemical equations.

> *Sample Problem: Refer back to Figure 4.25 (page 421), which illustrates the process of dissolving NaCl in one frame. Then, draw a five-panel sequence illustrating NaCl dissolving in water.*
>
> One way to visualize the sequence is described and illustrated below:
>
> - First panel: Sketch a beaker of water, with some water molecules drawn in. Beside the beaker, draw a 2-D representation of the NaCl crystal (alternating small and large circles, with charges drawn in).
>
> - Second panel: The solid is placed in the water. Draw the crystal at the bottom of the beaker.
>
> - Third panel: A few water molecules are interacting with a few ions, pulling the ions away from the surface of the crystal.
>
> - Fourth panel: More water molecules have carried away ions.
>
> - Fifth panel: No solid remains. All ions are surrounded by water molecules, and the crystal has dissolved.

1. Consider magnesium chloride, $MgCl_2$, another soluble ionic compound.

   a. Draw a five-panel sequence showing what happens when magnesium chloride is placed in water.

   b. Write a caption for your sequence, making specific reference to the attractive forces involved.

   c. Compare your magnesium chloride "dissolving movie" to the sodium chloride "dissolving movie" in the Sample Problem.

2. You determined whether chalk was soluble in water in Investigating Matter B.1, and Section B.2 also used chalk as an example.

   a. Draw a five-panel sequence showing what happens when chalk (calcium carbonate) is placed in water.

   b. Write a caption for your sequence, making specific reference to the attractive forces involved.

3. Another way to represent chemical and physical processes is by writing and interpreting chemical equations. For example, the following equation represents dissolving table salt in water:

$$NaCl(s) \longrightarrow Na^+(aq) + Cl^-(aq)$$

a. What do the symbols *(s)* and *(aq)* mean?

b. Write a sentence that interprets this chemical equation in words.

c. Water is not shown as a reactant or a product in this equation. How do you know, based upon this equation, that water is part of the process?

d. Review your results from Investigating Matter B.1 and write chemical equations for each of the ionic compounds that dissolved in water.

e. You classified some ionic compounds in Investigating Matter B.1 as insoluble. How would you write chemical equations for the interaction of these compounds with water? Use chalk $(CaCO_3)$ as an example. Write a chemical equation and write one or two sentences describing your reasoning.

4. Now model the dissolving process for molecular compounds.

a. Choose one soluble molecular compound from Investigating Matter B.1.

b. Draw a five-panel sequence illustrating what happens when this compound dissolves in water.

c. Write a caption for your sequence, making specific references to intermolecular forces.

d. Write a chemical equation that describes this process.

5. In Section B.2, you learned about factors that partly determine whether a specific ionic compound will dissolve in water. Suppose the following 2-D drawing represents ions in a sodium bromide (NaBr) crystal:

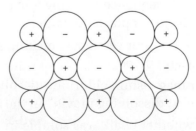

a. Magnesium cations (Mg^{2+}) are smaller than Na^+ ions, and sulfide (S^{2-}) ions are smaller than Br^- ions. Draw a 2-D representation of MgS that shows relative sizes and ion charges, compared to the NaBr crystal depicted above.

b. Would you predict MgS to be more or less soluble in water than NaBr? Use your model (and what you learned in Section B.2) to explain your prediction.

6. Consider the Riverwood fish kill that opened this unit. None of the newspaper articles have mentioned an oil spill (e.g., from a tanker truck) as a possible cause of the fish kill.

   a. Why is an oil spill an unlikely cause of the Riverwood fish kill? What would be observed at the macroscopic level? Draw a picture to illustrate your point.

   b. Discuss how intermolecular forces play a role in this scenario. Your discussion should include intermolecular forces within oil, within water, and between oil and water.

You are now familiar with why and how certain substances dissolve in water. You also know that solutions can contain different amounts of solute—think of putting a pinch of salt into a pot of boiling water versus adding a cup. Later in this unit, you will research and report concentrations of oxygen dissolved in water that are necessary for fish to survive. To interpret these data, we next explore how to describe and measure concentration.

B.5 SOLUTION CONCENTRATION

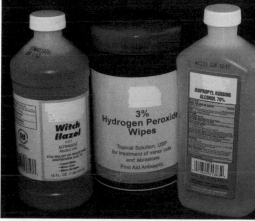

Solution concentration refers to how much solute is dissolved in a specific quantity of solvent or solution. There are several ways to report concentrations, and it is useful to be familiar with them. The simplest way to think about concentration is as the mass of a substance dissolved in a particular mass of water. Thus, if you made a solution by dissolving 5 g table salt (NaCl) in 100 g of water, you would report that as "5 g per 100 g."

Another way to express concentration is with percent values. For example, dissolving 5 g table salt in 95 g water produces 100-g solution with a 5% salt concentration (by mass). Here, the percent means parts solute per hundred total parts (solute plus solvent). So, a 5% salt solution could also be reported as a five-parts-per-hundred salt solution (5 pph salt). However, percent is much more commonly used. Similarly, solutions made by dissolving liquid molecular compounds are often reported using percent by volume. A common example is household rubbing alcohol solution, which is 70% isopropyl alcohol by volume. See Figure 4.28.

Figure 4.28 *Some common consumer products, including those pictured here, include percent composition on their labels.*

One of the most common units of concentration, and one you have encountered in many investigations, is molarity. The symbol for molarity is M, and it stands for moles of solute per liter of solution. If you wanted to make a liter of 0.10 M solution of NaCl, you would weigh out 0.10 moles (5.8 g) NaCl, dissolve it in a small quantity of water, and add more water until the total volume of the solution was 1.0 liter. If you only needed 500 mL of 0.10 M solution, you would dissolve 0.050 moles (2.9 g) NaCl in 0.50 L solution.

For solutions containing considerably smaller quantities of solute (as are found in many environmental water samples, including those from the Snake River), concentration units of *parts per million (ppm)* are sometimes useful. As you might expect, for very low concentrations, *parts per billion (ppb)* is often used. For example, the maximum concentration of nitrate allowed in drinking water is much too low to be conveniently written as a percent (it is 0.0010%). Instead, it is written as 10 ppm.

Nitrates are anions commonly found in natural waters as a result of fertilizer runoff. The EPA has established a maximum concentration limit of nitrates in drinking water at 10 ppm to prevent undesirable health effects resulting from elevated nitrate levels in the body.

The notion of concentration is part of daily life. For example, preparing beverages from premixed concentrate, adding antifreeze to water in an automobile radiator, and mixing pesticide or fertilizer solutions all require the use of solution concentrations. The following activity will help you review solution concentration, as well as gain experience with the chemist's use of this concept.

DEVELOPING SKILLS

B.6 DESCRIBING SOLUTION CONCENTRATION

Just as *percent* means "for every hundred," *per million* means "for every million." To express a concentration using parts per million (ppm), divide the mass of solute by the mass of solvent and multiply by 1 million.

Sample Problem 1: What is the concentration of a 1% salt solution expressed in ppm?

Because 1% of 1 million is 10 000, a 1% salt solution is 10 000 parts per million.

Sample Problem 2: A common intravenous saline solution used in medical practice contains 4.5 g NaCl dissolved in 495.5 g sterilized distilled water. What is the concentration of this solution, expressed as percent by mass?

Because this solution contains 4.5 g NaCl and 495.5 g water, its total mass is 500.0 g. Expressing a solution's concentration in percent by mass involves taking the mass of solute and dividing it by the mass of solution, then multiplying by 100. In the sample problem, that means dividing 4.5 g by 500.0 g and then multiplying by 100 to get 0.90%.

$$\text{Percent by mass} = \frac{4.5 \text{ g}}{500.0 \text{ g}} \times 100 = 0.90\%$$

Another way to look at it is to remember that *percent* means "per hundred." How many grams of NaCl do you have for each 100-g solution? Because there is a 500-g solution, we are dealing with 5 times as much material as in 100 g of solution. If there are 4.5 g NaCl in 500.0 g solution, there must be 1/5 that much NaCl in 1/5 of the solution—1/5 of 4.5 g NaCl is 0.90 g NaCl. That means there are 0.90 g NaCl per 100 g solution, or that the solution is 0.90% NaCl by mass, which is the same answer we arrived at previously.

> **Sample Problem 3:** *Suppose that the intravenous saline solution in Sample Problem 2 was made by dissolving 4.5 g NaCl in a total volume of 500 mL. What is the concentration of this solution, expressed as molarity (M)?*
>
> Molarity is moles of solute per liter of solution. To find moles of NaCl, we divide the mass of NaCl dissolved (4.5 g) by the molar mass of NaCl (58.44 g/mol). We then divide that by 0.500 L to find concentration:
>
> $$\text{mol NaCl} = 4.5 \ \cancel{\text{g NaCl}} \times \frac{1 \ \text{mol NaCl}}{58.44 \ \cancel{\text{g NaCl}}} = 0.077 \ \text{mol NaCl}$$
>
> $$\text{molarity} = \frac{0.077 \ \text{mol NaCl}}{0.500 \ \text{L}} = 0.15 \ \text{M NaCl}$$

Now answer the following questions:

1. One teaspoon of table sugar (sucrose, $C_{12}H_{22}O_{11}$) is dissolved in a cup of water. Identify

   a. the solute.   b. the solvent.

2. What is the concentration of each of these solutions expressed as percent sucrose by mass?

   a. 17 g sucrose is dissolved in 183 g water.

   b. 30.0 g sucrose is dissolved in 300.0 g water.

3. What is the concentration of each of these solutions expressed as ppm?

   a. 0.0020 g iron(III) ions dissolved in 500.0 g water.

   b. 0.25 g calcium ions dissolved in 850.0 g water.

4. Expressed in ppm, what is the concentration of each solution in Question 2?

5. At 60 °C, 100.0 g water can dissolve a maximum of 45 g KCl.

   a. What is the concentration of this solution, expressed as percent KCl by mass?

   b. What would be the new concentration if 155 g water were added?

6. You make a pitcher of flavored drink (Figure 4.29) using the instructions on a packet of unsweetened powder. It says to place the powder and 1 cup of sugar into a pitcher and then add water until you have a total volume of 2 quarts.

   a. One cup of sugar (sucrose, $C_{12}H_{22}O_{11}$) weighs 200 grams. How many moles of sugar did you use to make your flavored drink solution?

   b. One quart of liquid is equal to 0.95 L. What was the total volume of your solution in liters?

   c. Given your answers above, what is the concentration of sugar in your flavored drink, in units of mol/L?

Figure 4.29 *What would you need to know in order to calculate the concentration of sucrose in this pitcher of flavored drink?*

7. Your friend states that she prefers her flavored drink to be less sweet. She pours 200 mL of your flavored drink solution into a glass and adds another 200 mL of water.

   a. How many moles of sugar were in the 200 mL of flavored drink that your friend took from your original flavored drink solution?

   b. How many moles of sugar are in your friend's beverage after she adds water?

   c. What is the molar concentration of sugar in your friend's flavored drink? Support your answer either by showing a calculation or by explaining your reasoning.

   d. Your friend's beverage will not be as sweet as yours. How will the actual flavor compare? Describe your reasoning.

concept check 5

1. Consider two sugar solutions. One contains 0.20 moles sugar dissolved in 500 mL solution. The other contains 0.40 moles sugar dissolved in 1.5 L solution. Which solution is more concentrated? Describe your reasoning.
2. You have 100 mL of a 0.10-M NaCl solution. You need 200 mL of a 0.025-M NaCl solution. How would you use the first solution to make what you need? Use words and pictures to describe your procedure.

■ INVESTIGATING MATTER

B.7 MEASURING SOLUTION CONCENTRATION

Preparing to Investigate

The Riverwood fish kill may have been caused by something dissolved in the water. As you have read, scientists are conducting tests to determine what was dissolved in the water. Once dissolved substances have been identified, the next step often involves measuring the concentrations of those substances. How do chemists do this?

In this investigation, you will use one common technique for determining concentrations of a substance dissolved in water. This particular technique, **colorimetry**, is useful when the substance of interest is colored. Colorimetry is based on the principle that colored solutions absorb light. The higher the concentration, the more light that is absorbed by the solution (and less light is transmitted through to the other side).

You already use this principle in your everyday life, with your eyes as the instrument. Imagine looking at three different concentrations of orange-flavored drink. You can determine the relative concentrations by the darkness of each drink solution. The most concentrated solution will look darker—if you shined a flashlight through one side and looked at it on the other side, not as much light would pass through to your eye because it is absorbed by the many molecules dissolved in the orange drink. As the concentration decreases, fewer molecules are present to absorb the light, and the drink solution looks lighter.

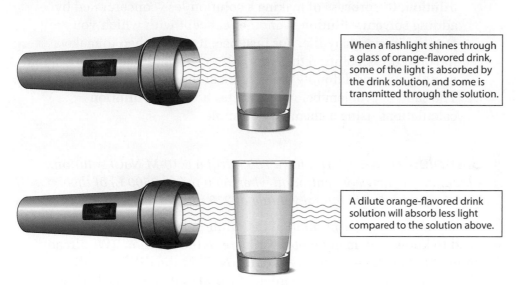

> When a flashlight shines through a glass of orange-flavored drink, some of the light is absorbed by the drink solution, and some is transmitted through the solution.

> A dilute orange-flavored drink solution will absorb less light compared to the solution above.

When chemists need to determine the concentration of a colored substance in a water sample, they first make up several solutions that contain the substance in known concentrations. They use an instrument to pass light through each of those solutions and measure how much of the original light is absorbed by the solution. The resulting measurement is called the **absorbance** of the solution. The relationship between concentration and absorbance is linear. With data from several solutions of known concentration (**standard solutions**), chemists construct a graph of concentration versus absorbance, as depicted in Figure 4.30, called a **calibration curve**. Then, the same measurement is made using the solution of unknown concentration. The absorbance of the unknown solution is located on the graph, and the corresponding concentration is found using the best fit line.

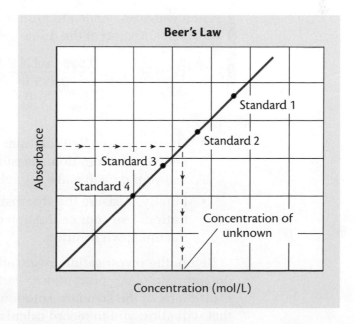

Figure 4.30 *A calibration curve showing the relationship between concentration and absorbance.*

You will use this same technique to determine the unknown concentration of a copper(II) sulfate pentahydrate ($CuSO_4 \cdot 5\ H_2O$) solution, which is blue in color. The basic steps are as follows:

- Obtain a stock solution from which all other standard solutions will be made. The stock solution is the most concentrated of the solutions that will be investigated. You will use 0.50 M $CuSO_4 \cdot 5\ H_2O$ as your stock solution.

- Prepare four more solutions of known concentration through **dilution**, the process of making a solution less concentrated by adding solvent. Dilution is another concept with which you are familiar in everyday life. For instance, if you wished to make a flavored drink solution that was half as concentrated as the original drink, you would double the volume of the solution by adding water. The sample problem below illustrates how to do dilution calculations using a chemistry example

Sample Problem: *You have 700 mL of a 0.10-M NaCl solution. Calculate the new concentration when 50.0 mL (0.0500 L) of this solution is diluted to a total volume of 275 mL.*

To calculate the concentration of the dilute solution in mol/L, we need to know how many moles NaCl are in the solution. (We already know the volume—275 mL.) All of the NaCl in the dilute solution originally come from the 50.0 mL (0.0500 L) of 0.10 M NaCl. Thus:

$$\text{mol NaCl} = 0.0500\ \cancel{\text{L solution}} \times \frac{0.10\ \text{mol NaCl}}{1.0\ \cancel{\text{L solution}}} = 0.005\ \text{mol NaCl}$$

Then water was added to produce a total volume of 0.275 L. Thus the concentration of the dilute solution is:

$$\frac{0.005\ \text{mol NaCl}}{0.275\ \text{L}} = 0.018\ \text{M NaCl}$$

- Once you have five solutions (including the stock solution) of known concentration, measure the absorbance of each solution and use your data to make a graph (calibration curve).

- Finally, measure the absorbance of the solution of unknown concentration. Use your calibration curve to determine the concentration of the unknown solution.

Before the investigation, read *Gathering Evidence* to learn what you will do and note safety precautions. Perform calculations to determine the concentrations of the standard solutions you will make. Construct a data table that will allow you to record calculated concentrations and measured absorbance (or % transmittance) readings for each of your standard solutions.

Gathering Evidence

Part I: Preparing the Standard Solutions

1. Before you begin, put on your goggles, and wear them properly throughout the investigation.

2. Label four clean, dry test tubes #1–4 and place them in a test tube rack. Also label four clean, dry Beral pipets #1–4.

3. Label an additional Beral pipet STOCK and use it only to transfer the stock 0.5 M copper(II) sulfate pentahydrate solution.

4. Label another Beral pipet WATER and use it only to transfer distilled water.

5. To make standard solution #1: Use the STOCK Beral pipet to transfer 2.0 mL of stock solution into a 10-mL graduated cylinder. Be sure the bottom of the meniscus is at exactly 2.0 mL. Use the Beral pipet labeled WATER to carefully fill the graduated cylinder, adding drops until the bottom of the meniscus sits exactly at the 10.00-mL mark. Mix the solution by drawing the solution into pipet #1 and emptying it into the graduated cylinder at least three times. Pour the solution into test tube #1.

6. Thoroughly clean and rinse the graduated cylinder.

7. Repeat Steps 5 and 6 for the remaining standard solutions, using 4.0 mL stock solution for #2, 6.0 mL stock solution for #3, and 8.0 mL stock solution for #4.

Part II: Colorimetry Measurements

Follow your teacher's instructions for measuring the absorbance of each of your standard solutions. Specific instructions will depend upon the instruments available for use in your classroom. However, expect to carry out these basic steps:

1. Obtain a clean, dry cuvette or other sample holder.

2. If your instrument allows you to select a wavelength, set it to a wavelength of 600 nm.

3. Calibrate the instrument for both "0% transmittance" and "100% transmittance" according to your teacher's instructions.

4. Measure and record either % transmittance or absorbance (depending upon your instrument) for the least concentrated solution. Rinse the cuvette with a small volume of your sample. Discard the rinsing. Then fill the cuvette or sample holder about ¾ full with your sample.

5. Make your measurement and record the value.

6. Measure the absorbance for each of the remaining standard solutions (stock and dilutions), always rinsing the cuvette with the desired solution before making the measurement.

7. Measure and record the absorbance of a solution of unknown concentration.

Analyzing Evidence

1. Construct a calibration curve, similar to Figure 4.30 (page 431), using absorbance and concentration data from your standard solutions. Follow these guidelines:

   a. If you recorded % transmittance values, you will first need to convert those readings to absorbance using this relationship:

   $$\text{Absorbance} = 2 - \log(\%T)$$

   b. Plot absorbance on the y-axis, and concentration on the x-axis.

   c. Draw a dot representing each data point. Then, with a ruler, draw a straight line that goes through the middle of your points. Don't simply connect the dots. Consider whether it makes sense for your line to go through the origin (0 concentration; 0 absorbance).

2. Use your calibration curve to report the concentration of the unknown sample.

Interpreting Evidence

1. Describe in words how you used your calibration curve to determine the concentration of the unknown sample.

Reflecting on the Investigation

2. Phosphate ions are colorless. However, they can be converted into a complex that is colored. Once converted, measurements similar to those you made are carried out to determine phosphate concentrations in natural waters. To the right are absorbance data from samples collected along the Godavari River in India.

Water Samples	Absorbance
G_1	0.122
G_2	0.130
G_3	0.135
G_4	0.141
G_5	0.162
G_6	0.166
G_7	0.169
G_8	0.192

   a. What patterns do you notice in the data?

   b. High phosphate concentrations at the most contaminated site along this river can be explained by large industrial waste from a local power station and waste containing molasses from a sugar factory. Which point along this river corresponds to this location? Describe your reasoning.

   c. High concentrations of phosphate often lead to lower concentrations of dissolved oxygen. Predict the pattern of dissolved oxygen concentrations from point G_1 down to point G_8 on this river.

B.8 FACTORS THAT AFFECT SOLUBILITY OF SOLIDS IN WATER

In solving the fish-kill mystery, you now know some substances that dissolve in water and how they dissolve. You can also describe and measure their concentrations. But what factors affect how much solute can dissolve in water? How do we describe the maximum amount of any substance that can dissolve under certain conditions?

Imagine preparing a water solution of potassium nitrate, KNO_3. As you stir the water, the solid, white crystals dissolve. The resulting solution remains colorless and clear. In this solution, water is the solvent, and potassium nitrate is the solute.

What will happen if you add a second scoopful of potassium nitrate crystals and stir? These crystals also may dissolve. However, if you continue adding potassium nitrate without adding more water, eventually some potassium nitrate crystals will remain undissolved—as a solid—on the bottom of the container, no matter how long you vigorously stir. At this point, the solution is said to be a **saturated solution**. The maximum quantity of a substance that will dissolve in a certain quantity of water (e.g., 100 g) at a specified temperature is called its **solubility** in water. In this example, the solubility of potassium nitrate might be expressed as "grams potassium nitrate per 100 g water" at a specified temperature.

From everyday experience, you probably know that both the size of the solute crystals and the vigor and duration of stirring affect how long it takes for a sample of solute to dissolve at a given temperature. But, with enough time and stirring, will even more potassium nitrate dissolve in water? It turns out that the solubility of a substance in water is a characteristic of the substance and cannot be changed by any extent of stirring over time.

Temperature and Solubility of Solids

So what does affect the actual quantity of solute that dissolves in a given amount of solvent? As you can see from Figure 4.31 (page 436), the mass of solid solute that will dissolve in 100 g water varies as the water temperature changes from 0 °C to 100 °C. The graphical representation of this relationship is called the solute's **solubility curve**.

Each point on the solubility curve indicates a solution in which the solvent contains as much dissolved solute as it normally can at that temperature. Such a solution is called a *saturated solution*. Thus, each point along the solubility curve indicates the conditions of a saturated solution. Look at the curve for potassium nitrate (KNO_3) in Figure 4.31 (page 436). Locate the intersection of the potassium nitrate curve with the vertical line representing 40 °C. Follow the horizontal line to the left and read the value. According to the solubility curve, the solubility of potassium nitrate in 40 °C water is 60 g KNO_3 per 100 g water.

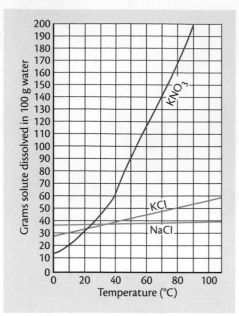

Figure 4.31 *Relationship between solute solubility in water and temperature.*

Sample Problem 1: At 50 °C, how much potassium nitrate will dissolve in 100 g water to form a saturated solution?

This value from Figure 4.31—80 g KNO_3 per 100 g water—represents the solubility of potassium nitrate in 50 °C water. By contrast, the solubility of potassium nitrate in cooler 20 °C water is only about 30 g KNO_3 per 100 g water. (Check the graph in Figure 4.31 to make sure that you can also find this 30-g value.)

Sample Problem 2: Consider a solution containing 80 g potassium nitrate in 100 g water at 60 °C. Locate this point on the graph. Locate where this temperature falls in relation to the solubility curve. What does this information tell you about the level of saturation of the solution?

Because each point on the solubility curve represents a saturated solution, any point on a graph below a solubility curve must represent an unsaturated solution. An **unsaturated solution** is a solution that contains less dissolved solute than the amount that the solvent can normally hold at that temperature.

DEVELOPING SKILLS

B.9 SOLUBILITY AND SOLUBILITY CURVES

Sample Problem 1: At what temperature will the solubility of potassium nitrate be 25 g per 100 g water?

Think of the space between 20 g and 30 g on the *y*-axis in Figure 4.31 as divided into two equal parts, then follow an imaginary horizontal line at "25 g/100 g" to its intersection with the curve. Follow a vertical line down to the *x*-axis. Because the line falls halfway between 10 and 20 °C, the desired temperature must be about 15 °C

As you have seen, the solubility curve is quite useful when you are working with 100 g water. For example, the solubility curve indicated that 60 g potassium nitrate will dissolve in 100 g water at 40 °C. But what happens when you work with other quantities of water?

> *Sample Problem 2:* How much potassium nitrate will dissolve in 150 g water at 40 °C?
>
> You can reason the answer in the following way. The quantity of solvent (water) has increased from 100 to 150 g—1.5 times as much solvent. That means that 1.5 times as much solute can be dissolved. Thus: 1.5×60 g $KNO_3 = 90$ g KNO_3.
>
> The answer can also be expressed as a simple proportion, which produces the same answer once the proportion is solved:
>
> $$\frac{60 \text{ g } KNO_3}{100 \text{ g } H_2O} = \frac{x \text{ g } KNO_3}{150 \text{ g } H_2O}$$
>
> $$150 \text{ g } H_2O \times 60 \text{ g } KNO_3 = 100 \text{ g } H_2O \times x \text{ g } KNO_3$$
>
> $$x \text{ g } KNO_3 = 150 \cancel{\text{ g } H_2O} \times \frac{60 \text{ g } KNO_3}{100 \cancel{\text{ g } H_2O}} = 90 \text{ g } KNO_3$$

Refer to Figure 4.31 in answering these questions.

1. What mass (in grams) of potassium nitrate (KNO_3) will dissolve in 100 g water at 60 °C?

2. What mass (in grams) of potassium chloride (KCl) will dissolve in 100 g water at 60 °C?

3. You dissolve 25 g potassium nitrate in 100 g water at 30 °C, producing an unsaturated solution. How much more potassium nitrate (in grams) must be added to form a saturated solution at 30 °C?

4. What is the minimum mass (in grams) of 30 °C water needed to dissolve 25 g potassium nitrate?

5. You place 50 g NaCl in 100 g water at 30 °C.

   a. Classify the solution as saturated or unsaturated.

   b. Of the 50 g NaCl, about what mass will dissolve?

   c. Describe what you will see in the beaker.

6. In Investigating Matter B.1, you classified several substances as soluble or insoluble in room-temperature water.

   a. Identify one substance that you classified as soluble.

   b. Do you think the solution containing this substance was saturated or unsaturated? Explain your reasoning.

   c. How could you test whether your solution was saturated or unsaturated?

B.10 FACTORS THAT AFFECT SOLUBILITY OF GASES IN WATER

You have noted that the solubility of ionic and molecular solids in water usually increases when the water temperature is increased. Does this phenomenon extend to gases dissolved in solution, and why is it important to consider dissolved gases in relation to the fish kill? As you may already know, fish and other aquatic organisms depend on oxygen dissolved in water to live.

To build an understanding of how gases dissolve, look at Figure 4.32, which shows the solubility curve for oxygen gas plotted as milligrams oxygen gas dissolved per 1000 g water. What is the solubility of oxygen gas in 20 °C water? In 40 °C water? As you can see, increasing the water temperature causes the gas to be less soluble! Note also the magnitude of the values for oxygen solubility. Compare these values with those for solid solutes, as shown in Figure 4.31 (page 436). At 20 °C, about 30 g potassium nitrate will dissolve in 100 g water. In contrast, only about 9 mg (0.009 g) oxygen gas will dissolve in 10 times more water (1000 g) at this temperature. It should be clear that most gases are far less soluble in water than are many ionic solids.

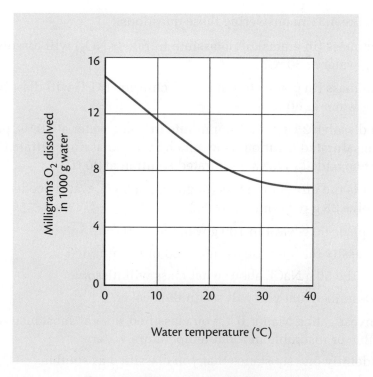

Figure 4.32 *Solubility curve for O₂ gas in water in contact with air.*

MODELING MATTER

B.11 SOLUBILITY AND CONCENTRATION

You have now learned about solubility, solubility curves, and the process of dissolving ionic and molecular compounds in water. As part of these discussions, visual models such as those presented in Figure 4.25 (page 421) and for which you drew five-panel illustrations (in Modeling Matter B.4, page 424) have helped to describe the process of dissolving. In this activity, you will combine these models with your knowledge of solubility curves to create new models of ions dissolved in water.

1. Suppose you dissolved 40 g potassium chloride (KCl) in 100 g water at 50 °C. You then let the solution cool to room temperature, about 25 °C.

   a. What changes would you see in the beaker as the solution cooled? See Figure 4.33.

   b. Draw models of what the contents in the beaker would look like at the particulate level at 50 °C, 35 °C, and 25 °C. Keep these points in mind: You should consider whether the sample at each temperature is *saturated* with excess solid present, or *unsaturated*, and draw the model accordingly. The solubility curve in Figure 4.33 and the particles depicted in Figure 4.25 (page 421) will be helpful. It is impossible to draw all the ions and molecules in this sample. The ions and molecules you draw will represent what happens on a larger scale.

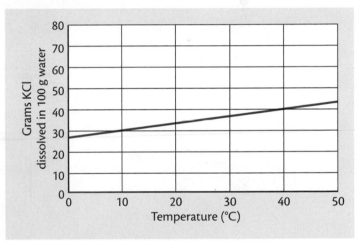

Figure 4.33 *Solubility curve for potassium chloride.*

2. An unsaturated solution will become more concentrated if you add more solute. Decreasing the total volume of water in the solution (such as by evaporation) also increases the solution's concentration. Consider a solution made by dissolving 20 g KCl in 100 g water at 40 °C.

   a. Draw a model of this solution.

   b. Suppose that while the solution was kept at 40 °C, one-fourth of the water evaporated.

      i. Draw a model of this final solution and describe how it differs from your model of the original solution.

      ii. How much water must evaporate at this temperature to create a saturated solution?

3. As you know from Investigating Matter B.7, a solution may be diluted by adding water.

   a. Draw a model of a solution containing 10.0 g KCl in 100 g water at 25 °C.

   b. Suppose you diluted this solution by adding another 100 g water with stirring (while keeping the temperature constant at 25 °C). Draw a model of the resulting solution.

   c. Compare your models in 3a and 3b. Describe key features that differ between the two models.

4. Recall Investigating Matter B.7, in which you used a colorimeter or spectrophotometer to measure concentrations of several $CuSO_4 \cdot 5\ H_2O$ solutions. Suppose the picture below represents the concentration of ions in 10 mL of stock solution: (*Note:* Cu^{2+} ions are represented by x's and SO_4^{2-} ions are represented by o's. Water molecules are not shown in this drawing.)

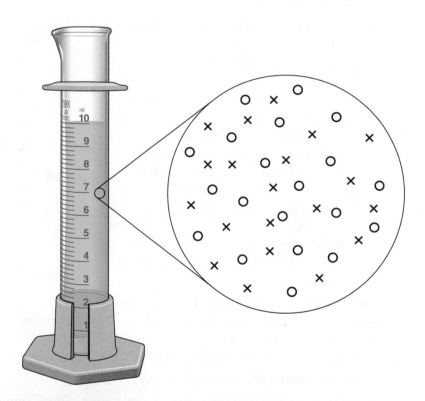

   a. You measured and transferred 1 mL of this stock solution to a 10-mL graduated cylinder. Draw a representation that shows the concentration of ions in this 1-mL sample.

   b. You then diluted the 1-mL sample to a total volume of 10 mL. Draw a representation that shows the concentration of ions in the diluted sample.

5. So far, you have represented ionic compounds dissolved in water. Now consider how another important substance, oxygen, could be represented as it dissolves in water.

a. When oxygen dissolves in water, does it break up into ions? Explain.

b. Refer to the solubility curves for potassium nitrate and for dissolved oxygen in Figures 4.31 (page 436) and 4.32 (page 438). Draw particulate-level pictures of a saturated solution of potassium nitrate (KNO_3) in 100 g of water at 0 °C and a saturated solution of dissolved oxygen in 100 g of water at 0 °C.

c. How do your drawings for 5b compare? Explain the similarities and differences.

B.12 TEMPERATURE, DISSOLVED OXYGEN, AND LIFE

Based on what you know about the effect of temperature on solubility, you may wonder if the temperature of the Snake River had something to do with the fish kill. As you just learned, water temperature affects how much oxygen gas can dissolve in the water. Various forms of aquatic life, including the many fish species, have different requirements for the concentration of dissolved oxygen needed to survive. Figure 4.34 provides this information for selected fish species.

How, then, does a change in the temperature of natural waters affect fish internally? Fish are *cold-blooded* animals. Their body temperatures rise and fall with the surrounding water temperature. If the water temperature rises, the body temperatures of fish also rise. This increase in body temperature, in turn, increases fish **metabolism**, a complex series of interrelated chemical reactions that keep fish alive. As these internal reaction rates speed up, the fish eat more, swim more, and require more dissolved oxygen. With rising temperatures, the rate of metabolism also increases for other aquatic organisms, such as aerobic bacteria, that compete with fish for dissolved oxygen.

Figure 4.34 *Dissolved-oxygen requirements of various species of fish.*

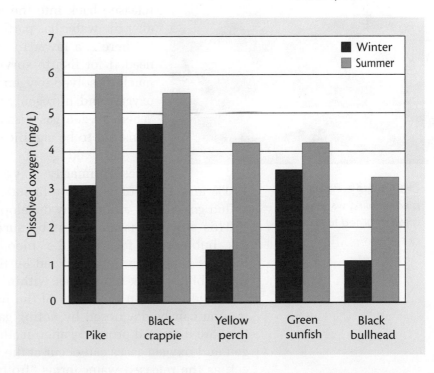

As you can see, an increase in water temperature affects fish by decreasing the amount of dissolved oxygen in the water and by increasing the oxygen consumption of fish. A long stretch of hot summer days sometimes results in large fish kills, where hundreds of fish suffocate because of a lack of dissolved oxygen. Table 4.3 summarizes the maximum water temperatures at which selected fish species can survive.

Table 4.3

Maximum Water Temperature Tolerance for Some Fish Species (24-hour exposure)		
	Maximum Temperature	
	°C	**°F**
Trout (brook, brown, rainbow)	24	75
Channel catfish	35	95
Lake herring (cisco)	25	77
Largemouth bass	34	93
Northern pike	30	86

Sometimes hot summer days are not to blame. Often, high lake or river temperatures can be traced to human activity. Many industries, such as electric-power generation, depend on natural bodies of water to cool heat-producing processes. Cool water is drawn from lakes or rivers into an industrial or power-generating plant, and devices called *heat exchangers* transfer thermal energy from the processing area to the cooling water. The heated water is then released back into the lakes or rivers, either immediately or after the water has partly cooled, as shown in Figure 4.35.

There is a lower limit on the amount of dissolved oxygen needed for fish to survive. Is there an upper limit? Can too much dissolved oxygen be a problem for fish? In nature, both oxygen and nitrogen gas are present in the air at all times. When oxygen gas dissolves, so does nitrogen gas. This fact turns out to be significant when considering the upper limit for dissolved gases. When the total amount of dissolved gases—primarily oxygen and nitrogen—reaches between 110% and 124% of saturation (a state called *supersaturation*), a condition called *gas-bubble trauma* may develop in fish.

Figure 4.35 *Power plant cooling water is often returned to lakes or streams.*

This situation is dangerous because the supersaturated solution causes gas bubbles to form in the blood and tissues of fish. Oxygen-gas bubbles can be partially utilized by fish during metabolism, but nitrogen-gas bubbles block capillaries within the circulatory system of fish. This blockage results in the death of the fish within hours or days. Gas-bubble trauma can be diagnosed by noting gas bubbles in the gills of dead fish if they are dissected promptly after death. Supersaturation of water with nitrogen and oxygen gases can occur at the base of a dam or a hydroelectric project, as the released water forms "froth," trapping large quantities of air, as shown in Figure 4.36.

Figure 4.36 *Froth at the base of a dam can trap large quantities of air. This can result in water that is supersaturated with dissolved oxygen and nitrogen gas.*

Back to the original question: *What caused the Riverwood fish kill?* You know from the newspaper articles that a likely cause is something dissolved in the water. Is it a case of concentrations of the substance that are too high for fish survival? Or in the case of dissolved oxygen, concentrations that are either too high or too low could be a culprit. For now, focus on dissolved oxygen. In Section C, you will investigate data related to other possible causes.

◼ MAKING DECISIONS

B.13 THE RIVERWOOD WATER MYSTERY: CONSIDERING DISSOLVED OXYGEN

In this activity, you will gather information about conditions that can lead to low or high dissolved oxygen levels and consider how such conditions could affect fish in the Snake River.

1. You learned in Sections B.10 and B.12 that water temperature affects the amount of dissolved oxygen in water.

    a. What is the relationship between water temperature and dissolved oxygen concentration?

    b. Describe a scenario in which a change in the temperature of Snake River water could lead to a fish kill in Riverwood.

2. In addition to changes in temperature, several scenarios exist that could lead to dissolved oxygen levels that would be either too high or too low for fish survival.

    a. Research and report two scenarios *not* related to temperature that can lead to low dissolved oxygen levels in rivers.

    b. Research and report one scenario *not* related to temperature that can lead to high dissolved oxygen levels in rivers.

3. Given your research for Question 2, which is a more likely cause for a fish kill: dissolved oxygen levels that are too high or too low? Explain.

SECTION B SUMMARY

Reviewing the Concepts

> The polarity of water helps explain its ability to dissolve many ionic solids.

1. What makes a water molecule polar?

2. Draw a model that shows how molecules in liquid water generally arrange themselves relative to one another.

3. Which region of a polar water molecule will be attracted to a

   a. K^+ ion?   b. Br^- ion?

4. Draw a particulate-level model of a KBr solution, showing how the water molecules are oriented toward the ions.

5. Describe the process of dissolving KCl

   a. using words. Be sure to discuss attractive forces involved in the process of dissolving.

   b. using a chemical equation.

> Whether a molecular substance dissolves in water depends on the relative strength of solute-water attractive forces, compared to competing solute–solute and water–water attractive forces.

6. Distinguish between polar and nonpolar molecules.

7. Would you select ethanol, water, or oil to dissolve a nonpolar molecular substance? Explain.

8. Explain the phrase "like dissolves like."

9. Explain why you cannot satisfactorily clean greasy dishes with plain water.

> In quantitative terms, the concentration of a solution expresses the relative quantities of solute and solvent in a particular solution.

10. A 35-g sample of ethanol is dissolved in 115 g water. What is the percent concentration of the ethanol, expressed as percent ethanol by mass?

11. Calculate the masses of water and sugar in a 55.0-g sugar solution that is labeled 20.0% sugar.

12. The EPA maximum standard for lead in drinking water is 0.015 mg/L. Express this value as parts per million (ppm).

13. How many moles of potassium hydroxide (KOH) are dissolved in 250 mL of a 0.15-M KOH solution?

14. Express the concentration, in mol/L, when 22 g $CuSO_4 \cdot 5\ H_2O$ is dissolved in 150 mL of solution.

> The solubility of a substance in water can be expressed as the quantity of that substance that will dissolve in a certain quantity of water at a specified temperature. Solubility curves are useful tools for determining whether a particular solution will be saturated or unsaturated.

15. Explain why three teaspoons of sugar will completely dissolve in a serving of hot tea, but will not dissolve in an equally sized serving of iced tea.

16. What is the maximum mass of potassium chloride (KCl) that will dissolve in 100.0 g water at 70 °C?

17. If the solubility of sugar (sucrose) in water is 2.0 g/mL at room temperature, what is the maximum amount of sugar that will dissolve in

    a. 100.0 mL water?

    b. 355 mL (12 oz) water?

    c. 946 mL (1 qt) water?

18. Rank the substances in Figure 4.31 (page 436) from most soluble to least soluble

    a. at 20 °C.   b. at 80 °C.

19. Distinguish between the terms *saturated* and *unsaturated*.

20. Using the graph on page 436, answer these questions about the solubility of potassium nitrate, KNO_3.

    a. What maximum mass of KNO_3 can dissolve in 100 g water if the water temperature is 20 °C?

    b. At 30 °C, 55 g KNO_3 is dissolved in 100 g water. Is this solution saturated, unsaturated, or supersaturated?

    c. A saturated solution of KNO_3 is formed in 100.0 g water at 75 °C. If some solute settles out as the saturated solution cools to 40 °C, what mass (in grams) of solid KNO_3 should form?

21. You are given a solution of KNO_3 of unknown concentration. What will happen when you add a crystal of KNO_3, if the solution is

    a. unsaturated?   b. saturated?

The solubility of a gaseous substance in water depends on the water temperature.

22. With increasing water temperature, how does the solubility of oxygen change?

23. How many milligrams of O_2 will dissolve in 1 000.0 g of water at 25 °C? (*Hint:* Use Figure 4.32 on page 438.)

24. Given your knowledge of gas solubility, explain why a bottle of warm cola produces more "fizz" when opened than does a bottle of cold cola.

25. Using the graph on page 441, determine how much more dissolved oxygen in water is required by yellow perch in the summer than in the winter.

Why do some substances readily dissolve while others do not?

In this section, you explored how ionic and molecular substances dissolve in water and how the quantities that dissolve are influenced by the structures of both solute and water molecules. You have also described and measured solution concentrations, used solubility curves, and considered the role of dissolved oxygen in aquatic life. Think about what you have learned, then answer the question in your own words in organized paragraphs. Your answer should demonstrate your understanding of the key ideas in this section.

Be sure to consider the following in your response: solute, solubility, electronegativity, polarity, and concentration.

Connecting the Concepts

26. Draw a model of a solution in which water is the solvent and oxygen gas (O_2) is the solute.

27. Many mechanics prefer to use waterless hand cleaners to clean their greasy hands. Explain

    a. what kind of materials are likely to be found in these cleaners.

    b. why using these cleaners is more effective than washing with water.

28. From each of these pairs, select the water source more likely to contain the higher concentration of dissolved oxygen. Give a reason for each choice.

    a. a river with rapids or a calm lake

    b. a lake in spring or the same lake in summer

    c. a lake containing only black crappie or a lake containing only black bullhead

29. Fluorine has the highest electronegativity of any element. Fluorine and hydrogen form a polar bond in hydrogen fluoride (HF). Which atom in HF would you expect to have a partial positive charge? Explain.

30. Suppose there are two identical fish tanks with identical conditions except for temperature. Tank A is maintained at a temperature 5 °C higher than is Tank B. Which tank could support the greater number of similar fish? Explain.

31. Using Figure 4.32 on page 438,

    a. determine the solubility of oxygen at 20 °C.

    b. express the answer you obtained in ppm.

32. Why does table salt (NaCl) dissolve in water but not in cooking oil?

Extending the Concepts

33. Describe how changes in solubility due to temperature could be used to separate two solid, water-soluble substances.

34. The continued health of an aquarium depends on the balance of the solubilities of several substances. Investigate how this balance is maintained in a freshwater aquarium.

SECTION C REACTIONS IN SOLUTION

How do we describe chemical behavior in aqueous solutions?

The challenge facing investigators of the Riverwood fish kill is to decide what was present—or absent—in the Snake River water that caused this crisis. In this section, you will learn about how dissolved substances in water solutions interact with one another. You will investigate the sources of some water contaminants and learn how scientists test for the presence of others. This background will ensure that you have the knowledge and skills to evaluate Snake River water data and, finally, to determine the cause of the fish kill.

GOALS

- Describe and write equations for precipitation reactions.
- Use solubility rules to predict the formation of a precipitate and design tests for dissolved ions.
- Distinguish among strong and weak acids and bases.
- Calculate and compare concentrations of hydronium and hydroxide ions and pH values in acidic, basic, and neutral aqueous solutions.
- Describe effects of pH changes on natural systems.
- Describe the composition and chemical behavior of a buffer.

concept check 6

1. Describe how the process of dissolving differs for ionic compounds compared to molecular compounds.
2. Which would you expect to have a greater influence on the effect of a solute in natural water, concentration or solubility? Explain.
3. What are some possible observable results of reactions between substances in aqueous solution?

■ INVESTIGATING MATTER

C.1 COMBINING SOLUTIONS

Preparing to Investigate

In this investigation, you will observe and consider the results of mixing aqueous solutions of ionic compounds. You will mix 0.1 M solutions of hydrochloric acid (HCl), silver nitrate ($AgNO_3$), sodium sulfate (Na_2SO_4), calcium chloride ($CaCl_2$), sodium hydrogen carbonate ($NaHCO_3$), barium nitrate ($Ba(NO_3)_2$), and sodium hydroxide (NaOH).

Before you begin, read *Gathering Evidence* to learn what you will need to do and note safety precautions. Plan for data collection and construct an appropriate data grid with blocks that indicate the pair of solutions to be mixed and provide room for observations and conclusions.

Gathering Evidence

1. Before you begin, put on your goggles, and wear them properly throughout the investigation.

2. Using pipets, place one drop each from the appropriate column and row onto the block on the acetate sheet. See Figure 4.37. Do not touch the pipets to the solutions on the sheet.

Figure 4.37 *Micropipets are used to dispense small quantities of solution onto the acetate sheet.*

3. Examine each combination. Record your observations and results.

4. Dispose of the mixtures on your acetate sheet as instructed by your teacher.

5. Wash your hands thoroughly before leaving the laboratory.

Analyzing Evidence

1. In your data grid, write the formulas of the possible products of each combination. Remember to balance positive and negative charges within each compound as you learned to do in Unit 1.

2. Write a plus sign (+) in each grid block that represents a combination that resulted in an observable reaction.

Interpreting Evidence and Making Claims

1. How did you know whether a reaction occurred when a pair of solutions was mixed?
2. What can you say about the solubility of the solid substances that were formed in some combinations? What evidence supports this claim?

Reflecting on the Investigation

3. What tests, besides visual observation, might be useful in deciding whether a reaction occurred?
4. What are some advantages and disadvantages of conducting this investigation on an acetate sheet instead of using test tubes or beakers?
5. What would you predict would happen if you mixed solutions of:
   a. sodium hydrogen carbonate and nitric acid (HNO$_3$)?
   b. silver nitrate and sodium chloride?
   c. barium nitrate and sodium chloride?

C.2 PRECIPITATION REACTIONS

In Investigating Matter C.1, several combinations of aqueous solutions led to the appearance of an insoluble material called a **precipitate** (Figure 4.38). You learned earlier in this unit that some substances are more soluble in water than others. In a process much like "dissolving in reverse," ions that make up poorly or moderately soluble compounds are attracted to one another in solution, forming groups of ions that eventually grow large enough to settle out of solution as a precipitate.

As an example, consider the combination of silver nitrate (AgNO$_3$) and sodium chloride (NaCl) solutions. Both of these ionic substances are soluble in water. When you combined them in Investigating Matter C.1, a white precipitate formed. Think about the possible identities

Figure 4.38 *A bright yellow precipitate forms when you add a solution of potassium iodide, KI, to a solution of lead(II) nitrate, Pb(NO$_3$)$_2$. The precipitate is lead(II) iodide, PbI$_2$.*

of this precipitate. Since both a cation and an anion are required for an ionic compound, the precipitate could be either sodium nitrate ($NaNO_3$) or silver chloride (AgCl). About 87 g $NaNO_3$ will dissolve in 100 g of 20 °C water. Under the same conditions, only 0.0002 g AgCl will dissolve. From this information, it is reasonable to conclude that the precipitate is AgCl. The following word equation describes the reaction that took place: Aqueous silver nitrate and aqueous sodium chloride react to form solid silver chloride and aqueous silver nitrate.

Translating to symbols gives:

$$AgNO_3(aq) + NaCl(aq) \longrightarrow AgCl(s) + NaNO_3(aq)$$

To more accurately reflect the form of the substances in solution, one could also write:

$$Ag^+(aq) + NO_3^-(aq) + Na^+(aq) + Cl^-(aq) \longrightarrow AgCl(s) + Na^+(aq) + NO_3^-(aq)$$

This type of equation is called a **total ionic equation**. Note that Na^+ and NO_3^- ions appear in the same form on both sides of the equation. That is, they do not participate in the reaction and are known as **spectator ions**. The equation can be rewritten without the spectator ions, resulting in a **net ionic equation**, as shown here:

$$Ag^+(aq) + Cl^-(aq) \longrightarrow AgCl(s)$$

In the preceding example, the precipitate was identified by finding the solubility of each possible product and comparing these values. Fortunately, ample solubility data are available for ionic compounds in water. A number of trends have been identified within these data that make it easier to identify a precipitate. These trends are known as **solubility rules**. Some general solubility rules are as follows:

- Ionic compounds containing group 1 or ammonium cations are soluble.
- Ionic compounds containing nitrate or acetate anions are soluble.
- Ionic compounds containing chloride, bromide, or iodide anions tend to be soluble unless they are combined with silver, mercury(I), copper(I), or lead(II).
- Ionic compounds containing carbonate, phosphate, or hydroxide ions tend to be insoluble or only slightly soluble unless they are combined with ammonium or a group 1 cation.

Solubility rules help to identify the products of a great variety of aqueous reactions. Remember that they are generalizations from solubility data, not absolute rules for memorization. When in doubt, or when faced with ions not addressed by these rules, refer to solubility data for individual compounds.

The next activity will provide practice in recognizing precipitates formed in aqueous reactions and representing these reactions with symbols, equations, and models.

Precipitation reactions are also known as *double-replacement reactions*.

Recall from Unit 1 that some elements, including mercury, copper, lead, and iron, can form cations with different charges. For this reason, Roman numerals are added to the name to indicate the charge on the ion.

DEVELOPING SKILLS

C.3 WRITING NET IONIC EQUATIONS

Sample Problem: Predict the identity of the precipitate formed in the reaction between aqueous copper(I) nitrate and potassium iodide solutions and write a net ionic equation for the process.

Possible precipitates include copper(I) iodide and potassium nitrate. Salts of both potassium and nitrate are soluble, so the product must be copper(I) iodide.

$$Cu^+(aq) + I^-(aq) \longrightarrow CuI(s)$$

1. Predict the identity of the precipitate formed in the reaction between each of the following pairs of aqueous solutions. Write both the name and the chemical formula of each precipitate and explain your conclusion.

    a. ammonium chloride and silver acetate

    b. sodium carbonate and calcium nitrate

    c. barium chloride and potassium hydroxide

2. Consider the reaction of aqueous lead(II) nitrate ($Pb(NO_3)_2$) and aqueous sodium iodide (NaI).

    a. What precipitate would form?

    b. How do you know?

    c. Write an equation for the reaction of these two solutions.

    d. Write a net ionic equation for this reaction.

    e. Draw three beakers representing:

        i. lead(II) nitrate before reaction.

        ii. sodium iodide before reaction.

        iii. the result of the reaction.

3. Choose two of the reactions in Investigating Matter C.1 that resulted in formation of precipitates. Identify the solutions that were mixed and write a net ionic equation for each reaction.

4. Think about the formation of calcium carbonate ($CaCO_3$) in a reaction between sodium carbonate (Na_2CO_3) and calcium chloride ($CaCl_2$). Recalling Modeling Matter B.4 and thinking about this process as the "opposite of dissolving," use a series of drawings to represent the reaction in a step-by-step progression.

INVESTIGATING MATTER

C.4 WATER TESTING

Asking Questions

Now that you have seen how ionic substances dissolve and what can happen when aqueous ionic solutions interact, think about how this knowledge could be helpful in analyzing natural water. How do chemists detect and identify certain substances or ions in water solutions? This investigation will allow you to use a method that chemists, including those investigating the Riverwood fish kill, use to detect the presence of specific ions in water solutions.

You will develop a procedure to test for the presence of iron(III) (Fe^{3+}) and calcium (Ca^{2+}) cations, as well as chloride (Cl^-) and sulfate (SO_4^{2-}) anions. These ions are often found in groundwater and tap water.

Given what you know about solubility and what you observed in Investigating Matter C.1, one test that you could conduct to determine the presence of an ion is known as a **confirming test**. That is, a positive test—one in which a precipitate forms or a change in solution color is noted—confirms that the ion in question is present. A negative test (no color or precipitate) means one of two things: either the ion is not present, or the ion is present in such low amounts that the chemical test involved is unable to produce a confirming result.

Such confirming tests are **qualitative tests**, tests that identify the presence or absence of a particular substance in a sample. In contrast, **quantitative tests** (like those you performed in Investigating Matter B.7) determine the *amount* of a specific substance present in a sample. Both types of tests would be used to determine the cause of the Snake River fish kill.

Think about the solutions that you could add to each of the ions listed above to create a positive result. In this investigation, you will be provided with aqueous solutions, including sodium carbonate, sodium hydroxide, silver nitrate, barium nitrate, and calcium chloride. You will also have a 24-well plate and pipets to dispense the solutions. With your lab partner, develop a plan to confirm the presence of each of the ions and write an equation for each test.

You will need to complete each confirming test on several different samples, including the following:

- a distilled-water **blank**—known not to contain any ions of interest.
- a **reference solution**—known to contain the sought ion.
- an unknown solution that may contain one or more of the ions.
- local tap water.
- natural (or other) water samples collected by you or your teacher.

Before you develop your complete procedure, draft a scientific question about the presence of one or more of these ions in a water source or sample of your choosing. Next, ask your teacher to approve your equations, then write out the steps you will take to complete each test for each type of sample.

> Technologies are available for detecting ions when their amounts are so small that a confirming test cannot produce a positive result.

Preparing to Investigate

After your teacher has approved your equations and procedure, prepare a data table or several data tables (one for each ion). Be sure to note the source of each natural-water sample. The following suggestions will help guide your ion analysis.

- If the ion is in tap or natural water, it will probably be present in a smaller amount than in the same volume of reference solution. Therefore, the quantity of precipitate or color produced in the tap-water or natural-water sample will be lower than in the reference solution.

- When completing an ion test, mix the contents of the well thoroughly, using a toothpick or small glass stirring rod. Do not use the same toothpick or stirring rod in other samples without first rinsing it and wiping it dry.

- In a confirming test based on color change, so few color-producing ions may be present that it is hard to tell if the reaction took place. Here are two ways to decide whether the expected color is present:
  - Place a sheet of white paper behind or under the well plate to make any color more visible.
  - Place the well plate on a black or dark surface to make a precipitate more visible.

Before you begin, read *Gathering Evidence* to note safety precautions.

Gathering Evidence

1. Before you begin, put on your goggles, and wear them properly throughout the investigation.

2. Conduct your confirming tests using pipets and well plates according to your developed procedure. See Figure 4.39.

3. Observe and record your results.

4. Dispose of the contents of the well plate as directed by your teacher.

5. Wash your hands thoroughly before leaving the laboratory.

Figure 4.39
Adding drops to a 24-well plate.

Interpreting Evidence

1. Consider the results of your tests.

   a. Provide a summary of the ions detected and in which samples they were detected.

   b. Did the results surprise you? Explain.

Making Claims

2. State and answer the scientific question that you developed. Support your answer with evidence.

3. Can you make a more general claim about the question you asked based upon this investigation? Why or why not?

Reflecting on the Investigation

4. Why were a reference solution and a blank used in each test?

5. What are some possible problems associated with the use of qualitative tests?

6. These tests cannot absolutely confirm the absence of an ion. Why?

7. How might your observations have changed if you had not cleaned your wells or stirring rods thoroughly after each test?

C.5 ACIDS AND BASES IN SOLUTION

You have seen that some reactions in aqueous solutions produce precipitates, a fact that can be applied to detecting the presence of some dissolved ions in solution. However, not all solution combinations in Investigating Matter C.1 produced insoluble solids. You probably noted combinations in which a gas was produced, as well as some in which it appeared that nothing happened. Consider the combination of hydrochloric acid (HCl) and sodium hydroxide (NaOH) solutions. You already learned in Unit 2 that HCl*(aq)* is an acid and NaOH*(aq)* is an example of a basic solution. The combination of these two solutions could be represented as:

$$HCl(aq) \;+\; NaOH(aq) \longrightarrow NaCl(aq) \;+\; HOH(aq)$$

The second product of this reaction, HOH*(aq)*, is more commonly known as $H_2O(l)$—ordinary water. The equation then becomes:

$$HCl(aq) \;+\; NaOH(aq) \longrightarrow NaCl(aq) \;+\; H_2O(l)$$

This is an example of a **neutralization** reaction. As is true for all neutralization reactions between compounds, this reaction produces water and a salt (an ionic compound). Such reactions are another important component of chemistry in aqueous solutions and can significantly affect the quality of natural waters such as the Snake River.

In 1887, Swedish chemist Svante Arrhenius defined an acid as any substance that generates hydrogen ions (H^+) when dissolved in water. He defined a base as a substance that generates hydroxide ions (OH^-) in water. In aqueous solutions, the hydrogen ion released by the acid is bonded to water. Such a combination can be represented as $H^+(aq)$, $H(H_2O)^+$ or, more commonly, as $H_3O^+(aq)$, the **hydronium ion**:

$$H^+(aq) \;+\; H_2O(l) \;\rightleftharpoons\; H_3O^+(aq)$$

Arrhenius also proposed that increases in atmospheric carbon dioxide may increase global temperatures.

Consider the example of carbon dioxide dissolving in water to produce carbonic acid, $H_2CO_3(aq)$:

$$H_2O(l) + CO_2(g) \rightleftharpoons H_2CO_3(aq)$$

The transfer of a hydrogen ion, H^+, from the carbonic acid molecule to a water molecule produces a hydrogen carbonate (bicarbonate) ion and a hydronium ion (H_3O^+), which is responsible for the solution's acidity:

$$H_2O(l) + H_2CO_3(aq) \rightleftharpoons H_3O^+(aq) + HCO_3^-(aq)$$

Water   Carbonic acid   Hydronium ion   Bicarbonate ion

Most common acids are substances containing one or more hydrogen atoms that can be released as $H^+(aq)$ in water. The remainder of the original acid molecule, with hydrogen's single electron still attached to it, becomes an anion. As shown in the following equations for two common acids, the dissolved acid produces hydronium ion and an anion.

Nitric acid: $H_2O(l) + HNO_3(aq) \longrightarrow H_3O^+(aq) + NO_3^-(aq)$

Sulfuric acid: $H_2O(l) + H_2SO_4(aq) \longrightarrow H_3O^+(aq) + HSO_4^-(aq)$

Many bases are composed of a cation and the hydroxide anion, OH^-. When such bases dissolve in water, the cations and hydroxide ions separate and disperse uniformly throughout the solution. The hydroxide ions (OH^-) give basic solutions (also referred to as alkaline solutions) their characteristic properties:

Potassium hydroxide: $KOH(s) \longrightarrow K^+(aq) + OH^-(aq)$

Barium hydroxide: $Ba(OH)_2(s) \longrightarrow Ba^{2+}(aq) + 2\,OH^-(aq)$

Acidic solutions have more $H_3O^+(aq)$ than $OH^-(aq)$, while alkaline solutions have more $OH^-(aq)$ than $H_3O^+(aq)$. Mixing equal amounts of $H_3O^+(aq)$ and $OH^-(aq)$ results in a neutralization reaction that destroys both acidic and basic characteristics, as summarized in this net ionic equation:

$$H_3O^+(aq) + OH^-(aq) \rightleftharpoons 2\,H_2O(l)$$

Pure (neutral) water is produced. It contains equal amounts of $H_3O^+(aq)$ and $OH^-(aq)$, as is true of all neutral substances.

Note the double arrow written in the equation. It is used to indicate that it is a **reversible reaction**; it proceeds at an equal rate in each direction, both to the right (the forward reaction) and to the left (the reverse reaction).

DEVELOPING SKILLS

C.6 ACIDS AND BASES

> *Sample Problem:* Name the base with formula NaOH, a
> compound often found in drain and oven cleaners, and write an
> equation showing the ions that are liberated when it dissolves
> in water.
>
> $$\text{Sodium hydroxide: NaOH}(s) \longrightarrow \text{Na}^+(aq) + \text{OH}^-(aq)$$

1. Hydrochloric acid (HCl) is an acid found in gastric juice (stomach fluid).

   a. Write an equation for the formation of ions as HCl dissolves in and reacts with water.

   b. Name each ion formed.

2. Separate equations for carbonic acid, nitric acid, and sulfuric acid reacting with water are shown on page 455. Note that the expression for carbonic acid's reaction also includes names and molecular models for each reactant and product. Complete the following steps for both nitric acid and sulfuric acid:

   a. Write the equation for the acid's reaction with water.

   b. Write the name of each reactant and product.

   c. Draw molecular models for each reactant and product. (*Hint:* Just as in carbonic acid, each hydrogen atom in nitric acid and sulfuric acid is connected directly to an oxygen atom.)

3. Each base in the following list has commercial and industrial uses. Name each base and write an equation showing the ions that are liberated when the base dissolves in or reacts with water:

   a. $Mg(OH)_2$, the active ingredient in milk of magnesia, an antacid.

   b. $Al(OH)_3$, a compound used to bind dyes to fabrics.

   c. NH_3, used as a cleaner and in synthesizing fertilizers and pharmaceuticals. (*Hint:* See Unit 2, page 251.)

4. A person consumes some antacid containing magnesium hydroxide, $Mg(OH)_2$, to relieve discomfort from excess stomach acid, HCl(*aq*).

   a. Write the overall equation for the neutralization of HCl(*aq*) with $Mg(OH)_2$(*aq*).

   b. Name each product shown in the equation you wrote in answer to Question 4a.

✓ concept check 7

1. How can you predict whether a precipitate will form when two aqueous solutions of ionic compounds are mixed?
2. Consider a neutralization reaction:
   a. What are the reactants?
   b. What happens to the properties of the reactants?
   c. What are the products?
3. Acids are often classified as either a "strong acid" or a "weak acid." What do you think chemists mean by the terms "strong acid" and "weak acid"? Why do you think this?

C.7 SOLUTION CONCENTRATION AND pH

Water and its solutions contain both hydronium ions (H_3O^+) and hydroxide ions (OH^-).

- In pure water and in **neutral solutions**, the concentrations of these ions are equal, but very small.

- In an **acidic solution**, the hydronium-ion concentration is larger than the hydroxide-ion concentration. In *very* acidic solutions the hydronium-ion concentration is much larger than the hydroxide-ion concentration.

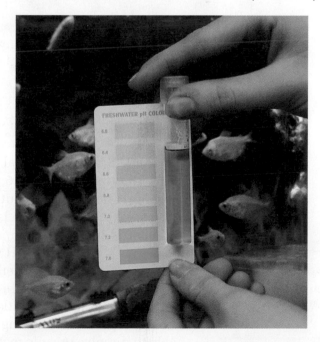

Figure 4.40 *Testing the pH of an aquarium.*

- In **basic solution**, the hydroxide-ion concentration is larger than the hydronium-ion concentration. In *very* basic solutions, the hydroxide-ion concentration is much larger than the hydronium-ion concentration.

Both acidic and basic solutions have effects on living organisms—effects that depend on their pH (Figure 4.40). The pH value of an aqueous solution is related to its hydronium-ion concentration (H_3O^+) present. The symbol *pH* stands for "power of hydronium ion," where "power" refers to the mathematical power (exponent) of 10 that expresses the hydronium ion's molar concentration.

Using similar reasoning, a solution with a pOH of 2.0 would contain 1×10^{-2} mol OH^- (0.01 mol OH^-) per liter.

For example, a solution containing 0.001 mol H_3O^+ (1×10^{-3} mol H_3O^+) per liter has an H_3O^+ concentration of 0.001 M (10^{-3} M) and thus a pH value of 3.0. The pH value can be interpreted as the exponent on the power of 10 with its sign changed. A solution with 0.000 000 01 mol H_3O^+ (1×10^{-8} mol H_3O^+) per liter (0.000 000 01 M) would have a pH of 8.0. Figure 4.41 shows how the pH of an aqueous solution is related to the molar concentration of H^+ (H_3O^+) and OH^-.

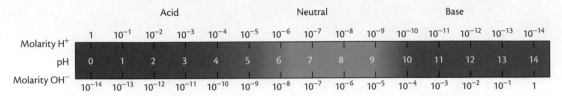

Figure 4.41 *The relationships among pH, molar concentration of H_3O^+, and molar concentration of OH^- at 25 °C.*

As you have just learned, the pH scale expresses acidity and alkalinity based on powers of 10. This is why, as you learned in Unit 2, a solution with pH 2.0 (such as lemon juice), is 10 times more acidic than is a soft drink at pH 3.0, which is 10 000 times more acidic than pure water, at pH 7.0 (four steps farther up the pH scale). See Figure 4.42. Similarly, a solution with pH 11.0 is 100 times more basic than is a solution with a pH of 9.0.

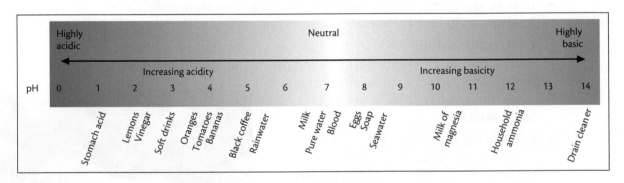

Figure 4.42 *The pH values of some common materials.*

The U.S. Environmental Protection Agency requires that drinking water be within the pH range 6.5 to 8.5. However, most fish can tolerate a slightly wider pH range, from about 5.0 to 9.0, in lake or river water. When the pH of rivers, lakes, and streams is too low (meaning high acidity), fish-egg development is impaired, hampering the ability of fish to reproduce. Bodies of water with low (acidic) pH values also tend to increase the concentrations of metal ions in natural waters by leaching metal ions from surrounding soil. These metal ions can include aluminum ions (Al^{3+}), which are toxic to fish when present in sufficiently high concentration. High pH (basic contamination) is a problem for living organisms mainly because alkaline solutions are able to dissolve organic materials, including skin and scales.

Expert freshwater anglers sometimes plan to fish in water between pH 6.5 and 8.2.

On a normal day, the pH of Snake River water in Riverwood ranges between 7.0 and 8.0, nearly optimal for freshwater fishing. Could the pH have changed abruptly, killing the fish? If so, was acidic or basic contamination responsible for the Riverwood crisis?

DEVELOPING SKILLS

C.8 INTERPRETING THE pH SCALE

Sample Problem: *Household vinegar (a solution of acetic acid, CH_3COOH) has a pH of about 3. Classify vinegar as acidic, basic, or neutral and estimate the hydronium- and hydroxide-ion concentrations in vinegar.*

> Vinegar is acidic (pH < 7)
>
> Vinegar with pH = 3 has:
>
> a hydronium-ion concentration of 10^{-3} mol/L.
>
> a hydroxide-ion concentration of 10^{-11} mol/L.

1. Using values shown in Figure 4.41, describe the mathematical relationship for aqueous solutions at 25 °C

   a. between pH values and H_3O^+ (H^+) concentrations.

   b. between H_3O^+ and OH^- concentrations.

2. Some common aqueous solutions and their typical pH values appear in the following list. Classify each solution as *acidic*, *basic*, or *neutral*.

   a. milk, pH = 6.7

   b. stomach fluid, pH = 1.3

   c. a solution of drain cleaner, pH = 14.0

3. Classify each of the following aqueous solutions as *acidic*, *basic*, or *neutral* and estimate the hydronium-ion and hydroxide-ion concentrations found in each.

   a. a cola drink, pH = 3.0

   b. sugar dissolved in pure water, pH = 7.0

   c. a solution of household ammonia, pH = 12.0

4. How many times more acidic is a cola drink than black coffee? *(Hint:* See the information provided in Figure 4.42.)

5. Clouds over Clingmans Dome (see Figure 4.43), a peak in Great Smoky Mountains National Park, have had pH levels as low as 2.0. Compared to rainfall at pH 5.5, estimate how many times more acidic the moisture has been in these clouds.

6. Products used to remove clogs from drains contain bases, such as sodium hydroxide (NaOH). If drain cleaner containing 10.0 g NaOH*(s)* is added to a clogged drain pipe filled with 2.5 L water,

   a. how many moles of NaOH were added?

   b. what is the molar concentration of NaOH in the resulting solution?

   c. use Figure 4.41 to find the pH of this drain-cleaner solution. (*Hint:* The concentration of OH^- is equal to the concentration of NaOH calculated in 6b.)

Figure 4.43 *Clouds over Clingmans Dome in the Great Smoky Mountains National Park.*

Figure 4.44 *This Colorado stream had a pH of about 3.0 before treatment with base. What could have caused this low pH?*

■ MAKING DECISIONS

C.9 SOURCES OF ACIDIC AND BASIC CONTAMINATION

Could acidic or basic contamination have caused the Riverwood fish kill? How do acids and bases get into natural waters such as the Snake River? (See Figure 4.44.) In this activity, you will gather information and explore these questions as you continue to investigate the incident in Riverwood.

Use the Internet and other resources provided or indicated by your teacher to answer the following questions. Be sure to consider the reliability of your sources.

1. Find information on fish kills caused by acidic conditions.
   a. List at least three different causes of overly acidic conditions leading to a fish kill.
   b. For each cause listed in 1a, identify a question that you could ask to help determine whether this cause could have been a factor in the Riverwood fish kill.

2. Find information on fish kills caused by basic conditions. (*Hint*: The terms *alkaline* or *caustic*, used to refer to solutions with high pH, may be useful in locating this information.)
   a. List at least one cause of overly basic conditions leading to a fish kill.
   b. For each cause listed in 2a, identify a question that you could ask to help determine whether this cause could have been a factor in the Riverwood fish kill.

3. List at least three effects of abnormally high or low pH on fish.

4. You learned about acid rain in Unit 2. Can acid rain cause fish kills? Explain and support your answer with evidence.

C.10 STRENGTHS OF ACIDS AND BASES

Figure 4.45 *What terms would you use to express the relative concentrations of these two tea beverages?*

In everyday discussions of solutions, *strong* tends to be associated with the idea of "concentrated," while *weak* usually means "diluted." (See Figure 4.45.) However, to a chemist, *strong* and *weak*—in relation to acid and base solutions—have very different meanings. Acids and bases are classified as strong or weak according to the extent to which they **ionize**, or form ions in solution. An acid is considered a **strong acid** if it ionizes completely, meaning that every dissolved acid molecule or unit reacts to produce a hydronium ion and an anion. A base is considered a **strong base** if every dissolved molecule or unit produces a hydroxide ion and a cation. The total dissolved molecules or ions does not matter. *Strong* signifies *complete* ionization, while *weak* implies only *partial* ionization.

When a strong acid dissolves in water, none of the original acid molecules remain. Nitric acid (HNO₃), found in acid rain, is a strong acid. The complete reaction within a nitric acid solution is represented this way:

$$H_2O(l) + HNO_3(aq) \longrightarrow H_3O^+(aq) + NO_3^-(aq)$$

All the original HNO_3 molecules have reacted with water, forming hydronium ions and nitrate ions.

In a **weak acid**, by contrast, only a relatively few dissolved acid molecules ionize to form hydronium ions and anions. That is, weak acids are only slightly ionized. Nitrous acid (HNO_2), sometimes also found in acid rain, is a typical weak acid. The partial ionization within aqueous nitrous acid solution is represented this way:

$$H_2O(l) + HNO_2(aq) \rightleftharpoons H_3O^+(aq) + NO_2^-(aq)$$

This is an example of a reversible reaction. An analogy is useful in illustrating this idea of reversible reactions: At a basketball game, if the number of people walking out of the arena to buy snack foods and drinks equals the number of people returning to their seats after buying their snacks and drinks, the system is described as in **dynamic equilibrium**. That is, two offsetting processes occur at equal rates, producing a state of balance where no net change is observed.

The forward and reverse *rates*—not the *amounts* of products and reactants—are *equal* in a system at equilibrium. In the basketball-game analogy, it is clear at any particular time that many more people remain seated watching the game than are moving to and from the snack counters. Likewise, at equilibrium, the concentration of HNO_2 is much larger than the concentrations of H_3O^+ and NO_2^-. However, a state of balance (dynamic equilibrium) is attained in both cases.

In addition to nitrous acid, acid rain may contain other weak acids, commonly sulfurous acid (H_2SO_3) and carbonic acid (H_2CO_3). Strong acids, such as sulfuric acid (H_2SO_4), may also be present.

Sodium hydroxide (NaOH) and potassium hydroxide (KOH) are two strong bases. Their aqueous solutions completely consist of sodium or potassium cations and OH^- ions in water. Even though the OH^- concentration may be limited by the solubility of a particular solid base in water, such a base is regarded as a strong base. Magnesium hydroxide, $Mg(OH)_2$, a commercial antacid ingredient, is an example of a strong (but sparingly soluble) base.

A weak base commonly found in the environment is the carbonate anion (CO_3^{2-}), a component of limestone and marble. The carbonate anion is considered basic, even though it has no hydroxide ions (OH^-) in its formula. Water reacts with carbonate anions to produce basic OH^- ions in an aqueous solution:

$$H_2O(l) + CO_3^{2-}(aq) \rightleftharpoons OH^-(aq) + HCO_3^-(aq)$$

Other anions that are basic, such as phosphate (PO_4^{3-}) and sulfite (SO_3^{2-}), also produce OH^- ions in water.

Strong acids are strongly ionized.

Weak acids are only *weakly* ionized.

Only 0.01 g $Mg(OH)_2(s)$ can dissolve in 1 L of water at room temperature, but magnesium hydroxide is still regarded as a strong base.

■ MODELING MATTER

C.11 STRONG VERSUS CONCENTRATED

In this activity, you will depict concentrated and dilute solutions of two imaginary acids—one strong (HSt) and the other weak (HWe).

In referring to an acidic solution, as you recall, the term *strong* indicates that all (or nearly all) acid molecules have reacted with the water to produce hydronium ions and anions. By contrast, *concentrated* indicates that there are a very large number of particles—molecules or ions—dissolved in a particular volume of solution.

Figure 4.46 depicts possible models for strong and weak acids. In this activity, you will construct models to illustrate the ideas you have just learned.

1. Draw four equal-sized rectangular boxes. Each box represents 1.0 L of solution. Use H^+ to represent hydrogen ions. Each H^+ symbol will represent 0.01 mol of dissolved hydrogen ions. Do *not* draw the solution's water molecules.

   Inside each box, draw between 2 and 20 acid molecules, or the appropriate numbers of hydrogen ions and anions, to illustrate the following conditions:

   a. In the first box, depict a *concentrated strong acid* (HSt) solution.

   b. In the second box, depict a *dilute strong acid* (HSt) solution.

   c. In the third box, depict a *concentrated weak acid* (HWe) solution.

   d. In the fourth box, depict a *dilute weak acid* (HWe) solution.

Use your depictions to answer the following questions.

2. Explain the difference between these pairs of words, used to describe a solution of an acid or base.

   a. weak, dilute   b. strong, concentrated

3. Calculate the molar concentration (molarity) of hydrogen ions (H+) depicted in each of your four drawings. Recall that each ion you drew represented 0.01 mol of particles.

4. How would the pH values for the four solutions that you depicted compare to one another?

5. a. Why is it easier to use a single symbol to represent 0.01 mol of particles, rather than trying to draw every particle in 0.01 mol of ions?

   b. How many total particles are actually in 0.01 mol of particles?

6. How might your model drawings mislead a viewer about what is actually present in each solution?

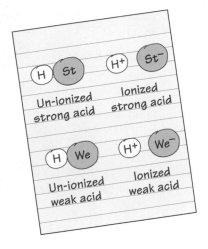

Figure 4.46 *Suggested models that can represent strong and weak acids in solution*

concept check 8

1. In your own words, explain how pH is related to the concentration of hydronium ions in an aqueous solution.

2. What types of particles would you expect to find in a concentrated solution of a strong base?

3. What type of substance might protect lakes and streams from the effects of acids, bases, and acidic precipitation?

C.12 BUFFERS

In Unit 2, you learned that acid precipitation causes damage to plants, fish eggs, and other aquatic life. One early mystery about such damage was the observation that the pH of some lakes and streams seemed unaffected by acidic precipitation. What protected these bodies of water from undergoing large changes in pH?

The key to solving this mystery came when researchers realized that bodies of water that suffered acidification due to acid rain (Figure 4.47) had two features in common. First, they were downwind from a dense array of power stations, metal-processing plants, or large cities, all of which produce nitrogen oxides and sulfur oxides. Second, these bodies of water were often surrounded by soils that cannot neutralize acid carried by the precipitation. If the soil cannot neutralize the acidic precipitation, the lake or stream into which the precipitation drains then becomes acidified.

By contrast, some bodies of water benefit from surrounding and underlying rock and soils that can neutralize the acidic precipitation, and consequently, they are not seriously affected by acid rain. In particular, the effects of acidic precipitation can be greatly reduced by limestone ($CaCO_3$), which reacts with acids to produce soluble calcium hydrogen carbonate:

Figure 4.47 *pH sampling of lake water.*

$$CaCO_3(s) + H_3O^+(aq) \longrightarrow Ca^{2+}(aq) + HCO_3^-(aq) + H_2O(l)$$

However, calcium carbonate can only prevent the pH from changing due to an added acid. If a base is added, there is nothing available to neutralize it, and water pH will increase. To keep the pH of a solution relatively steady—that is, neither increasing nor decreasing appreciably if small amounts of base or acid are added—a *buffer* is needed.

A **buffer** is a substance or combination of substances that can neutralize limited quantities of added acid (H_3O^+) or base (OH^-) without significantly altering its own pH value. Thus, a buffer must contain two components: (1) an acid to neutralize added base and (2) a base to neutralize added acid. Most buffer solutions contain relatively high concentrations of a weak acid and a salt of that acid—or a weak base and a salt of that base. For example, citric acid (a weak acid) and sodium citrate (a salt of citric acid) form a buffer combination that is often found in commercial food products.

Although calcium carbonate does not act as a true buffer—it cannot prevent added base from increasing the pH—the bicarbonate (HCO_3^-) anion produced serves as a buffer capable of neutralizing *either* added acid or base:

Neutralizing added acid: $H_3O^+(aq) + HCO_3^-(aq) \rightleftharpoons 2\,H_2O(l) + CO_2(g)$

Neutralizing added base: $OH^-(aq) + HCO_3^-(aq) \rightleftharpoons H_2O(l) + CO_3^{2-}(aq)$

In the next investigation, you will observe the unique behavior of buffers and gain a better understanding of how some bodies of water can withstand the effects of acid rain.

INVESTIGATING MATTER
C.13 ACIDS, BASES, AND BUFFERS

Preparing to Investigate

The acidity or alkalinity (basicity) of natural water is often monitored by measuring its pH in the field using test kits, pH or litmus paper, or a pH meter. Sometimes, however, it is important to know the exact concentration of acid or base in a sample. Measuring concentration requires a quantitative test, which can determine the amount of a substance that is present. For acids and bases, a common test used to analyze the amount of acid or base in a sample or solution is a titration. In an acid-base **titration**, a *standard* solution of a strong base (or acid) is slowly added to a known volume of acid (or base) solution of unknown concentration.

> A standard solution is a solution whose concentration is accurately known.

The solutions undergo a neutralization reaction (as in Section C.5, page 454) and an indicator or pH meter is used to determine when the neutralization is complete. The concentration of the unknown acid (or base) solution can then be calculated using the known volumes of standard and unknown solutions that reacted and the concentration of the standard solution.

In Part I of this investigation, you will perform a titration to determine the concentration of a sample of hydrochloric acid. In Part II, you will test a buffer solution, comparing the results of adding acid or base to water to the results of adding acid or base to a buffer solution. For Part II, your teacher will assign you to an "acid" group or a "base" group.

Before you begin, read *Gathering Evidence* to learn what you will need to do and note safety precautions. Plan for data collection and construct an appropriate data table. The table should provide space to record data (such as buret readings) and observations (including solution color) for each sample that you will test.

Gathering Evidence

Part I: Titration of Hydrochloric Acid of Unknown Concentration

1. Before you begin, put on your goggles, and wear them properly throughout the investigation.

2. Carefully add HCl to the "A" buret using a funnel.
   (*Caution: Hydrochloric acid is corrosive. If any hydrochloric acid accidentally spills on you, ask a classmate to notify your teacher immediately. Wash the affected area immediately with tap water and continue rinsing for several minutes.*)

3. Drain some HCl from the buret into a beaker to fill the buret tip. Ensure that the meniscus of the HCl is at the 0.00-mL mark (or below) with the tip filled.

4. Record the initial HCl buret reading on your data table to the nearest 0.01 mL.

5. Carefully add NaOH to the "B" buret using a funnel.
   (*Caution: Sodium hydroxide is corrosive. If any sodium hydroxide accidentally spills on you, ask a classmate to notify your teacher immediately. Wash the affected area immediately with tap water and continue rinsing for several minutes.*)

6. Drain some NaOH from the buret into a beaker to fill the buret tip. Ensure that the meniscus of the NaOH is at the 0.00-mL mark (or below) with the tip filled.

7. Record the initial NaOH buret reading on your data table to the nearest 0.01 mL.

8. Add ~10 mL of unknown HCl solution from the buret to a 125-mL Erlenmeyer flask. Record the final buret reading for the acid on your data table to the nearest 0.01 mL.

9. Add 2–3 drops of bromothymol blue indicator solution to the HCl solution in the Erlenmeyer flask.

10. Place the Erlenmeyer flask under the NaOH buret. Place a sheet of white paper under the flask.

11. Begin titrating: Carefully add ~1 mL of NaOH from the buret to the Erlenmeyer flask and swirl to mix (see Figure 4.48). Continue adding 1 mL portions until you begin to see a blue color that appears briefly and then quickly changes to green or yellow.

12. Once you have seen the blue color, begin adding the NaOH slowly and in smaller quantities. Remember to swirl the flask after each addition of NaOH. As you near the endpoint, the blue color should stay longer, but will still change to yellow when the flask is swirled. The longer the blue color stays, the smaller your next addition of NaOH should be. Eventually you will add NaOH one drop at a time until the blue color appears and stays.

Figure 4.48 *Swirling an Erlenmeyer flask under a buret.*

13. When the blue color appears and stays you have reached the endpoint. Record the buret reading to the nearest 0.01 mL.

14. Clean out the Erlenmeyer flask and repeat Steps 4–13 (the Erlenmeyer flask does not need to be dry, but must be clean). The solution in the Erlenmeyer flask can be poured down the drain. You will not need to add HCl or NaOH to the burets for the second trial.

15. Clean the Erlenmeyer flask and dispose of any remaining acid or base according to your teacher's instructions.

Part II: Investigation of Buffers

Acid Groups
Water Plus Acid

1. Refill the acid buret with HCl. If the meniscus is above the 0-mL mark, briefly open the valve, letting HCl run into a beaker until no air is left in the buret tip and the liquid surface is at or below the 0-mL mark. Record the volume.

2. Using a graduated cylinder, measure 40 mL of deionized (or distilled) water and transfer it into a clean 125-mL Erlenmeyer flask. Add 10 drops of universal indicator solution. Compare the color of the flask contents to the indicator scale. Record the pH.

3. Place the flask under the buret. Slowly open the buret valve and add five drops of acid to the flask. Swirl the flask.

4. Record the pH of this unbuffered acidic solution. Retain this flask as a color standard for comparison with the buffered solution.

Phosphate Buffer Plus Acid

5. Add 40 mL of buffer solution containing 0.1 M sodium hydrogen phosphate (Na_2HPO_4) and 0.1 M sodium dihydrogen phosphate (NaH_2PO_4) to a clean 125-mL Erlenmeyer flask.

6. Add 10 drops of universal indicator. Swirl the flask. Record the solution color and pH.

7. Record the initial buret volume.

8. Add five drops of 0.50 M HCl from the buret to the flask and swirl. Record the color and pH.

9. Continue to add HCl solution until the color and pH are identical to those in the color standard flask from the water plus acid procedure above.

10. Record the final buret volume.

11. Determine and record the volume of HCl solution added.

12. Dispose of the HCl solution in your buret according to your teacher's instructions.

13. Wash your hands thoroughly before leaving the laboratory.

Base Groups
Water Plus Base

1. Refill the base buret with NaOH. If the meniscus is above the 0-mL mark, briefly open the valve, letting NaOH run into a beaker until no air is left in the buret tip and the liquid surface is at or below the 0-mL mark. Record the volume.

2. Using a graduated cylinder, measure 40 mL of deionized (or distilled) water and transfer it into a clean 125-mL Erlenmeyer flask. Add 10 drops of universal indicator. Compare the color of the flask contents to the indicator scale. Record the pH.

3. Place the flask under the buret. Slowly open the buret valve and add five drops of base to the flask. Swirl the flask. See Figure 4.48.

4. Record the pH of this unbuffered basic solution. Retain this flask as a color standard for comparison with the buffered solution.

Phosphate Buffer Plus Base

5. Add 40 mL of buffer solution containing 0.1 M sodium hydrogen phosphate (Na_2HPO_4) and 0.1 M sodium dihydrogen phosphate (NaH_2PO_4) to a clean 125-mL Erlenmeyer flask.

6. Add 10 drops of universal indicator. Swirl the flask. Record the solution color and pH.

7. Record the initial buret volume.

8. Add five drops of base from the buret to the flask and swirl. Record the color and pH.

9. Continue to add NaOH solution until the color and pH are identical to those in the color standard flask from the water plus acid procedure above.

10. Record the final buret volume.

11. Determine and record the volume of NaOH solution added.

12. Dispose of the contents in the buret according to your teacher's instructions.

13. Wash your hands thoroughly before leaving the laboratory.

Analyzing Evidence

1. Write a balanced equation for the reaction in the Part I titration.

2. For each titration,

    a. calculate the amount (moles) of NaOH added from the buret to reach the endpoint.

    b. determine the amount (moles) of HCl in the Erlenmeyer flask.

    c. use the amount and volume of the HCl in the Erlenmeyer flask to determine the concentration of the unknown HCl solution.

3. Calculate the average concentration of the HCl solution.

Interpreting Evidence

1. How many drops of NaOH solution were needed to produce the same color (or pH value) in the phosphate buffer solution that was produced by adding 5 drops of NaOH to the distilled water? Explain any difference.

2. How many drops of HCl solution were needed to produce the same color (or pH value) in the phosphate buffer solution that was produced by adding 5 drops of HCl to the distilled water? Explain any difference.

Making Claims

3. What evidence suggests that the solution you used is actually a buffer?

4. The buffer used in this investigation included hydrogen phosphate ions ($HPO_4{}^{2-}$). Write an equation that shows how hydrogen phosphate ions could prevent the pH level of lake or river water from decreasing if limited quantities of acid (H_3O^+) were added to it.

Reflecting on the Investigation

5. Which one of these four pairs of substances would make the best buffer system? Why?

   - KCl and HCl
   - $NaC_2H_3O_2$ and $HC_2H_3O_2$
   - NaOH and H_2O
   - $NaNO_3$ and HNO_3

6. Why is a buret used in this investigation?

7. If a lake or river contained excess acid, could it be treated by adding base or buffer solution? Which solution would provide better results? Explain.

■ MAKING DECISIONS

C.14 ANALYZING WATER-QUALITY DATA

Snake River watershed data have been collected and monitored since the early 1900s. Although some measurements and methods have changed during that time, excellent data have been gathered, particularly in recent years (see Figure 4.49). Data are available for the following factors:

- water temperature
- dissolved oxygen
- rainfall
- water flow
- ammonia
- nitrate
- lead
- cadmium
- mercury
- arsenic
- pH
- total phosphates
- MTBE
- total suspended solids
- total organic carbon

Figure 4.49 *Researchers monitor water quality by measuring dissolved oxygen (left) and taking water samples both at the surface and several meters below the surface (right). Similar procedures provided the Snake River data.*

Your teacher will assign you to a group to study some of the data just listed. Each group will complete an analysis for its assigned data. After groups have finished their work, they will share their analyses and the whole class will discuss factors that might be related to the Riverwood fish kill.

The following guidelines will help you complete the analysis of data assigned to you by your teacher.

Data Analysis

Interpreting graphs of environmental data requires a slightly different approach from the one you used to interpret a solubility curve. Rather than seeking a predictable relationship, you will be looking for regularities or patterns among the values. Any major irregularity in the data may suggest a problem related to the contaminant or factor being evaluated. The following suggestions will help you prepare and interpret such graphs.

- Choose your scale so that the graph is large enough to fill most of the available space on the graph paper.

- Assign each regularly spaced division on the graph paper a convenient, constant value. The graph-paper line interval value should be easily divided by eye, such as 1, 2, 5, or 10, rather than awkward values such as 6, 7, 9, or 14.

- An axis scale does not have to start at zero, particularly if the plotted values cluster in a narrow range not near zero.

- Label each axis with the quantity and unit being plotted.

- Plot each point. Then draw a small symbol around each point, like this:

If you plot more than one set of data on the same graph, distinguish each by using a different color or small geometric shape to enclose the points, such as: ▣ or ◈ .

- Give your graph a title that will readily convey its meaning and purpose to readers.
- If you use a graphing calculator, computer software, or other technology to prepare your graphs, ensure that you follow the guidelines just given. Different devices and software have different ways to process data. Choose the appropriate type of graph (e.g., scatterplot or bar graph) for your data.

Next, follow the steps listed below when preparing and plotting your graphs, and answer the questions that follow.

1. Prepare a graph for each of your group's Snake River data sets. Label the x-axis (displaying the independent variable) with the consecutive months indicated in the data. Label the y-axis (dependent variable) with the water factor measured, accompanied by its units.

2. Plot each data point and connect the consecutive points with straight lines.

3. Is any pattern apparent in your group's plotted data?

4. Can you offer possible explanations for any pattern or irregularities that you detect?

5. What arguments might someone make to challenge your explanation?

6. How might fish be harmed by the pattern or patterns that your group has identified?

7. Do you think the data analyzed by your group might help account for the Snake River fish kill? Why? How?

8. Prepare to share your group's data-analysis findings in a class discussion. During the class discussion, take notes on key findings reported by each data-analysis group. Also note and record significant points raised in the data-analysis discussions.

Your class will reassemble several times during your study of Section D to discuss and consider implications of the water-analysis data you have just processed. In particular, guided by the patterns and irregularities found in your analysis of Snake River data, your class will seek an explanation that accounts for the observed data and for the resulting Riverwood fish kill. Good luck!

SECTION C SUMMARY

Reviewing the Concepts

> Mixing ionic solutions sometimes results in formation of a precipitate; evidence-based trends allow prediction of these reactions. Chemical tests for the presence of an ion in a solution can be developed based upon formation of a precipitate.

1. What is a precipitate?

2. List three ions that form compounds that are always soluble in water.

3. How is a "solubility rule" different from a "classroom rule"?

4. Write net ionic equations illustrating the result of each of the following combinations of ionic solutions.

   a. Ammonium carbonate and copper(II) chloride

   b. Sodium bromide and silver acetate

   c. Nickel(II) nitrate and potassium phosphate

5. Suppose that you wished to test for the presence of following ions in solution. Identify an ionic substance that could be added to test for each ion.

   a. Pb^{2+}

   b. CO_3^{2-}

   c. I^-

   d. Mg^{2+}

> Acids produce hydrogen (or hydronium) ions in water, and bases produce hydroxide ions. Strong acids and bases ionize completely; weak acids and bases ionize only partially.

6. Which of these compounds are acids? Which are bases? Which are neutral?

   a. LiOH

   b. $HC_2H_3O_2$

   c. $C_{12}H_{22}O_{11}$ (table sugar)

   d. H_2SO_3

7. Write an equation showing production of hydronium ion as these acids are mixed with water.

   a. Sulfuric acid (H_2SO_4), a strong acid used in automotive batteries.

   b. Hydrofluoric acid (HF), a weak acid sometimes used to etch glass.

   c. Acetic acid ($HC_2H_3O_2$), a weak acid that is the key ingredient in vinegar.

8. Complete these equations, depicting reactions that yield hydroxide ions.

   a. $??? \longrightarrow Ca^{2+}(aq) + 2\ OH^-\ (aq)$

   b. $NH_3(g) + H_2O(l) \rightleftharpoons ??? + ???$

9. What is the difference between a strong acid and a concentrated acid?

10. Give a specific example of an acidic solution that could be described as both strong and dilute.

11. What is the difference between a hydrogen ion and a hydronium ion?

pH is an expression of the molar concentration of hydronium ions in an aqueous solution. Acidic solutions contain a higher concentration of hydronium ions than hydroxide ions; basic solutions contain a higher concentration of hydroxide ions than hydronium ions.

12. Compare the concentration of hydroxide ions and hydronium ions in pure water.

13. What would happen to the concentration of hydronium ions in a solution of a strong acid if a solution containing an equal number of hydroxide ions were added to the solution?

14. A certain room-temperature solution contains 1×10^{-2} mol H_3O^+ and 1×10^{-12} mol OH^-. Is this solution acidic or basic? Explain.

15. Why do chemists often express the acidity of a solution as a pH value instead of a hydrogen ion concentration?

16. Oven cleaner has a pH of 13.0 and pure water a pH of 7.0. How many times more basic is oven cleaner than pure water?

17. A sample of acid rain has a pH of 3.0.

   a. What is its hydronium ion concentration?

   b. What is its hydroxide ion concentration?

18. Consider these three solutions:

   • lemon juice, containing 0.001 M H_3O^+.

   • stomach fluid, containing 0.1 M H_3O^+.

   • drain cleaner, NaOH, containing 0.1 M OH^-.

   a. What is the pH of each solution? See Figure 4.42, page 458.

   b. Which is most acidic?

   c. Which is most basic?

A buffered solution is capable of neutralizing limited amounts of either added acid or base, thus resisting changes in the solution's pH.

19. What are the general components of a buffer?

20. $H_2PO_4^-$ acts a buffer in cellular systems. Write an equation showing the reaction of $H_2PO_4^-$ ion with

   a. hydronium ion.   b. hydroxide ion.

21. Is it better to describe a buffered solution as resisting or preventing all changes in pH? Explain your answer.

22. How would a lake surrounded by limestone (a rock composed largely of calcium carbonate) respond to acid rain, compared to a lake that was surrounded by rock that does not react with acid rain?

23. Someone proposes that the acid-rain problem could be fixed by dissolving large quantities of basic substances in affected lakes. Do you agree? Explain.

How do we describe chemical behavior in aqueous solutions?

In this section, you have observed the possible results of combining ionic solution, used solubility rules to test for ions in water samples, studied the structure and behavior of acids and bases, and examined the role of buffers in solution. Think about what you have learned, then answer the question in your own words in organized paragraphs. Your answer should demonstrate your understanding of the key ideas in this section.

Be sure to consider the following in your response: precipitation reactions, net ionic equation, solubility rules, pH, weak acid/base, neutralization reactions, and buffers.

Connecting the Concepts

24. A methane molecule (CH_4) contains several hydrogen atoms. Describe a laboratory investigation that would allow you to decide whether or not methane behaves as an acid.

25. Explain how it might be possible for a strong-acid solution and weak-acid solution to have exactly identical pH values.

26. Why are shampoos and medicines often buffered?

27. In terms of their chemical reactions, compare antacids to buffers.

28. Most fish can live in water with pH values between 5.0 and 9.0.

a. What range of hydronium ion concentrations corresponds to this pH range?

b. By what factor does the hydronium ion concentration differ between these two pH values?

c. Explain, in terms of buffers, how it is possible for different species of fish to live in such different conditions.

d. Explain how acid rain might affect the distribution of fish species in an affected lake.

Extending the Concepts

29. What characteristics make the hydrogen ion uniquely reactive compared, for example, to a sodium or potassium cation with the same 1+ electrical charge?

30. Construct models and draw electron-dot diagrams of water and the hydronium ion. Compare their shapes, bonding, and interactions.

31. Examine the list of ingredients for several antacid products. Which ingredients are responsible for neutralizing acids?

32. Is it possible to have a solution with a pH value of zero? Does that mean that the solution has no measurable acidity? Explain your answer.

33. Sometimes the zone separating warring or hostile countries is called a "buffer zone." How does that particular use of the word *buffer* compare to a chemist's use of that term?

CHEMISTRY *AT WORK*
Q&A

You might think water quality is a matter of taste. But for millions of people around the globe, clean water is a matter of life and death. One of the places most affected by long-term water-quality issues is a region that spreads through the South Asian countries of India, Bangladesh, and Nepal. The only sources of drinking water for many of the residents are wells contaminated with arsenic. Over time, drinking this water produces a disease called arsenicosis, which causes skin discoloration and organ failure and, eventually, leads to cancer. Read on to see how one chemist's clear idea is making clean water flow for hundreds of thousands of people in South Asia.

Abul Hussam, Professor in the Department of Chemistry and Biochemistry at George Mason University in Fairfax, Virginia

Q. What is analytical chemistry, and why is it important for clean water?

A. Analytical chemistry is the study of components that make up natural and human-made materials, as well as how to determine those components. Chemical analysis is especially important for clean water, because the water we drink is usually composed of much more than just H_2O! Analytical chemists can help determine what other compounds might be dissolved in water—and if they're toxic, how to remove them.

Q. How did you get interested in chemistry, and why did you choose to focus your work on water quality?

A. My dad was a physician, and he encouraged my four brothers and me to get excited about science. When I was growing up in Bangladesh, he would bring home books with simple science experiments, and we would all give them a try. I became more and more interested in chemistry and came to the United States to get my doctorate in analytical chemistry. After I started working at George Mason University, I read an article in the *New York Times* about how people in Bangladesh and India were being poisoned by arsenic in their drinking water. My brother, a physician who still lives in our hometown, told me that he had seen people suffering from arsenicosis. I wanted to do something to help solve this problem, so I led a team of scientists to invent a filter we call the SONO Filter to remove arsenic from water.

There is no lack of water in Bangladesh, yet many people rely on tainted wells for drinking water.

Q. Why is there arsenic in drinking water in Bangladesh and India, and how serious is this problem?

A. People in this region get most of their drinking water from shallow wells. Arsenic is a natural part of the rocks and soil in these areas, and arsenic compounds dissolve in the groundwater. There is plenty of water in Bangladesh and India, but if water quality is bad, water quantity doesn't matter! Nearly 100 million people in these countries drink arsenic-contaminated water every day.

Q. How did you use chemistry to develop a solution to this problem?

A. In a way, the problem comes from the core of chemistry itself, because the chemical properties of arsenic compounds allow them to dissolve in water. Chemistry is also the way of resolving the problem, since we know that these same arsenic compounds bind to iron hydroxide. My colleagues and I invented a filter in which water passes through two buckets. The first bucket has a layer of iron and iron hydroxide sandwiched between two layers of sand, and the second has sand and activated charcoal. As contaminated water passes through the first layer, the arsenic binds to the iron hydroxide and remains there. The same layer also binds soluble iron, manganese, many toxic trace metals and some calcium. Sand and charcoal in both buckets remove microbes, particulate matter, and some dissolved organic compounds, which helps improve water's taste.

Use of a tube well, the source of the problems caused by arsenic.

Q. How successful is the SONO Filter, and what is your next step?

A. Since 2002, 170,000 SONO Filters have been manufactured and distributed to people in Bangladesh, Nepal, and India. These filters cost about $35, and by our calculations, they last about 11 years for a family of four. That's a good investment to protect a family's health. Right now, we're working on an arsenic filter that might work in a small water pitcher, like the ones many people use in the United States. We need to understand the science better, since a smaller filter will have to be comprised of materials that are even more efficient.

Q. How do I become a water quality chemist?

A. The most important thing is to get an understanding of basic chemistry. Water may look like a simple molecule, but when other compounds dissolve in water, their behavior gets very complex. Consider majoring in chemistry in college and taking analytical chemistry courses. Look for internships that put your laboratory knowledge to use solving real-life problems. Many experiments may work in the lab, but they don't work when you apply them to real problems. The lab and the field are two very different things, but lab experience helps you to solve real problems.

Cover

Empty space

Tap fitted with flow control junction

Coarse sand

CIM*

Coarse sand

Brick chips

Cover

Plastic pipe

Coarse sand

Wood charcoal

Fine sand

Brick chips

Tap fitted with flow control junction

Cross-section of the SONO filter.

SONO FILTER

Children at a primary school in Bangladesh collect filtered water.

RIVERWOOD NEWS

EDITORIAL

Attendance Urged at Town Council Meeting

A special town council meeting Wednesday will address four questions:

(1) What was responsible for the fish kill?

(2) Are Riverwood's current water treatment procedures adequate to ensure the safety of the town's water supply? If not, what immediate steps must be taken?

(3) How can similar fish kills be prevented in the future?

(4) Who, if anyone, should be responsible for the costs associated with the crisis?

Answers to these questions have practical and financial consequences for all Riverwood residents.

Those participating in next week's public meeting include representatives of industry and agriculture, U.S. Environmental Protection Agency (EPA) scientists who completed the river-water analyses, representatives from the county sanitation commission, and consulting engineers who have investigated some potential causes of the fish kill. Chamber of Commerce members representing Riverwood store owners and officials of the Riverwood Taxpayer Association also will make presentations at the meeting.

We urge you to attend and contribute to this meeting. Many questions remain. Was the fish kill an act of nature or was it due to human error? If a dissolved or suspended substance was responsible for the fish kill, what was the source of contamination? Is Riverwood's water supply safe? Should the town's business community be compensated, at least in part, for financial losses from the fish kill? If so, how should they be compensated and by whom? Who should pay for the drinking water brought to Riverwood?

> " *Many questions remain . . . Is Riverwood's water supply safe?* "

Was alternative drinking water necessary or was the water supply safe for human consumption? Can this entire situation be prevented in the future? If so, how, and at what expense? Who will pay for it?

The *Riverwood News* will set aside part of its Letters to the Editor column in the coming days for your comments on these questions and other matters related to the recent water crisis.

SECTION D WATER PURIFICATION AND TREATMENT

How is chemistry applied to produce safe drinking water?

GOALS

- Purify a sample of contaminated water.
- Describe the movement of water in Earth's hydrologic cycle including how water is purified by natural processes.
- Analyze steps of municipal water purification and identify contaminants removed by each process.
- Compare natural, municipal, and home water purification.
- Assess the risks and benefits of water disinfection methods.
- Evaluate the causes of and responses to the Riverwood fish kill.

✓ concept check 9

1. a. What is a precipitate?
   b. How are precipitates useful in water testing?
2. a. Why is the pH of water samples measured?
   b. What other substances would be affected by an elevated or depressed pH?
3. Why do many cities and towns add chlorine in some form to the drinking-water supply?

▪ MAKING DECISIONS

D.1 CONSIDERING CONTAMINANTS IN WATER SUPPLIES

In Making Decisions C.14, you graphed and analyzed data collected from the Snake River over the past weeks and years. A wide variety of phenomena and materials has been analyzed from the sample collected. Why is it important to track this information? What can these data tell us about the health and safety of the river—and more importantly, Riverwood's water supply? The EPA has developed drinking-water standards, summarized in Table 4.4. Using these standards along with information gathered from your research, answer the following questions. You will be expected to share your answers with the class in written form as directed by your teacher.

Table 4.4

Drinking Water Standards for Selected Contaminants

Substance	Maximum Contaminant Level (MCL)
pH	6.5–8.5
Total carbon	No total level set
Total lead	0.015 mg/L
Total cadmium	0.005 mg/L
Total mercury	0.002 mg/L
Total arsenic	0.05 mg/L
Nitrate (measured as nitrogen)	10 mg/L
Phosphate (measured as ortho-phosphate)	No total level set
Pesticides	See specific standards
PCBs	0.0005 mg/L

1. Why are we concerned about making measurements of the phenomenon or material that you graphed?

2. a. How is this phenomenon or material harmful and at what level?

   b. Is there a level at which it is beneficial?

3. a. What are some likely scenarios for how this material got into the river or how this phenomenon occurred?

   b. Would these be common or uncommon events? Explain.

4. Could this substance be removed from the waterway or could this phenomenon be reversed? If yes, how? If no, why not?

INVESTIGATING MATTER
D.2 FOUL WATER

Asking Questions

Your objective is to clean up a sample of foul water, producing as much "clean water" as possible, to a point where it could be used for hand washing. (**Caution:** *Do not test any water samples by drinking or tasting them.*)

Before you begin, think about what could make a sample of water foul. What contaminants might be present? How could you remove these contaminants? What methods are available to you for the purification?

To design your procedure, you will be expected to apply knowledge and skills you have developed during the entire course, not just in this unit. You may choose to use any or all of several different water-purification procedures described below: oil–water separation, sand filtration, and charcoal adsorption and filtration.

Preparing to Investigate

In this investigation, you will develop your own procedure for purifying foul water. To prepare, read through the descriptions of each technique and examine the available equipment. Recall that at the beginning of this unit (in Investigating Matter A.3), you performed investigations examining mixtures. Reviewing these investigations might be helpful as you prepare to develop a foul-water purification plan.

With your lab group, discuss the possible separation techniques and agree on which separation techniques and procedures you will use and in which order they will be used. Think about what data and observations you will record so that you can share the results of your purification. You will receive approximately 100 mL of foul water to purify. Consider and decide whether you will use the entire sample at once or instead use small portions of the sample to test techniques before committing the entire sample to the procedure.

Water Purification Techniques

Oil–Water Separation

As you probably know, if oil and water are mixed and left undisturbed, the oil and water do not noticeably dissolve in each other. Instead, two layers form. See Figure 4.50. In Section A.6, you learned that this type of mixture is a heterogeneous mixture.

So if your sample has two layers, you must decide which is the water layer and which is not aqueous. Then the layers must be separated and the

Figure 4.50 *Which layer in this carafe is mostly likely to contain oil?*

aqueous layer retained for further purification. Keep the following in mind as you decide how to determine the identity of each layer and how to separate them:

- Mixtures can be separated using physical methods.
- Which layer do you think will float on top of the other? (If you are unsure, how could you conduct a test to find out?)
- What role will the relative densities of the liquids play in deciding if the upper layer is aqueous or oil?

Sand Filtration

In **filtration**, solid particles are separated from a liquid by passing the mixture through a material that retains the solid particles and allows the liquid to pass through. The liquid collected after it has been filtered is called the **filtrate**. A sand filter traps and removes solid impurities—at least those particles too large to fit between sand grains—from a liquid. In Section A.6, you learned about mixtures that contain dissolved particles. They could be either heterogeneous or homogeneous. Mixtures with very large particle sizes were colloidal suspensions. Which of these types of mixtures could be separated by sand filtration?

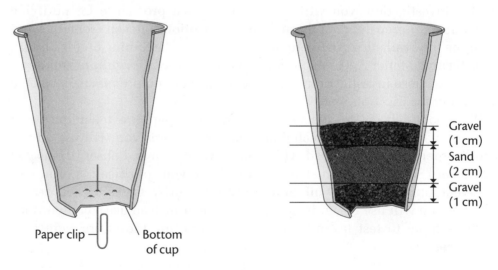

Figure 4.51 *Preparing a disposable cup (left) and layering gravel and sand (right) for sand filtration.*

As you prepare your filtration apparatus (see Figure 4.51) and design your procedure, think about the following question: Do all the particles in the "sand" have to be the same size? To be safe, you might want to always use the smallest grains of sand in this apparatus so that there is not much room between the grains and a large range of particles would be trapped. What would be possible drawbacks to this procedure?

Charcoal Adsorption and Filtration

Charcoal **adsorbs**, which means attracts and holds on its surface, many substances that could give water a bad taste, cloudy appearance, or an odor. In

Section A.4, you learned about cohesive forces between molecules in a substance. **Adhesive forces** are similar, except that they attract molecules of different substances to one another. Many substances that cause water to have a bad taste, a cloudy appearance, or an odor have stronger adhesive forces with charcoal than with water. Charcoal adsorption takes advantage of the fact that these contaminants stick to the charcoal and can be removed.

As you consider using this technique, keep in mind the very small size of the charcoal particles and how this affects the type of filtration setup that would be appropriate.

The pump system in a fish aquarium often includes a charcoal filter for adsorption of contaminants.

Gathering Evidence

Before you begin, put on your goggles, and wear them properly throughout the investigation.

As you prepare to purify your sample, assess what may be present in the sample that makes it "foul." Obtain approximately 100 mL of foul water from your teacher. Record the exact volume of the beginning sample. With your lab partner, discuss the evidence that this sample must be purified. Once you have made your observations, check in with your teacher before beginning the purification process.

Once your group has agreed on a procedure and had it approved by your instructor, follow your procedure to purify the foul-water sample. If necessary, revise your procedure as you proceed to improve your separation. Be sure to check with your teacher before attempting any revised procedure.

When you are finished, test the conductivity of your purified water.

Clean up your lab station, dispose of all materials as indicated by your teacher, and wash your hands thoroughly before leaving the laboratory.

Analyzing Evidence

1. What percent of your original foul water sample did you recover as purified water? This value is called the **percent recovery**.

2. What volume of liquid (in milliliters) did you lose during the entire purification process?

3. What percent of your original foul-water sample was lost during purification?

Making Claims

1. Is your purified water sample pure water? Provide evidence to support your answer.

2. How could you compare the quality of your final water sample with that obtained by other laboratory groups? That is, how should someone judge the success of each laboratory group? Remember that each group may have completed different purification steps or a different number of steps. Defend your answer using evidence from this investigation.

Reflecting on the Investigation

3. How could you improve the water-purification procedures you followed so that you could recover a higher percent of purified water?

4. a. Estimate the total time you spent purifying your water sample.

   b. In your opinion, did that time investment result in a large enough sample of sufficiently purified water? Explain.

5. It is sometimes said that "time is money."

   a. If you spent twice as much time purifying your sample, would that extra time investment pay off in higher-quality water?

   b. If you spent about 10 times as much time, would that extra investment pay off? Explain your reasoning.

6. Municipal water-treatment plants do not use distillation to purify water.

   a. Why?

   b. What advantages would distillation offer?

7. Are there types of contaminants that the procedures you used would not be capable of removing? Give examples and describe possible ways of removing them.

D.3 NATURAL WATER PURIFICATION

Riverwood's current concerns result from a crisis, but having access to water has always been an issue for human survival. Ensuring the quality of water for drinking has been a known concern for 6000 years.

Until the late 1800s, most Americans obtained water from local ponds, wells, and rainwater holding tanks. Wastewater and even human wastes were discarded into holes, dry wells, or leaching cesspools (pits lined with broken stone). Some wastewater was simply dumped on the ground.

By 1880, about one-quarter of U.S. urban households had flush toilets, and municipalities were constructing sewer systems. However, as recently as 1909, sewer wastes were often released without treatment into lakes and streams, from which water supplies were drawn at other locations. Many community leaders believed that natural waters would purify themselves indefinitely.

Waterborne diseases increased as the concentration of intestinal bacteria in drinking water rose. As a result, water filtering and chlorination soon began. However, municipal sewage—the combined waterborne wastes of a community—remained generally untreated. Today, sewage treatment is part of every U.S. municipality's water-processing procedures.

As you learned in Section A, there is a fixed quantity of water available on Earth. This water circulates through Earth's hydrologic cycle. Because materials are constantly added to water through natural and human processes, water must also be continually purified in order to be safe to use. The hydrologic cycle (see Figure 4.52) includes water-purification steps that address many potential threats to water quality. Thermal energy from the Sun causes water to

Environmentalists sometimes remark that "we are all downstream from someone else."

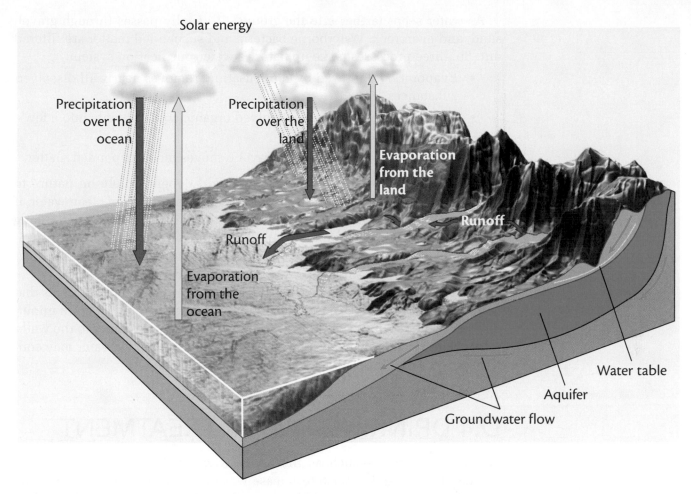

Solar energy

Precipitation over the ocean

Precipitation over the land

Evaporation from the land

Runoff

Runoff

Evaporation from the ocean

Water table

Aquifer

Groundwater flow

Figure 4.52 *Hydrologic cycle: Earth's water-purification* *system.* *The Sun provides energy for water to evaporate (yellow arrows). Evaporation leaves behind minerals and other dissolved substances. Water vapor condenses and falls as precipitation (red arrows), which runs off the land (blue arrows) to join surface water sources (lakes, streams, and rivers) or groundwater sources beneath Earth's surface. Both surface water and groundwater are sources of municipal and agricultural water. Any water—when returned to the surface—can evaporate and continue flowing through the hydrologic cycle. Throughout the cycle, evaporation, bacterial action, and filtration purify water.*

evaporate from oceans and other water sources. Dissolved heavy metals, minerals, or molecular substances do not evaporate and, thus, are left behind.

This natural process accomplishes many of the same results as distillation. Water vapor rises, condenses into tiny droplets in clouds, and—depending on the temperature—eventually falls as rain or snow. Raindrops and snowflakes are nature's purest form of water, containing only dissolved atmospheric gases. However, human activities release a number of gases into the air, making today's rain less pure than it used to be.

When raindrops strike soil, the rainwater collects natural and human-produced impurities. Organic substances deposited by living creatures become suspended or dissolved in the rainwater. A few centimeters below the soil surface, bacteria feed on these substances, converting them into carbon dioxide, water, and other simple compounds. Such bacteria help repurify the water.

As water seeps farther into the ground, it usually passes through gravel, sand, and even rock. Waterborne bacteria and suspended matter are filtered out. So, three processes make up nature's water-purification system:

- Evaporation, followed by condensation, removes nearly all dissolved substances.
- Bacterial action converts dissolved organic contaminants into a few simple compounds.
- Filtration through sand and gravel removes most suspended matter.

Given appropriate conditions, people could depend solely on nature to purify their water. Pure rainwater is the best natural supply of clean water. If water seeping through the ground encountered enough bacteria for long enough, all organic contaminants could be removed. Flowing through sufficient sand and gravel would remove suspended matter from the water. However, nature's system only works well if it is not overloaded.

If slightly acidic groundwater (pH less than 7.0) passes through rocks that contain slightly soluble compounds, such as magnesium and calcium minerals, chemical reactions with these minerals may add substances to the water rather than remove substances in the water. In this case, the water may contain an increased concentration of dissolved minerals.

D.4 DRINKING-WATER TREATMENT

Today, many rivers—such as the Snake River in Riverwood—are both a source of water and a place to release wastewater (sewage). Therefore, the water is cleaned twice, both before and after it is used. Pre-use purification, often called **drinking-water treatment,** takes place at a filtration and treatment plant. Figure 4.53 diagrams typical water-treatment steps, which are also described as follows:

- *Screening.* Intake water flows through metal screens that prevent fish, sticks, beverage containers, and other large objects from entering the water-treatment plant.
- *Pre-chlorination.* Chlorine, a powerful disinfecting agent, is added to kill disease-causing organisms.
- *Flocculation.* Crystals of alum—aluminum sulfate, $Al_2(SO_4)_3$— and slaked lime—calcium hydroxide, $Ca(OH)_2$—are added to remove suspended particles, such as colloidal clay, from the water. Suspended particles can give water an unpleasant, murky appearance. The alum and slaked lime react to form aluminum hydroxide, $Al(OH)_3$, a sticky, jellylike material that traps the suspended particles.
- *Settling.* The aluminum hydroxide (holding trapped particles from the water) and other solids remaining in the water are allowed to settle to the tank bottom.

- *Sand Filtration.* Any remaining suspended materials that do not settle out are removed by filtering the water through sand.

- *Post-chlorination.* The chlorine concentration in the water is adjusted to ensure that a low, but sufficient, concentration of chlorine remains in the water, thereby protecting the water from bacterial infestation.

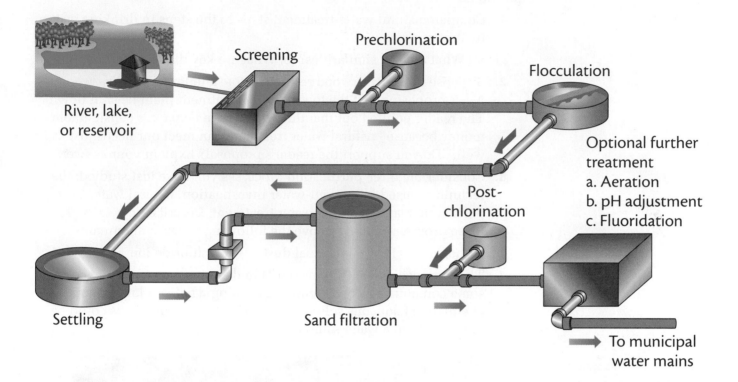

Figure 4.53 *Drinking-water treatment: A human-made water-purification system.* *Surface water is commonly cleaned at a water-treatment plant before being distributed to homes and businesses. Various steps in the cleaning process remove suspended materials, kill disease-causing organisms, and may remove odors or adjust pH levels.*

- *Optional Further Treatment.* One or more additional steps may take place, depending on community procedures.

  ○ *Aeration.* Some plants spray water into the air to remove odors and improve its taste.

  ○ *pH Adjustment.* Water may sometimes be acidic enough to slowly dissolve metallic water pipes. This process not only shortens pipe life, but may also cause copper (Cu^{2+}), as well as cadmium (Cd^{2+}) and other undesirable ions, to enter the home water supply. Lime—calcium oxide (CaO), a basic substance—may be added to neutralize such acidic water, thus raising its pH to a proper level.

  ○ *Fluoridation.* As much as about 1 ppm of fluoride ion (F^-) may be added to the treated water. Even at this low concentration, fluoride ions can reduce tooth decay.

DEVELOPING SKILLS

D.5 WATER PURIFICATION

Refer to Sections D.3 and D.4 (pages 482 and 484) to answer the following questions.

1. Compare natural water-treatment steps to the steps in drinking water treatment.

    a. What are key similarities? b. What are key differences?

2. Suppose that a Riverwood resident wrote a letter to the *Riverwood News* proposing that the town's water-treatment plant be shut down. The reader pointed out that this would save taxpayers considerable money because "natural water treatment can meet our needs just as well." Do you support the reader's proposal? Explain your answer.

3. Consider the three purification processes you have just studied: the techniques used in the foul-water investigation, natural water purification, and drinking-water treatment. Describe how each of these processes could remove the following water contaminants.

    a. hexane, C_6H_{14} b. coal dust   c. chloride ion

4. The presence of lead(II) ions (Pb^{2+}) in drinking water is of concern in some communities. See Figure 4.54. Using what you know about the chemistry of aqueous solutions, describe how lead(II) ions could be removed from a water sample.

Figure 4.54 *This system is used to monitor corrosion rates and lead release in the Washington Aqueduct. The Aqueduct provides drinking water to Washington, DC, where many lead service lines are still used to deliver drinking water.*

concept check 10

1. What substances would be easiest to remove from a contaminated sample of water? Explain.
2. What does it mean to disinfect drinking water?
3. What types of home water treatment are you aware of or have you used?

D.6 DISINFECTION OF DRINKING WATER

The single most common cause of human illness in the world is unhealthful water supplies. Very young children, particularly in developing countries, are most likely to be affected by organisms responsible for waterborne disease. Without a doubt, adding disinfection to drinking water treatment has helped save countless lives by controlling these organisms to provide access to safe water.

Chlorination

Disinfection of drinking water began in the United States only about 100 years ago with the addition of chlorine to water supplies in Jersey City, New Jersey. (See Figure 4.55.) Chlorination, the addition of chlorine to a water supply to kill harmful organisms, is still the most common method of disinfection. In water, chlorine kills disease-producing microorganisms.

In municipal water-treatment systems, chlorination can take place in several different ways:

- Chlorine gas, Cl_2, is bubbled into the water. Chlorine gas is not very soluble in water. It does react with water, however, to produce a water-soluble, chlorine-containing compound.

- An aqueous solution of sodium hypochlorite, NaOCl (the active ingredient in household bleach), is added to the water.

- Calcium hypochlorite, $Ca(OCl)_2$, is dissolved in the water. Available as both a powder and small pellets, calcium hypochlorite is often used in swimming pools. It is also a component of some household products sold as bleaching powder.

Figure 4.55 *Water treatment plant.*

Regardless of how chlorination takes place, chemists believe that chlorine's most active form in water is hypochlorous acid (HOCl). This substance forms whenever chlorine, sodium hypochlorite, or calcium hypochlorite dissolves in water.

Unfortunately, a potential problem is associated with adding chlorine to drinking water. Under some conditions, chlorine in water can react with organic compounds produced by decomposing animal and plant matter to form substances that, if in sufficiently high concentrations, can be harmful to human health. One group of such substances is known as the **trihalomethanes** (THMs). Many THMs, including chloroform ($CHCl_3$), are **carcinogens**—substances that are known to cause cancer.

Because of concern about the possible health risks associated with THMs, the EPA has placed a current limit of 80 parts per billion (ppb) on total THM concentration in water supply systems. Operators of water-treatment plants have several options for eliminating possible THM health risks. Possible risks associated with THMs must be balanced, of course, against the benefits of chlorinated water.

Alternative Disinfection Methods

Several other methods are commonly used for drinking-water disinfection. Two of these methods also use chlorine-based compounds: chloramines and chlorine dioxide. Chloramines are usually produced onsite (at the water treatment facility) by combining ammonia and chlorine in water. Monochloramine (NH_2Cl), the chloramine most commonly used for disinfection, is a weaker disinfectant than chlorine. It is used mainly for post-chlorination, because it results in formation of fewer THMs and is more stable (and thus longer lasting) in distribution pipes than elemental chlorine is.

Chlorine dioxide (ClO_2) is also generated onsite, by mixing sodium chlorite ($NaClO_2$) and elemental chlorine. This also reduces THM formation. It is a strong disinfectant and is not significantly affected by pH. However, it is costly, requires expertise to generate and monitor, and produces chlorite and chlorate byproducts.

Another commonly used chemical disinfectant is ozone (O_3). A very strong disinfectant, ozone must be generated onsite by passing dry air over high-voltage electrodes. Ozone is widely used in much of the world, but less commonly in the United States. Although no THMs are formed, some organic byproducts are produced. More importantly, it does not persist in water, so another disinfectant must be added to protect water in pipes. Also, ozone production and monitoring systems require skilled operators.

Drinking water can also be disinfected non-chemically through the use of ultraviolet radiation. As you know, UV radiation can damage the cells of organisms. Because, like ozone, it does not persist, another disinfectant is still required for distribution of water. UV radiation is effective against most known organisms and does not result in byproducts (see Figure 4.56). However, UV radiation is absorbed by solids, organic compounds, and colored substances in water, so it cannot be used in water that contains high levels of these materials.

Figure 4.56 *This UV disinfection system destroys waterborne microorganisms.*

DEVELOPING SKILLS

D.7 BENEFITS AND RISKS OF DISINFECTION

As you just learned, the production of THMs is a key disadvantage of the use of chlorination for drinking water disinfection. Other disinfection methods also have both benefits and risks. In this activity, you will evaluate the pros and cons of disinfection methods in several scenarios.

> *Sample Problem: One method for removal of THMs from drinking water is to pass the treated water through an activated charcoal filter. Activated charcoal can remove most organic compounds from water, including THMs. List one advantage and one disadvantage of this approach.*

Advantage: No additional substances are added to the treated water in the removal of THMs.

Disadvantage: Charcoal filters are expensive to install and operate. (Other possible answers include: disposal of filters is difficult because contaminants are hard to remove or filters must be replaced often.)

For each of the following scenarios, list two benefits of and two risks posed by the chosen approach to disinfection:

1. In a large municipal system, both pre-chlorination and post-chlorination are used to disinfect drinking water.

2. In a small, rural water system, ozone is chosen as the sole disinfectant for drinking water.

3. In a moderate-sized city water system, UV radiation is used for initial disinfection, followed by chloramine treatment before distribution.

List one advantage and one disadvantage of each of the following strategies for reducing THMs in drinking water:

4. Eliminate pre-chlorination. Chlorine is added only once, after filtering the water and removing much of the organic material.

5. Replace chlorine with chlorine dioxide.

CHEM**QUANDARY**

BOTTLED WATER VERSUS TAP WATER

Water that has been chlorinated sometimes has a bad taste. When people do not like the taste of tap water, think that the available water is unsafe to drink, do not have access to other sources of fresh water, or just want convenience, they may buy bottled water. This bottled water may come from a natural source, such as a spring, or it may be processed at the bottling plant.

Is this water any better for you than tap water that has been processed by a water treatment plant or that comes directly from ground or surface water? Could it possibly be harmful? What determines water quality? How can the risks and benefits of drinking water from various sources be assessed?

In your view, what two or three factors should you consider in deciding whether to drink tap water or bottled water? What factual information would be needed to establish the advantages and disadvantages of drinking bottled water rather than tap water?

D.8 HOME WATER TREATMENT

Although most water treatment in the United States is done through water supply networks, both in Riverwood and in other cities, certain water-quality issues are commonly addressed with in-home water treatment. Home water treatment is often used to improve the taste and cleaning properties of water that is already safe to drink. Households that obtain their water from wells also use home water treatment to supplement natural water purification and ensure a safe water supply.

Water Softening

Hard water causes some common household problems. It interferes with the cleaning action of soap. When soap mixes with soft water, it disperses to form a cloudy solution topped with a sudsy layer. In hard water, however, soap reacts with hard-water ions to form insoluble precipitates. These compounds appear as solid flakes or a sticky scum. The precipitated soap is no longer available for cleaning. Worse, soap curd can deposit on clothes, skin, and hair. The structural formula for this substance, the product of the reaction of soap with calcium ions, is shown in Figure 4.57.

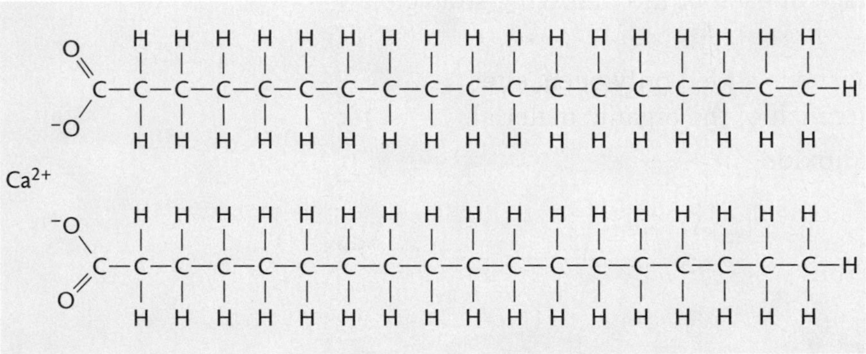

If hydrogen carbonate (bicarbonate, HCO_3^-) ions are present in hard water, boiling the water causes solid calcium carbonate ($CaCO_3$) to form. The reaction removes undesirable calcium ions and softens the water. However, the solid calcium carbonate can produce rocklike scale inside tea kettles, household water heaters, and even power-plant boilers. This scale (the same compound found in marble and limestone) acts as thermal insulation, partly blocking heat flow to the water. More time and energy are required to heat the water. Such deposits can also form in home water pipes. In older homes with this problem, water flow can be greatly reduced.

Fortunately, it is possible to soften hard water by removing some dissolved calcium, magnesium, or iron(III) ions. Adding sodium carbonate to hard water was an early method of softening water. Sodium carbonate (Na_2CO_3), known as *washing soda*, was commonly added to laundry water along with the clothes and soap. Hard-water ions, precipitated as calcium carbonate ($CaCO_3$) and magnesium carbonate ($MgCO_3$), were washed away in the rinse water. Water-softening substances in common use today include borax, trisodium phosphate, and sodium hexametaphosphate (as in Calgon). Calgon does not tie up hard-water ions as a precipitate, but rather as a new, soluble ion that does not react with soap.

Figure 4.57 *Structural formula of typical soap scum. This substance is calcium stearate.*

Scientific American Working Knowledge Illustration

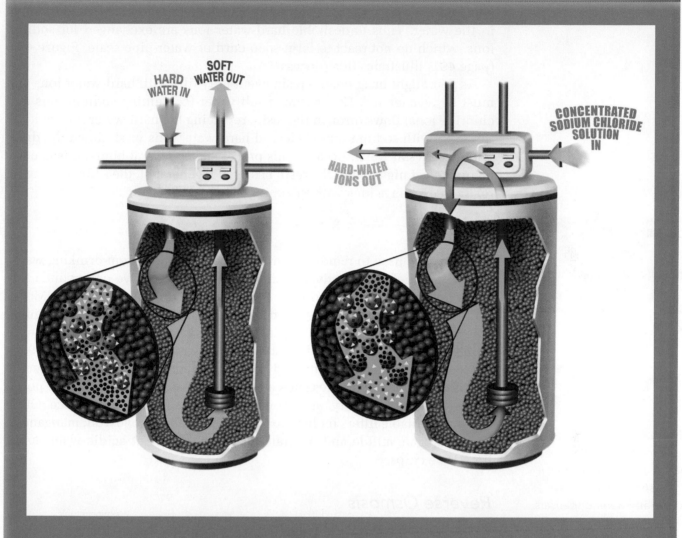

Figure 4.58 *Home water softener.* *Hard water that contains excess calcium (Ca^{2+}), magnesium (Mg^{2+}), or iron(III) (Fe^{3+}) cations prevents soap from lathering easily and can cause scale to build up inside pipes. Thus, where hard water poses a problem, many people install home water softeners. Water softeners contain ion-exchange resin, depicted here as orange beads. Left: Initially, sodium ions (Na$^+$, depicted as black dots in the magnified portion) are attached to the resin. As hard water flows into the tank, hard-water cations such as Mg^{2+} and Ca^{2+} (yellow triangles and red squares) become attached to the resin, releasing Na$^+$ ions into the water. This softened water flows through pipes to faucets, water heaters, and washing machines.*

Right: Eventually, all Na$^+$ ions are replaced by hard-water ions; no further ion exchange can occur. Then, the resin is "recharged" with Na$^+$ ions (black dots) by passing concentrated sodium chloride solution through the system. This displaces hard-water ions (yellow triangles and red squares) from the resin. They flow from the water softener into a drain. Thus, regenerated resin—with Na$^+$ ions again attached—is ready to soften hard water again.

If you live in a hard-water region, your home plumbing may include a water softener. Hard water flows through a tank containing an ion-exchange resin. Initially, this resin is filled with sodium cations (Na^+). Calcium and magnesium cations in the hard water are attracted to the resin and become attached to it. At the same time, sodium cations leave the resin and dissolve in the water. Thus undesirable hard-water ions are exchanged for sodium ions, which do not react to form soap curd or water-pipe scale. Figure 4.58 (page 491) illustrates this process.

As you might imagine, the resin eventually fills with hard-water ions and must be regenerated. Concentrated salt water (containing sodium ions and chloride ions) flows through the resin, replacing the hard-water ions held on the resin with sodium ions. Released hard-water ions wash down the drain with excess salt water. Because this process takes several hours, it is usually completed at night. After the resin has been regenerated, the softener is again ready to exchange ions with incoming hard water.

Filtration

Filters can be used to remove a variety of contaminants from drinking water. Many households use activated charcoal filters to remove compounds that cause undesirable odors and tastes, much as you did in Investigating Matter D.2. These filters also trap organic compounds, including THMs, and residual chlorine. Charcoal filters are available for point-of-entry (into the house) or point-of-use applications and must be regularly replaced. See Figure 4.59.

Mechanical filters, constructed of fabric, fiber, or ceramic, are used to remove suspended particles such as sand and clay. Also available for point-of-use and point-of-entry, mechanical filters require periodic service. Other types of available filters include oxidizing filters to remove iron, manganese, and hydrogen sulfide and neutralizing filters that treat acidic water using limestone chips.

Figure 4.59 *Point-of-use water treatment systems include pitchers with built-in filters (shown) and filters in refrigerator water dispensers.*

A semi-permeable membrane allows only particular molecules or ions to pass through it.

Reverse Osmosis

Reverse osmosis is a process that removes impurities from water by applying pressure to pass impure water through a semi-permeable membrane from a more concentrated region to a more dilute region (see Figure 4.60). Although this process removes many impurities, including inorganic compounds, microorganisms, and organic substances, it is often used in combination with filtration. Reverse osmosis can reduce the level of nitrates in water, but cannot remove them completely.

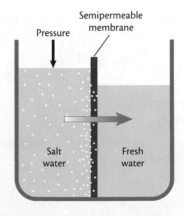

Figure 4.60 *Reverse osmosis.*

Disadvantages of reverse osmosis include production of large amounts of waste water and high initial and maintenance costs. Because of these limitations, reverse osmosis is generally used only to supply water for cooking and drinking.

Distillation

In Unit 3, you used the process of distillation to separate a mixture. You have also probably seen or purchased distilled water in the grocery store. In the process that prepares this product, water is heated to form steam, then the steam travels through a condenser where it is surrounded (but not in contact with) cold water. The steam condenses and the resulting water is collected in a separate container, leaving many contaminants behind. Although this method produces very pure water, some contaminants such as volatile organic compounds may be carried over with the water vapor.

Distillation is relatively inefficient—it requires significant quantities of impure water and energy to produce purified water. In addition, the equipment is expensive and requires regular maintenance. Finally, as you may know, distilled water lacks beneficial minerals and may taste "flat."

■ MAKING DECISIONS
D.9 SEARCHING FOR SOLUTIONS

Now that you have almost completed your analysis of the mystery of the Riverwood fish kill, you will be assigned a role to play in the town council meeting. Your teacher may also provide background material in addition to the data you have already analyzed within the unit.

This final Making Decisions activity will help you to prepare for the town council meeting by reviewing and synthesizing what you have learned throughout this unit and your chemistry course so far.

To begin, read the "Putting It All Together" on pages 498–501. Then answer the following questions in writing. These answers will not be shared with your classmates until the town council meeting takes place.

Part I: Background Information on Your Role

1. What constituency do you represent in your assigned role?
2. Do you represent a group that would rely mainly on facts or opinions?
3. What role would your group play in a meeting of this kind? When will your group present your information?

4. Is your group responsible for any of the substances or phenomena that were measured and analyzed? That is, did your group collect the data or introduce the substance into the watershed?

Part II: Preparing for the Meeting

5. Based on the data, what was the likely cause of the fish kill? Would that cause have also posed a problem for human consumption and use?

6. Would the substance have been removed from the water by normal treatment? If so, how? If not, how could it be removed?

7. Was the crisis managed properly? Should the water have been shut off? Why or why not?

8. Who should pay for the consequences of the fish kill and the water shut-off?

9. Depending upon your group's role, prepare one of the following: (a) an opening statement for the town council and a series of questions for the panel of experts, (b) a presentation of data and recommendations for the town council, or (c) a schedule and rules for the town council meeting, along with questions for the expert panel.

10. Create a visual aid to enhance your presentation.

SECTION D SUMMARY

Reviewing the Concepts

> Mixtures can be separated by physical methods.

1. What does it mean to "purify" water?
2. Identify at least three techniques for purifying water.
3. What types of impurities can be removed by each of the three techniques available to you in the foul-water investigation?
4. The procedures in the foul-water laboratory investigation could not convert seawater to water suitable for drinking.
   a. Explain why not.
   b. What additional purification steps would be needed to make seawater suitable for drinking?

> Water purification is required because contaminants are continually added to water through use, but the amount available in Earth's hydrologic cycle is fixed.

5. Make a diagram of the hydrologic cycle and label all stages.
6. Has the world's total water changed in the past 100 years? The past million years? Explain.
7. Consider this quotation: "Water, water, everywhere, nor any drop to drink." Describe a situation in which this would be true.
8. Explain what would happen to Earth's hydrologic cycle if water evaporation suddenly stopped.

9. One unique characteristic of water is that it is present in all three physical states (solid, liquid, and gas) in the range of temperatures found on Earth. How would the hydrologic cycle be different if this were not true?

> Water can be purified through the actions of the hydrologic cycle, municipal treatment, or home treatment.

10. List three major processes that occur in natural water purification and, for each, identify contaminants that the process removes.
11. How are the properties of aluminum hydroxide related to the process of flocculation?
12. Why is calcium oxide (CaO) sometimes added in the final steps of municipal water treatment?
13. Fluoride, an ingredient in many types of toothpaste, is sometimes added to municipal water supplies in the last stage of water treatment. How much fluoride is added, and what is its purpose?
14. What are two problems associated with use of hard water?
15. Identify three common hard-water ions.
16. Which water source in a given locality would probably have harder water, a well or a river? Explain.
17. Many homes use some sort of water treatment system. List two types of systems and identify the substances that each type of system removes.

> Water for human consumption is disinfected using a variety of methods; the most common disinfection method is chlorination.

18. What are advantages of chlorinated drinking water compared to untreated water?

19. Is there a disadvantage to using chlorinated water? Explain.

20. Water from a clear mountain stream may require chlorination to make it safe for drinking. Explain.

21. List two alternatives to the use of chlorination in municipal water treatment.

How is chemistry applied to produce safe drinking water?

In this section, you compared water purification through natural, drinking-water, and home systems, explored ways to clean a contaminated water sample, proposed methods to remove substances from water using knowledge acquired in previous sections, and examined different ways to disinfect water. Think about what you have learned, then answer the question in your own words in organized paragraphs. Your answer should demonstrate your understanding of the key ideas in this section.

Be sure to consider the following in your response: steps in natural water purification, steps in drinking-water treatment, risks and benefits of disinfection methods, and other ways water is treated.

Connecting the Concepts

22. The following are three water uses associated with water-purification techniques. Classify each as either a direct or an indirect water use. Explain your answers.
    a. manufacture of filter paper
    b. moistening of sand and gravel
    c. use of water to cool a distillation apparatus

23. Consider oil–water separation, sand filtration, charcoal adsorption and filtration, and distillation. Which purification procedure would be least practical to purify a city's water supply? Why?

24. A politician campaigning for election guarantees that "every household will have 100% pure water from every tap." Evaluate this promise and predict the likelihood of its success.

25. Explain why hard water can decrease the efficiency of a boiler in a steam-generated electric power plant.

26. Compare and contrast the use of activated charcoal and reverse osmosis in home water filtration systems.

27. Sketch and label two molecular-level representations of an ion-exchange resin bead—one bead before and one bead after treatment of hard water.

28. Why does the EPA limit the concentration of THMs to 80 ppb instead of requiring their total elimination from municipal water supplies?

29. Compare how the various processes used in the foul-water investigation (pages 479–482) are similar to steps in the natural purification of water.

30. Some physicians recommend consuming ~2 L of water daily. Municipal water supplies may contain up to 1 ppm fluoride. Assume that you drink 2 L of water per day. At 1 ppm fluoride, how many grams of fluoride ion would you consume in
    a. one day?   b. one week?   c. one year?

Extending the Concepts

31. You are marooned on a sandy island surrounded by ocean water. A stagnant, murky pond contains the only available water on the island. In your survival kit, you have the following items:

    - one nylon jacket
    - one plastic cup
    - two plastic bags
    - one length of rubber tubing
    - one knife
    - one 1-L bottle of liquid bleach
    - one 5-L glass bottle
    - one bag of salted peanuts

    Describe a plan to produce drinkable water, using only these items.

32. Charcoal-filter materials are available in various sizes—from briquettes to fine powder. List advantages and disadvantages of using either large charcoal pieces or small charcoal pieces for filtering.

33. Find out how much charcoal is used in a fish aquarium filter and how fast water flows through the filter. Estimate the volume of water that can be filtered by a kilogram of charcoal. What mass of charcoal would be needed to filter the daily water supply for Riverwood, population 19 500?

34. A group of friends is planning a four-day backpacking hike. Some members of the group favor carrying their own bottled water. Others wish to carry no water, buying bottled water in towns along the trail instead. Still others want to purchase and carry portable filters for use with stream water.

    a. What are advantages and disadvantages of each option?

    b. What additional information would the group need before deciding on a group plan?

35. How much water would you need to drink to get your minimum daily requirement of calcium from water that contains 300 ppm calcium carbonate?

36. Explain why we find hard water stains in old sinks around hot-water faucets more often than around cold-water faucets.

37. Research the sources and production of a brand of bottled water. Report its origin and identify substances that are removed and added before it is sold.

38. Discuss the health effects associated with sodium-based ion-exchange resins used in home water softening systems.

39. Describe and evaluate the practicality of two desalination processes for making seawater suitable for drinking.

PUTTING IT ALL TOGETHER

RIVERWOOD NEWS

TOMORROW'S
WEATHER:
partly cloudy

Breaking news at RiverwoodNewsLive.com

SPECIAL EDITION

Fish Kill Panel to Convene

Meeting Tonight

By Orlani O'Brien
RIVERWOOD NEWS STAFF REPORTER

Mayor Edward Cisko announced at a news conference earlier today that an expert panel will convene to discuss analyses of the Snake River fish kill at a special town council meeting tonight. The experts will present evidence and discuss the causes, effects, and implications of the recent fish kill crisis. The panel will make recommendations to the town council regarding:

- Actions to be taken immediately to ensure the safety of the Snake River and Riverwood's water supply.
- Preventive measures necessary to avoid future water problems and fish kills.
- Appropriate water testing and treatment for Riverwood.
- Management of water-quality crises.
- Responsibility for the Snake River fish kill.

Mayor Cisko announces tonight's special town council meeting.

Among the experts on tonight's panel is Dr. Steven Schmidt of the EPA, who accompanied Mayor Cisko at the press conference. Dr. Schmidt dissected fish taken from the river within a few hours of their death. He also directed the team that analyzed accumulated river-water data in their effort to determine the cause of the fish kill. His team will present the results of their investigation this evening.

Mayor Cisko refused to elaborate on the reasons for the incident. However, he invited the public to the special council meeting at 8 p.m. tonight at the town hall.

TOWN COUNCIL MEETING

Your teacher will assign you to one of the Riverwood groups that will participate in the special town council meeting (similar to Figure 4.61) and will provide some background information about your group's role in the fish kill situation.

Figure 4.61 *A town meeting in session. Open meetings such as this encourage citizen involvement in discussions of local issues and in related decision-making challenges.*

Your group's status as involved and affected, local or invited experts, or town council, will determine the role you will play in the town council meeting. The groups are as follows:

- Involved and Affected Groups
  - Agricultural cooperative representatives
  - Mining company representatives
  - Power company officials
  - Riverwood Chamber of Commerce members
  - Riverwood Taxpayer Association members

- Expert Panel Groups
  - County sanitation commission members
  - Consulting engineers
  - EPA scientists

- Government Group
  - Riverwood Town Council members

Each student team will prepare for the town council meeting as directed below and by your teacher. You will present your own views and develop questions for other groups. You will also make recommendations to the town council on at least some of the meeting action items.

Each student will also be responsible for an individual, written editorial or letter to the editor after the meeting, as described on page 501.

The town council meeting will begin with short presentations from the involved and affected groups, followed by the expert panel. The panel will present data, make recommendations, and answer questions from the audience. Finally, the council members will make and announce decisions for each of the agenda items.

Presentation

Involved and Affected Groups

- Each group should address the following questions:
  - How were we affected by the fish kill?
  - How do we as a group safeguard Riverwood's water?
  - Why did the fish kill happen? (Support this answer with evidence or reasoning.)
  - What should happen next to protect Riverwood's water and the Snake River?

- In addition, groups may choose to address preventive measures, management of the fish-kill crisis, appropriate water testing and treatment for Riverwood and the Snake River.

Expert Groups

- Each group should present data with respect to the recent fish kill as follows:
  - County Sanitation Commission: Historical river testing data and current water treatment protocols.
  - EPA Scientists: Recent river testing data and relevant safety data for fish and humans.
  - Consulting Engineers: Water flow patterns and disruptions, including activity at the power company dam and mining storage ponds.

- Each group should also make recommendations regarding:
  - Actions to be taken immediately to ensure the safety of the Snake River and Riverwood's water supply.
  - Preventive measures to avoid future water problems and fish kills.
  - Appropriate water testing and treatment for Riverwood.
  - Management of water-quality crises.
  - Responsibility for the Snake River fish kill.

Town Council

- Will listen to all presentations, then make, announce, and explain decisions about each of the following:

- ○ Actions to be taken immediately to ensure the safety of the Snake River and Riverwood's water supply.
- ○ Preventive measures to avoid future water problems and fish kills.
- ○ Appropriate water testing and treatment for Riverwood.
- ○ Management of water-quality crises.
- ○ Responsibility for the Snake River fish kill.

- Will be responsible for creating and enforcing the meeting schedule and time limits.

Guest Editorial or Letter to the Editor

Town Council Members

- Write a guest editorial explaining and justifying the decisions made in the town council meeting.
- Support all claims with scientific evidence and reasoning.
- Limit the editorial to three typed pages or the number of characters specified by your teacher.

All Other Group Members

- Write a letter to the editor supporting or disagreeing with the town council's recommendations.
- Support all claims with scientific evidence and reasoning.
- Limit the letter to one typed page or the number of characters specified by your teacher.

LOOKING BACK AND LOOKING AHEAD

The Riverwood water mystery is solved! In the end, scientific data and analysis provided the answer. Now, human ingenuity will provide strategies to prevent the recurrence of such a crisis. In the course of solving the problem, the citizens of Riverwood learned about the water that they take for granted — abundant, clean water flowing steadily from their taps—and gained a greater appreciation for it.

Although Riverwood and its citizens exist only on the pages of this textbook, their water-quality crisis could be real. The chemistry-related facts, principles, and procedures that clarified their problem and its solution have applications in your own home and community.

This completes the first four units of *Chemistry in the Community*. Many chemistry topics remain to be explored further, such as chemical equilibrium, biochemistry, and nuclear chemistry. The last three *ChemCom* units can be studied in any order desired. These three units build upon and apply the chemistry knowledge and skills you acquired through your study of the first four units.

INDUSTRY: APPLYING CHEMICAL REACTIONS

New Jobs May [...]

A Riverwood News Commentary
By Pak Jin-Woo   RIVERWOOD RESIDENT

Although no one can deny tha[...] Riverwood community needs [...] jobs, it's foolish to invite either [...] Nitrogen Products Company or [...] Battery Technology Corporatio[...] locate here to manufacture, re[...] tively, ammonia or lithium-ion[...] teries, without first considering all the consequences.[...]

The promise of 200 new jobs is alluring. Howev[...] of these jobs would be in only one company. Wha[...] happen if there's a decline in the market for ammon[...] lithium-ion batteries?

Wouldn't it be more prudent to distribute [...] 200 jobs among several different companies? Let's [...] from our recent experience with the city's former [...] employer, the now bankrupt Riverwood Corporatio[...]

Although long-term economic health is importan[...] must question the potential safety and environm[...] risks that each of these two chemical plants would po[...] our community.

For instance, ammonia manufacturing requires [...] high pressures and temperatures. Although accident[...] uncommon, the potential consequences of an expl[...] or spill are great. Several illnesses and even deat[...] workers have been documented at ammonia plants.

Battery production also likely involves risks t[...] workers. Now, most lithium-ion batteries are prod[...]

As you can infer from the newspaper commen[...] panies want to establish a chemical plant in Ri[...] tors acknowledge, either company would offer[...] nities to the Riverwood community. However[...] only factor to consider.

Later in this unit, you will help to decide w[...] (or neither company) to Riverwood. The che[...] will prepare you to make informed decisions a[...] operating such plants. Keep in mind the conce[...] as you learn about the chemistry involved [...] and batteries.

What role does nitrogen chemistry play in agriculture?

How is chemical energy transformed into electrical energy?

What chemical principles can be used in the responsible manufacture of ammonia?

What challenges must be met to optimize production and use of batteries?

?

Two large companies are interested in developing manufacturing facilities—and much-needed job opportunities—in Riverwood. What issues should citizens consider before inviting either company to locate in Riverwood?

Turn the page to learn about some chemical industries and the information the community needs to make a good decision.

More Job Option

A Riverwood News Commentary

By Greta Lederman   RIVERWOOD RESIDENT

Several weeks ago, the town coun; announced that two industrial firr; are interested in converting t empty Riverwood Corportion buil ing into a chemical manufacturir plant in Riverwood. Because of a la of employment opportunities in Ri erwood, many of your friends and neighbors have express relief at this news from EKS Nitrogen Products Compar and WYE Battery Technology Corporation.

When the Riverwood Corporation declared ban ruptcy last year, many residents lost their jobs. Althoug some have found new jobs, many have not. The unen ployment rate in Riverwood has hovered near 15% f most of the past year.

Can anything be done to boost our local econom Yes. We can allow one of these eager companies to build chemical plant in our town. New jobs would be create for chemists, chemical techni- cians, and office personnel. Each company claims it would need to hire about 200 new employees.

We need more job oppor- tunities; no one denies that. But some people may wonder whether a chemical plant is the best option for Riverwood. Aren't chemical plants danger- ous? Should our community wait for a different opportunity (whenever and whateve that might be)?

I admit I was concerned about the issue of safety; thu I started doing my own research. I've found that bot companies have strong safety and environmental records They both advocate green technologies. The fact tha both companies confirm plans to adapt the former River

> **"** *We ha opportu between that are locat*

In addition to studying the hydrologic and carbon cycles, you examined figures illustrating those processes. Such visual models help you organize related information, allowing you to easily trace interactions and connections.

You have just learned that nitrogen atoms also cycle among the atmosphere, soil, and organisms. However, a visual model depicting this nitrogen-cycle information has not been presented. In this activity, you will create that missing diagram.

Look at Figure 5.12 as well as the figures on pages 346 and 483. Notice how they depict the carbon and hydrologic cycles. Then review Section A.2 Fertilizer and the Nitrogen Cycle (pages 512–514) in this unit (and earlier textbook material if needed) to guide your completion of the following steps.

1. Construct your own diagram of the nitrogen cycle.
   a. Use arrows to show the direction of flow as nitrogen atoms cycle among the atmosphere, soil, and living organisms.
   b. Include chemical formulas and names for key molecules and ions at each cycle stage.
   c. Use pictures and color as needed to clarify details in your model.
   d. Make your model easy to follow. A classmate should be able to summarize the steps correctly by studying your diagram.
2. Exchange your model with a classmate's model.
3. Select an appropriate starting point on your classmate's diagram and trace nitrogen through its cycle.
4. Repeat Step 3, but use your classmate's diagram to write a description of the key steps in the nitrogen cycle. Your written description should be limited to information in your classmate's diagram, even if some features are different from those in your diagram.
5. Exchange diagrams and written descriptions with your classmate. You should now have the nitrogen-cycle diagram you originally drew and your classmate's written description based on it.
6. Compare the difficulty you experienced in completing these two tasks:
   a. transforming the book's description of the nitrogen cycle into a diagram
   b. transforming your classmate's diagram of the cycle into a written description
7. How closely did your classmate's written description reflect the actual structure and details of your diagram? Explain.
8. Compare the description that your classmate wrote about your diagram with the description of cycle steps on page 514.
   a. Compared to the textbook description, did your classmate's description omit or add any details or steps? Explain.
   b. Which description is more detailed? Explain.

9. Based on your classmate's description,

   a. how easy and convenient was your diagram to interpret and follow? Explain.

   b. how could you modify your diagram to improve its accuracy or clarity? Explain.

10. Considering your answers to Question 9, make any needed changes to your diagram so it more clearly illustrates the nitrogen cycle.

So far, you have learned about the components of fertilizer and how they are utilized by plants. What are some natural sources of fertilizer, and how does chemistry apply to the production of synthetic fertilizer?

concept check 2

1. Is the nitrogen cycle more similar to the carbon cycle or to the water cycle? Why?

2. What does it mean for a substance to be oxidized?

3. In a nitrate ion, are the electrons that make up the covalent bonds between N and O shared equally by the atoms? Explain.

A.5 FIXING NITROGEN BY OXIDATION–REDUCTION

Fertilizer Sources

Before modern methods of manufacturing ammonia were developed, nitrogen-containing fertilizers came from either animal waste or nitrate compounds. Large quantities of such compounds came from Chilean guano beds (Figure 5.13). At the turn of the 20th century, speculation arose that guano beds would be depleted by about 1930, raising fears of an agricultural crisis and the specter of world famine. Actually, the development of commercial ammonia production largely eliminated dependence on natural nitrate sources.

Using commercial ammonia for fertilizer has had a huge impact on agriculture and world food supplies. World ammonia production has increased dramatically over the last 65 years, as farmers have increased their use of fertilizer to meet the food needs of growing populations. The U.S. chemical industry produces about 20 billion pounds of ammonia every year. Most is dedicated to making fertilizer.

Figure 5.13 *Guano (seabird dung) deposits were one of the first commercial sources of fertilizer.*

In seeking ways to fix nitrogen gas artificially, scientists in 1780 first combined atmospheric nitrogen and oxygen by exposing them to an electric spark. However, the cost of electricity made this too expensive for commercial use. A less-expensive method, the **Haber–Bosch process**, replaced it. Fritz Haber and Karl Bosch first demonstrated this technique for making ammonia from hydrogen gas and nitrogen gas in Germany in 1909, according to this equation:

You will learn more about the Haber–Bosch process in Section B.

$$N_2(g) + 3 H_2(g) \rightleftharpoons 2 NH_3(g)$$

Oxidation–Reduction

In the Haber–Bosch reaction of nitrogen gas and hydrogen gas, electrons are involved in an oxidation–reduction reaction. As you know, atoms that apparently lose their share of electrons are involved in the process called *oxidation* (see page 102). For example, the conversion of metallic sodium atoms (Na) into sodium ions (Na^+) is oxidation because electrically neutral sodium atoms are oxidized to +1 sodium ions by loss of one electron per atom. Recall that the opposite process—the apparent gaining of electrons—is called *reduction*. The formation of chloride ions (Cl^-) from electrically neutral chlorine atoms is an example of reduction. Electrons can be transferred to or from particular atoms, molecules, or ions. The products of such oxidation–reduction reactions also include atoms, molecules, or ions.

Regardless of how the electron-transfer occurs, the same oxidation–reduction principles apply.

You can judge the relative tendency of a covalently bonded atom to attract electrons in compounds from that element's *electronegativity* (see page 401). Nonmetallic elements typically have higher electronegativities than do metallic elements. Figure 5.14 shows the electronegativity values for some common elements.

Increasing Electronegativity →

Increasing Electronegativity ↑

H 2.1																	
Li 1.0	Be 1.5												B 2.0	C 2.5	N 3.0	O 3.5	F 4.0
Na 0.9	Mg 1.2												Al 1.5	Si 1.8	P 2.1	S 2.5	Cl 3.0
K 0.8	Ca 1.0	Sc 1.3	Ti 1.5	V 1.6	Cr 1.6	Mn 1.5	Fe 1.8	Co 1.8	Ni 1.8	Cu 1.9	Zn 1.6	Ga 1.6	Ge 1.8	As 2.0	Se 2.4	Br 2.8
Rb 0.8	Sr 1.0	Y 1.2	Zr 1.4	Nb 1.6	Mo 1.8	Tc 1.9	Ru 2.2	Rh 2.2	Pd 2.2	Ag 1.9	Cd 1.7	In 1.7	Sn 1.8	Sb 1.9	Te 2.1	I 2.5
Cs 0.7	Ba 0.9	Lu 1.2	Hf 1.3	Ta 1.5	W 1.7	Re 1.9	Os 2.2	Ir 2.2	Pt 2.2	Au 2.4	Hg 1.9	Ti 1.8	Pb 1.8	Bi 1.9	Po 2.0	At 2.2
Fr 0.7	Ra 0.9															

Figure 5.14 *Electronegativity values of selected elements.*

Oxidation States

A convenient, yet arbitrary, way to express the degree of oxidation or reduction of particular atoms is by assigning each atom an **oxidation state**. The higher (more positive) the oxidation state becomes, the more an atom has become oxidized. The lower (less positive) the atom's oxidation state becomes, the more the atom has become reduced.

In binary compounds (compounds composed of two elements), we assign atoms of the element with the lower electronegativity a **positive oxidation state**, corresponding to an apparent loss of electrons. Likewise, we assign atoms of the more electronegative element a **negative oxidation state**, corresponding to an apparent gain of electrons.

Consider the key chemical change in the Haber–Bosch process, as depicted with electron-dot formulas and space-filling models:

Note that each nitrogen atom in N_2 shares six electrons with another nitrogen atom, resulting in a triple covalent bond. Both nitrogen atoms exert equal attraction for the shared electrons. Each nitrogen atom in N_2 has a **zero oxidation state**. (This is true of any atom of an element that is not bonded to any other element.) The two nitrogen atoms share their bonding electrons equally. There is no separation of electrical charge. As the reaction progresses, each nitrogen atom becomes covalently bonded to three hydrogen atoms. Each bonded nitrogen and hydrogen atom shares an electron pair, but they do not share the pair equally. Nitrogen atoms (electronegativity value = 3.0) have a greater attraction for these shared electrons than do hydrogen atoms (electronegativity value = 2.1). Thus, the nitrogen atom in each NH_3 molecule acquires a greater portion of shared electrons than does each hydrogen atom.

Consequently, the nitrogen atom in NH_3 is assigned a negative oxidation state. Likewise, each hydrogen atom in NH_3 has lost some share of bonding electrons due to the reaction, so hydrogen is assigned a positive oxidation state. Thus, the oxidation state of an atom in a particular substance depends on the identity of neighboring atoms to which it is covalently bonded.

The Haber–Bosch process converts difficult-to-use nitrogen gas molecules into more useful ammonia molecules, a form of fixed nitrogen. Once nitrogen is chemically combined with another element, it can be readily converted to other nitrogen-containing compounds. For example, under proper conditions, ammonia will react with oxygen gas, forming nitrogen dioxide:

$$4\ NH_3(g) + 7\ O_2(g) \longrightarrow 4\ NO_2(g) + 6\ H_2O(g)$$

This is an oxidation–reduction reaction. In forming NO_2, the nitrogen atom in ammonia has apparently been oxidized; it has lost part of its share of electrons. Why does this occur? Because oxygen is more electronegative than nitrogen (O = 3.5; N = 3.0), oxygen attracts bonding electrons more strongly than nitrogen does. Each oxygen atom is considered to be reduced. Each oxygen atom has gained more control of its bonding electrons than it originally had in O_2.

DEVELOPING SKILLS

A.6 DETERMINING OXIDATION STATES

Sample Problem 1: Which element in sulfur dioxide (SO_2) has a positive oxidation state?

Figure 5.14 (page 518) indicates that sulfur's electronegativity value is 2.5 and oxygen's is 3.5. In sulfur–oxygen covalent bonding, sulfur has the weaker electron-attracting ability. Therefore, sulfur is assigned a positive oxidation state in SO_2.

Sample Problem 2: The following equation represents the reaction of sulfur with oxygen gas to produce sulfur dioxide gas:

$$S_8(s) \ + \ 8\ O_2(g) \longrightarrow 8\ SO_2(g)$$

Why do chemists consider this reaction as an oxidation–reduction reaction in which sulfur is oxidized?

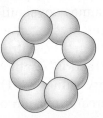

The element sulfur is found in eight-atom rings, S_8.

The oxidation state of sulfur changes from zero (in the pure element) to a positive value in the product, sulfur dioxide (see Sample Problem 1)—sulfur has been oxidized. By contrast, oxygen gas becomes reduced; it has become reduced from zero oxidation state to a negative oxidation state.

1.   Consider each of the following covalent compounds. Using electronegativity values from Figure 5.14, decide which element in each compound has (i) a positive oxidation state and which has (ii) a negative oxidation state:

   a. ammonia, NH_3

   b. hydrogen chloride, HCl

   c. hydrogen fluoride, HF

   d. oxygen difluoride, OF_2

   e. iodine trifluoride, IF_3

   f. phosphorus trifluoride, PF_3

2. Each of these compounds includes a metallic element and a nonmetallic element. Select the element in each compound possessing (i) a positive oxidation state and (ii) a negative oxidation state.

   a. sodium iodide, NaI

   b. lead(II) fluoride, PbF_2

   c. lead(II) sulfide, PbS

   d. potassium oxide, K_2O

   e. iron(III) chloride, $FeCl_3$

   f. sodium phosphide, Na_3P

   g. sodium chloride, NaCl

3. Consider your answers to Questions 1 and 2. What conclusions can you draw about the oxidation states of metals and nonmetals in binary compounds?

4. Consider this chemical equation: $8\ Ni\ +\ S_8\ \longrightarrow\ 8\ NiS$.

   a. Does this equation represent an oxidation–reduction reaction?

   b. If so, identify the element oxidized and the element reduced. If not, explain why.

5. The element iron is part of an essential system of energy transfer within human cells. In that system, Fe^{2+} ions are converted to Fe^{3+} ions. Does that change represent an oxidation or a reduction?

6. Within the nitrogen cycle (see page 514), nitrogen gas (N_2) undergoes particular chemical reactions in which it is oxidized and other reactions in which it is reduced. Identify, by name and formula, a nitrogen-cycle product that forms by nitrogen gas being

   a. oxidized.

   b. reduced.

A.7 PRODUCTS OF THE CHEMICAL INDUSTRY

By now, you know there are very few things that we use regularly that have not been modified in some way. This is the chemical industry's focus—transforming natural resources into useful products to meet a wide variety of needs and purposes. The chemical industry also creates new substances and materials as replacements for natural ones. For example, plastics often replace wood or metals, and synthetic fibers often replace cotton or wool.

Even though the chemical industry is a worldwide, multibillion-dollar enterprise that affects everyone's life daily through its products and economic impact, most people are not aware of what happens when new materials are produced. This raises questions about how chemical industries operate, how they manufacture new products, and what those products contain.

The modern chemical industry employs well over a million people world-wide. Over the past 80 years, it has grown through mergers of smaller companies and creation of new companies. During that time, the industry's focus has expanded from a limited range of basic products to more than 70 000 products. Hundreds of chemical companies form the third-largest manufacturing industry in the United States. Only industries that produce machinery and electrical equipment are larger. In fact, if we include the food and petroleum industries in the chemical-industry category, this represents the world's *largest* industry.

Most chemical products reach the public indirectly because they are used to produce other consumer materials. For instance, the automobile and home-construction industries use enormous supplies of industrial chemicals. They use paints and plastics for automobile body parts such as bumpers, dashboard panels, upholstery and carpeting, and synthetic rubber in tires. Home construction involves large quantities of plastics for carpeting and flooring, insulation, siding, window frames, piping, and appliances. It also involves using paints, metals, and air-conditioning coolants. Figure 5.15 shows a range of products from various chemical industries.

Figure 5.15 *The chemical industry produces many materials that have useful properties.*

In Riverwood, the two industrial companies under consideration manufacture products involving nitrogen and lithium-ion batteries. Among the products manufactured by EKS Nitrogen Products Company are nitric acid and ammonia, which are often used in chemical reactions that produce other materials. By contrast, the batteries produced by WYE Battery Technology Corporation are used in the assembly of electric and hybrid-electric vehicles directly in their manufactured form.

EKS is committed to producing high-quality nitrogen-based fertilizer in Riverwood at reasonable cost, using the best available technologies. The production and sale of fertilizer represents a multimillion-dollar business that employs thousands of people worldwide and affects the lives of nearly everyone, from farmers and gardeners to food producers and consumers. This is why some Riverwood residents are eager to invite EKS into their community.

■ MAKING DECISIONS

A.8 CHEMICAL PROCESSING IN YOUR LIFE

To become more aware of how pervasive the products of chemical processing are in everyday life, list five items or materials around you that have *not* been manufactured, processed, or altered from their natural form. Start by considering everyday items, such as clothes, objects in your home, modes of transportation, books, foods, communication devices such as phones and computers, and sports and recreation equipment—all things you routinely encounter.

Now answer the following questions. Be prepared to share and discuss your answers in class.

1. a. Which items on your list were wrapped, boxed, or shipped in materials that had been manufactured (Figure 5.16)? Explain.

   b. Is the packaging or shipping material necessary or simply a convenience? Why?

2. In what ways might each item or material on your list be better than or inferior to a manufactured, processed, or synthetic alternative? Consider factors such as cost, convenience, availability, and quality.

3. If a product is "100% natural," does that necessarily mean it was not involved in any processing or chemical or physical changes? Why? Support your answer with at least one example.

4. Many people avoid using synthetic fertilizers in their homes and gardens. Instead, they may use composted materials (food scraps and other plant and animal debris that have been decomposed).

   a. How does the practice of composting fit with the nitrogen cycle?

   b. Do you think that industrial composting could completely replace the use of synthetic fertilizers? Why or why not?

Figure 5.16 *Are the materials used to package these products manufactured? Is this packaging necessary? What other packaging options exist?*

SECTION A SUMMARY

Reviewing the Concepts

> Fertilizers provide nutrients necessary for plant growth.

1. List the three main elements in fertilizer.

2. What would the expression *20–10–15* mean if found on a fertilizer label?

3. Why is potassium content expressed as percent K_2O in fertilizer?

4. List each test performed in Investigating Matter A.1 (page 507), along with the ions that the test identified.

5. Describe the role in plant growth of each of the three key ingredients in a typical fertilizer.

6. Why is it useful for fertilizers to be available with different compositions, such as 7–7–7 and 20–10–10?

> Nitrogen is transformed chemically as it cycles through living systems and the physical environment.

7. Why do plants need nitrogen?

8. Given the fact that nitrogen is abundant in the atmosphere, why is it included in fertilizers?

9. What does it mean to *fix* nitrogen gas?

10. List three ways in which atmospheric nitrogen can be fixed.

11. List two nitrogen-containing ions that are useful to plant growth.

12. How do plants and animals differ in the ways they obtain
    a. nitrogen?   b. amino acids?

13. What is one role of denitrifying bacteria in the environment?

14. Summarize the steps of the nitrogen cycle.

> The tendency of an atom to attract electrons within a covalent chemical bond can be expressed by the electronegativity of that element.

15. Referring to Figure 5.14 (page 518), identify the element that is most electronegative.
    a. List the element's symbol and name.
    b. Is this element a metal or nonmetal?

16. Referring to Figure 5.14 (page 518), identify the element that is least electronegative.
    a. List the element's symbol and name.
    b. Is this element a metal or nonmetal?

17. Describe how electronegativity values for elements change as one moves
    a. from left to right across any period of the periodic table.
    b. down a group of the periodic table.

18. Arrange each of the following sets of elements in order of their increasing attraction for electrons within a bond:
    a. silicon, sodium, and sulfur
    b. nitrogen, phosphorus, and potassium
    c. bromine, fluorine, lithium, and potassium

Assigning oxidation states is a convenient way to identify oxidation or reduction of atoms in reactions.

19. What is the oxidation state of an atom in its elemental form (that is, not combined with an atom of another element)?

20. How does the oxidation state of an atom change when the atom is

    a. oxidized?   b. reduced?

21. How is it possible for the same element to be oxidized in one reaction and reduced in another?

22. What type of element (metal or nonmetal) is more often found in negative oxidation states when bonded to atoms of other elements?

23. Write the chemical formula for each of the following compounds and identify which element in each compound has a positive oxidation state and which has a negative oxidation state.

    a. water   c. carbon dioxide
    b. ammonia   d. magnesium chloride

The chemical industry transforms elements and compounds into other useful materials.

24. List four different ways that the chemical industry is involved in the production of a box of breakfast cereal.

25. Look around wherever you are sitting right now and identify four things you see that are products of the chemical industry.

26. How is most of the ammonia produced in the United States used?

27. What was the most common source of ammonia before the 20th century?

28. Write the chemical equation that represents the main reaction in the Haber–Bosch process.

What role does nitrogen chemistry play in agriculture?

In this section, you investigated nutrients that plants need to grow and thrive, constructed a model of the nitrogen cycle, and applied your understanding of oxidation–reduction reactions to ways in which nitrogen compounds are converted, both naturally and industrially, to form useful products for agriculture. Think about what you have learned, then answer the question in your own words in organized paragraphs. Your answer should demonstrate your understanding of the key ideas in this section.

Be sure to consider the following in your response: role of nitrogen compounds in plant growth, forms of nitrogen that are most useful to plants, and natural and industrial methods for converting nitrogen compounds.

Connecting the Concepts

29. Describe one advantage and one disadvantage of using commercial fertilizer instead of guano or other animal waste to fertilize crops.

30. The quantity of phosphorus in fertilizer is reported as percent P_2O_5.

    a. What is the percent phosphorus in P_2O_5?

    b. Fertilizers are burned (oxidized) for analysis and the burning produces P_2O_5. In what form is the phosphorus actually found in the original fertilizer?

31. Describe a procedure for determining whether a soil sample contains any fixed nitrogen.

32. Compare the fertilizer tests conducted in this unit to the ion tests in Unit 4 (page 452). In what ways are these tests similar and in what ways are they different?

33. A magazine article claims that "oxygen is needed for all oxidation reactions." Do you agree or disagree with that statement? Support your answer with evidence.

34. What does the electronegativity of an electrically neutral atom indicate about its tendency to become oxidized?

35. In general, how do the electronegativities of metallic and nonmetallic elements compare?

36. Consider the Haber–Bosch process.

    a. Write the Lewis-dot structure for each reactant and product involved.

    b. Using the concept of electronegativity, determine which atoms in that reaction have a positive oxidation state and which have a negative oxidation state.

37. How does the concept of a limiting reactant apply to the use of fertilizers?

Extending the Concepts

38. You have considered three major natural cycles: water, carbon, and nitrogen. Compare these cycles with respect to

    a. conservation of mass,

    b. types of chemical change, and

    c. participating organisms.

39. Why do some vegetarians claim that their diets make more economical use of resources than the diets of non-vegetarians?

40. Some historians claim that development of the Haber–Bosch process prolonged World War I. Explain.

41. a. Magnesium is a key component of chlorophyll in green plants. Explain why magnesium is generally not included in commercial fertilizers.

    b. Identify some other substances required by growing plants that are not included in fertilizers.

42. Review the list of ingredients in a multipurpose vitamin capsule for humans and compare this to the ingredients in a typical commercial fertilizer. Suggest reasons for the similarities and differences you find.

43. How does lightning fix nitrogen?

44. Research and report on denitrifying bacteria. Include their typical habitats and any unusual characteristics.

SECTION **B** # INDUSTRIAL PRODUCTION OF AMMONIA

What chemical principles can be used in the responsible manufacture of ammonia?

As you have learned, many industrial raw materials are extracted from Earth's crust (such as minerals, precious metals, sulfur, and petroleum), oceans (e.g., magnesium and bromine), and atmosphere. Nitrogen gas and oxygen gas, both obtained by low-temperature distillation from liquefied air, are valuable starting materials in the production of ammonia and nitric acid. As you will soon learn, producing ammonia also depends on understanding the implications of reversible reactions and chemical equilibrium.

GOALS

- Explain how temperature, reactant concentration(s), and the absence or presence of a catalyst affect reaction rates.
- Describe characteristics of a system in dynamic equilibrium.
- Use Le Châtelier's Principle to predict shifts in equilibria caused by perturbations to a system.
- Describe the Haber–Bosch process for industrial production of ammonia.
- Explain why many nitrogen-based compounds are effective chemical explosives.
- Describe how the production of ammonia can illustrate the goals and principles of green chemistry.

concept check 3

1. In some chemical equations a double arrow, $\rightleftharpoons$, is used. What does this notation represent?
2. According to collision theory, how does an increase in temperature affect collisions between gas particles?
3. Is the following statement an accurate description of a chemical system at equilibrium? "At equilibrium, all reactions have stopped." Explain.

B.1 KINETICS AND EQUILIBRIUM

Producing ammonia from nitrogen gas and hydrogen gas is a chemical challenge. As you learned in Section A, molecular nitrogen (N_2) is very stable. This means that nitrogen fixation, the chemical combination of nitrogen gas with other elements, has a substantial activation-energy barrier. As you learned in Unit 2 (page 211), a reaction with a large energy barrier requires either that the reactant particles have substantial kinetic energy or that a catalyst reduces the energy required to initiate the reaction.

The reaction of nitrogen gas with hydrogen gas is also difficult because some ammonia molecules decompose back to nitrogen gas and hydrogen gas during the synthesis reaction. As you learned in Unit 2 (page 251), this kind of reaction—one in which products re-form reactants at the same time that reactants form products—is known as a *reversible reaction*. The double arrows used below indicate that both forward and reverse reactions occur simultaneously:

$$N_2(g) + 3\,H_2(g) \rightleftharpoons 2\,NH_3(g)$$

How do chemists and chemical engineers, whether at EKS Nitrogen Products Company or elsewhere, overcome these obstacles to produce ammonia?

Kinetics: Producing More Ammonia in Less Time

Chemical kinetics is the study of how fast chemical reactions occur. The **reaction rate** expresses how fast a particular chemical change occurs. To determine a reaction rate requires an experiment in which the amount (often measured as concentration) of one reacting substance is measured as a function of time. For instance, in synthesizing ammonia, if you know how fast nitrogen is being used and how long the reaction has been going on, you can calculate the amount of ammonia produced.

For chemical reactions to occur, reactant molecules, atoms, or ions must collide with one another. According to **collision theory**, the reaction rate depends on the collision frequency and the energy involved in each collision (Figure 5.17). Increasing the concentrations of reactants speeds up reaction rates by increasing the number of collisions. High temperatures increase the reaction rate by providing more reacting molecules with sufficient energy to overcome the activation-energy barrier. Catalysts, on the other hand, increase the reaction rate by lowering the activation-energy barrier required for the reaction to occur, as shown in Figure 5.18.

> Any rate is defined as the change in some quantity divided by the time required for the change. A common example of a rate is speed: miles per hour.

> You first learned about collision theory in Unit 2. According to this theory, the reaction rate also depends on the orientation of the colliding molecules.

Figure 5.17 *For a chemical reaction to occur, reactant particles must collide with proper orientation and sufficient energy. What are the benefits and limitations of using billiard balls to model simple molecular collisions?*

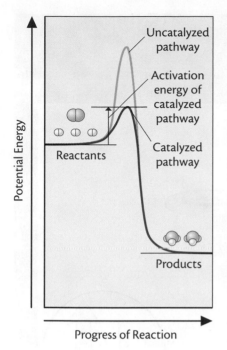

Figure 5.18 *A catalyst reduces the size of the activation energy barrier involved in synthesizing ammonia from its elements.*

Although a higher reaction temperature increases the average kinetic energies of the nitrogen and hydrogen molecules that react to form ammonia, ammonia itself becomes increasingly unstable at higher temperatures. The result is that ammonia decomposes back to nitrogen gas and hydrogen gas. If the reaction takes place at lower temperatures, fewer nitrogen and hydrogen molecules have enough energy to overcome the activation energy barrier, thus slowing the net rate of ammonia formation even though less ammonia decomposes at that lower temperature. So the formation of ammonia, the forward reaction, requires high temperatures, but high temperatures also increase the rate of decomposition of ammonia, the reverse reaction. What can we do to increase the rate of the forward reaction or decrease the rate of the reverse reaction, thus increasing the rate and yield of ammonia production?

The major breakthrough that led to profitable ammonia production was the discovery of a suitable catalyst. Catalysts made it possible to produce ammonia at lower temperatures (450–500 °C), thus slowing the rate of ammonia decomposition. Fritz Haber and his colleagues spent a great deal of time and energy in the early 1900s systematically searching for good catalysts. Today, the ammonia industry commonly employs iron as an ammonia synthesis catalyst.

Haber was awarded the 1918 Nobel Prize in Chemistry for his work on synthesizing ammonia.

Equilibrium: Favoring the Forward Reaction

Any reversible reaction appears to stop when the rate at which product forms equals the rate at which product reverts back to reactants—that is, when reactants and products attain **dynamic equilibrium**. At equilibrium, both the forward and the reverse reactions continue, but there is no further change in the amounts of reactants or products. At the point of dynamic equilibrium, the two opposing chemical changes are in exact balance, as modeled in Figure 5.19 and 5.20 (page 530).

Figure 5.19 *A system at dynamic equilibrium involves two ongoing processes acting in opposition to one another.*

Scientific American Conceptual Illustration

Figure 5.20 *Dynamic Equilibrium In the stoppered flask (left), we observe no overall change in the water level because its evaporation rate equals its rate of condensation. Water contained in the open flask (right) slowly escapes from the flask as water vapor. This happens because the rate of evaporation of water molecules is greater than the rate of condensation—this system is not in equilibrium. The closed system in the stoppered flask is, by contrast, an example of dynamic equilibrium.*

If the reverse reaction proceeds at an appreciable rate, then the amount of product that can be produced from the chemical reaction is decreased. Recall that in Unit 1 you learned how to predict the amount of product that would be produced by a known amount of reactant. When you made these calculations, you did not consider the possibility of a reverse reaction.

For many reactions, this is a valid simplification and it is not necessary to consider the reverse reaction. In these cases, the reverse reaction has a very slow rate, so we can assume that the forward reaction is the only important reaction.

However, the reverse reaction cannot be ignored in the synthesis of ammonia, so the net amount of ammonia formed from a given amount of nitrogen gas and hydrogen gas at a fixed temperature is limited by the competition between the forward and reverse reactions. One way to increase the amount of ammonia produced is to cool the ammonia as soon as it forms until it turns to a liquid and remove it from the reaction chamber. This prevents ammonia from decomposing back into nitrogen gas and hydrogen gas. If ammonia is continuously removed, the rate of the reverse reaction (decomposition of ammonia) is significantly decreased because there is less gaseous ammonia available to decompose. This causes the overall reaction (see equation on page 528) to favor the production of more ammonia because the rate of the forward reaction is greater than the rate of the reverse reaction.

> All reversible reactions in closed containers eventually reach equilibrium if conditions such as temperature remain constant.

Le Châtelier's Principle: Shifting the Equilibrium

The example above illustrates that a system at equilibrium can often be disturbed by changing the concentration of either reactants or products or by changing the temperature of the system. When one of these disturbances occurs, it causes either the forward or reverse reaction rate to become larger than the other and thus to be favored over the other. The reaction system will eventually reestablish dynamic equilibrium. However, the amounts of reactants and products present after equilibrium has been reestablished will be different from the amounts present before the disturbance.

For example, if more reactant was added to a system at equilibrium, the extra reactant will cause the forward reaction to increase, using up some of the added reactant, and produce more product. After the system returns to equilibrium, the concentration of the reactant that was added will be larger than the concentration of this reactant before the disturbance. The system has responded to the disturbance by undergoing a change that partially counteracts the initial effect of the disturbance. Thus, the initial equilibrium position is shifted. This effect, first described by the French chemist Henry Louis Le Châtelier in 1884, is commonly summarized in *Le Châtelier's principle*.

The external disturbance imposed on a system at equilibrium, sometimes called a *stress*, may be a change in the concentration of a particular reactant or product, a change in the temperature of the system, or (for a system including gases) a change in the total pressure. According to **Le Châtelier's principle**, the predicted shift in the equilibrium position is always in the direction that partially counteracts the imposed change in conditions. In the industrial production of ammonia, the removal of ammonia (a change in its concentration in the reaction vessel) results in the initial equilibrium position being shifted in favor of products. In the language of Le Châtelier's principle, the removal of ammonia is partially counteracted by the system, thus producing *more* ammonia.

> Shifting the position of an equilibrium system is often described by the direction of the favored reaction. If the disturbance causes the forward reaction to be favored, the equilibrium is shifted to the right side of the equation. If the reverse reaction is favored, the equilibrium is shifted to the left side of the equation.

Another external disturbance (stress) used to increase ammonia production is to add reactant molecules (nitrogen gas and hydrogen gas) continuously at high pressure. This higher pressure of reactant gases means the number of nitrogen and hydrogen gas molecules per unit volume is increased, thereby increasing the concentration of the reactants. The frequency of molecular collisions increases, which favors the forward reaction, and thus increases the amount of ammonia formed. This change can be viewed as partially counteracting the initial increased pressure because the total number of gas molecules has been decreased. (Four molecules of gas—three molecules H_2 and one molecule N_2—are replaced by two molecules of NH_3 gas.)

In many cases, changing the system's temperature can also cause an equilibrium system to shift. The direction of that effect can be predicted based on whether the forward reaction is exothermic or endothermic. For example, the synthesis of ammonia is exothermic:

$$N_2(g) \; + \; 3\,H_2(g) \; \rightleftharpoons \; 2\,NH_3(g) \; + \; \text{Thermal energy}$$

Thermal energy can be regarded as a product of the forward (left to right) reaction. Raising the temperature would tend to favor the reverse (right to left) reaction, a chemical change that absorbs thermal energy. Consequently, less ammonia is formed at equilibrium at conditions of higher temperatures. Remember, though, that the temperature must be high enough to provide the nitrogen and hydrogen molecules with adequate kinetic energy to react. A delicate balance is needed. The temperature must be high enough to produce significant amounts of ammonia, but not so high that it promotes an excessive rate of ammonia decomposition.

■INVESTIGATING MATTER
B.2 LE CHÂTELIER'S PRINCIPLE

Preparing to Investigate

In this investigation of a system at equilibrium, you will use what you have learned about Le Châtelier's principle to explore the effects of changing concentration and temperature on the position of an equilibrium system. The chemical system you will investigate is described by the following equilibrium equation:

Thermal energy + $[Co(H_2O)_6]^{2+}(aq) + 4\,Cl^-(aq) \rightleftharpoons [CoCl_4]^{2-}(aq) + 6\,H_2O(l)$

This system involves two complex ions. A **complex ion** is a chemical species composed of a single central atom or ion, usually a metal ion, to which other atoms, molecules, or ions are attached. One of your objectives in this investigation is to decide which complex ion above—$[CoCl_4]^{2-}$ or $[Co(H_2O)_6]^{2+}$—is blue and which is pink. See Figure 5.21.

Before you begin, read *Making Predictions* and *Gathering Evidence* to learn what you will need to do and note safety precautions. You will also need to create a data table that is appropriate for recording your data.

Figure 5.21 *Which complex ion is in this test tube?*

Making Predictions

Use the equation for this equilibrium system and what you know about solution chemistry to answer the following questions. Record the answers on your data sheet in a section labeled predictions.

1. How would the equilibrium system respond if chloride ion were removed from the system? Specifically, what would happen to the concentration of $(CoCl_4)^{2-}(aq)$?

2. How would the equilibrium system respond to an increase in temperature? Specifically, what would happen to the concentrations of $Cl^-(aq)$ and $(CoCl_4)^{2-}(aq)$?

3. In *Gathering Evidence*, you are told to add silver nitrate to the system. Which substance or substances present in the equilibrium will be affected by the addition of this reagent? Explain.

Gathering Evidence

1. Before you begin, put on your goggles, and wear them properly throughout the investigation.

2. Prepare a hot-water bath (60–70 °C) to use in Step 6.

3. Add 20 drops of 0.1 M cobalt(II) chloride $(CoCl_2)$ solution to a clean, dry test tube. Record the color.

4. Add 7 drops of 0.1 M silver nitrate $(AgNO_3)$ solution. (**Caution:** *$AgNO_3$ solution can stain skin and clothing. Handle with care.*)

5. Gently swirl the tube to ensure good mixing. Record the color.

6. Heat the tube in a hot-water bath for 30 seconds. Record the color.

7. Remove the tube from the hot-water bath. Add ~0.3 g sodium chloride (NaCl).

8. Gently agitate the tube. Heat the solution for 30 seconds in the hot-water bath. Record the color.

9. Place the test tube in a beaker containing ice water for 30 seconds. Record the color.

10. Reheat the test tube in the hot-water bath. Record the color.

11. Dispose of the mixture in the test tube as directed by your teacher.

12. Wash your hands thoroughly before leaving the laboratory.

Interpreting Evidence

To answer the following questions, refer to your observations and the equilibrium equation given in *Preparing to Investigate*.

1. What happened when you added $AgNO_3$ solution to the solution in the test tube?

2. What is the identity of the white precipitate that formed in Step 5? (*Hint:* Refer to Unit 4, page 448.)

3. Why did adding $AgNO_3$ solution affect the equilibrium, even though neither Ag^+ ions nor NO_3^- ions appear in the equilibrium equation? Write an equation to support your explanation.

4. Which way did the equilibrium shift after adding the $AgNO_3$ solution, toward reactants or products?

5. Which way did the equilibrium shift after adding NaCl in Step 7, toward reactants or products?

6. a. Which complex ion is pink: $[CoCl_4]^{2-}$ or $[Co(H_2O)_6]^{2+}$?

   b. Which complex ion is blue?

7. What was the effect of cooling the solution in Step 9?

8. Why did the solution's color change after heating in Step 8, but not in Step 6?

Making Claims

9. Review your results from Steps 5, 8, 9, and 10. Write a statement for each of these steps explaining how your results support or contradict Le Châtelier's principle.

Reflecting on the Investigation

10. Revisit the predictions you made before doing the experiment. Describe how well you were able to predict the behavior of the equilibrium system.

11. What would have been the effect if you had used hydrochloric acid in Step 7 instead of sodium chloride? Explain.

DEVELOPING SKILLS

B.3 CHEMICAL SYSTEMS AT EQUILIBRIUM

Sample Problem: For the following equilibrium system, describe three changes you could make to increase the formation of nitric oxide gas, NO(g), at equilibrium.

$$\text{Thermal energy} + 2\,NO_2(g) \rightleftharpoons 2\,NO(g) + O_2(g)$$

We need to identify factors that would cause the equilibrium to shift to the right, favoring NO production. Based on Le Châtelier's principle, any of the following changes would shift the equilibrium to the right: (i) increasing the temperature of the system, (ii) increasing the concentration of NO_2, or (iii) decreasing the concentration of either O_2 or NO.

1. For each equilibrium system, describe three changes you could make to favor the forward reaction.

   a. $2\,SO_2(g) + O_2(g) \rightleftharpoons 2\,SO_3(g) + \text{Thermal energy}$

   b. $H_2(g) + Cl_2(g) \rightleftharpoons 2\,HCl(g) + \text{Thermal energy}$

2. Examine the graph in Figure 5.22:

   a. What generalization can you make about the effect of temperature on the yield of ammonia?

   b. Do you think that generalization would remain valid for temperatures much lower than 400 °C? Explain your answer.

   c. What generalization can you make about the effect of total pressure on the yield of ammonia?

   d. What combination of temperature and pressure results in the highest ammonia yield?

3. For Questions 1a and 1b, describe three different changes you could make to favor the reverse reaction.

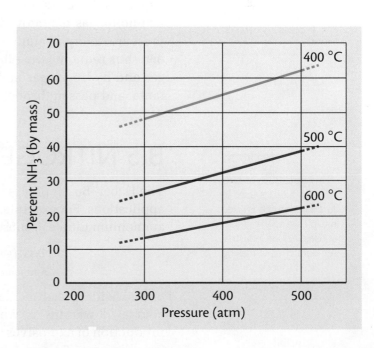

Figure 5.22 *Ammonia production in the system* $N_2(g) + 3\,H_2(g) \rightleftharpoons 2\,NH_3(g) + \text{Thermal energy}.$ *The graph depicts the effect of pressure and temperature changes on the percent NH_3 present at equilibrium.*

B.4 INDUSTRIAL SYNTHESIS OF AMMONIA

Figure 5.23 *This technician checks conditions within an industrial chemical storage system. What evidence suggests that safety is a priority?*

Large-scale ammonia production involves much more than allowing nitrogen gas and hydrogen gas to react in the presence of a catalyst. First, of course, the plant must obtain a continuous supply of the reactants. Nitrogen gas, which represents 78% of Earth's atmosphere, is liquefied from air through a series of steps involving cooling and compression. As you will soon learn, hydrogen gas can be obtained chemically from natural gas (mainly methane, CH_4).

To produce the hydrogen gas, chemical engineers first treat natural gas to remove sulfur compounds; then they allow methane to react with steam:

$$\text{Thermal energy} + CH_4(g) + H_2O(g) \rightleftharpoons 3\,H_2(g) + CO(g)$$

In modern ammonia plants, this endothermic reaction takes place at 200–600 °C and at pressures of 200–900 atm. Technicians, such as the one shown in Figure 5.23, must carefully control the ratio of methane to steam to prevent the formation of carbon compounds that would lower the yield of hydrogen gas.

Carbon monoxide, a product of the hydrogen-generating reaction shown above, is converted to carbon dioxide, which is accompanied by the production of additional hydrogen gas:

$$CO(g) + H_2O(g) \rightleftharpoons H_2(g) + CO_2(g)$$

All the hydrogen gas produced is then separated from carbon dioxide and from any unreacted methane.

In the Haber–Bosch process (see page 518), the reactants (hydrogen gas and nitrogen gas) are first compressed to high pressures (150–300 atm). Ammonia forms as the hot gases (at about 500 °C) flow over an iron catalyst. Ammonia gas is removed by converting it, by means of cooling and added pressure, to liquid ammonia, which is then removed from the reaction chamber, thus reducing the rate of ammonia decomposition. Nonreacted nitrogen gas and hydrogen gas are recycled, mixed with additional supplies of reactants, and passed through the reaction chamber again.

> The CO_2 formed in this reaction can be removed in several ways, including allowing it to react with calcium oxide (CaO or lime), which forms solid calcium carbonate ($CaCO_3$), or by dissolving the CO_2 gas at high pressure in water.

B.5 NITROGEN'S OTHER FACE

The Haber–Bosch process has provided fairly cheap ammonia for a variety of applications. For example, ammonia reacts directly with nitric acid to produce ammonium nitrate, a substitute for natural nitrates once used as fertilizers.

$$\underset{\text{Ammonia}}{NH_3(g)} + \underset{\text{Nitric acid}}{HNO_3(aq)} \longrightarrow \underset{\text{Ammonium nitrate}}{NH_4NO_3(aq)}$$

> About 80% of NH_3 produced is used in fertilizer and 5% is used in explosives. Ammonia's other uses include industrial refrigeration and making ice for hockey rinks.

The widespread availability of ammonia and nitrates has changed the course of warfare as well as agriculture. Ammonia is a reactant in the production of explosives, most of which are nitrogen-containing compounds (see Figures 5.24 and 5.25). The development of the Haber–Bosch process provided a convenient source of ammonia for making both fertilizers and

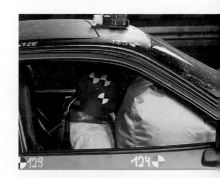

military munitions. This allowed Germany to continue fighting in World War I even after its shipping connections to Chilean nitrate deposits were cut off by the British Navy.

2,4,6-Trinitrotoluene (TNT)

NH_4NO_3
Ammonium nitrate

Nitroglycerin

Hexahydro-1,3,5-trinitro-1,3,5-triazine (RDX)

$Pb(—N=N=N)_2$
Lead azide

Figure 5.25 *Structural formulas of some common substances used as explosives. What similarities are shared by these chemical structures?*

Nitrogen-based explosives also have non-hostile, and even life-saving, uses. Air bags in automobiles are one such modern application. An air bag quickly inflates like a big pillow during a collision to reduce injuries to the driver and passengers. The uninflated air-bag assembly contains solid sodium azide, NaN_3. In a collision, sensors initiate a sequence of events that rapidly decompose the sodium azide to form a large volume of nitrogen gas:

$$3\ NaN_3(s) \longrightarrow Na_3N(s)\ +\ 4\ N_2(g)$$

Nitrogen gas quickly inflates the air bag (to about 50 L) within 50 ms (0.05 s) after the collision starts (see Figure 5.26).

The forces released by nitrogen-based explosives also blast road-cuts through solid rock during highway construction. To cut through the stone faces of hills and mountains, road crews drill holes, drop in explosive canisters, and then detonate the explosives.

Figure 5.26 *Automobile air bags deploy due to nitrogen gas released from a nitrogen-based explosive compound.*

Explosions in general result from the rapid formation of gaseous products from liquid or solid reactants. Detonating an explosive such as sodium azide, dynamite, or nitroglycerin produces gases that occupy more than a thousand times the volume of the original solid or liquid.

Many compounds used as explosives involve nitrogen atoms in a positive oxidation state and carbon in a negative oxidation state within the same reactant molecule. This creates conditions for the release of vast quantities of energy. The energy released in this type of explosive reaction is due in part to the formation of N_2, a highly stable molecule.

The powerful explosive nitroglycerin (see Figure 5.25, page 537) was invented in 1846. However, it was too sensitive (unstable) to be useful. Workers never knew when it was going to explode. The Nobel family built a laboratory in Stockholm to explore ways to control this unstable substance (Figure 5.27). Although the father and four sons were interested in explosives, one son, Alfred, was the most persistent experimenter.

> Chemical explosions are often rapid, exothermic oxidation–reduction reactions that produce large volumes of gaseous product.

Figure 5.27 *The production of nitroglycerin, devised by Alfred Nobel, was a dangerous process that needed careful monitoring. Note the one-legged stool that ensured that the attendant didn't fall asleep on the job.*

Carelessness led to accidental explosions. One killed Alfred's brother Emil. The city of Stockholm finally insisted that Alfred move his experimenting elsewhere. Determined to continue research to make nitroglycerin less unpredictable and dangerous, Alfred rented a barge and performed experiments in the middle of a lake.

Alfred finally discovered that mixing oily nitroglycerin with a finely divided solid (diatomaceous earth) caused nitroglycerin to become stable enough for safe transportation and storage. However, the nitroglycerin would still explode if a blasting cap activated it. This new, more stable form of nitroglycerin carried a new name—*dynamite*.

A new era in explosives had begun. At first, dynamite served peaceful uses in mining and in road and tunnel construction (see Figure 5.24, page 537). By the late 1800s, dynamite also found destructive use in warfare.

Military use of his invention caused Alfred Nobel considerable anguish and motivated him to use his fortune to benefit humanity. His will specified that his money be dedicated to annual international prizes for advances in physics, chemistry, physiology and medicine, literature, and peace. (The Swedish parliament later added economics as an award category.) Nobel Prizes, first awarded in 1901, are still regarded as the highest honors individuals can receive in these fields. Table 5.1 lists recent Nobel Prize recipients in chemistry, together with their contributions to chemical science.

Table 5.1 *Alfred Nobel's will established an annual award for those who "shall have made the most important chemical discovery or improvement."*

Nobel Laureates in Chemistry 2005–2010

Year	Awardees	Contributions
2010	Richard F. Heck, United States, University of Delaware; Ei-ichi Negishi, Japan, Purdue University; Akira Suzuki, Japan, Hokkaido University	Development of palladium-catalyzed cross couplings in organic synthesis
2009	Venkatraman Ramakrishnan, India, MRC Laboratory of Molecular Biology; Thomas A. Steitz, United States, Yale University, Howard Hughes Medical Institute; Ada E. Yonath, Israel, Weizmann Institute of Science	Studies of the structure and function of the ribosome
2008	Osamu Shimomura, Japan, Marine Biological Laboratory, Woods Hole, and Boston University Medical School; Martin Chalfie, United States, Columbia University; Roger Y. Tsien, United States, University of California, San Diego, Howard Hughes Medical Institute	Discovery and development of the green fluorescent protein, GFP
2007	Gerhard Ertl, Germany, Fritz-Haber-Institut der Max-Planck-Gesellschaft	Work on chemical processes on solid surfaces
2006	Roger D. Kornberg, United States, Stanford University	Studies of the molecular basis of eukaryotic transcription
2005	Yves Chauvin, France, Institut Français du Pétrole; Robert H. Grubbs, United States, California Institute of Technology; Richard R. Schrock, United States, Massachusetts Institute of Technology	Development of the metathesis method in organic chemistry, reducing potentially hazardous waste through green-chemistry synthesis methods

DEVELOPING SKILLS

B.6 EXPLOSIVE NITROGEN CHEMISTRY

Sample Problem: Another nitrogen-based explosive is trinitrotoluene (TNT). The following reaction describes the explosion of TNT.

$$2 \text{ } C_7H_5N_3O_6(s) \longrightarrow 6 \text{ } N_2(g) + 5 \text{ } H_2O(g) + 7 \text{ } CO(g) + 7 \text{ } C(s)$$

How many moles of CO are produced by the explosion of 1.5 moles of TNT?

$$1.5 \text{ } \cancel{\text{mol TNT}} \left(\frac{7 \text{ mol CO}}{2 \text{ } \cancel{\text{mol TNT}}} \right) = 5.3 \text{ mol CO}$$

The following equation describes the explosion of nitroglycerin:

$$4 \text{ } C_3H_5(NO_3)_3(l) \longrightarrow 12 \text{ } CO_2(g) + 6 \text{ } N_2(g) + 10 \text{ } H_2O(g) + O_2(g) + \text{ Energy}$$
Nitroglycerin

1. How many total moles of gaseous products form in the explosion of one mole of nitroglycerin?

2. One mole of gas at standard temperature and pressure occupies a volume of 22.4 L. One mole of liquid nitroglycerin occupies approximately 0.1 L. By what factor does the volume increase when one mole of nitroglycerin explodes? (Assume temperature remains constant.)

3. In fact, when nitroglycerin explodes, the temperature increase causes the gas volume to increase eight times more than the factor you just calculated in Question 2. By what combined factor does the total volume suddenly increase during an actual nitroglycerin explosion?

4. How do your answers to Questions 2 and 3 help explain the destructive power of a nitroglycerin explosion?

✓ concept check 4

1. What would be the effect of removing a product from a system at equilibrium? Explain your answer.
2. When water is left in a glass it evaporates completely. Is this a violation of Le Châtelier's principle? Why or why not?
3. One advantage of the Riverwood location for EKS is the availability of relatively inexpensive energy. Why is access to energy an important consideration for EKS when picking a site for ammonia production? What parts of the production of ammonia require energy?

B.7 FROM RAW MATERIALS TO PRODUCTS

Some chemical reactions you have observed in this course are essentially the same as reactions used in industry to synthesize chemical products. However, chemical reactions in industry must be scaled up to produce very large quantities of high quality products at low cost.

Four considerations become crucial in attempting to scale up chemical reactions: *engineering, profitability, waste,* and *safety*. Chemical engineers face many challenges in designing manufacturing systems for industry. One challenge is the management of energy, usually thermal energy. In a classroom laboratory, the energy released or taken in by a reaction is relatively small. When a reaction is scaled up to industrial quantities, the management of this energy transfer becomes crucial. If thermal energy is not removed appropriately, potentially dangerous, costly, and destructive situations can occur. In the case of ammonia production, many of the reactions are endothermic, so thermal energy must be added to the reaction system. If this energy input is not managed correctly, the efficiency of the reaction process will decrease, increasing costs and decreasing profitability. You will learn more about the energy requirements of ammonia production in the next section.

The chemical industry also faces challenges in dealing with unwanted materials that result from chemical processes. When reactions occur on an industrial scale, large quantities of waste can accumulate. The definition of waste is anything that is not the desired product of the reaction. So waste can refer to excess reactants, other substances (byproducts) produced by the reaction, or excess thermal energy.

Once laws were passed that regulated the handling of waste, many chemical industries discovered that, with some additional processing, they could turn some previously unwanted materials or products into valuable commodities. Some "waste" compounds can become intermediates in producing useful substances. Such wastes-turned-resources offer new sources of income. In the production of ammonia, carbon dioxide is produced as a waste product during the conversion of methane into hydrogen gas. However, this carbon dioxide can be used to produce urea, another key agricultural chemical. As a result, ammonia production facilities often manufacture urea as well.

$$2\,NH_3 + CO_2 \longrightarrow NH_2 {=} \overset{\displaystyle O}{\overset{\|}{C}} {=} NH_2 + H_2O$$
Urea

Additionally, catalysts and modified processes have allowed manufacturers to increase their efficiency and decrease the amounts of starting materials (reactants) they need. For example, refinements to the Haber–Bosch process have increased the efficiency of the process. This efficiency has come from improvements in reactor design, improved use of thermal energy transfer, and the development of new catalysts for each step in the process. In the next section, you will look at the role of energy in the production of ammonia. This information will help you decide whether EKS should be invited to open an ammonia production facility in Riverwood.

A major responsibility of the U.S. Environmental Protection Agency (EPA) is managing the cleanup of hundreds of chemical-waste sites in the United States.

The caffeine that is removed in the process of making decaffeinated coffee and teas is used in the manufacturing of caffeinated soft drinks and energy drinks.

B.8 AMMONIA AND ENERGY USE

By now, you know that the Haber–Bosch process, developed in the early 20th century, is still the dominant method of ammonia production. Even with improvements in efficiency, the production of ammonia and nitrogen-based fertilizers remains among the most energy intensive of common industrial processes. The fertilizer industry uses approximately 1.2% of all the human-generated power in the world. Production of ammonia (see Figure 5.28) represents 87% of this total.

As you learned in Section B.4, high temperatures and pressures are required for the successful reaction of hydrogen and nitrogen to form ammonia. In addition to this obvious use of energy, there are several less visible uses of energy in the overall ammonia production process. For instance, the preparation of each reactant requires energy before the actual reaction can take place. Nitrogen is widely available and would seem to be fairly economical to use, since it makes up about 80 percent of the air around us. However, nitrogen must be separated from air and purified before use in the Haber–Bosch process. This process requires a series of cooling and compression steps. Each of these steps requires energy.

Energy is also required to produce hydrogen. Of the several potential sources for hydrogen gas, methane or natural gas is most common. The first step in producing hydrogen from natural gas is the reaction of methane and steam.

Figure 5.28 *In ammonia production plants like the one shown here, large quantities of energy are required to maintain the high temperatures and pressures needed to synthesize ammonia.*

$$\text{Thermal energy} + CH_4(g) + H_2O(g) \rightleftharpoons 3\,H_2(g) + CO(g)$$

Note that this is an endothermic reaction, which means that energy must be added for this process to take place. Besides this energy input, this reaction consumes natural gas (methane), which is a fuel source. By using methane for this process, the methane is made unavailable for use as a fuel.

Reliance on methane as a source of hydrogen in ammonia production has another drawback. The cost associated with ammonia production is directly related to the price of natural gas. As natural gas prices have increased in Europe and the United States, the production of ammonia has decreased. Production has moved to areas of the world with more abundant supplies of, and lower prices for, natural gas.

As the price of natural gas continues to increase, interest in alternative sources of hydrogen for ammonia production grows. Other sources include coal gasification, biomass conversion, and electrolysis of water. However, none of these processes are as efficient as using methane.

One way to compensate for the increase in materials cost is to make the process more efficient. Recall that ammonia production facilities often are located with facilities that produce urea. This combination of production facilities is an example of how chemical industry can optimize the production process to minimize costs and waste generation. In the next section, you will learn more about the underlying principles of this type of optimization.

B.9 GREEN CHEMISTRY

Now that you know about the chemistry involved in the production of nitrogen-based fertilizers, you have part of the knowledge necessary to evaluate the EKS proposal to the town council. Another aspect to be considered is the relationship of the ammonia plant to the surrounding community.

An important global initiative, **green chemistry**, addresses the chemistry involved in an industrial process and its effects on humans and the environment. The goal is to make chemical products and production less hazardous to human health and to the environment. This initiative is sometimes termed *Benign by Design*. The word *benign* might be new to you. In medicine, something that is not harmful is called *benign*. In the context of industrial chemistry, benign by design means safety and environmental concerns are considered at the beginning of the development and design of an industrial process. Doing so might seem like common sense, but in years past, the harmful effects of chemical wastes and industrial processes were not as well understood as they are today. In meeting the objectives of green chemistry, companies also try to make their processes more efficient and profitable.

As you read the 12 Principles of Green Chemistry below, keep in mind what you have learned about how ammonia is produced and how these principles should be addressed in the EKS proposal to the Riverwood town council.

Principles guiding the green chemistry movement include these general points:

1. It is better to prevent waste than to treat it or clean it up after it forms.

2. Synthetic methods should be designed so that the desired product contains as much of the material used in the process as possible.

3. Whenever possible, reactants used and waste generated should be benign.

4. The products produced should be useful for their intended purpose but also as benign as possible.

5. Solvents and other potentially hazardous substances should be made unnecessary wherever possible and benign where not possible.

6. Energy requirements should be recognized for their environmental and economic impact and should be minimized wherever possible.

7. Raw materials (reactants) should be obtained from renewable resources wherever possible.

8. The industrial process used to make a product should contain as few steps as possible. The more steps involved, the more waste that can be generated.

9. Catalysts should be used whenever appropriate.

10. Chemical products should be designed so that if they decompose, the resulting products are not harmful.

11. Pollution detection should be incorporated into as many steps of a process as possible to detect and control the production of hazardous substances.

12. Chemical processes should be designed to reduce the chance for chemical accidents, including accidental releases, explosions, and fires.

The American Chemical Society's Green Chemistry Institute promotes education and research that encourages green-chemistry practices.

You have learned how the production of ammonia has changed over the years. Can you see how the changes in the production process of ammonia reflect the chemical industry's move toward green chemistry? In the next Making Decisions activity, you will use what you know about the production of ammonia and these ideas to identify which green chemistry principles will be most important when considering the EKS proposal.

■ MAKING DECISIONS

B.10 WHAT DOES RIVERWOOD WANT?

The first of several town meetings to discuss the possibility of a chemical plant in Riverwood is coming up soon. Representatives from EKS and WYE and town council members will attend. All interested local citizens are also encouraged to attend.

Some possible questions for the EKS representatives are listed below.

A. What are some wastes that EKS might generate in the manufacture of ammonia?

B. What is the efficiency of the process used by EKS to produce ammonia?

C. What hazards are associated with the substances that will be used by EKS?

D. Are there other options for starting materials that would more benign?

E. What are the contingency plans for accidental releases of materials?

F. What role will the price of energy have in the sustainability of the EKS facility in Riverwood?

G. What special safety training is needed by workers at the facility?

1. Identify the Principles of Green Chemistry that are addressed by each question.

2. Propose at least two other questions that could be used to evaluate EKS's commitment to the Principles of Green Chemistry.

3. It is not likely that one proposal will address all of these questions. Which questions (either in the list above or drafted by you or your classmates) must be addressed by the EKS proposal and which would be useful but not mandatory?

SECTION B SUMMARY

Reviewing the Concepts

> The rate of a particular reaction depends on temperature, reactant concentration(s), and the influence of a catalyst.

1. What is meant by the *rate* of a reaction?
2. What does *chemical kinetics* mean?
3. Explain why reactions tend to speed up
   a. with increased temperature.
   b. with increased concentration.
   c. when a suitable catalyst is added.

> When a system is in dynamic equilibrium, the rate of the forward reaction equals, and is thus balanced by, the rate of the reverse reaction.

4. What is equal about equilibrium?
5. What is dynamic about dynamic equilibrium?
6. How can you tell if chemical equilibrium is represented in a chemical equation?
7. What do *forward* and *reverse* mean when speaking about an equilibrium system?

> Le Châtelier's principle can be used to predict a shift in the equilibrium position of a reversible reaction.

8. Summarize Le Châtelier's principle.
9. List three types of stress that can be applied to an equilibrium system.
10. Consider the following system at equilibrium:

$$PCl_3(g) + Cl_2(g) \rightleftharpoons PCl_5(g) + \genfrac{}{}{0pt}{}{\text{Thermal}}{\text{energy}}$$

   What effect (if any) will each of the following changes have on the position of that equilibrium system?
   a. adding more Cl_2
   b. lowering the temperature
   c. removing some PCl_5 as it forms
   d. decreasing the total pressure

11. Consider the following reaction at equilibrium:

$$C(s) + H_2O(g) + \genfrac{}{}{0pt}{}{\text{Thermal}}{\text{energy}} \rightleftharpoons CO(g) + H_2(g)$$

   What effect (if any) will each of the following changes have on the position of that equilibrium system?
   a. lowering the temperature
   b. adding steam at constant volume to the equilibrium system
   c. adding a catalyst
   d. increasing the total pressure

12. What is the effect of removing some thermal energy from an exothermic reaction that is at equilibrium?

> Ammonia is commonly produced industrially by the Haber–Bosch process.

13. What are the sources of the reactants for the Haber–Bosch process?

14. The Haber–Bosch process works most effectively under particular conditions. What are the optimal conditions for

    a. pressure?

    b. temperature?

15. What is the advantage of removing ammonia from the reaction mixture as it forms in the Haber–Bosch process?

16. Why does the yield of ammonia decrease if the Haber–Bosch process is conducted at too high a temperature?

> Nitrogen compounds are commonly used in explosives.

17. What are the characteristics of a chemical explosion?

18. What two characteristics of nitrogen are responsible for the particular effectiveness of nitrogen-based explosives? Explain.

19. What were Alfred Nobel's contributions to chemistry?

20. Consider the equation for the principal reaction that causes an air bag to quickly inflate

$$3\ NaN_3(s) \longrightarrow Na_3N(s) + 4\ N_2(g)$$

What volume of nitrogen gas (at standard temperature and pressure, see Unit 2, page 188) would an air bag reaction involving 1.0 g NaN_3 produce?

> Modern management practices in the chemical industry stress conservation, safety, and pollution prevention in decisions about manufacturing, storing, transporting, and disposing of chemical materials.

21. "An ounce of prevention is worth a pound of cure." Explain how this relates to the green-chemistry initiative.

22. Provide an example of how following green chemistry principles can help a chemical-manufacturing company become more profitable.

What chemical principles can be used in the responsible manufacture of ammonia?

In this section, you learned about equilibrium and explored how it is affected by changes in a system, considered some nonagricultural uses of nitrogen, and studied the industrial production of ammonia. Think about what you have learned, then answer the question in your own words in organized paragraphs. Your answer should demonstrate your understanding of the key ideas in this section.

Be sure to consider the following in your response: kinetics, equilibrium, Le Châtelier's principle, energy use and management, scaling up, and green chemistry.

Connecting the Concepts

23. Consider the following equilibrium system:

$$2\,SO_2(g) + O_2(g) \longrightarrow 2\,SO_3(g) + \text{Thermal energy}$$

Explain how this equilibrium system would be affected by:

a. adding more molecules of oxygen gas at constant volume.

b. increasing the temperature.

c. increasing the volume of the reaction vessel.

d. increasing the total pressure on the system.

24. Scaling up a reaction to production levels involves many challenges not necessarily apparent at the lab scale. Discuss three such challenges faced by chemical engineers at EKS Nitrogen Products Company.

25. Refrigerated food lasts longer than food stored at room temperature. Explain.

26. Many explosive compounds contain nitrogen atoms in a positive oxidation state and carbon atoms in a negative oxidation state. What happens to the oxidation states of each type of atom if these compounds explode?

Extending the Concepts

27. Compare the nitrogen-supply crisis of the early 20th century to current concerns about petroleum supplies.

28. Before chemists can convert methane to nitrogen gas, they must remove sulfur-containing compounds. How is sulfur removed from natural gas? Why is this necessary? Include and explain relevant chemical equations as part of your answer.

29. In several situations, an industrial by-product has become a valuable commodity. Discuss, in detail, one example.

30. Describe some examples or analogies, other than chemical reactions, of dynamic equilibria.

31. The Toxics Release Inventory (TRI) is available from the EPA. Use the TRI to judge how much your state or metropolitan area has accomplished in reducing emissions from manufacturing facilities.

32. A pressure cooker reduces the total time needed to cook foods. Investigate the design of a pressure cooker and explain why it speeds up cooking times.

33. Select three green-chemistry principles and, for each, describe how you could apply and follow a comparable principle in your daily activities.

SECTION C GENERATING ELECTRICAL ENERGY FROM CHEMICAL REACTIONS

How is chemical energy transformed into electrical energy?

You have learned how EKS Nitrogen Products Company produces ammonia using the Haber–Bosch process. Such a chemical plant could affect Riverwood both positively and negatively. You will now learn more about WYE Battery Technology Corporation. Then you will be able to determine if EKS or WYE (or neither of the companies) should be invited to locate a chemical plant within the old Riverwood Corporation building.

In producing batteries for electric vehicles, WYE specializes in **electrochemistry**, chemical changes that produce or are caused by electrical energy. The following discussions and laboratory investigation provide background on electrochemistry principles. This information will help you understand how WYE's proposed new plant would operate.

GOALS

- Describe the design of a voltaic cell.
- Use the activity series of metals to predict the direction of electron flow within a particular voltaic cell.
- Use illustrations and half-reactions describing oxidation and reduction to explain chemical processes by which voltaic cells convert chemical energy to electrical energy.
- Apply the concepts of equilibrium and spontaneity to voltaic cells.

✓ concept check 5

1. What happens to an atom that is oxidized? To an atom that is reduced?
2. How can a metal activity series be used to predict chemical reactions?
3. How can energy produced by a voltaic cell be measured?

C.1 CONVERTING CHEMICAL ENERGY INTO ELECTRICITY

WYE Battery Technology Corporation proposes to manufacture batteries to be used for hybrid and electric vehicles (see Figure 5.29). To create the batteries, the company will need to harness the energy from oxidation–reduction (redox) reactions. You studied redox reactions at the beginning of the course in Unit 1. These are reactions that involve the gain and loss of electrons. You explored (in Unit 1, page 71) the reaction between a metal and the ionic solution of another metal. Through this investigation, you learned that some metals lose electrons (become oxidized) more readily than others; that is, some metals are more chemically active than others. The relative tendencies of metals to release electrons can be summarized in the activity series of metals (see Table 5.2). A metal that is higher in the activity series will give up electrons more readily than a metal that is lower. For example, according to Table 5.2, aluminum atoms are oxidized (lose electrons) more easily than iron atoms.

Figure 5.29 *The batteries produced by WYE will power electric vehicles much like this one.*

Table 5.2

Activity Series of Common Metals

Metal		Products of Metal Reactivity		
Li(s)	$\longrightarrow$	Li$^+$(aq)	+	e$^-$
Na(s)	$\longrightarrow$	Na$^+$(aq)	+	e$^-$
Mg(s)	$\longrightarrow$	Mg^{2+}(aq)	+	2 e$^-$
Al(s)	$\longrightarrow$	Al^{3+}(aq)	+	3 e$^-$
Mn(s)	$\longrightarrow$	Mn^{2+}(aq)	+	2 e$^-$
Zn(s)	$\longrightarrow$	Zn^{2+}(aq)	+	2 e$^-$
Cr(s)	$\longrightarrow$	Cr^{3+}(aq)	+	3 e$^-$
Fe(s)	$\longrightarrow$	Fe^{2+}(aq)	+	2 e$^-$
Ni(s)	$\longrightarrow$	Ni^{2+}(aq)	+	2 e$^-$
Sn(s)	$\longrightarrow$	Sn^{2+}(aq)	+	2 e$^-$
Pb(s)	$\longrightarrow$	Pb^{2+}(aq)	+	2 e$^-$
Cu(s)	$\longrightarrow$	Cu^{2+}(aq)	+	2 e$^-$
Ag(s)	$\longrightarrow$	Ag$^+$(aq)	+	e$^-$
Au(s)	$\longrightarrow$	Au^{3+}(aq)	+	3 e$^-$

Table 5.2 *The higher a metal is positioned in an activity series such as this, the more readily it gives up electrons.*

Connecting two metals from different positions on the activity series creates an electrical potential between the metals. Electrical potential (volts) is somewhat like water pressure in a pipe. Just as pressure causes water to flow in the pipe, **electrical potential** "pushes" electrons through the wire connecting the two metals. The greater the difference between the chemical activities of the two metals, the greater the electrical potential that the cell generates.

The differing tendency of metals to lose electrons allows an oxidation–reduction reaction to generate electrical energy. You can make a simple device called a **voltaic cell** from two half-cells connected in a circuit (see Figure 5.30). Each **half-cell** contains a metal partially immersed in a solution of ions of that metal. For example, one half-cell could contain copper metal immersed in a solution of Cu^{2+} ions. Another could contain a piece of zinc metal immersed in a solution of Zn^{2+} ions. You know from Unit 1, page 71, that the oxidation–reduction reaction between Cu^{2+} and Zn readily occurs. By separating these reactants into half-cells, the electrons are forced to flow through a wire and provide electrical energy to an external circuit.

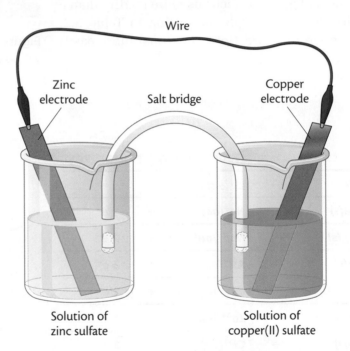

Figure 5.30 *A voltaic cell. Which metal is oxidized? Which metal is reduced? In which direction do electrons flow?*

A barrier prevents the solutions in the two half-cells from mixing. A wire connects the two metals, which act as **electrodes**, and allows electrons to flow between them, as shown in Figure 5.30. Such an electron flow constitutes an **electric current**. To complete the circuit and maintain a balance of electrical charges within the system, the two half-cells must be connected by a **salt bridge**. Dissolved ions in the salt bridge complete the internal electrical circuit by allowing ions to move freely between the half-cells, preventing build-up of electrical charge near the electrodes. Without the flow of ions, a positive electrical charge would build up in one half-cell and a negative charge would build up in the other half-cell. That situation would prevent any further flow of electrons in the wire.

■ INVESTIGATING MATTER
C.2 VOLTAIC CELLS

In this investigation, you will construct several voltaic cells and measure and compare the electrical potentials that they generate. You will also explore factors that may help determine the actual electrical potential a particular voltaic cell generates. You will set up the apparatus, formulate one or more testable questions, then design and perform an investigation to answer your question or questions.

Preparing to Investigate

1. Before you begin, put on your goggles, and wear them properly throughout the investigation.

2. Using scissors, make five notches in the filter paper. See Figure 5.31. Place the filter paper in the bottom of a Petri dish or other shallow container.

3. Label each "arm" with the symbol for the metal sample that will be placed on that arm.

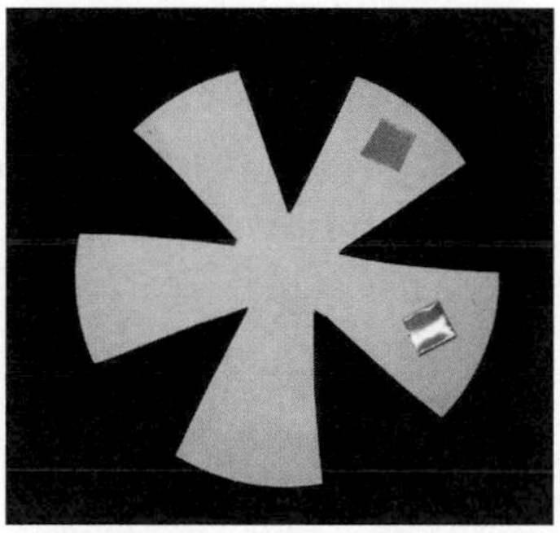

Figure 5.31 *The apparatus you will build is similar to the one depicted here. Voltaic cells result in spontaneous flow of electrons from the metal of higher activity to ions of the less-active metal.*

4. Saturate the center portion of the filter paper with KNO_3 solution from the dropper bottle.

5. Place 2–3 drops of $Cu(NO_3)_2$ on the end of one of the arms. The wet portions of the filter paper should meet.

6. Clean a copper strip with steel wool and place it on the area where the $Cu(NO_3)_2$ was dropped.

7. Repeat Steps 5 and 6 with the remaining metals and metal nitrates. You now have a completed apparatus.

8. Measure the electrical potential of each pair of metals by touching one electrode to each of the metal samples and reading the display. Record your results.

Asking Questions

Examine the apparatus and the other materials that your teacher provided. Think about what you learned about cell potential and electrodes in Section C.1 and develop one or more scientific questions that you can answer in this investigation. Share these questions with your laboratory group or classmates as directed by your teacher.

Gathering Evidence

Develop a procedure to answer your question or questions. Recall the discussion in Unit 2 (page 166) about experimental design and variables. Make sure that each investigation changes only one variable at a time. What variables will be controlled in all situations? Once the procedure has been approved by your teacher, carry out your investigations.

Interpreting Evidence

1. What changes in the apparatus provided the largest voltage? The smallest voltage?

2. List the cells (metal pairs) in order of lowest to highest potential.

3. How does the size of the electrode change the electrical potential of the cell?

Making Claims

4. Compare your results to the activity series in Table 5.2 (page 549). Do your results agree?

5. Would the electrical potential generated by cells composed of each of the following pairs of metals be larger or smaller than that of the Zn–Cu cell? Refer to Table 5.2 (page 549).

   a. Zn and Cr

   b. Zn and Ag

   c. Sn and Cu

6. How did changing the size of the zinc and copper electrodes affect the measured electrical potential? Explain and provide evidence supporting your answer.

7. Would an Ag–Au cell be a commercially feasible voltaic cell? Explain and provide evidence for your answer.

C.3 VOLTAIC CELLS AND HALF-REACTIONS

In the voltaic cells you constructed in the previous investigation, each metal sitting on filter paper soaked in a solution of its ions represented a half-cell. The activity series predicts that zinc is more likely to be oxidized (lose electrons) than copper. Thus, in the zinc–copper cell you investigated, oxidation (electron loss) occurred in the half-cell containing zinc metal immersed in zinc nitrate solution. Reduction (electron gain) took place in the half-cell consisting of copper metal in copper(II) nitrate solution. The half-reactions (individual electron-transfer steps) for that cell are:

Oxidation half-reaction: $Zn(s) \longrightarrow Zn^{2+}(aq) + 2\ e^-$

Reduction half-reaction: $Cu^{2+}(aq) + 2\ e^- \longrightarrow Cu(s)$

The electrode at which oxidation takes place is the **anode**. Reduction occurs at the **cathode**. (See Figure 5.32.)

The overall reaction in the zinc–copper voltaic cell is the sum of the two half-reactions, added so that the electrical charges balance—the total electrons lost and gained are the same—resulting in a net electrical charge of zero.

$$Zn(s) \longrightarrow Zn^{2+}(aq) + 2e^-$$
$$\underline{Cu^{2+}(aq) + 2e^- \longrightarrow Cu(s)}$$
$$Zn(s) + Cu^{2+}(aq) \longrightarrow Zn^{2+}(aq) + Cu(s)$$

Because a barrier separates the two reactants (Zn and Cu^{2+}) in the cell, the electrons released by zinc must travel through the wire to reach (and reduce) the copper ions.

The greater the difference in reactivity of the two metals in a voltaic cell, the greater the tendency for electron transfer to occur, and the greater the electrical potential (volts) of the cell. A zinc–gold voltaic cell, therefore, would generate a larger electrical potential than a zinc–copper cell. (See Table 5.2, page 549, to compare the placement of these metals in the activity series.)

One way to remember these electrode processes: Note that anode and its associated process (oxidation) both begin with vowels, whereas cathode and its process (reduction) both start with consonants.

Figure 5.32 REDuction always occurs at the **CAT**hode.

CHEM**QUANDARY**

ENERGY CELLS: ARE THEY ALL CREATED EQUAL?

You studied fuel cells in Unit 3 Section D (page 376). They are used to power cars, buses, and businesses. How do fuel cells differ from voltaic cells?

DEVELOPING SKILLS

C.4 GETTING CHARGED BY ELECTROCHEMISTRY

Each of the following questions deals with voltaic cells, their properties, and equations to describe them.

Sample Problem: Consider a voltaic cell containing lead metal (Pb) immersed in lead(II) nitrate solution, $Pb(NO_3)_2$, and a half-cell containing silver metal (Ag) in silver nitrate solution, $AgNO_3$.

   a. Predict the direction of electron flow in the wire connecting the two metals.

   b. Write equations for the two half-reactions and the overall reaction.

   c. Which metal is the anode and which is the cathode?

The answers are as follows:

   a. Table 5.2 (page 549) shows that lead is a more active metal than silver is. Therefore, lead will be oxidized, and electrons will flow from lead to silver.

   b. One half-reaction involves forming Pb^{2+} from Pb, as shown in Table 5.2. The other half-reaction produces Ag from Ag^+, which can be written by reversing the equation in Table 5.2 and doubling it so electrons lost and gained are the same:

$$Pb(s) \longrightarrow Pb^{2+}(aq) + 2e^-$$
$$2\,Ag^+(aq) + 2e^- \longrightarrow 2\,Ag(s)$$
$$\overline{Pb(s) + 2\,Ag^+(aq) \longrightarrow Pb^{2+}(aq) + 2\,Ag(s)}$$

   c. In the cell reaction, each Pb atom loses two electrons. Pb is thus oxidized, making it the anode. Each Ag^+ ion gains one electron, which is a reduction reaction. Because reduction takes place at the cathode, Ag must be the cathode.

> The total number of electrons consumed in reduction equals the total number of electrons liberated in oxidation. Overall, the voltaic cell based on these two half-reactions involves twice as many silver ions reduced as lead atoms oxidized.

Now answer the following questions.

1. Predict the direction of electron flow in a voltaic cell made from each specified pair of metals partially immersed in solutions of their ions.

   a. Al and Sn

   b. Pb and Mg

   c. Cu and Fe

2. A voltaic cell uses tin (Sn) and cadmium (Cd) as the electrodes. The overall equation for the cell reaction is

$$Sn^{2+}(aq) \; + \; Cd(s) \longrightarrow Cd^{2+}(aq) \; + \; Sn(s)$$

   a. Write the two half-reaction equations for this cell.

   b. Which metal, Sn or Cd, loses electrons more readily?

3. Sketch a voltaic cell composed of a Ni–Ni(NO$_3$)$_2$ half-cell linked to a Cu–Cu(NO$_3$)$_2$ half-cell.

4. For each voltaic cell designated below, identify the anode and the cathode. Assume that each voltaic cell uses appropriate ionic solutions.

   a. Cu–Zn cell

   b. Al–Zn cell

   c. Mg–Mn cell

   d. Au–Ni cell

concept check 6

1. Why won't a voltaic cell operate if the salt bridge is missing?
2. What factors affect a cell's electrical potential?
3. Consider the discussion of equilibrium in Section B. Write a statement about the relationship between equilibrium and a voltaic cell.

C.5 EQUILIBRIUM IN ELECTROCHEMICAL SYSTEMS

Voltaic cells create a flow of electrons by separating a pair of oxidation–reduction reactions and allowing both ions and electrons to move between the half-cells. When you built voltaic cells as part of Investigating Matter C.2, you compared the measured potential with the metal activity series. These reactions occurred without any input of additional information or stimuli. In chemical terms, the reactions were **spontaneous**. The direction that electrons flow is generally shown with an arrow. If the cell operates spontaneously, the arrow will be moving from anode toward the cathode. That means that the atoms of the anode will give up electrons, forming ions that enter the electrolyte solution. The concentration of ions will thus increase in this half-cell.

What would happen simultaneously to the electrolyte solution in the cathode half-cell? Consider what you learned about equilibrium in Section B.

A system is at equilibrium when the forward and reverse reactions occur at the same rate. For a voltaic cell to generate electricity, electrons must flow in only one direction. The voltaic cell will run only if the system is not at equilibrium. In fact, as the system gets closer to equilibrium, the cell's potential decreases.

The system can be reversed, but only if the electrons are forced to flow in a direction that is not spontaneous. When you conducted electroplating in Unit 1 (page 107), you forced electrons to flow in a direction that was not spontaneous. The energy to force the electrons to flow in reverse came from an outside source, in that case a 9-V battery. The electrons combined with metal ions to form elemental metal atoms, separating an ionic compound into the elemental components. This type of reaction transforms electrical energy into chemical energy and is called **electrolysis**.

Electrolysis can be used to extract alkali metals such as lithium from their ores by passing an electrical current through a molten mineral solution (an electrolyte), causing a chemical reaction. In this case, the flow of electrons reduces Li^+ ions in the electrolyte to lithium, $Li(s)$. This method of reducing metals is also called electrometallurgy. Electrolysis requires much electrical energy, so the cost of electricity is a factor in its use.

In the next activity, you will begin to visualize what is happening at a particulate level to each component in a voltaic cell.

> Electrolysis uses electrical energy to cause a non-spontaneous reaction. A voltaic cell generates electrical energy through a spontaneous reaction.

■ MODELING MATTER

C.6 VISUALIZING CHANGES WITHIN VOLTAIC CELLS

In Investigating Matter C.2 and sections following, you were asked to create and then draw voltaic cells. You identified the anode, cathode, electrolytic solutions, salt bridge, and direction of electron flow. In this activity, you will focus on the changes in each half-cell, including the flow of electrons and changes from atoms to ions and back.

1. Sketch a voltaic cell composed of a $Zn–Zn(NO_3)_2$ half-cell linked to a $Cu–Cu(NO_3)_2$ half-cell with a $NaNO_3$ salt bridge.
2. Consider the composition of each half-cell.
   a. What ions would be present in each of the electrolyte solutions?
   b. What ions are present in the salt bridge?
   c. Which way do electrons flow?
3. Consider the reaction at the cathode.
   a. Write the reaction that takes place at the cathode.
   b. What would happen to the cathode as the voltaic cell operates? Sketch three drawings of the cathode to represent
      i. Time 0: before the reaction begins
      ii. Time 1: after the cell has been operating for a short period of time
      iii. Time 2: at the point where the cell stops running

4. Now consider the reaction at the anode.

    a. Write the reaction that takes place at the anode.

    b. What would happen to the anode while the voltaic cell operates? Sketch three drawings of the anode to represent

        i. Time 0: before the reaction begins

        ii. Time 1: after the cell has been operating for a short period of time

        iii. Time 2: at the point where the cell stops running

5. What other term would describe the voltaic cell at Time 2, when the cell stops running?

6. You know that reduction occurs at the cathode. If the metal ions form metal atoms, how is charge conserved? In other words, what substances move into the cell? From where do these substances come?

7. What ions would be present in the cathode half-cell electrolyte solution at

    a. Time 0: before the reaction begins?

    b. Time 1: after the cell has been operating for a short period of time?

    c. Time 2: at the point where the cell stops running?

8. Answer question 7 for the anode half-cell electrolyte solution.

9. What would limit the length of time that the voltaic cell would operate? In other words, what is the limiting reactant or reactants?

10. Propose a way that the voltaic cell could operate again after it exhausts the limiting reactant.

11. Now that you have visualized what is happening at the atomic level in a voltaic cell, what do you think would happen to the electrodes if electricity were put into the system rather than taken out?

■ MAKING DECISIONS

C.7 WHAT DO WE NEED TO KNOW ABOUT WYE?

Green chemistry principles were listed and described in Section B.9 (page 543). Consider what you have just learned about voltaic cells, then answer the following questions.

1. Which of the 12 Green Chemistry Principles would be of most concern when considering manufacturing and use of voltaic cells?

2. Citizens of Riverwood must decide whether a new chemical plant will be invited to locate in the old Riverwood Corporation building. After reviewing the 12 principles, which 3 do you think are most important for making this decision?

3. What precautions must the Town Council implement to keep citizens of Riverwood safe if WYE is allowed to manufacture batteries in Riverwood?

SECTION C SUMMARY

Reviewing the Concepts

> Electrochemistry involves chemical changes that produce or are caused by electrical energy.

1. What is electrolysis?
2. What is a voltaic cell?
3. What is a half-cell?
4. Voltaic cells require a salt bridge.
   a. What is a salt bridge?
   b. Why is a salt bridge necessary for the operation of a voltaic cell?
   c. Describe one way to make a salt bridge.
5. Diagram a simple Ag–Cu voltaic cell. Label electrodes, solutions, and salt bridge.
6. Considering your results in Investigating Matter C.2 (page 551), does the electrical potential produced by a voltaic cell depend on the
   a. size of the electrodes? Explain.
   b. specific metals used? Explain.

> The activity series of metals can be used to predict the direction of electron flow within a particular voltaic cell.

7. In a voltaic cell, what process takes place at the
   a. anode?
   b. cathode?

8. Sketch a voltaic cell made from Ni and Zn in solutions of their ions. Label anode and cathode (identifying each metal) and show the direction of electron flow. See Table 5.2 (page 549).
9. Predict the direction of electron flow in a voltaic cell made from each of these metal pairs in solutions of their ions. See Table 5.2 (page 549).
   a. Ag and Sn
   b. Cr and Ag
   c. Cu and Pb

> Redox reactions can be used to convert chemical energy to electrical energy by harnessing the flow of electrons.

10. Using Table 5.2 (page 549), write the overall reaction and the two half-reactions for a voltaic cell made from each of these metal pairs in solutions of their ions.
    a. Ag and Sn
    b. Cr and Ag
    c. Cu and Pb
11. Sketch a voltaic cell made from Sn and Zn in solutions of their ions. Label anode and cathode (identifying each metal) and write a half-reaction for the process that occurs at each electrode. Explain, referring to the sketch, how electricity is generated by this system. See Table 5.2 (page 549).

> Voltaic cells operate spontaneously, leading to constant changes within the system.

12. Consider the anode and cathode in a voltaic cell.

   a. How does the anode change during operation?

   b. How does the cathode change during operation?

13. What changes occur in the solution on the anode side of a voltaic cell?

14. How does the concentration of the cathode cell solution change?

15. Assuming that the voltaic cell is constructed properly, when will it stop working?

How is chemical energy transformed into electrical energy?

In this section, you have constructed, diagrammed, and studied voltaic cells and their particulate level changes, written half-reactions, learned about equilibrium and voltaic cells, and related metal reactivity to voltaic cell operation. Think about what you have learned, then answer the question in your own words in organized paragraphs. Your answer should demonstrate your understanding of the key ideas in this section.

Be sure to consider the following in your response: anode, cathode, salt bridge, voltaic cell, electrolysis, half-reactions, equilibrium, and spontaneous reactions.

Connecting the Concepts

16. Compare Table 1.3 (page 75) and Table 5.2 (page 549). What is included in Table 5.2 that was not part of Table 1.3? Why is this information more important in the current unit?

17. Consider the discussion of Le Châtelier's principle in Section B.1 (page 531). Applying these ideas to a specified voltaic cell, how could you change the quantity of electrical energy produced without changing the identity of the electrodes?

18. In Unit 3, you learned that all energy transformations lead to some loss of energy. Where might some energy be lost in the transformation of chemical energy to electrical energy within a voltaic cell?

19. In everyday life, decisions made on an impulse may be referred to as *spontaneous*. How is this use of the word *spontaneous* similar to and different from how it is used to describe chemical reactions?

Extending the Concepts

20. Investigate the concept of *entropy* and discuss some of the challenges entropy presents in voltaic cells.

21. Recently, fuel cells have become available as an alternative to electrical energy generated in power plants and transmitted through wires. Investigate these cells and consider whether you would want them to be located in your neighborhood. Consider, for example, whether they are a "greener" alternative for electrical energy and what types of fuels they require and wastes they produce.

CHEMISTRY *AT WORK*
Q&A

Khalil Amine, Materials Scientist and Group Leader at Argonne National Laboratory in Argonne, Illinois

The small energy storehouses known as batteries have made it possible to power up electronic machines anyplace, not just near an electrical outlet. But as useful as batteries are, they're far from perfect. Take the battery in a car. Even though a car battery can charge up a vehicle's lights and start the engine, it doesn't provide enough energy to move the car very far—that's why cars still need gasoline. As Americans' interest in electric and hybrid gas–electric cars has grown, the need for better batteries has increased. Read on to see how one chemist's work is fueling the next generation of high-tech batteries for these cars.

An electric car developed recently in Norway.

Q. What is materials science, and why do you need it to develop new types of batteries?

A. Materials science is the study of matter and new ways to use matter to meet our needs. It's a field made up of many other fields, including physics and engineering, but it's mostly chemistry. To create new kinds of batteries, we have to invent new types of materials. We use materials science to try to understand everything about these materials: their structures, physical properties, electrochemical properties, and the impact that all these factors have on what we would like these materials to do, such as store energy.

Q. Why did you choose to go into materials science, and why did you focus on batteries?

A. When I was growing up in Morocco, I was always good at math and chemistry. I had to choose between these two fields before I left to go to college in France. Chemistry was much more interesting to me because there are so many different fields of chemistry to study. I earned a master's degree in chemistry and a doctoral degree in materials science, since it has such a heavy chemistry focus. I used my degree in materials science to get a job in Japan in 1990. In Japan, energy use is an important issue. The country has no oil, gas, or coal reserves, so it has to import almost all of its energy sources from overseas. Eventually, I started working on developing new types of batteries to fill this energy need. Now, I focus my own energy on the same kind of work for Argonne.

The typical lead-acid battery in a gasoline-powered car weighs more than thirty pounds.

Q. What's wrong with the batteries we already have in cars?

A. The batteries in gasoline-powered cars are lead-acid batteries. This type of technology is almost 90 years old, and these batteries are trustworthy, durable, and cheap. There's nothing wrong with using them the way we do now, to start our cars and power our headlights and radios. But cars that run only on electricity will need a battery that powers everything, including motion. Lead-acid batteries can't store enough energy for this purpose. A car would need 50 lead-acid batteries to run on electricity!

A lithium-ion cell-phone battery would not provide nearly enough energy to power a car.

Q. What types of batteries will replace lead-acid batteries in electric cars?

A. The best current candidate to replace lead-acid batteries is the lithium-ion battery. These batteries are the kind that we use in cell phones and computers now. The most common lithium-ion batteries have a cathode made of a lithium compound and an anode made of carbon. A solution of lithium ions in between allows current to pass from one end to the other. These batteries are small and light, so they are a good choice for cars.

Q. So, will we use cell-phone batteries for our cars in the future?

A. We won't use the same batteries because they still have a few kinks we'd like to work out. The electrolyte solution currently in use is flammable, so my colleagues and I are working on making a safer version. We're also experimenting with new materials for the anode. These new materials change the types of chemical reactions that happen in batteries, which can vastly increase the quantity of energy these batteries can store. Different types of metal alloys, such as nickel, manganese, and cobalt compounds, are all possibilities. We're especially interested in another kind of battery in development called a lithium-air battery. These batteries have a cathode made of porous carbon, which allows oxygen from the air to flow into the battery. The oxygen participates in chemical reactions that allow lithium-air batteries to store 12 times the quantity of energy that regular lithium-ion batteries can! These batteries might be powering our cars in a few decades.

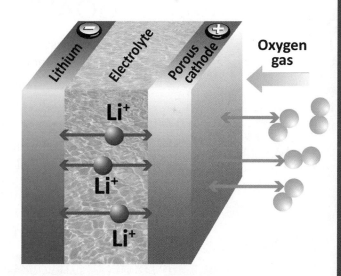

The lithium-air battery has a cathode of porous carbon (at right). The anode (at left) is usually nonporous lithium foil.

Q. What advice do you have for students interested in battery technology?

A. Being innovative and having a broad knowledge base is really important for this field. You should take classes in chemistry, electrochemistry, and materials science in college, along with other fields of science too—you never know what might spur an idea that will lead to a new or better battery. Right now, battery technology has electric potential for students with the right spark.

SECTION D INDUSTRIAL PRODUCTION OF BATTERIES

What challenges must be met to optimize production and use of batteries?

You have learned how voltaic cells produce electrical energy from chemical reactions. In this section, you will learn more about the batteries that WYE Battery Technology Corporation proposes to manufacture in Riverwood. Could these advanced vehicle batteries contribute to the economy of Riverwood and the United States? What risks are posed by manufacturing these batteries? Would EKS or WYE best suit the needs and capabilities of the Riverwood community? Read on to gather more evidence to support your decision.

GOALS

- Describe the function and composition of primary and secondary batteries.
- Calculate the electrical potential produced by a voltaic pile or battery.
- Analyze the life cycle of a battery.
- Use burden–benefit analysis in decisions about implementing chemical technologies.

concept check 7

1. How does a voltaic cell produce electrical energy?
2. In Investigating Matter C.2, you considered the effect of electrode size on a voltaic cell.
   a. Does the size of the electrode affect the electrical potential of a voltaic cell? Explain.
   b. What characteristics of the cell might be affected by the size of the electrode?
3. What type of substance would you expect to find in an alkaline battery?

■ INVESTIGATING MATTER

D.1 BUILDING A VOLTAIC PILE

Preparing to Investigate

In 1800, Alessandro Volta constructed a *voltaic pile* by stacking pairs of zinc and copper plates, each pair separated by cloth or cardboard that had been soaked in a solution of salt or sulfuric acid. Since concentrated sulfuric acid is corrosive, you will use a more benign electrolyte, copper(II) sulfate, in this investigation.

Gathering Evidence

Part I: Volta's Pile

1. Before you begin, put on your goggles, and wear them properly throughout the investigation.

2. Place each piece of zinc on top of a copper piece. Assemble four pairs of Cu–Zn pieces.

3. Prepare pieces of filter paper soaked in copper(II) sulfate as directed by your teacher.

4. Build a stack or pile of these pairs of discs as shown in Figure 5.33, with a piece of electrolyte soaked filter paper separating each pair from its neighbors. The sequence of the components is as follows: Cu, electrolyte, Zn, Cu, electrolyte, Zn, etc. (*Note:* Be careful that the electrolyte solution does not drip down the side of the stack as this can cause a short circuit between the elements of the pile.)

Figure 5.33 *Each pair of zinc and copper pieces is separated from the adjacent pair by a piece of filter paper soaked in electrolyte.*

5. When the device is arranged as specified, measure the voltage between the bottom-most copper piece and the top-most zinc piece. Record your data.

Part II: Constructing to Specifications

6. Based upon your results in Part I, build a pile that will produce an electrical potential of ~6.6 volts.

7. Describe the construction of your pile.

8. Try to power a small LCD electronic device such as a clock, thermometer, or calculator. Describe your results.

Interpreting Evidence

1. What was the measured potential of the pile in Part I?

2. What was the measured potential of your pile in Part II?

3. Were you able to power an electronic device in Part II? Explain.

Making Claims

4. What factors in the design of the pile would affect the potential generated? Explain.

Reflecting on the Investigation

5. How did you decide how to construct the pile in Part II?

6. How is the pile similar to commercial voltaic cells, i.e., batteries?

D.2 CELL POTENTIAL

Cell potential is sometimes informally referred to as voltage, since it is measured in volts.

As you observed, the cell potential of a voltaic pile varies with the number of "units" in the pile. You learned earlier that connecting two metals in a voltaic cell creates an electrical potential between the metals. In the voltaic pile, several cells are placed in series, so the total electrical potential is the sum of the individual cell potentials.

Volta used silver discs in place of copper in some of the voltaic piles that he constructed. Unfortunately for Volta, the only way he could compare the potential produced by different piles was by evaluating the shock he received when he completed the circuit. Fortunately for us, we can now accurately measure cell potential and know that silver–zinc cells produce a different potential than copper–zinc cells.

The potential of a cell reflects its tendency to move electrons through the cell. A cell with a small potential does not have much power to move electrons. A cell at equilibrium has no electron movement, and thus no potential.

Cell potentials for many electrochemical cells have been measured at standard conditions. Standard conditions refer to pure gases at 1 atm pressure, pure metals, and electrolyte solutions with concentrations of 1.0 mol/L. To make comparisons easier, these standard cell potentials are often split into contributions from each of their electrodes. These electrode potentials are known as standard potentials and denoted $E°$. Since it is not possible to measure the potential of an electrode in isolation, electrodes are compared against the standard hydrogen electrode ($2\,H^+(aq) + 2\,e^- \longrightarrow H_2(g)$), which has a potential defined as 0 V.

Electrode potentials are often organized within a table of standard reduction potentials. Given such a table, the standard potential of any cell can be calculated by first determining which metal will be reduced. The reduction half-reaction of the other electrode reaction (the oxidized species) is then reversed, and the sign of its potential is switched. The half-reaction equations and their potentials are added together to determine the overall standard cell potential. One or both of the half-reaction equations may need to be multiplied to ensure that the number of electrons lost and gained is equivalent, but this multiplication does *not* affect the standard potentials.

In the next activity, you will apply these ideas to some voltaic cells that you have investigated in Investigating Matter C.2 and D.1. As you look at and use Table 5.3, think about the metals listed and their everyday applications. What metals might you choose to use if you wanted to maximize electrical potential? Why?

Recall that the less active metal is more likely to be reduced, so Table 5.3 lists metals in the opposite order from a metal activity series.

Table 5.3

Standard Reduction Potentials

Species	Half-reaction	Standard Reduction Potential, $E°$ (volts)
Au^+/Au	$Au^+(aq) + e^- \longrightarrow Au(s)$	+1.61
Ag^+/Ag	$Ag^+(aq) + e^- \longrightarrow Ag(s)$	+0.80
Cu^{2+}/Cu	$Cu^{2+}(aq) + 2\,e^- \longrightarrow Cu(s)$	+0.34
H^+/H_2	$2\,H^+(aq) + 2\,e^- \longrightarrow H_2(g)$	0 (by definition)
Pb^{2+}/Pb	$Pb^{2+}(aq) + 2\,e^- \longrightarrow Pb(s)$	−0.13
Sn^{2+}/Sn	$Sn^{2+}(aq) + 2\,e^- \longrightarrow Sn(s)$	−0.14
Ni^{2+}/Ni	$Ni^{2+}(aq) + 2\,e^- \longrightarrow Ni(s)$	−0.23
Cd^{2+}/Cd	$Cd^{2+}(aq) + 2\,e^- \longrightarrow Cd(s)$	−0.40
Fe^{2+}/Fe	$Fe^{2+}(aq) + 2\,e^- \longrightarrow Fe(s)$	−0.44
Zn^{2+}/Zn	$Zn^{2+}(aq) + 2\,e^- \longrightarrow Zn(s)$	−0.76
Al^{3+}/Al	$Al^{3+}(aq) + 3\,e^- \longrightarrow Al(s)$	−1.66
Mg^{2+}/Mg	$Mg^{2+}(aq) + 2\,e^- \longrightarrow Mg(s)$	−2.36
Li^+/Li	$Li^+(aq) + e^- \longrightarrow Li(s)$	−3.05

DEVELOPING SKILLS

D.3 DETERMINING POTENTIAL IN ELECTROCHEMICAL CELLS

Sample Problem: Consider a voltaic cell containing lead metal (Pb) immersed in lead(II) nitrate solution, $Pb(NO_3)_2$, and a half-cell containing silver metal (Ag) in silver nitrate solution, $AgNO_3$.

   a. Determine which metal will be oxidized and which metal will be reduced.

   b. Write equations for the two half-reactions, including standard potentials.

   c. Write the equation for the overall reaction, including the standard potential.

The answers are as follows:

   a. Table 5.3 (page 565) shows that silver has a positive reduction potential (lies higher in the table). Therefore, silver will be reduced and lead will be oxidized.

   b. One half-reaction involves forming Pb^{2+} from Pb, which can be written by reversing the equation in Table 5.3. The other half-reaction produces Ag from Ag^+, as shown in Table 5.3, which must be doubled so that electrons lost and gained are the same. Standard potentials are not affected by this multiplication but do change signs when the equation is reversed.

$$Pb(s) \longrightarrow Pb^{2+}(aq) + 2e^- \qquad E° = + 0.13$$
$$2\,Ag^+(aq) + 2e^- \longrightarrow 2\,Ag(s) \qquad E° = + 0.80$$

   c. $Pb(s) + 2\,Ag^+(aq) \longrightarrow Pb^{2+}(aq) + 2\,Ag(s) \qquad E° = + 0.93$

1. How can the electrical potentials you measured in Investigating Matter D.1 be explained in terms of the values in Table 5.3 (page 565)?

2. A voltaic cell uses magnesium (Mg) and zinc (Zn) as the electrodes.
   a. Which metal will be the cathode?
   b. Write the cathode half-reaction, including the standard potential.
   c. Write the anode half-reaction, including the standard potential.
   d. Write the overall reaction equations for this cell, including the standard potential.

3. In Developing Skills C.4, you considered the flow of electrons and identified the anode and cathode in several voltaic cells. Now use Table 5.3 to calculate the expected electrical potential and identify the anode and cathode for each of the following voltaic cells. (Assume that each voltaic cell uses appropriate ionic solutions.)
   a. Cu–Fe cell
   b. Ni–Cd cell
   c. Pb–H_2 cell
   d. Li–Zn cell
   e. Al–Sn cell

4. Suppose you wanted to make a voltaic cell with a potential of at least 1.5 V. Use Table 5.3 (page 565) to propose one possible cell.

D.4 PRIMARY BATTERIES

Voltaic cells are a convenient way to convert chemical energy to electrical energy. Such cells can fit in small, portable containers. Commercial voltaic cells, called **batteries**, can be constructed from various combinations of metals and ions. In an ordinary zinc–carbon battery, zinc is the anode, and a graphite rod surrounded by a water-based paste of manganese(IV) oxide (MnO_2) and graphite serves as the cathode. A mixture of ammonium chloride and zinc chloride in an aqueous paste serves as the electrolyte. The fact that the electrolyte is in paste form, rather than liquid, accounts for other name of this type of battery—the **dry cell**.

Alkaline batteries, common in portable music players, have similar zinc and graphite–MnO_2 electrodes, but the electrolyte is an alkaline aqueous potassium hydroxide (KOH) paste. See Figure 5.34. Alkaline batteries are also considered dry cells, as are most modern consumer batteries. Zinc–carbon and alkaline batteries are **primary batteries**, that is, they are designed for a single use and cannot be recharged.

> The term *battery* once referred to two or more electrochemical cells connected together. In common use, though, a battery is any portable source of electrical energy from chemical energy.

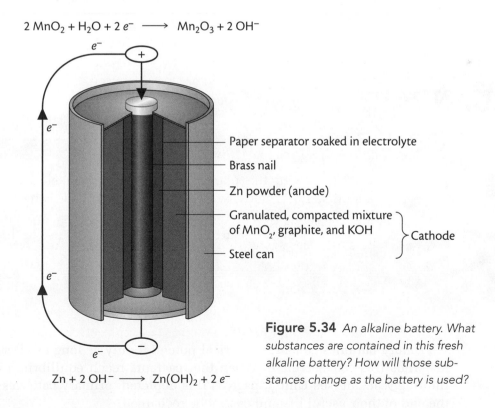

$$2\,MnO_2 + H_2O + 2\,e^- \longrightarrow Mn_2O_3 + 2\,OH^-$$

Paper separator soaked in electrolyte

Brass nail

Zn powder (anode)

Granulated, compacted mixture of MnO_2, graphite, and KOH $\Big\}$ Cathode

Steel can

$$Zn + 2\,OH^- \longrightarrow Zn(OH)_2 + 2\,e^-$$

Figure 5.34 *An alkaline battery. What substances are contained in this fresh alkaline battery? How will those substances change as the battery is used?*

Both zinc–carbon and alkaline batteries generate an electrical potential of 1.54 V. The following oxidation–reduction equations describe the chemical changes involved.

Zinc–Carbon Battery

Oxidation: $Zn(s) \longrightarrow Zn^{2+}(aq) + \cancel{2e^-}$

Reduction: $2\,MnO_2(s) + 2\,NH_4^+(aq) + \cancel{2e^-} \longrightarrow 2\,MnO(OH)(s) + 2\,NH_3(g)$

Overall: $Zn(s) + 2\,MnO_2(s) + 2\,NH_4^+(aq) \longrightarrow Zn^{2+}(aq) + 2\,MnO(OH)(s) + 2\,NH_3(g)$

Alkaline Battery

Oxidation: $Zn(s) + \cancel{2\,OH^-}(aq) \longrightarrow Zn(OH)_2(s) + \cancel{2e^-}$

Reduction: $2\,MnO_2(s) + H_2O(l) + \cancel{2e^-} \longrightarrow Mn_2O_3(s) + \cancel{2\,OH^-}(aq)$

Overall: $n(s) + 2\,MnO_2(s) + H_2O(l) \longrightarrow Zn(OH)_2(s) + Mn_2O_3(s)$

While zinc–manganese dioxide systems are the most common dry cells, there are several other types of primary dry cells in use. Lithium cells have an anode made of lithium foil and a cathode constructed from a metal oxide, commonly manganese dioxide. A lithium salt in an organic solvent serves as the electrolyte.

Zinc–air and zinc–silver oxide cells have the same anode (zinc powder) and electrolyte (KOH) as alkaline batteries, but use either air and carbon or silver oxide and carbon, as their cathodes. Zinc–air batteries are commonly used in hearing aids but suffer from a high self-discharge rate. Self-discharge is the natural loss of energy within stored batteries from chemical reactions within the cell. Zinc–silver oxide cells have fewer problems with self-discharge, but are more expensive.

CHEM**QUANDARY**

BATTERY SIZES

Each battery in the photo generates the same electrical potential, 1.5 V. So why does anyone need 1.5-V batteries larger than the smallest one? Wouldn't it save space, weight, and perhaps even resources to restrict consumer use to the smallest batteries?

Primary batteries generate electrical potential only as long as all starting materials (reactants) remain. When the reactants reach equilibrium or the limiting reactant (see Unit 3, page 343) is depleted, primary batteries are at the end of their useful life and cannot be recharged.

D.5 SECONDARY BATTERIES

Nickel-Based Batteries

Unlike primary batteries, secondary batteries are rechargeable. We can return their systems to their original states and reuse the batteries. Nickel–cadmium (NiCd) cells and lead-acid automobile batteries are common examples of such rechargeable batteries. The NiCd rechargeable battery is based on a nickel–iron battery that Thomas Edison developed in the early 1900s. The NiCd battery anode is cadmium, and the cathode is nickel oxide. These electrodes have a rolled design (Figure 5.35) that increases surface area and, thus, increases the generated current (electron flow).

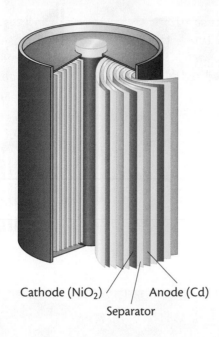

Cathode (NiO$_2$)   Anode (Cd)

Separator

Figure 5.35 *Cross-section of a NiCd battery.*

When this battery operates, Cd and NiO$_2$ are converted to Cd(OH)$_2$*(s)* and Ni(OH)$_2$*(s)*. These solid products cling to the electrodes, allowing them to convert back to reactants when the used battery is connected in a recharging circuit. In recharging, an external electrical potential causes electrons to flow in the opposite direction, thus reversing the reaction. Although you can recharge NiCd batteries many times, some of the same processes that prevent recharging of primary batteries can take their toll and eventually reduce the efficiency of recharging.

NiMH or nickel–metal hydride batteries (see Figure 5.36, page 570) use a metal–metal compound that absorbs hydrogen instead of the cadmium in NiCd batteries, improving both their performance and their environmental impact. Most gasoline–electric hybrid vehicles built before 2010 use NiMH cells for energy storage.

> NiCd batteries should be recycled to prevent the cadmium, which is toxic, from leaching into ground water.

Scientific American Working Knowledge Illustration

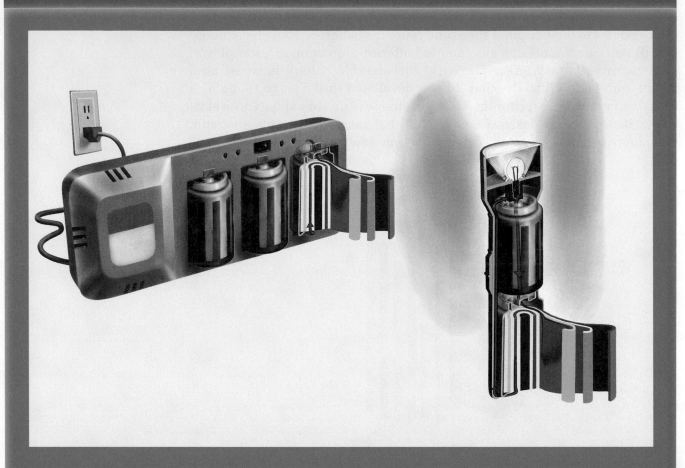

Figure 5.36 *Rechargeable batteries When a nickel–metal hydride (NiMH) battery is recharged, the charging circuit removes electrons (red) from the positive electrode (green) composed of nickel(II) hydroxide. This process oxidizes the Ni^{2+} to Ni^{3+}. Electrons are supplied to the negative electrode (blue). This causes a hydrogen-absorbing metal alloy to become reduced. When this battery is discharged during use, electron flow is reversed. The nickel(II) hydroxide gains electrons, while the metal hydride loses electrons.*

Lead-Acid Batteries

One everyday example of a rechargeable wet cell (a battery with a liquid electrolyte) is the 12-V automobile battery, known as a *lead-acid battery*. This battery, often used for starting, lighting, and ignition in a motor vehicle, is composed of a series of six electrochemical cells. As illustrated in Figure 5.37, each cell consists of an anode of uncoated lead (Pb) plates and a cathode of lead plates coated with lead(IV) oxide (PbO_2). The electrodes are immersed in a dilute solution of sulfuric acid, H_2SO_4, the electrolyte in this system.

When the vehicle's ignition is turned on, the electrical circuit is completed. As the electrons travel from one electrode to the other, they provide energy for the car's electrical systems. Metallic lead at the anode is oxidized to Pb^{2+}.

Lead-acid batteries are found in trucks, aircraft, motorcycles, self-starting lawnmowers, and motorized golf carts.

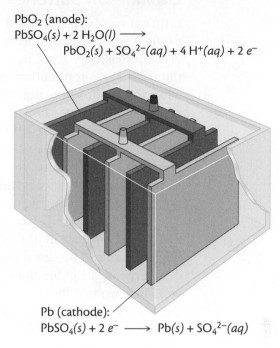

DISCHARGING

PbO$_2$ (cathode):
PbO$_2$(s) + SO$_4{}^{2-}$(aq) + 4 H$^+$(aq) + 2 e$^-$ $\longrightarrow$
PbSO$_4$(s) + 2 H$_2$O(l)

Pb (anode):
Pb(s) + SO$_4{}^{2-}$(aq) $\longrightarrow$ PbSO$_4$(s) + 2 e$^-$

CHARGING

PbO$_2$ (anode):
PbSO$_4$(s) + 2 H$_2$O(l) $\longrightarrow$
PbO$_2$(s) + SO$_4{}^{2-}$(aq) + 4 H$^+$(aq) + 2 e$^-$

Pb (cathode):
PbSO$_4$(s) + 2 e$^-$ $\longrightarrow$ Pb(s) + SO$_4{}^{2-}$(aq)

Figure 5.37 *Discharging and recharging an automobile battery. What furnishes the energy to recharge this battery?*

The freed electrons travel through the wire to the lead(IV) oxide cathode, reducing Pb^{4+} in PbO$_2$ to Pb^{2+}. Lead ions (Pb^{2+}) produced at both electrodes then form PbSO$_4$ by reacting with the electrolyte, as shown in the following equations.

$$\text{Oxidation: Pb}(s) + \text{SO}_4{}^{2-}(aq) \longrightarrow \text{PbSO}_4(s) + \cancel{2 e^-}$$
$$\underline{\text{Reduction: PbO}_2(s) + \text{SO}_4{}^{2-}(aq) + 4\,\text{H}^+(aq) + \cancel{2 e^-} \longrightarrow \text{PbSO}_4(s) + 2\,\text{H}_2\text{O}(l)}$$
$$\text{Overall: PbO}_2(s) + \text{Pb}(s) + 4\,\text{H}^+(aq) + 2\,\text{SO}_4{}^{2-}(aq) \longrightarrow 2\,\text{PbSO}_4(s) + 2\,\text{H}_2\text{O}(l)$$

If an automobile battery is used too long without being recharged, it runs down; that is, the redox reaction stops. Lead(II) sulfate eventually coats the electrodes, which reduces their ability to react and produce a current and electrical potential. In an automobile, recharging is accomplished by an alternator or generator, which converts some mechanical energy from the vehicle's engine into electrical energy that forces electrons to move in the opposite direction through the battery. This reverses the direction of the battery's chemical reactions (see Figure 5.37).

Lead-acid batteries can pose dangers if rapidly charged by an outside source of electrical energy. Hydrogen gas, formed by the reduction of H$^+$ in the acidic electrolyte, is released at the lead electrode. A spark or flame can ignite the hydrogen gas, causing an explosion. This is one reason you must be careful to avoid sparks when using jumper cables to start a vehicle with a dead battery.

You can sometimes recognize a run-down or dead car battery by observing layers of solid white PbSO$_4$ coating the electrodes.

Lithium-Ion Batteries

Lithium-ion batteries are a type of rechargeable batteries now commonly used in products such as laptops, cell phones, and MP3 players. They are also increasingly used in gasoline-electric vehicles and electric vehicles—the intended application for WYE's batteries. Unlike lithium (primary) batteries, which are not rechargeable, lithium-ion batteries use a carbon anode and a cathode constructed of a compound of lithium with oxygen and another metal. These lithium compounds include lithium cobalt oxide ($LiCoO_2$), lithium iron phosphate ($LiFePO_4$), and lithium manganese oxide ($LiMn_2O_4$). The electrolyte is usually an organic solvent and a lithium salt. Lithium ions move from the cathode, through the electrolyte, and attach to the carbon anode during charging. During use, this process is reversed.

Another type of lithium-ion battery, called a Li-poly or LiPo battery, uses a polymer gel instead of the liquid electrolyte. Lithium-ion batteries are easily customized to the devices that they power. They are slower to self-discharge and lighter than most other secondary batteries. However, these cells lose up to 20% of their capacity each year due to deposits that form in the electrolyte during use and charging. They are also subject to overheating and must contain safety features that increase production costs.

concept check 8

1. Distinguish between primary and secondary batteries.
2. How do lithium and lithium-ion cells differ?
3. What types of batteries are commonly recycled? Why do you think that these batteries (and not other types) are recycled?

D.6 MANUFACTURING AND RECYCLING BATTERIES

Battery Manufacturing

The manufacture of commercial batteries varies depending upon the type of battery, but all have several common features, including anode, cathode, electrolyte, and container. The composition of several types of batteries is summarized in Table 5.4.

Primary batteries are often constructed within a metallic (usually nickel-plated steel) can. For an alkaline battery, this can functions as part of the battery's cathode. Alkaline batteries are constructed by grinding the cathode materials, including manganese oxide and graphite, into small particles, then compacting this material into a hollow cylinder and inserting it into the steel can. A paper separator soaked in potassium hydroxide electrolyte is inserted inside the can/tablet assembly (see Figure 5.34, page 567), and a gel containing zinc powder is used to fill the inner portion of the battery. This zinc-containing gel acts as the anode. The battery is completed with a brass nail for current collection and a seal to contain the ingredients and extend battery life.

Lead-acid car batteries, by contrast, have containers made from polymers, often polypropene (polypropylene). These containers are divided into six sections, each of which will contain one cell. The electrodes are prepared on grids made from lead or a lead alloy, which are coated with pastes. The cathode paste contains lead oxide, sulfuric acid, and water, whereas the anode paste adds powdered sulfates to this mixture. A porous insulating material is used to separate the positive and negative plates. One element or cell (anode, cathode, and separator) is placed into each section of the container. The cells are connected with conductors, the container is filled with a mixture of sulfuric acid and water (battery acid), and the battery is sealed.

> Nearly 90% of U.S. used car batteries are recycled, leading to the recovery of lead metal, sulfuric acid, and battery-casing polymer.

Table 5.4

Composition of Common Primary and Secondary Batteries

		Principal Components		
	Type	**Anode**	**Electrolyte**	**Cathode**
PRIMARY	Zinc–carbon	Zn (sheet)	NH_4Cl or $ZnCl_2$	MnO_2/C
	Alkaline	Zn (powder)	KOH	MnO_2/C
	Lithium	Li (foil)	Organic solvent and Li salt	Metal oxide
	Zinc–air	Zn (powder)	KOH	Air/C
	Zinc–silver oxide	Zn (powder)	KOH	Ag_2O/C
SECONDARY	NiCd	Cd	KOH	NiO_2
	NiMH	Metal-metal compound	KOH	NiO_2
	Lead-acid	Pb	H_2SO_4	PbO_2
	Lithium ion	C	Organic solvent and Li salt	Li/metal/oxygen compound
	Li–poly or LiPo	C	Polymer gel and Li salt	Li/metal/oxygen compound

Adapted from Smith, M. J.; Gray, F. M. Batteries, from Cradle to Grave. *J. Chem. Ed.* 2010, 87, 162–167.

Lithium-ion batteries, like those WYE Battery Technology Corporation produces, are also constructed in individual cells. Each cell contains a graphite (carbon) anode and a manganese-based cathode, as well as a separator designed to allow for ion flow. The electrodes are made by creating a *slurry* (a suspension containing fine particles of an insoluble material) from a solvent and particles of active electrode materials and binders. The slurry is then applied to a surface to form electrodes, much like paste is applied to plates in the lead-acid battery construction process. The electrodes and separator are then assembled into a cell, an electrolyte consisting of an organic solvent and a lithium salt is added, and the cell is encased in a polymer-coated aluminum cover. Several cells are needed to construct an electric-vehicle battery.

Battery Recycling

Although recycling of lead-acid batteries is very common, recycling of other battery types is far less widespread in the United States. Recycling of lead-acid batteries makes sense both economically and ecologically. The lead can be reprocessed and reused, but it cannot be safely discarded. The situation is less clear cut for other batteries, particularly common primary alkaline and zinc–carbon cells. Since these batteries no longer contain mercury or other heavy metals, they are fairly safe for disposal. In addition, their recycling requires energy and time and results only in the recovery of common metals including zinc, steel, and manganese compounds.

The outlook is better for recycling of secondary batteries. Many of these batteries contain nickel, cadmium, or cobalt, which are scarcer than zinc and iron and may have sources located in areas of political unrest. While lithium supplies are also limited, the economic need to recover lithium does not yet play a large role in secondary battery recycling. The process of recycling any battery is energy-intensive. The combustible parts of the battery are burned off, then the remainder is cut up and thermal energy is added to liquefy the metals, which can then be separated.

Recycling is likely to be more practical for larger rechargeable batteries, such as those used in electric or hybrid-electric vehicles. Also, new batteries can be designed with the intention of recycling.

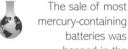

▮ MODELING MATTER
D.7 LIFE CYCLE OF A BATTERY

Earlier, in Modeling Matter A.4, you represented the nitrogen cycle using a diagram. Throughout this course, you have explored natural cycles, including the carbon and nitrogen cycles, as well as life cycles of human-made

---

Some companies are exploring the use of water to replace the more commonly used organic solvents in the slurry.

In Belgium and Austria, two of the world's leaders in battery recycling, more than half of all used batteries are collected.

The sale of most mercury-containing batteries was banned in the United States in 1996.

products. In many ways, because of the nature of atoms and conservation, these cycles are similar. In this activity, you will consider the life cycle of the product manufactured by WYE Battery Technology Corporation—the lithium-ion battery.

1. List the main components of a lithium-ion battery. Consider the electrodes, electrolyte, and casing or container materials.

2. What are the most common sources of the materials listed in Question 1?

3. Using Figure 1.49 on page 110 as a model, create a diagram representing the life cycle of a lithium-ion battery for an electric or hybrid-electric vehicle. Be sure to include at least one image for each of the following: materials acquisition, manufacturing, use and maintenance, and recycling or waste management.

4. Which steps in the life cycle of lithium-ion batteries consume energy? For each, explain the particular energy needs.

5. Which steps generate wastes or emissions? Explain.

6. What will happen to the materials when the batteries are depleted or the vehicle is removed from service?

CHEM**QUANDARY**

COMMODITY OR SPECIALTY

A commodity is a good or product with an identity that does not depend on its source or producer. Its price is a function of supply and demand, often set by global factors. On the other hand, a specialty product may have different characteristics depending upon its producer, or it may be available only from one or a few producers or sources. The price of a specialty product depends upon the desirability and availability of its features. The ammonia produced by EKS Nitrogen Products is an example of a commodity. Would the vehicle batteries produced by WYE Battery Technology Corporation be considered a specialty product or a commodity? Why? How, if at all, does this classification of products affect the decision facing Riverwood's citizens?

D.8 MOVING TOWARD GREENER METHODS AND PRODUCTS

As you learned earlier in this unit, industries, governments, and citizens in the United States and elsewhere have recognized their joint responsibility to ensure that chemical products are manufactured with maximum benefit and minimum risk to society. The EPA's Design for the Environment program, which includes green chemistry (pages 543–544), has already influenced the way chemical-processing, printing, and dry cleaning industries conduct business.

The chemical industry has developed new synthesis methods, based on using safe starting materials to replace toxic or environmentally unsafe substances. Green chemistry focuses on preventing environmental pollution directly at the point of manufacturing. In this approach, the chemical industry works as a social partner to sustain development and international trade without damaging the environment.

New green chemistry-based processes use more environmentally benign reactants than do traditional processes and create waste products that are less damaging to the air and water. For example, chemical processes can now use D-glucose, ordinary table sugar, to replace benzene, a known carcinogen. The D-glucose serves as a raw material for synthesizing other reactants that can be used to produce nylon and medicinal drugs. Also, in some processes, chemists can substitute nontoxic food dyes for catalysts composed of toxic metals such as lead, chromium, and cadmium.

Isocyanates are substances used to make polyurethanes, which in turn are used to produce seat cushions, insulation, and contact lenses. Carbon dioxide can replace phosgene, a toxic gas, in manufacturing isocyanates. Green chemistry has led to methods for synthesizing industrial chemicals using water as a solvent instead of toxic alternatives, whenever possible, and for employing materials that can be recycled and reused, thus significantly reducing waste-disposal problems.

The chemical industry has the responsibility to make products in ways that are as hazard-free as possible. The chemical industry must deal honestly with the public to ensure that people clearly understand risks and benefits of chemical operations. Chemistry companies also must assure consumers that their products are safe when used as intended.

Chemical industries must comply with relevant laws and regulations, as well as with voluntary standards set by many manufacturers themselves. Independent, outside organizations, such as the International Organization for Standardization in Geneva, Switzerland, encourage worldwide compliance.

No initiatives of the chemical industry or the government can eliminate all risks involved in manufacturing chemical substances, any more than we can completely eliminate the risks of automobile travel. However, knowing the risks, continuing to explore the sources of and alternatives to those risks, and making prudent decisions all remain essential.

> A chemical feedstock is a starting material from which we can make many other materials. For example, crude oil is a chemical feedstock for fuels and plastics.

As users of the industry's chemical products, all consumers share these responsibilities. Having studied some basic concepts about the manufacturing of chemical products, you can now turn to using such knowledge to weigh the risks and benefits of particular decisions—to you, the community, and the environment.

■ MAKING DECISIONS

D.9 ASSET OR LIABILITY?

If you were a Riverwood citizen, what would be your view? Should the town invite a chemical plant to locate in Riverwood? Which one? On what basis should such a decision be made?

To clarify this challenge, it is helpful to consider and evaluate both benefits and burdens—positive and negative factors—associated with any choice. Then you will be ready to address key questions that confront Riverwood citizens as they decide on their choices.

Positive Factors

- The Riverwood economy would improve. Either plant would employ about 200 residents. This would add about $8 million to Riverwood's economy each year. Also, each plant employee would indirectly provide jobs for another four people in local businesses. This is very desirable because 15% of Riverwood's labor force of 21 000 is currently unemployed.

- Farming costs might go down. Each year farmers in the Riverwood area spread 700 tons of fertilizer. The fertilizer comes from a fertilizer plant 200 miles away. Transport costs increase farmers' expenses by $15 for each ton of ammonia-based fertilizer. Local farmers thus stand to save about $10 000 each year in transportation costs.

- New local industries could result. With local access to lithium-ion batteries, small companies interested in developing and producing products that use these batteries may decide to locate in Riverwood. Already, a few local entrepreneurs have discussed the possibilities of producing electric delivery trucks and cargo vans that would be able to drive 100 miles on one charge (see Figure 5.38). The lithium iron phosphate batteries that they want to use could be produced within a special division at WYE. Such an expansion in battery production and creation of a new company that builds electric vehicles would create another 25–30 jobs, which would add another $600 000 per year to the local economy. Yet another possibility is the development of a lithium battery recycling plant in Riverwood.

Figure 5.38 *A local source of lithium-ion batteries might entice other manufacturers, such as makers of solar-electric trucks like this one, to build facilities in Riverwood.*

- As a commercial refrigerant, ammonia is used to produce large quantities of ice. Ready access to ammonia supplies could support a commercial ice-making plant in Riverwood. Such a company might employ 25 to 35 individuals, mainly drivers and some plant workers, which could add $500 000 each year to Riverwood's economy.

- The tax base would improve, as either company would contribute to the town's tax base. This will provide a large increase in revenues for the community.

Negative Factors

- Rates of worker-related injuries and accidents could increase. Ammonia is manufactured at high pressure and high temperature. At ordinary temperature and pressure, ammonia gas is extremely toxic at high concentrations. Large amounts of ammonia released due to an accident on the road or at the plant could injure or kill workers and other community members within the vicinity. Several cases of work-related injury or illness or even death in ammonia-based fertilizer plants are reported each year. Although few worker injuries have been reported in battery production plants, a few large fires have occurred. Such fires are difficult to fight, and they may add contaminants to the local air and water.

- Demand for lithium-ion batteries may decrease. Although recent government incentives and new technologies have increased public awareness and demand for electric and hybrid-electric vehicles, this may be short-lived. Electric/hybrid-electric vehicle demand may decrease, subsidies may end, or new technology may replace lithium-ion batteries. Hybrid-electric vehicle battery technology is still new—replacement or recycling issues may render these vehicles not competitive.

- Water quality might suffer. If ammonia, ammonia-bearing wastewater, or wastes from battery production leaked into the Snake River, the resulting water contaminants could threaten aquatic life.

- Specialty battery or ammonia markets might decline.

- Sources for raw materials may be uncertain. The United States does not have large natural deposits of lithium minerals and currently relies on Chile and Argentina for 97% of lithium supplies. These countries may decide it is more advantageous to use their lithium supplies to produce lithium-based batteries for export rather than export the raw materials to other countries.

- The fertilizer industry is among the largest consumers of ammonia. Current fertilizer-intensive agricultural methods have created controversy. In some cases, crop yields have declined despite use of increased quantities of synthetic fertilizer. Some farmers have elected to use less synthetic fertilizer (or none at all if they are attempting to be certified as "organic," see Figure 5.39), so ammonia demand may decline in coming years.

Figure 5.39 *An increase in organic farming could lead to decreased demand for synthetic fertilizers.*

Considering Burdens and Benefits

The poet William Wordsworth used the phrase, "Weighing the mischief with the promised gain . . ." in judging technological advancements (railroads, new in his time). Since the dawn of civilization, people have often accepted the burdens of new technologies to gain technology's benefits. Fire, one of civilization's earliest useful devices, gave people the ability to cook, warm themselves, and forge tools from metals. Yet fire out of control can destroy property and life. Every technology offers its benefits at a price.

One way to identify an acceptable new technology or venture is to evaluate it, finding an option that has a relatively low probability of producing harm—that is, delivers benefits that far outweigh the burdens. Unfortunately, benefit–burden analysis—weighing what Wordsworth called "mischief" against "promised gain"—is not an exact science.

For instance, some technologies may present high burdens immediately, while others may be associated with chronic, low-level risks for years or even decades. Many burdens are impossible to predict or assess with certainty. Individuals can control some potential burdens, but others must be addressed and controlled at regional or national levels. In short, it is quite difficult to conduct a thorough burden–benefit analysis. However, complete these activities:

> In many cases, electing not to make a decision is, in fact, also a decision—one accompanied by its own burdens and benefits.

1. Based on what you have learned in this unit, work in groups to create two lists for each chemical plant, one summarizing benefits and one summarizing burdens/risks.

2. Review your summaries of the burdens and risks. Are any negative factors completely unacceptable? If so, the plant associated with that risk or burden is probably not a viable option for Riverwood.

3. Within each list, mark the most valuable benefits with a plus symbol and the most serious burdens with a minus symbol.

4. Also consider the likelihood of occurrence of each benefit and burden. A burden that is fairly minor but almost certain to occur might merit more consideration than a burden that is more serious but extremely unlikely. Using a scale from 1 (highly unlikely) to 5 (extremely likely), rate each item on your lists in terms of its likelihood of occurrence.

5. Reread the opening commentaries.

   a. Identify the most convincing argument for a particular benefit or burden. Why do you find this argument convincing?

   b. Identify the least convincing argument for a particular benefit or burden. Why do you find this argument less convincing?

6. Based on your responses to Questions 3, 4, and 5, decide your own position on the question of locating a chemical plant near Riverwood. Discuss your view with your group.

7. As a class, discuss whether Riverwood should invite a chemical plant to the community. Be sure that the concerns of students who do not share the opinion of the majority are also heard and considered.

SECTION  SUMMARY

Reviewing the Concepts

> Batteries, which consist of one or more voltaic cells, provide convenient, portable ways to energize many common electrical devices.

1. What is the source of electrical energy in a battery?

2. Two types of voltaic cells are dry cells and alkaline batteries.

   a. List three similarities of these two cell types.

   b. How do these cells differ?

3. Write half-reaction equations for each of these oxidation-reduction processes. For each, identify (i) what is oxidized and (ii) what is reduced.

   a. $Pb(s) + Cu^{2+}(aq) \longrightarrow Pb^{2+}(aq) + Cu(s)$

   b. $Cr(s) + 3\ Ag^+(aq) \longrightarrow Cr^{3+}(aq) + 3\ Ag(s)$

4. Consider the following equation,

$$PbO_2(s) + Pb(s) + 4\ H^+(aq) + 2\ SO_4{}^{2-}\ (aq) \longrightarrow 2\ PbSO_4(s) + 2\ H_2O(l)$$

   a. In what type of battery does this reaction occur?

   b. What is (i) oxidized and (ii) reduced?

   c. Does this equation represent the charging or discharging of the battery?

   d. Identify the substance that you might observe as a white coating on battery electrodes.

   e. Under what conditions could this battery produce hydrogen gas?

> The electrical potential of a voltaic pile or battery depends upon the identity of its electrodes and can be calculated from standard reduction potentials.

5. Use Table 5.3 (page 565) to identify the anode and cathode and calculate the expected electrical potential for each of the following voltaic cells. (Assume that each voltaic cell uses appropriate ionic solutions.)

   a. Cu–Cd cell   c. Mg–H$_2$ cell   e. Au–Li cell

   b. Sn–Zn cell   d. Pb–Cd cell

6. Use Table 5.3 (page 565) to propose a voltaic cell that would generate each of the following potentials:

   a. 0.47 V   b. 1.60 V   c. 3.97 V   d. 0.32 V

> Life cycle analysis of products such as batteries allows designers and users to make informed choices about starting materials, manufacturing, use, and disposal.

7. Think about alkaline batteries.

   a. What are some common starting materials in the manufacture of these batteries?

   b. Are these readily available in the form that is required for manufacturing? Explain.

8. Consider primary and secondary batteries of the same size and electrical potential.

   a. How do the life cycles of these batteries differ?

   b. How might this affect which battery you choose? Explain.

9. Why shouldn't batteries be disposed of in landfills?

10. What are some challenges in recycling household batteries?

Burden–benefit analysis is useful in weighing both positive and negative consequences when making decisions.

11. List two positive and two negative aspects of the technologies involved in producing
   a. ammonia.   b. batteries.

12. Describe a specific change that the chemical industry has implemented in response to the principles of green chemistry.

What challenges must be met to optimize production and use of batteries?

In this section, you have constructed a voltaic pile, used half-reactions to determine the electrical potential of batteries, examined the composition of batteries, compared primary and secondary batteries, and used life-cycle analysis to study environmental challenges in the manufacturing and disposal of batteries. Think about what you learned, then answer the question in your own words in organized paragraphs. Your answer should demonstrate your understanding of the key ideas in this section.

Be sure to consider the following in your response: electrode potentials, primary and secondary batteries, life-cycle, and green chemistry principles.

Connecting the Concepts

13. Why do batteries eventually stop operating? Explain in terms of limiting reactants.

14. Why are some batteries rechargeable and some not?

15. Choose one of the three natural cycles that you have explored (carbon, nitrogen, and water) and compare it to the life cycle of a battery.

16. Identify one benefit and one burden or risk associated with each of the following:
   a. playing high-school basketball
   b. driving a car
   c. jogging
   d. receiving a dental X-ray
   e. applying fertilizers to garden plants
   f. drinking bottled water

Extending the Concepts

17. Consider voltaic cells and electrolytic cells (cells where electrolysis occurs).
   a. How are they similar?
   b. How do they differ?
   c. Sketch a diagram of each type of cell and highlight the differences.
   d. Are these processes opposites? Explain.

18. Explain how you can test the condition of a lead storage battery with a *hydrometer*, a device that measures liquid density. (*Hint:* Sulfuric acid solutions are more dense than liquid water.)

19. Identify several technologies in your community. Use the risk-assessment table below to assign an appropriate letter to each activity. Discuss your decisions and compare your rankings with those made by others.

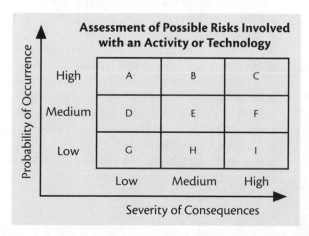

	Assessment of Possible Risks Involved with an Activity or Technology		
High	A	B	C
Medium	D	E	F
Low	G	H	I
	Low	Medium	High

Probability of Occurrence

Severity of Consequences

PUTTING IT ALL TOGETHER

RIVERWOOD NEWS

TODAY'S
WEATHER:
mostly sunny

Breaking news at RiverwoodNewsLive.com

MORNING EDITION

A Chemical Plant for Riverwood

Decision Nears: Is a Chemical Plant in Riverwood's Future?

By Gary Franzen
RIVERWOOD NEWS STAFF REPORTER

After months of study and discussion, the Riverwood Town Council is prepared to act on separate proposals from EKS Nitrogen Products Company and WYE Battery Technology Corporation to locate a plant near Riverwood in the old Riverwood Corporation building. At tonight's special meeting, the council will decide which, if either, plant to approve. The meeting starts at 7:30 p.m. in town hall.

Mayor Cisko remarked, "I'm very pleased with the turn-out we've had for the town meetings held on this issue. Tonight's council meeting is open to the public. I encourage all community members to attend and express their views about a chemical plant in Riverwood."

At the request of the Riverwood Town Council, both companies have prepared comprehensive summaries to inform citizens of their plans for a Riverwood plant. These summaries were circulated at previous town meetings and have been widely distributed in the community through newspapers, pamphlets, and on the town's Web site. Additional summaries can be obtained from the mayor's office.

Riverwood Corporation building

TOWN COUNCIL MEETING

Your class, with your teacher's guidance, will decide how to organize and conduct the town council meeting. Make sure that your format provides enough information so that Riverwood citizens will be able to consider benefits and burdens of each company's proposal and express their support for or opposition to each proposal. The meeting should conclude with a decision about whether a chemical plant will be allowed to operate in Riverwood, and, if so, which company will be invited to locate in Riverwood.

LOOKING BACK

Whether or not you decided to allow a chemical plant to locate in Riverwood, you have learned some valuable chemistry in the process of making that decision. You learned how key substances, such as ammonia, are produced and used, and how nitrogen cycles among the air, soil, and living organisms.

You also learned how chemists use electrochemical principles to harness chemical energy from spontaneous chemical reactions and also to provide energy enabling other reactions to occur. In addition, your acquired chemical knowledge and skills in analyzing burdens and benefits can help you to deal more effectively with future decision-making challenges.

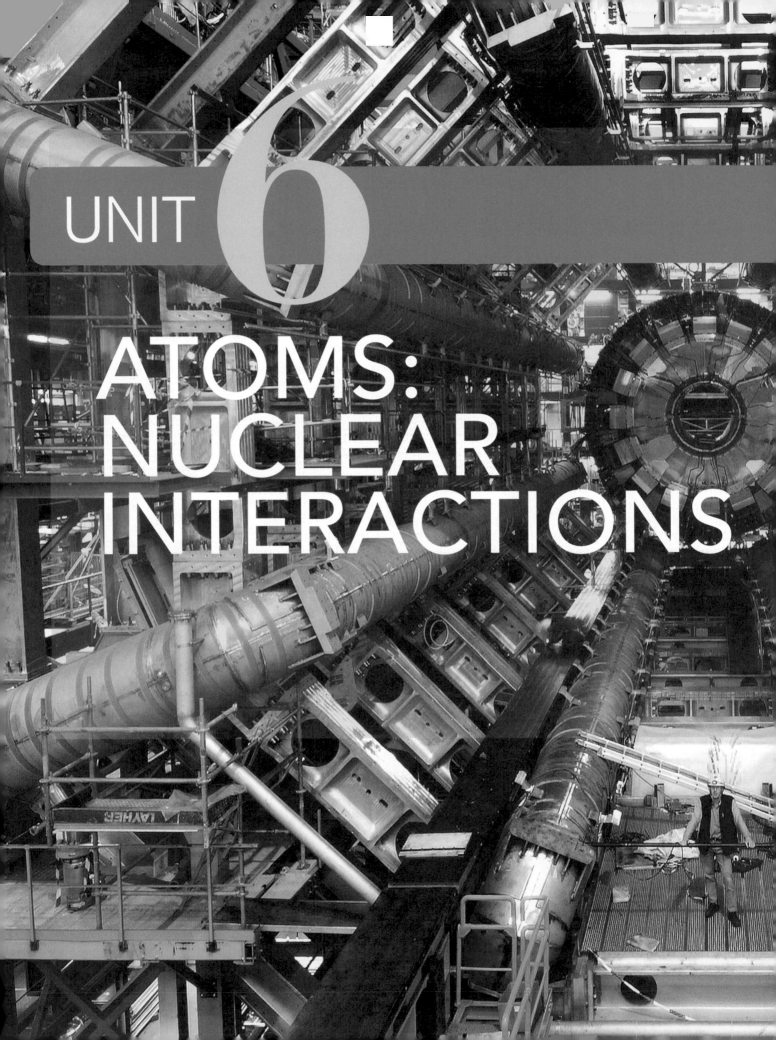

UNIT 6

ATOMS: NUCLEAR INTERACTIONS

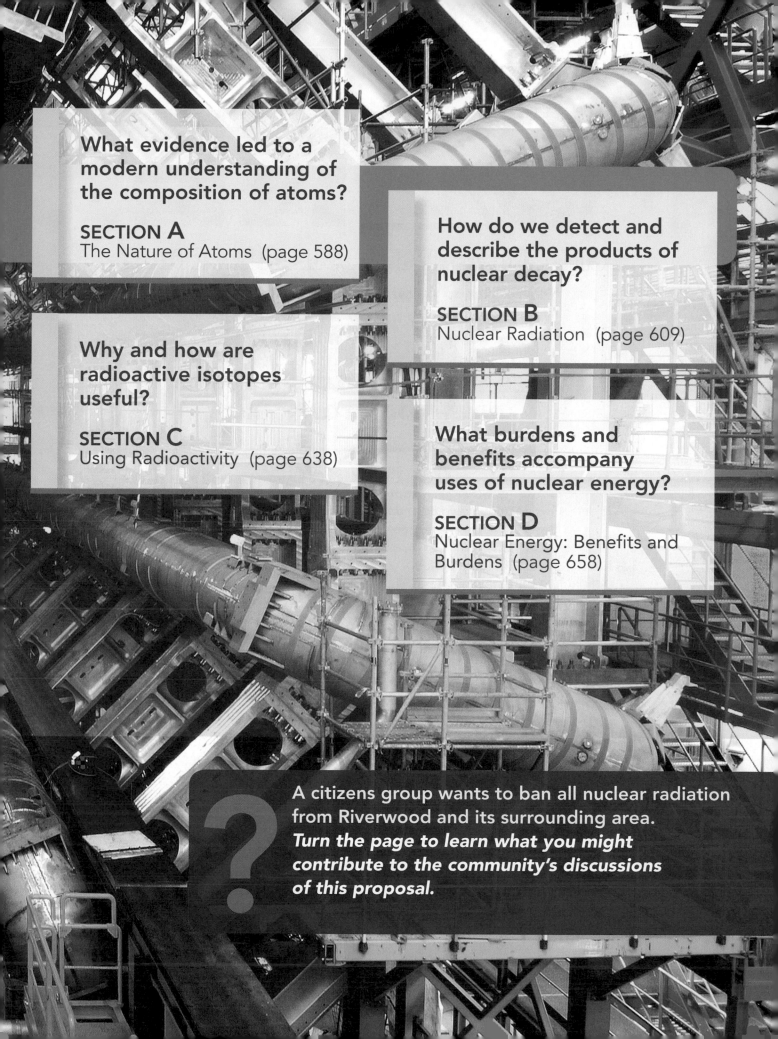

What evidence led to a modern understanding of the composition of atoms?

SECTION A
The Nature of Atoms (page 588)

How do we detect and describe the products of nuclear decay?

SECTION B
Nuclear Radiation (page 609)

Why and how are radioactive isotopes useful?

SECTION C
Using Radioactivity (page 638)

What burdens and benefits accompany uses of nuclear energy?

SECTION D
Nuclear Energy: Benefits and Burdens (page 658)

? A citizens group wants to ban all nuclear radiation from Riverwood and its surrounding area. *Turn the page to learn what you might contribute to the community's discussions of this proposal.*

C A N E

CITIZENS AGAINST NUCLEAR EXPOSURE

*invite you to attend an informative seminar
on banning the use of all nuclear energy
and materials in the Riverwood area.*

SEE YOU ON FRIDAY, 7:00 P.M., RIVERWOOD TOWN HALL

DO YOU KNOW
whether each statement below is true or false?
If not, you could be in danger of serious nuclear exposure!

1. Home smoke detectors contain radioactive materials.

2. Radioactive materials and radiation are unnatural—they did not exist on Earth until created by scientists.

3. All radiation causes cancer.

4. Human senses can detect radioactivity.

5. Individuals vary widely concerning how they are affected by exposure to radiation.

6. Small amounts of matter change to immense quantities of energy released by nuclear weapons.

7. Physicians can distinguish cancer caused by radiation exposure from cancer resulting from other causes.

8. Medical X-rays are dangerous.

9. Nuclear power plants create serious hazards to public health and to the environment.

10. An improperly operated nuclear power plant can explode like a nuclear weapon.

11. Some nuclear wastes must be stored for centuries to prevent dangerous radioactivity from escaping.

12. New, dangerous elements are being invented every day.

13. Nuclear power plants produce material that could be converted into nuclear weapons.

14. All nuclear medical techniques are highly dangerous.

A local Riverwood organization, Citizens Against Nuclear Exposure (CANE), has organized to prevent all uses of nuclear power, the disposal of nuclear waste, food irradiation, and nuclear medicine in the Riverwood area. The flyer reproduced on the opposite page is a sample of the organization's effort to communicate with Riverwood residents.

Some Riverwood citizens are wary of the restrictions CANE proposes because they fear that such restrictions would hinder access to some options in medical diagnosis and treatment. The grandmother of Ms. Lynn Paulson, a Riverwood High School chemistry teacher, is among these citizens. She asked Ms. Paulson whether any high-school chemistry students could provide background about nuclear science and technology to the senior citizens in the community. Several senior citizens plan to attend the announced CANE meeting, but they would like to acquire some background information to be ready to question CANE representatives about their proposal. Ms. Paulson has agreed to help. In this unit, you and your classmates will assume the role of Riverwood High School chemistry students preparing a community presentation.

The chemistry you have learned thus far involves chemical changes due to sharing or transferring outer-shell electrons among atoms. You will encounter much different changes in this unit—changes associated with nuclei (rather than electrons) of atoms. This unit examines nuclear radiation, radioactivity, and nuclear energy, plus implications of their use and development.

As you progress through this unit, record the ideas and applications you decide to share in your presentation. In particular, focus on the assertions contained in the CANE flyer. You may also want to investigate issues that will be of interest to your audience.

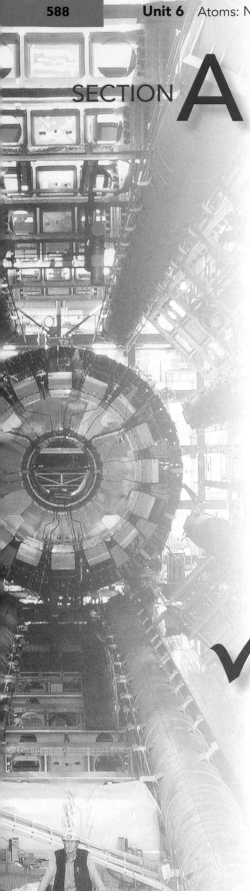

SECTION A

THE NATURE OF ATOMS

What evidence led to a modern understanding of the composition of atoms?

In this section, you will discover how both well-planned experiments and chance observations led to the current model of the atom's structure and to the discovery of radioactivity. Determining and describing the structure of atoms rank among the greatest scientific accomplishments of the past centuries. The stories of these great discoveries will help you understand atomic and nuclear chemistry and how researchers have used methods of science in their investigations.

GOALS

- Define and describe radioactivity.
- Distinguish between ionizing and non-ionizing radiation.
- Describe Rutherford's gold-foil experiment, including how its results led to a new model of the atom.
- Define isotope and radioisotope.
- Interpret a given isotope in terms of its atoms' protons and neutrons.
- Calculate the average molar mass of an element, given isotopic molar masses and abundance data.

concept check 1

1. Draw a representation of an atom of sodium. Label the components.
2. Are all atoms of sodium identical? Explain.
3. What are some sources of ionizing radiation?

■ INVESTIGATING MATTER
A.1 EXPLORING IONIZING RADIATION

You are being constantly bombarded by radiation. Whether you are at school, in an airplane, at a park, or inside your house, you are always exposed to multiple forms of radiation. In this investigation, you will examine a variety of everyday (and less common) objects using an instrument designed to detect ionizing forms of radiation.

Preparing to Investigate

Before you begin, read *Gathering Evidence* to learn what you will need to do and note safety precautions. Create a plan for data collection that will allow you to make some claims about materials that emit ionizing radiation.

Gathering Evidence

Part I: Background Radiation

1. Before you begin, put on your goggles and gloves, and wear them properly throughout the investigation.

2. Turn on your detector. Let the detector warm up for ~2 minutes. You will notice that your detector will indicate the presence of ionizing radiation even though it is not near a sample or pointed toward a source. This radiation is called **background radiation**. It is always present, and results from a variety of sources, some of which are natural and others that result from human activity.

3. Move to several places within the classroom or laboratory and take readings.

4. If your teacher approves and your detector is portable, take readings outside of the classroom.

5. Record the readings collected in your data table.

Part II: Measuring Ionizing Radiation from Samples

You will be provided with several objects and materials. Handle all samples with forceps. Use your detector to measure ionizing radiation emitted by each sample (see Figure 6.1). Record all readings in your data table.

Part III: Pinpointing Sources of Ionizing Radiation

Choose one or two samples that emit ionizing radiation. Handle all samples with forceps. Use your detector to identify the source of the radiation (an area or areas where the radiation seems to be concentrated). Record your findings in your data table.

Figure 6.1 *Some common objects emit ionizing radiation.*

Interpreting Evidence

1. Did background radiation vary from place to place? Explain.
2. Identify any locations inside or outside the classroom that had background radiation readings that were
   a. unusually high.
   b. unusually low.
3. Which samples emitted ionizing radiation?
4. Were the radiation levels emitted by the various samples similar or different? Support your answer with data.
5. Compare your results with those of your classmates.
   a. Did areas within and outside the classroom give consistent readings for background radiation from group to group?
   b. Did the same samples emit radiation?
   c. How do you explain any differences between your results and those of your classmates?

Making Claims

6. What statement can you make about the level of background radiation in and around your classroom?
7. Consider Part III of the investigation.
   a. Did all parts of the samples you examined emit ionizing radiation?
   b. If not, could you identify a purpose for the source of the ionizing radiation? Explain.

Reflecting on the Investigation

8. What safety precautions did you take? Would someone using this object every day use precautions?
9. Do you think you could predict, just by looking at it, whether a sample will emit ionizing radiation? Explain.

A.2 THE GREAT DISCOVERY

You learned about electromagnetic radiation in Unit 2 (see page 221).

The history of modern scientific investigation of the atom began with the study of radiation. Scientists have long been interested in light—because life would not exist without it—and other types of radiation. By the end of the 19th century, scientists had already studied many types of radiation from a variety of sources.

In 1895, a series of observations significantly broadened scientific understanding of radiation. The German physicist W. K. Roentgen was studying **fluorescence**, a phenomenon in which certain materials emit light when struck by radiant energy, such as ultraviolet rays (Figures 6.2 and 6.3).

Figure 6.2 *French physicist Henri Becquerel's investigations of fluorescence led to the discovery of radioactivity.*

Figure 6.3 *Ultraviolet light shining on particular objects, such as certain fabrics, produces fluorescence—visible light emitted by material exposed to such electromagnetic radiation.*

Roentgen found that certain materials fluoresced when exposed to beams of **cathode rays** emitted from the cathode when electricity passed through an evacuated glass tube (see Figure 6.4). A few years after Roentgen's work, cathode rays were identified as beams of electrons.

Roentgen was working with a cathode-ray tube covered by black cardboard. He observed an unexpected glow of light on a piece of paper across the room. The paper was coated with a fluorescent material, and Roentgen expected it to glow when exposed to radiation. However, visible radiation could not pass through the black cardboard covering the cathode-ray tube, and the fluorescent paper was not in the path of electrons from the tube. Roentgen hypothesized that some other radiation passing through the black cardboard had been emitted by the cathode-ray tube. He named the mysterious radiation *X-rays*, where *X* represented the unknown radiation. Scientists now know that **X-rays** are a form of high-energy electromagnetic radiation.

> Older computer monitors and TV screens were based on a version of the cathode-ray tube.

> In modern X-ray devices, the X-rays are generated when an electron beam strikes a metal target, which is often made of tungsten.

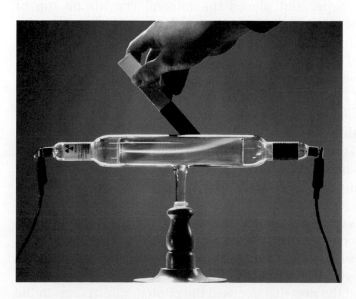

Figure 6.4 *A beam of electrons moves from the cathode (left) to the anode (right). The visible light emitted results from collisions of electrons with the fluorescent screen inside the tube. The deflection of the beam by a magnet indicates that the beam particles have a negative electrical charge. Collision of the electron beam with the glass or anode produces X-rays.*

Further experiments revealed that these X-rays could penetrate many materials, but could not easily pass through dense materials such as lead and bone. Scientists soon realized how useful X-rays could be in medicine. In fact, one early X-ray image Roentgen obtained was of his wife's hand. Figures 6.5 and 6.6 show some modern X-ray images.

Figure 6.5 *These X-ray images reveal a pair of normal human legs (left) and a pair of broken legs (right). Development of this useful medical diagnostic tool emerged from the study of fluorescence.*

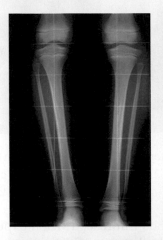
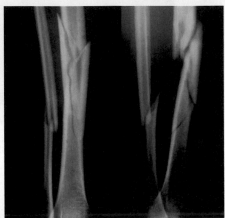

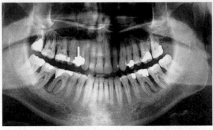

Figure 6.6 *An X-ray image of a human jaw. Such images help dentists detect cavities and other dental problems.*

Roentgen's discovery intrigued other scientists, including the French physicist Henri Becquerel. Because X-rays could produce fluorescence, Becquerel wondered if fluorescent minerals might give off X-rays as they fluoresce. In 1896, Becquerel placed in sunlight some crystals of a fluorescent mineral that contained uranium. He then wrapped an unexposed photographic plate in black paper and placed the mineral crystals on top of the wrapped plate. If the mineral did emit X-rays, they would penetrate the black paper and the exposed film would darken, even though it was shielded from light.

Cloudy weather prevented Becquerel from completing his experiments. He stored the wrapped photographic plates in a drawer with the uranium-containing mineral. After several days, he decided to develop some of the stored plates, thinking that perhaps some fluorescence might have persisted, causing some fogging of the photographic plates. When Becquerel developed the plates, he was astounded. Instead of faint fogging, the plates had been strongly exposed. Figure 6.7 illustrates the chain of events in Becquerel's investigation.

Fluorescence stops as soon as the external source of radiation (in this case, the Sun) is removed from the object. Thus, a fluorescent mineral in a dark drawer should not cause such an intense exposure. Scientists at that time could not offer a satisfactory explanation for Becquerel's observations. Becquerel suspected that the rays that exposed the photographic plates in the

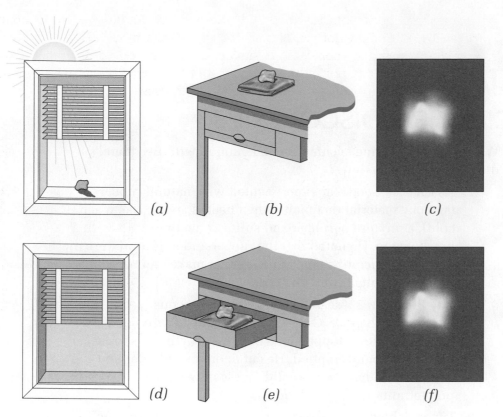

Figure 6.7 *Becquerel's investigation. Becquerel placed a fluorescent mineral in direct sunlight (a), then put it on an unexposed, wrapped photographic plate (b). Radiation exposed the plate (c). On a cloudy day (d), the wrapped plate was placed in a drawer with the mineral and kept from light (e); however, the mineral sample still exposed the photographic plate (f).*

drawer were more energetic and had much greater penetrating ability than X-rays. Thus, he interrupted his study of X-rays to investigate the mysterious radiation apparently given off by the uranium-containing mineral. Although he could not explain it, Becquerel had discovered **radioactivity**, which is now known to involve the spontaneous emission of particles and energy from atomic nuclei. This phenomenon is distinctly different from X-ray production or fluorescence (Figure 6.8).

Becquerel suggested that Marie Curie (Figure 6.9), a graduate student working with him, attempt to isolate the radioactive component of pitchblende, a uranium ore, for her research for her doctoral dissertation. Her preliminary work was successful. Her physicist husband, Pierre Curie, changed his research focus to join her on the pitchblende project. Working together, Marie and Pierre Curie discovered that the level of radioactivity in pitchblende was four to five times greater than expected from its known uranium content. The Curies suspected the presence of another radioactive element. After processing more than a thousand kilograms of pitchblende, they isolated tiny quantities (measured in milligrams) of two previously unknown radioactive elements. These elements later became known as polonium (Po) and radium (Ra).

Figure 6.8 *Many minerals exhibit fluorescence when exposed to ultraviolet light.*

Figure 6.9 *Marie Curie discovered two highly radioactive elements— radium (named for the radiation it emitted) and polonium (named for her native Poland). Her work earned her two Nobel Prizes.*

CHEM**QUANDARY**

SCIENTIFIC DISCOVERIES

What do the following events have in common with Becquerel's discovery of radioactivity?

1. As Charles Goodyear experimented with natural rubber (a sticky material that melts when heated and cracks when cold), a mixture of rubber and sulfur came in contact with a hot stovetop. He noted that the rubber-and-sulfur mixture did not melt. *Vulcanization*, a process that makes rubber more durable, resulted from this observation.

2. Roy Plunkett, a research chemist, used gaseous tetrafluoroethene ($F_2C{=}CF_2$) from a storage cylinder, but the gas flow stopped long before the cylinder should have completely emptied. He cut open the cylinder and discovered a new, white solid that is now known as polytetrafluoroethene, *Teflon*.

3. James Schlatter, a research chemist trying to produce an antiulcer drug, accidentally got some of the substance on his fingers. When he later licked his fingers to pick up a piece of paper, his fingers tasted very sweet, and he correctly linked the sweetness to the antiulcer drug. Instead of finding an antiulcer drug, he discovered *aspartame*, an artificial sweetener.

A.3 NUCLEAR RADIATION

The nuclei of unstable atoms are sources of nuclear radiation.

Many people respond with alarm and even panic when they hear the word *nuclear*. In addition, the general term *radiation*, which is sometimes used to refer particularly to nuclear radiation, can also cause anxiety. In reality, radiation falls into two general types: *ionizing radiation* and *non-ionizing radiation*.

Electromagnetic radiation in the visible and lower-energy regions of the spectrum (see Figure 6.10) is **non-ionizing** (long-wavelength) **radiation**. Non-ionizing radiation transfers its energy to matter, causing atoms or molecules to vibrate (infrared radiation), move their electrons to higher energy levels (visible radiation), or heat up (such as in microwave ovens). Although this radiation is generally considered safe, excessive exposure can be harmful. Sunburn, for example, results from an overexposure to non-ionizing radiation from the Sun (Figure 6.11). In fact, intense microwave and infrared radiation can cause lethal burns.

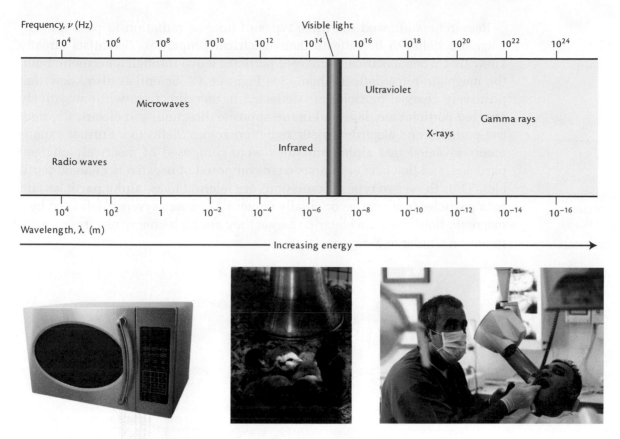

Frequency, ν (Hz)

Visible light

| 10^4 | 10^6 | 10^8 | 10^{10} | 10^{12} | 10^{14} | 10^{16} | 10^{18} | 10^{20} | 10^{22} | 10^{24} |

Ultraviolet

Microwaves

Gamma rays

X-rays

Infrared

Radio waves

| 10^4 | 10^2 | 1 | 10^{-2} | 10^{-4} | 10^{-6} | 10^{-8} | 10^{-10} | 10^{-12} | 10^{-14} | 10^{-16} |

Wavelength, λ (m)

⟶ Increasing energy ⟶

Figure 6.10 *The electromagnetic spectrum. Microwaves, infrared radiation (from heat lamps), and X-rays are all examples of electromagnetic radiation.*

Ionizing radiation, which includes high-energy electromagnetic radiation (short-wavelength ultraviolet radiation, X-rays, and gamma rays) and all nuclear radiation, carries more energy and potential for harm than non-ionizing radiation. Energy from ionizing radiation can eject electrons from atoms and molecules, forming molecular fragments and ions. These fragments and ions can be highly reactive. If formed within a living system, they can disrupt normal cellular chemistry, causing serious cell damage.

Nuclear radiation is a form of ionizing radiation that is caused by changes in the nuclei of atoms. In chemical reactions, the atomic number (number of protons) does not change. An atom of aluminum (13 protons) always remains an aluminum atom, and an iron atom (26 protons) always remains an iron atom. However, atoms with unstable nuclei—radioactive atoms—can spontaneously change their identities. A radioactive atom changes spontaneously through disintegration of its nucleus, which results in emission of high-speed particles and energy. When this happens, the identity of the radioactive atom often changes. An atom of a different element forms. This process is **radioactive decay**. The emitted particles and energy make up **nuclear radiation**.

Ernest Rutherford showed in 1899 that nuclear radiation included at least two different types of emissions, which he named *alpha rays* and *beta rays*. Shortly afterward, scientists discovered a third kind of nuclear radiation: *gamma rays*.

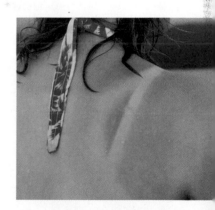

Figure 6.11 *Sunburn often results from overexposure of skin to ultraviolet radiation.*

Alpha (α), beta (β), and gamma (γ) are the first three letaters of the Greek alphabet.

Researchers allowed the three types of nuclear radiation to pass through magnetic fields to investigate their electrical properties. Scientists already knew that when electrically charged particles move through a magnetic field, the magnetic force deflects them. See Figure 6.12. Scientists also knew that positively charged particles are deflected in one direction, while negatively charged particles are deflected in the opposite direction, and electrically neutral particles and electromagnetic radiation are not deflected. Further experiments revealed that alpha emissions were composed of positively charged particles, and that beta emissions were composed of negatively charged particles. Thus these two types of emissions are referred to as **alpha particles** and **beta particles** (not rays, as originally named). **Gamma rays**, not deflected by a magnetic field, have no electric charge; they are high-energy electromagnetic radiation similar to X-rays.

> Alpha rays and beta rays are more commonly referred to as particles because they possess measurable mass.

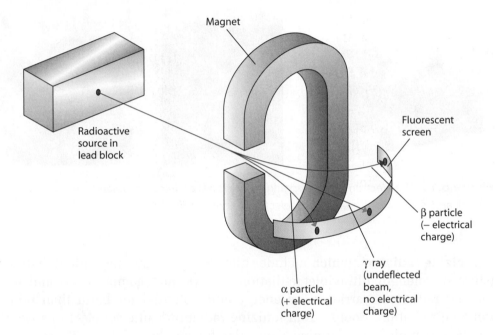

Figure 6.12 *Behavior of alpha (α) particles, beta (β) particles, and gamma (γ) rays passing through a magnetic field.*

By describing the nature of radioactivity, scientists toppled an old theory (a common event in scientific progress—new knowledge replaces an old body of knowledge). Once scientists knew about alpha particles, beta particles, and gamma rays, they became convinced that atoms, which were originally thought to be the smallest, most fundamental units of matter, must be composed of even smaller particles.

From the results of another experiment, the *gold-foil investigation*, Rutherford proposed a fundamental model of the atom that is still useful today. To do so, he developed an ingenious, indirect way to "look" at the structure of atoms.

A.4 THE GOLD-FOIL EXPERIMENT

Before Rutherford's research, scientists had tried to explain the arrangement of electrons and positively charged particles within atoms in several ways. In the most widely accepted model, an atom was viewed as a volume of positive electrical charge, with the negatively charged electrons embedded within, like peanuts in a candy bar. In the late 1800s this was known as the "plum pudding" model, because it resembled the distribution of raisins within that traditional English dessert.

About 1910, Rutherford decided to test the plum pudding model. Working in Rutherford's laboratory in Manchester, England, Hans Geiger and Ernest Marsden focused a beam of alpha particles—the most massive of the three types of nuclear radiation—at a thin sheet of gold foil only 0.000 04 cm (about 2000 atoms) thick (see Figure 6.13). Geiger and Marsden used a zinc sulfide-coated screen to detect the alpha particles after they passed through the gold foil (Figure 6.14). The screen emitted a flash of light where each alpha particle struck it. By observing the tiny light flashes at different positions with respect to the gold foil, Geiger and Marsden deduced the paths of the alpha particles as they interacted with the gold foil.

Figure 6.13 *A sheet of gold leaf is prepared for trimming. This very thin material was produced with machines in Germany, but gold foil or leaf was traditionally prepared by beating gold into sheets.*

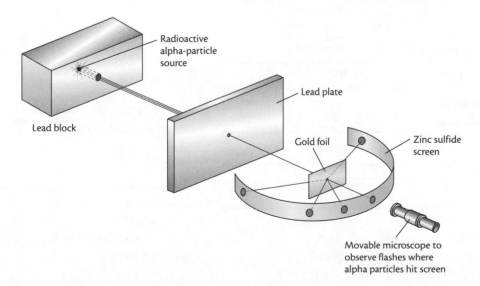

Figure 6.14 *The alpha-particle scattering experiment led Rutherford to conclude that the atom is largely empty space with an incredibly dense, positively charged nucleus at its center.*

Rutherford expected alpha particles to be scattered slightly as they were deflected by the gold atoms in the foil, producing a pattern similar to water being sprayed from a nozzle. However, he was in for quite a surprise. First, most of the alpha particles passed straight through the gold foil as if nothing were there (see Figure 6.15). This implied that most of the volume occupied by the gold atoms was empty space. But Rutherford was even more surprised that a few alpha particles, about 1 in every 20 000, bounced *back* toward the source. He described his astonishment this way: "It was almost as incredible as if you fired a 15-inch [artillery] shell at a piece of tissue paper and it came back and hit you."

Whatever repelled these deflected alpha particles must have been extremely small because most of the alpha particles went straight through the foil. Rutherford concluded, "On consideration, I realized that this scattering backward must be the result of a single collision, and when I made calculations I saw that it was impossible to get anything of that order of magnitude unless you took a system in which the greater part of the mass of the atom was concentrated in a minute nucleus. It was then that I had the idea of an atom with a minute massive centre, carrying a charge."

From these results, Rutherford developed the modern model of the nuclear atom. He named the tiny, dense, positively charged region at the center of the atom the *nucleus*. He envisioned that electrons orbited the nucleus, somewhat as planets orbit the Sun. Figure 6.15 illustrates how the nuclear model explains the results of the gold-foil experiment.

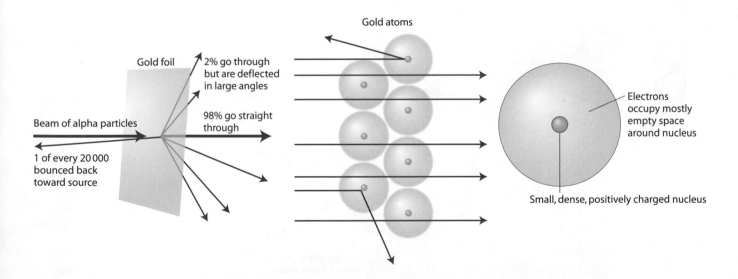

Figure 6.15 *Results of the alpha-particle scattering experiment. Most alpha particles passed through the foil but a few were deflected, some at very large angles (left). Proposed model (middle) to account for the results. Nuclear model of the atom (right).*

A.5 ARCHITECTURE OF ATOMS

Since Rutherford's time, scientific understanding of atomic structure has expanded and changed somewhat. Rutherford's model of a central, dense nucleus surrounded mostly by empty space is still valid. Later research has shown that the idea of orbiting electrons is incorrect. The general regions in which electrons are most likely to be found can be described, but particular movements or locations of electrons cannot be described.

Further research revealed that the nucleus is composed of two types of particles: neutrons, which are electrically neutral, and protons, which possess a positive charge. These particles, as well as electrons, are called **subatomic particles**. Protons and neutrons have about the same mass, 1.7×10^{-24} g. Although this mass is incredibly small, it is much greater than the electron's mass, which is 9.1×10^{-28} g. As shown in Table 6.1, one mole of electrons (6.02×10^{23} electrons) has a mass of only 0.0005 g. The same number of protons or neutrons would have a mass of about 1 g. In other words, a proton or a neutron is about 2000 times more massive than an electron. Thus, protons and neutrons account for nearly all the mass of every atom and also for nearly all of the total mass of every object that you encounter.

> Neutrons are actually slightly more massive than protons.

> Protons:
>   positive electrical charge
> Neutrons:
>   no electrical charge
> Electrons:
>   negative electrical charge

Table 6.1

Three Components of Atoms			
Particle	**Location**	**Charge**	**Molar Mass (g/mol)**
Proton	Nucleus	+1	1
Neutron	Nucleus	0	1
Electron	Outside of nucleus	−1	0.0005

The diameter of a typical atom is about 10^{-10} m, but an average nuclear diameter is 10^{-14} m, which is only one ten-thousandth ($10^{-14}/10^{-10}$) of the entire atom's diameter. Looking at this another way, the nucleus occupies only about one trillionth (10^{-12}) of an atom's total volume. Imagine that a billiard ball represents the diameter of an atom's nucleus. On that scale, electrons surrounding this billiard-ball nucleus would extend out in space more than one-half kilometer (about a third of a mile) away in all directions. This is consistent with the observation that most alpha particles, each the size of a helium nucleus, passed right through Rutherford's sheet of gold foil.

As you learned in Unit 1 (page 54), each atom of an element has the same number of protons in its nucleus, and each element has a unique number of protons. This number, called the *atomic number*, identifies the element. For example, each carbon atom nucleus contains six protons, Therefore, the atomic number of carbon is 6.

> In most periodic tables, the atomic number is written above the symbol of each element.

Yet all atoms of a given element do not necessarily have the same number of neutrons in their individual nuclei. Recall that atoms of the same element with different numbers of neutrons are called *isotopes* of that element. Naturally occurring carbon atoms, each containing six protons, may have six, seven, or even eight neutrons. The composition of these three carbon isotopes is summarized in Table 6.2. Figure 6.16 shows their nuclei.

Table 6.2

Three Carbon Isotopes				
Name	Total Protons (Atomic Number)	Total Neutrons	Mass Number	Total Electrons
Carbon-12	6	6	12	6
Carbon-13	6	7	13	6
Carbon-14	6	8	14	6

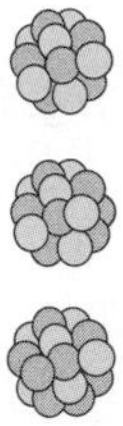

Figure 6.16 *Nuclei of three carbon isotopes, containing, respectively, 6, 7, and 8 neutrons (gray) and 6 protons (blue).*

Isotopes are distinguished by their different mass numbers. The mass number, as you learned in Unit 1 (page 54), represents the total number of protons and neutrons in an atom. The three carbon isotopes in Table 6.2 have mass numbers of 12, 13, and 14, respectively. To specify a particular isotope, the atomic number and the mass number are written in front of the element's symbol in a particular way. For example, an isotope of carbon (C) with an atomic number 6 and a mass number 12 is written this way:

$$^{12}_{6}\text{C}$$

An atom of strontium with an atomic number 38 and mass number 90 would be symbolized as follows:

$$^{90}_{38}\text{Sr}$$

Another way to identify a particular isotope is to write the name or symbol of the element followed by a hyphen and its mass number. For example, an isotope for carbon may be called carbon-12, C-12, or ^{12}C. The symbols, names, and nuclear composition of some isotopes are summarized in Table 6.3.

Table 6.3

Some Common Isotopes					
Symbol	Name	Total Protons (Atomic Number)	Total Neutrons	Mass Number	Total Electrons
Li	Lithium-7	3	4	7	3
O	Oxygen-16	8	8	16	8
Ne	Neon-20	10	10	20	10
Ga	Gallium-67	31	36	67	31
Ba	Barium-138	56	82	138	56
Pt	Platinum-195	78	117	195	78
Tl	Thallium-201	81	120	201	81
Pb	Lead-208	82	126	208	82

▮ DEVELOPING SKILLS

A.6 INTERPRETING ISOTOPIC NOTATION

> *Sample Problem:* Suppose you know that one product of a nuclear reaction is an isotope containing 85 protons and 120 neutrons. The atom therefore has a mass number of 205 (85 p + 120 n). What is the symbol of this element?
>
> $$^{205}_{85}?$$
>
> Consulting the periodic table, you find atomic number 85 represents the element astatine (At).
>
> $$^{205}_{85}\text{At}$$

1. Prepare a summary chart like Table 6.3 (previous page) for the following six isotopes. (Consult the periodic table for any needed information.)

   a. $^{12}_{?}\text{C}$
   b. $^{14}_{7}?$
   c. $^{16}_{?}\text{O}$
   d. $^{24}_{12}?$
   e. $^{202}_{?}\text{Hg}$
   f. $^{238}_{92}?$

2. Using Table 6.3 as a source of information, what general relationship do you note between the total number of protons and total number of neutrons for atoms of

   a. lighter elements with atomic numbers less than 20?

   b. heavier elements with atomic numbers greater than 50?

▮ MODELING MATTER

A.7 ISOTOPIC PENNIES

You learned earlier (Unit 1, page 34) that pre-1982 and post-1982 pennies have different compositions. As you might expect, these pennies also have different masses. In this activity, a mixture of pre- and post-1982 pennies will model or represent atoms of a naturally occurring mixture of two isotopes of the imaginary element *coinium*. Using the pennies, you will simulate one way scientists determine the relative amounts of different isotopes in a sample of an element.

You will receive a sealed container of 10 pennies that contains a mixture of pre-1982 and post-1982 pennies (Figure 6.17). Your container might hold any particular atomic mixture of these two isotopes. Your task is to determine the isotopic composition of *coinium* —*without opening the container.*

Figure 6.17 *How many pre- and post-1982 pennies are in your 10-coin sample of coinium?*

- Your teacher will give you some pre-1982 and post-1982 pennies, and a sealed container with a mixture of 10 pre- and post-1982 pennies, and will tell you the mass of the empty container. Record this information and the code number of your sealed container.

- Determine the isotopic composition of the element *coinium*. That is, find the percent pre-1982 and percent post-1982 pennies in your container. There is more than one way to find these answers.

Now answer these questions:

1. Describe the procedure that you followed to find the percent composition of coinium.

2. What property of the element *coinium* is different in its pre- and post-1982 forms?

3. Name at least one other familiar item that could serve as a model for isotopes.

4. You have examined and created models throughout the year. What are the limitations of this model in explaining isotopes?

concept check 2

1. Why was it necessary for Rutherford to know something about radiation before he could make a claim regarding atomic nuclei?
2. Describe the isotope symbolized $^{210}_{82}Pb$.
3. There are two naturally occurring isotopes of copper $^{63}_{26}Cu$ and $^{65}_{26}Cu$. If the average atomic mass of copper is 63.55 g/mol, which isotope occurs more frequently in nature? Explain.

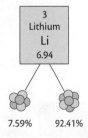

Figure 6.18 *Lithium's average atomic mass is a weighted average of the molar masses of its two isotopes, Li-6 and Li-7. Most lithium atoms (92.41%) are Li-7. Thus, lithium's average atomic mass is closer to 7 than to 6.*

A.8 ISOTOPES IN NATURE

Most elements in nature are mixtures of isotopes. Some isotopes of an element may be radioactive, whereas others are not. All isotopes of an element behave almost exactly the same way chemically, because they have the same electron distribution and differ only slightly in mass. If you consider chemical changes only, knowledge of isotopes is not particularly helpful. The average atomic mass of an element, as shown on the periodic table, is based on the relative natural abundances of isotopes of that element (see Figure 6.18 for lithium).

Marie Curie originally thought that only heavy elements were radioactive. It is true that naturally occurring **radioisotopes** (i.e., radioactive isotopes) are more common among the heavy elements. In fact, all naturally occurring isotopes of elements with atomic numbers greater than 83 (bismuth) are radio-

Some Natural Radioisotopes

Isotope	Abundance (%)
Hydrogen-3	0.00013
Carbon-14	Trace
Potassium-40	0.0012
Rubidium-87	27.8
Indium-115	95.8
Lanthanum-138	0.089
Neodymium-144	23.9
Samarium-147	15.1
Lutetium-176	2.60
Rhenium-187	62.9
Platinum-190	0.012
Thorium-232	100

Table 6.4

active. However, many natural radioisotopes are also found among lighter elements. Modern technology has made it possible to create a radioisotope of any element. Table 6.4 lists some naturally occurring radioisotopes and their isotopic abundances.

What is the relationship between an element's molar mass and the percent abundance of the element's istopes? To calculate the molar mass of an element, it is helpful to use the concept of a weighted average, as illustrated in Developing Skills A.9.

No stable (nonradioactive) isotopes have yet been found for elements with atomic numbers of 83 or greater.

DEVELOPING SKILLS

A.9 MOLAR MASS AND ISOTOPIC ABUNDANCE

Sample Problem 1: Consider an isotopic mixture of copper. Naturally occurring copper (Cu) consists of 69.1% copper-63 atoms and 30.9% copper-65 atoms. The molar masses of these two isotopes are:

$$Copper\text{-}63 = 62.93 \ g/mol$$
$$Copper\text{-}65 = 64.93 \ g/mol$$

What is the average molar mass of naturally occurring copper?

The equation for finding average molar masses is as follows:

$$\text{Molar mass} = \begin{pmatrix}\text{Fractional} \\ \text{abundance} \\ \text{of isotope 1}\end{pmatrix} \times \begin{pmatrix}\text{Molar mass} \\ \text{of isotope 1}\end{pmatrix} + \begin{pmatrix}\text{Fractional} \\ \text{abundance} \\ \text{of isotope 2}\end{pmatrix} \times \begin{pmatrix}\text{Molar mass} \\ \text{of isotope 2}\end{pmatrix} + \ldots \begin{pmatrix}\text{for each isotope} \\ \text{involved}\end{pmatrix}$$

Since there are two naturally occurring copper isotopes, the average molar mass of copper can be calculated as follows:

$$\text{Molar mass of Cu} = (0.691)(62.93 \ g/mol) + (0.309)(64.93 \ g/mol) = 63.5 \ g/mol$$

The decimal fractions must add up to 1. Why?

Sample Problem 2: For the coinium example, suppose that you found that the composition of the mixture was 0.4 (40%) pre-1982 pennies and 0.6 (60%) post-1982 pennies.

What is the average mass of a penny?

The equation setup is shown below, using the 40/60 coin mixture:

$$\text{Average penny mass} = (0.4) \times \begin{pmatrix}\text{Mass of} \\ \text{pre-1982 penny}\end{pmatrix} + (0.6) \times \begin{pmatrix}\text{Mass of} \\ \text{post-1982 penny}\end{pmatrix}$$

1. Calculate the average mass of a penny in your *coinium* mixture.

2. Calculate the average mass of a penny in your mixture another way: Divide the total mass of your entire penny sample by 10.

3. Compare the average masses that you calculated in Questions 1 and 2. These results should convince you that either calculation leads to the same result. If not, consult your teacher.

4. Naturally occurring boron (B) is a mixture of two isotopes. (Refer to the table below.)

   a. Do you expect the molar mass of naturally occurring boron to be closer to 10 or to 11? Why?

   b. Calculate the molar mass of naturally occurring boron.

5. Naturally occurring uranium (U) is a mixture of three isotopes. (Refer to the table below.)

   a. Do you expect the molar mass of naturally occurring uranium will be closest to 238, 235, or 234? Why?

   b. Calculate the molar mass of naturally occurring uranium.

Isotope Molar Mass and Abundance		
Isotope	**Molar Mass (g/mol)**	**% Natural Abundance**
Boron-10	10.0	19.90%
Boron-11	11.0	80.10%
Uranium-234	234.0	0.0054%
Uranium-235	235.0	0.71%
Uranium-238	238.1	99.28%

■ MAKING DECISIONS

A.10 FACT OR FICTION?

Look again at the statements at the start of this unit (page 586). Answer the following questions about the CANE flyer. This will help you start preparing your presentation for Riverwood senior citizens (see Figure 6.19).

1. Identify the specific statements on the flyer that you can now conclusively identify as either true or false? For each statement:

   a. list two pieces of evidence that helped you make your decision.

   b. list two public concerns about the statement that you plan to address in your presentation.

2. Choose one statement from the leaflet that you understand more completely now, but are still unable to confirm or deny. What else do you have to know before you can make a decision about that statement?

3. How helpful do you think it will be to discuss the history of some discoveries that you studied in this section when you talk to the Riverwood senior citizens? Explain your answer.

4. Think about the new terms that were introduced in this section.

   a. Which terms will you explain as part of your presentation?

   b. Select two terms from your answer to Question 4a. Prepare an explanation of each term that your audience will understand. Describe examples or real-world applications that you may use in your explanation.

The history of science is full of discoveries that build on earlier discoveries. The discovery of radioactivity was such an event. The investigations of Roentgen, Becquerel, the Curies, and Rutherford led to new knowledge and a better understanding of atomic structure. As you start to consider some current applications of nuclear radiation, think about the evidence and reasoning that supported these scientific advancements.

Figure 6.19 *What does your research lead you to conclude about the accuracy of statements made in the flyer? (See page 586.)*

SECTION A SUMMARY

Reviewing the Concepts

> Radioactive nuclei are unstable and undergo spontaneous changes in their structure.

1. Describe the sequence of events that led to Becquerel's discovery of radioactivity.

2. What three radioactive elements did the Curies find in pitchblende?

3. Define radioactivity.

4. List three types of nuclear radiation.

5. How did scientists determine that alpha particles have a positive electrical charge?

6. In what ways is gamma radiation different from alpha and beta radiation?

7. How did the idea that the atom is the smallest particle of matter change after the discovery of radioactivity?

> Radiation can be classified as either ionizing or non-ionizing, depending on the type of energy it transmits.

8. Define and give an example of
   a. ionizing radiation.
   b. non-ionizing radiation.

9. Why is ionizing radiation regarded as more dangerous than non-ionizing radiation?

10. Classify each of the following as ionizing or non-ionizing radiation.
    a. visible light   c. gamma rays
    b. X-rays   d. radio waves

11. How does ionizing radiation damage living cells?

> Rutherford's gold-foil experiment results led to a new model of the atom.

12. Describe Rutherford's gold-foil experiment.

13. a. What happened to most of the alpha particles observed in the gold-foil experiment?
    b. What did Rutherford conclude from this observation?

14. a. What happened to about 1 in every 20 000 alpha particles in the gold-foil experiment?
    b. What did Rutherford conclude from this observation?

15. What was the general structure of the atom that Rutherford proposed?

16. Sketch models to show the concept of an atom before and after Rutherford's gold-foil experiment.

17. What characteristic of alpha particles made them desirable as the beam in Rutherford's gold-foil experiment?

Most elements in nature are a mixture of isotopes. Some isotopes are radioactive.

18. Give the correct isotopic notation for copper-65.

19. Calculate the total neutrons in an atom of sulfur-34.

20. Copy the following table and find the value of each coded letter, *a* through *p*:

Symbol	Total protons	Total neutrons	Mass number
$^2_1 H$	*a*	1	*b*
$^{37}_c Cl$	*d*	*e*	*f*
$^g_h Tc$	43	56	*i*
$^{137}_j Cs$	*k*	*l*	*m*
$^n_o Ag$	47	60	*p*

21. How do the nuclei of carbon-12, carbon-13, and carbon-14 differ?

22. Consider the symbol $^{190}_{78} Pt$.

   a. What does the superscript 190 indicate?

   b. What does the subscript 78 indicate?

   c. How many neutrons are in a Pt-190 nucleus?

23. Neon (Ne) is composed of three isotopes with the following molar masses and relative abundances: Ne-20 (19.99 g/mol), 90.51%; Ne-21 (20.99 g/mol), 0.27%; and Ne-22 (21.99 g/mol), 9.22%.

   a. Based on these data, should neon's average atomic mass be closest to 20, 21, or 22? Why?

   b. Calculate the actual average atomic mass of neon. Show your calculations.

What evidence led to a modern understanding of the composition of atoms?

In this section, you have investigated ionizing radiation, read about the discoveries that led to Rutherford's model of the atom, and learned about isotope symbols, naturally occurring isotopes, and percent abundance. Think about what you have learned, then answer the question in your own words in organized paragraphs. Your answer should demonstrate your understanding of the key ideas in this section.

Be sure to consider the following in your response: key discoveries that led to the modern view of the atom (including the gold-foil experiment and the discovery of radioactivity), isotopes, and the structure of the atom.

Connecting the Concepts

24. A local politician proposes to ban all radiation in your community. Explain why this proposal has little chance of success.

25. Why is it possible to receive a suntan from ultraviolet radiation but not from radio waves?

26. How does the gold-foil experiment demonstrate the importance of evidence regarding events that cannot directly be seen?

27. Describe how the development of atomic theory illustrates the way scientific discoveries build on previous scientific knowledge and experiments.

28. In what way is fluorescence different from radioactivity?

29. In what way is a cathode-ray tube similar to a

    a. modern X-ray device?

    b. plasma lamp?

Extending the Concepts

30. Investigate the properties of gold and discuss why Rutherford probably selected that metal as an alpha-particle target.

31. Imagine that Rutherford proposed using beta particles rather than alpha particles in the gold-foil experiment. What result would you predict?

32. The neutron was discovered decades after the proton and electron were discovered. What made the discovering the neutron such a challenge?

33. Compare earlier models of the atom with

    a. Rutherford's model.

    b. the currently accepted model of the atom.

Figure 6.20 *Bricks are a common source of background ionizing radiation.*

SECTION B NUCLEAR RADIATION

How do we detect and describe the products of nuclear decay?

Of nearly 2000 known isotopes, there are more radioactive (unstable) isotopes than nonradioactive (stable) isotopes. Actually, most isotopes you encounter are not radioactive. However, naturally occurring radio-isotopes expose everyone to low levels of radiation. This radiation is from radioisotopes in building materials (such as brick and stone) in schools and homes (see Figure 6.20); in air, land, and sea; in foods you eat; and even within your own body. Because the human senses cannot detect nuclear radiation, various devices to detect and measure its intensity have been developed.

GOALS

- Identify sources of background radiation.
- Describe effects of ionizing radiation on human tissue and identify factors that determine the extent of damage.
- Distinguish among alpha particles, beta particles, and gamma rays and describe the effects of their emission on the composition of the nucleus.
- Write, complete, and balance nuclear equations.
- Describe methods for detecting and measuring ionizing radiation.

concept check 3

1. In a collection of 500 lithium atoms, would all atoms be identical in terms of electrons, neutrons, and protons? If not, describe what would be the same and what would be different.
2. a. What is your current definition of *background radiation*?
   b. Do you think it is possible to reduce your exposure to background radiation? Why or why not?
3. As you know, scientists use units to convey quantitative information such as distance (m), volume (L), concentration (mol/L), and density (g/mL). If you wanted to quantify the effects of ionizing radiation, what units might be useful? Propose one or two units of measurement and explain your reasoning.

B.1 EXPOSURE TO IONIZING RADIATION

When radioisotopes spontaneously decay, they usually emit alpha, beta, or gamma radiation. The type and intensity of radiation emitted helps to determine possible medical and industrial applications of particular radioisotopes. Each type of nuclear radiation also poses distinct hazards to human health.

Recall that a relatively constant level of radioactivity, called **background radiation**, is always present around and within you. Everyone receives background radiation at low levels from natural sources and from sources related to human activity. You will always experience at least some exposure to ionizing radiation.

Natural sources of background ionizing radiation include:

- High-energy particles from outer space that bombard Earth.
- Radioisotopes in rocks, soil, and groundwater: uranium (U-238 and U-235), thorium (Th-232), and the radioactive isotopes that form as they decay.
- Radioisotopes in the atmosphere: radon (Rn-222) and its decay products, including polonium (Po-210).
- Naturally occurring radioisotopes in foods and the environment, such as potassium-40 and carbon-14.

Advances in science and technology have created additional sources of background radiation, such as:

- Residual radioactive fallout from aboveground nuclear-weapon testing. (Most aboveground testing ended after the signing of the Limited Test Ban Treaty of 1963.)
- Increased exposure to radiation during high-altitude airplane flights (see Figure 6.21).

> The total quantity of radioisotopes released to the environment from fossil-fuel power plants is greater than the total quantity released from nuclear power plants.

Figure 6.21 *Air travel increases human exposure to ionizing radiation.*

- Radioisotopes released into the environment from both fossil fuel and nuclear power generation as well as other nuclear technologies.
- Radioisotopes released through the disturbance and use of rocks in mining and in making cement, concrete, and drywall.

Because of the effect of ionizing radiation on living tissue, it is important to monitor the quantity to which people are exposed over time. This quantity is referred to as the ionizing-radiation **dose**. The **gray** (Gy) is the SI unit that expresses the quantity of ionizing radiation absorbed by a particular sample, typically human tissue. An absorbed dose of one gray is defined as one joule of energy absorbed per kilogram of body tissue.

Not all forms of ionizing radiation, however, produce the same effect on living organisms. For example, alpha radiation will cause more harm internally to living organisms than will the same quantity of gamma radiation. The **sievert** (Sv) is the SI unit that expresses the ability of radiation—regardless of type or activity—to cause ionization in human tissue. Any exposure to radiation that produces the same detrimental effects as one gray of gamma rays represents one *sievert* of exposure. It is usually most convenient to express exposure in sieverts; this unit eases direct comparison across different types of ionizing radiation.

While the SI units for radiation exposure are the gray and the sievert, two other units have traditionally been used in the United States: the **rad** and the **rem** (see Table 6.5). The *rad* (like the gray) expresses the absorbed dose of radiation, and the *rem* (like the sievert) indicates ionizing effects on living organisms. Both the rad and the rem are one-hundredth of their corresponding SI units.

Table 6.5

Units of Radiation Dosage			
Unit	Absorbed Dose	Ionizing Effects	Definition
sievert (Sv)		X	1 Sv = radiation exposure that causes same effects as 1 Gy of gamma rays
rem		X	1 rem = 10^{-2} Sv
millirem (mrem)		X	1 mrem = 10^{-3} rem = 10^{-5} Sv
gray (Gy)	X		1 Gy = one joule of energy delivered to one kilogram of body tissue
rad	X		1 rad = 10^{-2} Gy

Even though a rem is only one-hundredth as large as a sievert, it is still much larger than typical exposures. Normal human exposures are so small that doses are expressed in units of *millirem* (mrem), where 1 mrem = 0.001 rem. One millirem of any type of radiation produces the same biological effects, whether the radiation is composed of alpha particles, beta particles, or gamma rays.

Some ionizing radiation comes from within your own body, as depicted in Figure 6.22. On average, people living in the U.S. receive about 360 mrem per person each year. About 300 mrem (83%) come from natural sources. Figure 6.23 shows the approximate proportion from each source.

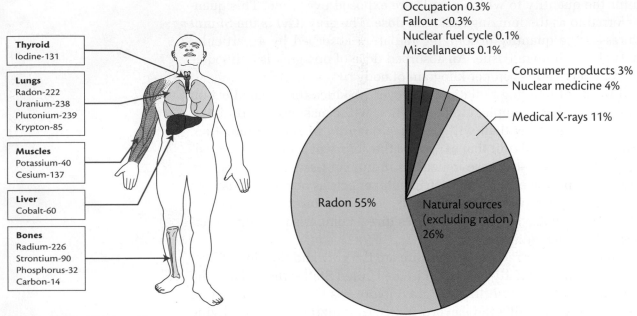

Thyroid
Iodine-131

Lungs
Radon-222
Uranium-238
Plutonium-239
Krypton-85

Muscles
Potassium-40
Cesium-137

Liver
Cobalt-60

Bones
Radium-226
Strontium-90
Phosphorus-32
Carbon-14

Other
Occupation 0.3%
Fallout <0.3%
Nuclear fuel cycle 0.1%
Miscellaneous 0.1%

Consumer products 3%
Nuclear medicine 4%

Medical X-rays 11%

Radon 55%

Natural sources (excluding radon) 26%

Figure 6.22 *All living things contain some radioactive isotopes, including these that are located in particular parts of the human body.*

Figure 6.23 *Sources of ionizing radiation in the United States.*

It is in your best interest to avoid unnecessary ionizing radiation exposure. What ionizing radiation level is considered reasonably safe? The U.S. government's background radiation limit for the general public is 500 mrem (0.5 rem) per year for any individual. The U.S. average exposure value of 360 mrem falls below this. The established U.S. limit for an individual's annual maximum safe exposure in the workplace is 5000 mrem (5 rem).

CHEM**QUANDARY**

RADIATION EXPOSURE STANDARDS

Why might radiation exposure standards for some individuals differ from those for the general public? Why do the standards differ for those who are exposed in their workplace?

■ MAKING DECISIONS

B.2 YOUR YEARLY IONIZING RADIATION DOSE

How does your yearly ionizing-radiation dose compare to the U.S. average of 360 mrem? Could you make decisions that would decrease your annual dose? On a sheet of paper, list the numbers and letters of each category in the following table. Then fill in the blanks on your sheet with appropriate quantities. Add all these quantities to estimate your annual ionizing-radiation dose.

Your Yearly Ionizing Radiation Dose

Common Sources of Radiation				Your Yearly Dose (mrem)
1. Where You Live				
a. Cosmic Radiation (from outer space)				
Your exposure depends on elevation. These are annual doses.				
sea level	**26 mrem**	4000–5000 ft	47 mrem	
0–1000 ft	**28 mrem**	5000–6000 ft	55 mrem	
1000–2000 ft	**31 mrem**	6000–7000 ft	66 mrem	
2000–3000 ft	**35 mrem**	7000–8000 ft	79 mrem	
3000–4000 ft	**41 mrem**	8000–9000 ft	96 mrem	_____ mrem
b. Terrestrial Radiation (from the ground)				
If you live in a state bordering Gulf or Atlantic coasts		23	mrem	
If you live in AZ, CO, NM, or UT		63	mrem	
If you live anywhere else in continental United States		46	mrem	_____ mrem
c. House Construction				
If you live in a stone, adobe, brick, or concrete building	7		mrem	_____ mrem
d. Power Plants				
If you live within 50 miles of a nuclear power plant		0.009	mrem	
If you live within 50 miles of a coal-fired power plant		0.03	mrem	_____ mrem
2. Food, Water, Air				
Internal Radiation (based on average values)				
a. from food (C-14, K-40) and water				
(radon dissolved in water)		40	mrem	40 mrem
b. from air (radon)		200	mrem	200 mrem

(Continued on next page)

Your Yearly Ionizing Radiation Dose

Common Sources of Radiation			Your Yearly Dose (mrem)	
3. How You Live				
Smoking cigarettes (1 pack / day)+	15	mrem	_____	mrem
Weapons test fallout*	1	mrem	_____	mrem
Travel by jet aircraft (per hour of flight)	0.5	mrem	_____	mrem
If you have porcelain crowns or false teeth	0.07	mrem	_____	mrem
If you wear a luminous (LCD) wristwatch	0.06	mrem	_____	mrem
If you go through airport security (each time)	0.002	mrem	_____	mrem
If you watch TV*	1	mrem	_____	mrem
If you use a video display (computer screen)*	1	mrem	_____	mrem
If you live in a dwelling with a smoke detector	0.008	mrem	_____	mrem
If you use a gas camping lantern with an old mantle	0.2	mrem	_____	mrem
If you wear a plutonium-powered pacemaker	100	mrem	_____	mrem
4. Medical Uses (radiation dose per procedure)				
X-rays: Extremity (arm, hand, foot, or leg)	1	mrem	_____	mrem
Dental	9	mrem	_____	mrem
Chest	6	mrem	_____	mrem
Pelvis/hip	65	mrem	_____	mrem
Skull/neck	20	mrem	_____	mrem
Barium enema	405	mrem	_____	mrem
Upper GI	245	mrem	_____	mrem
CT scan (head and body)	110	mrem	_____	mrem
Nuclear medicine (e.g., thyroid scan)	14	mrem	_____	mrem
Your Estimated Yearly Radiation Dose			_____	mrem

+This value represents a conservative estimate based upon information from the U.S. Department of Energy. Other estimates range from 20–1300 mrem.

* The value is less than 1 mrem, but adding that value would be reasonable.

Adapted from *Estimate your personal annual radiation dose*, American Nuclear Society, 2000.

1. Do any of the yearly dose values surprise you? Explain.
2. Compare your annual ionizing-radiation dose to the
   a. U.S. limit of 500 mrem.
   b. average background radiation (360 mrem).
3. Why is it useful to keep track of how many X-rays you receive each year?
4. How do geographic factors affect your annual ionizing-radiation dose?
5. a. What lifestyle changes could reduce a person's exposure to ionizing radiation?
   b. Would you decide to make those changes? Explain.

B.3 IONIZING RADIATION—HOW MUCH IS SAFE?

The two main factors that determine tissue damage due to ionizing radiation are radiation density (the number of ionizations within a given volume) and dose (the quantity of radiation received).

Gamma rays and X-rays are ionizing forms of electromagnetic radiation that penetrate deeply into human tissue. Ionizing radiation causes tissue damage by breaking bonds in molecules. At low levels of ionizing radiation, only a few molecules are damaged. In most low-dose cases, a body's systems can repair the damage. As the dose received increases, the total number of molecules affected by the radiation also increases. Generally, the damage to proteins and nucleic acids is of greatest concern because of their role in body structures and functions. Proteins form much of the body's soft tissue structure and compose enzymes, molecules that control the rates of cellular chemical reactions. If a large number of protein molecules are destroyed within a small region, too few functioning molecules may remain to enable the body to heal itself in a reasonable time (see Figure 6.24).

> See Unit 7 for more information on proteins and enzymes.

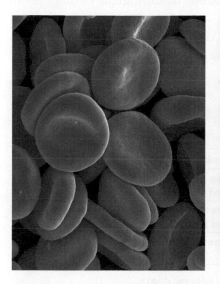 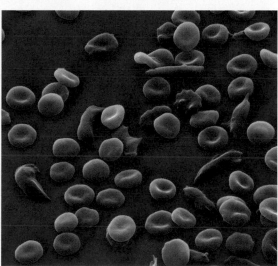

Figure 6.24 *Mutations are caused by damage to DNA, which can result from exposure to radiation. Compare the normal blood cells (left) to blood cells that are deformed (right) as a result of a mutation.*

Nucleic acids in DNA can be damaged by ionizing radiation. Minor damage causes **mutations**, which change the structure of DNA. Such mutations may result in the production of altered proteins and often kill the cell in which they occur. If the cell is a sperm or ovum, a mutation may lead to birth defects in offspring. Some mutations can lead to cancer, a disease in which cell growth and metabolism are out of control. When the DNA in many body cells is severely damaged, cells cannot synthesize new proteins to replace the damaged ones, and the organism or person dies.

> DNA molecules control cell reproduction and the synthesis of proteins.

Biological Damage from Radiation	
Factor	**Effect**
Dose	Most scientists assume that an increase in radiation dose produces a proportional increase in risk.
Exposure time	The more a given dose is spread out over time, the less harm it does.
Area exposed	The larger the body area exposed to a given radiation dose, the greater the damage.
Tissue type	Rapidly dividing cells, such as blood cells and sex cells, are more susceptible to radiation damage than are slowly dividing or non-dividing cells, such as nerve cells. Fetuses and children are more susceptible to radiation damage than are adults.

Table 6.6 *These four factors determine the actual effects of particular ionizing-radiation exposure.*

Table 6.6 lists factors determining the extent of biological damage from ionizing radiation. Table 6.7 summarizes the biological effects of large dosages of ionizing radiation. Because the values in Table 6.7 are so large, they are reported in rems, not millirems.

Becquerel observed a red spot on his chest after carrying a radium sample in his breast pocket. The dangers of ionizing radiation were unknown at that time.

Ionizing-Radiation Effects	
Dose (rem)	**Effect**
0–25	No immediate observable effects.
25–50	Small decreases in white blood cell count, causing lowered resistance to infections.
50–100	Marked decrease in white blood cell count. Development of lesions.
100–200	Radiation sickness—nausea, vomiting, hair loss. Blood cells die.
200–300	Hemorrhaging, ulcers, death.
300–500	Acute radiation sickness. 50% of those exposed die within a few weeks.
>700	100% die.

Table 6.7 *You can see how the consequences of radiation exposure change as dose increases.*

Large ionizing-radiation doses can have drastic effects on humans. Conclusive evidence that such doses produce increased cancer rates has been gathered from uranium miners and nuclear-accident victims. Some of the first cases of exposure to large doses of ionizing radiation occurred among workers who used radium compounds to paint numbers on watch dials that would glow in the dark (see Figure 6.25). The workers used their tongues to smooth the tips of their paint-brushes, and unknowingly swallowed small amounts of radio-active compounds. Later, these workers began to lose hair and became quite weak. Sometimes, this exposure even led to death.

Leukemia, a rapidly developing cancer of white blood cells, is commonly associated with exposure to high doses of ionizing radiation. Exposure also promotes other forms of cancer, anemia, heart problems, and cataracts (opaque spots on an eye lens).

Considerable controversy continues regarding whether very low doses of ionizing radiation, such as those from typical background sources, can cause cancer. Most of the data on cancer incidence have been based on human exposure to high doses of radiation. These data are extrapolated to much lower doses. Few studies have directly linked low radiation doses with cancer development. Most scientists agree that typical background levels of ionizing radiation are safe for most people. Some authorities argue that any increase above normal background levels increases the probability of developing cancer.

Figure 6.25 *These women, working in a factory in Orange, New Jersey, in the mid-1920s, painted radioactive radium onto watch dials so the watches could be read in the dark. Their exposure to ionizing radiation from radium often resulted in illness. Modern glow-in-the-dark watches do not contain radium.*

■ INVESTIGATING MATTER

B.4 ALPHA, BETA, AND GAMMA RADIATION

Preparing to Investigate

Investigating Matter A.1 provided an opportunity for you to make some initial ionizing radiation measurements and you now know more about its origin and effects on people. This investigation will allow you to learn more about specific properties of three types of ionizing radiation—alpha, beta, and gamma radiation.

To prepare for this investigation, review what you have already learned:

- Background radiation is always present (and can be detected and measured).

- Some objects emit ionizing radiation, which can be measured with the same devices used to measure background radiation.

- Alpha particles, beta particles, and gamma rays are all forms of ionizing radiation, but they have properties that differ from each other. For instance, you learned in Section A.2 (page 590) that alpha particles are positively charged, beta particles are negatively charged, and gamma rays have no electrical charge.

Also consider these two claims from Sections B.1 and B.3: "alpha radiation will cause more harm internally to living organisms than will the same quantity of gamma radiation" and "gamma rays . . . penetrate deeply into human tissue." Given these claims, you may want to know more about the likelihood of alpha particles penetrating human tissue—how does their penetrating ability compare to that of gamma rays? You might also wonder whether beta particles can penetrate skin or cause internal harm. After reading about doses and effects, perhaps you wonder how distance from a radiation source affects the dose you receive or how you can shield yourself from ionizing radiation.

This investigation will allow you to address these questions. Keep in mind that with proper handling, the radioactive materials in this investigation pose no danger to you. Nuclear materials are strictly regulated by state and federal laws. The radioactive sources you will use emit only very small quantities of radiation. Using them requires no special license. Nevertheless, you should handle all radioactive samples with great care, wearing protective gloves. Do not allow the radiation counter to come in direct contact with the radioactive material. Check your hands with a radiation monitor before you leave the laboratory.

Asking Questions

Read through *Gathering Evidence*. Write scientific questions that reflect the goals of Parts I and II. Think about a question that you might like to investigate for Part III. Write a draft of that question now (you will have the opportunity to revise this question after you complete Parts I and II). Construct a data table suitable for recording all relevant data that you will need to collect to answer your questions.

Gathering Evidence

Part I: Comparing the Penetrating Ability of Ionizing Radiation

1. Before you begin, put on your goggles, and wear them properly throughout the investigation.

2. Set up the apparatus shown in Figure 6.26. There should be space between the source and the detector for several sheets of glass or metal.

3. Turn on the counter. Allow it to warm up for at least 3 min. Determine the intensity of background radiation by counting the clicks for 1 min without any radioactive sources present. Record this background radiation value in counts per minute (cpm) in your data table.

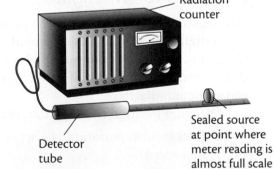

Figure 6.26 *You will need to determine the level of radiation detected in the absence of shielding.*

4. Put on protective gloves. Using forceps, place a gamma-ray source on the ruler at a point where it produces a nearly full-scale reading. Record the distance between the source and the detector.

5. Observe the meter for 30 s and estimate the number of counts per minute detected over this period. Record this approximate value. Subtract the background reading from that value and record the corrected results.

6. Without moving the radiation source, place a piece of cardboard (or an index card) between the detector and the source, as shown in Figure 6.27.

7. Observe the meter for 30 s. Record the typical reading. Then correct the reading for background radiation and record the corrected result in your data table.

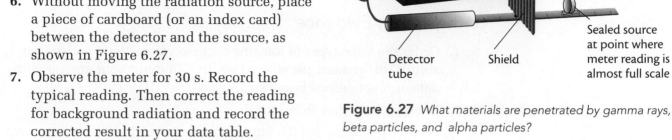

Detector tube   Shield   Sealed source at point where meter reading is almost full scale

Figure 6.27 *What materials are penetrated by gamma rays, beta particles, and alpha particles?*

8. Repeat Steps 6 and 7, replacing the cardboard with a glass or plastic sheet.

9. Repeat Steps 6 and 7, replacing cardboard with a lead sheet.

10. Repeat Steps 4 through 9, using a beta-particle source.

11. Repeat Steps 4 through 9, using an alpha-particle source.

Part II: Effect of Distance on Intensity

In Part I, you measured radiation intensity at a single distance from the source. What relationship would you expect between distance and radiation intensity? You will design your own procedure to investigate this relationship.

12. Write down your prediction about the relationship between distance and intensity. For instance, do you expect doubling the distance to lead to half the intensity reading?

13. Design a procedure to test your prediction. Think about how many different distances you should test and what data you will need to record. Write down your detailed procedure and obtain your teacher's approval before proceeding.

14. Carry out your investigation. Be sure to record all relevant data.

Part III: Further Investigations

15. In this part of the investigation, you will design, and possibly conduct, an additional investigation of your own design.

16. Note the materials available to conduct this investigation.

17. Think about one or two additional questions that relate to, or extend, the investigations you have already conducted using these ionizing radiation sources. Choose one question that you can address using the available equipment and materials. Write down the scientific question that you will investigate.

18. Propose and write down a detailed procedure for your investigation.

19. If you will carry out the investigation, first obtain your teacher's approval, then conduct your investigation. Be sure to record all relevant data.

Analyzing Evidence

1. Graph your data from Part II, plotting corrected cpm values on the *y*-axis and distances from the source to the detector (in cm) on the *x*-axis.

Interpreting Evidence

1. Order the three types of ionizing radiation from "least penetrating ability" to "greatest penetrating ability." Support your answer with data or observations from Part I.

2. Interpret your data from Part II.

   a. By what factor did the intensity of radiation (measured in counts per minute) change when the initial distance was doubled?

   b. Did this same factor apply when the distance was doubled again? Explain.

   c. State the mathematical relationship between distance and intensity.

3. It has been claimed that "alpha radiation will cause more harm internally to living organisms than will the same quantity of gamma radiation."

   a. Based on your investigations, which type of radiation is more likely to penetrate internal organs if the source of radiation is outside the body? Support your answer with data or observations.

   b. Describe what would need to happen for alpha particles to cause damage to internal organs.

> A *factor* is a number by which a value is multiplied to give a new value.

Making Claims

4. Of the shielding materials tested, which do you conclude is the

   a. most effective in blocking radiation? Cite supporting evidence.

   b. least effective in blocking radiation? Cite supporting evidence.

5. Based on your observations, what properties of a material appear to affect its ability to be penetrated by radiation?

6. Figures like the one on the next page (Figure 6.28) are often used to illustrate the relationship between distance from a radiation source and its intensity.

   a. Does the information in this figure fit with the data you collected in Part II?

   b. Write a caption for the figure that would help someone else interpret this diagram.

   c. What would the diagram look like at a total distance of 15 cm from the source? Explain your reasoning.

7. If you conducted an additional investigation in Part III:

   a. State the scientific question that you were addressing in your investigation.

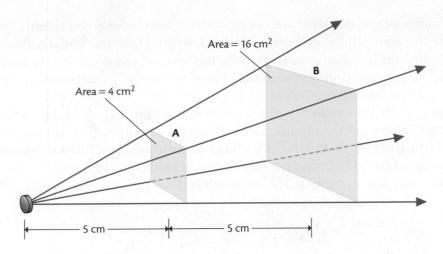

Area = 16 cm²

B

Area = 4 cm²

A

|← 5 cm →|← 5 cm →|

Figure 6.28

b. Summarize how you designed your experiment to address your question.

c. State your conclusions or claims. Support these conclusions or claims using data or observations from your investigation.

Reflecting on the Investigation

8. A patient receiving an X-ray is covered by a protective shield (Figure 6.29).

   a. What material would be a good choice for this apron?

   b. Why?

9. Whether or not you conducted an additional investigation, propose one additional question you have about alpha, beta, and gamma radiation.

You have found, through this investigation, that the three kinds of nuclear radiation differ greatly in their penetrating ability. You may wonder whether penetrating ability has any relationship to the structure of these nuclear emissions. In the next section, you will learn more about the composition of alpha, beta, and gamma radiation.

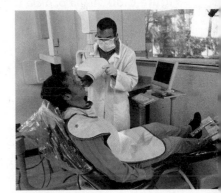

Figure 6.29 *What type of body shielding is used as this dental X-ray is obtained?*

B.5 NATURAL RADIOACTIVE DECAY

You know that alpha particles are positively charged. Why is this so? An alpha particle is composed of two protons and two neutrons. It is identical in composition to the nucleus of a helium-4 atom. Since alpha particles have no electrons, they have a double-positive electrical charge and are often symbolized as a doubly charged helium-4 atom, $_2^4 He^{2+}$. Alpha radiation (also called alpha emission) is released by many radioisotopes of elements with atomic numbers greater than 83.

Compared to a beta particle, an alpha particle has 5 to 50 times more energy and is more than 7300 times more massive. However, the larger and slower alpha particles are easy to stop when they are outside the body. Once inside the body, however, the electrical charge and energy of alpha particles can cause great damage to tissues. This damage occurs over very short distances (about 0.025 mm).

> Because of their relatively large mass, slower velocities, and large (2+) electrical charge, alpha particles lose most of their energy within a small distance.

Because alpha particles are very powerful tissue-damaging agents once inside the body, alpha emitters in air, food, or water are particularly dangerous to human life. Fortunately, outside the body, alpha particles are easy to block. As you noted in Investigating Matter B.4, alpha particles are stopped within a few centimeters by air.

Figure 6.30 illustrates a radium-226 nucleus emitting an alpha particle. During this process, the radium nucleus loses two protons, so its atomic number drops from 88 to 86, and it becomes an isotope of a different element, radon. In addition to losing two protons, radium-226 loses two neutrons, so its mass number drops by 4. The net result is the formation of radon-222. The decay process can be represented by the following nuclear equation:

$$\underset{\text{Radium-226}}{{}^{226}_{88}\text{Ra}} \longrightarrow \underset{\text{Alpha particle}}{{}^{4}_{2}\text{He}} + \underset{\text{Radon-222}}{{}^{222}_{86}\text{Rn}}$$

Atoms of two elements—helium and radon—have been formed from one atom of radium. (A radium compound is shown in Figure 6.31.) Note that atoms are *not* necessarily conserved in nuclear reactions, as they are in chemical reactions. Atoms of different elements can appear on both sides of a nuclear equation. Total mass numbers and atomic numbers, however, *are* conserved. In the above equation for radium-226 decay, the reactant mass number equals the sum of mass numbers of products ($226 = 4 + 222$). Also, the atomic number of radium-226 equals the sum of atomic numbers of products ($88 = 2 + 86$). Both relations hold true for all nuclear reactions.

Beta particles are fast-moving electrons emitted from a nucleus. Because they are so much lighter than alpha particles and travel at very high velocities, beta particles have much greater penetrating ability than do alpha particles. On the other hand, beta particles are not as damaging to living tissue.

During **beta decay**, a neutron in a nucleus decays to a proton and an electron. The proton remains in the nucleus, but the electron is ejected. The electron emitted from the nucleus is a beta particle. The following equation describes the process:

$$\underset{\text{Neutron}}{{}^{1}_{0}n} \longrightarrow \underset{\text{Proton}}{{}^{1}_{1}p} + \underset{\substack{\text{Beta particle}\\\text{(electron)}}}{{}^{0}_{-1}e}$$

Due to its negligible mass and negative electrical charge, a beta particle is assigned a mass number of 0 and an "atomic number" (nuclear charge) of −1. The net nuclear change due to beta emission is that a neutron is converted to a proton.

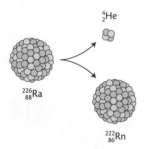

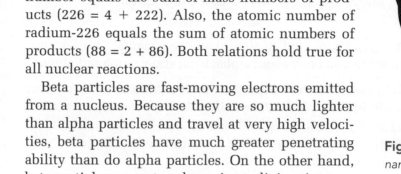

Figure 6.30 *Alpha-particle emission by radium-226. After emission, its mass number decreases by four (2 p + 2 n) and its atomic number decreases by two (2 p).*

Note that *n*, *p*, and *e* are symbols, respectively, for a neutron, a proton, and a beta particle. An emitted electron (beta particle) can also be symbolized by the Greek letter beta, β.

Figure 6.31 *Radium's name comes from the fact that small amounts of radium compounds, such as that on this antique watch face, emit enough radiation to glow in the dark. (See Figure 6.25, page 617.)*

The equation below shows a nucleus of lead-210 undergoing beta decay: The nucleus loses one neutron and gains one proton. The mass number remains unchanged at 210, but the atomic number increases from 82 to 83. The new nucleus formed is that of bismuth-210:

$$^{210}_{82}\text{Pb} \longrightarrow {}^{210}_{83}\text{Bi} + {}^{0}_{-1}e$$

| Lead-210 | Bismuth-210 | Beta particle (electron) |

Once again, the sum of all mass numbers remains the same in this nuclear reaction (210 on each side), and the sum of atomic numbers (the nuclear charge) remains unchanged (82 on each side).

Alpha and beta decay often leave nuclei in an energetically excited state. This type of excited state is described as **metastable** and is designated by the symbol m. For example, the symbol ^{99m}Tc represents a technetium isotope in a metastable or excited state. Energy from isotopes in such excited states is released as gamma rays—high-energy electromagnetic radiation that has as much or more energy than X-rays. Because gamma rays have neither mass nor charge, their release does not change the mass balance or charge balance in a nuclear equation. Table 6.8 summarizes general changes involved in natural radioactive decay.

Changes Resulting from Nuclear Decay				
Decay Type	**Symbol**	**Change in Atomic Number**	**Change in Total Neutrons**	**Change in Mass Number**
Alpha	$^{4}_{2}\text{He}$, α	Decreased by 2	Decreased by 2	Decreased by 4
Beta	$^{0}_{-1}e$, β	Increased by 1	Decreased by 1	No change
Gamma	γ	No change	No change	No change

Table 6.8
Summary of the results of alpha, beta, and gamma radioactive decay.

New isotopes produced by radioactive decay may also be radioactive, and therefore undergo further nuclear decay. Such a succession of decays is called a **decay series**. Uranium (U) (see Figure 6.32) and thorium (Th) are *parents* (reactants) in three natural decay series that start with U-238, U-235, and Th-232, respectively. Each decay series ends with formation of a stable isotope of lead (Pb). The decay series starting with uranium-238 contains 14 steps, as shown in Figure 6.33 (page 624).

Figure 6.32 *Uranium ore contains U-238 and U-235 isotopes, each of which participates in a decay series, producing several generations of radioactive decay products.*

Figure 6.33 *The uranium-238 decay series. Diagonal arrows show alpha decay. Horizontal arrows show beta decay. Here is how to interpret this chart: Locate radon-222 (Rn-222). The arrow pointing left shows that this isotope decays to polonium-218 by alpha (α) emission. This nuclear equation applies:*

$$^{222}_{86}Rn \longrightarrow ^{218}_{84}Po + ^{4}_{2}He$$

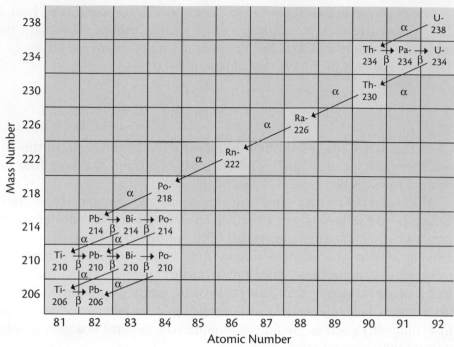

DEVELOPING SKILLS

B.6 NUCLEAR BALANCING ACT

The key to balancing nuclear equations is recognizing that both atomic numbers and mass numbers are conserved. Use the information in Table 6.8 (page 623) to complete the following exercises.

> *Sample Problem: Cobalt-60 is one source of ionizing radiation for medical therapy. Complete this equation for the beta decay of cobalt-60:*
>
> $$^{60}_{27}Co \longrightarrow ^{0}_{-1}e + \ ?$$
>
> Beta emission causes no change in mass number. Therefore, the new isotope will also have a mass number of 60. Thus, the unknown product can be written as ^{60}?. Because the atomic number increases by one during beta emission, the new isotope will have atomic number 28, one more than cobalt's atomic number. The periodic table indicates that atomic number 28 is nickel (Ni). The final equation is:
>
> $$^{60}_{27}Co \longrightarrow ^{0}_{-1}e + ^{60}_{28}Ni$$

Practice writing nuclear equations by completing these questions:

1. Write the appropriate symbol for the type of radiation given off in each reaction.

   a. The following decay process illustrates how archaeologists date the remains of ancient biological materials. Living organisms take in carbon-14 and maintain a relatively constant amount of it over their lifetimes. After death, no more carbon-14 is taken in, so the amount gradually decreases due to decay:

   $$^{14}_{6}C \longrightarrow ^{14}_{7}N + ?$$

   b. The following decay process takes place in some types of household smoke detectors:

   $$^{241}_{95}Am \longrightarrow ^{237}_{93}Np + ?$$

2. The two decay series beginning with Th-232 and U-238 are believed responsible for much of the thermal energy generated inside Earth. (Thermal contributions from the U-235 series are negligible. The natural abundance of U-235 is low.) Complete the following equations, which represent the first five steps in the Th-232 decay series. Identify the missing items A, B, C, D, and E. Each code letter represents a particular isotope or a type of radioactive emission. For example, in the first equation, Th-232 decays by emitting alpha radiation to form A. What is A?

   a. $^{232}_{90}Th \longrightarrow ^{4}_{2}He + A$

   b. $^{228}_{88}Ra \longrightarrow ^{0}_{-1}e + B$

   c. $^{228}_{89}Ac \longrightarrow C + ^{228}_{90}Th$

   d. $D \longrightarrow ^{4}_{2}He + ^{224}_{88}Ra$

   e. $^{224}_{88}Ra \longrightarrow E + ^{220}_{86}Rn$

> An alpha particle can be symbolized as either $^{4}_{2}He$ or $^{4}_{2}\alpha$.

concept check 4

1. Is it likely that a gamma ray—and nothing else—would be emitted from a nucleus? Explain.
2. How is writing nuclear equations different from writing chemical equations?
3. Radon is a noble gas, which means it is chemically unreactive. However, there is great concern over high levels of radon in homes. Why?

B.7 RADON

The gaseous element radon, which is the most massive of the noble gases, has always been a component of Earth's atmosphere. It is a radioactive decay product of uranium. In the 1980s, unusually high concentrations of radioactive radon gas found in some U.S. homes became a public health concern.

Radon is produced as the radioisotope uranium-238 decays in soil and building materials. (You can locate this radioactive decay product as Rn-222 in Figure 6.33, page 624.) Once it is emitted, radon is transported throughout the environment in many ways. Some radon produced in soil dissolves in groundwater. In other cases, radon gas seeps into houses through cracks in foundations and basement floors (see Figure 6.34).

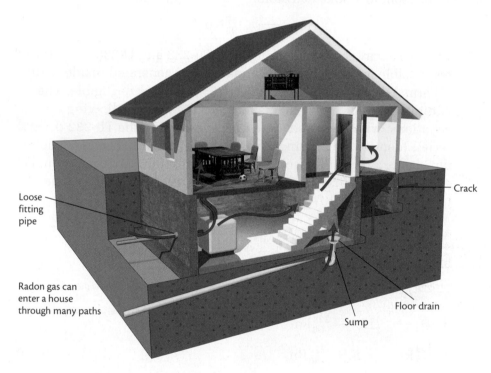

Loose fitting pipe

Crack

Radon gas can enter a house through many paths

Floor drain

Sump

Figure 6.34 *Radon gas is naturally released from some soils containing radioisotopes. Radon gas (red arrows) can enter a house by many paths.*

Figure 6.35 *Radon's high density and the fact that it enters a residence through foundation cracks means this radioactive gas can be removed by installing a venting system that withdraws air from the basement.*

In older houses, outdoor air entering through gaps in doors and windows can dilute the radon gas. However, to conserve energy, newer houses are built more airtight than are older houses. In a tightly sealed house, radon gas cannot mix freely with outdoor air or escape from the house. Consequently, radon gas concentrations may reach higher levels in newer homes. Remedies for high radon levels in houses include increasing ventilation and sealing cracks in floors (see Figure 6.35). Inexpensive radon test kits are available for home use.

The most serious danger of radon gas results from reactions that occur after it is inhaled. Radon decays to produce, in succession, radioactive isotopes of polonium (Po), bismuth (Bi), and lead (Pb). When radon gas is inhaled, it enters the body and is transformed, through radioactive decay, into these toxic heavy-metal ions, which cannot be exhaled as gases. These radioactive heavy-metal ions also emit potentially damaging alpha particles within the body.

Estimates indicate that about 6% of homes in the United States have radon levels higher than the exposure level recommended by the U.S. Environmental Protection Agency (EPA). See Figure 6.36. It is estimated that 10–15% of annual U.S. deaths from lung cancer are linked to the effects of indoor radon gas. These figures, although sobering, should be kept in perspective. About 80% of all U.S. lung-cancer deaths each year are attributed to cigarette smoking.

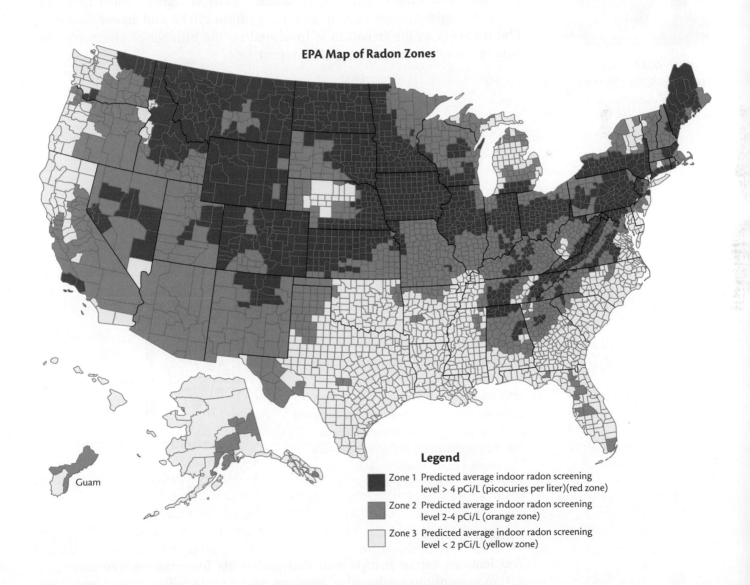

EPA Map of Radon Zones

Legend

Zone 1   Predicted average indoor radon screening level > 4 pCi/L (picocuries per liter)(red zone)

Zone 2   Predicted average indoor radon screening level 2-4 pCi/L (orange zone)

Zone 3   Predicted average indoor radon screening level < 2 pCi/L (yellow zone)

Figure 6.36 *Map highlighting U.S. areas by county that have experienced radon problems.*

B.8 NUCLEAR RADIATION DETECTORS

The only way to detect radioactive decay is to observe the results of nuclear radiation interacting with matter. In each of the following detection methods, visible and/or electronically detectable changes in matter enable a technician to determine when such radiation is present.

Geiger–Müller Tubes and Counters

An early device for detecting radioactivity still used today is the Geiger–Müller counter (Figure 6.37). This device contains argon gas that is ionized when radiation enters the tube. As illustrated in Figure 6.38, when the ionized gas strikes the detector, it produces electrical signals. Most radiation counters register these signals as both audible clicks and meter readings. The intensity of the radiation is indicated by the number of electronic signals or counts detected per minute (cpm).

Figure 6.37 *Using a Geiger–Müller counter to measure radiation dose.*

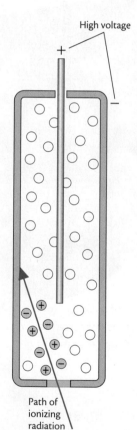

Figure 6.38 *As ionizing radiation passes into a Geiger–Müller detector tube, ions form in the gas inside. Positive ions are attracted to the tube's negatively charged outer wall, while negative ions are attracted to the positively charged center. These electrically charged particle movements constitute a pulse of electrical current. Each pulse is detected and counted.*

High voltage

Path of ionizing radiation

Film Badges

You learned earlier in this unit that emissions from radioactive materials will expose photographic film. Workers who handle radioisotopes wear film badges or other detection devices to monitor their exposure. Film badges are periodically collected and the film is developed and analyzed. The more ionizing radiation workers encounter, the greater the extent of exposure in their film badges. If workers receive doses in excess of federal limits, they are temporarily reassigned to jobs that minimize their future exposure to ionizing radiation.

Scintillation Counters

Another detection approach involves devices called **scintillation counters** that contain solid substances whose atoms emit light when they are excited by ionizing radiation. In modern scintillation counters, the flashes of light are detected electronically. The scintillation counter probe pictured in Figure 6.39 has a sodium iodide (NaI) detector that emits light when ionizing radiation strikes it.

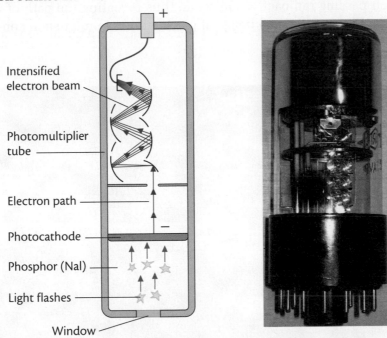

Figure 6.39 *A scintillation counter probe (left). Ionizing radiation causes flashes of light (scintillations) in the detector (NaI crystal). Each light flash is converted to an electron pulse that is increased many times as it moves through the photomultiplier tube (see close-up on right).*

Solid-State Detectors

In another technique, **solid-state detectors** monitor changes in the movement of electrons through silicon-based semiconductors as they are exposed to ionizing radiation. These detectors are often used in research laboratories.

Cloud Chambers

Ionizing radiation can also be detected in a cloud chamber. You will learn about this detection method in the following investigation.

■ INVESTIGATING MATTER

B.9 CLOUD CHAMBERS

Preparing to Investigate

A **cloud chamber** is a container filled with air saturated with water or another vapor, similar to saturated air on a very humid day. If cooled, this air becomes supersaturated with vapor. This is an unstable condition. When a radioactive source is placed near a cloud chamber filled with supersaturated air, the radiation passes through the chamber, ionizing gas molecules along its path. Vapor condenses to liquid on the ions that are formed, leaving a white trail along each passing radioactive emission, thus revealing the path of the particle or ray. Figure 6.40 is an image of particle tracks under such conditions.

Cloud-chamber trails resemble the vapor trails from high-flying aircraft.

Figure 6.40 *A cloud chamber, showing many particle trails created by ionizing radiation. The radioactive source is located on the left side.*

The cloud chamber you will use consists of a small plastic container and a felt band moistened with 2-propanol (isopropyl alcohol). The alcohol evaporates faster than water, so it saturates the enclosed air more readily. You will chill the cloud chamber with dry ice to promote supersaturation and cloud formation (Figure 6.41).

Figure 6.41
Cloud-chamber setup. Note the positions of the radiation source, dry ice, and flashlight.

Gathering Evidence

1. Before you begin, put on your goggles, and wear them properly throughout the investigation.
2. Fully moisten the felt band inside the cloud chamber with 2-propanol. Also place dark paper on the chamber bottom and moisten it with a small quantity of alcohol.
3. Using gloves and forceps, quickly place the radioactive source in the chamber.
4. To cool the chamber, place it on a flat surface of dry ice, ensuring that the chamber remains level.
5. Leave the chamber on the dry ice for 3–5 min.
6. Your teacher will adjust the room lighting. Focus a light source at an oblique angle (not straight down) through the container to illuminate the chamber base. If you do not observe any vapor trails, shine the light through the side of the chamber.
7. Observe the air inside the chamber near the radioactive source. Record your observations.

Interpreting Evidence

1. How would you describe your observations to someone who had never seen a cloud chamber?
2. What differences, if any, did you observe among the tracks?
3. Which type of radiation do you think would make the most visible tracks? Why?

Reflecting on the Investigation

4. What is the purpose of the dry ice?
5. Could a cloud chamber be used to detect non-ionizing radiation? Explain.

■ MAKING DECISIONS

B.10 ENSURING PUBLIC SAFETY

Consider again the statements found in the CANE flyer (page 586) and also your planned presentation to the senior citizens of Riverwood. With a small group of classmates, discuss the appropriateness of each of the following proposals for protecting the public from radiation hazards. For each proposal, identify statements in the CANE flyer that might encourage public groups to make such a proposal.

1. Because all radiation is bad, government regulators should ban exposure to all forms of radiation.

2. Although there are several different types of ionizing radiation (alpha, beta, and gamma), government standards for protecting the public against possible dangers should treat all radiation exposures identically.

3. Because nuclear radiation can be harmful, doctors should be required by law to inform patients of any medical procedure that involves nuclear radiation. Patients can then reject the treatment if they have concerns about radiation exposure (Figure 6.42).

4. The government should provide a waste-disposal site for permanently storing all wastes that have ever been identified as radioactive (Figure 6.43).

In this section, you studied the origins of ionizing radiation, sources of exposure to ionizing radiation, and some possible consequences of that exposure. Keep these ideas and concerns in mind as you explore uses of nuclear radiation in the next section.

Figure 6.42
Nuclear technology is widely used in medicine.

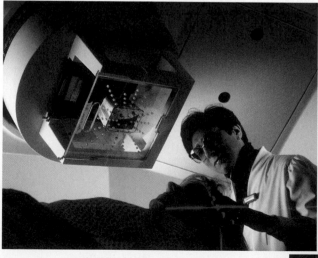

Figure 6.43 *Should the government be responsible for permanently storing all wastes that have ever been identified as radioactive?*

SECTION B SUMMARY

Reviewing the Concepts

> A relatively constant level of radioactivity, called background radiation, is always present and contributes to an individual's ionizing-radiation dose.

1. List five sources of background radiation.

2. Why does background radiation vary from one region to another?

3. Is it possible to eliminate background radiation? Explain.

4. What is the established U.S. background radiation limit for the general public expressed in

   a. mrem?

   b. Sv?

5. In what SI units is exposure to background radiation measured?

6. What radiation dose would you receive

   a. during a five-hour jet aircraft flight?

   b. from a CT scan and a chest X-ray?

   c. if you lived in a brick house 25 miles (40 km) from a coal-fired power plant for one year?

7. How does radon gas get into houses?

8. Why do radon levels tend to be higher in energy-efficient houses?

> Ionizing radiation has sufficient energy to break chemical bonds.

9. Why is the breaking of chemical bonds in living cells by ionizing radiation harmful?

10. What is a *mutation*?

11. How does ionizing radiation lead to an increase in mutation rates?

12. List four factors that determine the extent of biological damage from radiation.

13. What types of human tissue are most susceptible to radiation damage?

14. At what radiation dosage level would a person begin to experience nausea and hair loss? Express your answer in units of

    a. rem.   b. mrem.

> Alpha, beta, and gamma radiation have different properties that determine their effects on living tissues.

15. Which type or types of radiation would

    a. be stopped by a glass window pane?

    b. penetrate a cardboard box?

    c. penetrate a thin sheet of plastic but not a thin sheet of lead?

16. Explain why materials that emit alpha radiation are more dangerous inside the body than outside the body.

17. What is the relationship between radiation intensity and distance from the radiation source?

18. Suppose a beta source gives a corrected radiation reading of 640 cpm at a distance of 3 cm. Predict the corrected reading at a distance of

    a. 6 cm.   b. 12 cm.

> The emission of nuclear radiation changes the composition of the nucleus.

19. a. What is the composition of an alpha particle?

    b. List two symbols used to represent an alpha particle.

20. a. What is a beta particle?

    b. List two symbols used to represent a beta particle.

21. How is a beta particle formed?

22. Why does the emission of an alpha or beta particle create an atom of a different element?

23. What is a gamma ray?

24. Copy and complete the following nuclear equations.

    a. $^{6}_{3}\text{Li} + ^{1}_{0}n \longrightarrow ^{4}_{2}\text{He} + \underline{\quad}$

    b. $^{42}_{19}\text{K} \longrightarrow ^{0}_{-1}e + \underline{\quad}$

    c. $^{235}_{92}\text{U} \longrightarrow \underline{\quad} + ^{231}_{90}\text{Th}$

    d. $^{1}_{0}n + \underline{\quad} \longrightarrow ^{142}_{56}\text{Ba} + ^{91}_{36}\text{Kr} + 3\,^{1}_{0}n$

> Ionizing radiation may be detected by its interaction with matter using a variety of methods.

25. Describe how each of the following devices can detect the presence of ionizing radiation:

    a. Geiger–Müller tube and counter

    b. scintillation counter

    c. solid-state detector

    d. cloud chamber

    e. film badge

26. Would you expect alpha, beta, and gamma radiation to produce the same kinds of trails in a cloud chamber? Explain your answer.

How do we detect and describe the products of nuclear decay?

In this section, you investigated properties of alpha, beta, and gamma radiation, learned about sources of exposure to ionizing radiation and its possible health effects, considered nuclear decay, and explored how ionizing radiation can be detected. Think about what you have learned, then answer the question in your own words in organized paragraphs. Your answer should demonstrate your understanding of the key ideas in this section.

Be sure to consider the following in your response: background radiation; alpha, beta, and gamma radiation; penetrating power; dose; radioactive decay; and radiation detection methods.

Connecting the Concepts

27. High-level radioactive wastes are generally stored deep underground. Suggest two ways in which this method serves to keep people safe from excessive exposure to ionizing radiation.

28. In the 1940s and 1950s, some shoe stores invited customers to check the adequacy of shoe fit by X-raying their feet. Why did shoe stores discontinue that practice? (That shoe-store device was called a *fluoroscope*.)

29. Are the effects of shielding and distance the same for both ionizing and non-ionizing radiation? Explain your answer.

30. Radon is an inert noble gas that is fairly harmless to living things. Explain why its presence in homes constitutes a health hazard for occupants.

31. Describe fundamental differences between nuclear and chemical reactions.

32. A student sets up a cloud chamber and sees no white trails. What are some possible explanations for this result?

33. Two students live next door to each other. One receives three times more yearly radiation than the other. Explain how this could be possible.

34. Why do radiation detectors register counts even though no apparent source of radioactivity is near?

35. A heavy apron is provided for a patient who receives a dental X-ray.

   a. What element is probably used in the apron?

   b. What is the purpose of the apron?

   c. Why does the dentist or hygienist leave the room while the X-ray device is turned on?

Extending the Concepts

36. In terms of radiation, how is a sunburn different from a suntan? How do sunscreens work to prevent both?

37. Identify the five fastest-growing metropolitan areas in the United States. Rank them according to the average level of ionizing radiation exposure that inhabitants of similar houses receive.

38. Investigate and explain whether or not beta particles can be distinguished from electrons.

39. Why does ionizing radiation break some chemical bonds but not others?

CHEMISTRY *AT WORK*
Q&A

Dawn Shaughnessy, Nuclear Chemist at Lawrence Livermore National Laboratory in Livermore, California

With only a few exceptions, elements in the periodic table after polonium (Po, atomic number 84) are radioactive. Unlike polonium, radium, and a few other naturally occurring radioactive elements, almost all radioactive elements are synthesized, or created in a lab, including the "superheavy" elements—the ones with the largest atomic numbers. For many years, nuclear chemists at Lawrence Livermore National Laboratory have been leaders in synthesizing new superheavy elements. Read on to see how one chemist is focusing her efforts on extending the periodic table.

Q. What is nuclear chemistry, and why is it important for synthesizing new elements?

A. Nuclear chemistry is the study of radioactivity and nuclear reactions, including fission and fusion. Nuclear chemistry is central to creating new elements because we use fusion to synthesize these elements and radioactive signatures from their decay to detect their existence.

Q. How did you get into nuclear chemistry?

A. When I was in junior high, my parents got me a chemistry set. It sounds like a cliché, but it's true! I played with it all the time. Unfortunately, my high school didn't have much equipment, so we couldn't do many chemistry experiments there. But when I got to college, having access to a lab full of interesting tools made me realize that chemistry was what I wanted to do. One of my introductory chemistry professors was a nuclear chemist, and he spent a few weeks teaching us the basics of nuclear chemistry. I had never heard of nuclear chemistry, and I thought it was the coolest thing ever. When I finally got to take a nuclear chemistry class in college, I was hooked.

Q. Why did you end up working at Lawrence Livermore National Laboratory?

How element 118 was made. (top) Artist's conception of calcium ions traveling down the accelerator at a high velocity toward the rotating californium target. (bottom) The new element 118 travels through the accelerator to the detector.

A. There are lots of places that chemists can work. But if you want to work in nuclear chemistry, one of the U.S. Department of Energy's National Laboratories is the place to be. National labs do basic research with the specific goal of helping the country, and each one has a different focus. At Lawrence Livermore, one of our areas of focus is national security, which involves protecting the United States from nuclear weapons and disposing of radioactive waste safely. We also do basic nuclear chemistry research, such as discovering new elements. Scientists from Lawrence Livermore, including me, have played a part in discovering all the elements from 113 to 118.

Q. How do you go about discovering a new element?

A. We start by looking for two elements in the periodic table that have the right combination of protons and neutrons to add together to make the new element's atomic number. For example, if you want to make element 118 (also called ununoctium), you first need to find two elements with atomic numbers that add up to 118. You might try krypton (Kr) and lead (Pb), with atomic numbers 36 and 82. You then accelerate atoms of one of these elements to a very high velocity and slam them into atoms of the other element. In this process, the two elements fuse and a new element is created. This whole process happens inside a device called a particle accelerator.

Q. Lawrence Livermore doesn't have its own particle accelerator. How do you discover new elements without this device?

A. We have many partnerships, or collaborations, with nuclear chemists who work at particle accelerators around the world. We supply them with one of the starter elements. They supply the other starter element and conduct the reaction in their accelerator. We both get to bring something to the table. One of the labs we collaborate with most often is Dubna, a particle accelerator facility in Russia we've been working with since the 1980s. I fly to Russia to work there at least once a year.

The Joint Institute for Nuclear Research is in Dubna, near Moscow. This photo shows its synchrocyclotron.

Q. Superheavy elements can have very short half-lives. Why is it important for us to learn about them?

A. We don't know much about the bottom of the periodic table, including where it ends. Trying to create new elements will tell us if element 118 is truly the last element in the periodic table or if we can extend it even further. We also know very little about how these superheavy elements behave. Right now, we have them placed in groups based on their atomic number. But does element 116 really behave like other group 16 elements, such as oxygen? Creating these elements and trying to measure their properties is the only way to tell.

Experimenting with "atomic energy" at home is not safe, which is why modern chemistry sets do not claim to—unlike this vintage set. Yet a chemistry set is a good way to start investigating activity at the atomic level.

Q. What advice do you have for students interested in nuclear chemistry?

A. Take as many chemistry, physics, and math classes in high school as you can. When you get to college, think about doing a summer internship at a national lab like Lawrence Livermore. An internship is a great way to get hands-on experience in a wide variety of sciences, including nuclear chemistry—it can really help accelerate your career.

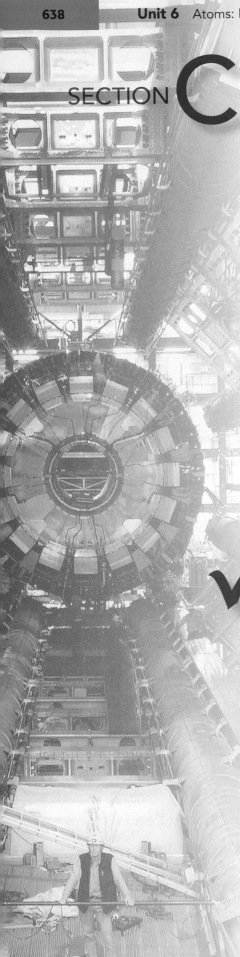

USING RADIOACTIVITY

Why and how are radioactive isotopes useful?

Each radioisotope decays and emits ionizing radiation at its own special rate. Scientists have devised convenient ways to measure, analyze, and report how rapidly (or slowly) particular radioisotopes decay. In this section, you will learn about radioactive decay rates, a characteristic that helps determine how useful or hazardous a radioisotope may be.

GOALS

- Define and describe half-life.
- Using its half-life, calculate the amount of a particular isotope that remains undecayed after a specified time.
- Describe how radioisotopes are used as diagnostic tracers in medicine.
- Explain how radioisotopes are used to kill cancerous cells.
- Describe the process of nuclear transmutation.
- Write, complete, and balance nuclear transmutation equations.

concept check 5

1. What does it mean to describe radiation as *ionizing radiation*?
2. What is balanced in a nuclear equation? How does this compare to a chemical equation?
3. What does half-life mean?

C.1 HALF-LIFE: A RADIOACTIVE CLOCK

How long does it take for a sample of radioactive material to decay? There is no simple answer to this question. However, by understanding how radioactive materials decay, scientists can predict how long a radioisotope used in a medical diagnostic test, for example, will remain active within the body; plan the long-term storage of hazardous nuclear wastes; and estimate the ages of old organisms, civilizations, or rocks.

The concept of **half-life** has several interpretations. The most common interpretation is that a half-life is the time it takes for one-half of the total radioactive atoms originally present in a sample to decay. Under normal conditions, the half-life for a particular radioisotope remains constant. This interpretation of half-life is useful when dealing with a sample that contains a large number of radioactive atoms. Because one gram of any element contains well over 10^{21} atoms, the typical sample size involves very large numbers of atoms and this definition of half-life is particularly useful.

Another interpretation, proposed by Ernest Rutherford in 1904, applies to an individual atom in a sample. In this view, half-life expresses the time within which a radioactive atom has a 50–50 chance to undergo radioactive decay. Table 6.9 lists the half-lives and decay reactions of several radioisotopes.

Table 6.9

Decay Equations and Half-Lives for Five Radioactive Isotopes

Radioisotope	Symbol	Half-Life
Hydrogen-3	$^{3}_{1}H \longrightarrow {}^{3}_{2}He + {}^{0}_{-1}e$	12.3 y
Carbon-14	$^{14}_{6}C \longrightarrow {}^{14}_{7}N + {}^{0}_{-1}e$	5.73×10^3 y
Phosphorus-32	$^{32}_{15}P \longrightarrow {}^{32}_{16}S + {}^{0}_{-1}e$	14.3 d
Potassium-40	$^{40}_{19}K \longrightarrow {}^{40}_{20}Ca + {}^{0}_{-1}e$	1.28×10^9 y
Radon-222	$^{222}_{86}Rn \longrightarrow {}^{218}_{84}Po + {}^{4}_{2}He$	3.82 d

Although it is not possible to predict when a particular radioactive atom will decay, each way of thinking about half-life is equally valid when working with the large numbers of atoms in typical chemical samples. The decay of carbon-14 is represented by this equation:

$$^{14}_{6}C \longrightarrow {}^{14}_{7}N + {}^{0}_{-1}e$$

All living matter contains carbon and, therefore, a small amount of radioactive C-14. Thus, all living organisms emit a small but constant level of radioactivity. In living matter, decaying C-14 atoms are constantly replaced by new carbon atoms. After death, no new carbon atoms are taken in. The radioactive C-14 atoms continue to decay, so the longer an organism has been dead, the fewer C-14 atoms it contains.

Table 6.9 indicates that C-14's half-life is 5730 years. If an organism contains 50.0 billion atoms of carbon-14 at the time of death, half of those atoms will have decayed after 5730 years pass, leaving 25.0 billion atoms of carbon-14. During the next 5730 years, another one-half of the atoms will decay, leaving 12.5 billion atoms of carbon-14. Therefore, if a sample from a previously living organism contains only one-fourth the number of C-14 atoms expected in a living organism, we can estimate that the sample is about 11 460 years old—that is, two half-lives must have passed since the organism died.

The relative abundance of C-14 has remained reasonably constant over thousands of years. C-14 is constantly being produced in the upper atmosphere by cosmic rays interacting with nitrogen atoms, while C-14 atoms are also undergoing radioactive decay.

Because the decay rate is directly related to the number of C-14 atoms present, scientists usually express the decay rate rather than the number of C-14 atoms.

In 2004, scientists reported that they could decrease the half-life of beryllium-7 by half a day (a 1% change) by trapping it in an electron-rich environment. The half-life of sodium-22 has also been decreased, in that case by implanting it in palladium (Pd) metal cooled to 12 K.

Every radioisotope has a specific half-life that is constant under normal circumstances. Half-lives of radioisotopes can be as short as a fraction of a second or as long as several billion years. For example, the half-life of polonium-212 is 3×10^{-7} seconds, while that of uranium-238 is 4.5 billion years. Thus, in one year, all atoms in a small sample of polonium-212 will probably have decayed, while well over 99% of the original uranium-238 atoms will still be present.

After 10 half-lives, only about 1/1000th or 0.1% of the original radioisotope atoms are still left to decay. (You can verify that statement with your own calculations.) That means that the rate of radioactive decay of the isotope has dropped to 0.1% of its initial level. This reduced level is often considered safe because it roughly approaches the level of normal background radiation.

Because there is no way to change the radioactive decay rate significantly for a particular isotope, radioactive waste disposal (or storage) can pose challenging problems, especially for radioisotopes with very long half-lives. You will examine that issue later in this unit.

In the following activity, you will model and explore the concept of half-life with heads-up and heads-down coins.

■ MODELING MATTER
C.2 UNDERSTANDING HALF-LIFE

In this activity you will receive 80 pennies and a box. Place all pennies heads up to represent the starting sample of *headsium*. Assume each heads-up penny represents an atom of the radioactive headsium. Its decay produces a tails-up penny—*tailsium*. Each shake of the closed box containing pennies represents one half-life. During this time a certain number of headsium nuclei will decay—flip over—to produce tailsium. You will investigate the relationship between the passage of time and the quantity of radioactive nuclei (heads-up pennies) that decay.

The following steps will lead you through this activity:

1. Prepare a data table for recording the undecayed headsium and decayed tailsium atoms after each of four half-lives. Include initial values for passage of 0 half-lives.

2. Place the 80 pennies heads up in the box.

3. Close the box and shake it vigorously.

4. Open the box. Remove all atoms that decayed into tailsium. Record the number of undecayed (headsium) and decayed (tailsium) atoms after this first half-life.

5. Repeat Steps 3 and 4 three more times. You will now have simulated the passage of four half-lives. Record your results for each half-life.

6. Follow your teacher's instructions to obtain pooled class data for total undecayed headsium atoms remaining after each half-life.

7. Using your own data and class-pooled data, prepare a graph by plotting the number of half-lives on the x-axis and the number of undecayed atoms remaining after each half-life on the y-axis. Plot and label two graph lines—one representing your own data and the other representing pooled class data.

Now answer these questions, based on your data:

1. a. Describe the appearance of your two graph lines. Are they straight or curved?

   b. Which set of data—yours or the pooled class data— provides a more convincing demonstration of half-life? Why?

2. About how many headsium nuclei would remain after three half-lives, if the initial sample had 600 headsium atoms?

3. If 190 headsium nuclei remain from an original sample of 3000 headsium nuclei, about how many half-lives must have passed?

4. Describe one similarity and one difference between your model based on pennies and actual radioactive decay. (*Hint:* Why did you pool the class data?)

5. How could you modify this model to demonstrate that different isotopes have different half-lives?

6. a. How many half-lives would be needed for one mole of a radioisotope to decay to 6.25% of the original number of atoms?

   b. Is it likely that any of the original radioactive atoms would still remain after

      i. 10 half-lives? Explain your answer.

      ii. 100 half-lives? Explain your answer.

7. a. In this simulation, can you predict when a particular headsium nucleus will "decay"?

   b. If you could follow the fate of an individual atom in a sample of radioactive material, could you predict when it would decay? Why or why not?

8. What other ideas could model the concept of half-life?

DEVELOPING SKILLS

C.3 APPLICATIONS OF HALF-LIVES

Use what you learned about half-life to answer the following questions. Figure 6.44 may help.

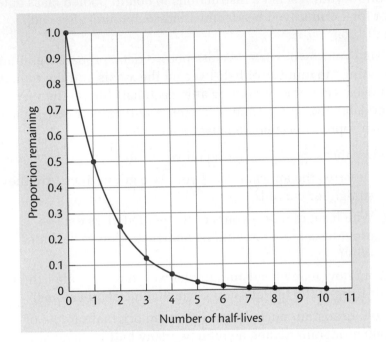

Figure 6.44 *This graph depicts the quantity of radioactive material remaining versus total half-lives that have passed. What proportion of material would remain after four half-lives?*

Sample Problem: *The half-life of O-15 is 2.0 min. Its radioactive decay produces N-15. How much O-15 will remain undecayed after 5.0 min, if the original sample contained 14.0 g O-15?*

First, determine the total half-lives that 5.0 min represents:

$$5.0 \ \cancel{\text{min}} \times \frac{1 \text{ half-life}}{2.0 \ \cancel{\text{min}}} = 2.5 \text{ half-lives}$$

Because the total number of half-lives is not an integer, use Figure 6.44 to estimate the proportion of O-15 remaining. Locate 2.5 half-lives on the *x*-axis. From that point, move directly upward on the graph until you touch the curved line. Then move left until you intersect the *y*-axis. Read this point from the graph: about 0.18. This means that 18% of the original O-15 sample remains:

$$(14.0 \text{ g O-15}) \times (0.18) = 2.5 \text{ g O-15 remain}$$

You can check to see if an answer of 2.5 g makes sense: After one half-life, 7.00 g O-15 would remain in the sample (the remaining mass represents N-15). After two half-lives, 3.50 g remain, and after three half-lives 1.75 g O-15 would remain undecayed. The answer we found, 2.5 g, is between 3.50 g and 1.75 g, which is what we would expect.

1   Suppose you received $1000 (see Figure 6.45) and could spend one-half in the first year, one-half of the balance in the second year, and so on. (*Note:* One year corresponds to one half-life in this analogy.)

   a. If you spent the maximum allowed annually, at the end of which year would you have $31.25 left?

   b. How much money would be left after 10 half-lives?

2. Potassium is a necessary nutrient for all living things and is the seventh most abundant element on Earth's surface, composing about 1.5% of its crust. About 0.01% of natural potassium atoms are the radioisotope potassium-40. K-40 has a half-life of nearly 1.3 billion (1.3×10^9) years.

   a. Assuming Earth is ~4.5 billion years old, how much of the K-40 at Earth's formation remained after one half-life?

   b. Roughly how many times more K-40 was present when Earth formed than is present now?

Figure 6.45 *If you initially had $1000 and spent half the money each year, how many years would it take until you had only $125?*

3. Strontium-90 is one of many radioisotopes generated by nuclear-weapon explosions. This isotope is especially dangerous if it enters our food supply. Strontium behaves chemically like calcium, because these elements are members of the same chemical family. Rather than passing through the body, radioactive strontium-90 is incorporated into calcium-based material, such as bone. A nuclear test-ban treaty in 1963 ended most aboveground weapons testing. But some Sr-90 released in previous testing is still present in the environment.

   a. Sr-90 has a half-life of 28.8 years. Track the decay of Sr-90 atoms that were present in the atmosphere in 1963 as follows:

      i. Using 1963 as year zero, when 100% of released Sr-90 was present, identify the years that represent completion of one, two, three, four, and five half-lives.

      ii. Calculate the percent of the original 1963 Sr-90 radioisotope present at the end of each half-life.

   b. Plot the percent of the original 1963 Sr-90 radioactivity level on the *y*-axis and the years 1963 to 2110 on the *x*-axis. Draw a smooth curve through the data points.

      i. What percent of Sr-90 formed in 1963 is present now?

      ii. What percent will remain in 2100?

   c. Compare your graph to that shown in Figure 6.44.

      i. How do the two graphs differ?

      ii. In what ways are the graphs similar?

CHEMQUANDARY

CARBON-14 DATING

It is not possible to determine the age of every artifact using carbon-14 dating. What kinds of materials might be good candidates for cabon-14 dating? What are some materials you could not date using carbon-14? Why does carbon-14 dating have a practical limit of about 50 000 years?

A conservator cuts a sample of a document for carbon-14 dating.

C.4 RADIOISOTOPES IN MEDICINE

During the late 19th and early 20th centuries, the rapid advancements in science fascinated the general public. This excitement about the "wonders of science" made people very susceptible to exaggerated claims about the healing powers of radiation. Many products were marketed as medicine with remarkable powers to cure. Some of these medicines advertised that they contained radium, claiming supposed benefits of that substance. Unfortunately, some actually contained radioactive materials and were quite hazardous. Radioactive medicines were sold in the United States as late as the 1930s.

The careful use of ionizing radiation and radioisotopes can be quite effective in medical diagnosis and treatment (see Figures 6.46 and 6.47). Such uses can be classified as either *diagnostic* or *therapeutic*. **Diagnostic** use helps doctors understand what is happening inside the body (see Figure 6.48), while **therapeutic** use involves treating a medical condition.

Figure 6.46 *Injection of radioactive tracer molecules allows noninvasive study of the metabolic function of internal organs.*

Figure 6.47 *Radioisotopes are prepared and used with knowledge of their half-lives, thus ensuring appropriate levels of activity as tracers. In this image, a lead-lined enclosure is used for protection as an injection of radioactive substance is prepared.*

A common diagnostic application uses radioisotope-tracer studies, based on detecting a radioisotope in particular parts of the body. Physicians know that certain elements collect in specific parts of the body (e.g., calcium in bones and teeth). They can investigate a given part of the body by using an appropriate radioisotope as a **tracer**. In a radioisotope-tracer study, the physician places radioisotopes with short half-lives in a patient's body. Such studies can identify cellular abnormalities and help physicians select appropriate therapy.

Tracers have properties that make them ideally suited to this task. First, radioisotopes have the same chemical properties as stable (non-radioactive) isotopes of the element. Researchers can apply a solution of an appropriate tracer isotope to the body, or they can feed or inject the patient with a biologically active compound containing the radioactive tracer element. A detection system, as shown in Figure 6.46, then allows a medical technician to track the tracer's location throughout the body.

For example, iodine-123 is used to diagnose problems of the thyroid gland, which is located in the neck. A patient drinks a tracer solution containing sodium iodide (NaI), in which some of the iodide ions are the radioactive isotope

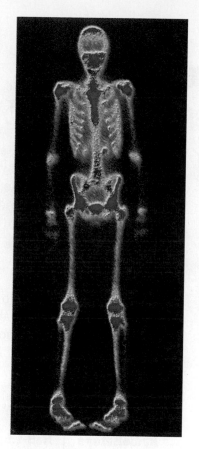

Figure 6.48 *Nuclear-medicine scan of bones of the entire human body.*

I-123. The physician, using a radiation detection system, monitors the rate at which this tracer is taken up by the thyroid. A healthy thyroid will incorporate a known amount of iodine. An overactive or underactive thyroid will take up, respectively, more or less iodine. The physician compares the measured rate of I-123 uptake by the patient to the normal rate for an individual of the same age, gender, and weight, then takes appropriate therapeutic action.

Technetium-99m (Tc-99m), a synthetic radioisotope, is the most widely used diagnostic radioisotope in medicine. It has replaced exploratory surgery as a way to locate tumors in the brain, thyroid, and kidneys. Tumors are areas of runaway cell growth; technetium concentrates where cell growth is fastest. A bank of radiation detectors around the patient's body can pinpoint the Tc-99m at the tumor's precise location.

Physicians use therapeutic radioisotopes because they emit radiation that carries enough energy to destroy living tissue. In some cancer treatments, doctors kill cancerous cells with ionizing radiation. For thyroid cancer, a patient takes a liquid or a pill that contains radioiodine. The radioiodine collects in and destroys the cancerous portion of the thyroid gland. In other cancer treatments, physicians may direct an external beam of ionizing radiation (from cobalt-60) at the cancerous spot. Physicians must administer treatments with great care. High radiation doses also kill normal cells.

Radioiodine is a way of designating the radioactive isotope of iodine.

CHEM**QUANDARY**

USING RADIOISOTOPES IN MEDICINE

One source of ionizing radiation is cobalt-60, which kills rapidly dividing cells. Consider two equally sized Co-60 samples shipped to two hospitals at the same time. At one hospital, the Co-60 is used to treat dozens of individuals, while at the other hospital it is used only once or twice. Why would *both* hospitals dispose of their Co-60 samples after five years?

Radiosodium is usually administered as a NaCl solution.

Other medical applications include the use of radiosodium (Na-24) to detect circulatory system abnormalities and radioxenon (Xe-133) to help locate lung embolisms (blood clots) and abnormalities. Table 6.10 summarizes medical uses for several radioisotopes.

Selected Medical Radioisotopes

Radioisotope	Half-Life	Use
Used as Tracers		
Technetium-99m	6.01 h	Measure cardiac output; locate strokes and brain and bone tumors.
Gallium-67	78.3 h	Diagnosis of Hodgkin's disease
Iron-59	44.5 d	Determine rate of red blood cell formation (these contain iron); anemia assessment
Chromium-51	27.7 d	Determine blood volume and lifespan of red blood cells
Hydrogen-3 (tritium)	12.3 y	Determine volume of the body's water; assess vitamin D use in body
Thallium-201	72.9 h	Assess coronary artery disease
Iodine-123	13.3 h	Diagnose diseases of the thyroid gland
Used for Radiation Therapy		
Cesium-137	30.1 y	Treat shallow tumors (external source)
Phosphorus-32	14.3 d	Treat leukemia, a bone cancer affecting white blood cells (internal source)
Iodine-131	8.0 d	Treat thyroid cancer (external source)
Cobalt-60	5.3 y	Treat shallow tumors (external source)
Yttrium-90	64.1 h	Treat pituitary gland cancer (internal source)

Table 6.10

You have learned that ionizing radiation, like that observed in a cloud chamber (page 630), is emitted by an unstable radioactive isotope as it decays, eventually to a stable nonradioactive isotope. Do you think it would be possible to reverse that process, converting a stable isotope into an unstable, radioactive isotope? Think about it. This question will be addressed later in this section.

C.5 NUCLEAR MEDICINE TECHNOLOGIES

Computers touch nearly all aspects of modern life. For example, health and medical science now employ two nuclear medicine technologies that rely heavily on computers to make sense of the large quantities of data obtained. These technologies are positron emission tomography (see Figure 6.49, page 648), which involves radioisotopes, and magnetic resonance imaging.

Positron emission tomography (PET) scans are based on a very unusual form of radioactive decay involving a few particular radioisotopes. Although most radioisotopes emit alpha, beta, or gamma radiation, a few radioisotopes emit radiation in the form of positrons. **Positrons** originate in the nucleus and have the same mass as beta particles (electrons). However, positrons differ from electrons in fundamental ways. Positrons have a positive electrical charge, while electrons are negatively charged.

Positrons are composed of antimatter. When a positron encounters an electron, both particles are annihilated (destroyed) and produce two gamma rays that are emitted in opposite directions. PET detects these gamma-ray pairs and, with the help of computers, determines where they originated. By observing a large number of such events, a computer-generated image gradually emerges.

The radioisotope tracer that emits positrons in PET scans is attached to a sugar molecule. Physicians can accurately determine the movement of each tagged sugar molecule as it progresses through the body. Because cancers grow faster than normal tissues, cancerous tissue metabolizes more sugar in a given time than does normal tissue. The radioisotope trace (sugar tag) eventually becomes more concentrated in regions of the body containing cancerous tissue. PET technology can thus detect and display metabolic activity. Doctors can use this information to investigate brain functioning without invasive surgery.

Magnetic resonance imaging (MRI) does not employ ionizing radiation. This technique relies on the properties of the protons in the nuclei of hydrogen atoms in large biomolecules. Because of its reliance on the properties of nuclear particles, this technique is classified as a nuclear medicine technique. MRI is an application of a laboratory process known as nuclear magnetic resonance (NMR), which was developed in the mid-20th century. NMR is a noninvasive technique that can identify atoms within a sample without altering and affecting the sample itself.

Tomography refers to producing a 3-D image of an internal object of interest.

Scientific American Working Knowledge Illustration

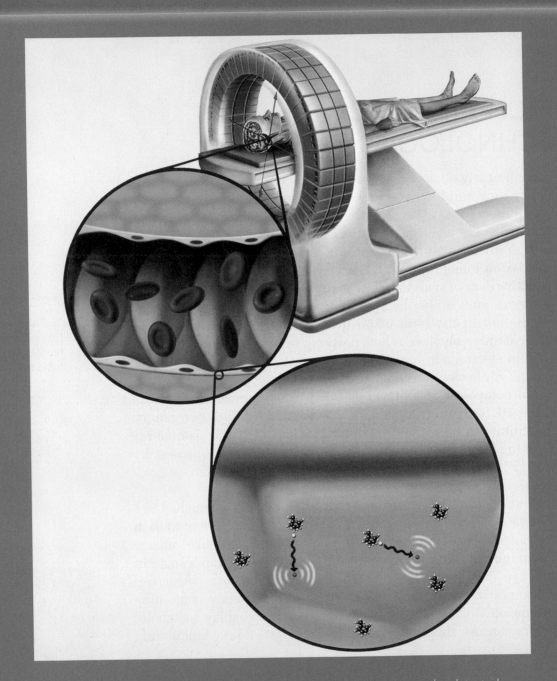

Figure 6.49 *Positron emission tomography (PET). Cancerous tissue can be detected using PET scans. In this illustration, a brain tumor is identified by detecting gamma radiation emitted when positrons collide with electrons. As depicted in the particulate-level representation (bottom image), sugar molecules are tagged with a radioisotope tracer that emits positrons (yellow spheres). When a positron collides with an electron (blue spheres) in the immediate vicinity, both particles are destroyed and produce two gamma rays (depicted as yellow waves). The gamma-ray pair is then detected.*

MRI imagery can produce useful images of soft tissues. A major benefit of MRI is that it does not rely on ionizing radiation. Unlike most other nuclear-medicine technologies, MRI uses radio waves of very low energies and involves no known health risks. Some patients were hesitant to undergo the procedure when it was called by its original name, nuclear magnetic resonance, due to fear evoked by the term nuclear. This unfounded fear prompted the name change to magnetic resonance imaging.

concept check 6

1. Would carbon-14 dating be useful for dating artifacts if the half-life of carbon-14 was not constant? Explain.
2. What specific characteristics of radioisotopes make them useful in
   a. medical diagnosis?
   b. treatment of cancer?
3. You are familiar with the synthesis of compounds. What does it mean to synthesize an element?

C.6 ARTIFICIAL RADIOACTIVITY

In 1919, Ernest Rutherford enclosed nitrogen gas in a glass tube and bombarded the sample with alpha particles. After analyzing the gas remaining in the tube, he found that some nitrogen atoms had been converted to an isotope of oxygen, according to the following equation:

$$\overset{4}{\underset{2}{}}\text{He} + \overset{14}{\underset{7}{}}\text{N} \longrightarrow \overset{17}{\underset{8}{}}\text{O} + \overset{1}{\underset{1}{}}\text{H}$$

| Helium-4 (alpha particle) | Nitrogen-14 | Oxygen-17 | Hydrogen-1 |

Rutherford had produced the first synthetic or artificial **transmutation** of an element, the first documented conversion of one element to another. He continued this work but was limited by the moderate energies of alpha particles then available. By 1930, scientists had developed particle accelerators that could produce highly energetic particles for bombardment reactions. Using these higher-energy particles, scientists created many other synthetic atoms, some of which were radioactive.

The first synthetic radioisotope (one not occurring in nature) was produced in 1934 by French physicists Irène and Frédéric Joliot-Curie (the daughter and son-in-law of Marie and Pierre Curie; see Figure 6.50). They bombarded aluminum atoms with alpha particles, producing radioactive phosphorus-30 and neutrons:

$$\overset{27}{\underset{13}{}}\text{Al} + \overset{4}{\underset{2}{}}\text{He} \longrightarrow \overset{30}{\underset{15}{}}\text{P} + \overset{1}{\underset{0}{}}n$$

Figure 6.50 *Irène and Frédéric Joliot-Curie produced the first synthetic radioisotope in 1934.*

Since then, researchers have accomplished many transformations of one element to another and have synthesized new radioactive isotopes of various elements. Many of the diagnostic radioisotopes noted in Table 6.10 (page 646) are synthetic. Technetium-99m, for example, is both a synthetic element and a radioisotope.

Most synthetic radioisotopes are produced by bombarding elements with neutrons, which are captured by target nuclei. This process requires less energy than many other bombardment reactions, because neutrons have no electrical charge and are not repelled by the positive charge of the nucleus. Such reactions produce radioactive nuclei that tend to emit beta particles, thus changing the atomic number and producing a different element.

The following examples show the formation of two synthetic radioisotopes often used as medical tracers, calcium-45 and iron-59.

$$^{44}_{20}\text{Ca} \ + \ ^{1}_{0}n \ \longrightarrow \ ^{45}_{20}\text{Ca}$$

$$^{58}_{26}\text{Fe} \ + \ ^{1}_{0}n \ \longrightarrow \ ^{59}_{26}\text{Fe}$$

Nuclear-bombardment reactions generally involve four particles:

- *Target nucleus:* the stable isotope that is bombarded.
- *Projectile particle (bullet):* the particle fired at the target nucleus.
- *Product nucleus:* the isotope produced in the reaction.
- *Ejected particle:* the lighter nucleus or particle emitted from the reaction.

> Reactions may release more than one ejected particle. For example, see the Cm-246 reaction on the next page.

For example, consider how the Joliot-Curies produced the first synthetic radioactive isotope, phosphorus-30. The four types of particles involved are identified as follows:

$$^{27}_{13}\text{Al} \ + \ ^{4}_{2}\text{He} \ \longrightarrow \ ^{30}_{15}\text{P} \ + \ ^{1}_{0}n$$

| Aluminum-27 target nucleus | Alpha particle projectile particle | Phosphorus-30 product nucleus | Neutron ejected particle |

CHEM**QUANDARY**

TRANSMUTATION OF ELEMENTS

Alchemists searched in vain for ways to transform (transmute) lead or iron into gold. Has such transmutation now become a reality? From what you know about nuclear reactions, do you think that lead, iron, or mercury atoms could be changed to gold? If so, try to write equations for the possible reactions.

DEVELOPING SKILLS

C.7 NUCLEAR-BOMBARDMENT REACTIONS

Sample Problem: *Nobelium (No) can be produced by bombarding curium (Cm) atoms with nuclei of a low-mass element. What element serves as the projectile particle in this reaction?*

$$^{246}_{96}Cm \ + \ ? \ \longrightarrow \ ^{254}_{102}No \ + \ 4\,^1_0n$$

Because the sum of product atomic numbers is 102, the projectile must have the atomic number 6, to make the sum of reactant atomic numbers also 102. Therefore, the projectile must be carbon (atomic number = 6). The total mass numbers of products is 258 (254 + 4), indicating that the projectile must have been carbon-12 (258 − 246 = 12).

The completed equation is as follows:

$$\underset{\text{target}\atop\text{nucleus}}{^{246}_{96}Cm} \ + \ \underset{\text{projectile}\atop\text{particle}}{^{12}_{6}C} \ \longrightarrow \ \underset{\text{product}\atop\text{nucleus}}{^{254}_{102}No} \ + \ \underset{\text{ejected}\atop\text{particle}}{4\,^1_0n}$$

> As you learned earlier, completing nuclear equations involves balancing atomic numbers and mass numbers.

Complete the following equations by supplying the missing numbers or symbols. Name each particle. Then identify the target nucleus, projectile particle, product nucleus, and ejected particle.

1. $^{59}_{27}? \ + \ ^{?}_{?}n \ \longrightarrow \ ^{60}_{?}?$ (Scientists produce most medically useful isotopes by bombarding stable isotopes with neutrons. This process converts the original nuclei to radioactive forms of the same element.)

2. $^{96}_{42}? \ + \ ^{?}_{?}H \ \longrightarrow \ ^{97}_{43}? \ + \ ^1_0?$ (Until it was synthesized in 1937, technetium was only an unfilled gap in the periodic table. All of its isotopes are radioactive. Any technetium originally on Earth has decayed. Technetium, the first element artificially produced, is now used in industry and medicine.)

3. $^{58}_{?}? \ + \ ^{209}_{?}Bi \ \longrightarrow \ ^{?}_{109}Mt \ + \ ^1_0?$ (In 1992, a research group in Darmstadt, Germany, created element 109 by bombarding bismuth-209 nuclei. The name meitnerium (Mt) honors Lise Meitner, the Austrian physicist who first proposed the idea of nuclear fission. (See page 659.)

Not only does the ability to transform one element into another provide new and powerful technological capabilities, it also has changed our view of elements.

C.8 EXTENDING THE PERIODIC TABLE

Since 1940, nearly 20 **transuranium** elements—with atomic numbers greater than the atomic number of uranium (92)—have been added to the periodic table. These elements have been synthesized in nuclear reactions, usually conducted in accelerators known as **cyclotrons** (see Figure 6.51). From 1940 to 1961, Glenn Seaborg and coworkers at the University of California–Berkeley synthesized and identified 10 new elements with atomic numbers 94 to 103, a prodigious feat.

> Seaborg and coworkers also identified over 100 new isotopes of various elements.

Figure 6.51 *Cyclotrons, pioneered by Ernest O. Lawrence, allow scientists to investigate high-energy bombardment of heavy nuclei with various particles. The images here illustrate the development of the cyclotron over several decades: (top left) an 11-inch chamber from 1932, (bottom left) the Berkeley Radiation Laboratory's 60-inch cyclotron, which was used by Glenn Seaborg, Albert Ghiorso, and their coworkers to discover several new elements, and (right) the world's largest cyclotron (~18 m in diameter) at Canada's National Laboratory for Particle and Nuclear Physics.*

None of those 10 elements occurs naturally. All were made by high-energy bombardment of heavy nuclei with various particles. For example, alpha-particle bombardment of plutonium-239 produced curium-242:

$$^{239}_{94}\text{Pu} + {}^4_2\text{He} \longrightarrow {}^{242}_{96}\text{Cm} + {}^1_0 n$$

Bombarding Pu-239 with neutrons yielded americium-241, a radioisotope now used in home smoke detectors:

$$^{239}_{94}\text{Pu} + 2\,{}^1_0 n \longrightarrow {}^{241}_{95}\text{Am} + {}^0_{-1} e$$

Seaborg was awarded the 1951 Nobel prize in chemistry for his work. Albert Ghiorso, a colleague of Seaborg, led the way in producing several new elements beyond lawrencium (element 103). One is element 106, produced by

Figure 6.52 *Glenn Seaborg and coworkers at the University of California–Berkeley synthesized and identified ten new elements beyond uranium. Can you identify any transuranium elements that they may have created?*

bombarding a californium-249 target with a beam of oxygen-18 nuclei, producing an isotope of element 106. To honor Seaborg's pioneering work, element 106 was named *seaborgium* (Sg). Glenn Seaborg, shown in Figure 6.52, has been called the father of the modern periodic table.

Traditionally, the discoverer of an element selects its name. For example, when Marie Curie first discovered element number 84 she named it *polonium* (Po) in honor of Poland, her home country.

Several scientific laboratories have claimed to have synthesized elements with atomic numbers greater than 92. For example, laboratories in both the former Soviet Union and the United States claimed the discovery of elements 104 and 105. Soviet scientists proposed naming them *kurchatovium* (Ku) and *dubnium* (Db), while U.S. scientists proposed the names *rutherfordium* (Ru) and *hahnium* (Ha). The International Union of Pure and Applied Chemistry (IUPAC) examined the claims for element discovery before recommending the official names. In 1997, the IUPAC approved the element names *rutherfordium* and *dubnium*.

At present, the IUPAC recognizes official names and symbols for the first 112 elements. Although claims for the discovery of other elements, 113 to 118, have been reported, the IUPAC has not yet recognized these discoveries or their official names. Scientists temporarily identify such unnamed elements by Latin prefixes indicating their atomic numbers. For example, element 113 is temporarily named ununtrium (un = 1, un = 1, tri = 3), which is symbolized as *Uut* until IUPAC officially recognizes the original discovery of that new element.

The name proposed by discoverers of element 111—roentgenium (Rg)—was approved by IUPAC in 2004. It honors W. K. Roentgen, who discovered X-rays. Roentgen was the first scientist you learned about in this unit. The process of confirmation and naming of new elements is very thorough and rigorous and takes a great deal of time. Evidence for element 112 was first reported in 1996. In 2010, the IUPAC officially recognized the discovery of and the name copernicium (Cn) for element 112.

The production of transuranium elements has enriched our understanding of the atomic nucleus. With the synthesis and identification of elements beyond atomic number 92, the periodic table has expanded to fill the actinide series as well as nearly all of Period 7.

Scientific research, like other human endeavors, can involve strong personalities, competition, and controversy.

IUPAC is a federation of national organizations that represent chemists around the world. It plays an important role in the international standardization of chemical nomenclature and terminology.

■ MAKING DECISIONS

C.9 OPINIONS ABOUT RADIOACTIVITY

Some older people tend to associate nuclear technologies with the use of atomic weapons at the close of World War II and with the atomic-weapon threats of the Cold War in the 1950s through the 1980s. Some of these people are likely to be in the audience when you speak to the Riverwood senior citizens.

The following opinions might be expressed by such community members. Decide how you would respond to each opinion, using knowledge you have gained in this unit (see Figure 6.53).

1. "I'm against having any isotopes in Riverwood. They're too dangerous."

2. "I don't want to live near anything that is radioactive."

3. "I don't know why scientists keep trying to make new elements. All the new ones are radioactive."

4. "I don't understand how cancer can be treated with radiation. I thought radiation *caused* cancer."

5. "We must outlaw radioactive material to eliminate the possibility of a nation using radioactive material to harm others."

Join with a classmate and share your responses to each opinion. How are your responses similar? How are they different?

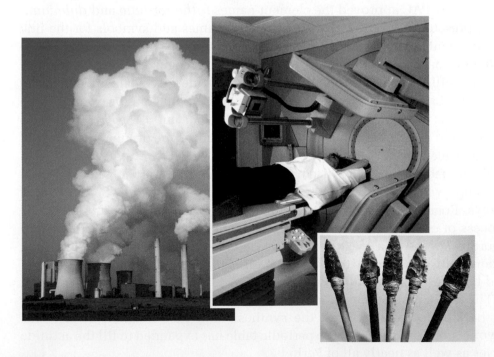

Figure 6.53 *How can each of these images suggest ways to respond to the concerns of Riverwood's citizens?*

SECTION C SUMMARY

Reviewing the Concepts

Half-life can be defined as the time required for half of the radioactive atoms in a sample to decay.

1. How can the concept of half-life be used to determine a material's age?

2. The half-life of carbon-14 is 5730 years. Provide a rough estimate of the percent of C-14 radioisotope that would be left after

   a. 24 hours.

   b. 100 years.

3. The half-life of astatine-209 is 5.4 h. Estimate the percent of At-209 that would be left after

   a. 24 hours.

   b. 100 years.

4. Phosphorus-32 has a half-life of 14.3 d. How long would it take for 1/8 of a sample of P-32 to remain unchanged?

5. Given a sample of 4.5 mol radon-222 (see Table 6.9, page 639), how many moles of Rn-222 would remain undecayed after

   a. 3.82 d?

   b. 15.3 d?

   c. 28.0 d?

Radioisotopes have properties that make them useful as tracers for diagnostic purposes.

6. How is metastable Tc-99m useful in medical diagnosis?

7. How is iron-59 used for diagnostic purposes?

8. Why do doctors use a radioisotope of iodine to detect thyroid problems?

9. Radioactive sodium chloride is appropriate for diagnosing circulatory problems, while radioactive xenon is helpful in searching for lung problems. Explain why each is used for its specific applications.

10. Medical personnel are selecting a radioisotope for diagnostic use. Why is each of the following considerations important in selecting the most suitable radioisotope?

    a. half-life

    b. mode of decay

    c. chemical properties of the element

Ionizing radiation emitted by some radioisotopes can be used for medical treatment.

11. How are cancer cells different from normal cells?

12. How do physicians use ionizing radiation to treat cancer?

13. What could happen if the source of ionizing radiation used to treat cancerous growth were

    a. too weak?

    b. too strong?

14. Name three radioisotopes that radiologists can use externally to treat cancer.

The conversion (transmutation) of one element to another can be accomplished by the high-energy bombardment of atomic nuclei with subatomic particles or other nuclei.

15. Define *transmutation*.

16. Why was Rutherford's ability to complete transmutations limited compared to the Curies' transmutation abilities?

17. Why are neutrons frequently used as projectiles in transmutation reactions?

18. Copy and complete each of the following transmutation equations:

   a. $\underline{\quad} \longrightarrow {}^{4}_{2}\text{He} + {}^{259}_{104}\text{Rf}$

   b. ${}^{238}_{92}\text{U} \longrightarrow \underline{\quad} + {}^{234}_{90}\text{Th}$

   c. ${}^{210}_{83}\text{Bi} \longrightarrow {}^{0}_{-1}\text{e} + \underline{\quad}$

   d. ${}^{95}_{40}\text{Zr} \longrightarrow \underline{\quad} + {}^{95}_{41}\text{Nb}$

   e. $\underline{\quad} \longrightarrow {}^{4}_{2}\text{He} + {}^{214}_{80}\text{Hg}$

Why and how are radio-active isotopes useful?

In this section, you explored the concept of half-life, learned about applications of radioisotopes in medicine, and considered transmutation and how it relates to the extension of the periodic table. Think about what you have learned, then answer the question in your own words in organized paragraphs. Your answer should demonstrate your understanding of the key ideas in this section.

Be sure to consider the following in your response: half-life, carbon dating, diagnostic and therapeutic uses of nuclear medicine, bombardment, and transmutation.

Connecting the Concepts

19. A student wrote this statement in a homework assignment: "After one half-life, half of the mass of a material has disappeared." Do you agree or disagree? Explain.

20. Would carbon dating help scientists determine the age of dinosaur remains? Explain.

21. Explain why an externally applied alpha emitter is an ineffective treatment for tumors deep within the body.

22. Graph the data in the following table and determine the

   a. half-life of thorium-234.

   b. minimum number of days required for thorium-234 to decay to 20% of its original activity.

Elapsed time (days)	% thorium-234 activity remaining
0	100
7	82
16	60
28	45
42	30
62	17
94	7

23. Scientists using carbon-14 dating generally do not use it to go back further than seven half-lives.

   a. Explain why.

   b. What is the maximum number of years a substance can be dated using carbon-14 dating?

24. A radiologist injects a patient's bloodstream with a radioisotope tracer sample registering 10 000 cpm (counts per minute) of radioactivity. Soon after, a technician draws 6.0 mL of blood, and the sample shows an activity of 10 cpm. What is the patient's total blood volume, assuming essentially no decrease in the activity of the radioisotope occurred during this clinical procedure?

Extending the Concepts

25. Compare PET and MRI in terms of

a. their method of operation.

b. data they provide for diagnosis.

c. radiation exposure for a patient.

26. Scientists originally were uncertain whether the oxygen gas produced during photosynthesis came from CO_2, H_2O, or both. How could radioisotope tracers be used to help settle that uncertainty?

27. A newly discovered element with an extremely short half-life is detected by analyzing its decay products. Explain how scientists can "work backward" to identify the original element.

28. Many gemstones are irradiated. Research and report on the reasons for this process.

29. Gold and other precious metals can be created from other metals by nuclear transformations.

a. So why don't commercial firms do this as a source of profit?

b. Is the synthetic precious metal distinguishable from the naturally occurring metal? Explain.

30. Research how the proton–neutron ratio is related to the stability of an atomic nucleus. Describe how you could use this ratio to predict the type of radioactive decay a particular atomic nucleus undergoes.

31. List foods that are currently irradiated with nuclear radiation before they are marketed for human consumption. Describe the process and evaluate its risks and benefits.

SECTION D NUCLEAR ENERGY: BENEFITS AND BURDENS

What burdens and benefits accompany uses of nuclear energy?

In the 1930s, a bombardment reaction involving uranium unlocked a new energy source and led to development of both nuclear power and nuclear weapons. This event marked the start of the nuclear age. How did scientists first unleash the enormous energy of the atom, and how have nuclear engineers harnessed atomic energy for both useful and destructive purposes?

GOALS

- Define and describe nuclear fission.
- Write and balance nuclear equations for fission reactions.
- Describe how nuclear energy is used to generate electricity in a nuclear power plant.
- Define and describe nuclear fusion.
- List and describe types of radioactive waste, including their sources.
- Describe and evaluate methods for disposal of radioactive wastes, including burdens and benefits of each.

✓ concept check 7

1. List at least three beneficial uses of nuclear technology or nuclear energy.
2. Describe how synthetic elements are created.
3. What is nuclear fission?

D.1 UNLEASHING NUCLEAR FORCES

Just before the start of World War II, German scientists Otto Hahn and Fritz Strassmann bombarded uranium with neutrons in the hope of creating a more massive nucleus and, thus, a new element. Much to their surprise, they found that one reaction product was atoms of barium, with only about half the atomic mass of the original target uranium atoms.

The first to understand what had happened was the Austrian physicist Lise Meitner (Figure 6.54), then living in Sweden, who had previously worked with Strassmann and Hahn. Meitner and her nephew Otto Frisch suggested that neutron bombardment had split the uranium atom into two parts of nearly equal mass. Other scientists quickly verified Meitner's explanation.

Hahn and Strassmann had actually triggered an array of related reactions. One of the reactions that produced barium is:

Figure 6.54 *Lise Meitner was first to suggest that nuclei might split due to neutron bombardment.*

$$\overset{1}{_{0}}n \;+\; \overset{235}{_{92}}U \;\longrightarrow\; \overset{140}{_{56}}Ba \;+\; \overset{93}{_{36}}Kr \;+\; 3\,\overset{1}{_{0}}n \;+\; \text{Energy}$$

Neutron   Uranium-235   Barium-140   Krypton-93   Neutrons

Splitting an atom into two smaller atoms is called **nuclear fission**. Scientists soon found that the uranium-235 nucleus can fission (split) into numerous pairs of smaller nuclei. The uranium usually did not split into two equal halves but into one element accounting for about 60% of uranium's mass (such as barium) and another element equivalent to about 40% of uranium's mass (such as krypton). Here is another example of a nuclear fission reaction involving uranium-235:

$$\overset{1}{_{0}}n \;+\; \overset{235}{_{92}}U \;\longrightarrow\; \overset{143}{_{54}}Xe \;+\; \overset{90}{_{38}}Sr \;+\; 3\,\overset{1}{_{0}}n \;+\; \text{Energy}$$

Neutron   Uranium-235   Xenon-143   Strontium-90   Neutrons

The nuclear fission of heavy atoms such as uranium releases a huge quantity of energy. Gram for gram, the released energy is at least a million times more than the energy of any chemical reaction. This is what makes nuclear explosions so devastating and nuclear energy so powerful.

Why does a nuclear reaction release much more energy than a chemical reaction? Recall what you know about chemical reactions, such as burning petroleum. Chemical reactions involve breaking chemical bonds in reactants and making new chemical bonds in products. When bonds are stronger in products than in reactants, energy is released, often as thermal energy. Thus, chemical energy is converted into thermal energy. There is no overall energy loss or gain. Similarly, mass is conserved in a chemical reaction. The nucleus of each atom, and thus its identity, remains intact in all chemical reactions. As a result, the number of atoms of each element remains unchanged. The atoms simply become rearranged. Balanced chemical equations illustrate this conservation of atoms and mass.

Lise Meitner fled to Sweden when Nazis assumed control of Austria and revoked her citizenship.

Fission of U-235 produces many other pairs of nuclei, such as Te-137 and Zr-97.

Not all nuclei are fissionable. U-235 is the only naturally occurring isotope that undergoes fission with lower-energy (thermal) neutrons. However, many synthetic nuclei (such as U-233, Pu-239, and Cf-252) also fission under neutron bombardment.

Nuclear reactions are also based on conserving energy and mass. However, during nuclear fission, very small quantities of mass are converted into measurable quantities of energy. Where does this energy originate?

The origin of nuclear energy lies in the force that holds protons and neutrons together in the nucleus. This force, called the **strong force**, is fundamentally different from the electrical forces that hold atoms and ions together in chemical bonds. It is also a thousand times stronger than the forces that result in chemical bonds. The strong force operates over very short distances, extending only across an atom's nucleus.

The forces holding nuclear particles together in the two atomic nuclei produced during U-235 fission are stronger than those in the nucleus of the uranium atom that was split. A small loss of mass results from forming two new nuclei and is converted into large quantities of released energy.

How much mass and energy are involved? The mass loss is very small, often less than 0.1% of the total mass of the fissioning atom. Even so, the conversion of these small quantities of mass into energy accounts for the vast power of nuclear reactions.

Albert Einstein's famous equation relates mass and energy: $E = mc^2$. This equation indicates that the energy released (E) equals the mass lost (m) multiplied by the speed of light (a very large number) squared (c^2). If one gram of matter were fully converted to energy, the energy released would equal that produced by burning 700 000 gallons of high-octane gasoline!

Such nuclear energy release has been harnessed by engineers to generate electricity (see Figure 6.55) and to create atomic weapons. However, the fission of one nucleus does not produce enough energy for practical use. How are fission reactions sustained to involve much larger quantities of nuclei?

> If one kilogram of U-235 fissions, a mass of about one gram is converted into energy.

> One gram of mass loss (1×10^{-3} kg) times the speed of light (3×10^8 m/s) squared equals 9×10^{13} J of energy.

Figure 6.55 *The core of a fission reactor, based on a nuclear chain reaction, emits visible light due to ionizing radiation released.*

Note from the equations on page 659 that another product of nuclear fission is the release of neutrons. These emitted neutrons can sustain the fission reaction by serving as reactants to split additional fissionable nuclei, which produce additional neutrons, which can split additional fissionable nuclei, and so on. The result is a **chain reaction** (see Figure 6.56).

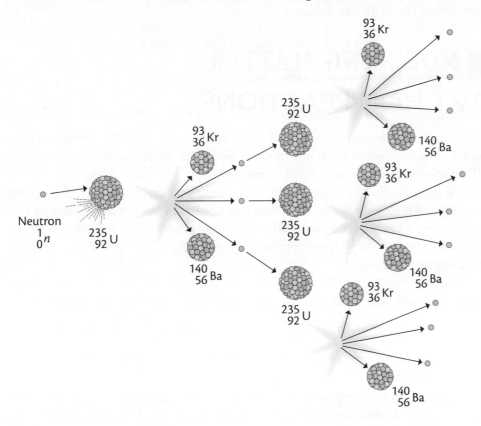

Figure 6.56 *A nuclear chain reaction. A neutron colliding with a uranium-235 nucleus initiates the reaction (left). The reaction continues and grows, as emitted neutrons encounter and split the nuclei of other fissionable atoms.*

Recall, however, that most of an atom is empty space. The probability that a neutron from a fission reaction will hit and split another fissionable nucleus depends on how much fissionable material is available. Unless a certain **critical mass** (minimum quantity) of fissionable material is present, the neutrons are not likely to encounter enough fissionable nuclei to sustain the reaction. However, if a critical mass of fissionable material is present, a chain reaction can occur, as depicted in Figure 6.56.

Shortly after the first fission reactions were explained in 1939, scientists recognized that they could employ large-scale nuclear reactions in military weapons. Germany and the United States soon initiated projects to build atomic bombs during World War II. In 1945, U.S. planes dropped two such bombs on Hiroshima and Nagasaki in Japan, which led quickly to the end of the war.

More recently, nuclear engineers have used the energy produced by nuclear fission chain reactions to generate electricity. They carefully monitor and control the rate of fission for such uses. Nuclear power plants harness the enormous energy produced by nuclear fission reactions, while also minimizing the risks of an uncontrolled chain reaction. You will soon learn more about those design features.

■ MODELING MATTER
D.2 CHAIN REACTIONS

Chain reactions sustain nuclear fission reactions in applications such as electrical power generation and atomic weapons. In this activity, dominoes will model some aspects of a chain reaction.

Each domino that falls represents a nucleus that has been split during fission. Figure 6.57 shows one way that you could set up the dominoes so that making one domino fall causes all other dominoes also to fall. Because you will model a specific fission reaction, your models will not match the one depicted in Figure 6.57.

The uranium-235 nucleus can fission into more than 100 different pairs of atoms. One way, for example, produces tellurium-137 and zirconium-97:

$$\,^{1}_{0}n \ + \ ^{235}_{92}U \ \longrightarrow \ ^{137}_{52}Te \ + \ ^{97}_{40}Zr \ + \ 2\,^{1}_{0}n \ + \ Energy$$

Figure 6.57 *Domino behavior can model key characteristics of a chain reaction.*

1. As this equation shows, splitting one U-235 nucleus releases two neutrons.
   a. Set up all the dominoes you receive from your teacher so that each falling domino will make two more erect dominoes fall.
   b. Sketch your setup.
   c. Push over the first domino and record what happens.
   d. Explain how this models the release of neutrons during the fission of U-235 as in the equation above.
   e. What aspects of the U-235 fission reaction are not modeled by the behavior of your domino setup?

2. Now you will assemble a critical mass of atoms (dominoes) to initiate a chain reaction.
   a. Set up the dominoes as in Step 1a.
   b. Assume that only dominoes (atoms) with seven total dots are fissionable.
   c. Remove all dominoes from your setup that do not have seven dots.
   d. Sketch your new setup.
   e. Push over the first domino and record what happens.

f. In what way does this model help clarify the idea of critical mass?

g. How can you ensure that you have a critical mass of "fissionable" dominoes in a particular setup?

3. Suppose you need to control the total neutrons emitted so that fission is just sustained.

 a. Set up the dominoes as in Step 1a.

 b. Devise a plan so that only half the dominoes will fall when you push over one domino. You should not remove any of the dominoes that are already set up.

 c. Describe and sketch your strategy.

 d. Try your plan and record what happens.

 e. What stopped the dominoes from falling?

 f. Use a domino model to describe how to control fission chain reactions.

4. Compare the domino arrangements in Question 1 and 3.

 a. Which is a better model of an atomic-weapon explosion? Explain.

 b. Which is a better model of fission in a nuclear power plant? Explain.

5. Propose another way to model a nuclear chain reaction.

 a. Explain how your model illustrates features of a nuclear chain reaction.

 b. Explain some limitations of your model.

D.3 NUCLEAR POWER PLANTS

The first nuclear reactors were designed and built during World War II. Since then, commercial companies have built many nuclear reactors to generate electricity, such as the nuclear power plant shown in Figure 6.58. In 2010, slightly more than 100 commercial nuclear reactors were generating electricity in the United States. Figure 6.59 (page 664) shows where these reactors are located. Globally, an estimated 438 nuclear reactors in 29 nations produce about 14% of the world's electricity.

Figure 6.58 *A nuclear power plant. Notice the reactor containment building with its domed roof.*

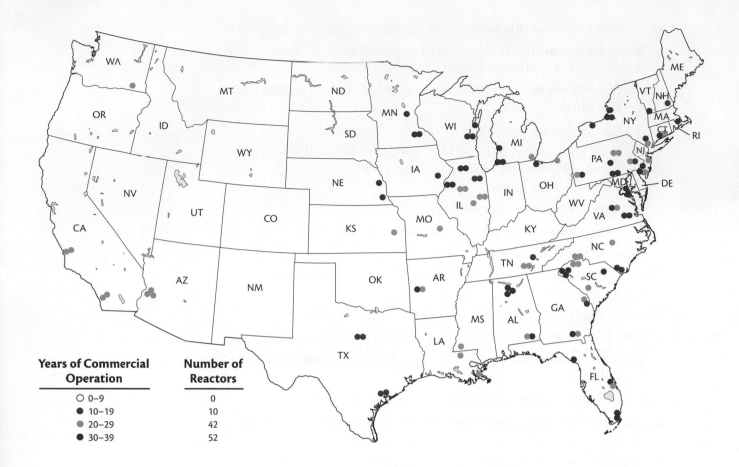

Figure 6.59 *Locations (dots) of all licensed U.S. nuclear power plants in 2010.*

Most conventional power plants generate electricity by burning fossil fuels to boil water and produce steam. A nuclear power plant operates in much the same way. However, instead of using fossil-fuel combustion to boil water, **nuclear power plants** use the thermal energy released from nuclear-fission reactions to heat water and produce steam. The steam spins turbines of giant generators, producing electricity.

The essential parts of a nuclear power plant, diagrammed in Figure 6.60, include the fuel rods, control rods, moderator, generator, and cooling system.

Fuel Rods

Coal-fired power plants burn thousands of tons of coal daily. By contrast, nuclear reactor fuel occupies a fraction of the volume needed for coal and is replenished only about once a year. Nuclear reactor fuel is small uranium dioxide (UO_2) pellets about the size and shape of short pieces of chalk. The energy in one uranium fuel pellet equals the energy in one ton of coal or 126 gal of gasoline. One nuclear power plant uses as many as 10 million fuel pellets. These pellets are arranged inside long, narrow steel cylinders—the fuel rods. The fission chain reaction occurs inside these rods (see Figure 6.60).

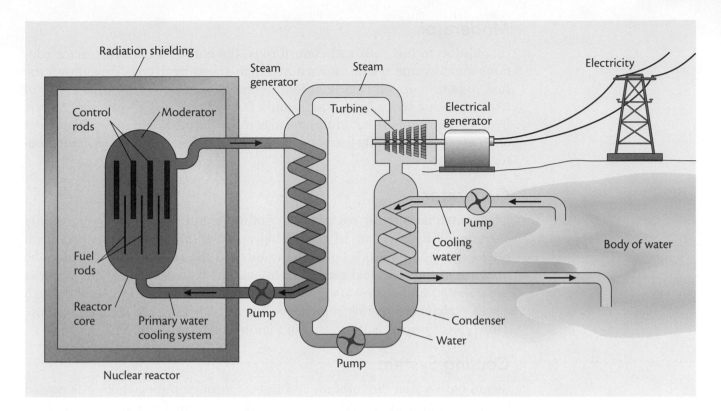

Figure 6.60 *Components of a nuclear power plant. Note the water flowing between the reactor and steam generator (red) and between the steam generator and outside body of water (blue). This water flow is important in controlling the power-plant operating temperature.*

Most uranium dioxide in fuel pellets is composed of the nonfissionable uranium-238 isotope. Only 0.7% of natural uranium is U-235, the fissionable isotope. In reactor fuel rods, U-235 composition has been enriched to about 3%, which is only a small fraction of total fuel-rod material. It is sufficient to sustain a chain reaction, but far less than enough to cause a nuclear explosion.

> Weapons-grade uranium usually contains 90% or more of the fissionable U-235 isotope.

Control Rods

The nuclei of some elements, such as boron or cadmium, can absorb neutrons very efficiently. Such materials are placed in control rods, which regulate the number of neutrons available for causing fission. Moving the control rods up or down (see Figure 6.61) between the fuel rods regulates the rate of the nuclear chain reaction. The reaction can be completely stopped by dropping the control rods all the way down between the fuel rods and absorbing nearly all neutrons released as U-235 fissions.

Figure 6.61 *A nuclear reactor core.*

Moderator

In addition to fuel rods and control rods, the core of a nuclear reactor contains a moderator, which slows down high-speed neutrons. This allows the fuel rods to more efficiently absorb neutrons, enhancing the probability of fission. Heavy water (with the formula D_2O, where D is the symbol for the hydrogen-2 isotope, deuterium), regular water (termed *light water* by nuclear engineers), and graphite (carbon) are the three most commonly used moderator materials.

Generator

In commercial nuclear reactors, the fuel rods and control rods are usually surrounded by a system of circulating water. In simpler reactors, the heat released by the fuel rods boils this water, and the resulting steam spins the turbines of the electrical generator. In another type of reactor, this water is superheated under pressure and does not boil. Instead, it circulates through a heat exchanger, where it boils the water contained in a second cooling loop. This type of reactor is illustrated in Figure 6.60 (page 665).

Cooling System

The white plumes seen rising from generating-plant cooling towers are condensed steam, not smoke.

Steam moves past the turbines and travels through pipes where it is cooled by water drawn from a nearby ocean, lake, or river. The cooled steam condenses to liquid water and circulates inside the generator. So much thermal energy is generated that some steam must also be released into the air. The largest and most prominent features of most nuclear power plants are tall, gracefully curved concrete cylinders, called cooling towers, where excess thermal energy is released. Some observers mistakenly assume that the cooling tower is the nuclear reactor.

A nuclear reactor is designed to prevent the full escape of radioactive material if a malfunction causes the release of radioactive material, including cooling water, within the reactor itself. The core of a nuclear reactor is surrounded by concrete walls two to four meters thick. Further protection is provided by enclosing the reactor in a building with thick walls of steel-reinforced concrete designed to withstand a chemical explosion or an earthquake. The reactor building is also capped by a domed roof that can withstand high internal pressure.

In the Chernobyl accident, one chemical reaction that caused trouble was high-temperature steam reacting with carbon from the moderator, producing CO and H_2. When H_2 mixed with air, it became explosive.

The well-known nuclear accident in 1986 at one of the four reactors at Ukraine's Chernobyl power plant occurred because too many control rods were withdrawn from the reactor and were not replaced fast enough. There was little control of the fission process, resulting in the buildup of much steam. The resulting explosion was not nuclear but was due to the buildup of high-temperature steam and to the chemical reactions that it triggered. Unfortunately, the plant had been built without a surrounding concrete containment building, unlike currently operating U.S. reactors. A large quantity of nuclear material from the reactor was also released directly into the environment.

Nuclear fission is not the only way to liberate nuclear energy. Soon you will learn about the kind of nuclear reaction that fuels the stars (and, indirectly, fuels all living matter).

D.4 NUCLEAR FUSION

In addition to releasing energy by splitting massive nuclei (fission), large quantities of nuclear energy can be generated by fusing, or combining, small nuclei. **Nuclear fusion** involves forcing two relatively small nuclei to combine into a new, more massive nucleus. As with fission, the energy released by nuclear fusion can be enormous, again due to the conversion of mass into energy. Gram for gram, nuclear fusion liberates even more energy than nuclear fission—between 3 and 10 times more energy (Figure 6.62).

Scientific American Conceptual Illustration

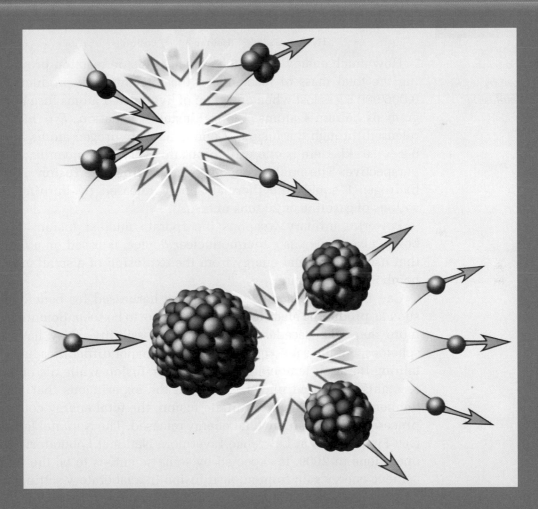

Figure 6.62 *Nuclear fission and fusion. Fusion (top) occurs when smaller nuclei (protons in blue, neutrons in gray) combine to form larger nuclei. Fission (bottom) occurs when a neutron of appropriate energy collides with particular large nuclei, creating two smaller nuclei. Both processes can release large quantities of energy, according to Einstein's famous equation, $E = mc^2$, where m is the mass converted into energy (E) and c is the speed of light. However, the total energy released decreases as the nuclei involved approach iron-56, the most stable nucleus.*

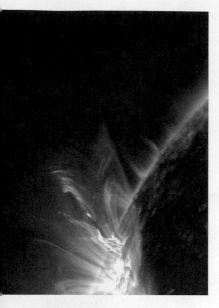

Figure 6.63 *The Sun's energy is produced by continuous nuclear-fusion reactions.*

Nuclear fusion powers the Sun and other stars (Figure 6.63). Scientists believe that the Sun formed when a huge quantity of interstellar gas, mostly hydrogen, condensed under the force of gravity. As the volume of gas decreased, its temperature increased to about 15 million °C, and hydrogen atoms began fusing into helium. The nuclei that fused together were all positively charged and tended to repel one another. The high temperature gave each nucleus considerable kinetic energy, which helped overcome the repulsions.

Once fusion was started, the Sun began to shine, converting nuclear energy into radiant energy. Scientists estimate that the Sun, believed to be about 4.5 billion years old, is about halfway through its life.

The nuclear-fusion reactions occurring in the Sun are rather complicated, but the result is the conversion of hydrogen nuclei into helium nuclei. The overall result can be summarized by this equation:

$$4\,{}^{1}_{1}\text{H} \longrightarrow {}^{4}_{2}\text{He} + 2\,{}^{0}_{+1}e + \text{Energy}$$

Hydrogen-1   Helium-4   Positrons

How much energy does such a nuclear fusion reaction produce? Comparing the total mass of reactants to the total mass of products reveals that 0.006 900 5 g is lost when one gram of hydrogen-1 atoms fuse to produce one gram of helium-4 atoms. Using Einstein's equation, $E = mc^2$, the energy released through the fusion of one gram of hydrogen atoms (one mol H) is 6.2×10^8 kJ. Here is one way to put this very large quantity of energy into perspective: The nuclear energy released from the fusion of one gram of hydrogen-1 equals the thermal energy released by burning nearly 5000 gallons of gasoline or 20 tons of coal.

Powerful military weapons incorporate nuclear fusion. The hydrogen bomb, also known as a thermonuclear device, is based on a fusion reaction that uses the thermal energy from the explosion of a small atomic (fission) bomb to initiate fusion.

Can the energy of nuclear fusion be harnessed for beneficial purposes, such as producing electricity? This remains to be seen. Scientists have spent more than five decades pursuing this possibility. They have tried many schemes, but have not yet succeeded. The major difficulties have been maintaining the high temperatures needed for fusion while also containing the reactants and fused nuclei. So far, in the experiments that have achieved temperatures sufficient to initiate fusion, the total energy consumed by the process is more than the total energy released. The National Ignition Facility (see Figure 6.64) at Lawrence Livermore National Laboratory, which began operations in 2009, is expected by some scientists to be the first facility to achieve energy gain from nuclear fusion in a laboratory setting.

If scientists finally succeed in controlling nuclear fusion in the laboratory, there is still no guarantee that fusion reactions will become a practical source of energy. Low-mass isotopes needed to fuel such reactors are plentiful and inexpensive, but confinement of the reaction could be very costly. Further, although the fusion reaction itself produces less radioactive waste than

Figure 6.64 *Scientists at the National Ignition Facility (NIF) aim to initiate nuclear fusion in the laboratory. This artist's rendering (right) shows a NIF target pellet (the white ball) inside a hollow capsule. Laser beams enter through openings on each end and compress and heat the target to conditions necessary for nuclear fusion. The target must be precisely centered inside the chamber before each experiment. A NIF technician (above) checks the target positioner, which serves as a reference to align the laser beams.*

nuclear fission, capturing the positrons and shielding the heat of the reaction could generate nearly as much radioactive waste as that produced now by fission-based power plants.

In splitting and fusing atoms, the nuclear energy that fuels the universe has been unleashed. Much good has arisen from it, but so have scientific, social, and ethical questions. Along with great benefits come great risks. How much risk is worth any potential benefits? In the next activity, you will analyze risks and benefits associated with deciding how to travel to see a friend.

> Electing not to make a decision is, in fact, also a decision—one with its own risks and benefits.

concept check 8

1. Consider the nuclear processes you have just studied.
  a. What is fission?
  b. What is fusion?
  c. How are they different?
  d. How are they similar?
2. What are some drawbacks to the use of nuclear energy?

■ MAKING DECISIONS

D.5 THE SAFEST JOURNEY

Suppose you want to visit a friend who lives 500 miles (800 km) away, using your safest means of transportation. Insurance companies publish reliable statistics on the safety of different methods of travel. Using Table 6.11, answer these questions:

Table 6.11

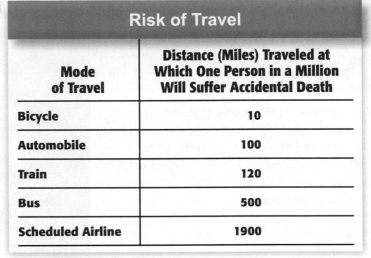

Risk of Travel	
Mode of Travel	**Distance (Miles) Traveled at Which One Person in a Million Will Suffer Accidental Death**
Bicycle	10
Automobile	100
Train	120
Bus	500
Scheduled Airline	1900

1. Assume there is a direct relationship between distance traveled and chance of accidental death (i.e., assume that doubling the distance doubles your risk of accidental death.)

   a. What is the risk factor (chance of accidental death) for traveling 500 miles by each mode of travel listed in the table? For example, Table 6.11 shows that the risk factor for biking increases by 0.000 001 for each 10 miles. Therefore, the bike-riding risk factor in visiting your friend would be

   $$\frac{500 \text{ miles} \times 0.000\ 001 \text{ risk factor}}{10 \text{ miles}} = 0.000\ 05 \text{ risk factor}$$

   b. What is the safest mode of travel (the one with the smallest risk factor)?

   c. Which is the riskiest mode of travel (the one with the largest risk factor)?

   d. Did your results surprise you? Explain.

2. The risk of accidental death is not the only factor to consider when choosing a mode of travel.

   a. List some benefits associated with each mode of transportation.

   b. List some risks associated with each mode of transportation.

   c. In your view, do benefits of riskier ways to travel outweigh their increased risks? Explain your reasoning.

   d. Identify some situations in which assumptions made in Question 1 would be invalid.

3. Do you think the statistics in Table 6.11 will apply 25 years from now? Explain.

4. What factor(s), beyond the risk to personal safety, would you include in a risk–benefit analysis before you decided how to travel?

5. Now use your results and reasoning to draw conclusions.

   a. Which mode of travel would you choose? Why?

   b. Would another person's risk–benefit analysis always lead to your decision? Why or why not?

CHEM**QUANDARY**

RISK-FREE TRAVEL?

Is there any way to travel to visit your friends or relatives that would be completely risk-free? Would it actually be safer not to visit them at all? Why?

D.6 RADIOACTIVE WASTE

Imagine that you live in a home that was once clean and comfortable, but now you have a major problem: You cannot throw away your garbage. The city forbids garbage removal because it has not decided what to do with the garbage. For about 50 years, your family has compacted, wrapped, and saved the garbage as efficiently as possible, but you are running out of room. Some bundles leak and are a health hazard. What can be done?

The U.S. nuclear power industry, the nuclear weapons industry, and medical and research facilities have a similar problem. Spent (used) nuclear fuel and radioactive waste products have been accumulating for about 50 years (see Figure 6.65). Some of these materials are still highly radioactive, while other materials—even initially—have low levels of radioactivity. It is uncertain where these materials, regardless of radioactivity levels, will be permanently stored or how soon.

Figure 6.65 *Low-level radioactive waste does not require the same disposal methods as spent fuel rods from nuclear reactors do. Yet improper disposal of such wastes poses serious hazards.*

Radioactive Wastes

Two broad categories of radioactive waste are high-level and low-level. **High-level radioactive wastes** are either (a) products of nuclear fission, such as those generated in a nuclear reactor or (b) transuranics, products formed when the original uranium-235 fuel absorbs neutrons. For example, plutonium-239 is a transuranic material. **Low-level radioactive wastes** have much lower levels of radioactivity. These wastes include used nuclear laboratory protective clothing, diagnostic radioisotopes, and air filters from nuclear power plants. Figure 6.66 illustrates the composition of radioactive wastes in the United States.

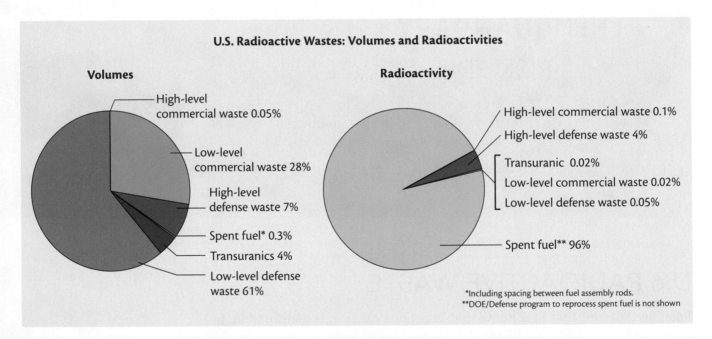

Figure 6.66 *Volume and radioactivity of U.S. radioactive wastes.*

Figure 6.66 emphasizes distinguishing waste volume from radioactivity level when considering radioactive wastes. High volume does not necessarily mean high radioactivity. For example, Figure 6.66 shows that military defense efforts produce the largest volume of radioactive wastes (61%), but this waste contributes slightly more than 4% of the total radioactivity in U.S. radioactive wastes. On the other hand, spent fuel from nuclear reactors occupies less than 1% of the volume of all radioactive wastes, yet it contributes 96% of the total radioactivity. Moreover, high-level radioactive wastes also can have extended half-lives, some as long as thousands of years.

Current Disposal Methods

Because high-level and low-level radioactive wastes have different charateristics, people dispose of them differently. We can put low-level wastes into sealed containers and bury the containers in lined trenches 20 feet deep.

Low-level military radioactive waste is disposed of at federal sites maintained by the U.S. Department of Energy (DOE). Since 1993, each state is responsible for disposal of its own commercial low-level radioactive wastes. Groups of states in several regions have formed compacts in which all members use a disposal site in one state for low-level wastes from all compact members.

High-level radioactive waste disposal requires a very different approach. Spent nuclear fuel presents the greatest challenge. A commercial nuclear reactor (power plant) typically produces about 30 tons of spent fuel annually. That means that the commercial nuclear power plants now operating in the United States generate roughly 3100 tons of waste annually. Approximately 60 000 tons of spent fuel are already stored in 34 states. Eventually, of course, the components of nuclear power-plant reactors will also become radioactive waste.

The U.S. nuclear weapons program contributed over 300 million liters of additional stored waste. The volume of military radioactive waste is much greater than the volume of commercial radioactive waste. This waste is in the form of a sludge—the waste product of extracting plutonium from spent fuel rods in military reactors. (Nuclear weapons are created from the plutonium.) As radioisotopes in the waste gradually decay, they emit radiation and thermal energy. In fact, without external cooling, such waste can become hot enough to boil. This makes military waste containment a continual challenge.

Operators must replace up to one-third of a nuclear reactor's fuel rods annually. This is necessary because the uranium-235 fuel becomes depleted as fissions occur and because accumulated fission products interfere with the fission process. The spent fuel rods are highly radioactive, with some isotopes continuing to decay for many thousands of years. Table 6.12 lists the half-lives of three radioisotopes produced by nuclear fission. All of these and many more are contained in spent fuel rods.

By federal law, nuclear reactor waste must be stored on-site, usually in waste storage tanks, until a permanent repository is created. Available storage space on-site is limited, however. The federal government is responsible for the final storage of high-level radioactive waste, but it has not yet opened permanent long-term disposal sites for high-level wastes. A site at Yucca Mountain, Nevada, was selected for this purpose and some construction to prepare for storage of high-level radioactive wastes was completed (see Figure 6.67). However, ongoing legal, political, and regulatory debate has made the future of the Yucca Mountain site uncertain.

Table 6.12

Some Radioisotopes Found in Spent Fuel Rods	
Isotope	**Half-Life**
Plutonium-239	24 110 y
Strontium-90	28.8 y
Barium-140	12.8 y

Figure 6.67 *Yucca Mountain, Nevada, was approved by the U.S. Congress as the site of a national repository for spent nuclear fuel and high-level radioactive waste. Ongoing controversy makes the future of this repository uncertain.*

Investigating Long-Term Disposal Methods

The method of long-term radioactive waste disposal favored by the U.S. government (and by many other nations) is mined geologic disposal. The radioactive waste would be buried at least one kilometer below Earth's surface and at least one kilometer from any water table in vaults that would presumably remain undisturbed.

To prepare radioactive waste for burial, spent fuel rods would first be allowed to cool for several decades in very large tanks of water. Over time, many of the radioisotopes would decay, lowering the radioactivity level to a point where the materials could be transferred to storage containers. These containers are currently stored on-site at many nuclear reactor facilities for transportation to a long-term disposal site when it becomes available.

Another option is to transform cooled radioactive wastes into glassy solids by application of heat, a process known as **vitrification** (see Figure 6.68). Although the encased material would still be highly radioactive, the waste would be much less likely to leak or leach into the environment because of the glasslike envelope encasing it. Technicians could seal the vitrified radioactive wastes in containers made of glass, stainless steel, or concrete. The Savannah River Site vitrification plant in South Carolina is the world's largest facility of its kind. France has used commercial vitrification for more than two decades (nuclear reactors produce about 78% of France's electrical energy).

Nowhere in the world, however, has radioactive waste been permanently buried. The challenge is to find technically, politically, and socially acceptable sites. The Japanese government has even considered deep-ocean burial.

Some geologic sites formerly assumed to be stable enough for radioactive waste disposal were later discovered to be unsafe. For example, technicians buried some plutonium at Maxey Flats, Kentucky, in a rocky formation that geologists believed would remain stable for thousands of years. Within a decade, however, some buried plutonium had moved dozens of meters away.

What are current plans to resolve our long-term radioactive waste disposal problems? By U.S. law, Congress selected two sites for permanent radioactive waste disposal from options provided by the DOE. These sites, located in regions of presumed geologic stability (see Figure 6.69), were the Waste Isolation Pilot Plant (WIPP) near Carlsbad, New Mexico, and Yucca Mountain, Nevada, an extinct volcanic ridge 100 miles northwest of Las Vegas.

> Each nuclear reactor site already uses this cooling process for wastes.

Figure 6.68 *One strategy to prevent radioactive waste from entering water supplies is to vitrify the waste—that is, convert it into a glassy solid.*

> The Hanford Vitrification Plant, which will be the world's largest when construction is completed in 2019, is designed to treat the 53 million gallons of radioactive and chemical waste stored near its site in eastern Washington State.

Figure 6.69 *Storing radioactive waste requires an underground site that will minimize contact with local water resources and remain geologically stable over many centuries. Siting poses great challenges. The waste will remain dangerously radioactive for thousands of years.*

Both proposals generated considerable debate and controversy about site locations and the means of transporting radioactive wastes to them. Although WIPP has been operating since 1999, the Yucca Mountain Repository is no longer being funded as of 2011, leaving future disposal options for U.S. radioactive waste uncertain.

DEVELOPING SKILLS

D.7 DISPOSING OF HIGH- AND LOW-LEVEL WASTES

Use Figure 6.66 (page 672) to answer the following questions:

1. Approximately what percent of radioactive waste is
   a. low-level waste?
   b. high-level (including transuranic) waste?

2. Which waste source accounts for the greatest volume of high-level radioactive waste?

3. Which two waste sources account for most of the low-level radioactive waste?

4. Should high-level or low-level radioactive wastes receive greater attention? Explain your answer.

5. For each of the following disposal strategies, identify at least one benefit and one risk:
   a. on-site storage
   b. vitrification (see Figure 6.70)
   c. mined geologic disposal

Figure 6.70 *A nuclear technician at La Hague, France, checks vitrified radioactive waste storage containers for radiation emissions.*

▮MAKING DECISIONS
D.8 DISCUSSING NUCLEAR ENERGY

Read the statements in the CANE flyer on page 586 once again, then complete the following activities and questions.

1. Identify the five statements on the flyer that are most closely connected to the ideas you studied in Section D.

2. Choose one of these statements that you now know to be false. Write a short rebuttal of the statement that would be clear to a fellow student who has not studied chemistry.

3. Select one of the statements that you now know to be true. Find a parent, grandparent, or other adult who is willing to discuss this statement with you. Ask the following questions, write down their answers, and respond to their questions to the best of your knowledge.

   a. Do you think this statement is true? Why?

   b. Does this statement cause you to be concerned? Why or why not?

   c. Would the ideas in this statement affect your opinion about whether a nuclear power plant should be located in your community? Why?

   d. What questions do you have about this statement?

4. Do you think that the person that you spoke to would support CANE's mission? Explain.

Whenever you prepare a written or oral presentation, it is important to be aware of your audience's knowledge, interests, and goals. Keep this in mind as you prepare for your presentation to Riverwood's senior citizens.

SECTION D SUMMARY

Reviewing the Concepts

> Some large nuclei, when bombarded by neutrons, undergo nuclear fission.

1. What is nuclear fission?

2. Name three isotopes that can undergo nuclear fission.

3. Write a balanced nuclear equation for

   a. the fission of U-235 by a neutron, producing Br-87, La-146, and several neutrons.

   b. the fission of U-235 by a neutron, producing Ba-144, Kr-90, and neutrons.

4. Why does a nuclear reaction release more energy than a chemical reaction?

5. State Einstein's mass–energy relationship and explain the meaning of each symbol.

6. Name the force that holds nuclear particles together and describe its characteristics.

7. Describe characteristics of a nuclear chain reaction.

8. Why is a critical mass of fissionable material needed to sustain a nuclear chain reaction?

> The electricity produced by nuclear power plants originates from the energy released by fission in controlled chain reactions.

9. Describe how most conventional (non-nuclear) power plants generate electricity.

10. State the equivalent quantities of coal and petroleum needed to produce the same total energy contained in one nuclear fuel pellet.

11. Why is U-235 used in nuclear power plants?

12. How does each of the following affect neutrons in a nuclear power plant?

    a. control rods

    b. moderator

13. Why is it impossible for the fuel in a nuclear power plant to cause a nuclear explosion?

14. List the three common moderators used in nuclear power plants.

15. The core of a nuclear reactor is surrounded by thick concrete walls. Give at least three reasons for these walls.

16. What is the composition of the white plumes often seen rising from nuclear power-plant towers?

> Nuclear fusion is the combination of two relatively small nuclei into a new, more massive nucleus.

17. Why are high pressures and temperatures needed to initiate fusion reactions?

18. How much more energy can nuclear fusion produce than nuclear fission?

19. Why isn't nuclear fusion currently practical for generating electricity in power plants?

20. State the equivalent quantities of coal and gasoline needed to produce the energy released by the fusion of one gram of hydrogen-1.

21. What is a thermonuclear weapon?

22. Explain why both of these statements are true:

    a. Nuclear fusion has not been used as an energy source on Earth.

    b. Nuclear fusion is Earth's main energy source.

> The permanent disposal of radioactive waste poses challenging problems and issues.

23. Radioactive waste is grouped into two major categories.

    a. List and describe these two categories.

    b. What is the major difference between the two categories?

24. In the United States, what are the two largest sources of radioactive waste? Refer to Figure 6.66 (page 672).

25. Refer to Figure 6.66 (page 672). Spent fuel makes up what percent of total U.S. radioactive wastes

    a. by volume?

    b. by radioactivity?

26. Why do technicians have to regularly replace nuclear fuel pellets even if the pellets are still radioactive?

27. Compare current methods for disposing of high and low-level radioactive wastes.

28. Why is vitrification a preferred method for handling and storing radioactive wastes?

29. a. Where are two potential U.S. sites for permanent radioactive waste disposal?

    b. Why were these particular sites selected?

What burdens and benefits accompany uses of nuclear energy?

In this section, you have considered the processes of nuclear fission and fusion and the energy released in these processes, learned how fission energy is harnessed in nuclear power plants, weighed risks of your personal decisions, and evaluated some options for disposal of radioactive waste. Think about what you have learned, then answer the question in your own words in organized paragraphs. Your answer should demonstrate your understanding of the key ideas in this section.

Be sure to consider the following in your response: nuclear fission, nuclear fusion, nuclear power plants, risk, and radioactive waste.

Connecting the Concepts

30. Explain the difference between nuclear fission and nuclear fusion.

31. Sometimes burning is a good way to dispose of some types of extremely toxic material. Why would burning be an unacceptable plan for destroying radioactive waste?

32. Explain why the disposal of radioactive waste is a challenging issue in many nations. Use the concepts of radioactive decay, half-life, and radiation shielding in your answer.

33. How does nuclear fusion compare to a chemical reaction in which two hydrogen atoms combine to form a hydrogen molecule, H_2?

34. Construct a diagram showing the energy transformations involved in producing electricity in a nuclear power plant. How would your diagram differ for a coal-fired power plant?

35. A simple way to minimize the need for long-term storage of radioactive waste might seem to be speeding up all radioactive decay rates involved. Why won't this plan work?

36. What are some factors complicating the cleanup of abandoned or improperly stored radioactive waste?

Extending the Concepts

37. Research how other countries dispose of their high-level radioactive waste and evaluate their methods in terms of risks and benefits.

38. When confronting problems surrounding high-level radioactive waste disposal, students sometimes propose loading the waste in a rocket and shooting it into the Sun. What are some risks and benefits of that plan?

39. Fusion reactions that produce iron and lighter elements can take place in the core of an ordinary star. However, elements with higher atomic numbers generally result from violent stellar explosions. Explain the difference.

40. What would be some advantages of nuclear fusion over nuclear fission for producing electricity?

41. Investigate conditions needed to sustain a critical-mass chain reaction in

a. a nuclear reactor.

b. a nuclear weapon.

PUTTING IT ALL TOGETHER

COMMUNICATING SCIENTIFIC AND TECHNOLOGICAL INFORMATION

Throughout your chemistry studies up to this point, you have used scientific ideas to evaluate and draw your own conclusions regarding published claims in the CANE flyer. Your final task in this unit is to prepare and present what you have learned to help some Riverwood residents draw their own conclusions about nuclear concerns.

As you already know, your audience is composed of senior citizens at a local community center. They are concerned that regulations proposed by CANE will affect their ability to receive a full range of medical care. They are also alarmed by some CANE claims about radioactive waste disposal.

The senior citizens have invited your class to address the scientific aspects of statements made in the flyer from CANE so that they can evaluate the group's message more completely.

INFORMING OTHERS ABOUT NUCLEAR SCIENCE AND TECHNOLOGY

Responding to the CANE Flyer

Each student group will be responsible for responding to one or more CANE statements. In preparing your presentation, coordinate with other groups working on statements that involve similar ideas.

For each statement, follow these steps:

1. Decide whether the flyer statement is true, false, or partially true.
2. Explain the science or technology that the statement involves. For example, if the statement is about nuclear power plants, you should review how such plants work. For statements about radiation, you should explain different types of radiation.
3. Prepare a replacement statement, if necessary, that more accurately and evenhandedly addresses the same topic/issue.

Targeting the Audience

Design your presentation to be understandable by adults with at least a high school education that most likely was completed many years ago. Review the concerns that were mentioned in the unit introduction and address these concerns whenever they are relevant to your topic.

Supporting the Presentation

When possible, include visual aids and everyday examples and applications. Also devise a way to evaluate the effect of your presentation. Try to assess whether or not the audience, after your presentation, is better able to make informed decisions about nuclear policies for Riverwood.

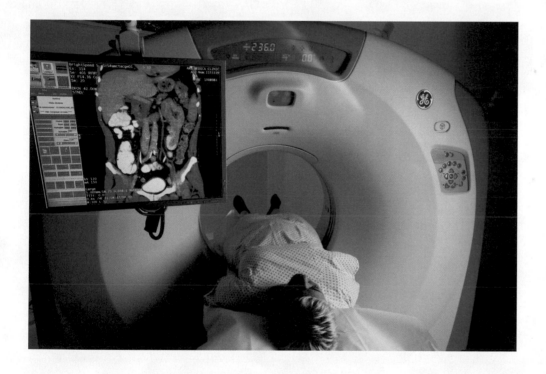

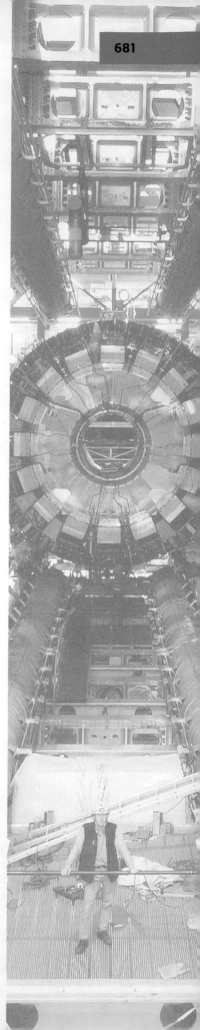

UNIT 7

FOOD: MATTER AND ENERGY FOR LIFE

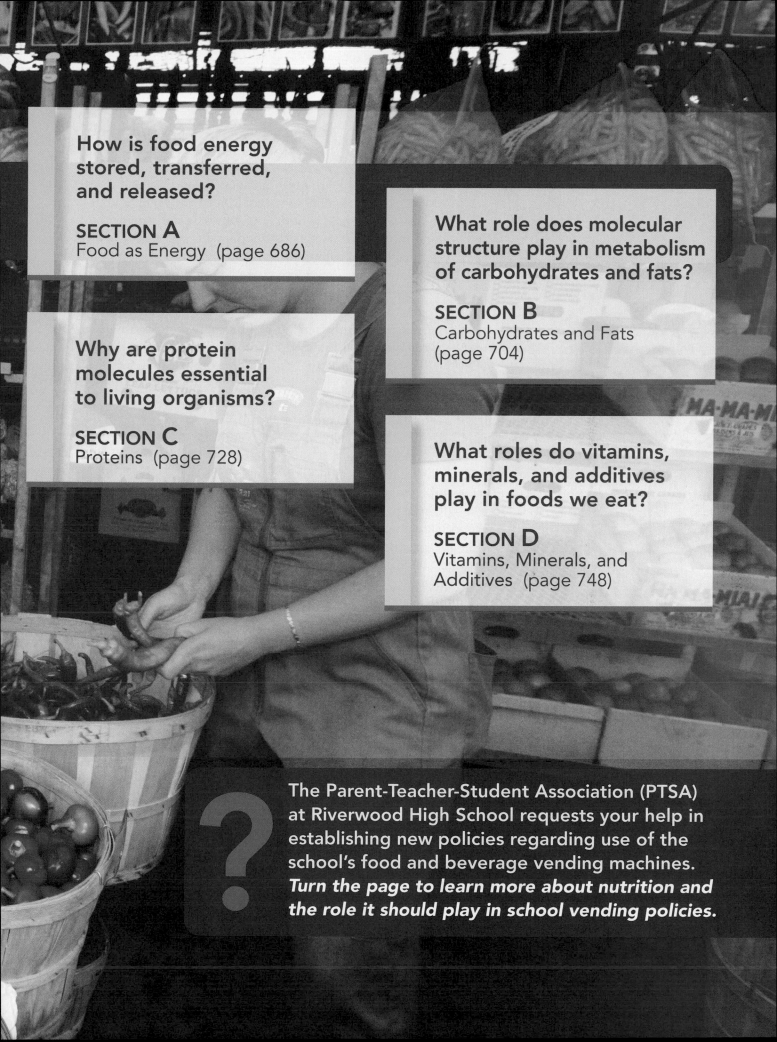

How is food energy stored, transferred, and released?

SECTION A
Food as Energy (page 686)

Why are protein molecules essential to living organisms?

SECTION C
Proteins (page 728)

What role does molecular structure play in metabolism of carbohydrates and fats?

SECTION B
Carbohydrates and Fats
(page 704)

What roles do vitamins, minerals, and additives play in foods we eat?

SECTION D
Vitamins, Minerals, and
Additives (page 748)

The Parent-Teacher-Student Association (PTSA) at Riverwood High School requests your help in establishing new policies regarding use of the school's food and beverage vending machines. *Turn the page to learn more about nutrition and the role it should play in school vending policies.*

Ervin Kostecky, a parent member of the Riverwood High School Parent-Teacher-Student Association (PTSA), recently read a report about the quality of food sold in U.S. middle school and high school vending machines. Mr. Kostecky is concerned that Riverwood High School's food and beverage vending machines do not meet energy and nutritional needs of students who use them. He has requested that a committee be appointed to recommend school policies regarding these vending machines.

In this unit, you will learn about the chemistry of foods. You will learn how energy contained in food is stored and released and how substances in food promote growth and repair in the body. You will also investigate the chemistry and the nutritional roles of fats, carbohydrates, proteins, vitamins, and minerals. This information will help you make sound policy recommendations to be implemented next year for the school's vending machines.

As a member of the Vending Machine Policy Planning Committee, you receive an inventory that documents a typical Riverwood High student's diet over three days. Together with other committee members (your classmates), you will conduct several analyses of this three-day food inventory. By the end of the unit, your results will provide insight into items that would complement a typical Riverwood High student's diet. You will then write a report summarizing your results and proposing recommended guidelines for selecting new items for the vending machines. This report will be used in the committee's discussions and, ultimately, in selecting future food and beverage items for school vending machines.

To aid your investigation, the PTSA has highlighted characteristics of a healthful diet found in the *Dietary Guidelines for Americans 2010*, issued by the U.S. Department of Health and Human Services.

Keep the following dietary advice as well as chemistry principles you have learned in this course in mind throughout this unit:

Characteristics of a healthful diet:

- Consume a variety of nutrient-dense foods and beverages within and among the basic food groups.

- Balance portion-controlled calories from the foods and beverages ingested with calories expended in physical activity.

- Choose foods that limit the intake of solid fats, added sugars, salt, and alcohol.

- Consume a sufficient quantity and variety of fruits and vegetables, while staying within energy needs.

Is this the vending machine of the future?

SECTION A   FOOD AS ENERGY

How is food energy stored, transferred, and released?

The term *diet* does not necessarily imply a weight-loss plan; it refers to the pattern of food and drink that one regularly consumes.

Where does the energy required for walking, running, playing sports, and even sleeping and studying come from? The answer is obvious: It comes from the food you eat. As you begin to analyze the three-day food inventory, you will learn the chemistry that will allow you to understand where that energy originates, how food energy is stored and used, and how decisions about eating can affect well being.

GOALS

- Use calorimetry to determine the energy stored in a particular food.
- Trace the flow of energy in food back to its sources and forward to its uses and dispositions.
- Analyze how the balance between energy intake and expenditure affects body mass and health.

✓ concept check 1

1. How can the amount of energy produced during a chemical reaction be determined?
2. What is a Calorie?
3. What are some components necessary for a healthful diet?

A.1 FOOD GROUPS

When considering foods, people often focus on particular categories, or **food groups**. A healthful diet includes foods from a variety of groups, such as those shown in Figure 7.1. Why? Figure 7.2 highlights the groups. What do you already know about the nutrient and energy contents of foods in each group?

Although it is healthful to consume foods from a variety of groups, the *Dietary Guidelines for Americans 2010* advise that people should increase their daily intake of fruits and vegetables, whole grains, and nonfat or low-fat milk and milk products. In addition, the *Dietary Guidelines* call for decreasing intake of foods rich in fats and sugars.

Some foods listed in the three-day inventory may be represented by categories in Figure 7.2, but others are not. Where should you classify foods not included in Figure 7.2? What range of food groups is represented in the diet you are analyzing? Find out in the next activity.

Figure 7.1 *An assortment of healthful foods.*

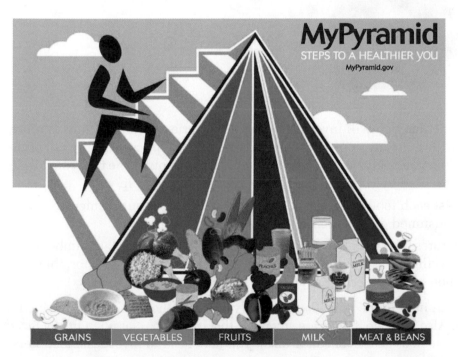

Figure 7.2 *The U.S. Department of Agriculture's food pyramid shows relative quantities of foods that should be consumed from each group, according to the Dietary Guidelines for Americans 2010. The small yellow "slice" represents fats, sugars, and salt, which should be limited in a healthful diet. See MyPyramid.gov for detailed individual recommendations.*

■ MAKING DECISIONS

A.2 DIET AND FOOD GROUPS

Your three-day food inventory analysis begins with the "big picture." That is, before analyzing the foods in terms of their characteristics—such as energy, fat, carbohydrate, protein, mineral, or vitamin content, you will verify what proportions of each food group the food inventory includes.

In particular, you will investigate whether a variety of food groups is represented within the inventory, or if one or two food groups constitute most of the food. When you analyze a food inventory, either your own, one that you devise, or one provided by your teacher, these suggestions will help you:

- List each food item, snack, beverage, and dietary supplement consumed during each of the three days.
- Express the quantity of each food item by estimating the number of servings, mass, or volume of the food or beverage. When possible, use food labels on each item to guide your quantity estimates.

For each item in the three-day inventory, indicate the food group to which it belongs. Also indicate if there are any items that cannot be classified in any listed category (such as snacks or soft drinks). You may need to investigate food items about which you are unsure. Then complete the following steps:

1. Construct a data table so that the total number of servings within each food group can be easily recorded. Include four columns, one for each day, and one for the daily average over the three days.

2. Record the quantity consumed within each food group for each day. Then calculate and enter the average number of daily servings for each group over the three days.

3. Using your data from Step 2, which food group or groups

   a. represent most of the foods within the diet?

   b. are minimally represented in the diet or not at all?

4. Overall, is there a reasonable balance of food groups in the inventory? If not, what dietary substitutions would lead to greater variety among food groups?

5. Complete the following steps for food items that were difficult to assign to a particular food group:

   a. List the item and the food group to which the item was assigned.

   b. Explain why it was difficult to assign this food item to a food group.

   c. Explain how you decided to which group to assign the item.

6. Is it better to analyze food consumed over three consecutive days or for only one day? Explain.

7. Under what circumstances would basing your food analysis on a particular three-day average be misleading?

■ INVESTIGATING MATTER

A.3 ENERGY CONTAINED IN A SNACK

Preparing to Investigate

Figure 7.3 shows part of a food label from a common snack food. Note the number of grams per serving. The label also shows the number of Calories (150). This is the quantity of food energy contained in one serving. In this investigation you will determine the energy contained in a sample of a snack food.

The quantity of energy contained in a particular food can be determined by carefully burning a known mass of the item under controlled conditions and measuring how much thermal energy is released. You completed similar investigations in Unit 3 when you determined the energy released during the combustion of several types of fuel, as well as a candle's heat of combustion (see Figure 3.41, page 331). This procedure is known as **calorimetry** and the measuring device is called a **calorimeter**.

Use the fuel-burning procedure (page 331) to guide you as you design a procedure to determine the energy contained in this particular snack sample. (*Caution: Do not use any nut-based snack food for this investigation; some students may be allergic to such items.*)

Figure 7.3 Part of the label from a popular snack food. Note the mass and total Calories contained in one serving.

Before starting the investigation, complete the following steps:

- Write a detailed procedure for determining the energy contained in the assigned snack food.
- Sketch your laboratory setup.
- Construct a data table to record all necessary measurements and observations for two separate trials.
- Have your teacher check and approve your procedure and data table.

Gathering Evidence

Before you begin, put on your goggles and wear them properly throughout the investigation. After your teacher approves your procedure, you may start the investigation. Conduct at least two trials.

Analyzing Evidence

If you know the mass of water and its temperature change, you can calculate the quantity of thermal energy that caused the temperature change. In this laboratory investigation, we assume that

Energy to heat water = Energy released by burning snack item

This assumes, of course, that no thermal energy is lost during the process.

> The precise specific heat capacity of water is 4.184 J/(g • °C).

Recall from Unit 3 that the specific heat capacity of liquid water is about 4.2 J/(g • °C). Thus, it takes about 4.2 J to raise the temperature of 1 g of water by 1 °C. If you know the mass of water heated and its temperature change, you can calculate the thermal energy absorbed by the water. If necessary, refer to page 338 to remind yourself how to complete this calculation.

Use data collected in this investigation to complete the following calculations:

1. Determine the mass (in grams) of the heated water.
2. Calculate the water's overall temperature change.
3. Calculate the total energy (in joules) required to heat the water.
4. The food **Calorie**, or *Cal* (written with an uppercase C), listed on food labels, is a much larger energy unit than the joule. One Calorie equals 4184 J, which can be conveniently rounded to 1 Cal = 4200 J.

> The *calorie* (cal), another common energy unit, is about four times the size of a joule (1 cal = 4.184 J). One Calorie (Cal), which is a unit still used in the U.S. for food-energy values, equals one kilocalorie (1 kcal) or 1000 cal.

   a. Calculate the total Calories used to heat the water.

   b. How many Calories were released by burning the item?

5. a. Calculate the energy released, expressed as Calories per gram (Cal/g) of the snack item burned, for each trial.

   b. Calculate the average Calories per gram for the snack item.

   c. Use reported values from the package label to calculate the declared Calories per gram for the snack item.

   d. Calculate the percent difference between the declared value found on the label and your average experimental value:

$$\% \text{ Difference} = \frac{|\text{Experimental value} - \text{Label value}|}{\text{Label value}} \times 100\%$$

Making Claims

1. Would you describe the snack food that you tested as energy-dense? Why or why not?

2. Compare your results with those of your classmates.

   a. Which snack food had the greatest variance among groups in calculated Cal/g?

   b. What do you think caused these differences?

Reflecting on the Investigation

3. Which aspects of your laboratory setup and procedure might account for any difference between your Cal/g value and the corresponding label value?

4. How could your laboratory setup and procedure be improved to increase the accuracy of your results?

5. Consider the snack item shown in Figure 7.4. What changes in setup or procedure would be necessary to find the Calories per gram in the snack item shown?

Figure 7.4 *Describe how you would determine the quantity of energy contained in these potato chips.*

concept check 2

1. How was the measurement of energy released in the Combustion investigation (Unit 3, page 337) similar to and different from the measurement of energy in snack foods?
2. Where does the energy in food originate?
3. How does your body convert food into usable energy?

A.4 ENERGY FLOW—FROM THE SUN TO YOU

The snack item you burned released enough energy to raise the temperature of a sample of water by several degrees Celsius. Where did that energy come from?

The ultimate answer is easy: All food energy originates from sunlight. Through **photosynthesis**, green plants capture and use solar energy to make large molecules from smaller, simpler ones (see Figure 7.5). Recall from Unit 3 (page 345) that green plants, through photosynthesis, use solar energy to convert water and carbon dioxide into carbohydrates and oxygen gas. Although a variety of carbohydrates are produced, an equation for photosynthesis usually depicts the production of glucose:

Photosynthesis: $6\,CO_2$ + $6\,H_2O$ + Solar energy $\longrightarrow$ $C_6H_{12}O_6$ + $6\,O_2$
Carbon   Water   Glucose   Oxygen
dioxide   gas

Figure 7.5 *Energy from the Sun was used to convert water and carbon dioxide into the molecules that constitute these onions. What type of energy conversion may happen next?*

For a discussion of energy involved in chemical reactions, see pages 333–336.

For this reaction to occur, bonds between carbon and oxygen atoms in carbon dioxide molecules and between oxygen and hydrogen atoms in water molecules must be broken. The atoms must then recombine in a different arrangement to form glucose and oxygen molecules. Recall from Unit 3 that breaking bonds always requires energy, whereas bond formation releases energy. In photosynthesis, the bonds in carbon dioxide and water molecules require more energy to break than is released when chemical bonds in glucose and oxygen molecules form. The energy needed to drive this endothermic reaction, as the photosynthesis equation indicates, comes from the Sun. The potential energy diagram in Figure 7.6 shows the energy relationships in this process.

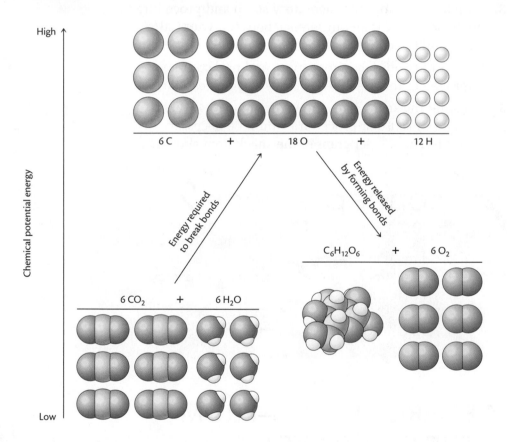

Figure 7.6 *Photosynthesis. Breaking bonds within carbon dioxide and water molecules requires energy. Energy is released when atoms combine to form glucose and oxygen molecules. The first step (bond breaking) requires more energy than the second step (bond forming) releases, so the overall process of photosynthesis is endothermic.*

How would a potential energy diagram for the process of metabolism look?

The Sun's radiant energy is converted to chemical energy stored within bonds of carbohydrate molecules. Living organisms release this chemical energy when they consume and metabolize carbohydrate molecules, converting them into lower-energy CO_2 and H_2O molecules. This chemical energy can be measured via calorimetry, as you just observed in Investigating Matter A.3.

Energy originally delivered as sunlight continues to flow through ecosystems as carnivores (meat-eating animals) consume plant-eating animals. Eventually, plants and animals die and decay; organisms that aid in decomposition use remaining stored-up energy. Thus, energy flows from the Sun to photosynthetic plants to herbivores and then to carnivores and decomposers.

The Sun's captured energy is dispersed and becomes less available as it is transferred from organism to organism. For example, as energy flows from one organism to another, some energy is dispersed into the environment as thermal energy. You may be surprised to learn that only a small fraction (about 10–15%) of food energy consumed by organisms is used for growth—for converting smaller chemical molecules to larger molecules that become part of an animal's structure. Over half the energy contained in consumed food is used to digest food molecules. The supply of useful energy declines as energy continues to transfer away from its original source—the Sun. Figure 7.7 depicts this decline in available energy.

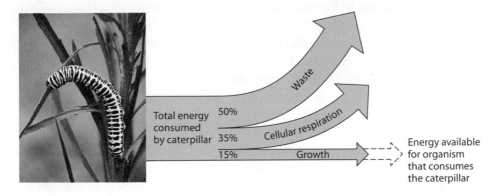

Figure 7.7 *The energy used by this caterpillar originated as solar energy. However, not all incident solar radiation is converted into usable energy. First, only a portion of the Sun's energy is actually stored as chemical energy in molecules within this plant. Next, only a fraction of that stored chemical energy is actually used by the caterpillar.*

CHEM**QUANDARY**

HOW DOES YOUR GARDEN GROW?

From a chemical viewpoint, consider this question: *As garden plants grow, what is the source of the material that causes their mass to increase?*

A.5 ENERGY RELEASE AND STORAGE

Some energy in the food you eat is used soon after you digest it (Figure 7.8). The rest of the energy may be stored for later needs. These available reserves of bodily energy are mainly in fats and, to a much smaller extent, in carbohydrates. Figure 7.9 shows some foods rich in carbohydrates and fats.

Whether your body uses energy from food recently ingested or from stored fat, the release of energy from these molecules depends on a series of chemical reactions inside your cells. In **cellular respiration**, plants and animals use oxygen to break down complex organic molecules into carbon dioxide and water molecules. The energy required to break bonds in the reactant molecules is less than the energy released when the bonds in carbon dioxide and water form. Thus energy is released in cellular respiration. You can think of that exothermic process as the reverse of the process shown in Figure 7.6 (page 692).

Figure 7.8 *A portion of the energy in food is expended soon after eating, particularly when engaging in energy-demanding activities such as bicycling. The remaining energy is then stored for later activity.*

> Carbon dioxide and water are also produced from burning petroleum, coal, and other fossil fuels. See page 345.

Organic compounds + Oxygen ⟶ Carbon dioxide + Water + Energy

Specialized structures within each cell use the energy released to carry out a variety of tasks such as energizing reactions, transporting molecules, disposing waste, storing genetic material, and synthesizing new molecules. That energy ultimately allows you to walk, talk, run, work, and think; it also provides energy to power your heart, lungs, brain, and other organs.

Carbohydrates, fats, and proteins can all be processed as fuel for your body. We will now examine cellular respiration through the oxidation of glucose, a carbohydrate. The overall equation, based on the oxidation of one mole of glucose, is as follows:

> This equation is the reverse of the equation for photosynthesis. See Figure 3.47 page 345.

$$C_6H_{12}O_6(aq) + 6\,O_2(g) \longrightarrow 6\,CO_2(g) + 6\,H_2O(l) + 686\ \text{Calories}$$

Figure 7.9 *Breads, grains, bagels, and pasta (left) contain significant portions of carbohydrates, whereas cooking oil, nuts, and avocado (right) are rich in fats.*

This equation actually summarizes a sequence of more than 20 integrated chemical reactions catalyzed by more than 20 different enzymes. Glucose (see Figure 7.10) is oxidized within cells throughout your body.

The energy needed to perform individual cellular functions is much less than the energy released by "burning" glucose molecules. Between its release from these energy-rich molecules and its later use in cells, energy is stored in biomolecules of ATP (adenosine triphosphate), shown in Figure 7.11.

Figure 7.10 *The icing and decorations on these cupcakes likely contain glucose, which is burned to provide energy or stored as fat in the body.*

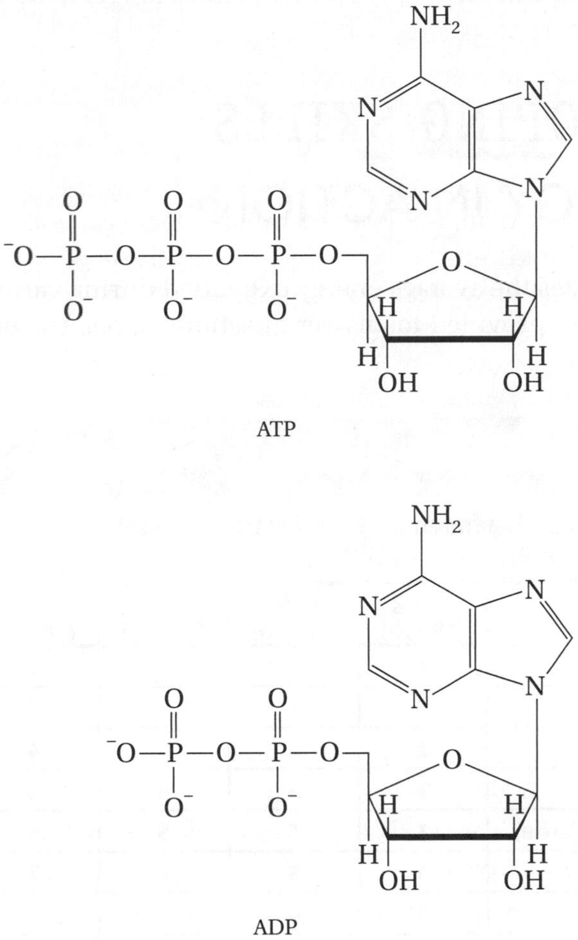

ATP

ADP

Figure 7.11 *The ionic forms of ATP and ADP. The colored atoms on ATP's left side are removed when ADP forms.*

As indicated by the following equation, more energy is required to break chemical bonds in adenosine diphosphate (ADP) and HPO_4^{2-} than is released when water and ATP form. The difference in those bond energies equals 7.3 Cal, which suggests that an ATP molecule can be regarded as an energy-storage site because energy is released when the reaction proceeds in the opposite direction.

$$7.3 \text{ Cal } + \text{ ADP}(aq) + \underset{\substack{\text{Hydrogen}\\\text{phosphate ion}}}{\text{HPO}_4^{2-}(aq)} \longrightarrow \text{H}_2\text{O}(l) + \text{ ATP}(aq)$$

When this reaction is reversed, each mole of ATP releases 7.3 Cal. This conveniently small quantity of energy, compared to the 686 Calories produced as one mole of glucose is oxidized, is used to energize particular cellular reactions. Some of these reactions require less energy than that supplied by a single ATP molecule, whereas other steps require the total energy released by several ATP molecules.

Oxidation of one mole of glucose produces enough energy to add 38 mol of ATP to short-term cellular energy storage. Each day, your body stores and later releases energy from at least 6.02×10^{25} molecules of ATP (100 mol ATP).

DEVELOPING SKILLS

A.6 ENERGY IN ACTION

Table 7.1 summarizes the average energy expended during various activities. Use the information provided to answer questions appearing on the following page.

Table 7.1

Energy Expended in Various Activities

Activity	Energy Expended (in Calories per Minute) for Individuals of Different Body Mass			
	46 kg (100 lb)	55 kg (120 lb)	68 kg (150 lb)	82 kg (180 lb)
Sleeping	1	1	1	1
Sitting	1	1	1	2
Playing volleyball	2	3	4	4
Weight lifting	2	3	4	4
Walking (17 min/mile)	3	4	5	6
Skateboarding	4	5	6	7
Swimming	5	6	7	9
Wrestling	5	6	7	9
Bicycling (moderate)	6	7	8	10
Playing soccer	6	7	8	10
Doing step aerobics	6	7	8	10
Playing tennis	6	7	8	10
Playing basketball	6	8	10	12
Playing football	6	8	10	12
Playing ice hockey	6	8	10	12
Doing martial arts	8	10	12	14
Running (8 min/mile)	10	12	15	18

Sample Problem: Table 7.1 indicates that a 55-kg (120-lb) person burns 5 Cal/min while skateboarding (Figure 7.12).

   a. This energy is supplied by ATP. If the oxidation of each mole of glucose ($C_6H_{12}O_6$) produces 38 mol ATP, how many moles of glucose would be needed to provide the energy for that one hour of skateboarding?

   b. What mass of glucose does your answer to Part a represent? The mass of one mole of glucose is 180 g.

To begin, find moles of ATP needed for one hour of skateboarding:

$$\frac{5 \text{ Cal}}{1 \text{ min}} \times \frac{60 \text{ min}}{1 \text{ h}} = 300 \text{ Cal/h}$$

$$\frac{300 \text{ Cal}}{1 \text{ h}} \times \frac{1 \text{ mol ATP}}{7.3 \text{ Cal}} = 41 \text{ mol ATP/h}$$

Because 38 mol ATP can be obtained from each mole of glucose, the skateboarder's need for 41 mol ATP can be met by slightly more than one mole of glucose:

$$41 \text{ mol ATP} \times \frac{1 \text{ mol glucose}}{38 \text{ mol ATP}} = 1.1 \text{ mol glucose}$$

What mass of glucose does that represent? Because the skateboarder needs more than one mole of glucose, that mass must be greater than 180 g:

$$1.1 \text{ mol glucose} \times \frac{180 \text{ g glucose}}{1 \text{ mol glucose}} = 200 \text{ g glucose}$$

Figure 7.12 *How much energy is expended in one hour of skateboarding?*

> More than 200 g glucose would actually be needed, since the energy-transfer steps are not 100% efficient.

Now answer the following questions:

  1. Assume your body produces approximately 100 mol ATP daily.

    a. How many moles of glucose are needed to produce that amount of ATP?

    b. What mass of glucose does your answer to Question 1a represent?

  2. One minute of muscle activity requires about 0.0010 mol ATP for each gram of muscle mass. How many moles of glucose must be oxidized to energize 454 g (1 lb) of muscle to dribble a basketball for one minute?

  3. a. How many hours do you typically sleep each night?

    b. How many Calories do you use during one night of sleep? Refer to Table 7.1.

    c. How many moles of ATP does this require?

    d. What kinds of activity require energy while you are sleeping?

A.7 ENERGY IN—ENERGY OUT

People follow various diets for different reasons. Some people may want to lose weight. Others may want to gain weight. Still others put little thought into what they eat. Instead, they merely eat for convenience or pleasure.

Some foods are needed to deliver important molecules to the body, regardless of their energy value. By contrast, other foods provide only energy. This latter category, which includes sugar-sweetened soft drinks, is sometimes described as furnishing "empty Calories."

The *Physical Activity Guidelines for Americans* encourages citizens of all ages to become more active. The report, issued in 2008 by the U.S. Department of Health and Human Services, cites research findings on the benefits of physical activity, including:

- improved cardiac, respiratory, and muscular fitness
- improved bone health
- improved cardiovascular and metabolic health biomarkers
- favorable body composition, and
- reduced symptoms of depression.

These benefits increase as you age. Research also shows that the total amount of physical activity is more important than its intensity, duration, or frequency. The guidelines measure physical activity in a unit called metabolic equivalent (MET). By contrast, the activities in this unit use Calories, so that you can compare energy consumed in your food to energy burned through your activities.

> If you eat 100 Cal per day more than you burn, you will gain about a pound each month.

If you want to lose weight, you must consume less energy than you expend. If your goal is to gain weight, you must do just the opposite. If losing weight or gaining muscle mass is desired, then regular physical activity, such as that shown in Figure 7.13, is needed in addition to the particular diet you follow. Now you will explore the interplay between personal activity and diet.

Figure 7.13 *Technology is available for monitoring total energy expenditure.*

■ DEVELOPING SKILLS

A.8 GAIN SOME, LOSE SOME

Look again at Table 7.1 on page 696. The quantity of energy expended depends on the duration of a given activity and the person's body mass. For example, during one hour of soccer playing, a 55-kg (120-lb) person expends 420 Cal, whereas a 68-kg (150-lb) person would expend 480 Cal in that same activity.

Use information in Table 7.1 (page 696) to answer the following questions.

> *Sample Problem:* *A 46-kg female usually takes 10.0 min to bike home from school. Would she burn more Calories by running home instead (at 8 min/mile for 15.0 minutes)?*
>
> Bicycling requires: Running requires:
> 6 Cal/~~min~~ × 10.0 ~~min~~ = 60 Cal   10 Cal/~~min~~ × 15.0 ~~min~~ = 150 Cal
>
> Thus, running home burns more energy by 90 Cal per trip.

1. Consider eating an ice-cream sundae (Figure 7.14). Assume that two scoops of your favorite ice cream contain 250 Cal; the topping adds 125 Cal more.

   a. Assume that your regular diet (without that ice-cream sundae) just maintains your current body weight. If you eat the ice-cream sundae and wish to "burn off" those extra Calories,

      i. for how many hours would you need to lift weights?

      ii. how far would you need to walk at 17 min/mile?

      iii. for how many hours would you need to swim?

   b. One pound of weight gain is equivalent to 3500 Cal. If you choose not to exercise, how much weight will you gain from eating the sundae?

   c. Now assume that you consumed a similar sundae once per week for 16 weeks. If you do not exercise to burn off the added Calories, how much weight will you gain?

2. Question 1 implies that eating an ice-cream sundae will cause weight gain unless you complete additional exercise.

   a. Can you think of a plan that would allow you to consume the sundae, do no additional exercise, and still *not* gain weight?

   b. Explain your answer.

   c. What concerns would you have about this plan?

3. Suppose you drank six glasses (250 g each) of ice water (0 °C) on a hot, summer day.

   a. Assume your body temperature is 37 °C. How many joules of thermal energy would your body use in heating that ice water to body temperature? Recall that the specific heat capacity of liquid water is about 4.2 J/(g • °C).

   b. How many Calories is this? Recall that 1 Calorie equals about 4200 J.

   c. A serving of french fries contains 240 Cal. How many glasses of ice water would you need to drink to "burn off" the Calories in one serving of french fries?

   d. Given your answer to Question 3c, does drinking large quantities of ice water seem like a reasonable strategy to lose weight? Explain.

4. Identify your favorite activity from Table 7.1 (page 696). How many minutes would you need to engage in that activity to burn off the Calories in one serving of french fries (see Question 3c)?

Figure 7.14 *How does an extra helping of ice cream affect your daily balance between energy intake and expenditure?*

■ MAKING DECISIONS

A.9 ENERGY INTAKE AND EXPENDITURE

Based on the chemical knowledge you have gained, you can find the total energy consumed in the three-day food inventory and estimate the energy expended based on typical activity levels. Using appropriate resources suggested by your teacher, calculate and record the food energy (Calories) contained in each item in the three-day inventory.

1. Calculate the total Calories consumed each day.

2. Calculate the average Calories consumed per day.

3. Using results from Making Decisions A.2 (pages 688–689), how many Calories per day (on average) are supplied by each Food Group?

4. Using Table 7.1 (page 696) and other resources provided by your teacher, develop a list of typical daily activities for a high school student and record how much energy is expended during each activity listed.

5. Based on the list you prepared in Question 4, calculate the total Calories expended over an average day. Calories are not only expended during the activities listed, but also by normal, resting-state body processes. Be sure to include this expenditure in your calculation.

6. a. Based only on Calories consumed and expended, decide whether the three-day inventory you are evaluating would be appropriate for weight maintenance (see Figure 7.15). Explain your answer.

    b. Identify and describe any limitations in your analysis that may affect your decision in Question 6a.

7. Based on your analysis thus far, what types of items would you recommend for the Riverwood High School vending machines in terms of the energy they provide?

Figure 7.15 *The balance between energy intake and expenditure affects one's body mass.*

SECTION A SUMMARY

Reviewing the Concepts

> Calorimetry can be used to determine the quantity of energy contained in a particular food sample.

1. List three common units used to express the energy content of food.

2. Sketch a simple calorimeter and label the purpose of each component.

3. Why is it important to know accurately the mass of water used in a calorimeter?

4. What is the mathematical relationship between a Calorie and a calorie?

5. Convert these energy quantities:
   a. 4375 cal to Cal
   b. 76 932 J to Cal
   c. 289 Cal to J
   d. 12 226 000 cal to kJ

6. a. How much thermal energy, in joules, is required to increase the temperature of a 115-g water sample by 10.0 °C?
   b. Suppose that a 2.0-g food sample was burned in a calorimeter to provide the thermal energy involved in Question 6a. Assume all thermal energy released was used to heat the water.
      i. How many calories per gram were present in the food sample?
      ii. How many Calories per gram were present in the food sample?

> All food energy originates from sunlight and is stored and released through a series of chemical reactions.

7. What process captures sunlight and transforms it into chemical energy?

8. Write a chemical equation for the production of glucose through photosynthesis.

9. Is photosynthesis an endothermic or exothermic chemical change? Explain.

10. Keeping in mind the law of conservation of energy, describe what happens to the food energy consumed by living creatures.

11. What is meant by an "energy rich" molecule?

12. What is the relationship between photosynthesis and cellular respiration? Write chemical equations that support your answer.

13. What is the difference, in terms of molecular composition and stored energy, between ADP and ATP?

14. How do ADP and ATP permit the controlled use of energy from glucose?

> The energy required for a physical activity depends on the particular activity, the total time involved, and the mass of the person engaged in the activity.

15. How much total energy (in Calories) is expended by a 68-kg (150-lb) person swimming for 35 minutes?

16. How much total energy (in joules) is expended by a 46-kg (100-lb) person walking for 56 minutes?

17. People expend energy even when they sit perfectly still. Why?

18. How many moles of ATP are required to provide energy for an 82-kg (180-lb) student to sit in class for 45 min?

19. Explain why two people performing exactly the same physical exercise may not burn the same total Calories.

20. Suppose a 55-kg (120-lb) man, whose regular diet and exercise just maintains his body weight, adds one chocolate candy bar containing 354 Cal to his diet each day for 30 days without increasing his level of exercise. Predict how much weight the person will gain in 30 days.

How is food energy stored, transferred, and released?

In this section, you began to analyze a three-day food and activity log in terms of quantity of Calories consumed and expended. You applied ideas that you learned in Unit 3 to food and food energy. Think about what you have learned in this unit and in previous units, and how these ideas are related, then answer the question in your own words in organized paragraphs. Your answer should demonstrate your understanding of the key ideas in this section.

Be sure to consider the following in your response: Calories, calorimetry, diet and exercise, ATP, ADP, and energy sources and pathways.

Connecting the Concepts

21. Compare the process of cellular respiration to combustion.

22. Does calorimetry directly measure the quantity of thermal energy liberated by a food or fuel? Explain.

23. Describe problems that could arise in calorimetry by using a food sample that is:

    a. very small.

    b. very large.

24. A student argues that eating a certain mass of chocolate or eating the same mass of apples results in the same gain in body weight. Explain why this is incorrect.

25. Figure 7.6 (page 692) shows an energy diagram for photosynthesis.

    a. Construct a similar energy diagram for cellular respiration.

    b. Compare the energy diagrams for photosynthesis and cellular respiration.

26. The term *respiration* is sometimes used to mean breathing. What is the relationship, if any, between cellular respiration and breathing?

27. In reference to a local seafood restaurant, someone remarked, "When you eat a pound of fish, you're eating ten pounds of flies." Aside from its questionable value as a meal-promoting strategy, how accurate is this message? Why?

28. Explain why low-Calorie food is sometimes described as *lite* or *light*, even though Calories are not a unit of mass.

29. You used similar calorimetry procedures in Unit 3 and in this unit to find the energy content of candles and food products, respectively. Does that imply that anything that can be burned to heat water can be used to fuel your body's metabolism? Explain.

Extending the Concepts

30. a. In what sense does the problem of world hunger involve an "energy crisis"?

    b. In what sense is it a "resource crisis"?

31. Research and report on the difference between *under*nourishment and *mal*nourishment. Could either term ever apply to an overweight person?

32. From an energy standpoint, are there advantages to eating low on the food chain? For example, is it energetically more favorable to use 100 lbs of grain or 100 lbs of beef as a food source?

33. Why is it not possible just to consume only pure ATP and thus eliminate some steps in metabolism?

34. Investigate the characteristics of a professionally-designed calorimeter. Make a sketch of its essential parts and explain its operation.

35. For hibernating animals, the storage of fat is critical for survival. Investigate a particular animal species to find out how it stores optimum quantities of fat for hibernation.

36. A student decides to lose some weight by not wearing a coat in cold winter weather. What knowledge of food energy might have inspired this idea? Does this plan have merit? Explain.

SECTION B CARBOHYDRATES AND FATS

What role does molecular structure play in metabolism of carbohydrates and fats?

You are now ready to explore how the chemical energy contained in foods, studied in Section A, is stored and consumed. Keep in mind the three-day inventory you are evaluating, and think about how this new knowledge may apply to its analysis.

GOALS

- Describe the general structure of a carbohydrate molecule.
- Describe the general structure and components of a fat molecule.
- Distinguish between saturated and unsaturated fats.
- Explain how structural differences in fat and carbohydrate molecules account for their different properties and energy content.
- Identify a limiting reactant, given the equation and reaction conditions.

✓ concept check 3

1. How is the process of cellular respiration similar to the process of combustion?
2. What is the source of the thermal energy released in an exothermic reaction?
3. What is the difference between unsaturated and saturated fat molecules?

B.1 CARBOHYDRATES: ONE WAY TO COMBINE C, H, AND O ATOMS

All **carbohydrate** molecules are composed of carbon, hydrogen, and oxygen. Glucose, which is the key energy-releasing carbohydrate in biological systems, has the molecular formula $C_6H_{12}O_6$. When such formulas were first established, chemists noted a 2:1 ratio of hydrogen atoms to oxygen atoms in carbohydrates, the same as in water. They were tempted to write the glucose formula as $C(H_2O)_6$, implying a chemical combination of carbon with six water molecules. Chemists even invented the term "carbohydrates" (water-containing carbon substances) for glucose and related compounds, although chemists later determined that carbohydrates contain no water molecules. However, like water, carbohydrate molecules *do* contain O—H bonds in their structures.

> Sugars, starch, and cellulose are examples of carbohydrates.

Carbohydrate molecules may be either simple sugars, such as glucose or fructose (see Figures 7.16 and 7.17), or chemical combinations of two or more simple sugar molecules. Simple sugars, called **monosaccharides**, are molecules containing five or six carbon atoms. Glucose (like most other monosaccharides) exists principally in a ring form: however, glucose can also exist in a chain form, as shown in Figure 7.17. Do both forms have the same molecular formula?

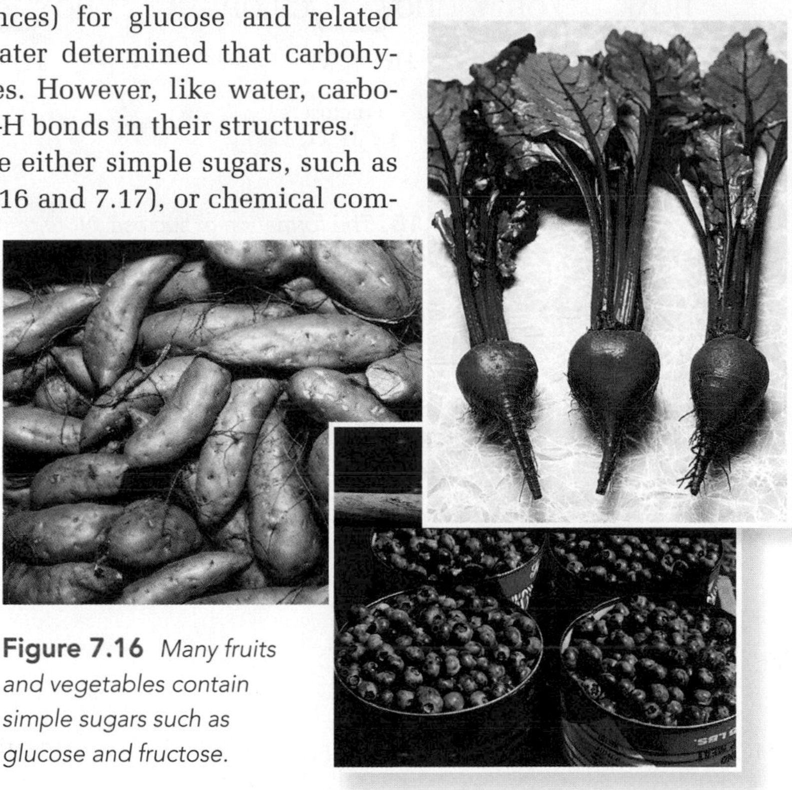

Figure 7.16 *Many fruits and vegetables contain simple sugars such as glucose and fructose.*

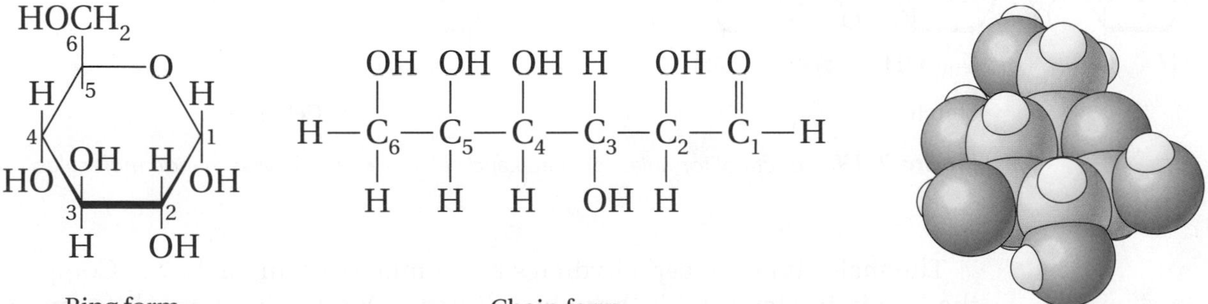

Ring form   Chain form

Figure 7.17 *Structural formulas for glucose. This carbohydrate, like most simple sugar and disaccharide molecules, is found primarily in ring form. How do ring and chain structures compare with one another?*

Condensation reactions were highlighted in Unit 3, pages 317–322, in the formation of esters and condensation polymers.

Sugar molecules composed of two monosaccharide units bonded together are called **disaccharides**. They are formed by a condensation reaction between two monosaccharides. Sucrose (table sugar, $C_{12}H_{22}O_{11}$) is a disaccharide composed of the ring forms of glucose and fructose, as illustrated in Figure 7.18.

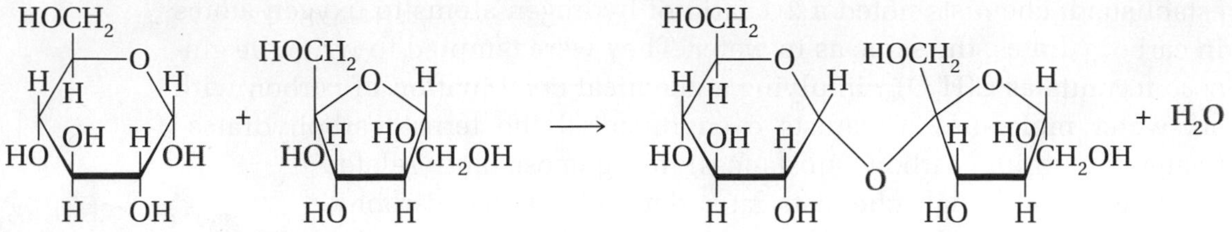

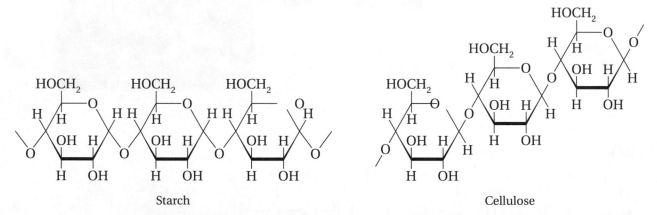

Figure 7.18 *The formation of sucrose. Note that particular —OH groups react (shown in red), resulting in elimination of one H_2O molecule.*

The reaction that forms disaccharides—a condensation reaction—can also cause monosaccharides to form polymers. Such polymers, not surprisingly, are called **polysaccharides** (see Figure 7.19). Starch, which is a major component of grains and many vegetables, is a polysaccharide composed of glucose units. Cellulose, which is the fibrous or woody material of plants and trees, is another polysaccharide formed from glucose.

All polysaccharides are polymers of monosaccharide molecules.

Starch

Cellulose

Figure 7.19 *Structural formulas for starch and cellulose, two polysaccharide molecules.*

The major types of carbohydrates are summarized in Table 7.2. Compare the bonds in starch with those in cellulose. Due to structural differences, starch is easily digested by the body (see Figure 7.20), whereas cellulose is indigestible by humans. Such indigestible types of carbohydrates are commonly called *fiber*.

The Composition of Common Carbohydrates

Table 7.2

Classification and Examples	Composition	Formula	Common Name or Source
Monosaccharides		$C_6H_{12}O_6$	
Glucose	–		Blood sugar
Fructose	–		Fruit sugar
Galactose	–		–
Disaccharides		$C_{12}H_{22}O_{11}$	
Sucrose	Fructose + glucose		Cane sugar
Lactose	Galactose + glucose		Milk sugar
Maltose	Glucose + glucose		Germinating seeds
Polysaccharides	Glucose polymers	$(C_6H_{10}O_5)_n$	
Starch			Plants
Glycogen			Animals
Cellulose			Plant fibers

Carbohydrates and fats are the primary high-energy substances in the human diet. One gram of a carbohydrate provides 4 Calories of food energy. Nutritionists recommend that about 45–65% of dietary Calories come from high-fiber carbohydrates. Worldwide, most people obtain carbohydrates by eating grains, such as rice, corn bread, wheat tortillas, bread, pasta, and beans. Other sources of carbohydrates include fruits, milk, and yogurt. Meats contain a small quantity of **glycogen**, the carbohydrate by which animals store glucose.

1 g carbohydrate = 4 Cal

Glycogen, similar in structure to plant starch, is used as a source of reserve energy by humans and animals.

Figure 7.20 *Breads and other foods contain starch, which is a digestible polysaccharide. Cellulose (found in wood, paper, and cotton) is a polysaccharide that humans cannot digest. What is a structural difference between digestible polysaccharides (such as starch) and cellulose? Which of these images show cellulose and which show starch?*

People in the United States tend to obtain more of their carbohydrates, on average, from wheat-based breads, potatoes, sugar-laden snacks, and desserts than is common in many other parts of the world. Currently, an average U.S. citizen consumes more than 82 kg (181 lb) of sugar annually, mainly from candy, desserts, and soft drinks.

B.2 FATS: ANOTHER WAY TO COMBINE C, H, AND O ATOMS

Unlike the terms *carbohydrate* and *protein*, *fat* has acquired its own general (and somewhat negative) meaning. From a chemical viewpoint, however, fats are just another major category of biomolecules with special characteristics and functions.

A significant part of a normal human diet, fats are present in meat, fish, poultry, oils, dairy products, nuts, and grains. When more food is consumed than is needed to satisfy energy requirements, much of the excess food energy is stored in the body as fat molecules. If food intake is not enough to meet the body's energy needs, then that stored fat is "burned" to release energy to make up the difference.

Fats are composed of carbon, hydrogen, and oxygen—the same three elements that compose carbohydrates. Fat molecules, however, contain fewer oxygen atoms and more carbon and hydrogen atoms. Thus, fats have a greater number of carbon–hydrogen bonds and fewer carbon–oxygen (and oxygen–hydrogen) bonds than do carbohydrate molecules. You can confirm this by comparing the structure of glyceryl tripalmitate, which is a typical fat molecule, shown in Figure 7.21, to that of glucose or starch molecules (Figure 7.17 on page 705 and Figure 7.19 on page 706, respectively).

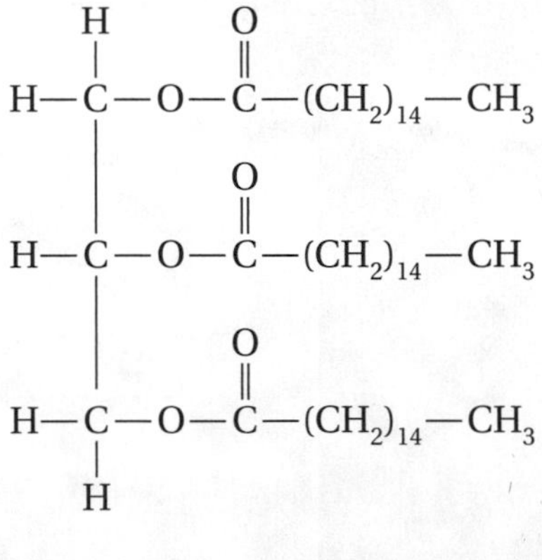

> When an organic substance is burned, it reacts with oxygen gas to form carbon dioxide, water, and thermal energy. Similarly, when fats, carbohydrates, and protein are "burned" in the body, they react with oxygen gas and are converted into carbon dioxide, water, and energy.

$$
\begin{array}{c}
\overset{\displaystyle H}{\underset{\displaystyle |}{|}}\overset{\displaystyle O}{\underset{\displaystyle \parallel}{}} \\
H-C-O-C-(CH_2)_{14}-CH_3 \\
| \overset{\displaystyle O}{\underset{\displaystyle \parallel}{}} \\
H-C-O-C-(CH_2)_{14}-CH_3 \\
| \overset{\displaystyle O}{\underset{\displaystyle \parallel}{}} \\
H-C-O-C-(CH_2)_{14}-CH_3 \\
| \\
H
\end{array}
$$

Figure 7.21 *The structural formula of glyceryl tripalmitate, a typical fat.*

Generally, fat molecules are nonpolar and only sparingly soluble in water. As you can see from Figure 7.21, fats have long hydrocarbon portions that prevent them from dissolving in polar solvents such as water. Their low water solubility and high energy-storing capacity give fat molecules, unlike carbohydrates, chemical properties similar to those of hydrocarbons.

The chemical properties of a fat molecule are due, in part, to the *fatty acid* groups in the molecule. **Fatty acids** are a class of organic compounds composed of long hydrocarbon chains with a carboxylic acid group (—COOH) at one end. Two fatty acids are shown in Figure 7.22.

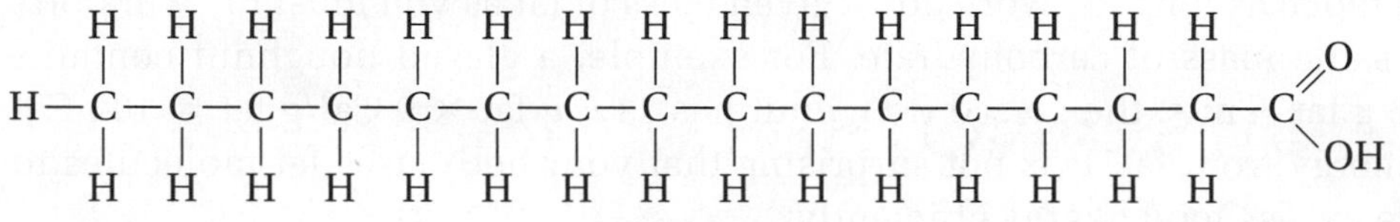

Palmitic acid

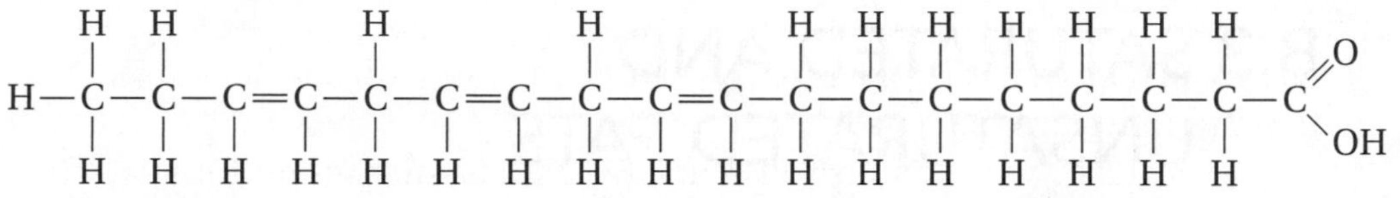

Linolenic acid

Figure 7.22 *Typical fatty acids.*

The typical fat molecule, a **triglyceride**, is a combination of a three-carbon molecule called *glycerol* and three fatty acid molecules.

| 1 Glycerol | + | 3 Fatty acid | ⟶ | 1 Triglyceride | + | 3 Water |
| molecule | | molecules | | (fat) molecule | | molecules |

In general, the terms "fat" and "triglyceride" are used interchangeably.

The formation of a triglyceride is shown in Figure 7.23. Each fatty acid molecule forms an ester linkage, as it reacts with an —OH group of glycerol. This produces one water molecule per ester linkage. The main product of this condensation reaction is a triglyceride composed of three ester groups. Although the three fatty acids depicted in Figure 7.23 are identical, fats can contain two or three different fatty acids.

The reaction that produces a fat molecule is similar to the reaction you observed in Unit 3 that produced an ester, methyl salicylate (page 319).

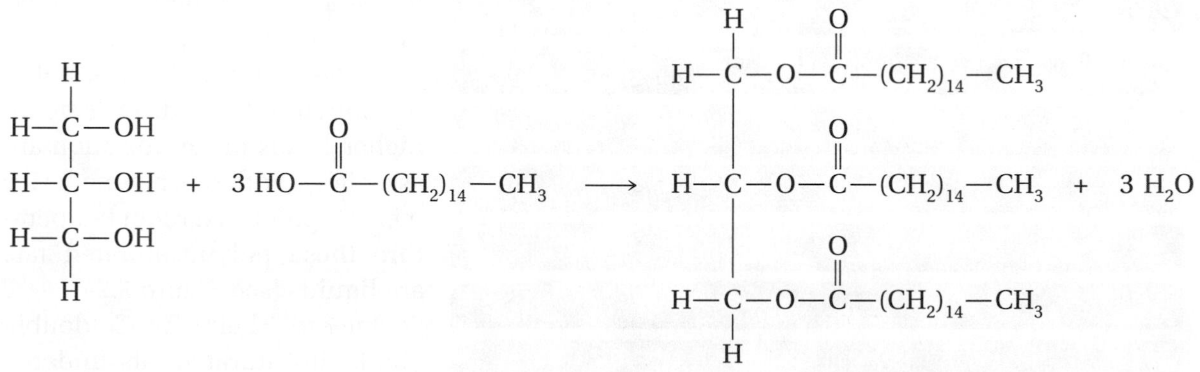

Glycerol   Palmitic acid   Glyceryl tripalmitate (a typical fat)   Water

Figure 7.23 *Formation of a typical fat molecule, a triglyceride. The colored atoms interact to form the fat molecule plus water.*

Because they have more carbon–hydrogen bonds than carbon–oxygen bonds, fats contain more stored energy per gram than do carbohydrates. In fact, one gram of fat contains over twice the energy stored in one gram of carbohydrate; 1 g fat is equivalent to 9 Cal, compared to 4 Cal for each gram of carbohydrate. Consequently, you must run more than twice as far or exercise twice as long to "work off" a given mass of fat as you must to "work off" the same mass of carbohydrate. For example, a glazed doughnut contains 11.6 g fat. Thus, the glazed doughnut has 11.6 g fat × 9 Cal/g fat = 104 Cal of energy from fat. It is not surprising that your body uses fat molecules to store excess food energy efficiently.

B.3 SATURATED AND UNSATURATED FATS

Recall from Unit 3 that hydrocarbons can be saturated (containing only single carbon–carbon bonds) or unsaturated (containing double or triple carbon–carbon bonds). Likewise, hydrocarbon chains in fatty acids are either saturated or unsaturated. Look again at the fatty acids in Figure 7.22 (page 709). Can you identify each as either saturated or unsaturated?

Fats containing saturated fatty acids are called **saturated fats**; those containing some unsaturated fatty acids are known as **unsaturated fats**. A **monounsaturated fat** contains just one carbon–carbon double bond in its fatty acid components. A **polyunsaturated fat** contains two or more C=C double bonds in the fatty acid portion of a triglyceride molecule. Based on these definitions, is the fat depicted in Figure 7.21 (page 708) saturated, monounsaturated, or polyunsaturated?

Triglycerides in animal fats are nearly all saturated and are solids at room temperature. However, fats from plant sources commonly are polyunsaturated or monounsaturated. In general, higher levels of unsaturation are associated with oils having lower melting points. At room temperature these polyunsaturated fats are liquids (see Figure 7.24).

Due to their C=C double bonds, unsaturated fats undergo addition reactions; saturated fats cannot undergo addition reactions. Thus, these two types of

> Oils are fats that are liquid at room temperature.

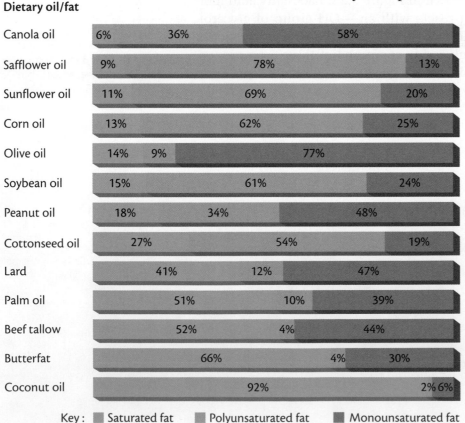

Dietary oil/fat

Canola oil	6%	36%	58%
Safflower oil	9%	78%	13%
Sunflower oil	11%	69%	20%
Corn oil	13%	62%	25%
Olive oil	14%	9%	77%
Soybean oil	15%	61%	24%
Peanut oil	18%	34%	48%
Cottonseed oil	27%	54%	19%
Lard	41%	12%	47%
Palm oil	51%	10%	39%
Beef tallow	52%	4%	44%
Butterfat	66%	4%	30%
Coconut oil	92%	2%	6%

Key : ■ Saturated fat   ■ Polyunsaturated fat   ■ Monounsaturated fat

Source : *Food Technology,* April 1989.

Figure 7.24 *The percent of polyunsaturated fat in fats from selected plant and animal sources.*

fats participate differently in body chemistry. Unsaturated fat molecules are much more chemically reactive. Increasing evidence suggests that saturated fats may contribute more to health problems than some unsaturated fats. Saturated fats are associated with formation of **arterial plaque**, which consists of fatty material that builds up in blood vessel walls (see Figure 7.25). The result is a condition commonly known as "hardening of the arteries," or **atherosclerosis**, which is a particular threat to coronary (heart) arteries and arteries leading to the brain.

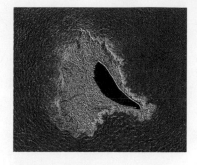

Figure 7.25 *An artery cross section, showing plaque (gray) on walls.*

DEVELOPING SKILLS

B.4 CALORIES FROM FAT

A sample of butter (Figure 7.26) provides the following nutritional values for one serving, which is defined as 1 tablespoon (tbsp) or 14 g:

Total fat	10.9 g	Polyunsaturated fat	0.4 g
Saturated fat	7.2 g	Calories	100

Figure 7.26 *What type of fat is primarily found in butter?*

> *Sample Problem: Calculate the percent of polyunsaturated fats contained in butterfat.*
>
> To find the answer, divide the mass of polyunsaturated fat by the mass of total fat and convert to a percent value:
>
> $$\frac{0.4\ g}{10.9\ g} \times 100\% = 4\%$$
>
> Thus, approximately 4% of butter's total fat is polyunsaturated.

1. Calculate the percent saturated fat in the total fat in butter.
2. The saturated and polyunsaturated fat percent values don't total 100%.
   a. What does the "missing" percent value represent?
   b. How many grams of fat does this represent?
3. Calculate the total percent fat in one serving of butter.
4. The *Dietary Guidelines for Americans 2010* suggest that from 20–35% of total Calories in human diets should come from fats. A way to evaluate this guideline for an individual food is to compare the total Calories from fat to the total Calories delivered in one serving of that food:
   a. Determine the total Calories from fat in 1 tbsp of butter.
   b. Calculate the percent of total Calories obtained from butter fat.
5. One serving (14 g) of margarine contains 10 g fat and 90 Cal.
   a. Find the percent of Calories obtained from margarine fat.
   b. Compare the percent of total Calories from margarine fat to the value from butter fat.

> Most fats, according to the *Dietary Guidelines*, should come from polyunsaturated and monounsaturated fatty acids, supplied by fish, nuts, and legumes.

> Recall that 1 g fat = 9 Calories.

B.5 HYDROGENATION

Partial **hydrogenation** adds hydrogen atoms to some C=C bonds in vegetable oil triglyceride molecules. The reaction with hydrogen converts a C=C double bond to a C—C single bond. Because this decreases the total unsaturated (C=C) sites, the partially hydrogenated product becomes more saturated, and the original liquid oil becomes semisolid. The ability to change oil from a liquid to a solid, depending on the extent of hydrogenation, allows food manufacturers to control the consistency and softness of their food products. Such partially hydrogenated fats are used in margarine, vegetable shortening, deep-fried foods, and snack foods.

The reaction between hydrogen and C=C double bonds in a polyunsaturated fat is typical of how C=C bonds in most alkenes react. Generally, C=C bonds in alkenes can be converted to C—C single bonds when they react with a variety of substances, including hydrogen.

You learned in Unit 3 (page 311) that a double bond in an alkene prevents adjacent carbon atoms in the double bond from rotating around the bond axis. The double-bonded carbon atoms align as shown in Figure 7.27. This inflexible arrangement between carbon atoms creates the possibility of *cis-trans isomerism*.

Ethene

Figure 7.27 *The structural formula for ethene.*

In **cis-trans isomerism**, two identical functional groups can be in one of two different molecular positions. Both groups can be on the same side of the double bond (the *cis* isomer), or they can be across the double bond from the other (the *trans* isomer), as illustrated in Figure 7.28.

cis-2-pentene   *trans*-2-pentene

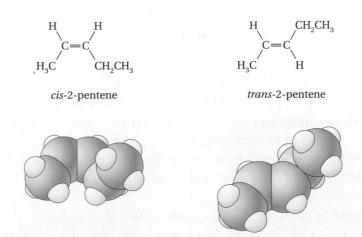

Figure 7.28 *A comparison of cis-2-pentene and trans-2-pentene. Note the positions of the functional groups.*

In the example of 2-pentene, each arrangement creates a different compound, even though the compounds have identical molecular formulas (C_5H_{10}). In *cis*-2-pentene, the molecule has the two hydrogen atoms on the same side (cis) of the double-bond plane (see Figure 7.28). In *trans*-2-pentene, the two hydrogen atoms are positioned on opposite sides (trans) of the double-bond plane.

Cis-trans isomerism, which was just described for an alkene, is also possible in unsaturated fats because fatty acid chains often contain C=C bonds. Unsaturated fatty acids in foods that have not been hydrogenated typically have their double bonds arranged in the cis arrangement, as shown in Figure 7.29. During hydrogenation, some cis double bonds break and reform in the trans arrangement. Although questions have arisen about nutritional safety of *trans*-fatty acids, studies addressing these concerns have not yet produced conclusive answers. However, the *Dietary Guidelines* advise that people should "keep *trans*-fatty acid consumption as low as possible."

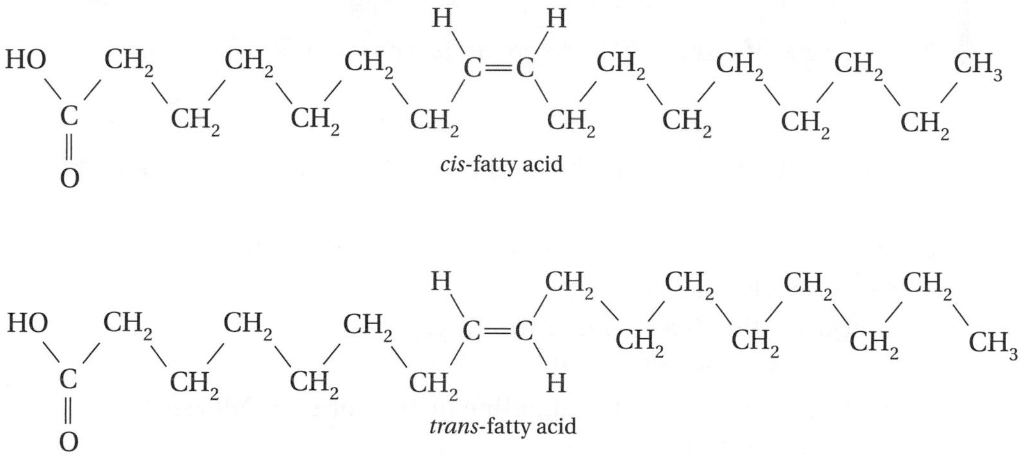

Figure 7.29 Cis- and trans-*fatty acids.*

Currently, most U.S. citizens obtain about 33% of their total food Calories from fats, within the upper limit of 35% recommended by *Dietary Guidelines*. Additionally, the *Guidelines* suggest that less than 7% of total food Calories should come from saturated fats.

High fat consumption is a factor in several modern health problems, including obesity and atherosclerosis. Most dietary fat consumed in the United States comes from processed meat, poultry, fish, and dairy products. Fast foods and deep-fried foods—such as hamburgers, french fries, fried chicken, and many snack items—add even more dietary fat. In addition, if your intake of food energy is higher than what you expend in physical activity, your body converts excess proteins and carbohydrates into fat for storage.

DEVELOPING SKILLS

B.6 FATS IN THE DIET

Your favorite ice-cream flavor just became available as frozen yogurt. The advertisement promotes frozen yogurt as a "reduced-fat alternative" to ice cream. Here are the nutritional data:

	Serving	Calories	Saturated Fat	Unsaturated Fat
Ice Cream	115 g	310	11 g	8 g
Frozen Yogurt	112 g	200	2.5 g	3.5 g

> **Sample Problem:** *Calculate the percent saturated fat by mass in a serving of ice cream.*
>
> To find the answer, divide the mass of saturated fat in a serving by the total mass of a serving and convert to a percent value:
>
> $$\frac{11\ g}{115\ g} \times 100\% = 9.6\%$$
>
> Thus, ~10% of a serving of ice cream is saturated fat.

> Recall that 1 g of fat contains 9 Cal of food energy.

1. Calculate and compare the total percent fat (by mass) contained in each dessert.

2. Another way to compare these desserts is to examine the percent of Calories from fat in each.

   a. Does either ice cream or frozen yogurt meet the guideline of 35% or less Calories from fat?

   b. Does either meet the guideline of 10% or less Calories from saturated fat?

CHEM**QUANDARY**

FAT-FREE FOOD?

You just compared ice cream to frozen yogurt in terms of their fat content. You might consider *fat-free* ice cream as another frozen dessert. Consider this ice cream label carefully. Is fat-free ice cream actually fat free? How do you know? What will happen to the food energy provided by fat-free ice cream if it is not immediately used by the body? Do you think it is wise to base a diet entirely on fat-free foods? Why or why not?

concept check 4

1. Describe structural similarities and differences between carbohydrate and fat molecules.
2. Explain why there are no cis or trans saturated fats.
3. Do all chemical reactions proceed until there are no reactants left? Explain.

B.7 LIMITING REACTANTS

Biochemical reactions in your body convert the fats, carbohydrates, and—under extreme conditions—proteins in foods you eat into energy. These biochemical reactions, like all chemical reactions, require the presence of a complete set of reactants to produce the desired product. Furthermore, the amount of product produced by a chemical reaction depends upon the amounts of reactants present.

Think about baking a cake and consider the following recipe:

2 cups flour	1 tablespoon baking powder
2 eggs	1 cup water
1 cup sugar	1/3 cup oil

The proper combination of these quantities (and some thermal energy) will produce one cake. What if you have 14 cups flour, 4 eggs, 9 cups sugar, 10 tablespoons baking powder, 10 cups water, and 3 1/3 cups oil. How many complete cakes can you bake?

Well, 14 cups flour is enough for 7 cakes (2 cups flour per cake). And there is enough sugar for 9 cakes (1 cup sugar per cake). The supplies of baking powder, water, and oil are sufficient for 10 cakes (confirm this with the recipe). However, it is not possible to make 10, 9, or even 7 cakes with available ingredients.

Why? Because only 4 eggs are available, which is enough for 2 cakes. The supply of eggs limits the number of cakes you can make. Excess quantities of other ingredients (flour, sugar, baking powder, water, and oil) remain unused. If you want more than two cakes, you will need more eggs.

In chemical terminology, the eggs in this cake-making analogy represent the *limiting reactant* (also called the *limiting reagent*). The limiting reactant is the starting material (reactant) that is used up entirely when a particular chemical reaction occurs. This starting material limits how much product can be formed.

The idea of limiting reactants applies equally well to living systems. A shortage of a key nutrient or reactant can severely affect the growth or health of plants and animals. In many biochemical processes, a product from one reaction becomes a reactant for other reactions. If a reaction stops because one substance is completely consumed (the limiting reactant), all reactions that follow it will also stop.

Recall that stoichiometry involves quantitative relationships among reactants and products in a reaction described by its chemical equation.

Limiting reactants were first introduced in Unit 3, page 343.

Fortunately, in some cases, alternate reaction pathways are available. If the body's glucose supply is depleted, for example, glucose metabolism cannot occur. One backup system oxidizes stored body fat in place of glucose. More drastically, under starvation conditions, structural proteins are broken down and used for energy. Producing glucose from protein is much less energy efficient than producing glucose from carbohydrates; if dietary glucose becomes available later, glucose metabolism reactions start up again.

Alternate reaction pathways are not a permanent solution. If the intake of a vital nutrient is consistently inadequate, that nutrient may become a limiting reactant in biochemical processes and affect personal health.

■ MODELING MATTER
B.8 LIMITING-REACTANT ANALOGIES

In Unit 2, you considered a container filled with super-bounce balls as an analogy for the kinetic molecular behavior of gases (see page 184). Another analogy, a cake recipe, introduced you to the idea of limiting reactants. Think about this and other limiting-reactant analogies (See Figure 7.30) as you answer the following questions.

1. Consider the cake-making analogy (page 715). Assume that you have 26 eggs and the quantities of all other ingredients as previously specified

   a. Which ingredient now limits the total cakes that you can make?

   b. How many total cakes can you make under the conditions specified?

   c. When the limiting reactant is fully consumed, how much of each other ingredient will be left over?

2. A restaurant prepares carryout lunch boxes. Each box consists of 1 sandwich, 3 cookies, 2 paper napkins, 1 milk carton, and 1 container. The current inventory is 60 sandwiches, 102 cookies, 38 napkins, 41 milk cartons, and 66 cardboard containers.

   a. As carryout lunch boxes are prepared, which item will be used up first?

   b. Which item is the limiting reactant?

   c. How many complete carryout lunch boxes can be assembled from this inventory?

3. Now consider the cellular respiration of glucose:

   $$C_6H_{12}O_6 + 6\,O_2 \longrightarrow 6\,CO_2 + 6\,H_2O + \text{thermal energy}$$

   a. Suppose that 2 glucose molecules are available for this reaction. How many oxygen molecules would be needed to completely react with these 2 glucose molecules?

b. Assume that 15 oxygen molecules are available to react. How many carbon dioxide and water molecules can be made from that amount of oxygen and unlimited glucose?
(*Hint:* Draw representations of all molecules available to react.)

4. Consider the cake-making and carryout lunch box analogies as they relate to chemical reactions.

a. Using pictures or symbols, write a balanced equation to represent each of these analogies. (*Note:* Your pictures should convey information comparable to that conveyed by a chemical equation.)

b. Describe one way in which these analogies help you to understand the notion of limiting reactants as applied to chemical reactions.

5. As you know, useful analogies also have limitations.

a. Consider the lunch-box analogy. Even if you run out of one item, can you use the other items to prepare carryout lunches that people would still buy and eat? Explain.

b. Consider the cake-making analogy. If you ran out of one ingredient, can you still use the other ingredients to make an edible cake? Explain.

c. Consider the glucose reaction. If one reactant is used up, or if you have less than the reaction requires (for instance, one glucose molecule but only three oxygen molecules), could any product be produced? Explain.

d. Considering your answers to Questions 5a through 5c, write a statement that summarizes a major limitation of both cake-making and lunch box analogies as they relate to actual chemical reactions.

6. Now it is your turn to devise a limiting-reactant analogy. Think about a limiting-reactant problem that you may encounter in everyday life.

a. Identify what item or product you are trying to create.

b. List all the reactants you will need.

c. Identify the limiting reactant and how many other reactants are left over when the reaction is completed.

Figure 7.30 *What key ingredients are needed to make this triple-stacker s'more? How could the "limiting reactant" concept apply here?*

B.9 LIMITING REACTANTS IN CHEMICAL REACTIONS

In chemical reactions, just as in baking, materials react in fixed ratios. These ratios are referred to as the *stoichiometry* of the reaction. The ratios are indicated in balanced chemical equations by the coefficients of each substance. Consider the equation for cellular respiration ("burning") of glucose:

$$C_6H_{12}O_6 \ + \ 6\,O_2 \ \longrightarrow \ 6\,CO_2 \ + \ 6\,H_2O \ + \ \text{thermal energy}$$

| 1 Glucose molecule | 6 Oxygen molecules | 6 Carbon dioxide molecules | 6 Water molecules |

Suppose you have 5 glucose molecules available to react with 60 oxygen molecules. Which substance will become the limiting reactant in this reaction? From the equation, you can see that 1 glucose molecule reacts with 6 oxygen molecules. That means that 5 glucose molecules would require 30 oxygen molecules to react completely. On the other hand, to use up all 60 oxygen molecules, 10 glucose molecules are required. Which of those two scenarios is actually possible in this reaction? In other words, which reactant—oxygen or glucose—will be used up completely in this reaction?

Because 60 oxygen molecules are available, all the glucose can react, with some oxygen left unreacted. Alternatively, to react completely, the 60 oxygen molecules would require 10 glucose molecules; however, only 5 glucose molecules are available to the reaction. Thus, glucose is the limiting reactant. Because the reaction stops after the 5 glucose molecules react with 30 oxygen molecules, 30 oxygen molecules will remain unreacted ("excess") at the end of the reaction.

A chemical equation can be interpreted not only in terms of molecules but also in terms of *moles* and *grams*. For example, the oxidation reaction for glucose can be interpreted in terms of moles of reactants and products:

$$C_6H_{12}O_6 \ + \ 6\,O_2 \ \longrightarrow \ 6\,CO_2 \ + \ 6\,H_2O \ + \ \text{thermal energy}$$

| 1 mol Glucose molecules | 6 mol Oxygen molecules | 6 mol Carbon dioxide molecules | 6 mol Water molecules |

Molar masses can be used to convert those molar amounts to grams:

$$C_6H_{12}O_6 \ + \ 6\,O_2 \ \longrightarrow \ 6\,CO_2 \ + \ 6\,H_2O \ + \ \text{thermal energy}$$

1 mol	6 mol	6 mol	6 mol
(1 × 180 g)	(6 × 32 g)	(6 × 44 g)	(6 × 18 g)
= 180 g	= 192 g	= 264 g	= 108 g

> The sum of reactant masses (180 g + 192 g) equals the sum of product masses (264 g + 108 g). Why must that be so?

This equation provides information about the stoichiometric relationships among the molecules that remain the same regardless of the reacting amounts involved. For example: If only 90.0 g (0.500 mol) glucose is oxidized (half the mass above), only 96.0 g (3.00 mol) O_2 is needed.

DEVELOPING SKILLS

B.10 USING CHEMICAL EQUATIONS TO FIND LIMITING REACTANTS

As illustrated by the following sample problems, chemical equations can help you identify the limiting reactant.

Sample Problem 1: *How much CO_2 can be produced if 100.0 g glucose reacts as completely as possible with 100.0 g O_2? Which is the limiting reactant—oxygen or glucose?*

First write the chemical equation:

$$C_6H_{12}O_6 + 6\,O_2 \longrightarrow 6\,CO_2 + 6\,H_2O + \text{thermal energy}$$

Next, calculate how much CO_2 could be produced if *each* reactant were fully consumed. The total amount of product formed depends on the amount of reactant that is used up first (the limiting reactant). When the limiting reactant is totally consumed, no more product can be produced. Calculating the amount of product that theoretically could be produced from each individual reactant makes it easy to identify the limiting reactant. That is, by comparing the results of these calculations, the *least* amount of product formed identifies the limiting reactant.

Consider how much CO_2 could be produced if glucose fully reacts (assume that sufficient oxygen is available):

$$100.0\ \text{g glucose} \times \frac{1\ \text{mol glucose}}{180.0\ \text{g glucose}} \times \frac{6\ \text{mol CO}_2}{1\ \text{mol glucose}} = 3.33\ \text{mol CO}_2$$

Determine how much CO_2 could be produced if the O_2 fully reacts (assume that sufficient glucose is available):

$$100.0\ \text{g O}_2 \times \frac{1\ \text{mol O}_2}{32.0\ \text{g O}_2} \times \frac{6\ \text{mol CO}_2}{6\ \text{mol O}_2} = 3.12\ \text{mol CO}_2$$

We find that 100.0 g O_2 produces less CO_2 than 100.0 g glucose would produce. That suggests that when 100.0 g O_2 combines with 100 g glucose, O_2 is the reactant that is used up first. When all the O_2 has reacted, the reaction stops, with some excess glucose still present.

Sample Problem 2: *How much glucose remains unreacted when the reaction just described finally stops?*

Here is one way to determine this. First find out how much glucose reacts with O_2:

$$100.0 \text{ g } O_2 \times \frac{1 \text{ mol } O_2}{32.0 \text{ g } O_2} \times \frac{1 \text{ mol glucose}}{6 \text{ mol } O_2} \times \frac{180 \text{ g glucose}}{1 \text{ mol glucose}} = 93.7 \text{ g glucose}$$

Then, find how much glucose remains unreacted:

(100 g glucose initially) − (93.7 g glucose reacted) = 6.3 g glucose left over.

Now it is your turn to check your understanding of this concept.

1. Lactose, milk sugar, is a disaccharide (see Table 7.2, page 707). In the human body, the enzyme lactase (see Figure 7.31) helps hydrolyze a lactose molecule into one molecule each of two different monosaccharides, glucose and galactose:

$$
\underset{\substack{\text{Lactose} \\ \text{342 g/mol}}}{C_{12}H_{22}O_{11}} + \underset{\substack{\text{Water} \\ \text{18 g/mol}}}{H_2O} \xrightarrow{\text{lactase}} \underset{\substack{\text{Glucose} \\ \text{180 g/mol}}}{C_6H_{12}O_6} + \underset{\substack{\text{Galactose} \\ \text{180 g/mol}}}{C_6H_{12}O_6}
$$

Suppose that you are investigating this reaction. During one trial, you combine 1.5 g lactose and 10.0 g water with a lactase enzyme.

a. Identify the limiting reactant in this trial.

b. How many moles of glucose will this reaction produce?

c. How many grams of glucose will be produced?

d. After the reaction stops, which reactant will be left over; that is, which reactant was in excess?

e. How many moles of that substance will be left unreacted?

2. Imagine you are also investigating how the body digests fat. The process starts with hydrolyzing the triglyceride. Each molecule reacts with three molecules of water to form three molecules of fatty acid and one molecule of glycerol:

$$
\underset{\substack{\text{Glyceryl tripalmitate} \\ \text{806 g/mol}}}{C_{51}H_{98}O_6} + \underset{\substack{\text{Water} \\ \text{18 g/mol}}}{3 \text{ } H_2O} \longrightarrow \underset{\substack{\text{Palmitic acid} \\ \text{256 g/mol}}}{3 \text{ } C_{16}H_{32}O_2} + \underset{\substack{\text{Glycerol} \\ \text{92 g/mol}}}{C_3H_8O_3}
$$

During one trial, you combine 4.0 g glyceryl tripalmitate with 1.0 g water.

a. Identify the limiting reactant in this trial.

b. How many moles of palmitic acid will this reaction produce?

Figure 7.31 *Some dairy products contain added lactase, an enzyme that promotes digestion of lactose, for "lactose-intolerant" individuals.*

   c. What mass (in grams) of palmitic acid will be produced?

   d. After the reaction stops, which reactant will be left over; that is, which reactant was initially in excess?

   e. How many moles of that substance will remain unreacted?

◼ MAKING DECISIONS

B.11 ANALYZING FATS AND CARBOHYDRATES

In Section A, you analyzed the energy content of a three-day diet. Using information in provided in this section, you can now identify the main sources of this energy.

1. Using appropriate resources, determine the mass of carbohydrates (in grams) contained in each item on the three-day food inventory. Then, calculate the percent of total Calories provided by carbohydrates. (*Note:* You calculated average Calories per day on page 700.) Report both sets of data in a suitable summary table.

2. If possible, identify the mass (in grams) of fiber in each item, as well. Record these data.

3. Using appropriate resources, determine the mass of fat (in grams) in each item in the inventory. If possible, identify the total mass of saturated and unsaturated fat in each item. Record these data.

4. Calculate the average mass of fat supplied daily by the three-day diet described in the food inventory.

5. a. Calculate the average daily energy (in Calories) supplied by fat.

   b. Based on the value calculated in Question 5a, is food energy from fat less than 30% of the total food energy supplied?

6. Identify possible ways to reduce the quantity of fat in the food inventory you are analyzing.

7. Now think of two possible Riverwood High School vending machine items that are *not* on your inventory.

   a. Find the total mass of carbohydrates (in grams) in each item. If possible, also report the total mass (in grams) of fiber in each item, as well. Record these data.

   b. Determine the mass of fat (in grams) in each item. If possible, report the total mass of saturated and unsaturated fat in each item. Record these data.

   c. Is the food energy from fat less than 30% of the total energy supplied by each item?

SECTION B SUMMARY

Reviewing the Concepts

> Carbohydrates, composed of carbon, hydrogen, and oxygen, include sugars, starches, and cellulose.

1. What is the origin of the term *carbohydrate*?

2. Using a shaded oval to represent a single monosaccharide molecule, sketch a model of a
   a. monosaccharide.
   b. disaccharide.
   c. polysaccharide.

3. Name two common monosaccharides and two common disaccharides.

4. All sugars are carbohydrates, but not all carbohydrates are sugars. Explain.

5. What kind of chemical reaction causes monosaccharides to link and produce more complex carbohydrates?

6. Why is starch classified as a polysaccharide?

7. How many Calories would be provided by 25 g carbohydrate?

> Fats and carbohydrates are both composed of carbon, hydrogen, and oxygen atoms, but they differ in structure and function.

8. What are fatty acids?

9. Consider the reaction that forms a typical fat molecule.
   a. List the substances that are combined in this reaction.
   b. How would you classify this reaction?
   c. Name the new functional group formed in this reaction.

10. Fats are examples of triglycerides. Why does that name provide an appropriate description of a fat molecule?

11. How many Calories would be provided by 10.6 g of olive oil? Olive oil is 100% fat.

12. What aspect of their structure makes fats sparingly soluble in water?

> Fats may be saturated or unsaturated. Unsaturated fats can be made more highly saturated through hydrogenation.

13. Define, in terms of their carbon—carbon bonds, the terms *saturated* fat and *unsaturated* fat.

14. Distinguish between monounsaturated and polyunsaturated fats.

15. With what is a saturated fat actually saturated?

16. How does the degree of saturation affect a fat molecule's properties?

17. The ingredient labels on some brands of margarine state that the oil is partially hydrogenated.
    a. What property does the product possess as a result of partial hydrogenation?
    b. Why do you think a manufacturer might decide against completely hydrogenating margarine?

18. What is the difference between cis and trans isomers?

19. Why do cis and trans isomers occur with a C=C double bond and not with a C—C single bond?

20. What are some natural sources of
    a. saturated fats?
    b. unsaturated fats?

21. Draw structural formulas for cis and trans isomers of 2-butene, $CH_3—CH{=}CH—CH_3$.

22. Draw a structural formula of a fat molecule containing both *cis*- and *trans*-fatty acids.

> Differences in structure between fat and carbohydrate molecules account for their different properties and energy content.

23. How are functional groups in fat molecules different from functional groups in carbohydrates? Use structural formulas to clarify your explanations.

24. From a chemical standpoint, explain why fat molecules contain more food energy per gram than do carbohydrate molecules.

25. List two chemical characteristics shared by all

    a. fats.   b. carbohydrates.

26. What property of fats makes them good energy-storage molecules?

> A chemical reaction involves substances interacting in specific, stoichiometric relationships. The limiting reactant determines how much product can be produced.

27. Is it ever possible to use up *all* reactants available for a chemical reaction? Explain.

28. How can a *limiting reactant* be identified?

29. Suppose you want to make a batch of s'mores. The "equation" for producing one s'more is as follows:

    2 graham crackers + 1 marshmallow
      + 3 chocolate pieces ⟶ 1 s'more

    If you have 12 graham crackers, 25 marshmallows, and 12 chocolate pieces:

    a. How many s'mores can you make?

    b. Which is the limiting "reactant"?

    c. How much of each excess reactant will be left over?

30. Consider the reaction of hydrogen gas and oxygen gas to produce water:

$$2\,H_2(g) + O_2(g) \longrightarrow 2\,H_2O(l)$$

    a. If 4.0 mol hydrogen gas are combined with 4.0 mol oxygen gas and the reaction described occurs, how many moles of water could be produced?

    b. What is the limiting reactant in this reaction?

    c. How much of the excess reactant will be left over?

31. One step in an early soap-making method was to allow fireplace wood ashes (containing potash, K_2O) to react with water, producing a highly basic potassium hydroxide (KOH) solution. The equation is as follows:

$$K_2O(s) + H_2O(l) \longrightarrow 2\,KOH(aq)$$

Assume that a 5.4-g sample of K_2O reacts as completely as possible with 9.0 g water.

    a. What will be the limiting reactant?

    b. What mass (in grams) of potassium hydroxide will be produced?

What role does molecular structure play in metabolism of carbohydrates and fats?

In this section, you have learned about the chemistry of nutrition and how your body uses the food you consume to sustain life. You have explored the chemical structures and properties of carbohydrate and fat molecules and the difference in energy available from the metabolism of these molecules. Think about what you have learned, then answer the question in your own words in organized paragraphs. Your answer should demonstrate your understanding of the key ideas in this section.

Be sure to consider the following in your response: structures of simple and complex carbohydrates; chemical composition and structures of fat molecules; the nutritional value of carbohydrates and fats; nutritional guidelines; and limiting reactants.

Connecting the Concepts

32. Why would it be unhelpful to label a container "sugar" in a chemical storeroom?

33. Explain how a relatively small number of subunits (fatty acids and monosaccharides) can create so many different kinds of fats and carbohydrates.

34. How do cis-trans isomers compare to the structural isomers you have learned about? (Recall that isomers were discussed in Unit 3, page 290.)

35. Polysaccharides are examples of naturally occurring polymers.
    a. How are they similar to synthetic polymers?
    b. How are they different from synthetic polymers?

36. Compare the chemical properties of fats to the chemical properties of alkanes.

37. Why is the term "burning fat" sometimes used to describe exercising, even though no actual burning is involved?

38. How are chemical equations similar to recipes used for cooking? How are they different?

39. In thinking about the reaction $HCl + NaOH \longrightarrow NaCl + H_2O$, one student concluded that 1.0 g HCl should react completely with 1.0 g NaOH. Explain why this reasoning is incorrect.

40. Consider the reaction described in Question 30. Specify a pair of starting masses (in grams) for hydrogen gas and oxygen gas so that *both* reactants would completely react in the chemical change.

Extending the Concepts

41. Research three different triglycerides and list the fatty acids in each.

42. Examine the ingredient listing on several food packages in your home. Identify the names of any specific carbohydrates or fats that you recognize.

43. Human consumption of *trans*-fatty acids may be associated with health risks. Investigate and summarize current research regarding this nutritional concern.

44. Fat-like products such as Olestra were designed and marketed as fat replacements in some foods. Investigate the chemistry of these fat substitutes and relate that information to health risks and benefits.

45. Assume that you currently consume 3000 Cal of food energy daily and want to lose 30 lb of body fat within two months (60 days).

   a. If you decide to lose that weight only by reducing your intake of food energy (with no extra exercise), how many food Calories, on average, would you need to remove from your daily diet?

   b. How many food Calories could you still consume daily?

   c. Would the scheme described in Questions 45a and 45b, if followed, be a sensible way to lose weight? Why?

46. Some well-publicized diet programs use terms such as *net carbs*, *counting carbs*, and *glycemic index*. Research the meanings of these terms and analyze their dietary usefulness based on the chemistry that you have learned.

CHEMISTRY *AT WORK*
Q&A

Mark Dewis, Flavor Chemist at International Flavors and Fragrances in Union Beach, New Jersey

Imagine a juicy steak or a ripe peach. These foods are much more than vehicles for calories and nutrients to get into our bodies. They're also tasty foods that make eating enjoyable. It's the job of flavor chemists to make sure that the foods we eat—even those that are changed from their original form, such as beef soup or peach yogurt—are delicious and taste like what we'd expect. Part of flavor chemistry is developing the flavors that we perceive in foods and making these flavors blend with the other ingredients in a food, whether it's yogurt, a soda, or chewing gum. Read on to see how one chemist makes what we eat a tasty way to get what we need.

Q. What is flavor chemistry?

A. At a basic and practical level, flavor chemistry is about understanding why food in its natural state tastes so good. If you take a tomato or grilled steak or wine or butter—or any natural food that people have been consuming for hundreds of years—the flavor chemistry industry tries to replicate this natural food experience. We try to figure out how nature makes food taste and how we can mimic the natural taste in a product.

Q. How did you become a flavor chemist?

A. When I was a high school student, I learned what I was good at and interested in so I wouldn't have to struggle in school. I'm lucky that chemistry happened to be one of those things. I majored in chemistry in college and went on to get a doctorate in synthetic organic chemistry. By the time I graduated, I had several options for careers, including pharmaceuticals, agricultural chemistry, and food and flavor chemistry. The last one was particularly fascinating—I found it compelling that you can do science to re-create food or improve taste.

Q. What are flavors, and where do they come from?

Umami comes from a Japanese word for flavorful. The taste of mushrooms is umami—not quite salty, sour, sweet, or bitter.

A. What the flavor industry and consumers call flavor is a combination of smell and taste. One aspect is aroma, which is kind of like an edible fragrance. We usually think of taste as one of the five basic tastes: bitter, salty, sour, sweet, and umami (or savory). We re-create a lot of flavors from natural sources, such as essential oils, extracts, and fermentation or enzymatic products from real food. We don't need all the flavor compounds in a food to re-create its flavor. For example, coffee has more than 500 flavor compounds, but we can re-create good coffee flavor from just a few of these. Some flavor compounds are made completely in the lab—the artificial flavors you might have seen on food packaging. Most ingredients in a flavor do come from natural sources, although if one ingredient is synthetic, the whole flavor is called artificial.

Q. What considerations go into making a new flavor?

A. First, you have to know what your target is because the same flavor may not work in different kinds of food and drinks. For example, the lemon-lime flavor that works in candy might not work in soda. A lemon-lime candy might be flavored with lemon and lime oils, which you isolate by pricking the fruits' skins. These are 90 percent limonene. Consumers are expecting a lemon-lime soda to be clear, so you can't use limonene because it's not soluble. To make a flavor realistic, you have to mix and match different flavor chemicals to give the product the flavor you want. For a strawberry flavor, is it a fresh strawberry or a candy strawberry? Different chemical combinations will give a different result.

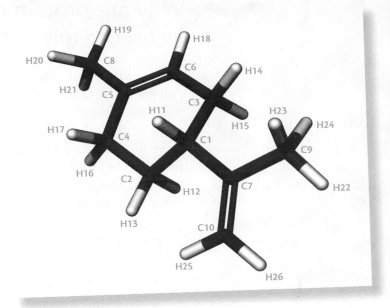

The limonene molecule gives oranges, lemons, and limes their fragrance. Limonene is a kind of hydrocarbon called a terpene.

Q. What advice do you have for students interested in flavor chemistry?

A. Take lots of chemistry classes, especially organic chemistry, because most flavor compounds tend to be organic compounds. It's important to take biology classes, too, including microbiology, biochemistry, and enzymology. You can get some good hands-on experience by getting internships in product-development labs at big food companies that make many different products. It gives you some insight into how flavors work in real life—for example, the different compounds that go into making apple-cinnamon flavors for cereal versus apple pie. You can also get some valuable experience right now. Flavor chemists taste each component that goes into a flavor to learn the set of ingredients they work with, and you can learn quite a bit by tasting different products that are supposed to have the same flavor. For example, chicken soup and cooked chicken, or chocolate and a chocolate milkshake. Once you start paying attention, you'll be surprised how different the "same" flavors taste.

A woman in Madagascar pollinates vanilla flowers by hand. The pods will be harvested and turned into a common flavor essence—vanilla extract.

C PROTEINS

Why are protein molecules essential to living organisms?

So far in this unit you have explored how carbohydrates and fats that you eat deliver energy that helps to keep you alive. The chemical bonds that store energy are broken, and energy is released in forming new, more strongly bonded molecules. In this section, you will investigate roles that *proteins* play in maintaining your well-being. For instance, proteins control the release of energy stored in foods and provide structure and support to maintain your cells and organs; in short, helping sustain life itself.

GOALS

- Describe the general structure of a protein molecule.
- Describe how amino acids in food are linked by peptide bonds to form proteins within the body.
- Describe how some protein molecules (enzymes) serve as catalysts.
- Explain why complete or complementary proteins are an essential part of the human diet.

concept check 5

1. Fats and carbohydrates are both made of carbon, hydrogen, and oxygen atoms. Why are the properties of fats and carbohydrates so different?
2. According to the collision theory for chemical reaction rates, what must happen for a chemical reaction to occur?
3. Could proteins be considered polymers? Explain.

C.1 PROTEINS—FUNCTION AND STRUCTURE

Whenever you look at a living creature, most of what you see is protein—skin, hair, feathers (Figure 7.32), eyeballs, fingernails, and claws. **Protein** also is a major structural component of human tissue. Inside your body, muscles, cartilage, tendons, and ligaments are composed of protein.

> The word *protein* is based on the Greek *proteios*, which means "of prime importance."

Figure 7.32 *This hummingbird feather is composed of many different proteins.*

At the cellular level, proteins help make it possible to transport materials into and out of cells. Your immune system depends on the ability of protein molecules to identify foreign substances. The rates of many chemical reactions that your cells require would be too slow were it not for special proteins called **enzymes**. Enzymes are specialized molecules that act as catalysts for biochemical reactions. Many proteins act as enzymes in the body. Your body contains tens of thousands of different proteins. Table 7.3 lists some major roles that proteins play in the human body.

Table 7.3

Major Groups of Proteins in the Human Body		
Type	**Function**	**Examples**
Structural proteins		
Muscle	Contraction, movement	Myosin
Connective tissue	Support, protection	Collagen, keratin
Chromosomal proteins	Part of chromosome structure	Histones
Membranes	Control influx and outflow, communication	Pore proteins, receptors
Transport proteins	Carrying of needed substances (such as O_2)	Hemoglobin
Regulatory proteins		
Fluid balance	Maintain pH, water, and salt content of body fluid	Serum albumin
Enzymes	Control metabolism	Proteases
Hormones	Regulate body functions	Insulin
Protective proteins	Attack foreign bodies	Antibodies (gamma globulin)

Proteins are polymers built from smaller molecules called **amino acids**. As Figure 7.33 shows, amino acid molecules contain carbon, hydrogen, oxygen, and nitrogen; a few, like cysteine, also contain sulfur. Just as monosaccharide molecules serve as building blocks for more complex carbohydrates, 20 different amino acids serve as the structural units of all proteins.

All amino acids have several structural features in common, but each amino acid is a unique molecule; four amino acids are shown in Figure 7.33. Every amino acid has two functional groups, the amino group ($-NH_2$) and the carboxylic acid group ($-COOH$).

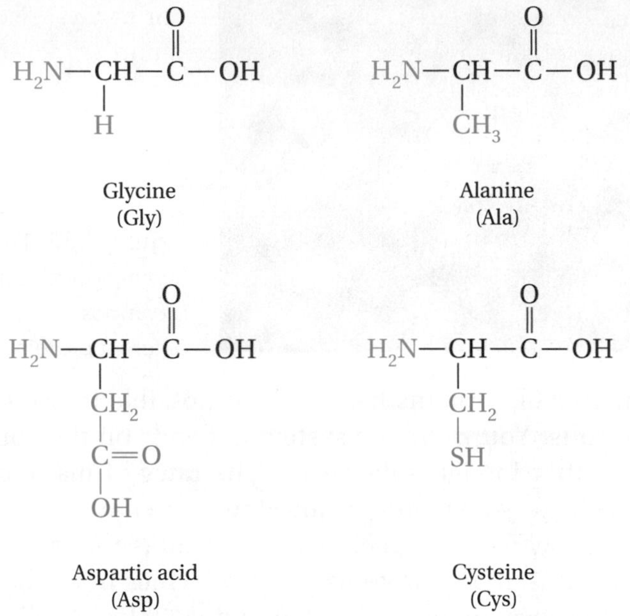

$$H_2N-CH-C-OH \quad\quad\quad H_2N-CH-C-OH$$

Glycine
(Gly)

Alanine
(Ala)

$$H_2N-CH-C-OH \quad\quad\quad H_2N-CH-C-OH$$

Aspartic acid
(Asp)

Cysteine
(Cys)

Figure 7.33 *Structural formulas of glycine, alanine, aspartic acid, and cysteine. The amino groups are highlighted in blue; carboxylic acid groups are in red; and green highlights unique side groups that distinguish amino acids from each other.*

Even though hundreds of amino acids have been identified in nature, the human body uses only 20 amino acids.

Like starch, nylon, and polyester, proteins are condensation polymers (see pages 321–322). The combination of two amino acid molecules with loss of one water molecule, as illustrated in Figure 7.34, is a typical condensation reaction. The bonds linking amino acid units together are **peptide bonds**. Proteins are chains of amino acids that vary in length, from ten to several thousand amino acids. Just as the 26 letters of the alphabet combine in different ways to form hundreds of thousands of words, the 20 amino acids can combine in a nearly infinite number of ways to form different proteins that meet many needs of living organisms.

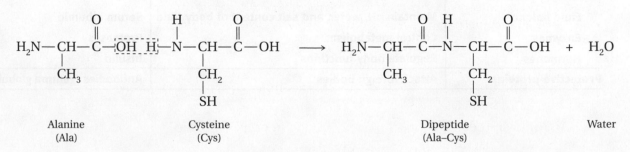

Alanine
(Ala)

Cysteine
(Cys)

Dipeptide
(Ala–Cys)

Water

Figure 7.34 *Formation of a dipeptide from two amino acids (Ala and Cys).*

■ MODELING MATTER

C.2 MOLECULAR STRUCTURE OF PROTEINS

Protein molecules differ from one another in the number and types of amino acids they contain and in the sequence in which the amino acids are bonded (see Figure 7.35). However, the way in which amino acids bond to one another to form these long polymers is the same for every protein molecule. You will now investigate how amino acids bond and how the sequence of amino acids affects a protein's structure.

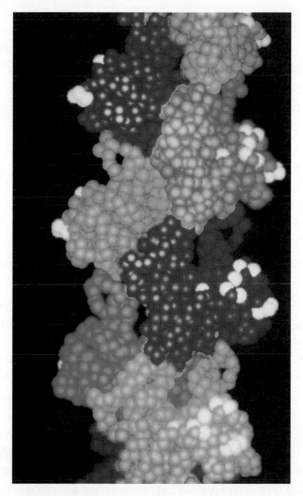

Figure 7.35 *Computer-generated space-filling model depicting a polymer of the protein actin. For clarity, each monomer (the repeated polymer unit) is shown in a different color.*

1. Draw structural formulas for glycine and alanine on a sheet of paper. Refer to Figure 7.33.

   a. Circle and identify the functional groups in each molecule.

   b. How are the two molecules alike?

   c. How do the two molecules differ?

2. Proteins are polymers of amino acids. Examine the equations in Figures 7.34 (page 730) and Figure 7.36 to see how two amino acids join. Notice that the amino group on one amino acid is bonded to the carboxylic acid group on another. This linkage, a peptide bond, is shown here:

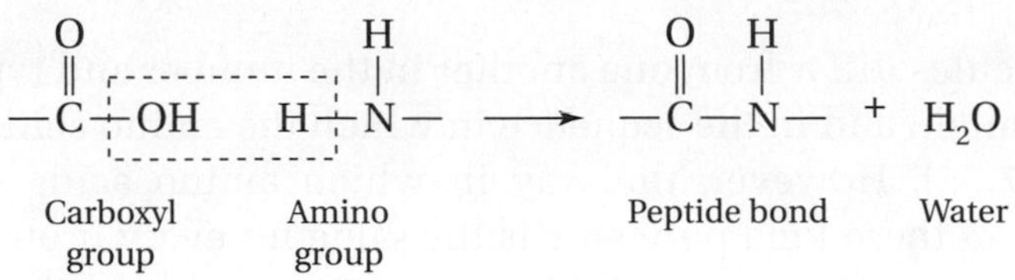

$$\underset{\substack{\text{Carboxyl} \\ \text{group}}}{-\overset{\overset{\displaystyle O}{\|}}{C}\!\!-\!\!OH} \quad \underset{\substack{\text{Amino} \\ \text{group}}}{H\!\!-\!\!\overset{\overset{\displaystyle H}{|}}{N}\!\!-} \quad \longrightarrow \quad \underset{\text{Peptide bond}}{-\overset{\overset{\displaystyle O}{\|}}{C}\!\!-\!\!\overset{\overset{\displaystyle H}{|}}{N}\!\!-} \quad + \quad \underset{\text{Water}}{H_2O}$$

Figure 7.36 *A carboxyl group combing with an amino group.*

When two amino acid units are linked by a peptide bond, the product is called a **dipeptide**. Because an amino acid contains at least one amino group and one carboxylic acid group, amino acids can form a peptide bond at either end.

3. Using structural formulas, write the equation for the reaction between two glycine molecules to form a dipeptide. Circle the peptide bond in the dipeptide.

4. Using structural formulas, write equations for possible reactions between a glycine molecule and an alanine molecule.

5. Examine structural formulas of the dipeptide products identified in Question 4. Note that each dipeptide still contains a reactive amino group and a reactive carboxylic acid group. These dipeptides can then react with other amino acids, forming even more peptide linkages.

6. Remembering that the carboxylic acid group of the first amino acid forms a peptide bond with the amino group of the second amino acid, explain why the amino acid sequence A-B-C is different from C-B-A.

7. Assuming that you have supplies of three different amino acid molecules—A, B, and C—and that each type of amino acid can be used only once, how many different tripeptides (three amino acids linked together) can be formed? Write all possible combinations.

8. Writing the sequence of amino acids by using their chemical names is a time-consuming and tedious process for most protein molecules. Consequently, chemists have devised three-letter (and even one-letter) abbreviations for each amino acid in proteins. For example, a tripeptide with a sequence of glycine–aspartic acid–cysteine is abbreviated Gly-Asp-Cys. Using three-letter abbreviations, write down all other possible sequences for these three amino acids.

9. How many tetrapeptides (four amino acids bonded together) could be formed from the four amino acids shown in Figure 7.33 (page 730)? Write the possible combinations by using their three-letter abbreviations.

10. The cells in your body build proteins from 20 different amino acids. If each protein can be as long as 10 000 amino acid units, would you estimate that the theoretical total of the different proteins that your cells could produce to be in the hundreds, thousands, or millions? Explain your answer.

11. How would a living organism benefit from its ability to synthesize so many different proteins? What potential problems do you envision?

C.3 PROTEIN IN YOUR DIET

What foods are the main protein sources in your diet? To which food groups shown in Figure 7.2 (page 687) do the protein-containing foods in Figure 7.37 belong?

When foods containing proteins reach your stomach and small intestine, peptide bonds break. The separated amino acid units then travel through the intestinal walls to the bloodstream, the liver, and then to the rest of the body. Individual cells can use these amino acids as building blocks for new proteins.

If you eat more protein than your body requires, or if your body needs to use protein because carbohydrates and fats are in short supply, amino acids are metabolized in the liver. Nitrogen atoms are removed and converted in the liver to urea (Figure 7.38), which is excreted through the kidneys in urine. The remaining portions of the amino acid molecules are either converted to glucose and oxidized, releasing 4 Cal of energy per gram, or stored as fat.

The human body can normally synthesize adequate supplies of 11 of the 20 required amino acids. The other nine, called **essential amino acids**, must be obtained from the diet. See Table 7.4 on pages 734–735.

Figure 7.37 *Foods, including plant sources, rich in protein.*

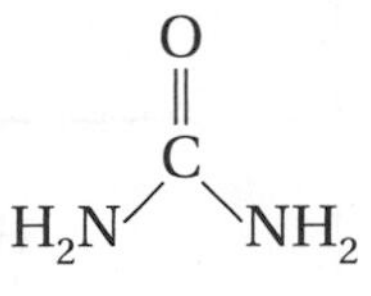

$$\begin{array}{c} O \\ \parallel \\ C \\ H_2N \diagup \quad \diagdown NH_2 \end{array}$$

Urea

Figure 7.38 *The structural formula for urea.*

Table 7.4

Amino Acids

Amino Acid	3-Letter Code	Structure	Amino Acid	3-Letter Code	Structure
Alanine	Ala		Glutamic acid	Glu	
Arginine	Arg		Glutamine	Gin	
Asparagine	Asn		Glycine	Gly	
Aspartic acid	Asp		*Histidine*	*His*	
Cysteine	Cys		*Isoleucine*	*Ile*	

NOTE: Amino acids in italics are essential.

Amino Acids

Amino Acid	3-Letter Code	Structure	Amino Acid	3-Letter Code	Structure
Leucine	*Leu*		Serine	Ser	
Lysine	*Lys*		*Threonine*	*Thr*	
Methionine	*Met*		*Tryptophan*	*Trp*	
Phenylalanine	*Phe*		Tyrosine	Tyr	
Proline	Pro		*Valine*	*Val*	

NOTE: Amino acids in italics are essential.

If an essential amino acid is in short supply in the diet, that amino acid can become a limiting reactant in building any protein containing that amino acid. When this happens, the only way the body can make that protein on its own is by breaking down one of its proteins containing that essential amino acid.

> Of course, the diet can be altered so that it includes that missing essential amino acid.

Any dietary source of protein that contains adequate amounts of all essential amino acids represents a source of **complete protein**. Most sources of animal protein contain all nine essential amino acids in quantities sufficient to meet dietary requirements. Plant protein and some sources of animal protein are incomplete, that is, they do not contain adequate amounts of all nine essential amino acids.

No single plant can provide adequate amounts of all essential amino acids, but certain combinations of plants can. These combinations of foods, which are said to contain **complementary proteins**, are part of many diets around the world (see Figure 7.39).

> Complementary proteins do not need to be consumed at every meal. However, sufficient complementary proteins must be consumed so that the daily protein intake is adequate.

Because the body cannot store amino acids, you must consume some every day. The recommended level of protein intake according to the *Dietary Guideline for Americans 2010* is 10–35% of total daily Calories consumed. The consumption of too much protein carries potential health risks. Excess protein causes stress on the two organs that metabolize amino acids: the liver and the kidneys. Excess protein also increases excretion of calcium ions (Ca^{2+}) that function in nerve transmission and help build bones and teeth. A diet with too much protein can even cause dehydration because more fluids are needed for the urinary excretion of urea. This problem is particularly relevant to athletes. Finally, consuming excess protein may lead to inadequate intake of other nutrients.

Figure 7.39 *Complementary proteins provide adequate amounts of all essential amino acids when consumed together in a single day.*

In 1997, the Food and Nutrition Board of the National Academy of Sciences created Dietary Reference Intakes (DRIs). The DRIs are quantitative estimates of nutrient intakes for planning and assessing the diets for healthy people. The DRIs for protein shown in Table 7.5, as well as for other nutrients not shown in the table, depend on a person's age and gender. DRIs are based on median heights and weights for a U.S. population of designated age and gender.

Food labels list nutritional information, as shown in Figure 7.40. Daily values are based on a daily intake of 2000 Calories and must be adjusted for an individual's body size, age, level of physical activity, and energy demands related to such exercise.

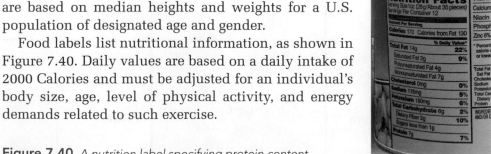

Figure 7.40 *A nutrition label specifying protein content.*

Table 7.5

Dietary Reference Intakes (DRIs) for Protein	
Life Stage Group	**RDA/AI* g/d**
Infants	
0–6 months	9.1*
7–12 months	11.0
Children	
1–3 years	13
4–8 years	19
Males	
9–13 years	34
14–18 years	52
19–70 years	56
> 70 years	56
Females	
9–13 years	34
14–70 years	46
> 70 years	46
Pregnant or Nursing	71

NOTE: Recommended Dietary Allowances (RDAs) are in **bold** type, Adequate Intakes (AIs) in ordinary type followed by an asterisk (*). RDAs and AIs may both be used as goals for individual intake. RDAs are set to meet the needs of almost all (97 to 98 percent) individuals in a group.

* Based on 1.5 g/kg/day for infants, 1.1 g/kg/day for 1–3 y, 0.95 g/kg/day for 4–13 y, 0.85 g/kg/day for 14–18 y, 0.8 g /kg/day for adults, and 1.1 g/kg/day for pregnant (using pre-pregnancy weight) and lactating women.

SOURCE: *Dietary Reference Intakes for Energy, Carbohydrate. Fiber, Fat, Fatty Acids, Cholesterol, Protein, and Amino Acids (2002/2005).* This report may be accessed via www.nap.edu.

DEVELOPING SKILLS

C.4 DAILY PROTEIN REQUIREMENTS

Sample Problem: What mass of protein should a woman over 70 consume each day, on average?

From Table 7.5, females > 70 years should consume an average of 46 g/d of protein.

Use information from Table 7.5 to answer the following questions:

1. What mass of protein should a person of your age and gender consume each day, on average?

2. For each kilogram of body mass, infants require more protein than adults do. Why should protein values per kilogram of body mass be highest for infants and become lower as a person ages (Figure 7.41)?

3. a. What food do infants consume that meets most of their relatively high protein needs?

    b. What evidence in Table 7.5 supports your answer? Explain.

Figure 7.41 *Why do an individual's age and developmental stage influence his or her protein needs?*

concept check 6

1. Amino acids have both an amino group and a carboxylic acid group. Why are both necessary for the formation of polypeptides?
2. Why does your body need a steady supply of amino acids from the protein in your diet?
3. Do catalysts exist within biological systems? What are they called and what is their function?

■ INVESTIGATING MATTER

C.5 ENZYMES

Preparing to Investigate

Think of a time you had to respond quickly, such as catching an object that fell unexpectedly or getting ready for school when you were late. How do your cells get the energy they need to respond quickly? You have already learned that food is "burned" to meet the body's continuous energy needs. The rate of "burning" can be adjusted from low to high, literally within a heartbeat.

What is the secret behind this impressive performance? It lies with biological catalysts called *enzymes*. Enzymes are able to speed up specific reactions without undergoing any lasting change themselves. In this investigation, you will monitor the rate of an enzyme-catalyzed reaction. In particular, you will investigate how the enzyme catalase affects the rate of decomposition of hydrogen peroxide (H_2O_2):

$$\text{2 } H_2O_2(aq) \xrightarrow{\text{catalase}} \text{2 } H_2O(l) \text{ + } O_2(g)$$

You will be assigned a particular food—apple, potato, or liver—to test for the presence of catalase. You will test a fresh piece of the food and one that has been boiled to investigate whether either material catalyzes the decomposition of hydrogen peroxide.

Making Predictions

1. What evidence will indicate that a gas is produced in the reactions?
2. Predict whether one or both pieces of the food, fresh or boiled, will contain catalase.

Gathering Evidence

1. Before you begin, put on your goggles and wear them properly throughout the investigation.

2. Obtain two pieces of your assigned food sample: one fresh and one boiled. (***Caution***: *Never taste or eat anything in the laboratory.*)

3. Label two 16 × 125-mm test tubes: one "fresh," the other "boiled."

4. Add 5 mL 3% hydrogen peroxide (H_2O_2) solution to each test tube. Do you see evidence that a reaction has taken place? Record your observations.

5. Add a portion of the fresh food sample to the appropriate test tube. Insert a stopper containing a segment of glass tubing into the mouth of the test tube, and arrange tubing as shown in Figure 7.42. Be sure the end of the glass tubing is submerged in the beaker of water.

6. Record the estimated number of bubbles formed per minute in the test tube for a total of 2 min.

7. Repeat Steps 5 and 6 with a sample of boiled food and the second test tube of hydrogen peroxide solution.

8. Discard and dispose of the samples and solutions as directed by your teacher.

9. Wash your hands thoroughly before you leave the laboratory.

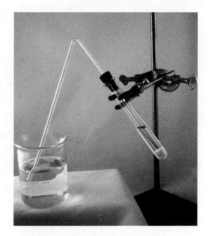

Figure 7.42 *Setup for enzyme catalysis investigation.*

Interpreting Evidence

1. Compare your experimental data with those of other class members who used

   a. the same material.   b. a different material.

2. What evidence for the presence of catalase do you see in your results?

Making Claims

3. What can you conclude about the enzyme's activity when comparing observations of the test with fresh material and boiled material? Cite evidence for your conclusion.

Reflecting on the Investigation

4. Why does commercial hydrogen peroxide contain preservatives?

5. Someone tells you that when hydrogen peroxide is put on a skin cut, the observed foaming shows that the cut is infected. Explain the actual chemical reason for this foaming.

If the same amount of original hydrogen peroxide solution were left undisturbed without adding any catalase-containing material, it would take days for the same extent of hydrogen peroxide decomposition to occur. How do enzymes work so well at catalyzing reactions? Read on.

C.6 HOW ENZYMES WORK

The speed of an enzyme-catalyzed reaction is hard to comprehend. In a single second, one molecule of the enzyme amylase helps to release 18 000 energy-rich glucose molecules from starch.

$$\text{-glucose-glucose-glucose-} + H_2O \xrightarrow{\text{amylase}} \text{-glucose-glucose-} + \text{glucose}$$

(portion of starch molecule) (remaining portion of starch molecule)

How do enzymes (or catalysts in general) work? According to the collision theory of reaction rates, chemical reactions can occur only when two reactant atoms or molecules collide with proper energy and orientation. When reactant molecules are large, the appropriate groups on the molecules that react must come together in the correct orientation. If molecules collide randomly, this is not likely to happen; thus the reaction rate will be very slow.

Catalysts cause reactions to occur more quickly by properly orienting reactant molecules. Reactions in living cells often involve very large molecules reacting at only one site, so catalysts are essential. Without enzymes or catalysts, many reactions would not occur at all (or at least not at reasonable rates at normal temperatures). Living systems manufacture a large variety of enzymes, each tailored to assist a particular cellular reaction (Figure 7.43).

In general, enzymes function as follows:

- A reactant molecule—known as the **substrate**—and the enzyme come together. The substrate molecule fits into the enzyme at an **active site**, where the substrate molecule's key functional groups are properly positioned. (Figure 7.44).

- The three-dimensional structure of the enzyme—how the amino acids in the protein polymer arrange themselves and interact—forms an active site. The shape of this site determines which substrate molecules can bind to the active site and what kind of reaction the enzyme is able to catalyze. The active site is not completely rigid and can change shape to accommodate different substrate molecules. Some enzymes have active sites that are fairly flexible and can accommodate a wide range of substrates. Other enzymes have rigid active sites, so only a few substrates can fit. This description of enzyme activity is known as the "induced-fit" model.

- The enzyme interacts with the substrate, weakening critical bonds and making the reaction more energetically favorable.

- The substrate is changed into a product or products as weakened bonds break and stronger ones form. One or more products then depart from the enzyme surface, freeing the enzyme to interact with other substrate molecules. Each enzyme molecule can participate in numerous reactions without any permanent change to its structure. Because of this regenerating (recycling), enzymes are needed in much smaller amounts than reactants.

The names of most enzymes include the suffix *-ase*. As with any catalyst, the enzyme name is included above the arrow in the equation representing the reaction. This shows that although the enzyme is involved in the reaction, it is not consumed by it.

The discussion of automobile catalytic converters on pages 211–213 describes how catalysts function; the reaction is speeded up because the catalyst lowers the activation-energy barrier.

Figure 7.43 *Without enzymes, this germinating avocado seed would be unable to develop into a mature plant. As the plant grows, it secretes enzymes that catalyze digestions of nutrients stored in the seed. The resulting smaller molecules are absorbed, thus promoting plant growth.*

Scientific American Conceptual Illustration

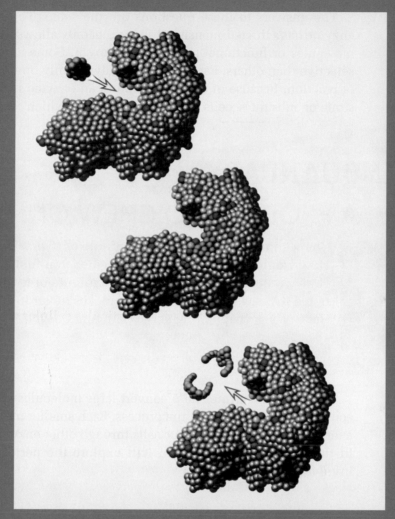

Figure 7.44 *An enzyme is a molecule produced by a living organism that catalyzes a specific chemical reaction. An active site within each enzyme accommodates (fits) only certain molecules, called substrates, depicted here in green. As the substrate binds to the enzyme's active site (upper left), the enzyme lowers the energy needed to break chemical bonds in the substrate and facilitates new bond formation. After a substrate is chemically altered, it detaches from the enzyme (lower right) and the enzyme is available to catalyze the transformation of more substrate molecules. Some enzymes can transform over 100 000 substrate molecules per second. The enzyme's entire three-dimensional structure maintains the shape and specificity of its active site. Thus, an enzyme's effectiveness depends upon careful control of surrounding factors that can alter its shape, such as pH and temperature. The shapes of most enzymes are significantly altered (denatured) under extreme conditions. In some cases, the enzyme becomes permanently denatured and is consequently unable to catalyze its specific reaction even if surrounding conditions later return to normal. While enzymes are not altered as they catalyze reactions, they are subject to degradation by certain reactions. Thus, they must be continually synthesized by cells that depend upon their role as biochemical catalysts.*

Each enzyme is as selective as it is fast. It catalyzes only certain reactions, even though substrates for many other reactions are also available. How does an enzyme "know" what to do? For example, why does amylase help to break starch molecules into thousands of glucose molecules instead of decomposing hydrogen peroxide molecules into water and oxygen gas?

The answers to these questions are the same: The active site of a given enzyme has a three-dimensional shape that only allows certain properly shaped molecules or functional groups to occupy it. Some enzymes are much more selective than others. Each enzyme catalyzes only one particular class or type of reaction. Because of the interaction of an enzyme's active site with its substrate or substrates, cells can precisely control which reactions take place.

CHEM**QUANDARY**

A PROBLEM TO CHEW ON

Chew an *unsalted* cracker for a minute or two before swallowing. How would you describe the taste when you first begin to chew? How would you describe the taste a minute or two later? If you noted a change, what causes the cracker flavor to change over time? What's happening at the particulate level?

Your body's enzymes help convert large molecules into many smaller molecules, releasing energy in the process. Each smaller molecule stores chemical energy that is released by your cells through other enzyme-catalyzed reactions. In the next investigation, you will explore the performance of an enzyme involved in this process.

■ INVESTIGATING MATTER

C.7 AMYLASE TESTS

Preparing to Investigate

In this investigation, you will explore how temperature and pH affect the performance of the enzyme amylase. Amylase contained in saliva helps to break down starch molecules into individual glucose molecules.

Glucose reacts with Benedict's reagent to produce a yellow-to-orange precipitate. The color and amount of precipitate formed is a direct indication of the glucose concentration generated by the enzyme-catalyzed reaction.

Your laboratory team will explore the performance of amylase either at room temperature or at a lower temperature. Before you begin, read *Gathering Evidence* and prepare a data table with appropriately labeled columns specifying data to be collected by each group member.

Making Predictions

1. Predict how temperature will affect the rate of the enzyme-catalyzed reaction.

2. Predict how pH will affect the rate of the enzyme-catalyzed reaction.

Gathering Evidence

Day 1. Preparing the Samples

1. Before you begin, put on your goggles, and wear them properly throughout the investigation.

2. Label five test tubes near the top with the temperature your group is assigned to investigate and with pH values of 2, 4, 7, 8, and 10. Also mark each label so that your test tubes can be distinguished from those of other groups (see Figure 7.45).

3. Using the pH solutions provided by your teacher, add 5 mL of pH 2, 4, 7, 8, and 10 solutions to the appropriate test tubes.

4. Add 2.5 mL of starch suspension to each tube.

5. Add 2.5 mL of 0.5% amylase solution to each tube.

6. Insert a stopper into each tube. Hold the stopper in place with your thumb or finger and shake each tube well for several seconds.

7. Leave the room-temperature test tubes in the laboratory overnight as directed by your teacher.

8. Give your teacher the test tubes that are to be refrigerated.

9. Wash your hands thoroughly before leaving the laboratory.

Figure 7.45 *Remember to label all test tubes, so that you can identify them later in the investigation.*

Day 2: Evaluating the Results

10. Before you begin, put on your goggles, and wear them properly throughout the investigation.

11. Prepare a hot-water bath by adding about 100 mL of tap water to a 250-mL beaker. Add a boiling chip. Warm the beaker on a hotplate. (Heat just to below boiling—hot, but not boiling.)

12. Add 5 mL of Benedict's reagent to each tube. Replace each stopper, being careful not to mix the stoppers. Hold the stopper in place with a thumb or finger and shake each tube well for several seconds.

13. Ensure that all test tubes are still clearly labeled. Remove the stoppers and place the test tubes into the hot-water bath.

14. Heat the test tubes in the hot-water bath until the solution in at least one tube has turned yellow or orange. Then continue heating for two to three more minutes.

15. Use tongs to remove the test tubes from the hot-water bath. Arrange them in a test-tube rack in order of increasing pH.

16. Observe and record the color of the contents of each tube.

17. Share your data with your classmates as directed by your teacher.

18. Wash your hands thoroughly before leaving the laboratory.

Benedict's reagent also provides a test for the presence of glucose in urine, a sympton of diabetes.

Interpreting Evidence

1. For each temperature, record the color of the solution in each test tube.

2. You know that Benedict's solution will react with glucose to form a yellow-orange precipitate. Based on your observations, rank the test tubes in order of least glucose present to most glucose present.

3. Using your ranking from Question 2, decide at which pH the amylase was the most effective. At which pH values (if any) was the amylase ineffective? Share your data with the class.

Making Claims

4. Make a scientific claim about how pH and temperature affect the ability of amylase to catalyze the decomposition of starch.

Reflecting on the Investigation

5. Many cells in the body only function properly within a certain pH range. For example, your blood is buffered to maintain a constant pH of about 7.4. Certain medical conditions can cause the pH of the blood to increase or decrease. Why do you think this is potentially dangerous?

6. What would be the effect on an individual if body temperature were to

   a. decrease significantly (as in hypothermia)?

   b. increase drastically (as with a severe fever)?

7. Do you think an enzyme that functions in the bloodstream would also function in the stomach? Explain.

Your cells synthesize many enzymes and other kinds of proteins to keep you alive. The amino acids used to synthesize these proteins are best obtained through a diet that provides appropriate amounts of protein. You will now decide whether a particular diet meets those protein needs.

■ MAKING DECISIONS

C.8 PROTEIN CONTENT

You have analyzed your three-day food inventory in terms of energy the diet provides and fat and carbohydrate molecules it delivers. Now consider whether the food provided supplies the recommended amounts of a key building block of living material—protein.

Use the food inventory to answer the following questions. Refer to Table 7.5 (page 737), if necessary.

1. What is the average total mass of protein (in grams) consumed daily?

2. What other information would you need to know regarding the person who follows this food-intake pattern to evaluate the appropriateness of protein supplied by these foods?

3. a. Does the person consume a good balance of essential amino acids? Explain your answer.

    b. What types of snack foods are high in protein (especially types that might be included in the vending machines at Riverwood High School)?

SECTION C SUMMARY

Reviewing the Concepts

> Proteins, major structural components of living creatures, fulfill many cellular roles.

1. Name three types of tissue in your body for which protein is the main structural component.

2. The name *protein* comes from a Greek word that means "of prime importance." Why is this name appropriate?

3. List five cellular functions where proteins are particularly important.

4. Name three food items composed primarily of protein.

5. Why are proteins considered polymers?

6. What chemical elements do proteins contain?

7. How many Calories would be provided by the metabolism of 3.6 g protein? (Recall that 1 g protein = 4 Cal.)

8. What is the chemical composition of an enzyme?

> Amino acids are the chemical subunits that make up proteins.

9. How does the relatively small number of different amino acids account for the vast variety of proteins found in nature?

10. What is a peptide bond? Use structural formulas to illustrate your answer.

11. How does the total number of amino acids vary within protein molecules?

12. What is the protein DRI value for

    a. a 4-month-old infant?

    b. a 36-year-old male of median height and weight?

13. Write structural formulas for the following molecules:

    a. A dipeptide of glycine and cysteine.

    b. A tripeptide abbreviated Asp-Ala-Cys.

14. Explain the meaning of the following terms:

    a. complete protein

    b. essential amino acid

    c. complementary proteins

15. On which two functional groups is the name "amino acid" based?

> Some protein molecules function as enzymes, that is, biological catalysts that speed up cellular reactions.

16. a. How are enzymes like other catalysts?

    b. How are enzymes different from other catalysts?

17. What would be the effect if all enzyme activity in the human body suddenly stopped? Explain.

18. Describe how high temperatures affect the ability of most enzymes to function.

19. Explain why enzymes can speed up only certain chemical reactions.

20. a. Explain the interaction of an active site and a substrate in terms of an induced-fit model.

    b. Describe at least one limitation to the induced-fit model for enzyme activity.

Why are protein molecules essential to living organisms?

In this section you have learned about the structure and function of proteins and the important role that proteins play in biochemical processes. Think about what you have learned and how these ideas are related, then answer the question in your own words in organized paragraphs. Your answer should demonstrate your understanding of the key ideas in this section.

Be sure to consider the following in your response: the structure of amino acids, peptide bonds, the role of proteins in the diet, and enzyme function.

Connecting the Concepts

21. a. In what ways are proteins similar to carbohydrates and fats?

    b. In what ways are proteins different?

22. If a steak is left on a barbecue grill too long, it turns black. What does that observation suggest about the chemical composition of protein in the meat?

23. Although many plants contain high levels of protein, vegetarians must be more concerned than non-vegetarians about including adequate protein in their diets.

    a. How can vegetarians ensure that they obtain all chemical building blocks needed to build required proteins?

    b. Explain your answer.

24. Using simple diagrams, including structural formulas, sketch how an enzyme might help form a peptide bond between two amino acids.

25. If a person followed a daily diet of 55 g fat, 75 g protein, and 85 g carbohydrate, how much energy would be provided through the metabolism of each? Show your calculations.

26. In Unit 4, you learned that fish species need a water pH range within which they can live. Given what you know about enzymes from the amylase tests in this section, explain why this is so.

27. Explain how the protein you actually eat—whether from beef, turkey, beans, nuts, or tofu—is transformed into human-body proteins.

Extending the Concepts

28. The genetic code in DNA carries blueprints for making proteins in the body. Explain how DNA helps determine the body's physical development and functioning.

29. The phrase "form follows function" is particularly applicable to enzymes. Explain how form and function are closely related for enzyme molecules.

30. Obtain information on the condition known as *ketosis*. What are its causes and effects? How can a high-protein diet lead to ketosis?

31. Explain why hydrogen peroxide is an effective antiseptic. How does your answer relate to enzyme action?

32. Research and report on chemical and physical properties of one or more amino acids used by the human body in building protein molecules.

33. What is a *zwitterion*? Under what conditions do amino acids become zwitterions?

SECTION D VITAMINS, MINERALS, AND ADDITIVES

What roles do vitamins, minerals, and additives play in foods we eat?

The focus of this unit so far has been on macronutrients in foods: proteins, carbohydrates, and fats. However, there are other substances in food, often in trace amounts: *vitamins, minerals,* and sometimes *additives.* Dietary guidelines imply that vitamins and minerals play vital roles within your body. These micronutrients occur naturally in many foods, and may also be added to foods, such as cereal or bread, to improve their quality as sources of various micronutrients. Food manufacturers use food additives for different reasons, as you will soon learn. What do vitamins and minerals do in your body? What purposes do food additives serve?

GOALS

- List key vitamins and describe their importance to the body.
- Explain why daily intake requirements of vitamins depend on the function and solubility properties of the vitamins as well as characteristics (such as age and gender) of individuals.
- List key minerals and explain their importance to health.
- List and describe main uses of food additives.
- Describe how titration and chromatography can be used to analyze food products.

✓ concept check 7

1. What are the primary roles of each of the following nutrients in a person's diet?
   a. carbohydrates
   b. fats
   c. proteins
2. a. Describe in your own words what an enzyme is and what it does.
   b. Then, speculate what a "co-enzyme" might be.
3. How does the mass of vitamins and minerals you need to consume each day compare to the mass of proteins, fats, and carbohydrates in your daily diet?

D.1 VITAMINS

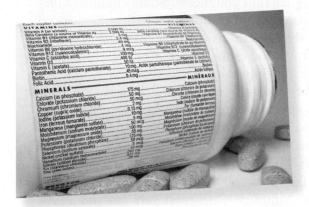

Figure 7.46 *Vitamin supplements are useful whenever sufficient vitamins are not provided in one's diet.*

Vitamins are biomolecules necessary for growth, reproduction, health, and life. Each vitamin is required in only a tiny amount. The total quantity of *all* vitamins required daily by an adult is only about 0.2 g; "a little goes a long way" with vitamins (see Figures 7.46 and 7.47).

How much is "enough"? That depends on age and gender, as is suggested by Table 7.6.

Figure 7.47 *Many foods such as leafy greens, liver, milk, eggs, and whole grains are rich in vitamins.*

Table 7.6

Age or Condition	Vit A (μg/d)	Vit D (μg/d)	Vit E (mg/d)	Vit K (μg/d)	Vit C (mg/d)	Vit B$_1$ (mg/d)	Vit B$_2$ (mg/d)	Vit B$_3$ (mg/d)	Vit B$_6$ (mg/d)	Vit B$_9$ (μg/d)	Vit B$_{12}$ (μg/d)
Males											
9–13 yrs	**600**	5*	**11**	60*	**45**	**0.9**	**0.9**	**12**	**1.0**	**300**	**1.8**
14–18 yrs	**900**	5*	**15**	75*	**75**	**1.2**	**1.3**	**16**	**1.3**	**400**	**2.4**
19–30 yrs	**900**	5*	**15**	120*	**90**	**1.2**	**1.3**	**16**	**1.3**	**400**	**2.4**
31–50 yrs	**900**	5*	**15**	120*	**90**	**1.2**	**1.3**	**16**	**1.3**	**400**	**2.4**
51–70 yrs	**900**	10*	**15**	120*	**90**	**1.2**	**1.3**	**16**	**1.7**	**400**	**2.4**
> 70 yrs	**900**	15*	**15**	120*	**90**	**1.2**	**1.3**	**16**	**1.7**	**400**	**2.4**
Females											
9–13 yrs	**600**	5*	**11**	60*	**45**	**0.9**	**0.9**	**12**	**1.0**	**300**	**1.8**
14–18 yrs	**700**	5*	**15**	75*	**65**	**1.0**	**1.0**	**14**	**1.2**	**400**	**2.4**
19–30 yrs	**700**	5*	**15**	90*	**75**	**1.1**	**1.1**	**14**	**1.3**	**400**	**2.4**
31–50 yrs	**700**	5*	**15**	90*	**75**	**1.1**	**1.1**	**14**	**1.3**	**400**	**2.4**
51–70 yrs	**700**	10*	**15**	90*	**75**	**1.1**	**1.1**	**14**	**1.5**	**400**	**2.4**
> 70 yrs	**700**	15*	**15**	90*	**75**	**1.1**	**1.1**	**14**	**1.5**	**400**	**2.4**
Pregnant	**770**	5*	**15**	90*	**85**	**1.4**	**1.4**	**18**	**1.9**	**600**	**2.6**
Nursing	**1300**	5*	**19**	90*	**120**	**1.6**	**1.6**	**17**	**2.0**	**500**	**2.8**

Dietary Reference Intakes (DRIs) for Selected Vitamins

Note: Recommended Dietary Allowances (RDAs) are in **bold** type. RDAs are established to meet the needs of almost all (97 to 98%) individuals in a group. Adequate Intakes (AIs) are followed by an asterisk (*). AIs are believed to cover the needs of all individuals in the group, but a lack of data prevent being able to specify with confidence the percent of individuals covered by this intake.

Source: Food and Nutrition Board, National Academy of Sciences: National Research Council, *Dietary Reference Intakes 2004.*

Vitamins perform very specialized tasks. Vitamin D, for example, helps move calcium ions from your intestines into the bloodstream. Without vitamin D, your body would not use much of the calcium you ingest. Some vitamins function as **coenzymes**, which are organic molecules that interact with enzymes and enhance their activity. For example, the B-vitamins act as coenzymes in releasing energy from food molecules. Figure 7.48 illustrates how a coenzyme functions.

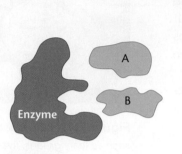

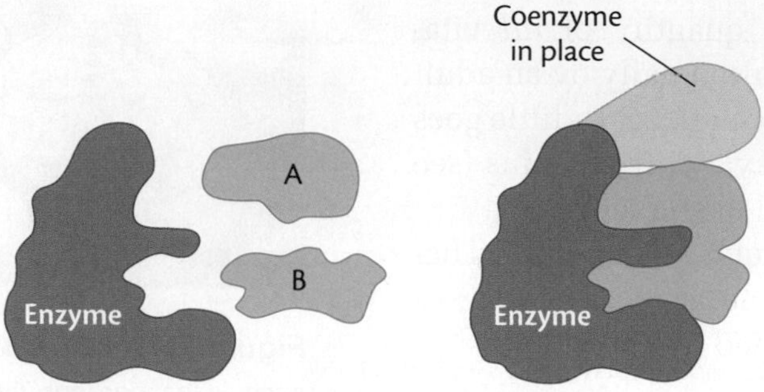

Figure 7.48 *A vitamin serving as a coenzyme.*

Long before the term *vitamin* was introduced early in the last century, people had discovered that small quantities of certain substances were necessary to maintain health. One example of vitamin deficiency is scurvy, once common among sailors; this condition is characterized by swollen joints, bleeding gums, and tender skin. Although early seafarers (in the 1700s and 1800s) did not know what caused scurvy, they commonly loaded citrus fruit on board, which they ate during long voyages to prevent scurvy. Scurvy is now known to be caused by vitamin C deficiency. Vitamin C is supplied by citrus fruit. In addition to vitamin C, about a dozen different vitamins have been identified over the past century, each critical to reactions occurring within the human body. Table 7.7 documents how some of those vitamins support human life.

Vitamins are classified as fat-soluble or water-soluble (see Table 7.7 and Figure 7.49). Water-soluble vitamins with polar functional groups pass directly into the bloodstream. They are not stored in the body; they must be ingested daily. Some water-soluble vitamins, including the B vitamins and vitamin C, are also destroyed by heat in cooking.

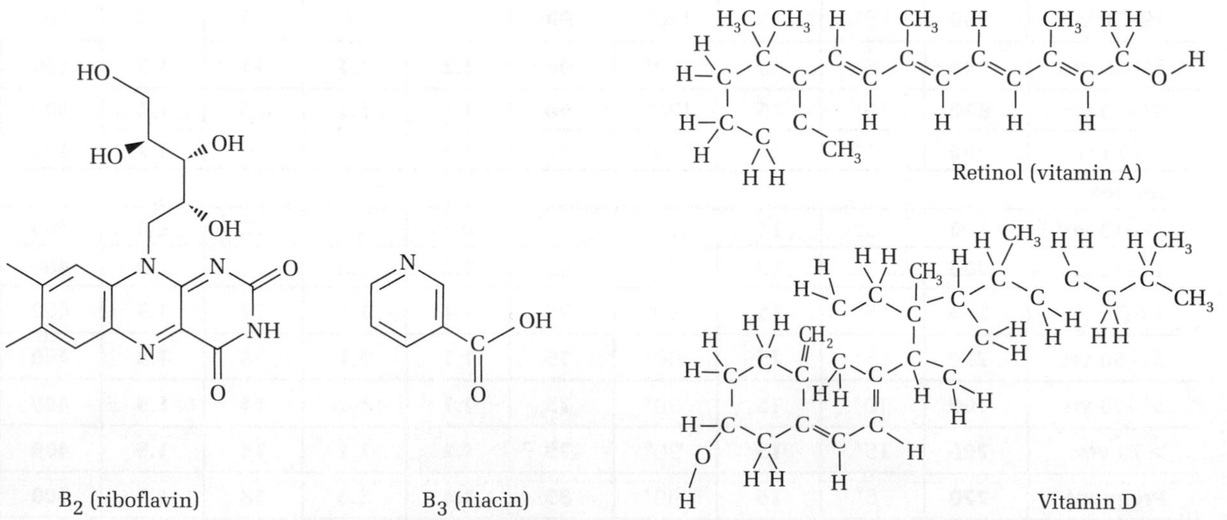

> Recall that "like dissolves like." See page 424.

Water-soluble vitamins:

B$_2$ (riboflavin)

B$_3$ (niacin)

Fat-soluble vitamins:

Retinol (vitamin A)

Vitamin D

Figure 7.49 *Structures of two water-soluble and two fat-soluble vitamins.*

Your body absorbs fat-soluble vitamins into the blood from the intestine with assistance from fats in the food you eat. Because the nonpolar structures of fat-soluble vitamins allow them to be stored in body fat, it is not necessary to consume fat-soluble vitamins daily. In fact, because fat-soluble vitamins accumulate within the body, they can build up to toxic levels if taken in excessively large quantities (megadoses).

Table 7.7

Vitamins by Category, Showing Sources and Deficiency Conditions

Vitamin (Name)	Main Sources	Deficiency Condition
Water-soluble		
B_1 (thiamine)	Liver, milk, pasta, bread, wheat germ, lima beans, nuts	Beriberi: nausea, severe exhaustion, paralysis
B_2 (riboflavin)	Red meat, milk, eggs, pasta, bread, beans, dark green vegetables, peas, mushrooms	Severe skin problems
B_3 (niacin)	Red meat, poultry, enriched or whole grains, beans, peas	Pellagra: weak muscles, no appetite, diarrhea, skin blotches
B_5 (pantothenic acid)	Liver, kidneys, yeast, egg yolk, broccoli, whole grains, yogurt, legumes, avocados, sweet potatoes	Anemia
B_6 (pyridoxine)	Muscle meats, liver, poultry, fish, whole grains	Depression, nausea, vomiting
B_7 (biotin)	Kidneys, liver, egg yolk, yeast, nuts	Dermatitis
B_9 (folic acid)	Kidneys, liver, leafy green vegetables, wheat germ, peas, beans	Anemia
B_{12} (cobalamin)	Red meat, liver, kidneys, fish, eggs, milk	Pernicious anemia, exhaustion
C (ascorbic acid)	Citrus fruits, melon, tomatoes, green peppers, strawberries	Scurvy: tender skin; weak, bleeding gums; swollen joints
Fat-soluble		
A (retinol)	Liver, eggs, butter, cheese, dark green and deep orange vegetables	Inflamed eye membranes, night blindness, scaling of skin, faulty teeth and bones
D (calciferol)	Fish-liver oils, fortified milk	Rickets: soft bones
E (tocopherol)	Liver, wheat germ, whole-grain cereals, margarine, vegetable oil, leafy green vegetables	Breakage of red blood cells in premature infants, oxidation of membranes
K (menaquinone)	Liver, cabbage, potatoes, peas, leafy green vegetables	Hemorrhage in newborns; anemia

DEVELOPING SKILLS

D.2 VITAMINS IN THE DIET

Sample Problem: Use Table 7.7 (page 751) to identify three water-soluble vitamins and one fat-soluble vitamin that are helpful in preventing anemia.

Referring to Table 7.7, water-soluble vitamins B_5, B_9, and B_{12}, and fat-soluble vitamin K have deficiency conditions that include forms of anemia.

1. Carefully planned vegetarian diets are nutritionally balanced. Individuals who follow a vegan diet do not consume any animal products, including eggs and milk. Because of their dietary limitations, vegans must ensure that they obtain the recommended daily allowances of two particular vitamins.

   a. Use Table 7.7 (page 751) to identify these two vitamins, and briefly describe the effect of their absence in the diet.

   b. How might individuals following a vegan diet avoid this problem?

2. Complete the following table about yourself, using data from Tables 7.6 (page 749) and 7.8.

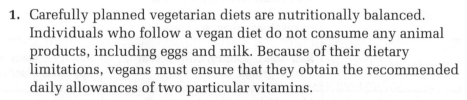

Vegetable (one-cup serving)	Your RDA		Total Serving to Supply Your RDA	
	B_1	C	B_1	C
Green peas				
Broccoli				

Figure 7.50 *How does the vitamin content of raw broccoli compare to that of steamed broccoli?*

   a. Would any of your entries change if you were of the opposite gender? If so, which entry or entries?

   b. Based on your completed table, why do you think variety is essential in a person's diet?

   c. Why might vitamin deficiencies pose problems even if people receive adequate supplies of food Calories?

3. Nutritionists recommend eating fresh fruit rather than canned fruit, and raw or steamed vegetables instead of canned or boiled vegetables (see Figures 7.50 and 7.51).

   a. What does food freshness have to do with vitamins?

   b. To what types of vitamins might nutritionists be referring?

   c. Is such advice sound? Explain.

Figure 7.51 *Why do dietary guidelines usually favor raw fruit over canned fruit?*

Table 7.8

Vitamin B₁ and C Content of Some Vegetables

Vegetable (one-cup serving)	Vitamin (in mg)	
	B₁ (thiamine)	C (ascorbic acid)
Green peas	0.387	58.4
Lima beans	0.238	17
Broccoli	0.058	82
Potatoes	0.15	30

4. Examine the structural formulas for the two water-soluble vitamins and two fat-soluble vitamins in Figure 7.49 (page 750).

   a. You investigated solubility characteristics in Unit 4 in an effort to make sense of the saying, "Like dissolves like." Identify particular features in the structures of the water-soluble vitamins that make them "like" water, and thus able to dissolve in water.

   b. What structural features of the fat-soluble vitamins make them "like" fats, and thus able to dissolve in fats? (Another way to think about this is to ask what structural features make them "unlike" water, and thus not very soluble in water.)

   c. How does the concept of intermolecular forces help to explain how a water-soluble vitamin dissolves in water but does not dissolve well in fat?

5. Below is the structural formula for one of the vitamins listed in Table 7.7 (page 751). This vitamin can be found in liver, peas, and green leafy vegetables, among other sources. Deficiency of this vitamin in a person's diet can result in anemia.

   a. Use Table 7.7 (page 751) to identify vitamins that match the description of this vitamin's sources and deficiency conditions.

   b. Based on its structure, do you think this vitamin is water-soluble or fat-soluble? Explain.

   c. Based on your answers to 5a and 5b, identify the vitamin.

INVESTIGATING MATTER

D.3 VITAMIN C

Preparing to Investigate

You may have heard vitamin C described as an "antioxidant." Based on the reaction with I_2 shown here, do you think that means that vitamin C oxidizes or reduces other substances?

Vitamin C, also called ascorbic acid, is a water-soluble vitamin. It is also among the least stable vitamins because it reacts readily with oxygen gas, and exposure to light or heat can decompose it. In this investigation, you will find out how much vitamin C is contained in some popular beverages, including fruit juices, milk, and soft drinks.

This investigation is based on a chemical reaction of ascorbic acid (vitamin C) with iodine (I_2). A colored solution of iodine (I_2) oxidizes ascorbic acid, forming the colorless products dehydroascorbic acid, hydrogen ions, and iodide ions:

$$I_2 \ + \ C_6H_8O_6 \ \longrightarrow \ C_6H_6O_6 \ + \ 2\,H^+ \ + \ 2\,I^-$$

| Iodine | Ascorbic acid (vitamin C) | Dehydroascorbic acid | Hydrogen ion | Iodide ion |

Figure 7.52 shows the structures of ascorbic acid and dehydroascorbic acid.

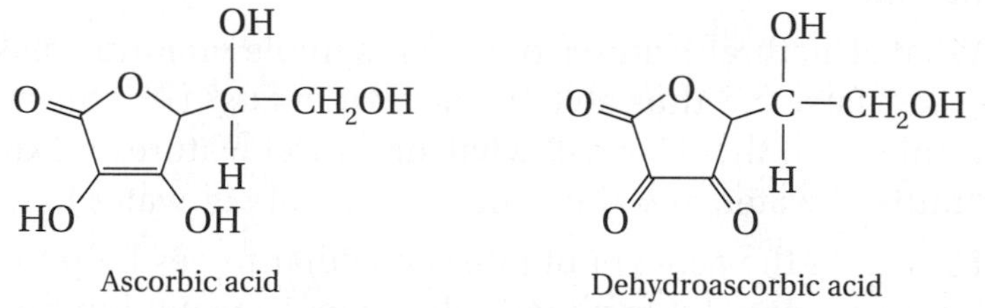

Figure 7.52 *Molecular structures of ascorbic acid and dehydroascorbic acid.*

You also conducted titrations in Unit 4, page 464.

You will conduct a titration (Figure 7.53), a common laboratory procedure to determine concentrations of substances in solution. This investigation involves adding a known amount of one reactant (iodine solution) slowly from a Beral pipet to a second reactant (ascorbic acid in the beverage) in a 24-well plate until just enough has been added for a complete reaction. The completion of the reaction, the **endpoint**, is signaled by a color change. Knowing the chemical equation for this reaction, you can then calculate the unknown amount of the second reactant (ascorbic acid) from the measured volume and concentration of the iodine-solution **titrant**.

The titration endpoint in the beverage-containing well of the well plate is the point where a dark blue-black color appears and does not disappear with additional stirring. This color is due to the reaction of excess iodine with starch. First, you add a starch indicator suspension to the beverage sample to be tested. Next, an iodine solution of known concentration is added drop by drop from a Beral pipet.

Figure 7.53 *Materials needed for fruit-juice titration.*

The chemical equation shows that as long as ascorbic acid is present, the iodine is quickly converted to colorless iodide ions; you will observe no bluish black iodine-starch product. When all the available ascorbic acid has been oxidized to colorless dehydroascorbic acid, the next drop of iodine solution added reacts with starch, producing the blue-black color, signaling the endpoint.

You will begin by completing a titration involving a solution of known vitamin C concentration. Once you have data from this titration, you can determine the mass of ascorbic acid that reacts with one drop of iodine solution. You can then calculate the mass (in milligrams) of vitamin C present in a 25-drop sample of each beverage. This information will allow you to rank the tested beverages in terms of the mass of vitamin C that each beverage contains.

Before starting this investigation, prepare a suitable data table. Leave room in the data table to record the total drops of vitamin C solution in 1.0 mL (Step 1) and the total drops of iodine needed to reach the endpoint (Step 5). Provide a horizontal row for each beverage that you will investigate.

Gathering Evidence

Part I. Standardizing the Iodine Solution

1. Before you begin, put on your goggles, and wear them properly throughout the investigation.

2. Fill a Beral pipet with vitamin C solution. Then determine how many drops of vitamin C solution delivered by that pipet represents a volume of 1.0-mL. (*Note:* The volume per drop may depend on the angle at which you hold the pipet. Keep the angle at which you hold your pipet consistent throughout the investigation.)

3. Fill a second Beral pipet with iodine solution.

4. Add 25 drops vitamin C solution into a well of a clean 24-well plate. The vitamin C solution has a known concentration of 1.0 mg vitamin C per milliliter of solution.

5. Add 1 drop of starch suspension to the same well.

6. Place a sheet of white paper underneath the well plate; it will help you detect the appearance of color.

7. Add iodine solution, one drop at a time, carefully counting drops, to the well containing starch and vitamin C mixture (see Figure 7.54). After adding each drop of iodine solution, use a toothpick to gently stir the resulting mixture.

8. Add and count the iodine solution drop by drop, with stirring, until the solution in the well remains bluish black for 20 seconds. If the color fades before 20 seconds have elapsed, add another drop of iodine solution.

Figure 7.54 *Proper technique for a well plate titration.*

9. Record the total drops of iodine solution needed to reach the endpoint (the appearance of the first "permanent" bluish black color).

10. Repeat Steps 3–9 two times and record your data.

Part II. Determining the Quantity of Vitamin C in Beverages

You will now use titration to determine the unknown quantity of vitamin C in beverages assigned by your teacher. Adapt the procedure used in Part I, considering these points:

- Colored beverages may not produce a true bluish black endpoint color. For example, red beverages may make the endpoint appear purple.

- You will need to conduct three trials and record appropriate data for each beverage analyzed. Construct an appropriate data table prior to beginning Part II.

Wash your hands thoroughly with soap and water before leaving the laboratory.

Analyzing Evidence

1. Determine the mass (in mg) of vitamin C that reacts with 1 drop of iodine solution. Use your data from Part I to perform the following calculations. The Sample Problem below illustrates how to set up the final calculation.

   a. What volume of standardized vitamin C solution did you use for each trial in Part I?

   b. How many milligrams of vitamin C did that volume of solution contain? The concentration of vitamin C solution used in Part I is 1.0 mg/mL.

   c. Determine the average number of drops of iodine used to react with the vitamin C in your three trials.

   d. Calculate the mass (in mg) of vitamin C that reacts with 1 drop of iodine solution. See the Sample Problem below.

Sample Problem: *Suppose you found that there were 30 drops of vitamin C solution in 1.0 mL and that it took 22 drops of iodine solution to reach the endpoint.*

These data lead to the calculated result that 0.038 mg vitamin C reacts with 1 drop iodine solution.

$$25 \text{ drops vitamin C} \times \frac{1 \text{ mL vitamin C}}{30 \text{ drops vitamin C}} \times \underbrace{\frac{1 \text{ mg vitamin C}}{1 \text{ mL vitamin C}}}_{(\text{step 1})} \times \underbrace{\frac{1}{22 \text{ drops I}_2}}_{(\text{step 5})}$$

= 0.038 mg vitamin C per drop I_2 solution

2. Determine the mass of vitamin C (in mg) contained in 25 drops of each beverage you analyzed in Part II.

Interpreting Evidence

1. Rank the tested beverages in terms of how much vitamin C each contains, from the highest quantity to the lowest.

2. Among the beverages tested, in your opinion, were any vitamin C levels

    a. unexpectedly low? If so, explain.

    b. unexpectedly high? If so, explain.

3. Imagine that you added too many drops of iodine solution during the titration and missed the true endpoint. Will this procedural error increase or decrease your calculated milligrams of vitamin C in the sample? Explain.

Reflecting on the Investigation

4. Describe how the concept of limiting reactants applies to titration as an analytical technique. In the titrations you performed, what was the limiting reactant?

5. Supposed you heated the beverages, let them cool, and then performed the titrations in Part II.

    a. Would you expect the number of drops of I_2 solution required to titrate each beverage to be higher or lower than the data you recorded?

    b. Explain.

D.4 MINERALS: ESSENTIAL WITHIN ALL DIETS

Of the more than 100 known elements, only 32 are believed essential to support human life. In nutrional terms, **minerals** refer to these elements (including calcium, sodium, chlorine, selenium, and zinc) in their ionic forms. For convenience, essential minerals are divided into **major minerals** (also called **macrominerals**), and **trace minerals** (also called **microminerals**). Some are quite common; others are likely to be found in large quantities only on research-laboratory shelves.

Minerals have several functions in the body. Some minerals become part of the body's structural material, such as bones and teeth. Others help enzymes do their jobs. Still others help maintain the health of the heart and other organs. The thyroid gland, for example, uses only a miniscule quantity of iodine (only millionths of a gram daily) to produce the vital hormone thyroxine (Figure 7.55). The field of bioinorganic chemistry explores how minerals function within living systems.

Figure 7.55 *The thyroid gland does not function properly without proper levels of iodine in one's body. Iodine deficiency causes goiter (depicted here).*

Table 7.9

	Dietary Minerals	
Mineral	**Typical Food Sources**	**Deficiency Condition**
Macrominerals		
Calcium (Ca)	Milk, dairy products, canned fish	Rickets in children; osteomalacia and osteoporosis in adults
Chlorine (Cl)	Table salt, meat, salt-processed foods	—
Magnesium (Mg)	Seafood, cereal grains, nuts, dark green vegetables, cocoa	Heart spasms, anxiety, disorientation
Phosphorus (P)	Meat, dairy products, nuts, seeds, beans	Blood cell disorders, gastrointestinal tract and renal dysfunction
Potassium (K)	Orange juice, bananas, dried fruits, potatoes	Poor nerve function, irregular heartbeat, sudden death during fasting
Sodium (Na)	Table salt, meat, salt-processed food	Headache, weakness, thirst, poor memory, appetite loss
Sulfur (S)	Protein (e.g., meat, eggs, legumes)	Conditions related to deficiencies in sulfur-containing essential amino acids
Trace minerals		
Chromium (Cr)	Animal and plant tissue, liver	Loss of insulin efficiency with age
Cobalt (Co)	Animal protein, liver	Conditions related to deficiencies in cobalt-containing vitamin B_{12}
Copper (Cu)	Egg yolk, whole grains, liver, kidney	Anemia in malnourished children
Fluorine (F)	Seafood, fluoridated water	Dental decay
Iodine (I)	Seafood, iodized salt	Goiter
Iron (Fe)	Meat, green leafy vegetables, whole grains, liver	Anemia; tiredness and apathy
Manganese (Mn)	Whole grains, legumes, nuts, tea, leafy vegetables, liver	—
Molybdenum (Mo)	Whole grains, legumes, leafy vegetables, liver, kidney	Weight loss, dermatitis, headache, nausea, disorientation
Selenium (Se)	Meat, liver, organ meats, grains, vegetables	Muscle weakness, Keshan disease (heart-muscle disease)
Zinc (Zn)	Shellfish, meat, wheat germ, legumes, liver	Anemia, growth retardation

Your body contains rather large quantities, at least 100 mg per kilogram of body mass, of each of the seven macrominerals. Each trace mineral is present in relatively small quantities, less than 100 mg per kilogram of body mass in an average adult. However, trace minerals are just as essential in a human diet as macrominerals. Any essential mineral, whether macromineral or trace mineral, can become a limiting reactant if it is not present in sufficient quantity.

The essential minerals and their dietary sources and deficiency conditions are listed in Table 7.9. Several other minerals, including arsenic (As), cadmium (Cd), and tin (Sn), are known to be needed by laboratory test animals. These minerals and perhaps other trace minerals may be essential to human life. You may be surprised to learn that the widely known poison arsenic might be an essential mineral. In fact, many substances beneficial in low doses become toxic in higher doses. Table 7.10 summarizes recommended daily doses for several macrominerals and trace minerals.

Table 7.10

Dietary Reference Intakes (DRIs) for Selected Minerals

Age or Condition	Calcium (mg/d)	Phosphorus (mg/d)	Magnesium (mg/d)	Iron (mg/d)	Zinc (mg/d)	Iodine (μg/d)
Males						
9–13 yrs	1300*	**1250**	**240**	**8**	**8**	**120**
14–18 yrs	1300*	**1250**	**410**	**11**	**11**	**150**
19–30 yrs	1000*	**700**	**400**	**8**	**11**	**150**
31–50 yrs	1000*	**700**	**420**	**8**	**11**	**150**
51–70 yrs	1200*	**700**	**420**	**8**	**11**	**150**
> 70 yrs	1200*	**700**	**420**	**8**	**11**	**150**
Females						
9–13 yrs	1300*	**1250**	**240**	**8**	**8**	**120**
14–18 yrs	1300*	**1250**	**360**	**15**	**9**	**150**
19–30 yrs	1000*	**700**	**310**	**18**	**8**	**150**
31–50 yrs	1000*	**700**	**320**	**18**	**8**	**150**
51–70 yrs	1200*	**700**	**320**	**8**	**8**	**150**
> 70 yrs	1200*	**700**	**320**	**8**	**8**	**150**
Pregnant	1000*	**700**	**350**	**27**	**11**	**220**
Nursing	1000*	**700**	**310**	**9**	**12**	**290**

Note: Recommended Dietary Allowances (RDAs) are in **bold** type. RDAs are established to meet the needs of almost all (97 to 98%) individuals in a group. Adequate Intakes (AIs) are followed by an asterisk (*). AIs are believed to cover the needs of all individuals in the group, but a lack of data prevent being able to specify with confidence the percent of individuals covered by this intake.

Source: Food and Nutrition Board, National Academy of Sciences: National Research Council, *Dietary Reference Intakes 2004.*

DEVELOPING SKILLS

D.5 MINERALS IN THE DIET

Use the values in Table 7.10 (page 759) to answer these questions.

> *Sample Problem:* How many cups of broccoli would a 16-year-old female need to eat each day to reach her daily iron allowance? *(1 cup broccoli = 1.1 mg iron)* Assume that broccoli is her only dietary source of iron.
>
> The iron DRI for a 16-year-old female is 15 mg:
>
> $$15 \, \text{mg iron} \times \frac{1 \text{ cup broccoli}}{1.1 \, \text{mg iron}} = 14 \text{ cups broccoli}$$

Figure 7.56 *Which foods are good sources of minerals such as calcium?*

1. One slice of whole wheat bread contains 0.8 mg iron.
   a. How many slices of whole-wheat bread would supply your daily iron allowance? (Assume that this is your only dietary source of iron.)
   b. Predict health consequences of consuming an inadequate quantity of iron.

2. One cup of whole milk contains 288 mg calcium. How much milk would you need to drink daily to meet your daily allowance for that mineral, if milk were your only dietary source of calcium?

3. One medium pancake (Figure 7.56) contains about 27 mg calcium and 0.4 mg iron.
   a. Does a pancake provide a greater percent of your DRI for calcium or for iron?
   b. Explain your answer.

4. The following questions focus on calcium and phosphorous.
   a. What total mass of each of these minerals do you need to consume daily?
      i. calcium.   ii. phosphorus.
   b. Why are the values in the answer to Question 4a higher than DRI values for other listed essential minerals? (*Hint:* Consider how calcium and phosphorus are used in the body.)
   c. List several good dietary sources of
      i. calcium.   ii. phosphorus.
   d. Predict the health consequences of a deficiency of
      i. calcium.   ii. phosphorus.
   e. Would a particular age group or gender be especially affected by the consequences you listed in 4d? If so, which and why?

5. Most table salt, sodium chloride (NaCl), includes a small amount of added potassium iodide (KI).
   a. Why do you think KI is added to table salt?
   b. If you decide not to use iodized salt, what other kinds of food could you use as sources of iodine?

"Iodized salt" refers to such products.

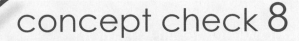

concept check 8

1. Consider minerals and vitamins as nutrients.
   a. Which of these nutrients is usually found in ionic form?
   b. Which of these nutrients is usually found in covalent form?
   c. How else would you describe similarities or differences between vitamins and minerals to a friend?
2. Vitamins and minerals occur naturally in foods. Why then are some foods labeled as "enriched" with particular vitamins or minerals?
3. Beyond minerals and vitamins, identify one food additive with which you are familiar. Why might that additive be included in a food product?

D.6 FOOD ADDITIVES

Small amounts of vitamins and minerals occur naturally in food. Some foods, especially processed foods such as packaged snacks or frozen entrees, also contain small amounts of **food additives**. Manufacturers add those substances during processing to increase the nutritive value of foods or to enhance their storage life, visual appeal (see Figure 7.57), flavor, or ease of production. A food label might provide this ingredient information:

Sugar, bleached flour (enriched with niacin, iron, thiamine, and riboflavin), semisweet chocolate, animal and/or vegetable shortening, dextrose, wheat starch, monocalcium phosphate, baking soda, egg white, modified corn starch, salt, nonfat milk, cellulose gum, soy lecithin, xanthan gum, mono- and diglycerides, BHA, BHT.

This list shows quite a collection of ingredients! You probably recognize the major ingredients, such as sugar, flour, shortening, and baking soda, and some additives such as vitamins (niacin, thiamine, and riboflavin) and minerals (iron and monocalcium phosphate). However, you probably do not recognize the food additives xanthan gum (an emulsifier that helps produce uniform, non-separating water–oil mixtures (see Figure 7.58, page 762) and BHA and BHT (butylated hydroxyanisole and butylated hydroxytoluene, which are antioxidants that act as preservatives).

Figure 7.57 *Food-coloring additives are often used to make food items more attractive and appealing.*

Scientific American Working Knowledge Illustration

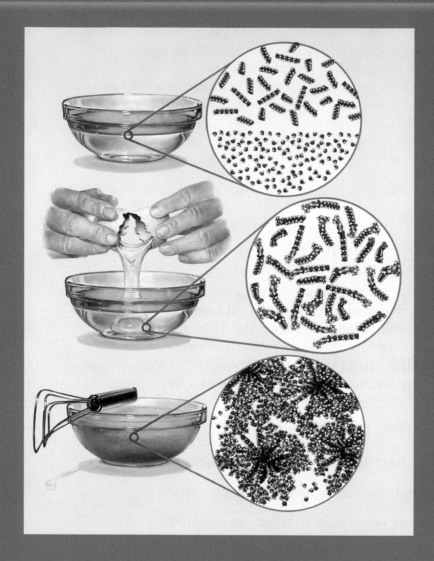

Figure 7.58 *Emulsifier action.* *These illustrations depict how lecithin, a molecule contained in egg yolk, emulsifies (mixes together) an oil–water system. In the top view, oil is layered over water. The particulate-level enlargement reveals that oil molecules are long, nonpolar carbon chains quite different in structure and polarity from the small, polar, water molecules, also shown. Water–water intermolecular forces are much stronger than are oil–water intermolecular forces. Hence these two liquids do not mix. Each lecithin molecule, the emulsifying agent (middle view), has a long, nonpolar carbon chain and a smaller, polar region involving phosphorous, oxygen, nitrogen, and hydrogen atoms. Lecithin molecules orient themselves so that they simultaneously interact with both oil and water molecules. Once an egg is cracked open and added and the entire mixture is whisked, micelles form (bottom view). Each micelle is a spherical arrangement of oil molecules surrounded by and attracted to the nonpolar region of lecithin molecules. The polar regions of lecithin molecules align at the outside of a micelle, causing the entire sphere to intermingle among polar water molecules. This stable arrangement permits oil molecules to remain evenly dispersed through water. Emulsification is used to make numerous food products, including mayonnaise and ice cream.*

Food additives have been used since ancient times. For example, salt has been used for centuries to preserve foods, and spices helped disguise the flavor of food that was no longer fresh. To make foods easier and less expensive to distribute and store, most manufacturers, especially those of processed foods, rely on food-preservation additives. Table 7.11 summarizes the major categories of food additives. The structural formulas of two common additives are shown in Figure 7.59.

Color and taste additives often enhance the commercial appeal of food products. In the following investigation, you will analyze several commonly used food-coloring agents (dyes).

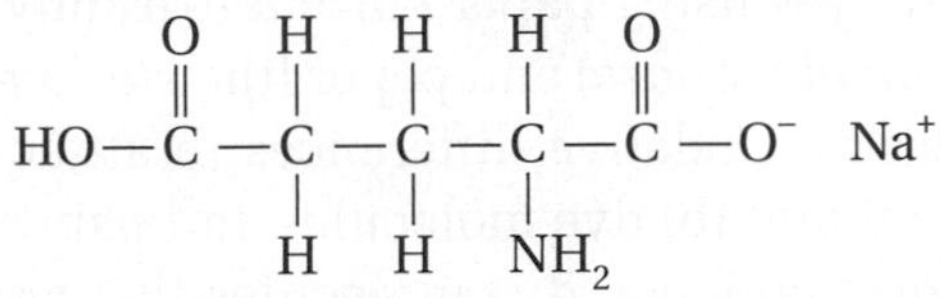

Monosodium glutamate (MSG) 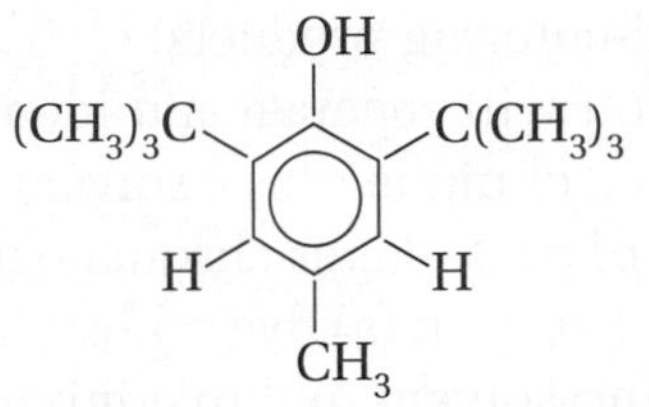   Butylated hydroxytoluene (BHT)

Figure 7.59 *The molecular structures of MSG and BHT.*

Table 7.11

Food Additive Types, Purposes, and Examples		
Additive Type	**Purpose**	**Examples**
Anticaking agents	Keep foods free-flowing	Sodium ferrocyanide
Antioxidants	Prevent fat rancidity	BHA and BHT
Bleaches	Whiten foods (flour, cheese); hasten cheese maturing	Sulfur dioxide, SO_2
Coloring agents	Increase visual appeal	Carotene (natural yellow color); synthetic dyes
Emulsifiers	Improve texture, smoothness; stabilize oil-water mixtures	Cellulose gums, dextrins
Flavoring agents	Add or enhance flavor	Salt, monosodium glutamate (MSG), spices
Humectants	Retain moisture	Glycerin
Leavening agents	Give foods light texture	Baking powder, baking soda
Nutrients	Improve nutritive value	Vitamins, minerals
Preservatives and antimycotic agents (growth inhibitors)	Prevent spoilage, microbial growth	Propionic acid, sorbic acid, benzoic acid, salt
Sweeteners	Impart sweet taste	Sugar (sucrose), dextrin, fructose, aspartame, sorbitol, mannitol

■ INVESTIGATING MATTER

D.7 ANALYZING FOOD-COLORING ADDITIVES

Preparing to Investigate

Many candies contain artificial coloring agents to increase their visual appeal. Colorless candies would be quite dull! In this investigation, you will analyze the food dyes in two commercial candies and compare them with dyes in food-coloring products.

You will separate and identify the food dyes using **paper chromatography**. This technique uses a solvent (called the *mobile phase*) and paper (the *stationary phase*). Paper chromatography is based on relative differences in attraction between (a) dye molecules and solvent and (b) dye molecules and paper. As the solvent and dye mixture moves up the paper, dye molecules that are more strongly attracted to paper will more readily leave the solvent and separate out onto the paper, leaving a colored spot on the paper. Dye molecules less attracted to paper (and more attracted to solvent) will separate out onto the paper later. Thus, characteristic areas representing different dye molecules will appear on the paper.

To analyze the results of this chromatography investigation, you will calculate the R_f value for each spot. The R_f value, as illustrated in Figure 7.60, is a ratio involving the distance traveled by each dye compared to the distance the solvent has moved up the paper. The actual distance that the dye and solvent travel might vary from trial to trial, but the R_f value will remain constant for a particular dye. Thus chemists use the R_f value, not the actual distance traveled, to identify the molecules.

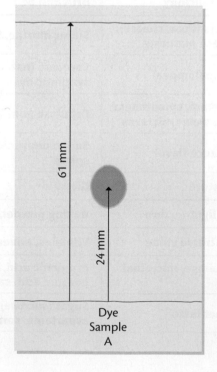

61 mm

24 mm

Dye
Sample
A

Calculating R_f for Dye Sample A
$$R_f = \frac{24 \text{ mm}}{61 \text{ mm}} = 0.39$$

Figure 7.60 *Paper chromatograph and calculation of R_f values.*

Read through *Gathering Evidence* and prepare an appropriate data table for this investigation.

Gathering Evidence

1. Before you begin, put on your goggles, and wear them properly throughout the investigation.

2. Obtain one piece each of two different commercial candies from your teacher, as well as a food-coloring sample. The food coloring and candies should all be the same color.

3. Put each candy into a separate well of a well plate. Note which candy is in each well. Add 5–10 drops of water to each well. Stir the mixture in each well with a separate toothpick until the colored coating completely dissolves from the candy. Add 3 to 4 drops of the food coloring sample to a third well. Observe and record the initial colors of each sample.

4. Obtain a strip of chromatography paper, handling it only by its edges. With a pencil (do not use a pen), draw a horizontal line 2 cm from the bottom of the paper strip and another horizontal line 3 cm from the top.

Figure 7.61 *Spotting a sample on chromatography paper.*

5. Next, place a spot of dye solution on the bottom line; the spot should not be large. Place a drop of the first candy's colored solution with a toothpick, as shown in Figure 7.61.

6. Allow the drop of solution to sit until the spot it makes stops spreading out on the paper. Then apply a second drop of the same sample on top of the spot.

7. Obtain a graduated cylinder to be used as a chromatography chamber. Fill the chamber/cylinder with the solvent (water) to a depth of 1 cm.

8. Lower the spotted chromatography paper's bottom edge into the chromatography chamber until the edge rests evenly in the solvent (water). Be sure the colored spots remain above the solvent surface. Cover the chromatography chamber. See Figure 7.62.

9. Repeat this procedure (Steps 4–8) for the second candy's colored solution and also for the food-coloring sample.

10. Allow the solvent in each cylinder to move up the paper until solvent reaches the top penciled line (3 cm from the top). Then remove the paper from the chambers and, using a pencil, mark the farthest point of solvent travel. Allow the paper to air-dry overnight.

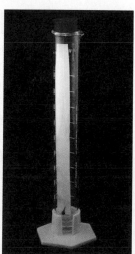

11. After the paper has dried, record the colors observed for the dye sample and candy solutions.

12. Measure the distance (in cm) from the initial pencil line where you placed the spots to the center of each dye spot. Record these distances in your data table.

Figure 7.62 *Prepared chromatography paper in a graduated cylinder chromatography chamber. Note that the sample spot is above the solvent level.*

13. Measure and record the distance (in cm) that the solvent moved.

Analyzing Evidence

1. Calculate the R_f value of each dye spot you investigated.
2. Which sample solution, if any, created a single spot rather than several spots?

Making Claims

1. Which dye in the sample solutions had the greatest attraction for the paper? Use evidence to support your claim.
2. Which dye in the sample solutions had the greatest attraction for the solvent? Use evidence to support your claim.
3. Based on your data, do any of the three samples contain the same dyes? Use evidence to support your claims.

Reflecting on the Investigation

4. The candy and food-coloring packages list each dye that they contain. Compare this information with your experimental results.
   a. What similarities did you find?
   b. What differences did you find?
   c. If you found differences, what are some possible reasons for them?
5. Why was it important to use a pencil rather than a pen to mark lines on the chromatography paper in Step 3?

D.8 REGULATING ADDITIVES

Both processed and unprocessed foods may contain contaminants that were not deliberately added, such as pesticides, mold, antibiotics used to treat animals, insect fragments, food-packaging materials, or dirt. We presume that food purchased in grocery stores and restaurants is safe to eat. In most cases, that is true. Nonetheless, in the past, some food additives and contaminants were identified as or suspected of posing hazards to human health.

Safe food in the United States is required by law—the Federal Food, Drug and Cosmetic Act of 1938. This act authorized the Food and Drug Administration (FDA) to monitor food's safety, purity, and wholesomeness. This act has been amended to address concerns about pesticide residues; artificial colors and food dyes; potential cancer-causing agents (carcinogens); and **mutagens**, which are agents that cause mutations, or changes in DNA. Food manufacturers must complete a battery of tests and provide extensive evidence about the safety of any proposed food product or additive. Any new food product must earn FDA approval before it is marketed.

According to the amended Federal Food, Drug and Cosmetic Act, ingredients that are known not to be hazardous and were in use for a long time prior to the act were exempted from testing. These substances, rather than legally defined as additives, constitute the "generally recognized as safe" (GRAS) list. The GRAS list, periodically reviewed in light of new findings, includes items such as salt, sugar, vinegar, vitamins (for example, vitamin C and riboflavin), and some minerals.

In accord with the Delaney Clause, which was added to the Act in the 1950s, every proposed new food additive must be tested on laboratory animals (usually mice). The Delaney Clause specifies that "no additive shall be deemed to be safe if it is found to induce cancer when it is ingested by man or animal." Thus, approval of a proposed additive is denied if it causes cancer in the test animals.

Since the 1950s, great advances have occurred in science and technology. Improvements in chemical-analysis techniques permit scientists to detect even smaller amounts of potentially harmful substances. Consequently, scientists can now detect and study food contaminants that have always been present but previously were undetected.

This new information has resulted in greater understanding of potential risks associated with human exposure to particular food additives. People now recognize that amounts of additives comparable to those causing cancer in test animals are often vastly greater than would ever be encountered in a human diet. In light of this, Congress passed the Food Quality and Protection Act in 1996. This legislation, which replaced the Delaney Clause, states that manufacturers may use food additives that present a "negligible risk."

Many concerns about food additives remain. Sodium nitrite ($NaNO_2$), for instance, is a color stabilizer and spoilage inhibitor used in many cured meats, such as hot dogs and lunch meats. Nitrites are particularly effective in inhibiting growth of the bacterium *Clostridium botulinum*, which produces botulin toxin. This toxin is the cause of botulism, an often fatal disease. Sodium nitrite, however, may be a carcinogen. In the stomach, nitrites are converted to nitrous acid:

$$NaNO_2(aq) \ + \ HCl(aq) \ \longrightarrow \ HNO_2(aq) \ + \ NaCl(aq)$$

| Sodium nitrite | Hydrochloric acid | Nitrous acid | Sodium chloride |

Nitrous acid can then react with compounds formed during protein digestion, producing nitrosoamines, known as potent carcinogens. An example of this reaction is shown below. The concentration of carcinogenic compounds produced, however, is generally low, below the toxic threshold.

$$HNO_2 + R-\underset{\underset{R}{|}}{N}-H \ \longrightarrow \ R-\underset{\underset{R}{|}}{N}-N=O + H_2O$$

Testing a new food additive usually requires millions of dollars and years of research.

Figure 7.63 *Nitrites are used to prevent spoilage and preserve flavor in some meat products.*

CHEM**QUANDARY**

NITRITE ADDITIVES

Consider these benefits and risks of using nitrites to preserve meats:

	Using Nitrites	**Eliminating Nitrites**
Benefit	Decreases risk of formation of botulin toxin	Decreases risk of formation of possible carcinogens
Risk	Increases risk of formation of possible carcinogens	Increases risk of formation of botulin toxin

Based on this information, do you think people should keep using nitrites to preserve meats?

D.9 ARTIFICIAL SWEETENERS

Because many people believe they should reduce the amount of sugar in their diets, many use low-Calorie sweeteners. Saccharin (found in Sweet 'N' Low™) was the first sugar substitute used extensively in the United States. There has been some controversy over saccharin use because early investigations conducted with rats suggested that massive amounts of saccharin may cause cancer. Further investigation indicated that a link between saccharin and cancer was very weak. Currently, it is generally believed that saccharin is safe for human consumption. Today, saccharin is used in many candies, baked goods, jellies, and jams.

The sugar-substitute aspartame (NutraSweet™ and Equal™) is an ingredient in many diet beverages and thousands of other food products. Aspartame (Figure 7.64) is a chemical combination of two natural amino acids, aspartic acid and phenylalanine, neither of which, by itself, tastes sweet. One gram of aspartame contains roughly the same food energy as one gram of table sugar (4 Cal), but aspartame tastes 200 times sweeter than sugar. Because we need smaller quantities of aspartame to sweeten a product, it as a "low-Calorie" alternative to sugar—one that is also safe for people with diabetes. Annually, thousands of tons of aspartame are used in the United States to sweeten diet drinks and foods. Aspartame decomposes at cooking temperatures. However, Splenda, an artificial sweetener based on sucralose (a chlorinated carbohydrate; see Figure 7.64), can be used directly in cooking.

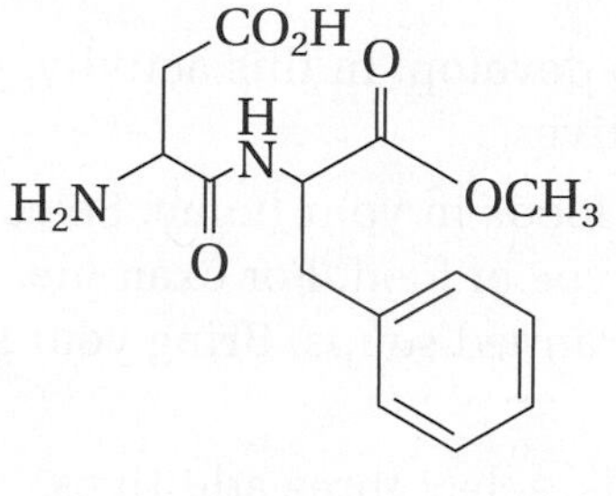

Aspartame

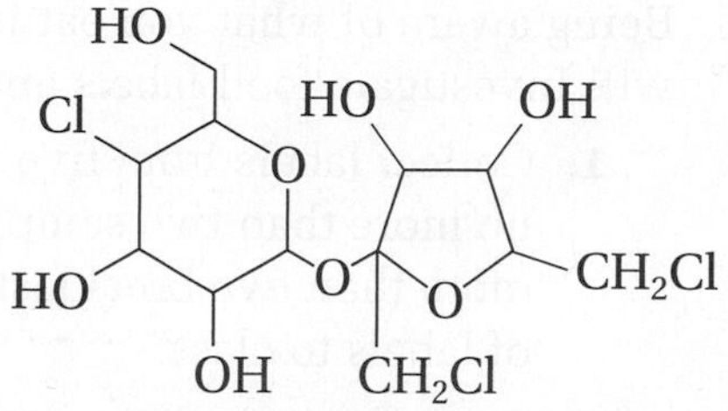

Sucralose

Figure 7.64 *Molecular structures of aspartame and sucralose.*

Although no serious warning regarding aspartame has been issued for the general population, aspartame does pose a health hazard to phenylketonurics—individuals who cannot properly metabolize phenylalanine. An FDA-required warning, *Phenylketonurics: Contains Phenylalanine,* on the label of foods with aspartame highlights the potential risk (see Figure 7.65).

Individuals with specific medical conditions may need to avoid certain foods and food additives. For instance, people with diabetes (see Figure 7.66) must regulate their intake of carbohydrates, including sugars. People with high blood pressure (hypertension) must avoid excess sodium. Some people have food allergies. If such restrictions apply to you, you must always read food labels. Sometimes a new ingredient will be used in a food product that you have used safely in the past; this new ingredient may put you at risk.

Figure 7.65 *The warning to phenylketonurics found on a can of diet drink containing aspartame.*

Figure 7.66 *A diabetic individual checks her blood glucose level.*

DEVELOPING SKILLS

D.10 FOOD ADDITIVE SURVEY

Being aware of what you eat is a wise habit to develop. In this activity, you will investigate food labels and consider additives.

1. Collect labels from five or six packaged foods in your home. Select no more than two samples of the same type of food. For example, no more than two breakfast cereals or two canned soups. Bring your set of labels to class.

2. From the ingredient listings on the labels, select three additives.

3. Complete a summary table with this format:

   a. List the three additives in a vertical column along the left side.

   b. Make four vertical columns to the right of the additives with these headings: "Food Product Where Found," "Purpose of Additive (if known)," "Chemical Formula," and "Other Information."

4. Use Table 7.11 (page 763) to review the purposes of particular food additives. Then answer the following question for each food additive in your summary table: Why do you think this additive is included in this food?

5. What alternatives to food additives can you propose to prevent food spoilage?

MAKING DECISIONS

D.11 ANALYZING VITAMINS AND MINERALS

So far, you have analyzed the three-day food inventory in terms of fats, carbohydrates, and proteins. Now you will decide whether the inventoried foods provide adequate quantities of vitamins and minerals.

1. Your teacher will specify two vitamins and two minerals for you to analyze.

2. Using references provided by your teacher, analyze each food item in your three-day inventory for the presence of those two vitamins and two minerals. Record these values.

3. Determine the average intake of those two vitamins and two minerals over the three days covered in the three-day inventory. Record these data.

4. **a.** Does your analyzed food inventory provide enough of each specified vitamin and mineral?

   b. If not, what types of foods can be added to increase intake of these vitamins and minerals? Identify the food groups to which these food types belong.

5. Some people believe that eating a well-balanced diet provides all the vitamins a person needs to remain healthy. Others suggest that vitamin supplements can enhance personal health. Proponents of both opinions have engaged in considerable advertising to promote their viewpoints.

   a. After considering your data, which viewpoint do your results support?

   b. What could you do to help people make more informed decisions about this issue?

SECTION D SUMMARY

Reviewing the Concepts

> Vitamins are organic molecules necessary for basic life functions.

1. Vitamins are not considered foods, yet they are vital to healthful diets. Explain.

2. What is a coenzyme?

3. Give an example of a function served by a vitamin.

4. List some typical symptoms of dietary deficiency of
   a. vitamin B_{12}.
   b. vitamin A.
   c. vitamin D.

> The required daily intake of a vitamin depends on properties of the vitamin as well as characteristics of individuals.

5. Vitamins are micronutrients. What does this mean about the quantity of vitamins required daily?

6. What molecular properties determine whether a vitamin is fat-soluble or water-soluble?

7. Why should people take water-soluble vitamins more regularly than fat-soluble ones?

8. Give two examples of how DRIs for vitamins and minerals vary for individuals, based on age and gender.

> Minerals are elements that are essential for human life.

9. Name three essential minerals.

10. Of the more than 100 known elements, how many are believed essential for human life?

11. What is the difference between a *macromineral* and a *trace mineral*?

12. What roles do minerals play in the body?

13. List a macromineral and a trace mineral and describe the deficiency condition for each.

14. What are some dietary sources of
   a. magnesium?   c. iron?
   b. potassium?   d. molybdenum?

15. Regarding minerals in the diet, some people believe that "if a little is good, more must be better." Evaluate this idea, particularly considering such minerals as arsenic (As) and cadmium (Cd).

Food additives are used to increase the nutritive value, storage life, visual appeal, taste, or ease of producing foods.

16. What determines whether a substance is considered an additive or a basic component of a food?

17. List two typical food additives and their functions.

18. Give an example of a food additive used to
    a. increase nutritive value.
    b. improve storage life.
    c. enhance visual appeal.
    d. ease production.

19. a. What is a carcinogen?
    b. What is a mutagen?

20. How does the Food Quality and Protection Act differ from the Delaney Clause?

21. Aspartame's energy value is the same as that of sugar, 4 Cal/g. Why, then, is aspartame useful as a low-Calorie artificial sweetener?

22. Explain why aspartame can't be used in baking.

Titration is a common laboratory procedure used to determine the amount of solute in a particular solution. Paper chromatography is a technique for separating and identifying the components of a solution.

23. Describe the key steps in a vitamin C titration, and explain how this process allows you to determine the concentration of vitamin C in a beverage.

24. What is meant by a titration's *endpoint*?

25. What reactions allowed you to detect the presence of vitamin C (usually colorless) in solution in Investigating Matter D.3?

26. Suppose you complete a vitamin-C titration and determine that 0.035 mg vitamin C reacts with 1 drop iodine solution. How much vitamin C is in a particular sample that turns bluish-black after adding
    a. 1 drop of iodine?
    b. 12 drops of iodine?

27. For what juices would the vitamin-C titration endpoint be difficult to observe? Explain.

28. a. How is an R_f value determined?
    b. What does its R_f value indicate about a particular solute?

29. In a paper-chromatography investigation, the solvent moves 5.8 cm, and a food-dye component moves 3.9 cm. What is the R_f value for the food-dye? Show calculations to support your answer.

30. What properties allow chromatography to separate components of a solution?

31. A student tests two samples of dye solution with paper chromatography. In A, the solvent moves 6.2 cm, while the dye moves 4.1 cm. In B, the solvent moves 5.3 cm, and the dye moves 4.1 cm. Do these data support the idea that A and B are the same dye? Explain.

What roles do vitamins, minerals, and additives play in the foods we eat?

In this section you have studied the structure, sources, and functions of vitamins and minerals, considered reasons for adding substances to foods, and used quantitative techniques to investigate the presence of vitamins and additives in food products. Think about what you have learned, then answer the question in your own words in organized paragraphs. Your answer should demonstrate (and perhaps clarify!) your understanding of the key ideas in this section.

Be sure to consider the following in your response: vitamins, minerals, food additives, and methods for detecting and quantifying components of food products.

Connecting the Concepts

32. Although high temperatures can destroy some vitamins in foods, they rarely affect the quality of minerals in foods. Explain.

33. Why do food standards use the term "generally recognized as safe" (GRAS), rather than "always safe" or "100% guaranteed safe" with certain food additives?

34. Would paper chromatography provide any useful information if one or more sample components were colorless? Explain.

35. How could you modify Investigating Matter D.3 to find out how much vitamin C is lost during cooking a food item?

36. One cup of spinach provides 51 mg calcium and 1.7 mg iron. Does spinach provide a greater percent of your DRI for calcium or for iron? Show calculations to support your answer.

37. Suppose a government banned all food additives. Overall, do you think the ban would produce positive or negative results? Explain.

38. Suppose one component in a chromatography sample was quite volatile, tending to evaporate when placed on paper. What effect might this property have on the experimental results?

39. All substances described in this unit consist of atoms, molecules, or ions. With that in mind, why do you think some people describe food additives as "chemicals," but do not describe carbohydrates and fats also as chemicals?

40. In a paper chromatography investigation, a sample failed to move up the paper when water is used as a solvent. How could the investigation be modified to address this problem?

Extending the Concepts

41. Investigate some additional types of chromatography, such as gas chromatography and column chromatography. Explain how each works.

42. Modern food additives are generally safe and healthful. That was not always the case in the past. Research the history of food additives and the unexpected impact of some early additives on human health.

43. When administered in large doses, a prospective food additive is found to cause cancer in laboratory rats. Should that evidence be used to withhold the approval for its use in human foods? Explain your answer.

44. Research some food-preparation methods that tend to preserve vitamin content. Discuss the chemical concepts that can account for the effectiveness of such methods.

45. Liver is a rich source of many trace minerals. Explain this based on liver physiology.

46. Research *goiter*. When and where has it been most prevalent? Why? What specific product makes it much less common today?

47. Some historians claim that the most significant contribution to the success of the British Navy in the 1700s was adding sauerkraut to shipboard food supplies. Investigate and explain the possible connection.

PUTTING IT ALL TOGETHER

GUIDING THE PTSA'S DECISIONS

Now that you have completed a series of analyses of the three-day food and activity inventory list, it is time to meet with other members of the Vending Machine Policy Planning Committee to consider next year's Riverwood High School food vending machine policies and to convey recommendations to the PTSA. To guide your writing, review the results of your analyses and think of points you would like to raise about students' eating habits. Use guidelines for a healthful diet provided by the PTSA at the start of this unit (page 685) to decide on the content and presentation of your report.

As a group, the team (your class) should make recommendations regarding the types or classes of foods that should be supplied in Riverwood High School food vending machines. Focus on generalized recommendations, such as: "The vending machines should offer food options high in calcium." In your discussion, be sure to support proposed recommendations with evidence and/or data from your three-day food inventory analyses.

Once your team has prepared its list of recommendations, each team member should write a report to submit to the PTSA. Here are guidelines for writing your report:

Thinking About Your Audience

As you studied this unit, you learned much about food chemistry, so you may decide to include some technical terminology in your report. If you do, make sure you explain these terms, so that PTSA members can understand your findings and recommendations. Many of them have studied chemistry (they know about bonding and other fundamental chemistry concepts), but are unfamiliar with food chemistry.

Writing

Your written report should include these sections:

1. *Introduction*—a brief summary highlighting features of the three-day food inventory. In one or two paragraphs, provide readers with an overview of what you found in your food-inventory analysis. In addition, list your top three vending machine policy recommendations. Hold the details of your study and detailed explanations of your recommendations until later.

2. *Background Information*—information that PTSA members need to understand your report. Here are some questions you may need to answer for readers:

- What are fats, carbohydrates, and proteins? You may include information about each substance's molecular structure, energy content, and function in the human body.

- Why do people need fats, carbohydrates, and proteins in their diets?

- What roles do vitamins and minerals play in the diet?

- You may include structural formulas and other visual aids, but remember that you need to guide your readers in what to look for in any visuals.

3. *Data Analysis*—the body of your report. In this section, you bring together detailed data you gathered concerning energy, fats, carbohydrates, protein, and vitamins and minerals. Organize and present the information so that it will make sense to readers. Include graphs and tables as appropriate. You may wish to address questions such as these:

- What is the average daily Calorie intake? How reasonable is this value? Does the average accurately represent each day in the inventory, or does the Calorie intake vary widely from day to day?

- How does daily Calorie intake compare with daily Calorie expenditure?

- What percent of total Calories are supplied by carbohydrates, fats, and protein? Are these percent values in line with USDA recommendations?

- What percent of fats are saturated and unsaturated? Do these levels meet USDA recommendations?

- Which vitamins and minerals are present at adequate levels? Which are lacking?

4. *Conclusions*—here you will provide further details regarding recommendations you highlighted at the start of your report. Discuss each recommendation in light of your findings and the guidelines for a healthful diet summarized at the start of this unit. List representative vending-machine food items that (a) meet one or more of your proposed recommendations, and (b) do not meet your recommendations, and explain how you decided on the particular items. Keep in mind that nutrition is not the only consideration—taste, appearance, and whether students would actually eat the items are also considerations.

LOOKING BACK

This unit focused on food, a mixture of chemical substances that you necessarily must encounter daily. You can now explain how food that you eat provides energy for daily living and structural components for growth. You can attach deeper chemical meaning to terms such as *carbohydrate*, and can better evaluate consequences associated with deciding to consume or not consume certain foods. The next time you hear someone remark, "You are what you eat," smile and remark that you know some of the chemistry behind the true meaning of their words!

The Scientific Method versus Scientific Methods

Scientists deepen their knowledge and understanding of the natural world by observing and manipulating their environment. The inquiry approach used by scientists to solve problems and seek knowledge has led to vast increases in understanding how nature works. Many efforts have been made to formalize and list the steps that scientists use to generate and test new knowledge. You may have been asked to learn the "steps" of the Scientific Method, such as Make Observations, Define the Problem, and so on. But no one comes to an understanding of how scientific inquiry works by simply learning a list of steps or definitions of words.

Although you may not go on to become a research scientist, it is important that all students acquire the ability to conduct scientific inquiry. Why? Everyone is confronted daily with endless streams of facts and claims. What should be accepted as true? What should be discarded? Having well-developed ways to evaluate and test claims is essential in deciding between valid and deceptive information.

What abilities are needed to conduct scientific inquiry? According to the National Science Education Standards (NSES), they include the abilities to

- identify questions and concepts that guide scientific investigations.
- design and conduct scientific investigations.
- use technology and mathematics to improve investigations and communications.
- formulate and revise scientific explanations and models using logic and evidence.
- recognize and analyze alternative explanations and models.
- communicate and defend a scientific argument.

These skills are necessary for both doing and learning science. They are also important skills to evaluate information in daily living. You can only acquire these skills through practice—by doing exercises and investigations such as those contained in this textbook.

Even with all the abilities listed above, doing science is a complex behavior. The NSES outlines the ideas that all students should know and understand about the practices of science:

- Scientists usually inquire about how physical, living, or designed systems function.
- Scientists conduct investigations for a wide variety of reasons.
- Scientists rely on technology to enhance the gathering and manipulation of data.
- Mathematics is essential in scientific inquiry.
- Scientific explanations must adhere to criteria such as: a proposed explanation must be logically consistent; it must abide by the rules of evidence; it must be open to questions and possible modification; and it must be based on historical and current scientific knowledge.
- Results of scientific inquiry—new knowledge and methods—emerge from different types of investigations and public communication among scientists.

The last statement above acknowledges that—despite generalizations that can be listed—there are many paths to gaining new scientific knowledge. That is what makes studying scientific processes so important. Learning how to acquire the abilities to do and understand scientific inquiry may be the most important and useful thing you learn in this course.

Numbers in Chemistry

Chemistry is a quantitative science. Most chemistry investigations involve not only measuring but also a search for the meaning among the measurements. Chemists learn how to interpret as well as perform calculations using these measurements.

Scientific Notation

Chemists often deal with very small and very large numbers. Instead of using many zeros to express very large or very small numbers, they often use scientific notation. In scientific notation, a number can be rewritten as the product of a number between 1 and 10 and an exponential term—10^n, where n is a whole number. The exponential term is the number of times 10 would have to be multiplied or divided by itself to yield the appropriate number of digits in the number. For instance, 10^3 is $10 \times 10 \times 10$, or 1000; $3.5 \times 10^3 = 3500$.

> **Sample Problem 1:** *Express the distance between New York City and San Francisco, 4 741 000 meters, using scientific notation.*
>
> $4\ 741\ 000$ m $= (4.741 \times 1\ 000\ 000)$ m or $\mathbf{4.741 \times 10^6}$ m

> **Sample Problem 2:** *Express the amount of ranitidine hydrochloride in a Zantac tablet, 0.000 479 mol, using scientific notation.*
>
> $0.000\ 479$ mol $= 4.79 \times 0.0001$ mol or $\mathbf{4.79 \times 10^{-4}}$ mol

It is easier to assess magnitude and to perform operations with numbers written in scientific notation than with numbers fully written out. As you will see, it is also easier to communicate the precision of the measurements involved.

Rules for *adding and subtracting* using scientific notation:

Step 1.   Convert the numbers to the same power of 10.

Step 2.   Add (subtract) the non-exponential portion of the numbers. *The power of ten remains the same.*

> **Sample Problem 3:** *Add $(1.00 \times 10^4) + (2.30 \times 10^5)$.*
>
> **Step 1.**   A good rule to follow is to express all numbers in the problem to the highest power of ten. Convert 1.00×10^4 to 0.100×10^5.
>
> **Step 2.** $(0.100 \times 10^5) + (2.30 \times 10^5) = 2.40 \times 10^5$

Rules for *multiplying* using scientific notation:

Step 1. Multiply the nonexponential numbers.

Step 2. Add the exponents.

Step 3. Convert the answer to scientific notation.

> *Sample Problem 4: Multiply (4.24 × 10²) by (5.78 × 10⁴).*
>
> **Steps 1 and 2.** $(4.24 \times 5.78) \times (10^{2+4}) = 24.5 \times 10^6$
>
> **Step 3.** Convert to scientific notation $= 2.45 \times 10^7$

Rules for *dividing* using scientific notation:

Step 1. Divide the nonexponential numbers.

Step 2. Subtract the denominator exponent from the numerator exponent.

Step 3. Express the answer in scientific notation.

> *Sample Problem 5: Divide (3.78 × 10⁵) by (6.2 × 10⁸).*
>
> **Steps 1 and 2.** $\dfrac{3.78}{6.2} \times (10^{5-8}) = 0.61 \times 10^{-3}$
>
> **Step 3.** Convert to scientific notation $= 6.1 \times 10^{-4}$

Practice Problems*

1. Convert the following numbers to scientific notation.

   a. 0.000 036 9

   b. 0.0452

   c. 4 520 000

   d. 365 000

2. Carry out the following operations:

   a. $(1.62 \times 10^3) + (3.4 \times 10^2)$

   b. $(1.75 \times 10^{-1}) - (4.6 \times 10^{-2})$

   c. $\dfrac{6.02 \times 10^{23}}{12.0}$

   d. $\dfrac{(6.63 \times 10^{-34}\,\text{J·s}) \times (3.00 \times 10^8\,\text{m s}^{-1})}{4.6 \times 10^{-9}\,\text{m}}$

*Answers to odd-numbered problems can be found on page ANS-1.

Dimensional Analysis

Dimensional analysis, also called the *factor-label method*, is used by scientists to keep track of units in calculations and to help guide their work in solving problems. The method is helpful in setting up problems and also in checking work.

Dimensional analysis consists of three basic steps:

Step 1. Identify equivalence relationships in order to create suitable conversion factors.

Step 2. Identify the given unit(s) and the new unit(s) desired.

Step 3. Arrange each conversion factor so that each unit to be converted can be divided by itself (and thus cancelled).

Sample Problem 1: *In an exercise to determine the volume of a rectangular object, your laboratory partner measured the object's length as 12.2 in (inches). However, measurements of the object's width and height were recorded in centimeters. In order to calculate the object's volume, convert the object's measured length to centimeters.*

Step 1. Find the equivalence relating centimeters and inches.
$$2.54 \text{ cm} = 1 \text{ in}$$

Step 2. Identify the given unit and the new unit.
Given unit = in   new unit = cm

Step 3. Create a fraction so that the "given" unit (in) can be divided and thus cancelled.
$$12.2 \text{ in} \times \frac{2.54 \text{ cm}}{1 \text{ in}} = 31.0 \text{ cm}$$

Sample Problem 2: *How many seconds are in 24 hours?*

Step 1. Identify the equivalence:
1 hr = 60 min   1 min = 60 s

Step 2. Given unit: hr   new unit: s

Step 3. Arrange for the given unit to cancel and progress to the desired unit:
$$24 \text{ hr} \times \frac{60 \text{ min}}{1 \text{ hr}} \times \frac{60 \text{ s}}{1 \text{ min}} = 86\,400 \text{ s}$$

Practice Problems*

3. The distance between two European cities is 4 741 000 m. That may sound impressive, but to put all those digits on a car odometer is slightly inconvenient. Kilometers are a better choice for measuring distance in this case. Convert the distance to kilometers.

4. The density of aluminum is 2.70 g/cm³. What is the mass of 234 cm³ of aluminum?

Significant Figures

Significant figures are all the digits in a measured or calculated value that are known with certainty plus the first uncertain digit. Numerical measurements have some inherent uncertainty. This uncertainty comes from the measurement device as well as from the human making the measurement. No measurement is exact. When you use a measuring device in the laboratory, read and record each measurement to one digit beyond the smallest marking interval on the scale.

Guidelines for Determining Significant Figures

Step 1.   All digits recorded from a laboratory measurement are called significant figures.

The measurement of 4.75 cm has three significant figures.

Note: If you use a measuring device that has a digital readout, such as a balance, you should record the measurement just as it appears on the display.

Measurement	Number of Significant Figures
123 g	3
46.54 mL	4
0.33 cm	2
3 300 000 nm	2
0.033 g	2

Step 2.   All non-zero digits are considered significant.

Step 3.   There are special rules for zeros. Zeros in a measurement or calculation fall into three types: middle zeros, leading zeros, and trailing zeros.

Middle zeros are always significant.

303 mm   a middle zero—always significant. This measurement has three significant figures.

A leading zero is never significant. It is only a placeholder, not a part of the actual measurement.

0.0123 kg   two leading zeros—never significant. This measurement has three significant figures.

A trailing zero is significant when it is to the right of a decimal point. This is not a placeholder. It is a part of the actual measurement.

23.20 mL   a trailing zero—significant to the right of a decimal point. This measurement has four significant figures.

The most common errors concerning significant figures are (1) reporting all digits found on a calculator readout, (2) failing to include significant trailing zeros (14.150 g), and (3) considering leading zeros to be significant—0.002 g has only one significant figure, not three.

Practice Problem*

5. How many significant figures are in each of the following?

  a. 451 000 m
  b. 4056 V
  c. 6.626×10^{-34} J·s
  d. 0.0065 g
  e. 0.0540 mL

Using Significant Figures in Calculations

Addition and subtraction: The number of *decimal places* in the answer should be the same as in the measured quantity with the smallest number of *decimal places*.

> *Sample Problem 1:* Add the following measured values and express the answer to the correct number of significant figures.
>
> $$\begin{array}{r} 1259.1 \ \text{g} \\ 2.365 \ \text{g} \\ +\ 15.34 \ \text{g} \\ \hline 1276.805 \ \text{g} \end{array} \quad = \textbf{1276.8 g}$$

Multiplication and division: The number of *significant figures* in the answer should be the same as in the measured quantity with the smallest number of *significant figures*.

> *Sample Problem 2:* Divide the following measured values and express the answer to the correct number of significant figures.
>
> $$\frac{13.356 \ \text{g}}{10.42 \ \text{mL}} = 1.2817658 \ \text{g/mL} = \textbf{1.282 g/mL}$$

Practice Problem

6. Report the answer to each of these using the correct number of significant figures.

   a. 16.27 g + 0.463 g + 32.1 g

   b. 42.04 mL − 3.5 mL

   c. 15.1 km × 0.032 km

   d. $\dfrac{13.36 \ \text{cm}^3}{0.0468 \ \text{cm}^3}$

Note: Only measurements resulting from scale readings or digital readouts carry a limited number of significant figures. However, values arising from direct counting (such as 25 students in a classroom) or from definitions (such as 100 cm = 1 m, or 1 dozen = 12 things) carry unlimited significant figures. Such "counted" or "defined" values are regarded as exact.

Interpreting Graphs

Graphs are of four basic types: pie charts, bar graphs, line graphs, and *x-y* plots. The type chosen depends on the characteristics of the data displayed.

Pie charts show the relationship of the parts to the whole. This presentation helps a reader visualize the magnitude of difference among various parts. They are made by taking a 360° circle and dividing it into wedges according to the percent of the whole represented by each part.

Bar graphs and **line graphs** compare values within a category or among categories. The horizontal axis (*x*-axis) is used for the quantity that can be controlled or adjusted. This quantity is the **independent variable.** The vertical axis (*y*-axis) is used for the quantity that is influenced by the changes in the quantity on the *x*-axis. This quantity is the **dependent variable.** For example, a bar graph could present a visual comparison of the fat content (dependent variable on the *y*-axis) of types of cheese (independent variable on the *x*-axis). Such a graph would make it easy to choose a cheese snack with a low-fat content. Bar graphs can also be useful in studying trends over time.

Graphs involving ***x-y* plots** are commonly used in scientific work. Sometimes it is difficult to decide if a graph is a line graph or an *x-y* plot. In an *x-y* plot, it is possible to determine a mathematical relationship between the variables. Sometimes the relationship is the equation for a straight line ($y = mx + b$), but other times it is more complex and may require transformation of the data to produce a simpler graphical relationship. The first example below refers to a straight-line or direct relationship.

Sample Problem 1: A group of entrepreneurs was considering investing in a mine that was said to contain gold. To verify this claim, they gave several small, irregular mined particles to a chemist, who was told to use nondestructive methods to analyze the samples.

The chemist decided to determine the density of the small samples. The chemist found the volume of each particle and determined its mass. The data collected are shown in the table below.

Particle	Volume (mL)	Mass (g)
1	0.006	0.116
2	0.012	0.251
3	0.015	0.290
4	0.018	0.347
5	0.021	0.386

Use the x-y *plot of these data to evaluate whether the particles are gold.*

Gold Particle Data

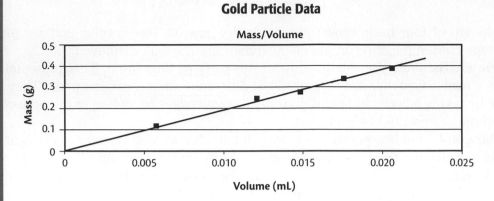

Since the *x-y* plot is linear, the sample materials are likely the same. If point (0, 0) is included (a sample of zero volume has zero mass), the slope is about 19 g/mL, close to gold's density (19.3 g/mL), so these are likely gold particles.

Sample Problem 2: The graph below is a plot of data gathered at constant temperature involving the volume of 2.00 mol of ammonia (NH_3) gas measured at various pressures. Determine whether the measurements are related by a simple mathematical relationship.

Volume of 2.00 mol NH_3 at Different Pressures

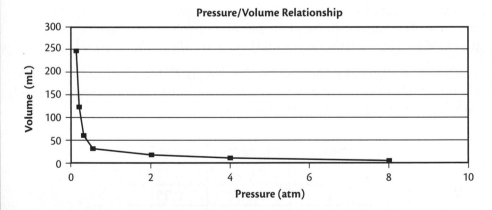

When the best smooth graph line is not a straight line (or direct relationship), the data can be manipulated to see if any other simple mathematical relationship is possible. Several of the most common types of mathematical relationships are inverse, exponential, and logarithmic. Each relationship has unique characteristics that can often be identified from the graphical presentation. Knowing the mathematical relationship allows scientists to interpret the data. In this case, it appears that as the pressure increases the volume decreases, which is a characteristic of an inverse relationship. In testing this theory, the value of 1/V (the inverse of the volume) can be calculated, recorded in another column of the table, and plotted versus the pressure.

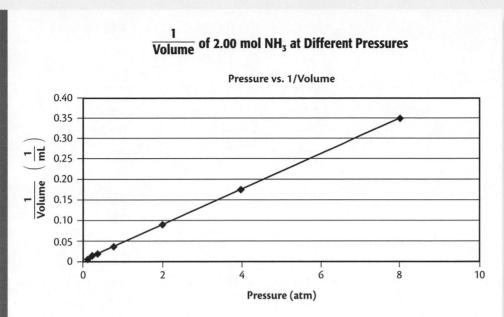

$\dfrac{1}{\text{Volume}}$ **of 2.00 mol NH$_3$ at Different Pressures**

Pressure vs. 1/Volume

This graph exhibits a straight line showing that pressure is directly related to the inverse of the volume. This leads to the mathematical result that pressure and volume are inversely related. If this mathematical manipulation had not resulted in a straight line, some other reasonable relationships might have been considered and tested.

Equations, Moles, and Stoichiometry

Moles and Molar Mass

The **mole** is regarded as a counting number–a number used to specify a certain number of objects. **Pair** and **dozen** are other examples of counting numbers. A mole equals 6.02×10^{23} objects. Most often, things that are counted in units of moles are very small—atoms, molecules, or electrons.

The modern definition of a mole specifies that one mole is equal to the number of atoms contained in exactly 12 g of the carbon-12 isotope. This number is named after Amedeo Avogadro, who proposed the idea, but never determined the number. At least four different types of experiments have accurately determined the value of Avogadro's number—the number of units in one mole. Avogadro's number is known to eight significant figures, but three will be enough for most of your calculations—6.02×10^{23}.

The modern atomic weight scale is also based on C-12. Compared to C-12 atoms with a defined atomic mass of exactly 12, hydrogen atoms have a relative mass of 1.008. Therefore, one mole of hydrogen atoms has a mass of 1.008 g. One mole of oxygen atoms equals 15.9994 g—also called the molar mass of oxygen. The total mass of one mole of a compound is found by adding the atomic weights of all of the atoms in the formula and expressing the sum in units of grams.

> **Sample Problem:** *Calculate the mass of one mole of water.*
>
> $$2 \text{ mol H} \times \frac{1.008 \text{ g H}}{1 \text{ mol H}} = 2.016 \text{ g H}$$
>
> $$1 \text{ mol O} \times \frac{16.00 \text{ g H}}{1 \text{ mol O}} = 16.00 \text{ g O}$$
>
> molar mass of water = 2.016 g + 16.00 g = 18.02 g

Practice Problems*

Find the molar mass (in units of g/mol) for each of the following:

1. Acetic acid, CH_3COOH

2. Formaldehyde, $HCHO$

3. Glucose, $C_6H_{12}O_6$

4. 2-Dodecanol, $CH_3(CH_2)_9CH(OH)CH_3$

*Answers to odd-numbered problems can be found on page ANS-1.

Gram–Mole Conversions

Conversions between grams and moles can be readily accomplished by using the technique of dimensional analysis (see Appendix B).

> *Sample Problem 1:* What mass (in grams) of water contains 0.25 mol H_2O?
>
> The mass of one mole of water (18.02 g) is found as illustrated earlier. Two factors can be written, based on that relationship:
>
> $$\frac{1 \text{ mol } H_2O}{18.02 \text{ g } H_2O} \text{ and } \frac{18.02 \text{ g } H_2O}{1 \text{ mol } H_2O}$$
>
> The second conversion factor is chosen so that each unit to be converted is divided by itself and thus cancelled. Units for the answer are g H_2O, as expected.
>
> $$0.25 \text{ mol } H_2O \times \frac{18.02 \text{ g } H_2O}{1 \text{ mol } H_2O} = 4.5 \text{ g } H_2O$$

> *Sample Problem 2:* How many moles of water molecules are present in a 1.00-kg sample of water?
>
> $$1.00 \text{ kg } H_2O \times \frac{1000 \text{ g } H_2O}{1 \text{ kg } H_2O} \times \frac{1 \text{ mol } H_2O}{18.02 \text{ g } H_2O} = 55.5 \text{ mol } H_2O$$

Practice Problems*

5. Acetic acid, CH_3COOH, and salicylic acid, $C_7H_6O_3$, can be chemically combined to form aspirin. If a chemist uses 5.00 g salicylic acid and 10.53 g acetic acid, how many moles of each compound are involved?

6. Calcium chloride hexahydrate, $CaCl_2 \cdot 6\ H_2O$, can be sprinkled on sidewalks to melt ice and snow. How many moles of that compound are in a 5.0-kg sack of that substance?

A Quantitative Understanding of Chemical Formulas

Calculating percent composition from a formula

The percent by mass of each component found in a sample of material is called its **percent composition.** To find the percent of an element in a particular compound, first calculate the molar mass of the compound. Then find the total mass of the element contained in one mole of the compound. Then divide the mass of the element by the molar mass of the compound and multiply the result by 100%.

> *Sample Problem 1:* Calculate the molar mass of sucrose, $C_{12}H_{22}O_{11}$
>
> | | | |
> |---|---|---|
> | 12 mol C (12.01 g/mol) | = | 144.1 g |
> | 22 mol H (1.008 g/mol) | = | 22.18 g |
> | 11 mol O (16.00 g/mol) | = | 176.0 g |
> | Molar mass of $C_{12}H_{22}O_{11}$ | = | 342.3 g |

*Answers to odd-numbered problems can be found on page ANS-1.

Sample Problem 2: *Find the mass percent of each element in sucrose (rounded to significant figures).*

$$\% \text{ C} = \frac{\text{mass C}}{\text{mass C}_{12}\text{H}_{22}\text{O}_{11}} \times 100\% = \frac{141.1 \text{ g C}}{342.3 \text{ g C}_{12}\text{H}_{22}\text{O}_{11}} \times 100\% = 42.10\% \text{ C}$$

$$\% \text{ H} = \frac{\text{mass H}}{\text{mass C}_{12}\text{H}_{22}\text{O}_{11}} \times 100\% = \frac{22.18 \text{ g H}}{342.3 \text{ g C}_{12}\text{H}_{22}\text{O}_{11}} \times 100\% = 6.48\% \text{ H}$$

$$\% \text{ O} = \frac{\text{mass O}}{\text{mass C}_{12}\text{H}_{22}\text{O}_{11}} \times 100\% = \frac{176.0 \text{ g O}}{342.3 \text{ g C}_{12}\text{H}_{22}\text{O}_{11}} \times 100\% = 51.42\% \text{ O}$$

Practice Problems*

7. Barium sulfate, $BaSO_4$, is commonly used to detect gastrointestinal tract abnormalities; it is administered as a water suspension by mouth prior to X-ray imaging. Find the mass percent of each element in barium sulfate.

8. Sodium acetate is a common component in commercial thermal packs. Find the mass percent of each element in sodium acetate, $NaCH_3COO$.

Deriving formulas from percent composition data

An **empirical formula** gives the relative numbers of each element in a substance, using the smallest whole numbers for subscripts. The empirical formula of a compound can be calculated from percent composition data. A formula requires the relative numbers of moles of each element, so percent values must be converted to grams and grams to moles. It is easiest to assume 100 g of compound. Then each percent value is equal to the total grams of that element.

Sample Problem 1: *A hydrocarbon consists of 85.7% carbon and 14.3% hydrogen by mass. What is its empirical formula?*

Step 1. Assume 100 g of compound—thus 85.7 g are carbon and 14.3 g are hydrogen.

Step 2. Use dimensional analysis to find the total moles of each element in the compound.

$$85.7 \text{ g C} \times \frac{1 \text{ mol C}}{12.01 \text{ g C}} = 7.14 \text{ mol C}$$

$$14.3 \text{ g H} \times \frac{1 \text{ mol H}}{1.008 \text{ g H}} = 14.2 \text{ mol H}$$

Step 3. Determine the smallest whole-number ratio of moles of elements by dividing all mole values by the smallest value.

$$\frac{7.14 \text{ mol C}}{7.14} = 1 \text{ mol C}$$

$$\frac{14.2 \text{ mol H}}{7.14} = 1.99 \text{ mol H}$$

The ratio of moles C to moles H = 1:2, so the empirical formula for the compound must be CH_2.

*Answers to odd-numbered problems can be found on page ANS-1.

> *Sample Problem 2:* *The percent composition of one of the oxides of nitrogen is 74.07% oxygen and 25.93% nitrogen. What is the empirical formula of that compound?*
>
> **Step 1.** 100 g of compound consists of 74.07 g oxygen and 25.93 g nitrogen.
>
> **Step 2.** $25.93 \ \cancel{g\,N} \times \dfrac{1 \ \text{mol N}}{14.01 \ \cancel{g\,N}} = 1.851 \ \text{mol N}$
>
> $74.07 \ \cancel{g\,O} \times \dfrac{1 \ \text{mol O}}{16.00 \ \cancel{g\,O}} = 4.629 \ \text{mol O}$
>
> **Step 3.** $\dfrac{1.851 \ \text{mol N}}{1.851} = 1 \ \text{mol N}$
>
> $\dfrac{4.629 \ \text{mol O}}{1.851} = 2.50 \ \text{mol O}$
>
> **Step 4.** This ratio (1:2.5) does not consist entirely of whole numbers. Thus all the numbers in the ratio must be multiplied by a number that converts the decimal to a whole number. In this case the number is 2, and the empirical formula becomes N_2O_5.

Practice Problems*

9. The percent composition by mass of an industrially important substance is 2.04% H, 32.72% S, and 65.24% O. What is the formula of this compound?

10. Determine the empirical formula of a compound that contains (by mass) 38.71% C, 9.71% H, and 51.58% O.

Mass Relationships in Chemical Reactions

Since the total number of atoms is conserved in a chemical reaction, their masses must also be conserved, as expected from the law of conservation of mass. In the equation for the formation of water from the elements hydrogen and oxygen, $2H_2(g) + O_2(g) \longrightarrow 2H_2O(l)$, 2 molecules of hydrogen gas and 1 molecule of oxygen gas combine to form 2 molecules of water. One could also interpret the equation this way: 2 mol hydrogen gas react with 1 mol oxygen gas to form 2 mol water. Using the molar mass of each substance, the mass relationships in the table below can be determined. The ratio of moles of hydrogen gas to moles of oxygen gas in forming water will be 2:1. If 10 mol hydrogen gas are available, 5 mol oxygen gas are required.

$2 H_2(g)$ +	$O_2(g)$ $\longrightarrow$	$2 H_2O(l)$
2 molecules	1 molecule	2 molecules
2 mol	1 mol	2 mol
2 mol × (2.02 g/mol)	1 mol × (32.00 g/mol)	2 mol × (18.02 g/mol)
4.04 g	32.00 g	36.04 g

Solving problems involving the masses of products and/or reactants is conveniently accomplished by dimensional analysis. And remember, all numerical problems involving chemical reactions involve a correctly balanced equation.

*Answers to odd-numbered problems can be found on page ANS-1.

Sample Problem: *Find the mass of water formed when 10.0 g hydrogen gas completely reacts with oxygen gas.*

$$2\ H_2(g)\ +\ O_2(g) \longrightarrow 2\ H_2O(l)$$

Step 1.   Find the moles of hydrogen gas represented by 10.0 g, using the molar mass of H_2.

$$10.0\ \cancel{g\ H_2} \times \frac{1\ mol\ H_2}{2.02\ \cancel{g\ H_2}} = 4.95\ mol\ H_2$$

Step 2.   Find the total moles of H_2O produced by 4.95 mol H_2. From the balanced equation, you know that for every 2 mol H_2, 2 mol H_2O are produced.

$$4.95\ \cancel{mol\ H_2} \times \frac{2\ mol\ H_2O}{2\ \cancel{mol\ H_2}} = 4.95\ mol\ H_2O$$

Step 3.   Find the mass of H_2O that contains 4.95 mol H_2O by using the molar mass of water.

$$4.95\ \cancel{mol\ H_2O} \times \frac{18.02\ g\ H_2O}{1\ \cancel{mol\ H_2O}} = 89.2\ g\ H_2O$$

Most chemistry students find it is more convenient to set up all three steps in one extended calculation. Assure yourself that all units divide and cancel except for grams of water (g H_2O)—an appropriate way to express the answer sought in the problem.

$$10.0\ \cancel{g\ H_2} \times \underbrace{\frac{1\ \cancel{mol\ H_2}}{2.02\ \cancel{g\ H_2}}}_{\substack{\text{Molar mass} \\ \text{of } H_2}} \times \underbrace{\frac{2\ \cancel{mol\ H_2O}}{2\ \cancel{mol\ H_2}}}_{\substack{\text{Coefficients} \\ \text{in equation}}} \times \underbrace{\frac{18.02\ g\ H_2O}{1\ \cancel{mol\ H_2O}}}_{\substack{\text{Molar mass} \\ \text{of } H_2O}} = 89.2\ g\ H_2O$$

Practice Problems*

11. Find the mass of copper(II) oxide formed if 2.0 g copper metal completely reacts with oxygen gas.

$$2\ Cu(s) + O_2(g) \longrightarrow 2\ CuO(s)$$

12. What mass of water is produced when 25.0 g methane gas reacts completely with oxygen gas?

$$CH_4(g) + 2\ O_2(g) \longrightarrow 2\ H_2O(g) + CO_2(g)$$

*Answers to odd-numbered problems can be found on page ANS-1.

Answers to Practice Problems

Appendix B

1. a. 3.69×10^{-5}
   b. 4.52×10^{-2}
   c. 4.52×10^{6}
   d. 3.65×10^{5}
3. 4741 km
5. a. 3
   b. 4
   c. 4
   d. 2
   e. 3

Appendix D

1. 60.05 g/mol
3. 180.16 g/mol
5. 0.1754 mol acetic acid, 0.0362 mol salicylic acid
7. 58.84% Ba, 13.74% S, 27.42% O
9. H_2SO_4
11. 2.5 g CuO

Glossary

A

absolute zero
the lowest temperature theoretically obtainable, which is −273 °C or 0 K (zero kelvin)

absorbance
a measure of the amount of light that is absorbed by a particular substance/solution

accuracy
extent to which a measurement represents its corresponding actual values

acid precipitation
see acid rain

acid rain
fog, sleet, snow, or rain with a pH lower than about 5.6 due to dissolved gases such as SO_2, SO_3, and NO_2

acidic solution
an aqueous solution with a pH less than 7; it turns litmus from blue to red

acid
ion or compound that produces hydrogen ions, H^+ (or hydronium ions, H_3O^+), when dissolved in water

activation energy
the minimum energy required for the successful collision of reactant particles in a chemical reaction

active site
the location on an enzyme where a substrate molecule becomes positioned for a reaction

activity series
the ranking of elements in order of chemical reactivity

addition polymer
a polymer formed by repeated addition reactions at double or triple bonds within monomer units

addition reaction
a reaction at the double or triple bond within an organic molecule

adhesive force
force that causes molecules of different substances to be attracted to one another

adsorb
to take up or hold molecules or particles to the surface of a material

alcohol
a nonaromatic organic compound containing one or more —OH groups

alkali metal family
the group of elements consisting of lithium, sodium, potassium, rubidium, cesium, and francium

alkaline
a basic solution containing an excess of hydroxide ions (OH^-)

alkane
a hydrocarbon containing only single covalent bonds

alkene
a hydrocarbon containing one or more double covalent bonds

alkyne
a hydrocarbon containing one or more triple covalent bonds

alloy
a solid solution consisting of atoms of two or more metals

alpha particle (α)
high-speed, positively charged particle emitted during the decay of some radioactive elements; consists of a helium nucleus, $^4_2He^{2+}$

amino acid
an organic molecule containing a carboxylic acid group and an amine group; serves as a protein building block

anion
a negatively charged ion

anode
an electrode in an electrochemical cell at which oxidation occurs

aqueous solution
a solution in which water is the solvent

aquifer
a structure of porous rock, sand, or gravel that holds water beneath Earth's surface

area
the total space that makes up the surface of an object

aromatic compound
a ring-like compound, such as benzene, that can be represented as having alternating double and single bonds between carbon atoms

arterial plaque
build-up in blood vessel walls that results from fatty material

atherosclerosis
condition commonly known as "hardening of the arteries"

atmosphere
(a) a unit of gas pressure (atm); (b) the gaseous envelope surrounding Earth and composed of four layers: troposphere, stratosphere, mesosphere, and thermosphere

atom
the smallest particle possessing the properties of an element

atomic number
the number of protons in an atom; this value distinguishes atoms of different elements

Avogadro's law
equal volumes of all gases, measured at the same temperature and pressure, contain the same number of gas molecules

B

background radiation
the relatively constant level of natural radioactivity that is always present

balanced
in a chemical equation, when the total number of each type of atom is the same for both the reactants and products

balanced chemical equation
see chemical equation

ball-and-stick model
molecular model where each ball represents an atom, and each stick represents a pair of shared electrons (a single covalent bond) connecting two atoms

base
ion or compound that produces OH^- ions when dissolved in water

base unit
an SI unit that expresses a fundamental physical quantity (such as temperature, length, or mass)

basic solution
an aqueous solution with a pH greater than 7; it turns litmus from red to blue

battery
a device composed of one or more connected voltaic cells that supplies electrical current

beta decay
the radioactive decay of a nucleus accompanied by the emission of a beta particle

beta particle (β)
negatively charged particle emitted during the decay of some radioactive elements; high-speed electron

bias
a tendency to a particular belief or perspective

biodiesel
an alternative fuel or fuel additive for diesel engines made from various materials such as new or recycled vegetable oils and animal fats

biomolecule
large organic molecule found in living systems

blank
a solution or substance known not to contain any ions or molecules of interest

bottoms
components from petroleum found in the lower trays of a fractionating tower after distillation

Boyle's law
the pressure and volume of a gas sample at constant temperature are inversely proportional; $PV = k$

branched-chain alkane
an alkane in which at least one carbon atom is bonded to three or four other carbon atoms

branched polymer
a polymer formed by reactions that create numerous side chains rather than linear chains

brittle
a property of a material that causes it to shatter under pressure

buffer
a substance or combination of dissolved substances capable of resisting changes in pH when limited quantities of either acid or base are added

C

calibration curve
a graph constructed from data collected on solutions of known concentration

Calorie (Cal)
a unit of energy; thermal energy required to raise the temperature of one kilogram of water by one degree Celsius; commonly used to express quantity of food energy; informally called *food calorie*; 1 Cal = 1 000 cal; 1 kJ = 4.184 Cal

calorimeter
measuring device to determine energy released from the burning of a substance or other chemical reactions

calorimetry
procedure to determine the energy released during the combustion of several types of fuel or from other chemical reactions

carbohydrate
substance such as sugar or starch that is composed of carbon, hydrogen, and oxygen atoms; a main source of energy in foods

carbon chain
carbon atoms chemically linked to one another, forming a chainlike molecular structure

carbon cycle
the movement of carbon atoms within Earth's ecosystems, from carbon storage as plant and animal matter, through release as carbon dioxide due to cellular respiration, combustion, and decay, to reacquisition by plants

carbon footprint
the quantity of greenhouse gases emitted based upon individual activities; measured in kilograms of carbon dioxide (CO_2)

carboxylic acid
an organic compound containing the —COOH group

carcinogen
substance known to cause cancer

catalyst
a substance that speeds up a chemical reaction but is itself unchanged

catalytic convertor
the reaction chamber in an auto exhaust system designed to accelerate the conversion of potentially harmful exhaust gases to nitrogen gas, carbon dioxide, and water vapor

cathode
an electrode in an electrochemical cell at which reduction occurs

cathode ray
a beam of electrons emitted from a cathode when electricity is passed through an evacuated tube

cation
a positively charged ion

cellular respiration
the process that is used by organisms to convert complex organic molecules into carbon dioxide and water molecules, with an overall release of energy

chain reaction
in nuclear fission, a reaction that is sustained because it produces enough neutrons to collide with and split additional fissionable nuclei

Charles' law
the volume of a gas sample at constant pressure is directly proportional to its kelvin temperature; $V = kT$

chemical bond
the attractive force that holds atoms or ions together; *see also* covalent bonds; ionic bonds

chemical change
an interaction of matter that results in the formation of one or more new substances

chemical energy
a form of potential energy stored in chemical compounds

chemical equation
a symbolic expression summarizing a chemical reaction, such as $2 H_2(g) + O_2(g) \longrightarrow 2 H_2O(g)$

chemical formula
a symbolic expression representing the elements contained in a substance, together with subscripts that indicate the relative numbers of atoms of each element, such as H_2O

chemical kinetics
study of the rate of chemical reactions

chemical properties
properties only observed or measured by changing the chemical identity of a sample of matter

chemical reaction
the process of forming new substances from reactants that involves the breaking and forming of chemical bonds

chemical symbol
an abbreviation of an element's name, such as N for nitrogen or Fe for iron

cis-trans isomerism
isomers based on arrangement about a double bond in a molecule

claim
one- or two-sentence statement summarizing an important result of an investigation

climate
the average or prevailing weather conditions in a region

cloud chamber
a container filled with supersaturated air that, when cooled and exposed to ionizing radiation, produces visible trails of condensation, tracing paths taken by radioactive emissions

coefficient
a number in a chemical equation that indicates the relative number of units of a reactant or product involved in the reaction

coenzyme
an organic molecule that interacts with an enzyme to facilitate or enhance its activity

cohesive force
attractive force between molecules in a substance, especially in a liquid

collision theory
for a reaction to occur, reactant molecules must collide in proper orientation with sufficient kinetic energy

colloid
a mixture containing solid particles small enough to remain suspended and not settle out

colorimetry
a chemical analysis method that uses color intensity to determine solution concentration

combustion
a chemical reaction with oxygen gas that produces heat and light; burning

complementary proteins
multiple protein sources that provide adequate amounts of all essential amino acids when consumed together

complete protein
a protein source for humans containing adequate amounts of all essential amino acids

complex ion
a single central atom or ion, usually a metal ion, to which other atoms, molecules, or ions are attached

compound
a substance composed of two or more elements bonded together in fixed proportions; a compound cannot be broken down into simpler substances by physical means

compressed natural gas (CNG)
natural gas condensed under high pressure (160–240 atm) and stored in metal cylinders; CNG can serve as a substitute for gasoline or diesel fuel

concentration
see solution concentration

condensation
converting a substance from a gaseous state to a liquid state

condensation polymer
a polymer formed by repeated condensation reactions of one or more monomers

condensation reaction
the chemical combination of two organic molecules, accompanied by the loss of water or other small molecules

condensed formula
a chemical formula that provides additional information about bonding; for example, the condensed formula for propane, C_3H_8, is CH_3—CH_2—CH_3 or $CH_3CH_2CH_3$

conductor
a material that allows electricity (or thermal energy) to flow through it

confirming test
a laboratory test giving a positive result if a particular chemical species is present

control
in an experiment, a trial that duplicates all conditions except for the variable under investigation

covalent bond
a linkage between two atoms involving the sharing of one pair (single bond), two pairs (double bond), or three pairs (triple bond) of electrons

cracking
the process in which hydrocarbon molecules from petroleum are converted to smaller molecules, using thermal energy and a catalyst

criteria pollutant
an Environmental Protection Agency classification for a pollutant commonly found throughout the United States and detrimental to human health or the environment, such as carbon monoxide (CO), sulfur oxides (SO_x), nitrogen oxides (NO_x), ozone (O_3), lead (Pb), and particulate matter (PM)

critical mass
the minimum mass of fissionable material needed to sustain a nuclear chain reaction

cross-linking
polymer chains interconnected by chemical bonds; causes polymer rigidity

crude oil
unrefined liquid petroleum as it is pumped from the ground by oil wells

crystal
A solid 3-D network with a regular arrangement of anions and cations

currency
circulating money, includes both coins and bills (banknotes)

cycloalkane
a saturated hydrocarbon containing carbon atoms joined in a ring

cyclotron
accelerator used by scientists to conduct nuclear reactions, particularly bombardment reactions to produce new elements

D

data
objective pieces of information, such as information gathered in a laboratory investigation

decay series
the decay of a particular radioisotope, which yields a different radioisotope that, in turn, decays; this sequential decay process may continue through several more radioisotopes until a stable nucleus is produced

density
the mass per unit volume of a given material that is often expressed as g/cm^3

dependent variable
measured or observed variable in an experiment that is used to draw conclusions about effects of changes to the independent variable

deposit
naturally occurring collection of ores in the lithosphere

derived unit
an SI unit formed by mathematically combining two or more base units

diagnostic
a test that helps doctors understand what is happening inside the body

diatomic molecule
a molecule made up of two atoms, such as chlorine gas, Cl_2, or carbon monoxide, CO

dilution
process of making a solution less concentrated by adding solvent

dimer
a molecule composed of two monomers

dipeptide
a molecule consisting of two amino acids bonded together

direct water use
water consumed by an end user

disaccharide
a sugar molecule (such as sucrose) composed of two monosaccharide units bonded through a condensation reaction

distillate
the condensed products of distillation

distillation
a process that separates liquid substances based on differences in their boiling points; *see also* fractional distillation

dose
quantity of ionizing radiation that people are exposed to over time

dot structure
see electron-dot structure

double covalent bond
a bond in which four electrons are shared between two adjacent atoms

drinking-water treatment
pre-use purification of water that occurs at a filtration and treatment plant

dry cell
battery in which the electrolyte is in paste form, rather than liquid

ductile (ductility)
a property of a material that permits it to be stretched into a wire without breaking

dynamic equilibrium
see equilibrium

E

elastic collision
molecular collision in which there is no gain or loss in total kinetic energy

electric current
the flow of electrons, as through a wire connecting electrodes in a voltaic cell

electrical potential
the tendency for electrical charge to move through an electrochemical cell (based on an element's relative tendency to lose electrons when in contact with a solution of its ions); it is measured in volts (V)

electrochemistry
the study of chemical changes that produce or are caused by electrical energy

electrode
a strip of metal or other conductor serving as a contact between an ionic solution and the external circuit in an electrochemical cell

electrolysis
the process in which a chemical reaction is caused by passing an electrical current through an ionic solution

electromagnetic radiation
radiation ranging from low-energy radio waves to high-energy X-rays and gamma rays; includes visible light

electromagnetic spectrum
comprising the full range of electromagnetic radiation frequencies; *see also* electromagnetic radiation

electron
a particle possessing a negative electrical charge; electrons surround the nuclei of atoms

electron-dot formula
see Lewis dot structure

electron-dot structure
a structure of a substance or ion in which dots represent the valence electrons in each atom

electronegativity
an expression of the tendency of an atom to attract shared electrons within a chemical bond

electroplating
the deposition of a thin layer of metal on a surface by an electrical process involving oxidation–reduction reactions

elements
the fundamental chemical substances from which all other substances are made

endothermic
a process that requires the addition of energy

endpoint
point where a titration is stopped, usually because an appropriate indicator just changes color

energy efficiency
the use of smaller quantities of energy to achieve the same effect

enzyme
biological catalyst

equilibrium
the point in a reversible reaction where the rate of products forming from reactants is equal to the rate of reactants forming from products; also called *dynamic equilibrium*

essential amino acid
amino acid not synthesized in adequate quantities by the human body and that must be obtained from protein in the diet

ester
an organic compound containing the —COOR group, where R represents any stable arrangement of bonded carbon and hydrogen atoms

evidence
qualitative observations or quantitative data; experimental support for claims; should be used to answer the questions, "How do I know what I know?" and "Why am I making this claim?"; it should be used in presentable format along with an explanation

exothermic
a process that involves the release of energy

experimental design
relying on making measurements on one variable while changing another variable for investigations

extrapolation
the process of estimating a value beyond a known range of data points

F

family (periodic table)
see group

fat
energy-storage molecule composed of carbon, hydrogen, and oxygen

fatty acid
organic compound made up of long hydrocarbon chains with carboxylic acid groups at one end

filtrate
the liquid collected after filtration

filtration
the process of separating solid particles from a liquid by passing the mixture through a material that retains the solid particles

fixed
the process in which nitrogen is combined with other elements to produce nitrogen-containing compounds that plants can use chemically

fluorescence
the emission of visible light when exposed to radiant energy (usually ultraviolet radiation)

food additive
a substance added during food processing that is intended to increase nutritive value or enhance storage life, visual appeal, or ease of production

food group
one of several categories for foods that humans eat, typically grains, vegetables, fruits, milk, meats, and beans

force
a push or pull exerted on an object; expressed by the newton (N), an SI unit

formula unit
a group of atoms or ions represented by a compound's chemical formula; simplest unit of an ionic compound

fossil fuel
a fuel (such as coal, petroleum, or natural gas) believed to be formed from plant or animal remains that were buried under Earth's surface for millions of years

fraction
(a) a mixture of petroleum-based substances with similar boiling points and other properties; (b) one of the substances collected during distillation

fractional distillation
a process of separating a mixture into its components by boiling and condensing the components; *see also* distillation

frequency (υ)
the number of waves that pass a given point each second; in other words, the rate of oscillation; for electromagnetic radiation, the product of frequency and wavelength equals the speed of light

fuel cell
a device for directly converting chemical energy into electrical energy by chemically combining a fuel (such as hydrogen gas) with oxygen gas; does not involve combustion

functional group
an atom or a group of atoms that imparts characteristic properties to an organic compound

G

gamma ray (γ)
high-energy electromagnetic radiation emitted during the decay of some radioactive elements

global warming
the observed and predicted increases in average global surface temperatures

glycogen
the carbohydrate by which animals store glucose

gray (Gy)
the unit expressing the quantity of ionizing radiation delivered to tissue; it equals one joule absorbed per kilogram of body tissue

Green Chemistry
the design of chemical products and processes that require fewer resources and less energy and that reduces or eliminates reliance on and generation of hazardous substances

greenhouse effect
the trapping and returning of infrared radiation to Earth's surface by atmospheric substances such as water and carbon dioxide

greenhouse gas
atmospheric substance that absorbs infrared radiation, such as CO_2, N_2O, and CH_4

groundwater
water from an aquifer or other underground source

group (periodic table)
a vertical column of elements in the periodic table; also called a *family*; group members share similar properties

H

Haber–Bosch process
the industrial synthesis of ammonia from hydrogen and nitrogen gases, which involves high pressure and temperature accompanied by a suitable catalyst

half-cell
a metal (or other electrode material) in contact with a solution of ions to form one half of a voltaic cell

half-life
the time for a radioactive substance to lose half of its radioactivity from decay; at the end of one half-life, 50% of the original radioisotope remains

half-reaction
a chemical equation explicitly showing either a loss of electrons (oxidation) or a gain of electrons (reduction); any oxidation–reduction reaction can be expressed as the sum of two half-reactions

halogen
see halogen family

halogen family
the group of elements consisting of fluorine, chlorine, bromine, iodine, and astatine

hazardous air pollutant
substance in air that is known or suspected to cause cancer or other serious health effects or adverse environmental effects; air toxic

heat
see thermal energy

heat of combustion
the quantity of thermal energy released when a specific quantity of a material burns

heterogeneous mixture
a mixture that is not uniform throughout

high-level radioactive waste
(a) products of nuclear fission, such as those generated in a nuclear reactor; (b) transuranics, products formed when the original uranium-235 fuel absorbs neutrons

histogram
a graph indicating the frequency or number of instances of particular values (or value ranges) within a set of related data

homogeneous mixture
a mixture that is uniform throughout; a solution is a homogeneous mixture

hybrid vehicle
a vehicle that combines two or more power sources; the combination of gasoline and electric power is the most common design

hydrocarbon
a molecular compound composed only of carbon and hydrogen atoms

hydrogen bond
strong intermolecular force in compounds in which a hydrogen atom is bonded directly to an atom of oxygen, nitrogen, or fluorine

hydrogenation
a chemical reaction that adds hydrogen atoms to an organic molecule

hydrologic cycle
see water cycle

hydronium ion
in aqueous solutions, the ion formed when the hydrogen ion released by the acid is bonded to water, commonly represented as $H_3O^+(aq)$

hydrosphere
all parts of Earth where water is found, including oceans, clouds, ice caps, glaciers, lakes, rivers, and underground water supplies

I

ideal gas
a gas that behaves under all conditions as described by the kinetic molecular theory or by the ideal gas law

ideal gas law
a mathematical relationship that describes the behavior of an ideal gas sample, $PV = nRT$, where P = gas pressure; V = gas volume; n = moles of gas; R = gas constant, 0.0821 L • atm/(mol • K); and T = kelvin temperature

independent variable
variable that is manipulated by the investigator during an experiment

indirect water use
water consumed in the preparation, production, or delivery of goods and services

inference
a conclusion based on analysis of data and observations

infrared (IR)
electromagnetic radiation just beyond the red (low-energy) end of the visible spectrum

intermolecular force
force of attraction among molecules

International System of Units (SI)
the modernized metric system

ion
electrically charged atom or group of atoms; negative ions are called *anions*, and positive ions are called *cations*

ionic compound
a substance composed of positive and negative ions

ionize
to convert to ions

ionizing radiation
nuclear radiation and high-energy electromagnetic radiation with sufficient energy to produce ions by ejecting electrons from atoms and molecules

isomer
a molecule that has the same formula as another molecule and differs from it only by the arrangement of atoms or bonds

isomerization
a chemical change involving the rearrangement of atoms or bonds within a molecule without changing its molecular formula

isotope
atom of the same element with differing numbers of neutrons

K

kelvin temperature scale (K)
the absolute temperature scale, where zero kelvins (0 K) represents the theoretical lowest possible temperature; 0 °C = 273 K

kinetic energy
energy associated with the motion of an object

kinetic molecular theory (KMT)
observed gas behavior and gas laws explained by rapidly moving particles (gas molecules) that are relatively far apart and change direction only through collisions with each other or the container walls

kinetics
see chemical kinetics

L

law of conservation of energy
energy can change form but cannot be created or destroyed in any chemical reaction or physical change

law of conservation of matter
matter is neither created nor destroyed in any chemical reaction or physical change

Le Châtelier's principle
the predicted shift in the equilibrium position that partially counteracts the imposed change in conditions

Lewis dot structure
the representation of atoms, ions, and molecules where valence-electron dots surround each atom's symbol; it is useful for indicating covalent bonding; also called *electron-dot structure*

Lewis structure
see Lewis dot structure

life cycle
the sequence of steps that a material or product undergoes from raw materials to product to final disposal

limiting reactant
a starting substance that is used up first in a chemical reaction; sometimes called the *limiting reagent*

liquefied petroleum gas (LPG)
petroleum-based gaseous substance fuel, propane (C_3H_8)

lithosphere
the solid outer layer of Earth, which also includes land areas under oceans and other bodies of water

low-level radioactive waste
nuclear laboratory protective clothing, diagnostic radioisotopes, and air filters from nuclear power plants

luster
the reflection of light from the surface of a material

M

macromineral
mineral essential to human life and occurring in relatively large quantities (at least 5 g) in the body; also called a *major mineral*

macroscopic
large enough to be seen by an unaided eye

magnetic resonance imaging (MRI)
a non-invasive computerized method of imaging soft human tissues by use of a powerful magnet and radio wave

major mineral
see macromineral

malleable
a property of a material that permits it to be flattened without shattering

mass number
the sum of the number of protons and neutrons in the nucleus of an atom of a particular isotope

material's life cycle
several distinct stages, which include aquisition, manufacturing, use/reuse/maintenance, and recycle/ waste management

matter
anything that has mass and occupies space

mean
an expression of central tendency obtained by dividing the sum of a set of values by the number of values in the set; also known as the *average value*

median
within an ascending or descending set of values, the number that represents the middle value with an equal number of values above and below it

metabolism
complex series of interrelated chemical reactions that keep organisms alive

metal
a material possessing properties such as luster, ductility, conductivity, and malleability

metalloid
a material with properties intermediate between those of metals and nonmetals

metastable
nuclei in an energetically excited state; designated by the symbol m

meter (m)
the SI base unit of length

micromineral
mineral essential to human life and occurring in relatively small quantities (less than 5 g) in the body; also called a *trace mineral*

mineral
(a) a naturally occurring solid substance commonly removed from ores to obtain a particular element of interest or value; (b) an inorganic substance needed by the body to maintain good health

mixture
a combination of materials in which each material retains its separate identity

model
tool to understand and interpret natural phenomena that can be physical, mathematical, or conceptual and represent, explain, or predict observed behavior; models are particularly helpful in chemistry to account for molecular-level interactions

molar concentration (M)
see molarity

molar heat of combustion
the quantity of thermal energy released from burning one mole of a substance

molar mass
the mass (usually in grams) of one mole of a substance

molar volume
the volume occupied by one mole of a substance; at 0 °C and 1 atm, the molar volume of any gas is 22.4 L

molarity (M)
the concentration determined by dividing the total moles of solute by the solution volume (expressed in liters); also known as *molar concentration*

mole (mol)
the SI unit for amount of a substance, equal to 6.02×10^{23} units, where the unit may be any specified entity; it is the chemist's "counting" unit

molecular formula
a chemical expression indicating the total atoms of each element contained in one molecule of a particular substance

molecule
the smallest particle of a substance retaining all the properties of that substance

monatomic ion
ion composed of only one atom, such as chloride (Cl^-)

monomer
a compound whose molecules can react to form the repeating units of a polymer

monosaccharide
a simple sugar (such as glucose or fructose) that cannot be hydrolyzed to produce other sugars

monounsaturated fat
a fat molecule containing one carbon–carbon double bond

mutagen
a material that causes mutations in DNA

mutation
changes in the structure of DNA that may result in production of altered protein material

N

negative oxidation state
the negative number assigned to an atom in a compound when that atom has greater control of bonding electrons than the control exerted by one or more atoms to which it is bonded

net ionic equation
equation written without the spectator ions, such as $Ag^+(aq) + Cl^-(aq) \longrightarrow AgCl(s)$

neutral solution
a water solution in which H^+ (H_3O^+) and OH^- concentrations are equal (pH = 7)

neutralization
combining an acid and a base in amounts that result in the elimination of all excess acid or base

neutron
a particle without electrical charge; found in the nucleus of an atom

newton (N)
the SI unit of force that is roughly equal to the force exerted by a mass of 100 g at Earth's surface

nitrogen cycle
the movement of atmospheric nitrogen atoms through Earth's ecosystems via collection by bacteria, conversion into ammonia or ammonium ions, conversion into nitrate ions, uptake by plants, passage through the food chain, release as ammonia or ammonium ions, and conversion back into atmospheric nitrogen

noble gas family
an unreactive element belonging to the last (right-most) group on the periodic table

nonconductor
a material that does not allow electrical current (or thermal energy) to flow through it

nonionizing radiation
electromagnetic radiation in the visible and lower-energy regions of the electromagnetic spectrum with insufficient energy to form ions when it transfers energy to matter

nonmetal
a material possessing properties such as brittleness, lack of luster, and nonconductivity; nonmetals are often insulators

nonpolar molecule
a molecule that has an even distribution of electrical charge with no regions of partial positive and negative charge

nonrenewable resource
a resource in limited supply that cannot be replenished by natural processes over the time frame of human experience

nuclear fission
splitting an atom into two smaller nuclei

nuclear fusion
the combination of two nuclei to form a new, more massive nucleus

nuclear power plant
a facility where thermal energy generated by the controlled fission of nuclear fuel drives a steam turbine, producing electrical power; in other words, a facility that converts nuclear energy into electricity

nuclear radiation
a form of ionizing radiation that results from changes in the nuclei of atoms

nucleus, atomic
the dense, positively charged central region of an atom that contains protons and neutrons

O

observation
data/information that you can collect with your senses; what you see, hear, feel, or smell

octane rating
a measure of the combustion quality of gasoline compared to the combustion quality of isooctane; the higher the number, the higher the octane rating; also called *octane number*

oil sands
source of petroleum that contains bitumen, a viscous, heavy crude oil

oil shale
sedimentary rock containing a material (kerogen) that can be converted to crude oil

ore
a rock or other solid material from which it is profitable to recover a mineral containing a metal or other useful substances

organic chemistry
a branch of chemistry dealing with hydrocarbons and their derivatives

oxidation
any process in which one or more electrons can be considered as lost by a chemical species

oxidation state
the apparent state of oxidation of an atom; also called *oxidation number*

oxidation–reduction (redox) reaction
a chemical reaction in which oxidation and reduction simultaneously occur

oxidized
see oxidation

oxidizing agent
a species that causes another atom, molecule, or ion to become oxidized; the oxidizing agent becomes reduced in this process

oxygenated fuel
a fuel with oxygen-containing additives, such as methanol, that increase the octane rating and reduce harmful emissions

P

paper chromatography
a method for separating substances that relies on solution components having different attractions to the solvent (mobile phase) and paper (stationary phase)

particulate level
the realm of unseen atoms, molecules, and ions in contrast to the observable macroscopic entities

particulate pollutant
microscopic particle that enters the air from either human activities or natural processes

pascal (Pa)
the SI pressure unit; equal to one newton of force applied per square meter, $1\ Pa = 1\ N/m^2$

peptide bond
the chemical bond that links amino acids together in peptides and proteins

percent composition
the percent by mass of each component in a material; or, specifically, the percent by mass of each element within a compound

percent recovery
the proportion of sought material recovered in a process

period (periodic table)
a horizontal row of elements in the periodic table

periodic properties
chemical or physical properties that vary among elements according to trends that repeat as atomic number increases

periodic relationship
regular patterns among chemical and physical properties of elements arrayed on a periodic table

periodic table of the elements
an arrangement of elements in order of increasing atomic number, such that elements with similar properties are located in the same vertical column (group)

petrochemical
any organic compound produced from petroleum or natural gas

pH scale
method used as a convenient way to measure and report acidic, basic, or chemically neutral character of a solution; pH is based on a solution's hydrogen ion (H^+ or H_3O^+) molar concentration

photochemical smog
a potentially hazardous mixture of secondary pollutants formed by solar irradiation of certain primary pollutants in the presence of oxygen

photon
an energy bundle of electromagnetic radiation that travels at the speed of light

photosynthesis
the process by which green plants and some microorganisms use solar energy to convert water and carbon dioxide to carbohydrates (stored chemical energy)

physical change
a change in matter in which the identity of the material involved does not change

physical property
a property that can be observed or measured without changing the identity of the sample of matter

polar molecule
a molecule with regions of partial positive and negative charge resulting from the uneven distribution of electrical charge

pollutant
an undesirable contaminant that adversely affects the chemical, physical, or biological characteristics of the environment

polyatomic ion
an ion composed of two or more atoms, such as the ammonium cation, NH_4^+, or acetate anion, $C_2H_3O_2^-$

polymer
a molecule composed of very large numbers of identical repeating units

polysaccharide
a polymer composed of many monosaccharide units

polyunsaturated fat
a fat molecule containing two or more carbon–carbon double bonds

positive oxidation state
the positive number assigned to an atom in a compound when that atom has less control of its electrons than it has as a free element

positron
a positively charged subatomic particle with the same mass as an electron; the antimatter counterpart to the electron

positron emission tomography (PET)
a technique for examining metabolic activity in tissues (particularly in the brain) by measuring blood flow containing tracers that emit positrons

postulate
an accepted statement used as the basis for developing an argument or explanation; also called an *axiom*

potential energy
energy associated with position

precipitate
an insoluble solid substance that has separated from a solution

precision
describes how closely repeated measurements cluster around the same value

pressure
force applied per unit area; in SI, pressure is expressed in pascals (Pa)

primary air pollutant
a contaminant that directly enters the atmosphere; it is not initially formed by reactions of airborne substances

primary battery
battery designed for a single use and that cannot be recharged

product
a substance formed in a chemical reaction

protein
a major structural component of living tissue made from many linked amino acids

proton
a particle possessing a positive electrical charge that is found in the nuclei of all atoms; the total protons in an element's atom equals its atomic number

Q

qualitative test
a chemical test indicating the presence or absence of an element, ion, or compound in a sample

quantitative test
a chemical test indicating the amount or concentration of an element, ion, or compound in a sample

R

rad
a unit that expresses the quantity of ionizing radiation absorbed by tissue; 1 rad = 0.01 Gy (Gray)

radiation
energy emitted in the form of electromagnetic waves or high-speed particles; refers to both ionizing radiation and nonionizing radiation

radioactive decay
a change in an atom's nucleus due to the spontaneous emission of alpha, beta, or gamma radiation

radioactivity
the spontaneous emission of nuclear radiation

radioisotope
radioactive isotope

range
the difference between the highest and lowest values in a data set

reactant
a starting material in a chemical reaction

reaction rate
an expression of how fast a particular chemical change occurs

redox reaction
see oxidation–reduction reaction

reduced
see reduction

reducing agent
a species that causes another atom, molecule, or ion to become reduced; the reducing agent, in turn, becomes oxidized in this process

reduction
any process in which one or more electrons can be considered as gained by a chemical species

reference solution
a solution of known composition used as a comparison in chemical tests

refined
removal of impurities from a desired material

rem
a unit that expresses the ability of radiation to cause ionization in human tissue; 1 rem = 0.01 Sv (sievert)

renewable resource
a resource that can be replenished by natural processes over the time frame of human experience

reversible reaction
a chemical reaction in which products form reactants at the same time that reactants form products

S

salt bridge
a connection that allows a voltaic cell's two half-cells to be in electrical contact without mixing; specifically, a tube containing an electrolyte (such as potassium chloride solution) that completes the internal circuit of a voltaic cell

saturated fat
a fat molecule containing only single carbon–carbon bonds within its fatty acid components

saturated hydrocarbon
a hydrocarbon consisting of molecules in which each carbon atom is bonded to four other atoms

saturated solution
a solution in which the solvent has dissolved as much solute as it can retain stably at a specified temperature

scientific model
a representation of either a part of the natural world or of a scientific theory

scientific question
question that provides a framework for gathering and analyzing data that will ultimately result in being able to describe, explain, or predict natural phenomena

scientific theory
a coherent set of ideas that explains many related observations or events in the natural world and offer "how"-type explanations of phenomena in the natural world

scintillation counter
a detector of ionizing radiation that measures light emitted by atoms that have been excited by ionizing radiation

secondary air pollutant
a contaminant generated in the atmosphere by chemical reactions between primary air pollutants and natural components of air

shell (electron)
energy level surrounding an atom's nucleus within which one or more electrons reside; outer-shell electrons are commonly called *valence electrons*

sievert (Sv)
an SI unit that expresses the dose equivalent of absorbed radiation that causes the same biological effects as one gray of gamma rays

single covalent bond
a bond in which two electrons are shared by the two bonded atoms

smog
the potentially hazardous combination of smoke and fog; *see also* photochemical smog

solid-state detector
a device used to monitor changes in the movement of electrons through semiconductors as they are exposed to ionizing radiation

solubility
the quantity of a substance that will dissolve in a given quantity of solvent to form a saturated solution at a particular temperature

solubility curve
a graph indicating the solubility of a particular solute at different temperatures

solubility rules
trends that have been identified within known data that make it easier to identify a precipitate

solute
the dissolved species in a solution; the solute is usually the smaller component of a solution

solution
a homogeneous mixture of two or more substances

solution concentration
the quantity of solute dissolved in a specific quantity of solvent or solution

solvent
the dissolving agent in a solution; the solvent is usually the larger component in a solution

space-filling model
model that depicts atoms in contact with each other

species
a general name used in chemistry for atoms, molecules, ions, free radicals, or other well-defined entities

specific heat capacity
the quantity of thermal energy needed to raise the temperature of 1 g of a material by 1 °C; the expression commonly has units of J/(g °C)

spectator ion
ion that does not participate in the reaction

spontaneous
reactions occurring without any input of additional stimuli; the overall energy of products is lower than the overall energy of reactants

standard solution
a solution of known concentration

standard temperature and pressure (STP)
conditions of 0 °C and 1 atm

stoichiometry
the relationships by which quantities of substances involved in a chemical reaction are linked and calculated

straight-chain alkane
an alkane consisting of molecules in which each carbon atom is linked to no more than two other carbon atoms

strong acid
an acid that fully ionizes in solution to liberate H^+ (H_3O^+); no molecular form of the acid remains

strong base
a base that fully liberates OH^- in solution

strong force
the force that holds protons and neutrons together in an atom's nucleus

structural formula
a chemical formula showing the arrangement of atoms and covalent bonds in a molecule, in which each electron pair in a covalent bond is represented by a line between the symbols of two atoms

structural isomers
substances involving rearrangement of atoms or bonds within a molecule but sharing a common molecular formula

subatomic particle
particle smaller than an atom; commonly regarded as electrons, protons, neutrons

subscript
the number printed below the line of type indicating the total atoms of a given element in a chemical formula; in H_2O, for example, the subscript 2 specifies the total H atoms

substance
an element or a compound; that is, a material with a uniform, definite composition and distinct properties

substituted hydrocarbon
carbon-backbone hydrocarbon with other elements substituted for one or more hydrogen atoms

substrate
a molecule that interacts with an enzyme and undergoes a reaction

superconductivity
the ability of a material to conduct an electrical current with zero electrical resistance; with present technology, operating superconductors must be extremely cold

surface water
water found on Earth's surface, such as oceans, rivers, and lakes

suspension
a mixture containing large, dispersed solid particles that can settle out or be separated by filtration

sustainability
present-day activities that preserve the ability of future generations to thrive and meet their resource needs on a habitable Earth

synergistic interaction
an interaction where the combined effect of several factors is greater than the sum of their separate effects

T

temperature inversion
an atmospheric condition where a cool air mass is trapped beneath a less dense warm air mass; it most frequently occurs in a valley or over a city

tetrahedron
a regular triangular pyramid; the four bonds of each carbon atom in an alkane point to the corners of a tetrahedron

therapeutic
treating a medical condition

thermal energy
the energy a material possesses due to its temperature; also known as *heat*

titrant
in a titration, the solution of known concentration that is added until an endpoint is reached

titration
a laboratory procedure for determining the concentrations of dissolved substances

total ionic equation
equation that accurately reflects the form of all substances in solution, such as $Ag^+(aq) + NO_3^-(aq) + Na^+(aq) + Cl^-(aq) \longrightarrow AgCl(s) + Na^+(aq) + NO_3^-(aq)$

trace mineral
see micromineral

tracer
a readily-identifiable material, such as a radioisotope, used to diagnose disease or to determine how the body is responding to treatment

transmutation
the conversion of one element to another either naturally or artificially

transuranium
any element with an atomic number greater than 92 (uranium)

triglyceride
a fat molecule composed of a simple three-carbon alcohol (glycerol) and three fatty acid molecules

trihalomethane (THM)
substance that in sufficiently high concentrations can be harmful to human health; a methane molecule with 3 hydrogen atoms substituted with halogen atoms

troposphere
the layer of the atmosphere closest to Earth's surface where most clouds and weather are located

Tyndall effect
the scattering of a beam of light caused by reflection from suspended particles

U

unsaturated fat
a fat molecule containing one or more carbon–carbon double bonds; it is monounsaturated if each fat molecule has a single double bond and polyunsaturated if it has two or more double bonds

unsaturated hydrocarbon
a hydrocarbon molecule containing one or more double or triple bonds

unsaturated solution
a solution containing a lower concentration of solute than a saturated solution contains at a specified temperature

V

valence electron
electron in the outermost shell of an atom; these relatively loosely held electrons often participate in bonding with other atoms or molecules

vaporization
the phase change that occurs when a substance changes from a liquid state to a gaseous state

viscosity
resistance to flow

vitamin
a biomolecule necessary for growth, reproduction, health, and life

vitrification
the conversion of material into a glassy solid by the application of high temperatures

volatile organic compound (VOC)
reactive carbon-containing substance that readily evaporates into air, such as components of gasoline and organic solvents

voltaic cell
an electrochemical cell in which a spontaneous chemical reaction produces electricity

W

water cycle
repetitive processes of rainfall (or other precipitation), run-off, evaporation, and condensation that circulate water within Earth's crust and atmosphere; also called the *hydrologic cycle*

wavelength (λ)
the distance between corresponding points of two consecutive waves; for electromagnetic radiation, the product of frequency and wavelength equals the speed of light

weak acid
an acid that does not fully ionize in solution to liberate H^+ (H_3O^+) but remains primarily in molecular form

X

X-ray
high-energy electromagnetic radiation that cannot penetrate dense materials such as bone or lead but can penetrate less dense materials

Z

zero oxidation state
the value of the oxidation state of an element's atoms when not chemically combined with any other element

Photo Credits

104 (right) Paul Degginger/Getty Images

108 ACS

109 (top) Michael Newman/PhotoEdit

109 (bottom) David Barber/PhotoEdit

110 (top row) Peter Johnson/Corbis; Corbis; Corbis

110 (middle) Corbis

110 (bottom) Kevin Fleming/Corbis

118 Richard Megna/Fundamental Photographs, NYC

120 John Cancalosi/Peter Arnold, Inc.

121 Alan Schein/Alamy

122 Brand X Pictures/Getty Images

123 Richard Megna/Fundamental Photographs, NYC

125 Tomd/Dreamstime.com

127 NASA

128 The Canadian Press/Darryl Dyck

129 Brand X Pictures/Getty Images

133 Michael Ciesielski Photo/ACS

134 (left) Designer: Joe Fitzgerald, Engraver: Don Everhart

134 (right) Designer: Jamie Franki, Engraver: Norman E. Nemeth

138 (counterclockwise from top right) Ramón Rivera-Moret; Image Finders/Index Stock; David Young-Wolff/PhotoEdit; David Young-Wolff/PhotoEdit; Gary Gladstone, Inc./The Image Bank/Getty Images; Ball Corporation

139 EcoStar LLC

143 (top) DNY59/istockphoto.com

143 (bottom) Peter Sprio/istockphoto.com

145 Judy A. Mosby

UNIT 2

146 Mary Van de Ven/Pacific Stock/Photolibrary

148 (top) U.S. Air Force photo by Tech. Sgt. Polly Bennett

148 (middle) Amy Walters/istockphoto

148 (bottom) Tim Graham/Alamy

149 (top) AP Photo/Charlie Riedel

149 (middle) Natalia Bratslavsky/Dreamstime.com

149 (bottom) Jenny Hager/The Image Works

153 (top) Ramón Rivera-Moret

153 (bottom) Michael Ciesielski Photo/ACS

155 (both) Michael Ciesielski Photo/ACS

156 (left) Corbis

156 (right) PhotoDisc/Getty Images

157 (left) ACS

157 (right) Sonbeam/Dreamstime.com

159 The Photo Works

160 PhotoDisc/Getty Images

161 (top) Bert Bolle

161 (bottom) Michael Ciesielski Photo/ACS

171 Sofiaworld/Dreamstime.com

174 (top) Arnulf Husmo/Stone/Getty Images

174 (bottom) Mark Conlin/Alamy

175 Michael Ciesielski Photo/ACS

181 (left) Russ Lappa/Photo Researchers

181 (right) Richard Megna/Fundamental Photographs, NYC

184 Nicky Blade/istockphoto.com

188 (top) medobear/istockphoto.com

188 (bottom) ACS

191 ACS

193 ACS

200 (top) Jason Church

200 (middle) Ted Foxx/Alamy

200 (bottom) Chris Selby/Alamy

201 (top) Richard Megna/Fundamental Photographs, NYC

201 (bottom) Newscom

204 NASA

207 Roger Ressmeyer/Corbis

212 Raymond Reuter/Corbis

213 Tony Freeman/PhotoEdit

217 (all) ACS

218 ACS

219 ACS

221 (top left) NASA/JPL/USGS

221 (top right) NASA/Goddard/SDO AIA Team

221 (bottom row) Blend/Fotostock/Fotosearch; ImageState/Alamy; Craig Sillitoe/The Age (Melbourne)/Fairfaxphotos

223 Cathy Melloan/PhotoEdir

224 (left) Dennis Flaherty/Photo Researchers

224 (right) John Beatty/Tony Stone

UNIT 4

601 Michael Ciesielski Photo/ACS

605 Photodisc/Getty Images

609 Natal'ya Bondarenko/istockphoto

610 © Dan Barnes/istockphoto

612 © 1999 Richard Megna/Fundamental Photographs

613 Corbis/Agefotostock

614 (top) AP Photo/M. Spencer Green

614 (bottom) Mauro Fermariello/Photo Researchers, Inc.

615 (left) Andrew Syred/Photo Researchers, Inc.

615 (right) Oliver Meckes/Nicole Ottawa/Photo Researchers, Inc.

617 General Records of the Department of Energy, Argonne National Laboratory/National Archives and Records Administration—Great Lakes Region

621 Brand X Pictures/Photolibrary

622 Richard Megna/Fundamental Photographs

623 Arthur Hill/Visuals Unlimited

626 Ted Spiegel/Corbis

628 Hank Morgan/Photo Researchers

629 Poil 22:18, 26 Apr 2005 (UTC)

630 CERN/Science Photo Library/Photo Researchers

631 (both) ACS

632 (left) Peter Arnold, Inc./Alamy

632 (right) U.S. NRC

636 (top) Jacqueline McBride/Lawrence Livermore National Laboratory

637 (top) RIA Novosti/TopFoto/The Image Works

637 (bottom) Max Whitby and Theodore Gray

643 Jesus Jauregui/istockphoto

644 (top) Kenneth Garrett/National Geographic Stock

644 (bottom left) Philippe Garo/Photo Researchers, Inc.

644 (bottom right) Belmonte/Agefotostock

645 ISM/Phototake

649 Bettmann/Corbis

650 The Granger Collection, New York

652 (left, both) Images Courtesy of Fermilab

652 (right) Courtesy of TRIUMF. Photo from Timothy I. Meyer, Ph.D.

653 Corbis

654 (left to right) Esemelwe/istockphoto; PRNewFoto/Elekta, Inc./Publicity photo via AP; Werner Forman/Art Resource, NY

659 Corbis

660 H. Mark Weidman Photography/Alamy

662 Michael Ciesielski Photo/ACS

663 Naturfoto-Online/Alamy

665 Carlos Humberto T.D.C./Contact Press Images/ PictureQuest

668 NASA/Goddard Space Flight Center

669 (left) Lawrence Livermore National Laboratory

669 (right) Lawrence Livermore National Laboratory National Ignition Facility

670 (top to bottom) Jeff Greenberg/Agefotostock; Dragan Cvetanovic/Dreamstime.com; G. Vockel/Agefotostock; Luminis/istockphoto; AP Photo/Mel Evans

671 (top) Photonica/Getty Images

671 (bottom) Roger Ressmeyer/Corbis

673 Courtesy of U.S. Department of Energy

674 (both) U.S. Department of Energy

675 Roger Ressmeyer/Corbis

681 Mauro Fermariello/Photo Researchers, Inc.

UNIT 7

682 Chris Cheadle/Photolibrary

684 (top) © Mark Hunt/Agefotostock

684 (bottom left) David Young-Wolff/Photo Edit

684 (bottom right) Glowimages RF

685 Rubberball/istockphoto

687 (top) Maximilian Stock Ltd./Photo Researchers, Inc.

687 (bottom) mypyramid.gov

688 Graça Victoria/istockphoto

689 Michael Ciesielski Photo/ACS

691 (top) Michael Ciesielski Photo/ACS

691 (bottom) StockByte/Getty Images

693 (middle) PhotoDisc/Getty Images

693 (bottom) desuza communications/istockphoto

694 (top) Digital Vision/Getty Images

694 (bottom left) Charles D. Winters/Photo Researchers

694 (bottom right) FoodPix/Getty Images

695 redmonkey8/istockphoto

Project Credits

CHEMCOM Project Credits

ChemCom is the product of teamwork involving individuals from all over the United States over more than twenty-five years. The American Chemical Society is pleased to recognize all who contributed to *ChemCom*. The team responsible for the sixth edition of *ChemCom* is listed on the copyright page. Individuals who contributed to the initial development of *ChemCom*—for the first edition in 1988, the second and third editions in 1993 and 1998, respectively, the fourth edition in 2002, and the fifth edition in 2006—are listed below.

Principal Investigator: W. T. Lippincott

Project Manager: Sylvia Ware

Chief Editor: Henry Heikkinen & Conrad L. Stanitski

Contributing Editor: Mary Castellion

Assistant to Contributing Editor: Arnold Diamond

Editor of Teacher's Guide: Thomas O'Brien & Patricia J. Smith

Revision Team: Diane Bunce, Gregory Crosby, David Holzman, Thomas O'Brien, Joan Senyk, Thomas Wysocki

Editorial Advisory Board: Joseph Breen, Glenn Crosby, James DeRose, I. Dwaine Eubanks, Lucy Pryde Eubanks, Regis Goode, Henry Heikkinen (chair), Mary Kochansky, Ivan Legg, W. T. Lippincott (ex officio), Steven Long, Nina McClelland, Lucy McCorkle, Carlo Parravano, Robert Patrizi, Max Rodel, K. Michael Shea, Patricia Smith, Susan Snyder, Conrad Stanitski, Jeanne Vaughn, Sylvia Ware (ex officio)

Writing Team: Rosa Balaco, James Banks, Joan Beardsley, William Bleam, Kenneth Brody, Ronald Brown, Diane Bunce, Becky Chambers, Alan DeGennaro, Patricia Eckfeldt, Dwaine Eubanks (dir.), Henry Heikkinen (dir.), Bruce Jarvis (dir.), Dan Kallus, Jerry Kent, Grace McGuffie, David Newton (dir.), Thomas O'Brien, Andrew Pogan, David Robson, Amado Sandoval, Joseph Schmuckler (dir.), Richard Shelly, Patricia Smith, Tamar Susskind, Joseph Tarello, Thomas Warren, Robert Wistort, Thomas Wysocki

Steering Committee: Alan Cairncross, William Cook, Derek Davenport, James DeRose, Anna Harrison (ch.), W. T. Lippincott (ex officio), Lucy McCorkle, Donald McCurdy, William Mooney, Moses Passer, Martha Sager, Glenn Seaborg, John Truxall, Jeanne Vaughn

Consultants: Alan Cairncross, Michael Doyle, Donald Fenton, Conrad Fernelius, Victor Fratalli, Peter Girardot, Glen Gordon, Dudley Herron, John Hill, Chester Holmlund, John Holman, Kenneth Kolb, E. N. Kresge, David Lavallee, Charles Lewis, Wayne Marchant, Joseph Moore, Richard Millis, Kenneth Mossman, Herschel Porter, Glenn Seaborg, Victor Viola, William West, John Whitaker

Synthesis Committee: Diane Bunce, Dwaine Eubanks, Anna Harrison, Henry Heikkinen, John Hill, Stanley Kirschner, W. T. Lippincott (ex officio), Lucy McCorkle, Thomas O'Brien, Ronald Perkins, Sylvia Ware (ex officio), Thomas Wysocki

Evaluation Team: Ronald Anderson, Matthew Bruce, Frank Sutman (dir.)

Field Test Coordinator: Sylvia Ware

Field Test Workshops: Dwaine Eubanks

Field Test Directors: Keith Berry, Fitzgerald Bramwell, Mamie Moy, William Nevill, Michael Pavelich, Lucy Pryde, Conrad Stanitski

Pilot Test Teachers: Howard Baldwin, Donald Belanger, Navarro Bharat, Ellen Byrne, Eugene Cashour, Karen Cotter, Joseph Deangelis, Virginia Denney, Diane Doepken, Donald Fritz, Andrew Gettes, Mary Gromko, Robert Haigler, Anna Helms, Allen Hummel, Charlotte Hutton, Elaine Kilbourne, Joseph Linker, Larry Lipton, Grace McGuffie, Nancy Miller, Gloria Mumford, Beverly Nelson, Kathy Nirei, Elliott Nires, Polly Parent, Mary Parker, Dicie Petree, Ellen Pitts, Ruth Rand, Kathy Ravano, Steven Rischling, Charles Ross, Jr., David Roudebush, Joseph Rozaik, Susan Rutherland, George Smeller, Cheryl Snyder, Jade Snyder, Samuel Taylor, Ronald Tempest, Thomas Van Egeren, Gabrielle Vereecke, Howard White, Thomas Wysocki, Joseph Zisk

Field Test Teachers: Vincent Bono, Allison Booth, Naomi Brodsky, Mary D. Brower, Lydia Brown, George Bulovsky, Kay Burrough, Gene Cashour, Frank Cox, Bobbie Craven, Pat Criswell, Jim Davis, Nancy Dickman, Dave W. Gammon, Frank Gibson, Grace Giglio, Theodis Gorre, Margaret Guess, Yvette Hayes, Lu Hensen, Kenn Heydrick, Gary Hurst, Don Holderread, Michael Ironsmith, Lucy Jache, Larry Jerdal, Ed Johnson, Grant Johnson, Robert Kennedy, Anne Kenney, Joyce Knox, Leanne Kogler, Dave Kolquist, Sherman Kopelson, Jon Malmin, Douglas Mandt, Jay Maness, Patricia Martin, Mary Monroe, Mike Morris, Phyllis Murray, Silas Nelson, Larry Nelson, Bill Rademaker, Willie Reed, Jay Rubin, Bill Rudd, David Ruscus, Richard Scheele, Paul Shank, Dawn Smith, John Southworth, Mitzi Swift, Steve Ufer, Bob Van Zant, Daniel Vandercar, Bob Volzer, Terri Wahlberg, Tammy Weatherly, Lee Weaver, Joyce Willis, Belinda Wolfe

Field Test Schools: California: Chula Vista High, Chula Vista; Gompers Secondary School, San Diego; Montgomery High, San Diego; Point Loma High, San Diego; Serra Junior-Senior High, San Diego; Southwest High, San Diego. Colorado: Bear Creek Senior High, Lakewood; Evergreen Senior High, Evergreen; Green Mountain Senior High, Lakewood; Golden Senior High, Golden; Lakewood Senior High, Lakewood; Wheat Ridge Senior High, Wheat Ridge. Hawaii: University of Hawaii Laboratory School, Honolulu. Illinois: Project Individual Education High, Oak Lawn. Iowa: Linn-Mar High, Marion. Louisiana: Booker T. Washington High, Shreveport; Byrd High, Shreveport; Caddo Magnet High, Shreveport; Captain Shreve High, Shreveport; Fair Park High, Shreveport; Green Oaks High, Shreveport; Huntington High, Shreveport; North Caddo High, Vivian; Northwood High, Shreveport. Maryland: Charles Smith Jewish Day School, Rockville; Owings Mills Junior-Senior High, Owings Mills; Parkville High, Baltimore; Sparrows Point Middle-Senior High, Baltimore; Woodlawn High, Baltimore. New Jersey: School No. 10, Patterson. New York: New Dorp High, Staten Island. Texas: Clements High, Sugar Land; Cy-Fair High, Houston.

Virginia: Armstrong High, Richmond; Freeman High, Richmond; Henrico High, Richmond; Highland Springs High, Highland Springs; Marymount School, Richmond; Midlothian High, Midlothian; St. Gertrude's High, Richmond; Thomas Dale High, Chester; Thomas Jefferson High, Richmond; Tucker High, Richmond; Varina High, Richmond. Wisconsin: James Madison High, Madison; Thomas More High, Milwaukee. Washington: Bethel High, Spanaway; Chief Sealth High, Seattle; Clover Park High, Tacoma; Foss Senior High, Tacoma; Hazen High, Renton; Lakes High, Tacoma; Peninsula High, Gig Harbor; Rogers High, Puyallup; Sumner Senior High, Sumner; Washington High, Tacoma; Wilson High, Tacoma.

Safety Consultant: Stanley Pine & William H. Breazeale, Jr.

Social Science Consultants: Ross Eshelman, Judith Gillespie

Art: Rabina Fisher, Pat Hoetmer, Alan Kahan (dir.), Kelley Richard, Sharon Wolfgang

Copy Editor: Martha Polkey

Production Consultant: Marcia Vogel

Administrative Assistant: Carolyn Avery

ACS Staff: Rebecca Mason Simmons, Martha K. Turckes

Student Aides: Paul Drago, Stephanie French, Patricia Teleska

Second Edition Revision Team

Project Manager: Keith Michael Shea & Ted Dresie

Chief Editor: Henry Heikkinen

Assistant to Chief Editor: Wilbur Bergquist

Editor of Teacher's Guide: Jon Malmin

Second Edition Editorial Advisory Board: Diane Bunce, Henry Heikkinen (ex officio), S. Allen Heininger, Donald Jones (chair), Jon Malmin, Paul Mazzocchi, Bradley Moore, Carolyn Morse, Keith Michael Shea (ex officio), Sylvia Ware (ex officio)

Teacher Reviewers of First Edition: Vincent Bono, New Dorp High School, New York; Charles Butterfield, Brattle Union High School, Vermont; Regis Goode, Spring Valley High School, South Carolina; George Gross, Union High School, New Jersey; C. Leonard Himes, Edgewater High School, Florida; Gary Hurst, Standley Lake High School, Colorado; Jon Malmin, Peninsula High School, Washington; Maureen Murphy, Essex Junction Educational Center, Vermont; Keith Michael Shea, Hinsdale Central High School, Illinois; Betsy Ross Uhing, Grand Island Senior High School, Nebraska; Jane Voth-Palisi, Concord High School, New Hampshire; Terri Wahlberg, Golden High School, Colorado.

Teacher Reviewers of Second Edition: Michael Clemente, Carlson High School, Gibraltar, MI; Steven Long, Rogers High School, Rogers, AR; William Penker, Neillsville High School, Neillsville, WI; Audrey Mandel, Connetquot High School, Bohemia, NY; Barbara Sitzman, Chatsworth High School, Chatsworth, CA; Kathleen Voorhees, Shore Regional High School, West Long Branch, NJ; Debra Compton, Cy-Fair High School, Houston, TX; Christ Forte, York Community High School, Elmhurst, IL; Gwyneth D. Sharp, Cape Henlopen High School, Lewes, DE; Louis Dittami, Dover-Sherborn High School, Dover, MA; Sandra Mueller, John Burroughs School, St. Louis, MO; Kirk Soule, Sunset High School, Beaverton, OR; Sigrid Wiolkinson, Athens Area High School, Athens,

PA; Millie McDowell, Clayton High School, Clayton, MO; Leslie A. Roughley, Steward School, Richmond, VA; Robert Houle, Bacon Academy, Colchester, CT; Robert Storch, Bishop Ireton High School, Alexandria, VA; Michael Smolarek, Neenah High School, Neenah, WI; Fred Nozawa, Timpview High School, Provo, UT; Michael Sixtus, Mar Vista High School, Imperial Beach, CA

Safety Consultant: Stanley Pine

Editorial: The Stone Cottage

Design: Bonnie Baumann & P.C. & F., Inc.

Art: Additional art for this edition by Seda Sookias Maurer

ACS also offers thanks to the National Science Foundation for its support of the initial development of *ChemCom*, and to NSF project officers Mary Ann Ryan and John Thorpe for their comments, suggestions, and unfailing support.

Fourth Edition Credits

Chief Editor: Henry Heikkinen

Revision Team: Laurie Langdon, Robert Milne, Angela Powers, Christine Gaudinski, Courtney Willis

Revision Assistants: Cassie McClure, Seth Willis

Teacher Edition: Lear Willis, Joseph Zisk

Ancillary Materials: Regis Goode, Mike Clemente, Ruth Leonard

Fourth Edition Editorial Advisory Board: Conrad L. Stanitski (Chair), Boris Berenfeld, Jack Collette, Robert Dayton, Ruth Leonard, Nina I. McClelland, George Miller, Adele Mouakad, Carlo Parravano, Kirk Soulé, Maria Walsh, Sylvia A. Ware (ex officio), Henry Heikkinen (ex officio)

ACS: Sylvia Ware, Janet Boese, Michael Tinnesand, Guy Belleman, Patti Galvan, Helen Herlocker, Beverly DeAngelo

ACS Safety Committee: Henry Clayton Ramsey (Chair), Wayne Wolsey, Kevin Joseph Edgar, Herbert Bryce

Technical Reviewers: Steve Cawthron, Kenneth Hughes, Susan C. Karr, Mary Kirchhoff, David Miller, Charles Poukish, Mary Ann Ryan, Tracy Williamson

Teacher Edition Reviewers: Drew Lanthrum, Karen Morris

Safety Consultants: Stanley Pine, Herbert Bryce

Fifth Edition Credits

Chief Editor: Henry Heikkinen

Project Manager: Angela Powers

Revision Team: Laurie Langdon, Robert Milne, Wendy Naughton

ACS: Sylvia Ware, Michael Tinnesand, Terri Taylor, Helen Herlocker, Jodi Wesemann

ACS Project Editor: Rebecca Strehlow

Technical Reviewer: Conrad Stanitski

Technical Contributors: Janet Cohen, Susan Heikkinen, Jack A. Ladson, COLOR Science Consultancy

Fourth Edition Reviewers: Kirsten Almo, Mark Boehlor, Susan Berrend, Martin Besant, Regis Goode, Gary Jackson, Jane Meadows, Walt Shacklett, Terri Taylor

Teacher Contributors: Pat Chriswell, Robert Dayton, Regis Goode, Drew Lanthrum, Joelle Lastica, Steve Long, Cece Schwennsen, Barbara Sitzman, Terri Taylor

Index

Chart of the Elements

Element	Symbol	Atomic Number	Average Atomic	Element	Symbol	Atomic Number	Average Atomic
Actinium	Ac	89	[227]	Erbium	Er	68	167.26
Aluminum	Al	13	26.98	Europium	Eu	63	151.96
Americium	Am	95	[243]	Fermium	Fm	100	[257]
Antimony	Sb	51	121.76	Fluorine	F	9	19.00
Argon	Ar	18	39.95	Francium	Fr	87	[223]
Arsenic	As	33	74.92	Gadolinium	Gd	64	157.25
Astatine	At	85	[210]	Gallium	Ga	31	69.72
Barium	Ba	56	137.33	Germanium	Ge	32	72.64
Berkelium	Bk	97	[247]	Gold	Au	79	196.97
Beryllium	Be	4	9.01	Hafnium	Hf	72	178.49
Bismuth	Bi	83	208.98	Hassium	Hs	108	[277]
Bohrium	Bh	107	[272]	Helium	He	2	4.003
Boron	B	5	10.81	Holmium	Ho	67	164.93
Bromine	Br	35	79.90	Hydrogen	H	1	1.008
Cadmium	Cd	48	112.41	Indium	In	49	114.82
Calcium	Ca	20	40.08	Iodine	I	53	126.90
Californium	Cf	98	[251]	Iridium	Ir	77	192.22
Carbon	C	6	12.01	Iron	Fe	26	55.85
Cerium	Ce	58	140.12	Krypton	Kr	36	83.80
Cesium	Cs	55	132.91	Lanthanum	La	57	138.91
Chlorine	Cl	17	35.45	Lawrencium	Lr	103	[262]
Chromium	Cr	24	52.00	Lead	Pb	82	207.2
Cobalt	Co	27	58.93	Lithium	Li	3	6.94
Copernicium	Cn	112	[285]	Lutetium	Lu	71	174.97
Copper	Cu	29	63.55	Magnesium	Mg	12	24.31
Curium	Cm	96	[247]	Manganese	Mn	25	54.94
Darmstadtium	Ds	110	[281]	Meitnerium	Mt	109	[276]
Dubnium	Db	105	[268]	Mendelevium	Md	101	[258]
Dysprosium	Dy	66	162.50	Mercury	Hg	80	200.59
Einsteinium	Es	99	[252]	Molybdenum	Mo	42	95.96

Chart of the Elements

Element	Symbol	Atomic Number	Average Atomic	Element	Symbol	Atomic Number	Average Atomic
Neodymium	Nd	60	144.24	Silicon	Si	14	28.09
Neon	Ne	10	20.18	Silver	Ag	47	107.87
Neptunium	Np	93	[237]	Sodium	Na	11	22.99
Nickel	Ni	28	58.69	Strontium	Sr	38	87.62
Niobium	Nb	41	92.91	Sulfur	S	16	32.07
Nitrogen	N	7	14.01	Tantalum	Ta	73	180.95
Nobelium	No	102	[259]	Technetium	Tc	43	[98]
Osmium	Os	76	190.23	Tellurium	Te	52	127.60
Oxygen	O	8	16.00	Terbium	Tb	65	158.93
Palladium	Pd	46	106.42	Thallium	Tl	81	204.38
Phosphorus	P	15	30.97	Thorium	Th	90	232.04
Platinum	Pt	78	195.08	Thulium	Tm	69	168.93
Plutonium	Pu	94	[244]	Tin	Sn	50	118.71
Polonium	Po	84	[209]	Titanium	Ti	22	47.87
Potassium	K	19	39.10	Tungsten	W	74	183.84
Praseodymium	Pr	59	140.91	Ununhexium	Uuh	116	[293]
Promethium	Pm	61	[145]	Ununoctium	Uuo	118	[294]
Protactinium	Pa	91	231.04	Ununpentium	Uup	115	[288]
Radium	Ra	88	[226]	Ununquadium	Uuq	114	[289]
Radon	Rn	86	[222]	Ununtrium	Uut	113	[284]
Rhenium	Re	75	186.21	Uranium	U	92	238.03
Rhodium	Rh	45	102.91	Vanadium	V	23	50.94
Roentgenium	Rg	111	[280]	Xenon	Xe	54	131.29
Rubidium	Rb	37	85.47	Ytterbium	Yb	70	173.05
Ruthenium	Ru	44	101.07	Yttrium	Y	39	88.91
Rutherfordium	Rf	104	[265]	Zinc	Zn	30	65.38
Samarium	Sm	62	150.36	Zirconium	Zr	40	91.22
Scandium	Sc	21	44.96				
Seaborgium	Sg	106	[271]				
Selenium	Se	34	78.96				